FLORA OF TROPICAL EAST AFRICA

APOCYNACEAE (Part 2)

DAVID GOYDER[1], TIMOTHY HARRIS[1], SIRO MASINDE[2],
ULRICH MEVE[3] & JOHAN VENTER[4]

For the family description, see part 1. Notes on floral structures within the more derived subfamilies of Apocynaceae: the five stamens are fused apically to the expanded stylar head and together form a compound structure called the gynostegium. The ovary is therefore almost entirely concealed within a ring of stamens (the staminal column) whose filaments are usually fused into a tube, but free in subfamily Periplocoideae. Only the sterile apex of the stylar head remains visible, level with or extending beyond the anthers; the receptive parts of the stylar head are on its underside behind the frequently sclerified margins of the anthers which form a chamber for the deposition of pollen. The pollen transfer apparatus is formed by secretions from the stylar head and, in periplocoid genera, generally consists of a spatulate translator onto which pollen tetrads, or more rarely pollen masses (pollinia), from adjacent anthers are shed; in the two remaining subfamilies, the secretions form a central corpusculum linked by caudicles or translator arms to 2 or 4 pollinia of adjacent anthers – the unit is transferred in its entirety from one flower to another and is called a pollinarium.

The corona presents many diagnostic characters for generic and specific recognition. Following the system devised by Liede & Kunze in Pl. Syst. Evol. 185: 275–284 (1993), it may be corolline (derived from the corolla) or gynostegial (from the staminal column). Corolline coronas can be divided into those occuring in the corolla lobe sinuses and those forming an annulus in the corolla tube. Gynostegial coronas again have two basic elements, a staminal corona attached dorsally to the stamens, and an interstaminal corona, and these elements can be combined in a number of ways: e.g. staminal corona lobes only; a fused ring of staminal and interstaminal lobes; fused staminal and interstaminal corona with additional staminal lobes.

[1] Royal Botanic Gardens, Kew: David Goyder **Secamonoideae** (genus 46, with Tim Harris); **Asclepiadoideae**: tribes Fockeeae (genera 47, 48); Marsdenieae (genera 49–51); Ceropegieae (Leptadeniinae; Anisotominae, in part) (genera 52–54) & Asclepiadeae (genera 67–87)
[2] East African Herbarium, National Museums of Kenya: **Asclepiadoideae**: tribe Ceropegieae (Anisotominae, in part; Stapeliinae, in part) (genera 55–57)
[3] Lehrstuhl für Pflanzensystematik, Universität Bayreuth: **Asclepiadoideae**: tribe Ceropegieae (Stapeliinae, in part) (genera 58–66)
[4] Department of Plant Sciences, University of the Free State, Bloemfontein: **Periplocoideae** (genera 32–45)

Key to the derived subfamilies of Apocynaceae (formerly Asclepiadaceae) – see Fig. 36

1. Staminal filaments free; anther lacking sclerified margins; pollen in tetrads and appearing granular, or occasionally aggregated into pollinia, deposited onto a spatulate pollen carrier or translator situated on the stylar head, between adjacent anthers . . . . **Periplocoideae** (p. 117)

   Staminal filaments, when present, united into a tube or annulus; anthers with sclerified margins (anther wings or guide rails), adjacent anther wings adnate to each other, with a groove between them; pollen of each anther cell aggregated into 1 or 2 pollinia, the pollinia from adjacent anther locules attached directly or indirectly to a corpusculum situated on the stylar head, at the top of the guide rails 2

2. Corpusculum pale, minute; pollinia 4 to each corpusculum, attached directly or on short stalks **Secamonoideae** (p. 167)

   Corpusculum hard, horny, mostly dark brown or black; pollinia 2 per corpusculum, attached by variously structured translator arms . . . . **Asclepiadoideae** (p. 177)

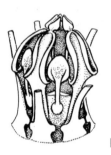

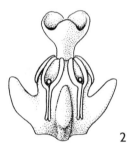
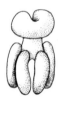

1

2

3

FIG. 36. Stylised gynostegia and pollen transport arrangements (to be used in conjunction with the key to subfamilies): 1, PERIPLOCOIDEAE: corona lobes cut away to reveal stamens depositing granular pollen onto the spoon-like translator; 2, SECAMONOIDEAE: adjacent anther wing margins sclerified and closely adnate, with a groove between, pollinarium with four pollinia attached ± directly to the minute soft corpusculum; 3, ASCLEPIADOIDEAE: gynostegium with two corona lobes removed to expose anthers, again with sclerified margins to the anther wings and a groove between, pollinarium of two pollinia attached indirectly to the hard corpusculum by translator arms (but note Fockeeae — see fig. 55: 1, p. 179 (two pollinia attached direct to the corpusculum)). Drawn by M. Tebbs.

subfam. PERIPLOCOIDEAE

Periplocoideae (Asclepiadaceae) K. Schum. in E. & P. Pf. 4(2): 209 (1895); Venter & R.L.Verhoeven in Taxon 46: 705 (1997)

33 genera and approximately 190 species occur in Africa, Madagascar, Europe, Asia and Australia. Approximately 100 species, and 16 of the 19 genera found in Africa, are endemic to the continent. 14 genera and 38 species are present in the Flora area.

The sequence of genera follows the evolutionary relationships elucidated by Ionta & Judd in Ann. Missouri Bot. Gard. 94: 360–375 (2007).

1. Plants epiphytic (anchored on trees) or rupicolous (anchored on rocks) in high rainfall and mist areas 2
   Plants anchored in soil under various climatic conditions 3
2. Corona lobes densely white-pubescent; translator apex split; pollen in pollinia . . . . 45. **Epistemma** (p. 166)
   Corona lobes glabrous; translator apex not split, pollen in tetrads . . . . 44. **Sarcorrhiza** (p. 164)
3. Corolla tube campanulate, cylindrical or reflexed 4
   Corolla rotate with tube shallow and saucer-shaped 11
4. Corolla tube campanulate or cylindrical 5
   Corolla tube reflexed 10
5. Stamens arise at base of corolla tube or from inner base of coronal feet 6
   Stamens arise at mouth of corolla tube 8
6. Flowers large, 3–6 cm long and 5–9 cm diameter . 33. **Cryptostegia** (p. 120)
   Flowers small, less than 2 cm long and less than 3 cm diameter 7
7. Inside of corolla tube and inside of corolla lobes papillate; lower corona of vertical ridges with hooded apices radiating into corolla tube cavity; no upper corona present . . . . 34. **Parquetina** (p. 122)
   Inside of corolla tube glabrous or villous, corolla lobes glabrous; lower corona of clavate to acicular lobes; upper corona, when present pocket-like or pocket-like with lobes from pocket rim . . . . 35. **Cryptolepis** (p. 126)
8. Pollen in pollinia; lower corona lobes basally fused into a collar around the stamens; style terete, glabrous . . . . 41. **Schlechterella** (p. 157)
   Pollen in tetrads; lower corona lobes basally free, if fused into a collar style will be ribbed and hairy 9
9. Woody perennial climbers with numerous root tubers per plant . . . . 37. **Chlorocyathus** (p. 138)
   Suffrutescent herbaceous erect, prostrate or climbing plants with only one taproot tuber per plant . . . . 40. **Raphionacme** (p. 146)
10. Coronal collar of deltoid coronal feet fused to lobular interstaminal nectaries; stamens glabrous 42. **Baseonema** (p. 161)
    Corona lobes deeply bifid to tetrafid; stamens hirsute . . . . 34. **Parquetina** (p. 122)

11. Interpetiolar collar of green frills; flowers large,
    15–25 mm diameter .... **39. Mondia** (p. 143)
    Interpetiolar lines or dentate ridges without frills;
    flowers less than 10 mm diameter 12
12. Erect shrubs; flowers very small, 2–3 mm long and
    less than 5 mm diameter .... **36. Sacleuxia** (p. 135)
    Climbers; flowers 5–20 mm long and 8–30 mm
    diameter 13
13. Stamens and inside of corolla lobes hairy .... **32. Periploca** (p. 118)
    Stamens and inside of corolla lobes glabrous 14
14. Inflorescences inconspicuous, few-flowered;
    peduncles less than 10 mm long; corolla lobes
    with glandulars swellings at their inner bases ... **43. Buckollia** (p. 161)
    Inflorescences conspicuous, paniculate, many-
    flowered; peduncles long, usually from 10–60 mm
    long; corolla lobes without any glandular
    swellings at their inner bases .... **38. Tacazzea** (p. 140)

32. PERIPLOCA

L., Sp. Pl.: 211 (1753); N.E. Br. in F.T.A. 4(1): 256 (1902); Browicz in Arb. Korn. 11:
5–104 (1966); Venter in S. Afr. Journ. Bot. 63: 123–128 (1997)

Shrub, scrambler or liana with white latex. Leaves persistent or soon falling, opposite, broadly elliptic, linear-elliptic, broadly ovate or linear-ovate, sometimes bracteate. Inflorescences terminal, compact to lax, few to many-flowered. Sepals mostly with paired colleters at inner base. Corolla tube saucer-shaped; lobes glabrous to villous, inside with or without dark coloured glandular centres and densely puberulous white spots. Corolline corona inserted at mouth of the corolla tube, lobes simple, undivided or bi- or tri-segmented; undivided lobes filiform, linear or ovate, apically simple, bi-fid or tetra-fid; bi-segmented lobes subulate; tri-segmented lobes with central segment filiform or linear, apically simple or bifid, lateral segments fleshy, ovoid-deltoid and fused to inside base of corolla lobes, basally fused with staminal filaments. Gynostegial corona absent. Stamens inserted directly below corona lobes, anthers villous or hirsute, pollen in tetrads. Interstaminal nectaries fused with staminal filaments, lobular. Style-head ovoid to broadly ovoid, translators with receptacle, stalk and sticky disc. Gynostegium exposed from corolla. Follicles paired, linear-ovoid to very narrowly ovoid, narrowly to horizontally divergent.

Periploca includes 12 species which are widely distributed over Africa, Europe and Asia.

Periploca linearifolia *Quart.-Dill. & A. Rich.* in Ann. Sci. Nat.-Bot. ser. 2, 14: 263 (1840); F.P.S. 2: 413 (1952); Malaisse in Fl. Rwanda 3: 109 (1985); U.K.W.F. ed. 2: 175 (1994); K.T.S.L.: 496 (1994); Venter in Fl. Eth. 4(1): 103 (2003); Burrows in Pl. Nyika Plateau: 73 (2005). Type: Ethiopia, Adowa, *Quartin-Dillon & Petit* s.n. (P!, holo.; K!, iso.)

Large liana; stems up to 20 m long or more, glabrous, bark greyish-brown. Leaves shortly petiolate, leathery; blade linear to very narrowly ovate, 3–7(–9) × 0.2–0.5 cm, long attenuate, base cuneate or obtuse, glabrous. Inflorescences panicle-like cymes, usually many-flowered; peduncles 1–2 cm long; bracts ovate, ± 1 mm long; pedicels 2–6 mm long. Sepals broadly ovate to sub-orbicular, ± 1 × 1 mm, apex obtuse. Corolla sub-herbaceous; tube ± 0.5 mm long; lobes violet with dark glabrous centre above, linear-ovate to very narrowly ovate, 3–4 × 1–2 mm, obtuse to retuse, white-

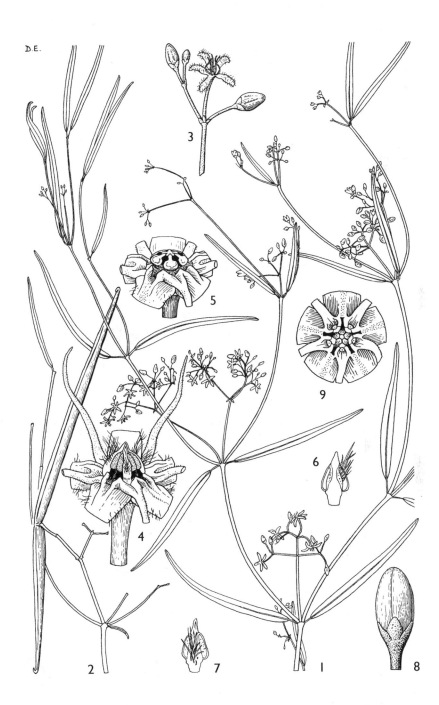

FIG. 37. *PERIPLOCA LINEARIFOLIA* — **1**, stem with flowers, × 1; **2**, stem with follicles, × 1; **3**, part of inflorescence, × 3; **4** & **5**, flower, × 15; **6**, adaxial view of anther, × 25; **7**, abaxial view of anther with hairs, × 25; **8**, young bud, × 8; **9**, flower from above (style removed) × 8. Drawn by D. Erasmus.

villous on margin and apex. Corona lobes violet, puberulous, trisegmented; central segments 3–4 mm long, filiform, puberulous, lateral segments fleshy ovoid-deltoid and fused to inner base of corolla lobes. Stamens ± 1 mm long. Style ± 0.5 mm long, style-head ± 1 mm long. Follicles horizontally divergent, cylindrical-ovoid, 6–12(–16) cm long, 4–5 mm diameter. Seeds black, 7–9 mm long, coma 2.5–3 cm long. Fig. 37, p. 119.

UGANDA. Karamoja District: Mt Moroto, 11 June 1970, *Katende & Lye* 392!; Kigezi District: Rubanda County, Echuya Forest Reserve, 25 Apr. 1970, *Katende* 218! & Kinkizi County, Bwindi Forest, Byumba, 15 Nov. 1997, *Eilu* 212!
KENYA. Naivasha District: Lake Naivasha, 21 Aug. 1972, *E. Polhill* 151!; Kiambu District: Escarpment, 18 Jan. 1970, *Kokwaro* 2190!; Kisumu-Londiani District: Tinderet Forest Reserve, 14 June 1949, *Maas Geesteranus* 4921!
TANZANIA. Arusha District: Ketumbeine Forest Reserve, 2 Apr. 2000, *Kindeketa et al.* 265!; Ufipa District: Mbizi Forest, 26 Nov. 1958, *Napper* 1088!; Njombe District: Livingstone Mts, E-facing slope of Ligala Mt, 15 Feb. 1991, *Gereau & Kayombo* 4057
DISTR. U 1–3; K 1, 3–6; T 2–4, 6–8; also known from South Sudan, Ethiopia, Congo-Kinshasa, Burundi, Malawi and Zambia
HAB. Bamboo forest, podocarp forest, riparian scrub forest, open dry montane forest and semi-deciduous dry forest; 700–2800 m
USES. The latex is fed to cows to enhance lactation; very strong twine, rope and fish nets are woven from the stem bark; the Masai brew a tea from the roots

SYN. *Periploca linearis* Hochst. in Flora 24, Intelligentblatt 2: 25 (1841), type as for *P. linearifolia*
   P. linearifolia Quart.-Dill. & A. Rich. var. *gracilis* Browicz in Arb. Korn. 11: 75 (1966). Type: South Sudan, Lado, Yei River, *Sillitoe* 260 (K!, holo.)
   P. refractifolia Gilli in Ann. Naturhist. Mus. Wien 77: 17 (1973). Type: Tanzania, Njombe District: Livingstone Mountains, Madunda, *Gilli* 396 (W!, holo.)

33. CRYPTOSTEGIA

R. Br., Bot. Reg. 5: t. 435 (1819); Marohasy & P.I. Forst. in Austr. Syst. Bot. 4: 571–577 (1991); Klackenberg in Adansonia, ser. 3, 23: 205–218 (2001)

Liana or scrambling sub-shrub with white latex; stems twining. Leaves leathery. Inflorescences terminal, few-flowered. Corolla pinkish to white, glabrous; tube campanulate to funnelform; lobes narrowly ovate to ovate, obtuse. Corolline corona of 5 entire or bifid lobes from about middle of corolla tube; each lobe with basal papillose pad above stamens. Gynostegial corona absent. Stamens from base of corolla tube; anthers narrowly ovate with acuminate apex, completely fertile, pollen in tetrads. Interstaminal nectar pockets at base of corolla tube. Style terete, style-head narrowly ovoid; translators of elliptically concave receptacle on sticky disc. Gynostegium concealed within corolla-tube. Follicles paired, widely divergent, narrowly ovoid, triangular, apex apiculate. Seeds oblong, with a hairy coma.

A genus of two species endemic to Madagascar. Introduced into Africa, probably as ornamental garden plant.

Cryptostegia madagascariensis *Decne.* in DC., Prodr. 8: 492 (1844); Marohasy & P.I. Forst. in Austr. Syst. Bot. 4: 571 (1991); Klackenberg in Adansonia, ser. 3, 23: 205–218 (2001); Thulin in Fl. Somalia 3: 142 (2006). Type: Madagascar, cult. in Mauritius, *Bojer* s.n. (G-DC!, holo.)

Liana or sub-shrub, several metres tall; stems sparsely lenticellate, interpetiolar ridges with hair-like colleters, axilliary colleters acicular, reddish. Leaf petiole 3–10 mm long; blade oblong-elliptic, elliptic, ovate or sub-orbicular, (2–)4–9 × 2–5 cm, acuminate, base round, glabrous or pubescent on both sides or only beneath.

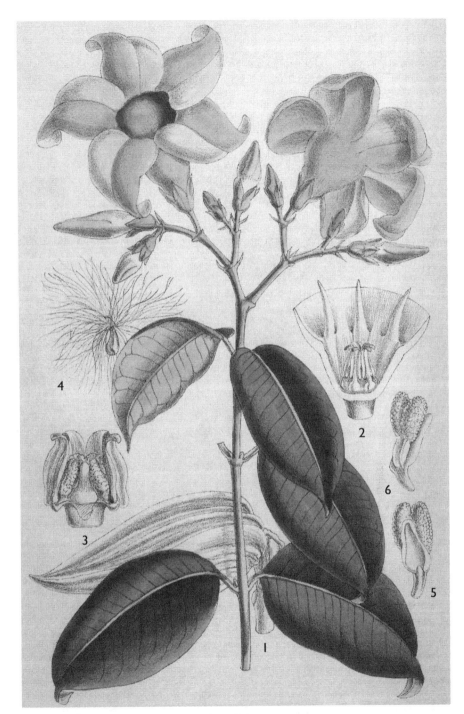

Fig. 38. *CRYPTOSTEGIA MADAGASCARIENSIS* — **1**, flowering shoot, with one of the two follicles shown behind; **2**, basal portion of corolla tube cut away to show corolline corona lobes and gynostegium; **3**, gynostegium with one stamen removed to expose stylar head; **4**, seed with coma of hairs; **5 & 6**, posterior and anterior views of translator with pollen tetrads. Reproduced from Bot. Mag. 130: t. 7984 (1904).

Inflorescences with peduncles stout, 8–10 mm long, glabrous; pedicels 6–8 mm long, pubescent; bracts narrowly triangular, 6–8 mm long, pubescent on outside. Flower buds ovoid with narrowly deltoid apex. Sepals narrowly ovate to narrowly triangular, 12–14 mm long, attenuate, pubescent on outside. Corolla pink, pubescent on outside, tube 15–20 mm long, lobes 20–40 mm long. Corona lobes acicular, entire, 6–9 mm long. Staminal filaments 2–3 mm long, anthers 3–4 mm long. Ovaries 2–3 mm long, style ± 1 mm long, style-head 3–4 mm long. Follicles 6–10 cm long, 2–4 cm diameter. Seeds brown, obliquely obovate to obliquely elliptic, 6–9 mm long, with black base, warty; coma white, 20–25 mm long. Fig. 38, p. 121.

KENYA. Nairobi District: Coryndon Museum Snake Park, 16 Mar. 1962, *Kirrika* 446; Mombasa District: Shanzu Beach, 28 Nov. 1974, *E. Polhill* 442!; Kwale District: El Capricchio Beach Cottages, 20 km S of Mombasa, 25 Dec. 1972, *Robertson & Greathead* 1807
TANZANIA. Lushoto District: Mangula, near Pentecostal Church, 23 Feb. 1983, *Kisena* 45 & Kizigu[?], cult. Amani, 29 Apr. 1947, *May* s.n.; Tanga District: Mtotohovu, 8 Dec. 1935, *Greenway* 4245
DISTR. **K** 4, 7; **T** 3, 6, 8; **Z**; native to Madagascar, and widely cultivated in tropical Africa
HAB. Gardens near the coast, escape in coastal grassland and woodland; altitude 0–100 m on the coast, ± 1600 m in Nairobi
USES. *Greenway* 4245 reports "originally introduced as a rubber vine and now found in most gardens on the coast as an ornamental shrub."

NOTE. *Cryptostegia madagascariensis* seems to have escaped into the wild in Kenya and Tanzania. *C. grandiflora* R. Br., a closely related species that differs most notably in its bifid corona lobes, is widely cultivated in other African countries and has become naturalised in Ethiopia. There are no records from the F.T.E.A. region, but this species may have been overlooked. U.O.P.Z.: 220 (1949) lists *C. grandiflora* occuring on Zanzibar – this dubious record probably derives from a single fruiting specimen (*Greenway* ADZ9) from Zanzibar originally named as *C. grandiflora*, which Klackenberg subsequently determined as *C.* cf. *madagascariensis*

34. **PARQUETINA**

Baill. in Bull. Soc. Linn. Paris 2: 806 (1889) & Hist. Pl. 10: 294 (1891); K. Schum. in E. & P. Pf. 4(2): 218 (1895); Venter in S. Afr. Journ. Bot. 75: 557–559 (2009)

Omphalogonus Baill. in Bull. Soc. Linn. Paris 2: 812 (1890); N.E. Br. in F.T.A. 4(1): 256 (1902)

Lianas with white latex; stems woody, twining. Leaves glabrous, petiolate, leathery. Inflorescences axillary, many-flowered, glabrous. Sepals glabrous, with colleters at their inner bases. Corolla coriaceous, deep crimson, purplish or pink. Corolline corona near base of corolla tube or on inversion of corolla tube, coronal feet fused to stamens and nectaries forming a collar. Stamens arising from inner base of coronal feet; pollen in tetrads. Gynostegial corona absent. Style terete; style-head deltoid, obtuse; translator with receptacle, stalk and sticky disc. Follicles paired, horizontally opposite.

A genus of two species, endemic to tropical Africa.

Note: Young stems, leaves, peduncles and flowers of both species turn black when drying. The flowers are so different in morphology that the two species were separated in different genera (Venter & Verhoeven in S. Afr. Journ. Bot. 62: 23–30 (1996)), but DNA-analysis proves them to be sister species (Ionta & Judd in Ann. Missouri Bot. Gard. 94: 372 (2007)).

Corolla lobes spreading, 6–7 mm long, inner surface uniformly
    pink to deep violet; gynostegium and corona lobes enclosed
    within the corolla tube; corona lobes fused along their length
    to corolla tube, channelled and hooded, ± 2 mm long . . . . . .   1. *P. calophylla*
Corolla tube and lobes reflexed; lobes 10–12 mm long, within
    velvety and violet at base turning brown towards apex;
    gynostegium and corona exserted from corolla mouth; corona
    lobes linear to filiform and segmented towards their apices,
    5–7 mm long . . . .   2. *P. nigrescens*

1. **Parquetina calophylla** (*Baill.*) *Venter* in S. Afr. Journ. Bot. 75: 558 (2009). Type:
Tanzania, Zanzibar, *Boivin* 1008A (P!, holo.)

Large liana; roots unknown; stems up to 20 m long, pale brown, glabrous, verrucose,
flaky. Leaf petiole (2–)5–7 cm long; blade broadly ovate, 9–12(–14) × 5–11 cm,
obtuse to cuspidate, base cordate, veins pale green above, dark green beneath.
Inflorescences 10–30-flowered; peduncles 1–3 cm long; bracts acicular, ± 2 mm long;
pedicels 4–5 mm long. Sepals sub-orbicular, 2 × 2 mm. Corolla glabrous outside; tube
campanulate, 5–6 mm long, inside of mouth papillate; lobes spreading, broadly
elliptic to broadly ovate, 6–7 × 6–7 mm, obtuse, inside papillate, pink, maroon, dark
red or violet to deep violet. Coronal collar borne in lower half of corolla tube; lobes
violet to maroon, 2 mm long and fused to corolla tube's inner face for ± 2 mm,
radiating into corolla tube cavity, radiating processes channelled, upper part
bisegmented, outer segment subulate to obtuse, inner segment concavely hood-
shaped. Stamens green; filaments terete, ± 0.5 mm long, glabrous; anthers narrowly
hastate-ovate, ± 0.5 mm long, pubescent to rarely glabrous on back. Interstaminal
nectaries sub-orbicular, erect around style. Style terete, ± 1 mm long, style-head ±
1 mm long; translators ± 1 mm long. Gynostegium concealed in corolla tube.
Follicles very narrowly deltoid-ellipsoid to deltoid-ovoid, 2-edged, obtuse-acute,
11–25 cm long, 2–4 cm diameter, glabrous. Seeds dark brown, narrowly elliptic to
obliquely narrowly elliptic, 6–8 mm long, reticulate to warty; coma 3–4 cm long.
Fig. 39, p. 124.

UGANDA. Mengo District: Entebbe, 10 Oct. 1902, *Mahon* 7!
KENYA. Northern Frontier/Tana River Districts: Garissa, 26 Dec. 1942, *Bally* 1995; Machakos
    District: Kibwezi, 15 Sep. 1961, *Polhill & Paulo* 463!; Tana River District: Tana River National
    Primate Reserve, Baomo Lodge road, 3 July 1988, *Medley* 364!
TANZANIA. Pare District: Muheza, Kuze Kibago Village, 29 June 2000, *Mwangoka & Haji* 1411!
    & Lutindi Forest Reserve, NW of Mkole Village, 11 May 1987, *Iversen et al.* 87418!; Lindi
    District: 32 km on Mnazimoja–Mtwara road, 2 Mar. 1991, *Bidgood, Abdallah & Vollesen* 1762!
DISTR. U 4; K 1, 4, 7; T 2, 3, 6–8; Z; Senegal and Mali to Central African Republic and South
    Sudan
HAB. Swampy forest, riverine forest and coastal thicket to grassland; 0–1150 m
USES. The roots are used as an aphrodisiac. The flossy cortical fibres are spun into fish nets and
    bow strings

SYN. *Omphalogonus calophyllus* Baill. in Bull. Soc. Linn. Paris 2: 812 (1890) & Hist. Pl. 10: 300
    (1891); N.E. Br. in F.T.A. 4(1): 256 (1902); T.T.C.L.: 67 (1949); Venter & R.L. Verh. in S.
    Afr. Journ. Bot. 62: 24 (1996)
    O. nigritanus N.E. Br. in K.B. 1912: 279 (1912); F.P.S. 2: 411 (1952). Type: Southern
    Nigeria, *A.S. Thomas* 1011 (K!, holo.)
    Periploca calophylla (Baill.) Roberty in Bull. Inst. Franc. Afr. Noire 15: 1429 (1953)
    Parquetina nigrescens sensu K.B. 15: 205 (1961), *non* (Afzel.) Bullock, partly

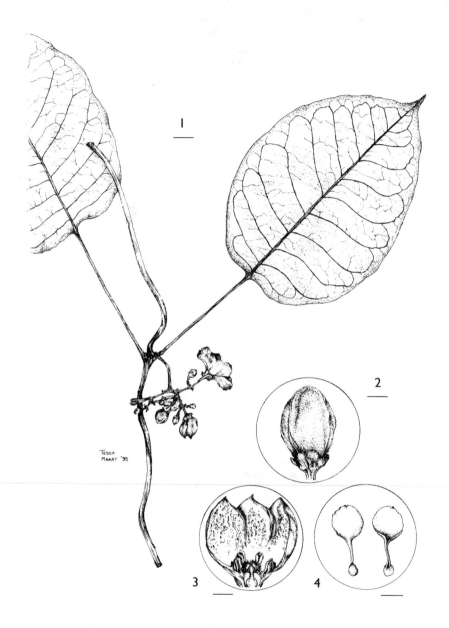

Fig. 39. *PARQUETINA CALOPHYLLA* — **1**, stem with leaves and flowers; **2**, bud; **3**, flower opened showing the corona lobes, hairy stamens and pistil; **4**, translators, posterior and anterior view. Scale bars: 1 = 10 mm, 2 & 3 = 2.5 mm, 4 = 0.3 mm. Reproduced by permission of the Editor, South African Journal of Botany 62: 26, fig. 3.

2. **Parquetina nigrescens** (*Afzel.*) *Bullock* in K.B. 15: 205 (1961); U.K.W.F. ed. 2: 175 (1994); K.T.S.L.: 495 (1994); Venter in S. Afr. Journ. Bot. 75: 558 (2009). Type: locality unknown, *anonymous* s.n. (BM!, holo.)

Liana with copious latex; roots tuberous; stems twining, 10 m or more long; twigs green to purple, older stems brown, rough, scaly. Leaf petiole 2–5(–10) cm long, purplish; blade mostly elliptic, sometimes broadly ovate to narrowly ovate, (7–)13–14(–16) × (3–)5–8(–11) cm, cuspidate or acuminate, base cordate, glossy, bright green above, pale green beneath, veins pale green to purplish beneath. Inflorescences 10–30-flowered, pale green, glabrous; peduncles 1–4 cm long; bracts acicular, 1–2 mm long; pedicels 2–5 mm long. Sepals broadly ovate, 1–2 mm long, 2 mm wide. Corolla pale green, creamy green or white outside, glabrous; tube 2–4 mm long, halfway reflexed, partly hispid; lobes elliptic to ovate, rounded, reflexed, 10–12 mm long, 4–5 mm wide, fleshy coriaceous, inside velvety and deep crimson, deep violet or black-violet at base turning brown to dark brown towards apex. Coronal collar borne at corolla inversion; lobes 5–7 mm long, filiform to linear, 2–4-segmented above middle with upper two segments tortuous, rarely undivided, greenish-white, pale green or pale yellow. Stamens green to brown with white hispid hairs on filaments and backs of anthers; filaments terete, 1–2 mm long; anthers narrowly ovate, 3–4 mm long. Style terete, 1–2 mm long; style-head deltoid, truncate, 3 mm long; translators 2.5–3 mm long. Gynostegium exserted from corolla. Follicles paired, horizontally divergent, cylindrical-ovoid, 12–21 cm long, 1–2 cm diameter, glabrous. Seeds brown, elliptic, 3–4 mm long, warty; coma white, 24–25 mm long.

UGANDA. Mengo District: Kipayo Estate, June 1914, *Dummer* 955!
DISTR. U 4; across West Africa, Cameroon, Gabon, Equatorial Guinea, Central African Republic, Congo-Kinshasa, Angola
HAB. Thickets, gallery forest; ± 1100 m
USES. An antidote against intestinal worms. Its latex is used for skin troubles

SYN. *Periploca nigrescens* Afzel., Stirpium Guinea 1: 2 (1817); Venter & R.L. Verh. in S. Afr. Journ. Bot. 262: 27 (1996)
  P. afzelii G. Don, Gen. Syst. 4: 163 (1837). Type: Sierra Leone, collector unknown (no type located, synonymy by N.E. Br. (1902))
  Parquetina gabonica Baill. in Bull. Soc. Linn. Paris 2: 806 (1889). Type: Gabon, *Duparquet* 1864, no. 1 (P!, holo.)
  Periploca preussii K. Schum. in E.J. 17: 117 (1893). Types: Angola, Cuanza Norte, Golungo Alto, Sobato de Mussengue, *Welwitsch* 4232 (K!, lecto., designated here, see note below; ?B†, BM!, COI!, MO!, isolecto.); Cameroon, between Mokonye and Kumba Ninga, *Preuss* 151 (B†, paralecto.); Gabon, Sibange Farm, *Soyaux* 406 (B†, K! paralecto.), *Buttner* 402 (B†, paralecto.); Lagoa Chinchoxa, *Soyaux* 46 (B†, paralecto.)
  P. gabonica (Baill.) A. Chev. in Rev. Bot. Appl. 31: 251 (1951)

NOTES. *Verdcourt* 3598A, a fruiting specimen from **K** 7 (Voi Hotel, Teita District), has been determined as this species, but in the absence of flowers its identity is questionable.
  Venter & Verhoeven in S. Afr. Journ. Bot. 62: 27 (1996) stated that *Preuss* 151 (B†) was the holotype of *Periploca preusii* K. Schum. This was in effect a lectotypification, as several other collections had been cited in the protologue. Extant duplicates of two of the original syntypes have now surfaced – indeed they were cited by Venter & Verhoeven (1996), but not recognised as syntypes at the time. We therefore propose to overturn the earlier lectotypification, and designate the Kew duplicate of *Welwitsch* 4232 as the lectotype of this name – duplicates of this collection are distributed in a number of other herbaria. D.J. Goyder & H.J.T. Venter.

35. **CRYPTOLEPIS**

R. Br., Asclepiadeae: 58 (1810); N.E. Br. in F.T.A. 4(1): 242 (1902) & in Fl. Cap.
4(1): 526 (1907)

Leposma Blume, Bijdr.: 1049 (1826)
Lepistoma Blume, Fl. Javae: 7 (1828), *nom. illeg.*
Cryptolobus Steud., Nom. Bot., ed. 2: 450 (1840), *non* Spreng. (1818)
Ectadiopsis Benth. in Benth. & Hook.f., Gen. Pl. 2: 741 (1876)

Mesophytic climbers or sclerophyllous shrubs, scramblers or small trees with white
or occasionally yellow or orange latex; interpetiolar ridges and leaf axils with reddish
dentate colleters. Leaves opposite on normal shoots, sub-fascicled on stunted shoots,
petiolate to sessile; blade herbaceous, leathery or succulent. Inflorescences few to
many-flowered or flowers solitary. Flower buds with corolla apex helically twisted and
deltoid to apiculate. Corolla tube campanulate and shorter than lobes, rarely
cylindrical and longer than lobes, mostly with elliptic papillose spot below every
corona lobe; lobes linear to ovate. Corolline corona 1- or 2-whorled; lower whorl of
lobes inserted around middle of corolla tube, mostly concealed; lobes clavate,
acicular, peg-like or filiform, glabrous, free; upper whorl, when present, of simple
pockets or lobed pockets in corolla lobe sinuses. Gynostegial corona absent. Stamens
inserted near base of corolla tube; filaments short, mostly dilated; anthers deltoid to
hastate, glabrous or rarely hairy; pollen in tetrads. Insterstaminal nectifery pouches
at base of corolla tube. Style-head deltoid, broadly deltoid, ovoid or broadly ovoid,
translators with receptacle concavely narrowly elliptic and sessile on sticky disc.
Gynostegium concealed at bottom of corolla tube. Follicles paired, widely to
narrowly divergent, cylindrical-ovoid to narrowly ovoid.

A genus of 27 species in Africa, 3 in Asia and 1 in Arabia.

1. Leaves violet spotted or blotched, beneath pale greyish-
   green or glaucous 2
   Leaves not violet spotted or blotched, beneath green or
   pale green 3
2. Leaf petiolate, lamina base obtuse to cuneate, pale greyish-
   green beneath; inflorescence bracts acicular; corolla
   creamy white, tube villous on inside near base; anthers
   villous . . . .   1. *C. africana*
   Leaves sessile with their bases cordate or obtuse, glaucous
   beneath; inflorescence bracts broadly ovate; corolla
   purplish-pink, glabrous inside; anthers glabrous . . . . . .   3. *C. hypoglauca*
3. Erect, procumbent or scrambling shrubs; flower buds
   with conical or narrowly conical apices 4
   Climbers; flower buds with apiculate to long-apiculate
   apices 5
4. Decumbent, erect or scrambling shrub up to 1.25 m high;
   corolla lobes ovate, 2–4 mm long, 1 mm wide . . . .   6. *C. oblongifolia*
   Erect to decumbent shrublet up to 0.8 m high; corolla
   lobes oblong-ovate, 4–8 mm long, 2 mm wide . . . .   8. *C. producta*
5. Leaf apices obtuse mucronate, glabrous . . . .   7. *C. obtusa*
   Leaf apices with apiculate or acuminate drip-tips 6
6. Stems, leaves and peduncles scabridulous to hirsute, or
   obtuse and scabridulous to hirsute . . . .   5. *C. microphylla*
   Stems, leaves and peduncles glabrous 7

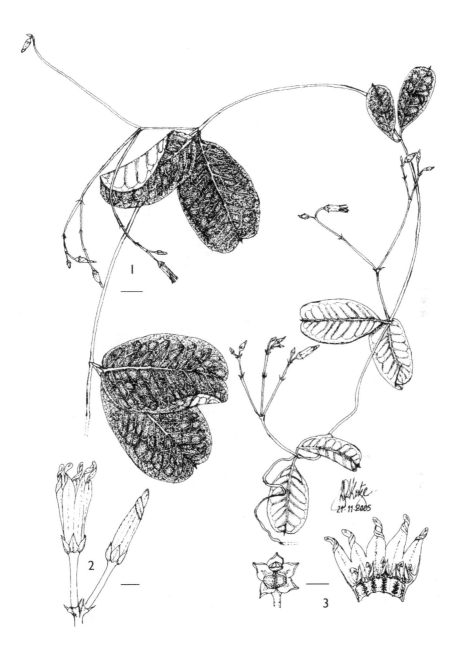

FIG. 40. *CRYPTOLEPIS AFRICANA* — **1**, stem with leaves and flowers; **2**, inflorescence; **3**, flower with corolla opened, calyx and pistil separated from it. Scale bars: 1 = 10 mm, 2 and 3 = 2 mm. 1 & 2 from *Luke & Mbinda* 5978; 3 from *Drummond & Helmsley* 3836. Reproduced by permission of the Editor, South African Journal of Botany 73: 41, fig. 2.

7. Milksap yellow-orange; inflorescence bracts oblong-ovate
    to narrowly elliptic, 5–7 mm long; corolla 2–3 cm long   9. *C. sanguinolenta*
    Milksap white; inflorescence bracts acicular to narrowly
    ovate, ± 1 mm long; corolla 0.7–2 cm long 8
8. Corolla 1.5–2 cm long, tube glabrous on inside; no
    corona in corolla lobe sinuses . . . .   2. *C. apiculata*
    Corolla 0.7–0.8 cm long, tube villous on inside; bifid
    corona lobes in corolla lobe sinuses . . . .   4. *C. ibayensis*

1. **Cryptolepis africana** (*Bullock*) *Venter & R.L. Verh.* in S. Afr. Journ. Bot. 73: 42
(2007). Type: Kenya, Kwale District: Buda Mafisini Forest, 13 km WSW of Gazi,
Drummond & Hemsley 3836 (K!, holo. (2 sheets and spirit collection); EA, iso.)

Climber 2–3 m high; roots unknown; stems slender, twining, brown, papillose to
minutely scabridulous, nodes enlarged. Leaves sub-coriaceous; petiole 0.5–1 cm
long; blade elliptic to oblong-elliptic, 6–8 × 4–5 cm, retuse-mucronate, mucro
reddish, base obtuse to sometimes cuneate, green above, beneath pale grey-green,
violet spotted and blotched, glabrous. Inflorescences axillary, few-flowered,
glabrous; peduncles slender, primary 1.5–2.5 cm long, secondary 2–3.5 cm long;
pedicels ± 5 mm long; bracts acicular, ± 1 mm long. Buds oblong-ovoid, long-
apiculate. Sepals narrowly triangular to narrowly ovate-triangular, ± 2 × 0.7 mm,
glabrous, acute. Corolla creamy-white; tube campanulate, 2–3 mm long, outside
glabrous, inside villous below stamens; lobes linear to linear ovate, 7–8 × ± 1 mm,
round to acute, glabrous. Corona of lower whorl only, lobes clavate, ± 0.5 mm long.
Stamens villous; filaments ± 0.5 mm long; anthers creamy yellow, ovate-triangular,
± 1 mm long. Styles free, terete, ± 1 mm long, glabrous; style-head broadly ovoid.
Fig. 40, p. 127.

KENYA. Kwale District: Buda Mafisini Forest, 13 km WSW of Gazi, 16 Aug. 1953, *Drummond &
    Hemsley* 3836!; Kilifi District: Kaya Ribe, 2 Sep. 1999, *Luke & Mbinda* 5978!
DISTR. **K** 7; not known elsewhere
HAB. Coastal lowland forest; 0–100 m

SYN. *Cryptolepis sinensis* Merr. subsp. *africana* Bullock in K.B. 10: 282 (1955)

2. **Cryptolepis apiculata** *K. Schum.* in P.O.A. C: 320 (1895); N.E. Br. in F.T.A. 4(1):
244 (1902); T.T.C.L.: 64 (1949); K.T.S.L.: 490 (1994). Type: Tanzania, Tanga District:
Amboni, *Holst* 2564 (B†, holo.; M!, lecto., designated here; K!, COI!, iso.)

Many-stemmed climber; roots unknown; stems twining, up to 3 m long, 2–4 mm
diameter, brown, glabrous, sparsely lenticellate. Leaves leathery; petiole 0.5–3 cm
long; blade narrowly elliptic or oblong-elliptic, (5–)7–13(–17) × (2–)3–5(–7) cm,
obtuse-attenuate, base obtuse to cuneate, glossy dark green above, pale green
beneath, glabrous. Inflorescences axillary, open, few-flowered, glabrous; peduncles
sturdy, primary 3–5 cm long, secondaries 3–6 cm long; pedicels sturdy, 0.8–1 cm
long; bracts narrowly ovate, ± 1 mm long. Bud ovoid, long-apiculate. Flowers
glabrous. Sepals narrowly ovate to narrowly triangular, 2–3 × 1 mm. Corolla white,
creamy yellow or pale greenish-yellow; tube campanulate, 3–5 mm long, 2–4 mm
diameter, with papillose spots; lobes linear, 12–16 × 1–2 mm, rounded. Corona: lower
whorl of clavate or acicular lobes, ± 1 mm long; upper whorl of sinus pockets.
Staminal filaments ± 1 mm long; anthers pale yellow, narrowly triangular, ± 1 mm
long, attenuate. Style terete, ± 0.5 mm long; style-head broadly angular-deltoid, ±
1 mm long, attenuate. Follicles paired, horizontally opposite, brown, cylindrical-
ellipsoid, 15–16 cm long, 0.8–1 cm diameter, attenuate, glabrous. Seeds obliquely
narrowly elliptic, 8 × 2 mm; coma white, 3–4 cm long.

KENYA. Kwale District: Kaya Bombo, 3 Jan. 1993, *Luke* 3490!; Kilifi District: Mangea Hill (Sita), 24 Mar. 1989, *Luke & Robertson* 1784! & Mida, June 1928, *Graham* 1526!
TANZANIA. Tanga District: Kange Sisal Estate, 23 Jan. 1952, *Faulkner* 240! & Pongwe Sisal Estate, 12 Apr. 1968, *Faulkner* 4097! & 8 km SE of Ngomeni, 29 July 1953, *Drummond & Hemsley* 3517!
DISTR. **K** 7; **T** 3; also recorded from Congo-Kinshasa, Mozambique and Zimbabwe
HAB. Lowland forest and shrubby thickets in grassland; 30–350 m

3. **Cryptolepis hypoglauca** *K. Schum.* in P.O.A. C: 320 (1895); T.T.C.L.: 64 (1949); K.T.S.L.: 491 (1994). Type: Tanzania, Tanga District: Amboni, *Holst* 2728 (B†, holo.; K!, lecto., designated here)

Many-stemmed climber with white latex; roots unknown; stems twining, pale brown to reddish brown, up to 6 m long, 2–5 mm diameter, glabrous, lenticellate. Leaves sessile or sub-sessile, leathery; blade elliptic, broadly elliptic, broadly obovate or sub-orbicular, (6–)8–10(–16) × (4–)5–8(–10) cm, retuse or obtuse, mucronate or apiculate, base cordate to obtuse, green above, glaucous beneath, violet spotted or blotched, glabrous. Inflorescences axillary, rarely terminal, open, monochasial branches 3–6-flowered, glabrous; peduncles sturdy, reddish brown, primary 2–3(–10) cm long, secondaries 1–3 cm long, densely packed with floral bracts and scars; bracts broadly ovate to triangular, clasping, ± 1 mm long; pedicels sturdy, reddish brown, 1–5 mm long. Buds sub-globose, short-apiculate. Sepals green, reddish blotched, narrowly ovate to narrowly triangular, ± 2 × 0.5 mm, glabrous, margins fimbriate. Corolla dull purple, brownish-red, pinkish-purple or pink, glabrous; tube broadly campanulate, 2.5–3 mm long, with papillose spots; lobes ovate to broadly ovate, 3–4 × 1–2.5 mm, acute to obtuse. Corona: lower whorl of pale green to yellow ovoid lobes with curved apex, ± 1 mm long; upper whorl of sinus pockets. Stamens glabrous, filaments ± 0.5 mm long, anthers whitish, triangular, ± 1mm long. Style ± 0.5 mm long; style-head conical, long-attenuate, ± 1.5 mm long. Follicles paired, horizontally opposite, dark reddish-brown, angular cylindrical-ovoid, 25–30 cm long, 5–8 mm diameter. Seeds brown, very narrowly sub-elliptic to very narrowly sub-ovate, 10 × 2 mm; coma white, 3–5 cm long.

KENYA. Meru District: Ngaia Forest Reserve, 6 May 2001, *Luke & Luke* 7434!; Kwale District: Shimoni, 20 Aug. 1953, *Drummond & Hemsley* 3917!; Kilifi District: Sabaki River, May 1960, *Rawlins* 892!
TANZANIA. Tanga District: Sawu, 9 July 1976, *Faulkner* 4914!; Ulanga District: Udzungwa Mountain National Park, Mwaya–Mwanihana Route, 5 Nov. 1997, *Luke & Luke* 4883!; Masasi District: Ndanda Mission, 19 March 1991, *Bidgood, Abdallah & Vollesen* 2086!; Zanzibar, Chwaka, 22 May 1961, *Faulkner* 2837!
DISTR. **K** 4, 7; **T** 3, 6-8; **Z**; Cameroon, Central African Republic and N Mozambique
HAB. Moist evergreen lowland and coastal forest on deep humic soil over coral, or sub-montane forest; 0–1100 m

4. **Cryptolepis ibayensis** *L. Joubert & Venter* in J.L.S. 160: 355 (2009), replacement name for *C. bifida* L. Joubert & Venter, *non* (Blume) P.I. Forst. Type: Tanzania, Pare District: Mkomazi Game Reserve, Ibaya stream, *Abdallah & Mboya* 3979 (K!, holo.; EA!, iso.)

A climber with white latex; roots unknown; stems slender, twining, purplish-brown, ± 4 mm diameter, verrucose, glabrous. Leaves leathery; petiole 5–6 mm long; blade recurved, elliptic, 5–7 × 2–3 cm, acuminate, cuspidate or drip-tipped, base truncate, green above, pale green beneath, glabrous. Inflorescences terminal and axillary, sub-compact, cymose with 3–7 compact monochasiums, few-flowered, glabrous; primary peduncles 1–2 cm long, secondary 1–2 cm long; pedicels ± 3 mm long; bracts densely packed, acicular, ± 1 mm long. Flower buds oblong, apiculate. Sepals free, ovate, ± 1 mm long, glabrous, rounded. Corolla yellowish-green; tube oblong-campanulate, ±

3 mm long, outside glabrous, inside villous below stamens; lobes linear to linear-ovate, 4–5 × ± 1 mm, rounded, glabrous. Corona: lower whorl with lobes clavate, ± 0.8 mm long; upper whorl of pockets with linear, deeply cleft lobes of ± 1 mm long from pocket rim. Stamens sub-sessile; anthers hastate, ± 0.5 mm long, attenuate, glabrous. Style terete, ± 0.3 mm long; style-head angular-deltoid, ± 0.5 mm diameter. Fruit and seed unknown.

KENYA. Machakos District: Mbuinzau [Mbinzao] Hill, 29 Jan. 1942, *Bally* 1749!
TANZANIA. Pare District: Mkomazi Game Reserve, Ibaya stream, 22 Jan. 1996, *Abdallah & Mboya* 3979!
DISTR. **K** 4; **T** 3; not known elsewhere
HAB. Upland forest; ± 1200 m

SYN. *Cryptolepis bifida* L. Joubert & Venter in J.L.S. 157: 344 (2008), *non* (Blume) P.I. Forst. (1993). Type as above

NOTE. Dimensions of floral parts may be incorrectly small as the two known collections on which the measurements were made have only buds, no mature flowers.

5. **Cryptolepis microphylla** *Baill.* in Bull. Soc. Linn. Paris 2: 804 (1889); N.E. Br. in F.T.A. 4(1): 245 (1902). Type: Angola, Queta Mountains, *Welwitsch* 5940 (P?, holo.; G!, lecto., designated here; BM!, K!, iso.)

Slender climber; roots unknown; stems twining, brown, up to 3 m long, 3–5 mm diameter, glabrous, scabrous or hirsute, lenticellate. Leaves herbaceous; petioles slender, 2–3 mm long, scabridulous to hirsute; blade oblong-ovate, oblong-elliptic, ovate or elliptic, 2–4(–7) × 0.7–1.5 cm, obtuse-mucronate, acuminate or long-attenuate, base obtuse, above bright green, sparsely scabridulous to hirsute, beneath pale green, scabridulous to hirsute, dense on prominent midrib. Inflorescences few-flowered, open, hirsute; peduncle slender, 1–1.2 cm long; bracts linear, 2–3 mm long; pedicels slender, 1.8–2 cm long. Buds ovoid, long-apiculate. Sepals very narrowly triangular to linear-ovate, 4–5 × 1 mm, attenuate, scabridulous to hirsute. Corolla creamy yellow, glabrous; tube campanulate, 2–3 mm long, ± 1 mm diameter; lobes linear, 10–12 × 1–2 mm, obtuse. Corona glabrous; lower whorl of peg-like lobes, acute, ± 1 mm long; upper whorl of sinus pockets. Stamens glabrous, filaments ± 0.5 mm long; anthers narrowly triangular, ± 1 mm long, apiculate. Style ± 0.5 mm long; style-head purplish-red, deltoid, ± 1 mm long. Follicles and seeds unknown.

TANZANIA. Tanga District: Kilulu Hill, 26 July 2001, *Luke & Chidzinga* 7546!; Rufiji District: Selous Game Reserve, source of Nahomba River, 10 Dec. 1978, *Vollesen* 4920!
DISTR. **T** 3, 8; Congo-Brazzaville, Congo-Kinshasa, Central African Republic, Angola
HAB. Lowland forest and riverine thicket; 250–650 m

6. **Cryptolepis oblongifolia** (*Meisn.*) *Schltr.* in J.B. 34: 315 (1896); N.E. Br. in F.T.A. 4(1): 248 (1902) & in Fl. Cap. 4(1): 529 (1907); T.T.C.L.: 64 (1949); F.P.S. 2: 406 (1952); Malaisse in Fl. Rwanda 3: 97 (1985); Burrows in Pl. Nyika Plateau: 70 (2005). Type: South Africa, KwaZulu-Natal, Umgeni, *Krauss* 132 (K–herb. Bentham!, lecto., designated here; K–herb. Hooker!, BM!, MO!, iso.)

Erect, procumbent or scrambling shrub of up to 1.25 m high; roots woody, hemicryptophytic; stems reddish-brown, up to 1.5 m long, 4–7 mm diameter, scabridulous, sometimes lenticellate. Leaves leathery, sessile or sub-sessile; blade linear-ovate to broadly ovate, narrowly obovate, obovate, linear elliptic or elliptic, (2–)4–9(–12) × (0.2–)1–1.5(–2.4) cm, obtuse-mucronate, acute or acuminate, base cuneate to obtuse, above dark green, beneath pale green to glaucous, glabrous. Inflorescences axillary and terminal, compact, few to many-flowered, glabrous; peduncles 1–15 mm long; bracts very narrowly ovate to very narrowly triangular, ± 1.5 mm long; pedicels 2–5 mm long. Buds ovoid with apex conical. Flowers sweet-

D.E.

FIG. 41. *CRYPTOLEPIS OBLONGIFOLIA* — **1**, flowering shoot, × ²/₃; **2**, fruiting shoot, × ²/₃; **3–6**, variation in leaf shape, × 1; **7**, inflorescence, × 4; **8**, half-flower, × 8; **9**, flower from above, × 4; **10 & 11**, stamen from beneath and from above, × 27; **12 & 13**, translators with pollen deposited, × 38. 1, 7–13 from *Thomas* 3909; 2 from *Purseglove* 2240; 3 & 5 from *Milne-Redhead & Taylor* 9760; 4 from *Richards* 195; 6 from *Michel* 4643. Drawn by D. Erasmus.

scented. Sepals ovate to broadly ovate, 1.5–2.5 mm long, 1 mm wide, green tipped purple, glabrous with margins fimbriate. Corolla creamy white to pale lemon-yellow inside, pale brownish-green to purplish outside, glabrous; tube campanulate, 1.5–2 mm long, with papillose spots; lobes oblong-ovate to obliquely ovate, obtuse, 2–4 × ± 1 mm, glabrous. Corona double: lower whorl of clavate lobes, ± 1 mm long; upper whorl of sinus pockets, pocket-rim sometimes lobed. Stamens glabrous, filaments ± 0.5 mm long; anthers hastate, ± 1 mm long, acuminate. Style terete, ± 0.5 mm long; style-head conical, obtuse to bifid, ± 1 mm long. Follicles single or paired, when paired ± 45° to 135° divergent, narrowly ovoid, (4–)8–12(–16) cm long, 5–10 mm diameter, ribbed, glabrous. Seeds dark brown, obliquely ovate, 2–4 mm long, warty; coma whitish, 3–4 cm long. Fig. 41, p. 131.

UGANDA. Mengo District: Wabusana, Bulemzi, 2 June 1935, *A.S. Thomas* 1269!; Mubende District: 10 km NW of Katera, 16 Mar. 1969, *Lye* 2310!; Ankole District: Bunyaruguru, Kasunju Hill, 5 Jan. 1953, *Osmaston* 2762!
KENYA. Trans-Nzoia District: Kitale, May 1964, *Tweedie* 2802!; North Kavirondo District: Kakamega Forest, 7 Jan. 1968, *Perdue & Kibuwa* 9484!; South Kavirondo District: Kisii, Bana, Sep. 1933, *Napier* 5333!
TANZANIA. Ufipa District: Sakalilo, 1 Dec. 1954, *Richards* 3504!; Iringa District: Mafinga, 20 Oct. 1979, *Mwasumbi* 11938; Masasi District: 29 km from Newala on Masasi road, 6 Mar. 1963, *Richards* 17756!
DISTR. U 1–4; K 3–5; T 4, 6–8; widely distributed across tropical and southern Africa
HAB. Open rocky miombo woodland, *Uapaca* woodland on grey sandy loam, riverbank woodland, grassland; 800–2300 m
USES. Chimpanzees eat this plant apparently when feeling unwell. The woody roots are used as aphrodisiac; a root decoction is drunk for coughs and tuberculosis

SYN. *Ectadium oblongifolium* Meisn. in Hook. in London Journ. Bot. 2: 542 (1843)
  Ectadiopsis oblongifolia (Meisn.) B.D. Jacks. in Ind. Kew. 1(2): 822 (1893)
  Ectadiopsis oblongifolia (Meisn.) Schltr. in E.J. 20, Beibl. 51: 10 (1895), *nom. illeg.*; U.K.W.F. ed. 2: 174 (1994)
  Cryptolepis elliottii Schltr. in J.B. 33: 300 (1895). Type: Burundi, Urundu Hills, *Scott Elliot* 8372 (BM!, holo.; K!, iso.)
  Ectadiopsis suffruticosa K. Schum. in E.J. 28: 453 (1900). Type: Tanzania, Iringa, Uhehe, Weru Area, *Goetze* 665 (B†, holo.)
  Cryptolepis suffruticosa (K. Schum.) N.E. Br. in F.T.A. 4(1): 251 (1902); T.T.C.L.: 64 (1949)
  C. buxifolia Chiov., Racc. Bot. Miss. Consol. Kenya: 80 (1935). Type: Kenya, Mt Kenya NE, Meru, *Balbo* 16 (FT!, lecto., designated here; FT!, iso.)

7. **Cryptolepis obtusa** *N.E. Br.* in K.B. 1895: 110 (Apr./May 1895) & in F.T.A. 4(1): 246 (1902) & in Fl. Cap. 4(1): 528 (1907); K.T.S.L.: 491 (1994). Types: Mozambique, between Tete and sea coast, Mar. 1860, *Kirk* s.n. (K!, lecto., designated by Bullock in K.B. 10: 281 (1955))

Climber with white latex; roots unknown; stems slender, twining, brown, 3–5 m long, 2–4 mm diameter, glabrous. Leaves herbaceous; petioles 4–10 mm long; blade oblong-elliptic, oblong-obovate, elliptic or obovate, 3–7(–9) × 0.2–0.4 cm, obtuse-mucronate, mucro 1–2 mm long, base truncate, bright green above, pale green beneath, glabrous. Inflorescences axillary, lax to sub-compact, with 3–7 monochasia, few to many-flowered, glabrous; peduncles frail, primary 5–15 mm long, secondaries 5–10 mm long; pedicels 2–4 mm long; bracts acicular, ± 1 mm long. Buds ovoid to ellipsoid, apiculate. Flowers glabrous. Sepals broadly ovate, 1.5–2 × 1–1.5 mm. Corolla creamy yellow; tube campanulate, 1.5–2 mm long, with papillose spots; lobes linear to linear-ovate, 6–8 × ± 1 mm, obtuse. Corona double; lower whorl of clavate lobes, ± 1 mm long; upper whorl of sinus pockets. Stamens sub-sessile; anthers hastate, ± 0.8 mm long, attenuate, puberulous. Style terete, ± 0.5 mm long; style-head broadly conical, 0.5–1 mm diameter. Follicles horizontally opposite, cylindrical-ovoid, 8–15 cm long, 4–7 mm diameter. Seeds dark brown, obliquely narrowly ovate, 5–7 mm long; coma 30–40 mm long. Fig. 42, p. 133.

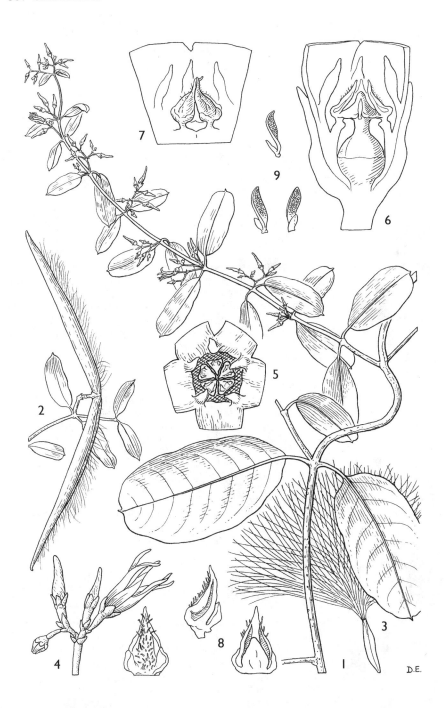

Fig. 42 *CRYPTOLEPIS OBTUSA* — **1**, flowering shoot, × ²/₃; **2**, fruiting shoot, × ²/₃; **3**, seed with coma of hairs, × 2; **4**, inflorescence, × 3; **5**, centre of flower from above, × 8; **6**, half-flower, × 12; **7**, portion of flower showing gynostegium and 3 of the 5 corolline corona lobes, × 12; **8**, stamens in dorsal, lateral and ventral view, × 20; **9**, translators, × 27. 1, 4–9 from *Balsinhas & Marrime* 385; 2 & 3 from *Faulkner* Kew 460. Drawn by D. Erasmus.

TANZANIA. Uzaramo District: Dar es Salaam, 20 Aug. 1969, *Harris* 3150!; Rungwe District: Lufiro River, 21 May 1957, *Richards* 9858!; Kilwa District: Mwera, 22 July 1970, *Ruffo* 388!
DISTR. **T** 6–8; Zambia, Malawi, Mozambique, Zimbabwe and South Africa
HAB. Open savanna, grassland, wasteland and low shrubby vegetation; 0–500 m
USES. Medicinal plant

SYN. *Cryptolobus obtusa* K. Schum. in P.O.A. C: 320 & 424 (Aug. 1895). Type: East Africa, *Stuhlmann* 7827 (B†?). Synonymy after N.E. Br.: 246 (1902)

8. **Cryptolepis producta** *N.E. Br.* in F.T.A. 4(1): 247 (1902). Type: Angola, Amboella, Kubango River above Kinimarva, *Baum* 457 (K!, holo.; BM!, W!, iso.)

Shrublet of 0.3–0.8 m high with milky latex; woodstock perennial; stems erect, spreading or decumbent, brown, 3–5 mm diameter, scabridulous, sometimes lenticellate. Leaves leathery; petiole 1–5 mm long, scabridulous; blade sub-globose to narrowly ovate or oblong-ovate, 2–5(–7) × (0.5–)1–2(–3) cm, obtuse-mucronate, acute or acuminate, base obtuse to cuneate, above dark green, beneath pale green, scabridulous or scabrous all over or on veins only. Inflorescences axillary and terminal, compact or sometimes lax, up to 10-flowered, scabrous; peduncles 0.2–1(–4) cm long; pedicels 2–6 mm long; bracts acicular, ± 1 mm long. Buds narrowly ovoid, apex narrowly conical. Sepals ovate to narrowly triangular, 1.5–2 × ± 1 mm, glabrous or scabridulous, margins fimbriate. Corolla dull yellow or creamy yellow on inside, yellow tinged red or purplish to brownish-purple on outside, glabrous; tube campanulate, 3–4 mm long, with papillose spots below corona lobes; lobes oblong-ovate, (4–)6–8 × 1.5–2 mm. Corona double, glabrous; lower whorl of clavate, ± 1 mm long lobes; upper of sinus pockets with acicular or fan-shaped lobes. Stamens fused with papillose spots; filaments 0–0.5 mm long; anthers hastate, attenuate, 1.5–2 mm long and slightly hairy. Style ± 0.5 mm long, terete; style-head broadly conical, bifid, ± 1 mm diameter. Follicles single or paired, when paired ± 45° to 180° divergent, narrowly ovoid-conical to cigar-shaped, 5–8 cm long, 1–1.3 cm diameter, ribbed, acute to blunt-acute, glabrous. Seeds dark brown, ovate to oblong-ovate, 5–7 mm long, warty; coma 2–3 cm long.

TANZANIA. Ufipa District: Sumbawanga, 10 Nov. 1956, *Richards* 6960!; Rungwe District: Ulambaya, Mlale, 25 Oct. 1976, *Leedal* 3908!; Songea District: Songea Airfield, 4 Apr. 1956, *Milne-Redhead & Taylor* 8125!
DISTR. **T** 4, 7–8; Congo-Kinshasa, Burundi, Angola, Zambia, Malawi, Zimbabwe and Botswana
HAB. Sandy dry grassland with scattered bushes, miombo woodland; common on disturbed ground; 1000–2400 m

SYN. *Ectadiopsis producta* (N.E. Br.) Bullock in K.B. 10: 278 (1955)

9. **Cryptolepis sanguinolenta** (*Lindl.*) *Schltr.* in Westafr. Kautschuk-Exped.: 308 (1900). Type: Sierra Leone, 1822, *Don* s.n. (K!, holo.)

Climber with yellow-orange to orange latex; roots unknown; stems twining, dark brown, up to 15 m long, glabrous, lenticellate. Leaves sub-leathery; petiole 5–7 mm long; blade narrowly ovate, ovate or narrowly elliptic to elliptic, (4–)6–9.5 × 2–3.5 cm, acuminate to long attenuate, base rounded, deep green above, paler green beneath, glabrous, midvein prominent beneath. Inflorescences axillary or terminal, few-flowered, lax, glabrous; peduncles 2–3.5 cm long, slender; pedicels 1–2 cm long, slender; bracts oblong-ovate to narrowly elliptic, 5–7 × 1–2 mm, obtuse. Bud narrowly ovoid, long-apiculate. Flower glabrous. Sepals narrowly ovate to ovate, 3–6 × 1–2 mm wide, obtuse, glabrous. Corolla greenish-yellow, yellow or yellow-orange; tube narrowly campanulate, (2–)4–6 mm long, 1–2 mm wide, with papillose spots; lobes linear or oblong ovate, (11–)20–27 × 1–3 mm, obtuse. Corona sometimes double;

lower whorl of clavate or arrow-like lobes, 1–2 mm long, fused to papillose spots; upper lobes, when present, peg-like from sinus pocket rims. Staminal filaments ± 0.5 mm long; anthers hastate, 1–2 mm long, acuminate. Style ± 0.5 mm long, terete; style-head broadly conical, acute, ± 1 mm long. Follicles paired, horizontally opposite, cylindrically ovoid, attenuate, 19–24 cm long, 4–6 mm diameter. Seeds brown, obliquely oblong to obliquely narrowly ovate, 8–10 mm long, 2–3 mm wide; coma white, 2–2.5 cm long.

UGANDA. Toro District: Kibale Forest, Ngogo, 15 June 1997, *Eilu* 112!; Kigezi District: Kinkizi County, Bwindi Forest, Isasha Gorge, 17 Feb. 1997, *Eilu* 255!; Mengo District: Nakasongola, Buruli, 4 Sep. 1943, *A.S. Thomas* 3763!
TANZANIA. Kigoma District: Tubira Forest, 1 Apr. 1994, *Bidgood & Vollesen* 3018! & Gombe Stream Reserve, 6 May 1992, *Mbago & Lyanga* 1115!; Buha District: Kasakela Reserve, Melinda Stream, 19 Nov. 1962, *Verdcourt* 3372!
DISTR. U 2, 4; T 4; West Africa, Cameroon, Central African Republic, Congo-Kinshasa and Angola
HAB. Moist semi-deciduous or evergreen forest, thicket woodland, lake-beach and stream bank forest; 500–1800 m
USES. Extract administered as antidote to malaria

SYN. *Pergularia sanguinolenta* Lindl. in Bot. Mag. 52: t. 2532 (1825)
    Cryptolepis triangularis N.E. Br. in J.L.S. 30: 92 (1894) & in F.T.A. 4(1): 245 (1902). Types: Nigeria, Nupe, *Barter* 1333 (K!, lecto., designated here, P! isolecto.); Nigeria, Abbeohuta, *Barter* 3359 (K!, L! (Herb. Lugd. Bat. 908-335-719) syn.); Angola, *Welwitsch* 5993 (BM!, BR!, COI!, G!, K!, syn.)
    Strophanthus radcliffei S. Moore in J.L.S. 37: 180 (1906). Type: Uganda, Coast of Victoria Nyanza, *Bagshawe* 589 (BM!, holo.)

36. SACLEUXIA

Baill., Hist. Pl. 10: 265 (1890); K. Schum. in E. & P. Pf. 4(2): 226 (1895) & in E.J. 23: 232 (1896); Bullock in K.B. 15: 393 (1962)

Gymnolaema Benth. in G.P. 2: 740 (1876); K. Schum. in E. & P. Pf. 4(2): 211 (1895); N.E. Br. in F.T.A. 4(1): 241 (1902), non *Gymnoleima* Decne. (1844)
Macropelma K. Schum. in P.O.A. C: 321 (1895)

Virgate, erect, evergreen, shrubs; latex white or clear; tubers numerous on lateral roots. Leaves opposite. Inflorescences axillary. Corolla broadly campanulate, glabrous inside, glabrous or pubescent outside. Corolline corona arising at base of corolla tube; coronal feet columnar, fleshy, fused laterally with interstaminal nectaries into coronal annulus; upper part of lobes, when present, oblong or narrowly ovate with bifid apices or tiara-like. Gynostegial corona absent. Stamens sessile or sub-sessile from inner apex of coronal feet, anthers fully fertile, pollen in tetrads. Interstaminal nectaries opposite petals, lobular from rim of coronal annulus or absent. Gynostegium exposed above corolla mouth. Style-head sub-sessile on ovaries; translator receptacle broadly ovate to broadly angular-ovate, stype cuneate. Follicles paired, divergent.

A small genus of two species, endemic to the F.T.E.A. region. Although readily distinguished morphologically, a molecular marker could not be found to support this.

Inside of corolla lobes yellow, creamy yellow or white; primary peduncles slender and usually long, (1–)4–13 cm long . . . .   1. *S. newii*
Inside of corolla lobes deep purple-brown or crimson; primary peduncles stout and short, 0.2–0.4 cm long . . . .   2. *S. tuberosa*

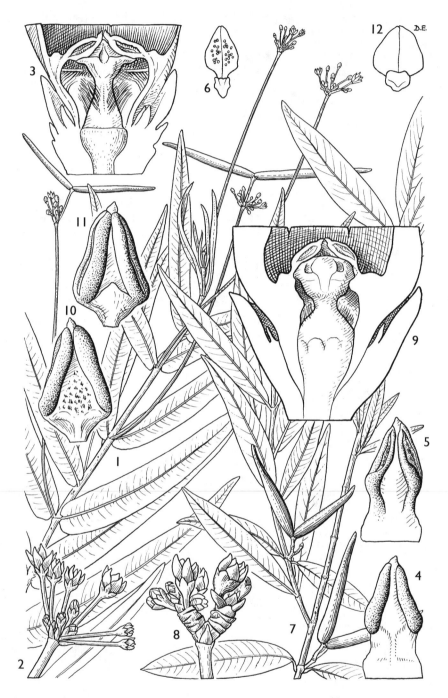

FIG. 43. *SACLEUXIA NEWII* — **1**, habit, × ²⁄₃; **2**, inflorescence, × 2; **3**, half-flower, × 16; **4 & 5**, stamens in dorsal and ventral view, × 37; **6**, translator, × 37. *SACLEUXIA TUBEROSA* — **7**, habit, × ²⁄₃; **8**, inflorescence, × 2; **9**, half-flower, × 16; **10 & 11**, stamens in dorsal and ventral view, × 37; **12**, translator × 37. 1–6 from *Drummond & Hemsley* 3364; 7–12 from *Tanner* 360. Drawn by D. Erasmus.

1. **Sacleuxia newii** (*Benth.*) *Bullock* in K.B. 15: 394 (1962); U.K.W.F. ed. 2: 174 (1994); K.T.S.L.: 496 (1994). Type: Tanzania, Kilimanjaro, Aug. 1871, *New* s.n. (K!, holo.)

Erect or straggling shrub of up to 3 m high; latex white; root tubers sub-spherical, 3–15 cm diameter; stems purplish-brown, glabrous to sparsely puberulous. Leaves sessile to sub-sessile, petiole 0–3 mm long; blade narrowly ovate or oblong ovate, (3–)6–11(–16) × 1.1–2 cm, attenuate, base cordate; or leaves less often with petiole 1–7 mm long and fluted and blade sometimes falcate, linear to narrowly lanceolate, (6–)11–18 × 0.2–1.6 cm, attenuate, base cuneate to obtuse; young leaves pubescent, mature leaves papery to herbaceous, above mid-green, glossy, glabrous with impressed pubescent crimson midrib, beneath pale green, pubescent, puberulous or glabrous with main vein pubescent. Inflorescences puberulous; peduncles slender, (1–)4–13 cm long, with fascicle of flowers at apex; bracts ovate, 1–2 mm long; pedicels slender, 2–5 mm long. Sepals triangular, ± 1.5 × 1 mm, obtuse, sub-glabrous. Corolla yellow or creamy yellow, glabrous to sparsely pubescent on outside; tube ± 1 mm long; lobes ovate, 1.2–2 × ± 1 mm, obtuse. Corona lobes, when present, creamy yellow, narrowly ovate or oblong, apices bifid or not, ± 0.6 mm long. Anthers whitish to creamy yellow, narrowly triangular, ± 0.5 mm long, attenuate, full length fertile. Ovaries glabrous or puberulous; style terete, ± 0.5 mm long; style-head very broadly ovoid, blunt, ± 0.5 mm long, ± 0.8 mm diameter. Follicles horizontally divergent, cylindrical-ovoid, sinuate, 4–6 cm long, 3–4 mm diameter, long-attenuate, glabrous or puberulent. Seeds obliquely narrowly obovate, 4–7 mm long; coma 1–2.5 cm long. Fig. 43/1–6, p. 136.

KENYA. Machakos District: Tsavo National Park East, Mzinga Hill, 19 Apr. 1966, *Gillett* 17237! & Chyulu Plains, Soitpus Hill, 30 July 2000, *Luke & Luke* 6408!; Teita District: E slopes of Mt Kasigau, pipeline path from Makwasingyi, 13 Feb. 1996, *Goyder et al.* 4030!
TANZANIA. Lushoto District: Usambaras, Mt Bomolo, Amani, 10 May 1950, *Verdcourt* 188!; Handeni District: Kwa Mkono, 5 May 1981, *Archbold* 2842!; Kilosa District: Magubike Village, 5 Feb. 1988, *Pócs* SAU–88012!
DISTR. **K** 4, 6, 7; **T** 3, 6; not known elsewhere
HAB. Miombo woodland or shrubby grassland; always associated with rock crevices, rocky ledges, rocky hilltops, rock slabs, sometimes locally common; 600–1800 m

SYN. *Gymnolaema newii* Benth. in Hooker's Icon. Pl. 12: t. 1186 (1876); T.T.C.L.: 66 (1949)
 Sacleuxia salicina Baill., Hist. pl. 10: 265 (1891). Type: Tanzania, Morogoro District [Zanzibar], Nguru, *Sacleux* 758 (P!, holo.)
 Macropelma angustifolia K. Schum. in P.O.A. C: 321 (1895). Type: Tanzania, Handeni District: Kwa Mberue [Merue], *Fischer* 383 (B†, holo.). Synonymy based on Schumann's description

2. **Sacleuxia tuberosa** (*E.A. Bruce*) *Bullock* in K.B. 15: 394 (1962); U.K.W.F. ed. 2: 174 (1994); K.T.S.L.: 496 (1994). Type: Tanzania, Mwanza District: Lake Victoria, coast of Speke Gulf, *B.D. Burtt* 2475 (K!, holo., K!, iso.)

A virgate evergreen shrub of up to 4 m high; latex white or colourless; root tubers potato-like; stems erect, spindly, sparsely branched; bark purplish to dark brown, puberulent to sparsely puberulent. Leaves in clusters at apices of stems, petiolate; petiole 2–10 mm long, puberulent above; blade coriaceous, narrowly ovate, 8–9(–14) × 1–2.5 cm, acute to attenuate, base obtuse to cuneate; above dark green, glossy, glabrous but puberulent on midrib and margin, beneath dull green, tinged magenta, puberulous, midrib and secondary veins prominent. Inflorescences terminal and axillary, pubescent to puberulent, many-flowered, flowers densely clustered; peduncles sturdy, 2–4 mm long; pedicels 2–3 mm long, bracts widely triangular-ovate, 1 mm long. Sepals broadly triangular-ovate, ± 1 mm long. Corolla tube ± 1 mm long; lobes inside deep purple-brown or crimson, broadly triangular, 0.6–1.1 × 0.7–0.8 mm, outside green, pubescent. Coronal annulus with nectary lobes triangular, ± 0.2 mm long, with pouches between corona and corolla. Anthers triangular, ± 1 mm long, apices attenuate. Ovaries ± 1 mm long, puberulous; style ± 0.5 mm long, style-head ovoid, ± 0.5 mm long, bluntly bifid. Translator receptacle broadly ovate. Follicles parallel to 90° divergent, narrowly ovoid, 3.5–5 cm long, 7–10 mm diameter, apiculate, puberulent. Seeds brown, cuneately obovate, 4–5 mm long; coma white, 8–10 mm long. Fig. 43/7–12, p. 136.

KENYA. Naivasha District: Njoroa Gorge, Hell's Gate Steam Jets, 21 Oct. 1962, *Verdcourt* 3289! & Ol Longonot Estate, 9 Dec. 1961, *Kerfoot* 3395!; Masai District: Mt Suswa, 7 Oct. 1962, *Glover & Samuel* 3309!
TANZANIA. Mwanza District: Barika, 19 July 1951, *Tanner* 360! & Mwanza, Juma Island, Jan. 1953, *Procter* 136!; Masai District: 80 km W of Endulen, 27 July 1957, *Bally* 11605!
DISTR. **K** 3, 6; **T** 1, 2; not known elsewhere
HAB. Miombo woodland and grassland in crevices or on shallow soils of rocky outcrops, sometimes locally common; 1200–2000 m
USES. The Wasukuma use the branches for arrow shafts. The tubers are pounded up green and this mash is then rubbed into scabies.

SYN. *Gymnolaema tuberosa* E.A. Bruce in K.B. 1934: 304 (1934); T.T.C.L.: 66 (1949)

37. CHLOROCYATHUS

Oliv. in Hooker's Icon. Pl. 16: t. 1557 (1887); Venter in S. Afr. Journ. Bot. 74: 288–294 (2008)

Kappia Venter, A.P. Dold & R.L. Verh. in S. Afr. Journ. Bot. 72: 530 (2006)

Perennial liana with white latex; roots branched, tubers numerous; stems twining, nodes swollen. Interpetiolar stipules fleshy, sub-spherical, dentate. Leaves opposite, petiolate, margins undulate. Inflorescences many-flowered dichasia and/or monochasia. Corolla greenish, semi-succulent. Corolline corona inserted in corolla mouth, 5-lobed; lobes broadly obcordate or obtriangular, laterally fused pocket-like to corolla lobes. Gynostegial corona absent. Stamens inserted directly beneath corona lobes and fused to their inner bases; pollen in tetrads, grains 4–16-porate. Interstaminal nectaries pocket/disc-like near base of corolla tube. Gynostegium exposed above corolla mouth. Follicles paired, divergent, narrowly ovoid or narrowly ellipsoid. Seeds narrowly ovate to oblong-ovate, flattened, concavo-convex, with distal coma of hairs.

A genus of two species, endemic to Africa.

Chlorocyathus monteiroae *Oliv.* in Hooker's Icon. Pl. 16: t. 1557 (1887); Venter in S. Afr. Journ. Bot. 74: 290 (2008). Type: Mozambique, Maputo [Delagoa Bay], 1882, *Monteiro* s.n. (K!, holo.)

Perennial climber of up to 8 m high, few-stemmed; tubers numerous from lateral roots, globoid to cylindrical-ovoid, 10–30 cm long, 3–7 cm diameter; aerial stems woody, twining, branching lateral, pubescent. Leaf petiole 1–10 mm long; blade semi-succulent, narrowly ovate, ovate, elliptic, narrowly obovate, obovate or suborbicular, 1–6 × 0.5–3 cm, acute to obtuse, base cuneate to obtuse, above dark green, glossy, sparsely pubescent, beneath pale green, pubescent. Inflorescences raceme-like, 2–5-flowered, pubescent; peduncles 0.5–2 cm long; bracts ovate, 1–3 mm long, pubescent; pedicels 0.2–1 cm long. Sepals ovate, 2–4 × 1–2 mm, acute, outside pubescent. Corolla bright green to light green, outside pubescent; tube cylindrical-campanulate, 4–6 mm long; lobes spreading, ovate to narrowly ovate, 6–10 × 3–5 mm, obtuse to acute. Corona lobes trisegmented; foot obtriangular, laterally fused to corolla lobes, succulent; central segment creamy white to white,

FIG. 44. *CHLOROCYATHUS MONTEIROAE* — **1**, flowering stem; **2**, flower; **3**, corolla opened showing corona lobes, stamens and style with style-head. Scale bars: 1 = 10 mm, 2 & 3 = 2 mm. All from *Faulkner* 1479. Reproduced by permission of the Editor, South African Journal of Botany 74: 290, fig. 2.

filiform to subulate, 2–3 mm long; lateral segments pinkish to purplish, ± 1 mm long, corniculate. Staminal filaments filiform, 1–2 mm long; anthers narrowly ovate, 2–3 mm long, acute. Style terete, 2–4 mm long; style-head ovoid, obtuse. Follicles horizontally paired, very narrowly ovoid, tapering, 7–11 cm long, 7 mm diameter. Seeds narrowly ovate or rhomboid, 7 mm long; coma white, 3 cm long. Fig. 44, p. 139.

KENYA. Kwale District: 8 km E of Mackinnon road, 9 Sep. 1953, *Drummond & Hemsley* 4227! & Mwachi Forest Reserve, 16 May 1990, *Robertson & Luke* 6167! & cult. ex Malindi Road, 13 Dec. 1975, *Bock* H 16047
TANZANIA. Handeni District: Handeni, 1 Oct. 1954, *Faulkner* 1479!
DISTR. **K** 7; **T** 3; Angola, Mozambique, Zimbabwe, Namibia and South Africa
HAB. Succulent thicket in dry thorn bushland; 300–800 m
NOTE. *R. monteiroae* is uncommon in Kenya and Tanzania and its flowers are smaller than in the southern Africa specimens.

SYN. *Raphionacme loandae* Schltr. & Rendle in Hiern, Cat. Afr. Pl. Welw. 1: 679 (1898). Type: Angola, Boa Vista, *Welwitsch* 4274 (BM!, holo.)
    R. monteiroae (Oliv.) N.E. Br. in Fl. Cap. 4(1): 533 (1907)

38. **TACAZZEA**

Decne. in DC in Prodr. 8: 492 (1844); Benth. in Gen. pl. 2: 745 (1876); K. Schum. in E. & P. Pf. 4(2): 215 (1895); N.E. Br. in F.T.A. 4(1): 260 (1902) & in Fl. Cap. 4(1): 540 (1907); Bullock in K.B. 9: 350 (1954); H. Huber in Prodr. Fl. SW.-Afr. 113: 7 (1967); Venter *et al.* in S. Afr. Journ. Bot. 56: 95 (1990)

Large lianas or erect shrubs with white latex; stems woody, bark often verrucose. Leaves mostly opposite, sometimes alternate, rarely whorled, petiolate, interpetiolar ridges with reddish colleters. Inflorescences terminal and sub-terminal cymes with dichasial and monochasial branches, many-flowered, bracteate. Sepals with ciliate margin, with pairs of colleters at inner bases. Corolla rotate, tube saucer-shaped, lobes ovate or oblong. Corolline corona arising from corolla mouth, coronal feet, staminal filaments and nectaries fused into a wavy annulus, lobes filiform or narrowly ovate. Gynostegial corona absent. Stamens from inner apex of coronal feet, glabrous; filaments free, anthers glabrous, pollen in tetrads. Interstaminal nectaries lobular, cone-shaped around style. Style-head broadly ovoid, translators spatula-like. Gynostegium exposed above corolla. Follicles paired, divergent, cylindrical-ovoid to ovoid.

Leaf with apex obtuse-acuminate to obtuse or emarginate, mucronate and base rounded to cordate; inflorescences with peduncles and pedicels slender and frail; follicles 180° divergent, narrowly ovoid with apex long-apiculate, 3–9 cm long . . . . 1. *T. apiculata*
Leaf with apex acute to acuminate and base cuneate or obtuse-tapering; inflorescences with peduncles and pedicels sturdy; follicles slightly divergent, cylindrical-ovoid with apices obtuse-acute, 7–20 cm long . . . . 2. *T. conferta*

1. **Tacazzea apiculata** *Oliv.* in Trans. Linn. Soc., Bot. 29: 108 (1875); K. Schum. in P.O.A. C: 320 (1895); N.E. Br. in F.T.A. 4(1): 267 (1902); T.T.C.L.: 68 (1949); F.P.S. 2: 418 (1952); Malaisse in Fl. Rwanda 3: 117 (1985); U.K.W.F. ed. 2: 175 (1994); K.T.S.L.: 498 (1994); Venter in Fl. Eth. 4(1): 103 (2003) & in Fl. Somalia 3: 141 (2006). Type: Uganda, Madi Stream, Dec. 1862, *Speke & Grant* 711 (K!, lecto., designated by Venter *et al.* in S. Afr. Journ. Bot. 56: 93–112 (1990))

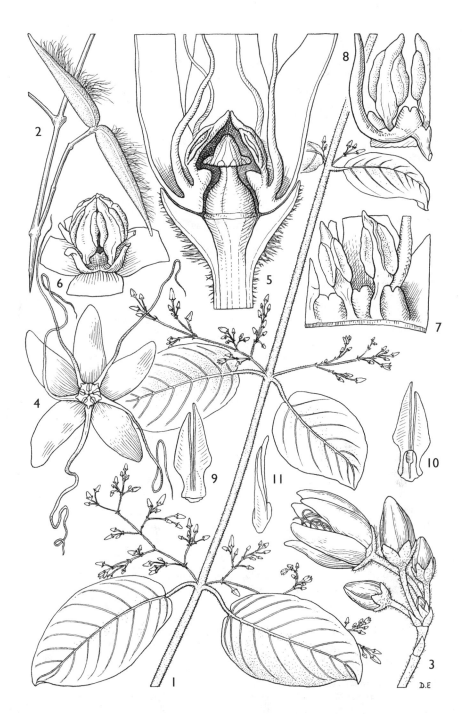

Fig. 45. *TACAZZEA APICULATA* — **1**, flowering shoot, × ⅔; **2**, paired follicles, × ⅔; **3**, part of inflorescence, × 4; **4**, flower from above, × 4; **5**, half-flower, × 10; **6**, centre of flower with corona lobes cut off to show gynostegium, × 8; **7 & 8**, portions of corolla showing relative positions of stamens, coronal feet and interstaminal nectaries, × 10; **9–11**, translators, × 23. 1, 3–11 from *Drummond & Hemsley* 4263; 2 from *Jarrett* 461. Drawn by D. Erasmus.

Giant liana; stems up to 20 m long, bark reddish brown, glabrous. Leaves semi-coriaceous; petiole 1–6 mm long; blade ovate to broadly ovate or elliptic to broadly elliptic, 3–22 × 2–12 cm, obtuse-acuminate to obtuse or emarginate, mucronate, base rounded to cordate, glabrous to sparsely tomentose above, tertiary veins conspicuously netted, beneath whitish to greyish tomentose. Inflorescences frail; peduncles 3–60 mm long, whitish tomentose; bracts ovate, ± 1 mm long, puberulous; pedicels 2–15 mm long, puberulous. Sepals broadly ovate, 1.5 mm long, pale green to brownish-red. Corolla pale green to pale yellow or reddish, glabrous, tube 0.5–1 mm long; lobes ovate, 5–7 × 2 mm, obtuse to acute. Coronal feet ± 1 mm long, lobes yellowish or reddish, 5–11 mm long, filiform with apex sometimes 2- or 3-fid. Stamens 1.5–2 mm long, filaments terete; anthers broadly oblong. Interstaminal nectaries sub-quadrate, emarginate or bifid. Style-head sub-sessile, broadly ovoid; translator receptacle rhomboidal with apex deeply split. Follicles horizontally divergent, narrowly ovoid, 3–9 cm long, 0.5–1.5 cm diameter, long-apiculate, hirsute or tomentose to softly puberulous or glabrous. Seeds obliquely ovate, 3–9 mm long; coma 2–5 cm long. Fig. 45, p. 141.

UGANDA. Ankole District: Mitooma, 1 Feb. 1991, *Rwaburindore* 3158! & Ruizi River, 17 Feb. 1951, *Jarrett* 461!; Mengo District: Kabaka's Lake, Mar. 1936, *Chandler & Hancock* 142!
KENYA. Northern Frontier District: Samburu, Ndoto Mountains between Nkurnit and Manmanet Ridge, 25 Oct. 1995, *Bytebier & Kirika* 45!; Machakos District: Mbeere, Kindaruma Dam, 4 Oct. 2000, KSCP/PGRWG 003/94/2000!; Tana River District: Tana River National Primate Reserve, Mulondi Swamp, 22 Mar. 1990, *Luke et al.* 778!
TANZANIA. Handeni District: Turiani, 22 Dec. 1969, *Harris* 3826!; Uzaramo District: Dar es Salaam, Ubongo, 27 Nov. 1968, *Mwasumbi* 10414!; Lindi District: Litipo Forest Reserve, eastern side of Lake Lutamba, 26 Feb. 1991, *Bidgood, Abdallah & Vollesen* 1732!
DISTR. U 2, 4; K 1–4, 7; T 1, 3–8; widely distributed across tropical and southern Africa
HAB. Swamp, stream and lake bank forests; 0–2000 m
USES. *Tacazzea apiculata* is supposed to be very poisonous

SYN. *Tacazzea thollonii* Baill. in Bull. Soc. Linn. Paris 2: 807 (1889). Type: Congo-Brazzaville, Ogôone, *Savorgon* 507 (P!, holo.)
    T. kirkii N.E. Br. in K.B. 1895: 248 (1895); T.T.C.L.: 68 (1949). Type: Mozambique, Zambesi Region, Lupata, March/June 1859, *Kirk* s.n. (K!, lecto., designated by Venter in S. Afr. Journ. Bot. 56: 93 (1990))
    T. bagshawei S. Moore in J.B. 44: 88 (1906). Type: Uganda, Mengo District: Entebbe, *Bagshawe* 745 (BM!, holo.)

2. **Tacazzea conferta** *N.E. Br.* in K.B. 1895: 247 (1895) & in F.T.A. 4(1): 265 (1902); T.T.C.L.: 68 (1949); Venter in Fl. Eth. 4(1): 103 (2003); Burrows in Pl. Nyika Plateau: 75 (2005). Type: Ethiopia, Efat, *Roth* 407 (K!, holo.).

Giant liana; stems up to 20 m long, young stems tomentose, older stems glabrous, verrucose. Leaves petiolate; petiole 1–4 cm long, puberulous, adaxially with 2 fleshy colleters at apex; blade elliptic to oblong or oblong-ovate, 6–16 × 3–10 cm, acute to acuminate or drip-tipped-acuminate, base cuneate to obtuse, glabrous, or above glabrous to sparsely puberulous and beneath puberulous. Inflorescences puberulous; peduncles sturdy, 0.5–4 cm long; pedicels 0.4–2 cm long; bracts broadly ovate to ovate, 1–3 mm long. Sepals reddish-brown, broadly ovate, ± 1.5 mm long, outside puberulous, margin ciliate. Corolla pale green to pale yellow or reddish; tube ± 1 mm long, sparsely puberulous outside; lobes broadly oblong, 5–6 × 2–3 mm, obtuse. Coronal feet ± 1 mm long, lobes 4–8 mm long, filiform with apex helically twisted and sometimes bifid. Stamens 2–3 mm long, filaments terete, anthers ovate, acute. Interstaminal nectaries lobular, ± 1 mm long. Style-head sub-sessile, broadly ovoid; translator receptacle broadly ovate with apex emarginate. Follicles slightly divergent, pendulous, cylindrical-ovoid, 7–20 cm long, 0.7–1.6 cm diameter, obtuse-acute. Seeds obliquely ovate, 0.8–1.4 cm long; coma 1.5–5 cm long.

UGANDA. Toro District: Ruwenzori, 1905, *Bauer* 666!; Mbale District: Sebei, 12 Dec. 1938, *A.S. Thomas* 2606!; Kigezi District: Virunga Mts, north foot of Muhavura, 26 Nov. 1954, *Stauffer* 972!

KENYA. Nakuru District: Mount Londiani, 1 Dec. 1967, *Perdue & Kibuwa* 9213!; Fort Hall District: Kimakia Forest Reserve, 12 Feb. 1965, *Gillett & Kabuye* 16649!; Kericho District: Mau Forest Reserve, 23 Nov. 1971, *Magogo* 1492!

TANZANIA. Arusha District: Ngorongoro, Empakaai Crater, 20 Nov. 1973, *Frame* 275!; Ufipa District: Nsanga Forest, 8 Aug. 1960, *Richards* 13015!; Kilosa District: Ukaguru Mountains, N slope of Myera Ridge, 13 Feb. 1988, *Pocs, Minja & Persson* 88019!

DISTR. U 2, 3; **K** 1, 3–7; **T** 2–8; Ivory Coast, Central African Republic, Congo-Kinshasa, Rwanda, Burundi, Ethiopia, Zambia, Malawi

HAB. Bamboo or mixed *Podocarpus* afromontane forests, moist bushland, riparian woodland; 1500–3000 m

USES. The roots which are strongly vanilla-scented are used as milk enhancer for both humans and cattle

SYN. *Tacazzea floribunda* K. Schum. in E.J. 30: 381 (1902); T.T.C.L.: 68 (1949). Type: Tanzania, Mbeya/Rungwe District: Poroto Mountains, Ngozi [Ngosi], *Goetze* 1289 (B†; K!, lecto., designated by Venter in S. Afr. Journ. Bot. 56: 93 (1990))

T. galactagoga Bullock in K.B. 9: 358 (1954); U.K.W.F. ed. 2: 175 (1994); K.T.S.L.: 498 (1994). Type: Tanzania, Morogoro District: Tanana in Uluguru Mts, *E.M. Bruce* 757 (K!, holo.; BM!, iso.)

39. **MONDIA**

Skeels in U.S.D.A. Bur. Pl. Industr. Bull. 223: 45 (1911); Bullock in K.B. 15: 203 (1961)

Chlorocodon Hook.f. in Bot. Mag. 97: t. 5898 (1871), *non* (DC.) Fourr. (1869); K. Schum. in E. & P. Pf. 4(2): 215 (1895); N.E. Br. in F.T.A. 4(1): 254 (1902) & in Fl. Cap. 4(1): 541 (1907)

Climbers with white latex; stems twining; interpetiolar stipules frill-like. Leaves opposite, herbaceous to sub-leathery; petiole fluted, with reddish colleters; lamina with main vein sunken above, with reddish colleters towards base, prominent beneath; secondary veins looping. Inflorescences of lax axillary paniculate cymes, bracts acicular. Flower buds ovoid-conical. Corolla rotate, coriaceous; tube saucer-shaped; lobes obliquely narrowly ovate to sub-orbicular, reddish, maroon or purplish on inside. Corolline corona lobes free, fleshy, tri- or bi-segmented, central segment corniculate or ligulate and absent or present, lateral segments flap-like with apices rounded. Gynostegial corona absent. Stamens sub-sessile from inner bases of coronal feet, anthers broadly hastate, whitish, acuminate; pollen in tetrads. Interstaminal nectary lobes scoop-shaped, laterally fused with inner bases of coronal feet, arranged cone-like around styles. Style-head very broadly ovoid, dark coloured. Follicles paired, narrowly ovoid. Seeds with coma of hairs.

A genus of two species, endemic to Africa.

Corona lobes with two segments (only flap-like lateral lobules
    present, and corniculate or ligulate central lobule absent);
    sepals broadly elliptic to orbicular with obtuse apex . . . .   1. *M. ecornuta*
Corona lobes with three segments (a central corniculate or
    ligulate lobule and two flap-like lateral lobules); sepals ovate
    to elliptic with acute to acuminate apex . . . .   2. *M. whitei*

1. **Mondia ecornuta** (*N.E. Br.*) *Bullock* in K.B. 15: 203 (1961); K.T.S.L.: 495 (1994). Type: Kenya, Kilifi District: Ribe, near Mombasa, May 1880, *Wakefield* s.n. (K!, holo.)

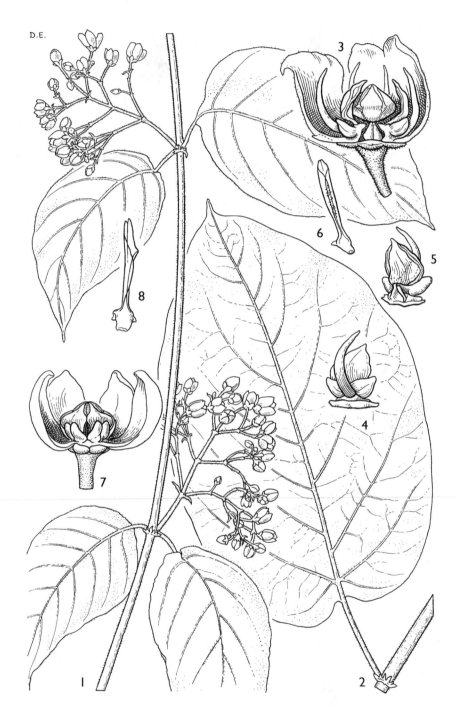

FIG. 46. *MONDIA WHITEI* — **1 & 2**, habit, × ²/₃; **3**, half-flower, × 3; **4 & 5**, stamen in dorsal and ventral view, showing attachment of corona lobe, × 3; **6**, translator, × 8. *MONDIA ECORNUTA* — **7**, flower with one corolla lobe removed to expose gynostegium, × 2; **8**, translator, × 8. 1 & 2 from *Drummond & Hemsley* 1432; 3–6 from *Dawkins* 706; 7 & 8 from *Faulkner* 558. Drawn by D. Erasmus.

Climber with whitish latex; roots unknown; stems 5+ m long. Leaves thin-coriaceous; petiole 10–35 mm long; lamina elliptic to ovate, 9–12(–15) × 4–6(–10) cm, acuminate to cuspidate, sometimes recurved, base obtuse to cordate, bright green above, paler green beneath. Inflorescences few-flowered, glabrous; peduncles 1–3 cm long; bracts ± 2 mm long; pedicels 1–4 cm long. Sepals broadly elliptic to orbicular, 2–3 × 2–3 mm, obtuse, glabrous, margin sparsely fimbriate. Corolla glabrous; tube ± 0.5 mm long; lobes purple, dark maroon, maroon or brownish, 7–10 × 4–5 mm, inside papillose, outside glabrous, pale green to greenish-yellow, sometimes green-maroon spotted. Corona lobes bi-segmented, flap-like lobules 1–2 × 1–2 mm. Anthers 4–5 × 3–4 mm. Interstaminal nectary lobes with margins entire, agglutinated together into a hollow cone around styles and ovaries, small apertures left at inner base of coronal feet. Styles ± 0.5 mm long; style-head ± 3 × 3 mm. Follicles 10–12 cm long, 2–2.5 cm diameter. Fig. 46/7–8, p. 144.

KENYA. Kilifi District: Cha Simba, 14 Aug. 1989, *Luke & Robertson* 1881!; Kwale District: 4 km W of Shimoni, 30 May 1990, *Luke & Robertson* 2352! & Mrima Hill, 3 Feb. 1989, *NMK Mrima-Dzombo exped.* in MDE 25!
TANZANIA. Lushoto District: Amani Nature Reserve, Muheza, 17 Nov. 1998, *Hizza* 212!; Pangani District: Bushiri Estate, 6 May 1950, *Faulkner* 558!; Zanzibar: Dimani District, 16 Nov. 1999, *Fakih* 417!
DISTR. **K** 7; **T** 3, 6; **Z**; Mozambique
HAB. Coastal bushland, woodland, dense forest or mangrove forest edge on coral or limestone outcrops, also more inland; 0–700 m
USES. Root extract drunk as antimalarial and anti-oxyuriasis remedy; root "skin" macerated and given orally against bilharziasis.

SYN. *Chlorocodon ecornutus* N.E. Br. in K.B. 1895: 111 (1895) & F.T.A. 4(1): 254 (1902), as *ecornuta*; T.T.C.L.: 64 (1949)

2. **Mondia whitei** (*Hook.f.*) *Skeels* in U.S.D.A. Bur. Pl. Industr. Bull. 223: 45 (1911); U.K.W.F. ed. 2: 175 (1994); K.T.S.L.: 495 (1994), as *whytei*. Type: South Africa, KwaZulu-Natal, Mfundisweni, May 1880, *White* s.n. (K!, holo.; K!, iso.)

Large liana; roots aromatic; stems up to 20 m long. Leaves herbaceous; petiole 2–7 cm long, glabrous to puberulent; lamina ovate, broadly ovate, elliptic, broadly elliptic or sub-orbicular, (6–)12–20(–28) × (3–)6–14(–20) cm, acuminate to cuspidate, sometimes recurved, base obtuse to cordate, glabrous to sparsely puberulent or sparsely puberulent on veins only, bright green above, paler green beneath. Inflorescences 10–20-flowered, glabrous to puberulent; flower odour most unpleasant; peduncles 2–4 cm long; bracts ± 2 mm long; pedicels ± 1 cm long. Sepals ovate to elliptic, 2–3 × 1–2 mm, acute to acuminate, outside glabrous to puberulent, margin fimbriate. Corolla glabrous; tube 2–3 mm long; lobes 9–11 × 4–6 mm, inside violet, maroon, wine red, mauve red, glabrous, outside pale green, glabrous. Corona lobes tri-segmented; lateral flaplike lobules green-yellow or creamy yellow, 1–2 × 1–2 mm; central corniculate or ligulate lobule darker coloured, 5–8 mm long. Anthers 4–5 × 3–4 mm. Interstaminal nectary lobes with margins dentate, not agglutinated. Styles 0.5–1 mm long; style-head ± 3 × 3 mm, acute. Follicles 8–12 cm long, 2–4 cm diameter. Seeds obliquely ovate, 8–10 mm long; coma 2–2.5 cm long. Fig. 46/1–6, p. 144.

UGANDA. Toro District: km 16 on Fort Portal road, 28 Feb. 1959, *Lind* 2374!; Busoga District: 13 km W of Kamuli on road to Namsagali (Nansagazi?), 29 Apr. 1953, *G.H.S. Wood* 794!; Mengo District: Nansagazi, Nakiza Forest, 24 Jan. 1951, *Dawkins* 706!
KENYA. Machakos District: Chyulu Hills, road to Kwibezi, 27 Dec. 2001, *Luke & Luke* 8241!; Kiambu District: cult. Nairobi, 26 Mar. 1952, *Bell* 1!; North Kavirondo District: Kakamega, May 1935, *Dale* 3431!
TANZANIA. Arusha District: Arusha National Park, Ngurdoto, 22 Apr. 1998, *Mkeya & Lema* 1035; Kigoma District: Mt Livandabe, 6 June 1997, *Bidgood et al.* 4328!; Songea District: Gumbiro, Mutandazi River, 9 May 1956, *Milne-Redhead & Taylor* 10138!

DISTR. U 1–4; **K** 4–6; **T** 2–4, 6–8; widely distributed across tropical and southern Africa

HAB. Moist habitats ranging from swamp forest, swampy shrubby grassland and riverine forest to disturbed forest; 0–1800 m

USES. The ginger-scented roots of *Mondia whitei* are used to alleviate flatulence, to settle the stomach, and as a tonic and an aphrodisiac.

SYN. *Chlorocodon whitei* Hook.f. in Bot. Mag. 97: t. 5898 (1871); N.E. Br. in F.T.A. 4(1): 255 (1902) & Fl. Cap. 4: 542 (1907); T.T.C.L.: 64 (1949)

    Periploca latifolia K. Schum. in P.O.A. C: 321 (1895) & in E.J. 23: 232 (1896). Type: Burundi, without locality, *Stuhlmann* 1619 (B†, holo.); synonymy based on N.E. Br. in F.T.A. 4(1): 255 (1902) and description of protologue

    Tacazzea amplifolia S. Moore in J.B. 50: 337 (1912). Type: Angola, Cazengo, *Gossweiler* 616 (BM!, holo.; K!, P!, iso.)

    T. viridis A. Chev., Expl. Bot. Afr. Occ. Fr.: 429 (1920) *nomen nudum,* ex Hutch. & Dalziel, F.W.T.A. 2: 52 (1931); Bullock in K.B. 1937: 339 (1937). Type: Ivory Coast, Mankono, between Dialakoro and Kènègonè, *Chevalier* 21975 (P!, holo.; K!, iso.)

40. **RAPHIONACME**

Harv. in London Journ. Bot. 1: 22 (1842); Benth. in G.P. 2 (2): 745 (1876); N.E. Br. in F.T.A. 4(1): 268 (1902) & in Fl. Cap. 4(1): 532 (1907); Venter in S. Afr. Journ. Bot. 75: 292–350 (2009)

Zuchellia Decne. in DC., Prodr. 8: 492 (1844)
Pentagonanthus Bullock in Hooker's Icon. Pl.: t. 3585 (1962)

Geophytic suffrutescent herbs or herbaceous (rarely woody) climbers; latex white; root tuber turnip-shaped or cylindrical; underground stems perennial, erect, one or few from crown of tuber; aerial stems annual, seldom perennial, twining, erect or procumbent, branching dichotomous or lateral, interpetiolar ridges with dentate colleters. Leaves opposite, sessile or petiolate; blade herbaceous to leathery, secondary veins arching or divaricate. Inflorescences terminal and/or axillary, racemose, plumose or globose, few to many-flowered, glabrous or hairy. Corolla tube campanulate to cylindrical, inner face fluted; lobes ovate, obovate or triangular, spreading to reflexed, outside glabrous or hairy, inside glabrous. Corolline corona arising from mouth of corolla tube; coronal feet inconspicuous or conspicuous and columnar or cone-shaped, free or seldom fused collar-like; lobes simple or basally tri-segmented, apices simple or variously incised, glabrous. Gynostegial corona absent. Stamens arising from inner base of coronal feet, filaments free; anthers ovate, triangular or hastate, glabrous; pollen in tetrads. Nectary pockets from base of corolla tube, fused to coronal feet. Gynostegium just protruding from to elevated above corolla mouth. Styles terete or fluted, glabrous or hairy; style-head, broadly ovoid, oblong-ovoid or conical, translators with receptacle, stalk and sticky disc. Follicles paired or single, erect or divaricate, narrowly ovoid to cylindrical. Seeds ovate with coma or ring of hairs.

With the exception of one species that occurs in Arabia, all the other 36 species of *Raphionacme* are endemic to and widely distributed over Africa, the winter rainfall areas excluded. Most *Raphionacme* species are from grassland or savanna, but a few inhabit semi-desert or true desert, a few swampy habitat and one forest habitat.

1. Corona lobes entire or segmented at their apices only 2
   Corona lobes basally segmented or parted 10
2. Coronal feet fused into a collar with fleshy lobules,
    lobules opposite the corolla lobes; style hairy and
    fluted . . . . 11. *R. splendens* (p. 154)
   Coronal feet free from one another; style glabrous
    and terete or narrowly obconical 3

1. **Raphionacme borenensis** *Venter & M.G. Gilbert* in J.L.S. 99: 401 (1989); Venter in Fl. Eth. 4(1): 108 (2003) & in Fl. Somalia 3: 140 (2006). Type: Ethiopia, Bale Region, track from Negele to Biderra, *Gilbert & Jones* 66 (K!, holo.)

Erect herb; tuber massive, turnip-like; aerial stems erect, branching lateral. Leaf petiole 1–2 mm long; blade ovate to broadly ovate to obovate, 2.5–7.5 × 1.5–3 cm, acute to auminate, base cuneate, above glabrous to puberulous, green with lustre, beneath puberulous, dull green to green flushed purple, margin wavy. Inflorescences terminal to sub-terminal, 2–3-flowered, sparsely to densely puberulous; peduncle 2–10 mm long; pedicels 3–4 mm long. Sepals ovate to triangular, 1–2 × ± 1 mm, puberulous. Corolla tube campanulate, 4 mm long, outer side puberulous; lobes spreading, purple, ovate, 5–7 × 2–4 mm, acute to obtuse. Corona lobes whitish, unsegmented, upper part filiform, 5–12 mm long, feet dilated. Staminal filaments broadly semi-deltoid, 1–2 mm long, basally fleshy, terminally shortly filiform; anthers narrowly ovate, ± 2 mm long. Style 2–3 mm long; style-head broadly ovoid, 2–3 mm long; translators 2 mm long. Gynostegium base in corolla mouth. Follicles slender, fusiform, the pair horizontal. Seeds not known.

DISTR. **K** 1?; Ethiopia, Somalia
HAB. *Acacia-Commiphora* bushland on limestone, in full sun or partial shade, usually in rather bare, stony ground; 750–1600 m

NOTE. *Raphionacme borenensis* is known from just N of the Kenya/Ethiopia border and could well be present in Kenya.

2. **Raphionacme brownii** *Scott-Elliot* in J.L.S. 30: 91 (1895); N.E.Br. in F.T.A. 4: 273 (1902). Type: South Sudan, Falaba, *Scott Elliot* 5179 (K!, holo.; BM!, iso.)

Erect herb; tuber ovoid, up to 10 × 6 cm; aerial stems up to 25 cm long, branching lateral, sparsely to densely scabrous and siliceous. Leaves mostly absent till after flowering, sessile, sub-sessile to shortly petiolate; blade linear or linear-ovate, 8–16 × 0.3–0.8 cm, acuminate, base cuneate, conduplicate, glabrous to sparsely siliceous above, sparsely to densely siliceous beneath, midvein prominent, secondary veins arching to apex. Inflorescences terminal, lax cymes of several 3–6-flowered branches, scabrous or siliceous; peduncles frail, 10–50 mm long; bracts narrowly triangular, 1–2 mm long; pedicels frail, 6–10 mm long. Sepals triangular to ovate, 1–2 × 0.5–1 mm, acute to obtuse, scabrous to siliceous. Corolla violet, mauve, pink or greenish-pink or tube pale green with lobes lilac, outside sparsely scabrous or siliceous, inside glabrous; tube cylindrical, 3–4 mm long; lobes ovate to narrowly ovate, 4–6 × 1.5–2 mm, acute to rounded. Corona lobes white, green or yellow, filiform, 4–6 mm long, apex entire or bifid, tortuous, papillose. Staminal filaments at base laterally dilated and fused to corolla tube, upper part filiform, 0.5 mm long; anthers greenish to pinkish-white, ovate, 1–1.5 mm long, mucronate. Style very narrowly obconical, 3 mm long; style-head ovoid to broadly ovoid, 1 × 1–2 mm. Gynostegium base in corolla mouth. Follicles erect, solitary or paired, very narrowly obovoid, 5–28 × 0.5 cm, acuminate. Seeds oblong-ovate, 5–6 × 1 mm; coma 4 cm long.

DISTR. **U** 1?; Senegal, Sierra Leone, Guinea, Guinea Bissau, Mali, Burkina Faso, Ivory Coast, Ghana, Togo, Benin, Nigeria, Chad, Central African Republic and South Sudan
HAB. Dry open savanna or short grassland on shallow soils on laterite, schistose rock and shales, or in rock pans; 0–1000 m
USES. Tuber and follicles used as food

SYN. *Raphionacme baguirmienis* A. Chev. in Etud. Fl. Afr. Centr. Franc. 1: 197 (1913). Type: Chad, Moyen-Chari, region of Lake Iro, Fort Archambault, Koulfe, Kaba-Mara, *Chevalier* 8995 (P!, lecto. designated by Venter in S. Afr. Journ. Bot. 75: 301 (2009))

NOTES. *Raphionacme brownii* is a pyrophyte and can be found flowering two weeks after a fire. *R. brownii* is known from just north of the Uganda/South Sudan border, its absence from the FTEA area is probably due to poor collecting. A U.K.W.F. ed. 2: 175 (1994) record from **K** 4, Langata, has not been substantiated.

3. **Raphionacme flanaganii** *Schltr.* in E.J. 18, Beibl. 45: 2 (1894); Gordon-Gray in Fl. Pl. Afr. 40: t. 1599 (1970). Type: South Africa, Eastern Cape Province, Komgha, near Kabousie River, *Flanagan* 118 (B†, holo.; PRE!, lecto., designated by Venter in S. Afr. Journ. Bot. 69: 212–213 (2003); GRA!, NBG!, iso.)

Perennial climber with one to few main stems; tuber spindle-shaped, up to 30 cm diameter; aerial stems up to at least 2 m long, branching lateral, velutinous. Leaf petiole 1–15 mm long; blade obovate, narrowly obovate or ovate, 3–11 × 0.5–5 cm, acute, obtuse or truncate-mucronate, above velutinous to glabrous, beneath velutinous, primary vein prominent. Inflorescences subterminal, 7–30-flowered, velutinous; peduncles 5–25 mm long; bracts narrowly ovate, 2–3 mm long; pedicels 2–6 mm long. Sepals ovate, 2–4 × 1–2 mm, acute to obtuse, outside velutinous. Corolla green, outside velutinous; tube campanulate, 2–3 mm long; lobes reflexed, yellow-green with reddish to purple triangular spot at base, oblong-ovate to ovate, 5–7 × 2–3 mm, obtuse to acute. Corona lobes tripartite, white; central segment filiform with dilated base, apex tortuose, 4–10 mm long, papillose; lateral segments corniculate, 2–3 mm long. Staminal filaments filiform, ± 2 mm long; anthers ovate, ± 2 mm long, white. Style terete, ± 1.5 mm long; style-head ovoid, ± 1.5 mm long, obtuse. Gynostegium base in corolla mouth. Follicles paired, seldom solitary, pendulous, cylindrical-ovoid, 5–10 cm long, ± 1 cm diameter, puberulous. Seeds 6–7 mm long; coma 1–2 cm long.

KENYA. Kitui District: Mutomo Hill, 3 Mar. 1960, *Bally* 12096!; Masai District: Chyulu Plains, Oleeilelu Hill, 16 Feb. 2001, *Luke & Luke* 7343! & Emali Hill, 7 Mar. 1940, *V.G.L. van Someren* 31/229!
DISTR. **K** 4, 6; Mozambique and South Africa
HAB. Dry bushland, dry forest or streambank forest; 1000–1800 m

SYN. *Raphionacme scandens* N.E. Br. in K.B. 1895: 111 (1895). Type: South Africa, KwaZulu-Natal, Tugela, *Gerrard* 1312 (K!, holo.; BM!, NH!, W!, iso.)

NOTE. It is surprising to find *R. flanaganii* in Kenya while it is absent from Tanzania and just about so from Mozambique, but the absence is probably due to poor collecting. The plants found in Kenya have smaller flowers than those in South Africa.

4. **Raphionacme globosa** *K. Schum.* in E.J. 17: 118 (1893); N.E. Br. in F.T.A. 4: 271 (1902). Type: Angola, Malange, *von Mechow* 327 (B†; Z!, lecto.; K!, iso. designated by Venter in S. Afr. Journ. Bot. 75: 309, 2009)

An erect, single to few-stemmed herb; tuber conical to spherical, up to 5–10 cm diameter; aerial stems up to 0.5 m long, branching lateral, glabrous, sparsely hirsute or scabrous. Leaf petiole 2–5 mm long; blade linear to very narrowly ovate to less often narrowly ovate, 5–19 × 0.2–2 cm, acute to attenuate, base cuneate, above dark green and sometimes densely dotted, beneath pale green and mostly densely dotted, margin reddish, glabrous with midrib and margin hirsute or scabrous, or sparsely hirsute to scabrous all over; secondary veins mostly visible, arching. Inflorescences terminal, rarely also subterminal, many-flowered globose cyme, densely hirsute to scabrous; peduncles 1–4.5 cm long; bracts subulate, 2–3 mm long; pedicels ± 2 mm long. Sepals subulate to narrowly triangular, 2–3 × 0.8–1 mm, hirsute or scabrous. Corolla hirsute or scabrous on outside; tube campanulate, 2–3 mm long; lobes white, whitish-green, or creamy-yellow, tips purple, pink or red, sometimes tinged violet, brown or red, ovate to triangular, 3–4 × 1–2 mm. Corona pale green tinged purple; lobes trisegmented; central segment filiform, 2–3 mm long, papillose, apex twirled; lateral segments subulate, ± 1.5 mm long, glabrous. Staminal filaments sub-sessile; anthers white, narrowly angular-ovate, connective reddish, attenuate. Style reddish, terete, 1–2 mm long; style-head ovoid, obtuse,1–2 × 0.5 mm. Gynostegium base in corolla mouth. Follicles solitary, erect, cylindrical-ovoid, ± 7.5 cm long, ± 7 mm diameter, tapering into a blunt tip, smooth, glabrous. Seeds ± 5 mm long; coma 2–2.5 cm long.

TANZANIA. Mbeya District: Songwe Valley, 20 km W of Mbeya, 30 Nov. 1994, *Goyder et al.* 3866!; Songea District: 80 km on Songea–Tundura road, 30 Jan. 1991, *Bidgood, Abdallah & Vollesen* 1323! & Mkukira River, 19 Feb. 1956, *Milne-Redhead & Taylor* 8270!
DISTR. **T** 7, 8; Congo-Kinshasa, Angola, Zambia, Malawi
HAB. Wooded grassland, grassland and forest margins on rocky shallow soils to deeper loamy soils; 900–1400 m

SYN. *Raphionacme virgultorum* S. Moore in J.B. 50: 338 (1912). Type: Angola, Kubango, near Fort Princeza Amelia, *Gossweiler* 2267 (BM!, lecto., designated by Venter in S. Afr. Journ. Bot. 75: 309 (2009))

5. **Raphionacme grandiflora** *N.E. Br.* in K.B. 1895: 111 (1895) & in F.T.A. 4(1): 270 (1902); Bullock in K.B. 8: 353 (1953). Type: Zambia, Niomkolo, *Carson* 5 (K!, lecto., designated by Bullock in K.B. 8: 353 (1953))

Herb, erect, single or few-stemmed; tuber conical to spherical, up to 8 cm diameter; aerial stems erect, up to 80 cm long, branching lateral, sparsely to densely pubescent or velutinous. Leaf petiole 0–2.5 cm long; blade obovate, ovate, narrowly obovate, narrowly ovate or linear, (2–)10–21 × (0.3–)1–6 cm, often narrowing from base to apex of stem, above dark green, glabrous, pubescent or velutinous and often densely dotted, beneath pale green and glabrous with veins pubescent or velutinous or hairy all over; secondary veins arching. Inflorescences 1–3-flowered, terminal and from upper 2–3 nodes; indumentum as on stem; peduncles rarely present; bracts subulate to narrowly triangular, 1–5 mm long; pedicels 1.5–5 cm long. Sepals narrowly ovate, 5–6 × ± 2 mm, acuminate, margin fimbriate. Corolla glabrous; tube campanulate, 5–9 mm long, pentangular, basally 5-saccate; lobes ovate to triangular, 115–20 × 8–10 mm, inside deep violet, deep violet-blue or deep red, sometimes with green tips. Corona white, whitish flecked violet, violet or greenish; lobes fused to corolla lobe bases, fleshy, rectangular, 4–5 × 4–5 mm, apex trisegmented, central segment filiform, 5–7 mm long, lateral segments corniculate to subulate, 1–2 mm long. Staminal filaments filiform, 1–2 mm long; anthers creamy-white with connective purplish, narrowly ovate-hastate, acute, 7–8 mm long, upper half with pollen, lower of two thickened horn-like callosities. Style terete, 5–6 mm long; style-head broadly oblong-ovoid, obtuse, 5–9 × 4–8 mm. Gynostegium base in corolla mouth. Follicles paired, erect, linear-ovoid, 15–18 cm long, 8–10 mm diameter, attenuate, glabrous. Seeds narrowly ovate, 8–10 mm long; coma 3 cm long. Fig. 47, p. 151.

TANZANIA. Ufipa District: Sumbawanga, 30 Dec. 1956, *Richards* 7397!; Songea District: 95 km from Songea on Njombe road, 29 Jan. 1991, *Bidgood, Abdallah & Vollesen* 1318! & near Lumecha Bridge, 23 Jan. & 17 Mar. 1956, *Milne-Redhead & Taylor* 8394 & 8394B!
DISTR. **T** 4, 8; Rwanda, Zambia, Malawi, Mozambique and Zimbabwe
HAB. Miombo woodland and grassland on alluvium, sand or laterite, often on stony soil among rocks; 900–1400 m

SYN. *Raphionacme grandiflora* N.E. Br. subsp. *glabrescens* Bullock in K.B. 8: 353 (1953). Type: Mozambique, Moebede, Lugela, Mocuba, *Faulkner* 7 (K!, holo.; EA, PRE!, iso.)
Pentagonanthus grandiflorus (N.E. Br.) Bullock in Hooker's Icon. Pl. 36: t. 3583 (1962)
P. grandiflorus (N.E. Br.) Bullock subsp. *glabrescens* (Bullock) Bullock in Hooker's Icon. Pl. 36: t. 3583 (1962)

6. **Raphionacme longifolia** *N.E. Br.* in K.B. 1895: 110 (1895). Type: Mozambique, Lower Shire Valley, Maramballa, 18 Jan. 1863, *Kirk* s.n. (K!, lecto., designated by Venter in S. Afr. Journ. Bot. 75: 322 (2009)

FIG. 47. *RAPHIONACME GRANDIFLORA* — **1**, leafy shoot with tuber, × ¹/₃; **2**, stem with leaves and flowers, × 1; **3**, stem with leaves and fruit, × ¹/₃; **4**, sepal, × 6; **5**, section of flower to show corolla lobe and corona lobe, × 1; **6**, corona lobe from outside, × 3; **7**, pistil, × 6; **8, 9, 10**, translator from outside, inside and laterally, respectively, × 6; **11**, seed, × 1.5. Drawn by Margaret Stones, and reproduced with permission from Hooker's Icon. pl. 36: t. 3583.

Herb up to 75 cm long, erect, terminally twining, single- or few-stemmed, with white latex; tuber ovoid to flattened-globose, up to 10 cm diameter; aerial stems brownish, branching lateral, puberulous. Leaves sessile; blade linear to very narrowly ovate, (5–)12–15(–20) × (0.4–)0.88–1(–2.2) cm, attenuate, base tapering, often involute, above green, puberulent, beneath pale green, puberulent, midvein prominent, whitish, secondary veins looping. Inflorescences axillary and terminal, fascicles 5–10-flowered, puberulous; peduncles 2–10 mm long; bracts subulate, 2–3 mm long; pedicels 3–6 mm long. Sepals triangular to ovate, ± 2 × 1 mm, densely puberulent. Corolla outside densely puberulent; tube campanulate, ± 2 mm long; lobes oblong-ovate to triangular, 4–6 × 2–3 mm, inside greenish, greenish-brown, greenish-purple, maroon-brown or dull purple-red. Corona whitish to purple; coronal feet columnar, ± 1 mm long; lobes from outer apex of coronal feet, trisegmented, central segment filiform, 5–6 mm long; lateral segments subulate, ± 1 mm long. Staminal filaments from inner apex of coronal feet, filiform, 1–2 mm long; anthers white, oblong-ovate, 1–2 mm long. Style terete, ± 2 mm long, glabrous; style-head ovoid, ± 2 × 1 mm. Gynostegium base in corolla mouth. Follicles paired, erect, cylindrical-ovoid, ± 10 cm long. Seeds unknown.

TANZANIA. Morogoro District: 35 km along the Morogoro road to Dodoma, 29 Dec. 1970, *Botany students* DSM 2124!; Rufiji District: Selous Game Reserve, Kivuko Hill, 13 Dec. 1998, *Luke & Luke* 5608!; Iringa District: lower slopes of Mt Image, N side of Kitongo Gorge, 60 km E of Iringa, 9 Dec. 1994, *Goyder et al.* 3923!
DISTR. **T** 6–8; Zambia, Malawi, Mozambique, Zimbabwe and Botswana
HAB. Miombo grassland; 500–1600 m
USES. Tuber used as food

SYN. *Raphionacme decolor* Schltr. in Bot. Untersuch. 1: 265 (1916). Type: Zambia, Kalambo, between Mbala and Kasanga, *R.E. Fries* 1396 (UPS, holo.; BLFU!, photo)

7. **Raphionacme longituba** *E.A. Bruce* in K.B. 1937: 419 (1937). Type: Tanzania, Tabora District: Kakoma, *H.M. Lloyd* 45 (K!, holo.; EA, iso.).

A spreading herb, up to 0.5 m tall; tuber ovoid to compressed-globose, 12–22 cm diameter; aerial stems spreading to decumbent, dichotomously branched, 15–20 cm long, hispid. Leaf petiole ± 3 mm long, hispid, with reddish colleters above; blade often succulent, elliptic to ovate, 5–7 × 2–4 cm, obtuse-mucronate, base obtuse, sparsely hispid, dotted, secondary veins divaricate to patent, margin thickened, undulating, densely hispid. Inflorescences axillary, 3–4-flowered, hispid; peduncle 1–3 cm long; bracts subulate, 3–7 mm long; pedicels 5–6 mm long. Sepals narrowly triangular, 4–6 × 1–2 mm, outside hispid. Corolla green tinged maroon or purplish to brown outside, white, white tinged pink to purplish inside; tube cylindrical-campanulate, 7–10 mm long, outside sparsely hispid; lobes spreading, oblong to narrowly ovate, 9–12 × 3–5 mm, obtuse to acute. Corona conniving dome-like over gynostegium; lobes fleshy, white, greenish-white to creamy-white, narrowly ovate to ovate, acuminate or bifid, concave, 5–8 × 2–3 mm. Staminal filaments linear, ± 1 mm long; anthers oblong ovate to narrrowly ovate, ± 3 mm long, acuminate, upper half with pollen, lower a white callosity. Styles terete, 7–8 mm long, apically fused; style-head ovoid, acute, ± 2 mm long. Gynostegium base in corolla mouth. Follicles solitary, erect, very narrowly ovoid, 17–18 cm long, 1–1.5 cm diameter, tapering. Seeds ± 9 mm long; coma ± 3 cm long.

TANZANIA. Ufipa District: Tatanda Mission, 26 Feb. 1994, *Bidgood, Mbago & Vollesen* 2508!; Manyoni District: S of Itigi Station on Chunya road, 21 Apr. 1964, *Greenway & Polhill* 11677!; Mbeya District: Hot Springs in Songwe River Gorge, 27 May 1990, *Carter, Abdallah & Newton* 2493!
DISTR. **T** 4, 5, 7; Zambia and Malawi
HAB. Grassland or seasonally inundated grassland on limestone or other soils; 1200–1800 m

SYN. *Raphionacme ernstiana* Meve in Bradleya 18: 71 (2000). Type: Tanzania, Singida District: 51 km SW of Itigi on road to Rungwa, *E. & M. Specks* 1038 (K, holo.; UBT!, iso.)

8. **Raphionacme madiensis** *S. Moore* in J.B. 46: 294 (1908); Malaisse in Fl. Rwanda 3: 110 (1985); U.K.W.F. ed. 2: 175 (1994). Type: Uganda, West Nile District: Madi, near Nimuli, *Bagshawe* 1611 (BM!, holo.)

Herb, up to 0.2 m tall, profusely lactiferous; tuber sub-spherical, up to 16 cm diameter; aerial stems erect to spreading, branching lateral, 10–16 cm long, puberulent to scabridulous. Leaf petiole 5–20 mm long, puberulent; blade ovate, narrowly ovate, elliptic, narrowly elliptic, obovate or narrowly obovate, (5–)8–15 × (1–)3–5 cm, acute, base tapering, crinkled, above green, puberulent, veins pale and densely puberulous, beneath pale green, puberulent, veins prominent and densely puberulous, margin wavy. Inflorescences terminal and axillary, compact, many-flowered, densely puberulous to scabridulous; peduncles 0.5–1 cm long; bracts subulate, 3–5 mm long; pedicels 5–7 mm long. Sepals very narrowly triangular or narrrowly ovate, 2–3 mm long. Corolla tube campanulate, ± 2 mm long, puberulous outside; lobes oblong-ovate, oblong-triangular, narrowly ovate or narrowly triangular, 5–7 × ± 2 mm, outside puberulent, green, inside glabrous, green or dull green to whitish with brown, pink, magenta or violet streaks, and/or tips and/or margins. Coronal feet columnar, ± 1 mm long, fused laterally to inner base of corolla lobes, apically tri-segmented; central segment filiform, 5–7 mm long, apex bifid or not, tortuous, papillose, lateral segments filiform, 2–3 mm long, glabrous or papillose. Staminal filaments filiform, curving inwards, ± 1 mm long; anthers white, ovate, ± 2 mm long, apices fused together over style-head. Style terete, ± 1 mm long; style-head ovoid, acute. Gynostegium base in corolla mouth. Follicles solitary, erect, ovoid, 4.5–6 cm long, 1.2–1.5 cm diameter, with tapering apex. Seeds unknown.

UGANDA. Karamoja District: Lokales, Moroto, May 1971, *J. Wilson* 2067!; West Nile District: Arua Hill, Mar. 1938, *Hazel* 460!; Teso District: Teso, Apr. 1932, *Chandler* 604!
KENYA. Trans-Nzoia District: NE Mt Elgon, May 1971, *Tweedie* 1119!
TANZANIA. Musoma District: SW of Klein's Camp, 23 Dec. 1969, *Greenway* 13895!; Mbeya District: Pwizi River, 3 km from Mjele Village on road from Utengele to Galula, 9 Oct. 1976, *Leedal* 4048! & Sao Hill, Madibira Road, 8 Dec. 1962, *Richards* 17355!
DISTR. U 1, 3; K 3; T 1, 4, 7; Congo-Kinshasa, Rwanda, Burundi, Angola, Zambia
HAB. Wooded grassland and grassland on rocky outcrops and in shallow rock-basins; 850–2050 m
USES. The tuber is said to be edible.

SYN. *Raphionacme wilczekiana* Germain in B.J.B.B. 22: 74 (1952). Type: Congo-Kinshasa, Plain of Ruzizi, Kabunambo to River Ruzizi, *Germain* 5567 (BR!, holo.; SRGH! iso.)

9. **Raphionacme michelii** *De Wild.* in Ann. Mus. Congo, ser. 5, 1: 181 (1904); U.K.W.F. ed. 2: 175 (1994). Types: Congo-Kinshasa, Kimbele, *Cabra & Michel* 47 (BR!, lecto., designated by Venter in S. Afr. Journ. Bot. 75: 327 (2009); K!, photo.)

An erect or spreading herb of up to 15 cm high; tuber sub-spherical, 7–10 cm diameter; aerial stems erect to spreading, 3–15 cm long, branching lateral, puberulent to pubescent. Leaves with petiole 0–1 cm long, sometimes stem-clasping, puberulent to pubescent; blade very narrowly obovate to obovate, 4–6 × 1–2 cm, acute, margin wavy, base tapering, above glabrous or pubescent, spotted, below puberulent to pubescent, denser on prominent veins; veins arching to apex. Inflorescences terminal, open, few-flowered, puberulent to pubescent; peduncles 0.2–1.7 cm long; bracts subulate, 2–4 mm long; pedicels 0.7–1.8 cm long. Sepals narrowly triangular to narrowly ovate, 2–3 × ± 1 mm, attenuate, outside puberulent to pubescent. Corolla tube campanulate, 3–5 mm long, outside glabrous to pubescent; lobes spreading, oblong-ovate, 12–20 × 3–5 mm, obtuse to sub-acute, outside glabrous to pubescent, inside mauve. Corona lobes filiform with dilated base, 1–2 cm long, papillose or glabrous, apex tortuous. Staminal filament bases semi-ovoid, ± 1 mm long, succulent; filaments filiform, 2–4 mm long; anthers whitish, oblong-ovate, 3–4 mm long, fully fertile. Style terete, 3–4 mm long; style-head ovoid, 3–4 × ± 2 mm, acute. Gynostegium elevated above corolla mouth. Follicles and seed unknown.

KENYA. Nyanza District: 3 km E of Kindu Bay, 15 Mar. 1989, *Hartmann & Newton* 28593!
DISTR. **K** 5; Congo-Kinshasa, Angola and Zambia
HAB. Rocky grassland; 1000–1300 m

SYN. *Raphionacme gossweileri* S. Moore in J.B. 46: 294 (1908). Type: Angola, Kuiriri, east of Kossuogo, *Gossweiler* 3273 (BM!, holo.; K!, iso.)

10. **Raphionacme moyalica** *Venter & R.L. Verh.* in S. Afr. Journ. Bot. 63: 101 (1997); Venter in Fl. Eth. 4(1): 110 (2003). Type: Kenya, Northern Frontier District, Moyale, *Gillett* 14021 (K!, holo.)

Erect herb of ± 10 cm high; tuber turnip-shaped, ± 13 cm long, 6.5 cm diameter; aerial stems erect, branching lateral, scabridulous. Leaves sessile to subsessile; blade very narrowly ovate to very narrowly elliptic or obovate, 55–65 × 8–10 mm, acute, scabridulous. Inflorescences terminal and axillary, branched cymes, each branch ± 5-flowered, densely scabridulous. Corolla tube campanulate, 2–3 mm long, outside scabridulous; lobes spreading, maroon-violet, ovate, 3–4 × 1–2 mm, outside densely scabridulous. Corona lobes maroon-violet, unsegmented; feet fleshy, laterally dilated, concave; upper part filiform, 7–15 mm long, terminally tortuous. Staminal filaments ± 0.5 mm long, basally fleshy, broadly sub-conical, terminally filiform; anthers ovate, ± 1 mm long, acute, connective violet. Style terete, 1–2 mm long; style-head ovoid, acute, ± 1 mm long. Gynostegium base in corolla mouth. Follicles and seeds unknown.

KENYA. Northern Frontier District: Moyale, 10 Oct. 1952, *Gillett* 14021!; Meru District: Meru National Park, 21 Nov. 2000, *Luke et al.* 7093!
DISTR. **K** 1, 4; Ethiopia
HAB. Mountain bushland and grassland; 500–1200 m

11. **Raphionacme splendens** *Schltr.* in J.B. 33: 301 (1895); Malaisse in Fl. Rwanda 3: 110 (1985); Meve, Willke & Albers in Illustr. Handb. Succ. Pl.: Asclepiadaceae: 227 (2002). Type: Uganda, Ruwenzori, *Scott Elliot* s.n. (BM!, holo.; K!, iso.)

Erect herb of up to 50 cm high, 1–3-stemmed; tuber narrowly ovoid to obconical, 8–20 cm long, 2–6 cm diameter; aerial stems erect, branching lateral, sparsely puberulent, sparsely scabrous or pubescent. Leaves often only develop after flowering; petioles 0–3 mm long; blade linear, linear-ovate, narrowly ovate, ovate, narrowly obovate, obovate or elliptic, 3–12 × 1–4 cm, acute, attenuate, acuminate or obtuse-mucronate, base cuneate to obtuse, above glabrous to sparsely puberulous or sparsely scabridulous, beneath sparsely puberulent, sparsely scabridulous, densely pubescent or scabrous; midrib prominent, secondary veins looping towards apex, margin often wavy. Inflorescences terminal and axillary, lax, 1–10-flowered, puberulous to pubescent; peduncles 1–15 mm long; bracts ovate to narrowly triangular, 1–2 mm long; pedicels 5–30 mm long. Sepals ovate to triangular, 2–4 × 1–2 mm, outside sparsely puberulent to pubescent, reddish-green. Corolla tube campanulate or narrowly urceolate to broadly urceolate, 2–4 mm long, outside glabrous to sparsely pubescent or puberulent, greenish; lobes reflexed, narrowly ovate, narrowly elliptic, narrowly obovate or narrowly triangular, 10–30 × 2–9 mm, obtuse to acute, outside brownish-pink, dark brown or greenish, sparsely pubescent or hirsute, inside whitish-pink, pink, mauve or purple to blueish. Coronal feet columnar, exserted, 1–2 mm long, basally fused into an annulus with fleshy lobules; upper part purple to white, filiform, (4–)7–12(–20) mm long, tortuous. Stamens from inner apex of coronal columns, filaments purple, 2–4 mm long, semi-erect; anthers greenish, ovate to narrowly ovate, acute, 2–4 mm long. Style 4–8 mm long, terete; central part swollen cone-like or oblong-like, vertically furrowed with fine hairs; style-head greenish, oblong-ovoid, acute, (2–)4–6(–8) mm long. Gynostegium elevated above corolla mouth. Follicles single, rarely paired, erect, very narrowly ovoid, 5–11 cm long, 4–15 mm diameter, glabrous, puberulous or hirsute. Seeds narrowly ovate, 5–8 mm long; coma 25–30 mm long.

A widespread pyrophytic species occuring across tropical Africa. A single subspecies is found within the Flora area – the second, subsp. *bingeri* (A. Chev.) Venter, occurs in W Africa.

subsp. **splendens**

Stems straight, leaves linear to less often narrowly ovate or narrowly obovate.

UGANDA. Bunyoro District: Lake Albert, Kibiro, 19 Feb. 1906, *Bagshawe* 910!; Ruwenzori Mountains, 1893–94, *Scott Elliot* s.n.!; Toro District: Mpanga River, Toro, *Bagshawe* 1060!
KENYA. Kwale District: Kaya Lunguma, 29 May 1996, *Luke et al.* 4494! & near Kiwegua, 27 May 1990, *Luke & Robertson* 2331!; Lamu District: Bolaa [Boula], 6 Nov. 1946, *Adamson* 220 in *Bally* 5916!
TANZANIA. Biharamulo District: 42 km on Biharamulo–Muleba road, 1 July 2000, *Bidgood, Leliyo & Vollesen* 4910!; Tanga District: Machui on Tanga–Pangani road, 26 Nov. 1955, *Faulkner* 1956!; Buha District: 52 km on Kasula – Kibondo road, 20 June 2000, *Bidgood, Leliyo & Vollesen* 4758!
DISTR. U 2, 4; K 1, 7; T 1–4, 6, 8; Cameroon, Central African Republic, Sudan, South Sudan, Ethiopia, Rwanda, Burundi, eastern parts of Congo-Kinshasa, Zambia, Zimbabwe, Malawi and Mozambique
HAB. Open dry forest, *Acacia* and Miombo woodland, wooded grassland and grassland on soils varying from red laterite loam to clay or sand, sometimes in seasonal marshes; 0–1500 m
USES. A decoction of the leaves is used for conjunctivitis. The tuber is edible and tastes like sweet potato

SYN. *Raphionacme excisa* Schltr. in J.B. 33: 301 (1895); Venter in Fl. Eth. 4(1): 109 (2003). Type: Uganda, Ruwenzori, *Scott Elliot* s.n. (BM!, holo.; K!, iso.)
    Brachystelma bingeri A. Chev. in Rev. Cult. Col. 8: 65 (1901). Type: Mali, Zomblara and Tieviana, *Chevalier* 992 (P!, holo.)
    Raphionacme jurensis N.E. Br. in F.T.A. 4(1): 272 (1902); F.P.S. 2: 413 (1952). Type: South Sudan, Jur, between Jur and Wau Rivers, *Broun* s.n. (K!, holo.)
    R. bagshawii S. Moore in J.B. 45: 50 (1907). Type: Uganda, Toro District: Mpanga River, *Bagshawe* 1060 (BM!, holo.)
    R. daronii Berhaut in Bull. Soc. Bot. France Sci: 374 (1955). Type: Senegal, Thies–Ngazobil, Velor, *Berhaut* 1638 (P!, holo.; K!, iso.)
    R. bingeri (A. Chev.) Lebrun & Stork in Bull. Mus. Hist. Nat., Paris 6: 225 (1984)

12. **Raphionacme welwitschii** *Schltr. & Rendle* in J.B. 34: 97 (1896); Cribb & Leedal, Mt Fl. S. Tanz.: 103, t. 23c (1982); Meve, Willke & Albers in Illustr. Handb. Succ. Pl.: Asclepiadaceae: 228 (2002). Type: Angola, Ambaca, between Halo and Zamba, *Welwitsch* 4234 (BM!, holo.; K!, iso.)

Suffrutescent herbaceous climber; tuber hemi-spherical, up 20 cm diameter; aerial stems annual, branching lateral, twining, up to 1 m long, velutinous. Leaf petiole (0.7–)1.5–2.5(–4) cm long, velutinous; blade herbaceous to coriaceous, narrowly ovate, ovate, elliptic, narrowly obovate or obovate, (4–)5–11(–21) × (1.5–)3–7(–9) cm, acute to obtuse-mucronate, base cuneate to obtuse, dark green above, pale green beneath, velutinous to tomentose, lateral veins semi-divaricate. Inflorescences axillary, compact, 10–15-flowered fascicle of monochasia, velutinous; peduncles 3–15 mm long; bracts broadly ovate, 1–2 mm long; pedicels 2–6 mm long. Sepals

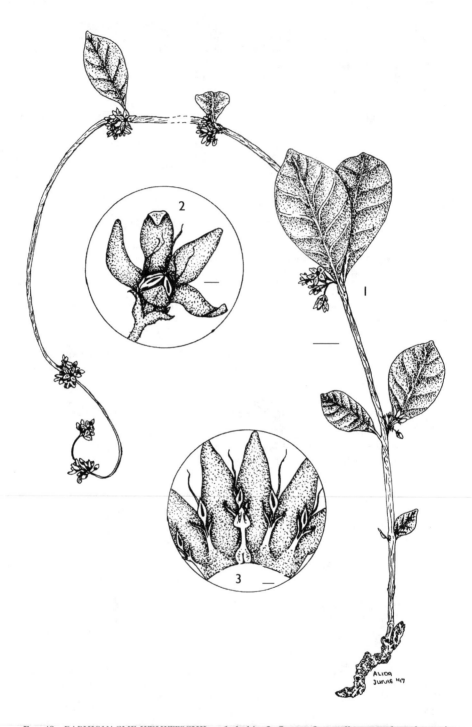

Fig. 48. *RAPHIONACME WELWITSCHII* — **1**, habit; **2**, flower; **3**, corolla opened to show tri-segmented corona lobes, stamens and gynoecium. Scale bars: 1 = 10 mm, 2 & 3 = 2 mm. 1 & 2 from *Luke & Mbinda* 5978; 3 from *Drummond & Helmsley* 3836. Reproduced by permission of the Editor, South African Journal of Botany 75: 344, fig. 33.

broadly triangular, broadly ovate or ovate, 2–3 × 1–2 mm, acute to obtuse, outside velutinous, green tinged red. Corolla tube campanulate, 2–4 mm long, greenish, glabrous; lobes reflexed, greenish-yellow to green with brown, reddish-brown or purplish-brown markings near base of lobe, ovate, 5–6(–12) × 3–4 mm, acute to obtuse, outside tomentose. Corona pale yellow, pale green or green, glabrous, lobes trisegmented to tripartite; central segment filiform to acicular, 2–5 mm long, apex simple or bifid; lateral segments acicular to claw-shaped. Staminal filaments filiform, 1–2 mm long, base dilated; anthers greenish to white, narrowly ovate, 2 mm long, acute to acuminate. Style 2–3 mm long, terete; style-head ovoid, ± 2 mm long, acute. Gynostegium elevated above corolla mouth. Follicles solitary or rarely paired, narrowly ovoid, tapering, 4–6 cm long, 1–1.3 cm wide; seeds brown, obliquely ovate, 7 mm long; coma coppery coloured, 2.5 cm long. Fig. 48, p. 156.

TANZANIA. Kigoma District: Uvinza–Mpanda road, km 42, 23 Nov. 1962, *Verdcourt* 3441!; Mbeya District: Songwe valley, 20 km W of Mbeya, 30 Nov. 1994, *Goyder et al.* 3864!; Masasi District: Ndanda Mission, 17 Mar. 1991, *Bidgood, Abdallah & Vollesen* 2053!
DISTR. **T** 4, 6–8; Angola, Malawi, Mozambique, Congo-Kinshasa, Zambia, Zimbabwe
HAB. Miombo woodland and grassland; 400–1600 m

SYN. *Zuchellia angolensis* Decne. in DC., Prodr. 8: 492 (1844), non *Raphionacme angolensis* (Baill.) N.E. Br. (1902). Type: Angola, *anonymous* s.n. (P!, holo.)
   Raphionacme denticulata N.E. Br. in F.T.A. 4(1): 275 (1902). Type: Malawi, *Mahon* s.n. (K!, holo.)
   R. verdickii De Wild. in Ann. Mus. Congo, ser. 5 (1): 182 (1904). Type: Congo-Kinshasa, Katanga, Lufira, *Verdick* 283 (BR!, holo.; K! photo)
   Chlorocyathus welwitschii (Schltr. & Rendle) Bullock in Cavaco, Contrib. Etude Fl. Lunda Recolt. Gossw. 1946–48: 130 (1959)

41. SCHLECHTERELLA

K. Schum. in E. & P. Pf., Index: 462 (1899); Venter & R.L. Verh. in S. Afr. Journ. Bot. 64: 350–355 (1998)

Pleurostelma Schltr. in J.B. 33: 303, t. 351 (1895), *non* Baill. (1890)
Triodoglossum Bullock in Hooker's Icon. Pl. 36: t. 3584 (1962)

Perennial suffrutescent climber with white latex; root tubers turnip-shaped or cylindrical. Stems twining. Leaves opposite or fascicled; interpetiolar ridges with dentate colleters. Inflorescences terminal and axillary. Corolla tube campanulate to shallowly campanulate; lobes oblong-ovate or obovate. Corolline corona inserted in mouth of the corolla tube, coronal feet fused with staminal filaments and nectaries forming an annulus, lobes ligulate or cylindrical, 2-, 3-, or 5-fid. Gynostegial corona absent. Stamens arise from inner base of coronal feet; filaments filiform; anthers ovate to ovate-hastiform, base with white callosities, pollen in pollinia. Interstaminal nectaries lobular, fused laterally with coronal feet. Style-head broadly ovoid, translators spatulate. Gynostegium just protruding from corolla mouth. Follicles paired, widely divergent, cylindrical-ovoid. Seeds not known.

An African genus of two species found from Ethiopia and Somalia in the north-east to Mozambique in the south-east.

Corolla up to 4 mm long, lobes obovate; corona lobes ligulate, terminally 3–5-fid, up to 1.5 mm long; plants scabridulous; leaves opposite and narrowly obovate to obovate . . . .   1. *S. abyssinica*
Corolla 6–12 mm long, linear to oblong-ovate; corona lobes cylindrical, 4–7 mm long, 2–3-fid with one segment usually longer; plants glabrous; leaves clustered or opposite and linear to very narrowly ovate or very narrowly elliptic . . . .   2. *S. africana*

Fig. 49. *SCHLECHTERELLA AFRICANA* — **1**, flowering stem, × ²⁄₃; **2**, stems showing fascicles of
leaves on stunted lateral shoots, × ²⁄₃; **3**, fruiting specimen, × ²⁄₃; **4**, partial inflorescence, ×
2; **5**, centre of flower showing corona and gynostegium, × 8; **6**, half-flower, × 8; **7**, stamen,
× 12; **8**, translator, × 12. 1, 5–8 from *Newbould* 3242; 2 & 3 from *Greenway* 9546. Drawn by
D. Erasmus.

1. **Schlechterella abyssinica** (*Chiov.*) *Venter & R.L. Verh.* in S. Afr. Journ. Bot. 64: 353 (1998); Venter in Fl. Eth. 4(1): 110 (2003) & in Fl. Somalia 3: 140 (2006). Type: Ethiopia, Borena, *Cufodontis* 165 (FT!, lecto.; designated by Venter & R.L. Verh. in S. Afr. Journ. Bot. 64: 350 (1998))

Climber, up to 1.5 m long, slender, scabridulous; tuber turnip-shaped to cylindrical, 10–25 cm long, 6–10 cm diameter; aerial stems annual or perennial, twining. Leaves opposite; petiole 3–4 mm long; blade recurved, narrowly obovate to obovate, 5–10 × 2–3.5 cm, acute to obtuse-mucronate, base cuneate, dull green, dotted. Inflorescences axillary and terminal, lax, with 3–5 monochasial branches of ± 5 flowers each; peduncles 2–14 mm long; bracts triangular, ± 1 mm long; pedicels 1–3 mm long. Sepals acicular to triangular, 1–1.2 × 0.6–0.7 mm. Corolla glabrous; tube 1.7–2 mm long, campanulate; lobes spreading, obovate, 1.7–2 × 0.8–1 mm, obtuse, dark violet, violet, magenta or mottled green. Coronal annulus ± 0.2 mm high; lobes connivent over gynostegium, white to greenish-white to pale green, 1.5 mm long, ligulate with apices 3–5-fid. Stamens ± 1 mm long. Follicles brown, very narrowly ovoid, 6 cm long, 8 mm diameter, tapering, glabrous or sparsely scabridulous.

UGANDA. West Nile District: Madi, Leya River, 25 Mar. 1945, *Greenway & Eggeling* 7248!
KENYA. Northern Frontier District: Mt Furrole, 15 Sep.1952, *Gillett* 13912! & Marsabit, Jibisa, *Newton & Powys* 3680!
DISTR. U 1; K 1; Ethiopia, Somalia
HAB. Scattered tree grassland on stony soil with *Terminalia* and *Lophira* or *Commiphora* and *Acacia*; 1300–1700 m

SYN. *Raphionacme abyssinica* Chiov. in Miss. Biol. Borana, Racc. Bot. 4: 161 (1939)
    Triodoglossum abyssinicum (Chiov.) Bullock in Hooker's Icon. Pl. 36: t. 3584 (1962)

2. **Schlechterella africana** (*Schltr.*) *K. Schum.* in E. & P. Pf. Nachträge 2: 60 (1900); U.K.W.F. ed. 2: 175 (1994); Venter in Fl. Eth. 4(1): 112 (2003) & in Fl. Somalia 3: 140 (2006). Type: Kenya, Machakos District, Ngomeni, *Scott Elliott* 6175 (BM!, holo.; K! iso.)

Woody climber, glabrous; root tuber cylindrical; aerial stems up to 1.5 m long, lateral shoots sometimes stunted. Leaves opposite on normal shoots, fascicled in verticillate clusters on stunted shoots, sessile to shortly petiolate; blade linear to very narrowly ovate, 4.5–8.5 × 0.2–1 cm, obtuse to attenuate, margin revolute and undulate, base cuneate to rounded. Inflorescences lax, frail, with few-flowered monochasial branches; peduncles 1–2 cm long; bracts triangular, ± 1 mm long; pedicels 3–5 mm long. Sepals ovate-triangular, 1 × 0.5 mm, acute. Corolla tube shallowly campanulate, 0.5–1 mm long; lobes linear to oblong-ovate, 6–11 × 2–4 mm, acute, slightly reflexed, pale greenish outside, inside creamy yellow, white flushed mauve-pink to mauve-purple or pale mauve. Corona creamy yellow, annulus ± 0.5 mm high, lobes cylindrical, 2–3 mm long; terminal part 2- or 3-fid, inner segment longest, filiform, 2–4 mm long. Follicles reddish brown, cylindrical-ovoid, 8–9 cm long, ± 5 mm diameter, blunt-acute. Fig. 49, p. 158.

KENYA. Northern Frontier District: Isiolo to Matthews Range, 16 Dec. 1958, *Newbould* 3242!; Machakos District: Mtito Andei, *Greenway* 9546!; Teita District: Ndara Ranch, 7 Nov. 1992, *Harvey, Mwachala & Vollesen* 37!
TANZANIA. Pare District: near River Ruvu NW of Pare Mountains, 4 Nov. 1955, *Milne-Redhead & Taylor* 7229! & Kisangiro, Oct. 1927, *Haarer* 913
DISTR. K 1, 3, 4, 7; T 3; Ethiopia, Somalia and Mozambique
HAB. *Acacia-Commiphora-Lannea-Sterculia* woodland or semi-desert scrub; 300–1400 m
USES. The tubers are used as a source of water by local people, and as food for their camels

SYN. *Pleurostelma africanum* Schltr. in J.B. 33: 303, t. 351 (1895)
    Tacazzea africana (Schltr.) N.E. Br. in F.T.A. 4(1): 261 (1902)

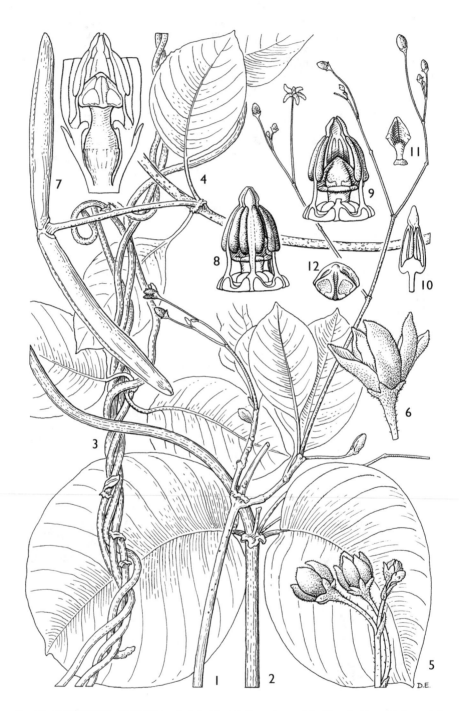

FIG. 50. *BASEONEMA GREGORII* — **1–4**, habit with flowering and fruiting shoots, × ²/₃; **5**, partial
inflorescence, × 2; **6**, flower, × 4; **7**, half-flower, × 8; **8**, gynostegium, × 8; **9**, gynostegium with
one stamen removed to expose stylar head, × 8; **10**, stamen, ventral view, × 8; **11**, translator,
× 8; **12**, top of stylar head, × 8. 1, 5 & 6 from *Bally* 8125; 2, 7–12 from *Bally* 8745; 3 from
Gardner 3612; 4 from *van Someren* 184. Drawn by D. Erasmus.

42. BASEONEMA

Schltr. & Rendle in J.B. 34: 97 (1896); N.E. Br. in F.T.A. 4(1): 259 (1902); Thonner, Flowering Plants of Africa: 442 (1915); Bullock in K.B. 8: 64 (1953); Venter & R.L. Verh. in S.Afr. J. Bot. 75: 445–455 (2009)

Climber with white latex; stems twining, pubescent. Leaves opposite, petiolate, interpetiolar stipules forming a swollen ridge with dentate margin; blade with veins patent in lower half of blade, divaricate in upper half. Inflorescences axillary, lax. Corolla tube reflexed, lobes outside velutinous, inside glabrous, green to yellowish-green. Corolline corona annular on corolla inversion, coronal feet massive. Gynostegial corona absent. Staminal filaments fused to coronal feet, pollen in tetrads. Interstaminal nectaries lobular, fused into an annulus with coronal feet. Gynostegium exposed. Follicles paired, very narrowly oblong-ovoid.

A monotypic genus, endemic to dry savanna in Kenya and Tanzania.

Baseonema gregorii *Schltr. & Rendle* in J.B. 34: 97 (1896); U.K.W.F. ed. 2: 174 (1994); K.T.S.L.: 489 (1994). Type: Kenya, Machakos District, Kenani & Ongalea Mountains, *Gregory* 14 (BM!, holo.)

Climber with white latex; roots unknown; stems woody, bark brown, young stems pubescent, older sparsely pubescent. Leaves pubescent to velutinous; petiole 0.5–1 cm long; blade broadly ovate to broadly elliptic to sub-orbicular, 2–4 × 2–4 cm, acute to acuminate, recurved, margin undulate, base obtuse, dark green above, pale green beneath. Inflorescences few-flowered, sparsely pubescent, purplish, nodes woolly; primary peduncles 2–4 (–7) cm long, secondary 1–5 cm long; bracts acicular, 1–2 mm long, outside tomentose; pedicels 1–4 cm long. Sepals ovate to triangular, 1–2 × 1 mm, outside velutinous. Corolla lime green to yellow-green mottled maroon, velutinous on outside, inside glabrous; tube 1 mm long, reflexed at coronal feet; lobes oblong, 4–5 × ± 2 mm, apex obtuse, main vein prominent. Coronal feet ovoid-deltoid, ± 1 mm long, corona lobes absent; interstaminal nectary lobes ligulate, acute to bifid, ± 1 mm long. Style ± 1 mm long; style-head conical-ovoid, translator receptacle ovate with linear stalk and sub-spherical sticky disc. Follicles 5–10 cm long, 7–11 mm diameter, puberulous to sub-glabrous. Seeds 10–11 × 2 mm, compressed; coma 2–3 cm long. Fig. 50, p. 160.

KENYA. Machakos District: 1 km S of Hunters Lodge, 11 Feb. 1996, *Goyder et al.* 4006 & 237 km from Mombasa to Nairobi, 15 Apr. 1960, *Verdcourt & Polhill* 2695!; Teita District: Mt Kasigau, 4 km S of Rukanga, 2 June 1981, *M. & C. Gilbert* 6122!
TANZANIA. Masai District: Naberera, Olimosori waterhole, W of Okuto Hill, 14 Nov. 1983, *Vincent* 202; Pare District: Lembeni, 2 Apr. 1952, *Bally* 8125!
DISTR. **K** 1, 4, 6, 7; **T** 2, 3; not known elsewhere
HAB. Dry bushland or evergreen forest on rocky outcrops; 600–1500 m

43. BUCKOLLIA

Venter & R.L. Verh. in S. Afr. Journ. Bot. 60: 97 (1994)

Woody climbers with white latex; stems twining, bark verrucose. Leaves opposite or fascicled, petiolate; blade hairy, secondary veins divaricate and parallel. Inflorescences terminal on short lateral shoots, or axillary, monochasial or dichasial with two monochasial branches, each branch bearing up to 10 flowers. Corolla tube saucer-shaped; lobes with glandular swellings at inner bases. Corolline corona arising at the mouth of the corolla tube; lobes simple, filiform, distinct. Gynostegial corona absent. Stamens from inner bases of coronal feet, glabrous; filaments filiform;

anthers ovate, apex acute, pollen in tetrads. Interstaminal nectaries lobular, distinct. Gynostegium exposed. Follicles paired, divergent to horizontal, long cylindrical-ovoid, tapering to blunt apex, puberulent and verrucose.

A genus of two species, endemic to dry parts of north-eastern and eastern Africa.

All leaves opposite, ovate to broadly ovate, discolorous, beneath
    densely white to rusty tomentose; corolla lobes white tomentose
    on outside, inside puberulous; follicles sparsely verrucose . . . . .   1. *B. tomentosa*
Most leaves clustered on short lateral shoots, narrowly to broadly
    obovate, green-pubescent; corolla lobe glabrous; follicles
    densely verrucose . . . .   2. *B. volubilis*

1. **Buckollia tomentosa** (*E.A. Bruce*) *Venter & R.L. Verh.* in S. Afr. Journ. Bot. 60: 98 (1994); Venter in Fl. Eth. 4(1): 103 (2003). Type: Uganda, Acholi District, Agoro, *Eggeling* 1704 (K!, holo.)

Woody climber with white latex; roots unknown; stems sparsely verrucose, densely tomentose in young stems. Leaves opposite; petiole 1–2 cm long, rusty-tomentose; blade ovate to broadly ovate, 2–5 × 1–2 cm, acute to acuminate, margin undulate, base obtuse, discolorous, dark green and puberulent above, densely white-tomentose with midrib rusty-tomentose beneath. Inflorescences densely tomentose to pubescent; peduncles 7–10 mm long; bracts ovate, 1–2 mm long, acuminate; pedicels 4–5 mm long. Sepals triangular to ovate, acute, ± 2 mm long, white-tomentose outside. Corolla greenish to creamy-yellow; tube ± 0.5 mm long; lobes ovate to oblong-ovate, 5–6 × 3 mm, obtuse, tomentose outside, sparsely puberulent within, glandular swelling violet-brown. Corona lobes filiform, 3–4 mm long, puberulent. Staminal filaments ± 1 mm long, anthers 1–2 mm long. Style-head sub-sessile, angular-ovoid, apex bifid, translator receptacle broadly obovate and subsessile on trifid sticky disc. Follicles brown, 13–18 cm long, ± 9 mm diameter, sparsely verrucose. Seeds unknown.

UGANDA. Karamoja District: near Loyoro, Feb. 1960, *J. Wilson* 844!
DISTR. U 1; Ethiopia, Sudan
HAB. 'Scrub savanna'; 1300–1400 m

SYN. *Tacazzea tomentosa* E.A. Bruce in K.B. 1936: 477 (1936); Bullock in K.B. 9: 354 (1954)

2. **Buckollia volubilis** (*Schltr.*) *Venter & R.L. Verh.* in S. Afr. Journ. Bot. 60: 97 (1994); Venter in Fl. Eth. 4(1): 103 (2003) & in Fl. Somalia 3: 135 (2006). Type: 'Central Africa', *Scott Elliot* s.n. (BM!, holo.; K!, iso.)

Woody climber with white latex; roots tuberous; stems up to 4 m long, scrambling, branches divergent to horizontal, lateral shoots mostly stunted. Leaves fascicled on stunted shoots or opposite on normal shoots; petiole 1–3 mm long; blade narrowly to broadly ovate, 3–7 × 1–3 cm, acute to obtuse or retuse, base cuneate, green, pubescent. Inflorescences white-tomentose; peduncles 2–8 mm long; pedicels 1–2 mm long. Flowers aromatic. Sepals ovate to broadly ovate, 1–2 × 1–2 mm. Corolla green to yellowish-green, glabrous; tube ± 1 mm long; lobes narrowly triangular to narrowly ovate, 5–7 × 2–3 mm, acute to obtuse, glandular swelling globular, dull red-brown. Corona lobes pink, 4–9 mm long, base yellow to brownish, glabrous to sparsely puberulous. Staminal filaments 1 mm long, anthers 1–2 mm long. Style columnar, ± 1 mm long; style-head ovoid, obtuse to concave; translator with receptacle broadly elliptic, stalk linear, sticky disc hemispherical. Follicles greyish-brown, 14–19 cm long, 6–8 mm diameter, densely verrucose. Seeds narrowly ovate, 5–7 × 2 mm, smooth; coma whitish, 2–3 cm long. Fig. 51, p. 163.

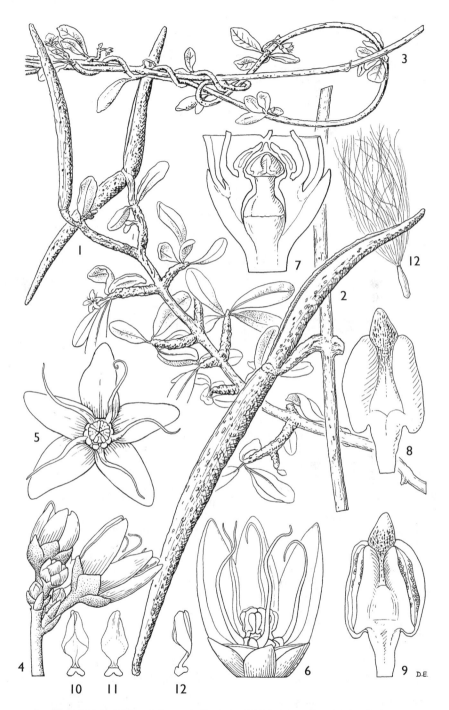

Fig. 51. *BUCKOLLEA VOLUBILIS* — **1–3**, flowering and fruiting stems, × ²/₃; **4**, inflorescence, × 3; **5**, flower from above, × 3; **6**, flower with two corolla lobes removed to expose corona and gynostegium, × 6; **7**, half-flower, × 8; **8 & 9**, stamen in dorsal and ventral view, × 27; **10–12**, translators, × 27; **13**, seed and coma of hairs, × 1¹/₃. 1, 4–12 from *Drummond & Hemsley* 2307; 2 from *Gillett* 12613; 3 from *Tanner* 1274. Drawn by D. Erasmus.

UGANDA. Toro District: Ruwenzori, anno 1893–94, *Scott Elliot* s.n.!
KENYA. Northern Frontier District: Moyale, 20 Oct. 1952, *Gillett* 14068! & Dandu, 22 Mar. 1952,
   Gillett 12613!; Teita District: between Mwatate and Voi, 18 Oct. 1961, *Archer* 291!
TANZANIA. Masai District: Minjingu, 107 km S of Arusha, 4 Jan. 1962, *Polhill & Paulo* 1050!;
   Lushoto District: Buiko, 30 Apr. 1953, *Drummond & Hemsley* 2307!; Pare District: Mkomazi, 7
   km towards Tia Panda from Lembeni, 11 Dec. 2000, *Bruyns* 8685!
DISTR. **U** 2; **K** 1, 3, 6, 7; **T** 1–3; Ethiopia, Somalia
HAB. Dry bushland on granite, basement screes or quartz outcrops; 750–1400 m
USES. Pounded roots in drinking water are used to enhance lactation.

SYN. *Raphionacme volubilis* Schltr. in J.B. 33: 302 (1895)
   Tacazzea volubilis (Schltr.) N.E. Br. in F.T.A. 4(1): 262 (1902)
   Curroria volubilis (Schltr.) Bullock in K.B. 9: 359 (1954); U.K.W.F. ed. 2: 174 (1994);
   K.T.S.L.: 491 (1994)

44. SARCORRHIZA

Bullock in Hooker's Icon. Pl. t. 3585 (1962)

Perennial climbing epiphyte; latex colour not recorded; root-tubers oblong-ovoid.
Stems contorted, glabrous, succulent. Leaves petiolate, blade coriaceous.
Inflorescences terminal and axillary, few-flowered. Corolla rotate, lobes spreading,
bicolored, green and yellowish-red. Corolline corona lobes filiform. Gynostegial
corona absent. Stamens fused to base of corona lobes, pollen in tetrads.
Gynostegium exposed. Follicles paired.

A monotypic genus.

Sarcorrhiza epiphytica *Bullock* in Hooker's Icon. Pl.: t. 3585 (1962). Type: Tanzania,
Lushoto District, path to Mt Bomole, near Amani, *Verdcourt* 1725 (K!, holo.)

Climbing epiphyte; root tubers oblong-ovoid, purple-brown, 8–9 × 3–4 cm
diameter; stems contorted, branched; bark reddish, sub-fleshy, glabrous, shiny.
Leaves glabrous; petiole 2–5 mm long, dark crimson, glabrous, puberulous or
pubescent; blade oblong-elliptic to oblong-ovate, 6.7–10 × 1.8–3.3 cm, acute to
acuminate, base obtuse to cordate, main vein glabrous or puberulous above, lateral
veins looping towards blade tip. Inflorescences terminal and axillary, 2–4- flowered,
puberulous or pubescent; peduncles ± 2 mm long; bracts triangular or ovate, ± 1 mm
long, margin fimbriate, overlapping; pedicels ± 3 mm long. Sepals ovate to
triangular-ovate, ± 1 × 0.7 mm, sparsely puberulous. Corolla glabrous; tube
campanulate, ± 1 mm long; lobes spreading, ovate, 4–6 × 3–4 mm, obtuse-acute,
green or yellowish-green outside, inside with outer edge yellow or greenish, middle
zone dark crimson-brown. Corona creamy yellow; coronal feet dilated, fused with
stamens and interstaminal nectaries into an annulus; lobes from apices of coronal
feet, filiform, 2–3 mm long, glabrous. Stamens from inner base of coronal feet;
filaments filiform, ± 1 mm long; anthers ovate, ± 1 mm long, apiculate; pollen in
tetrads. Interstaminal nectaries lobular, erect around style. Style-head sessile on
ovaries, ovoid, translator receptacle ovate. Gynostegium exposed from corolla
mouth. Follicles paired, horizontally divergent, falcate, cylindrical, 10–15 cm long,
5–10 mm diameter, puberulous or pubescent. Seeds not known. Fig. 52, p. 165.

TANZANIA. Lushoto District: Kwamkoro, 15 Dec. 1959, *Semsei* 2957! & Amani, 2 Jan. 1934,
   Greenway 3679!; Morogoro District: Uluguru Mountains, 8 Nov. 1932, *Schlieben* 2939!
DISTR. **T** 3, 6; Ivory Coast, Congo-Kinshasa
HAB. Epiphyte on trees in evergreen moist and mist forest, also a shrubby creeper over rock;
   1000–1900 m

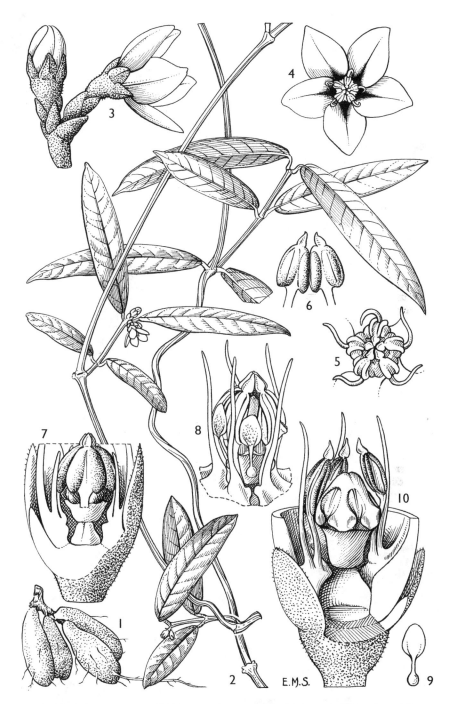

FIG. 52. *SARCORRHIZA EPIPHYTICA* — **1**, tuberous roots, × ¹/₃; **2**, stem with leaves and flowers, × ²/₃; **3**, part of inflorescence, showing bracts and flowers, × 4; **4**, flower from above, × 4; **5**, flower centre from above, × 8; **6**, stamens from inside, × 12; **7**, section of flower bud, × 12; **8**, corona lobes and gynostegium, × 12; **9**, translator, 12; **10**, section of flower to show corona, stamens and style-head, × 12. All from *Verdcourt* 1725. Drawn by Margaret Stones, and reproduced with permission from Hooker's Icon. Pl. 36: t. 3585

45. **EPISTEMMA**

D.V. Field and J.B. Hall in K.B. 37: 117 (1982)

Epiphytic or rupicolous climber; latex white. Stems succulent, nodes swollen. Leaves opposite, petiolate, blade coriaceous or subsucculent. Inflorescences terminal or subterminal, several- to many-flowered; peduncles bracteate. Calyx lobes fused at extreme base. Corolla rotate; tube shallowly bowl-shaped, with or without radiating ridges and alternating hollows; lobes broadly ovate to ovate, yellowish-green to purple or reddish-brown. Corolline corona arising at corolla lobe sinuses directly

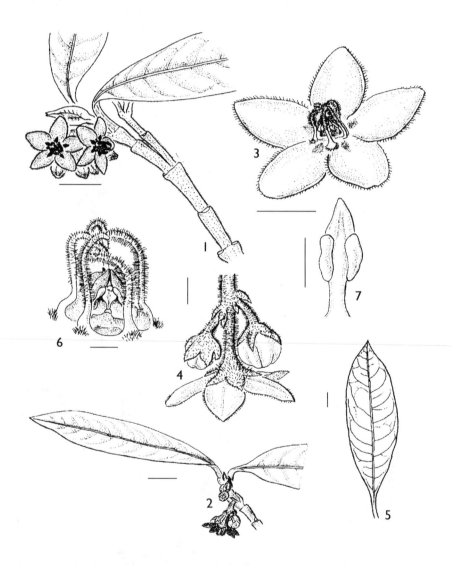

FIG. 53. *EPISTEMMA NEUERBERGII* — **1 & 2**, flowering branches; **3**, flower; **4**, inflorescence; **5**, leaf; **6**, gynostegium and corolline corona lobes; **7**, stamen. Scale bars: 1, 2 & 5 = 10 mm; 3 = 5 mm; 4 = 2 mm; 7 = 1 mm. All from *Fischer & Biedinger* 227. Drawn by E. Fischer and reproduced by permission of the Editor, South African Journal of Botany 77: 681, fig. 1.

above stamens, lobes filiform, distinct, sometimes club-shaped at obtuse tips, basally fused with staminal filaments and nectaries. Gynostegial corona absent. Stamens with pollen in pollinia. Interstaminal nectaries fused laterally with coronal feet. Style-head broadly obovoid; translators broadly spatulate, apically split. Gynostegium exposed from corolla. Follicles paired, fusiform, widely divergent.

4 species distributed across Central and West Africa.

Epistemma neuerburgii *Eb. Fisch.* & *Venter* in S. Afr. Journ. Bot. 76: 681 (2011). Type: Rwanda, Nyungwe National Park, between Karamba and Gisakura, *Fischer & Biedinger* 227 (BR, holo.; BOL, KOBL, iso.)

Epiphytic climber; stems twining, up to 1.5 m long, bark greenish-brown, glossy, pubescent. Leaf petiole up to 2.5 cm long; lamina lanceolate, 12 × 3.5 cm, acuminate, coriaceous, pubescent. Inflorescences terminal and sub-terminal, 4–8-flowered, pubescent; peduncle up to 7 mm long; bracts ovate-acuminate; pedicel up to 7 mm long. Calyx lobes ovate, 2–2.5 mm long, margin ciliate, without pubescent. Corolla up to 1.4 cm in diameter; tube purple, 1–1.5 mm long, outside glabrous, inside with few hairs near bases of corona lobes; lobes ovate, acute to acute-obtuse, 7–8 × 3–4 mm, within reddish-purple becoming darker towards base, glossy, margins ciliate. Corona lobes incumbent on anthers, filiform, 4–5 mm long, densely white-pubescent. Stamens arise from inner base of corona, filaments terete, 1 mm long; anthers greenish white, ovoid with acute connective, 2 mm long, cells ± 1 mm long; pollinia ovoid, ± 1 mm long. Nectaries lobular, fitting around style. Styles ± 0.5 mm long; style-head broadly obovoid with obtuse apex, 1 × 1.5 mm; translator receptacle broadly ovate and deeply split, ± 1 × 0.8 mm. Gynostegium exposed from corolla tube. Fruit and seeds not known. Fig. 53, p. 166.

UGANDA. Kigezi District: Ihihizo, Bwindi Impenetrable National Park, no date, *Hafashimana* 721! DISTR. U 2; Rwanda
HAB. Moist forest; ± 1500 m

subfam. SECAMONOIDEAE

The smallest of the subfamilies of Apocynaceae. This group is most diverse on Madagascar, both in numbers of species and genera represented. Approximately 7 genera are recognised, with ± 150 species. The single genus *Secamone* occurs in continental Africa, and just 10 species are to be found within the Flora area.

46. **SECAMONE**

R. Br. in Mem. Wern. Soc. 1: 55 (1809); Goyder in K.B. 47: 437–474 (1992); Klackenberg in Adansonia 23: 317–335 (2001)

Rhynchostigma Benth. in Hooker's Ic. Pl. 12: 77 (1876) & in Gen. Pl. 2: 771 (1876)

Lianes and twining shrubs with white latex. Young shoots and inflorescences glabrous or puberulent. Leaves opposite, petiolate. Inflorescences terminal or axillary (extra-axillary is the norm in Asclepiadoideae), cymose or composed of an aggregation of cymes. Calyx of 5 ± free, with ciliate margins. Corolla rotate or campanulate, 1–8 mm long, glabrous or hairy on the inner surface, 5-lobed to halfway or beyond; corolline corona present or occasionally absent, formed of 5, usually paired, nonvascularised fleshy ridges running from the base of the corolla tube to the corolla lobe sinus, sometimes extending up the inner face of the corolla lobes or forming a pocket obscuring the sinus. Gynostegial corona staminal, of 5 laterally or more rarely dorsally compressed or subulate lobes. Staminal filaments entirely free or connate at the base; anthers with entire or fimbriate appendages. Pollinaria minute, with 4 pollinia attached directly or by extremely short, flattened caudicles to the pale, porous corpusculum. Stylar head appendage terete, lobed or dilated, exserted beyond the top of the staminal column or not. Follicles paired but sometimes only one developing, smooth or shallowly striate. Seeds ovate, compressed, with a coma of white or ivory hairs.

A genus of ± 90 species distributed across the paleotropics, with its centre of diversity in Madagascar, and with 21 species in continental Africa.

1. Corona lobes dorsiventrally flattened and about as long as the staminal column or longer; stylar head appendage exserted from staminal column for more than 3 mm .... 10. *S. racemosa* (p. 176)
   Corona lobes laterally compressed or subulate or reduced; stylar head appendage exserted from staminal column for 1 mm or less 2
2. Corolla lobes pubescent adaxially 3
   Corolla lobes glabrous 4
3. Adaxial face of corolla pubescent only at the base of the lobe; corolline corona of 5 pairs of fleshy ridges stretching from the base of the corolla tube to halfway along lobes; stylar head appendage clavate .... 9. *S. africana* (p. 175)
   Adaxial face of corolla lobes pubescent except on the margin; corolline corona absent; stylar head appendage not dilated .... 6. *S. alpini* (p. 173)
4. Stylar head appendage obconic, clavate or with two divergent lobes 5
   Stylar head appendage not dilated 10
5. Mature leaves pubescent with rusty spreading hairs 6
   Mature leaves glabrous or with few sparsely scattered hairs 7
6. Corolla 1.5–3 mm long; cymes terminal and axillary, up to 6-flowered; lower leaf surface with dense tuberculate papillae in addition to indumentum 7. *S. stuhlmannii* (p. 174)
   Corolla 3–4 mm long; cymes mostly terminal and always single flowered; lower leaf surface ± smooth with red hairs restricted to upper and lower midrib .... 8. *S. clavistyla* (p. 175)
7. Corolline corona formed of 5 membranous pockets obscuring corolla lobe sinus adaxially; corolla tube 1 mm long excluding coronal pockets, at least 1.5 mm long including the coronal pockets ..... 1. *S. gracilis* (p. 169)
   Corolline corona pockets absent; corolla tube less than 1 mm long 8
8. Corolla robust, usually 3–4.5 mm long; inflorescences regularly divided into straight, robust branches; pedicels ± 0.4 mm thick .... 9. *S. africana* (p. 175)
   Corolla thin, 1.5–3 mm long; inflorescences subumbelliform or irregularly branched, the branches more slender, often flexuous; pedicels usually 0.1–0.3 mm thick 9

9. Inflorescences lax, 2–3 times dichotomously
   branched; abaxial leaf surface minutely but
   densely tuberculate papillate . . . .   2. *S. retusa* (p. 169)
   Inflorescences congested, with few branches;
   abaxial leaf surface without tuberculate papillae
   (viewed under at least × 25) . . . .   3. *S. punctulata* (p. 170)
10. Corolla ± 1.5 mm long; staminal corona lobes
    triangular or quadrate, ± half the length of the
    staminal column or less; leaves lanceolate . . . . . .   4. *S. attenuifolia* (p. 171)
    Corolla 2–2.5 mm long; staminal corona lobes falcate,
    half to as long as the staminal column; leaves mostly
    ovate to suborbicular, but occasionally lanceolate .   5. *S. parvifolia* (p. 171)

1. **Secamone gracilis** *N.E. Br.* in K.B. 1895: 248 (1895) & in F.T.A. 4(1): 278 (1902).
Type: Kenya, Coast Province, Mombasa, *Wakefield* s.n. (K!, holo.)

Slender climber, glabrous except for a few red hairs at the nodes of young
shoots. Leaves with petiole 0.4–1(–1.3) mm long; lamina oblong or elliptic, 0.6–2.8
× 0.2–1.2 cm, apex obtuse or rounded, often minutely apiculate, base rounded to
subcuneate, thin, lower surface minutely but densely tuberculate-papillate. Flowers
solitary or in few-flowered, dichotomously branched axillary cymes; peduncles
0.5–3 mm long; pedicels 3–8 mm long. Calyx lobes ovate or oblong, 0.5–1 × 0.3–0.5 mm,
apex generally obtuse. Corolla white, 3–5 mm long, divided to halfway or beyond;
tube ± 1 mm long excluding the coronal pockets, campanulate; lobes oblong, 2–4 ×
0.8–1 mm, obtuse. Corolline corona of 5 membranous pockets on the inner surface
of the corolla obscuring the corolla lobe sinus for 0.6–0.8 mm above the mouth of
the corolla tube, the pockets each with a narrow flange running to the base of the
corolla tube. Staminal corona lobes slender, erect, ± ²/₃ length to as along as the
staminal column, falcate or linear. Stylar head appendage obconic, shallowly 2-
lobed, verruculose, exserted for ± 0.2 mm from staminal column. Follicles and
seeds unknown.

KENYA. Mombasa District: Mombasa, Dec. 1884, *Wakefield* s.n.!; Kilifi District: Kambe Kaya, 25
    Mar. 1981, *Hawthorne* 191!
TANZANIA. Handeni District: Kwa Mkono, 13 km E of Handeni, Oct. 1976, *Archbold* 2212!
DISTR. **K** 7; **T** 3; known only from these three collections
HAB. Coastal forest; 0–500 m

2. **Secamone retusa** *N.E. Br.* in K.B. 1895: 248 (1895) & in F.T.A. 4(1): 279 (1902).
Type: Tanzania, Zanzibar, *Kirk* s.n. (K!, holo.)

Woody climber; stems twining, glabrous except for a minute pubescence at the
joints of the inflorescence and a few scattered hairs on undersides of leaves. Leaves
with petiole (1–)2–4.5(–7) mm long; lamina coriaceous, oblong or obovate,
2.7–6.5(–7.7) × 0.8–2.8 cm, apex obtuse or rounded, occasionally retuse, apiculate,
base rounded to cuneate, the margins commonly inrolled; veins prominent
underneath, lower surface paler than the upper, lower surface minutely but densely
tuberculate-papillate. Inflorescence terminal and axillary, lax, 15–30(–40) mm long,
2–3 times dichotomously branched, the branches slender, spreading, often at ± 90°;
pedicels slender, 3–6(–8) mm long. Calyx lobes broadly ovate or oblong, 0.5–1 ×
0.4–0.7 mm, obtuse, margins ciliate. Corolla yellow or greenish white, drying orange-
brown with paler margins, 2–3 mm long, united for about ¹/₄ of its length; tube ± 0.5 mm
long, campanulate; lobes oblong, 1.5–2.5 × ± 1 mm, obtuse or rounded at the apex.
Corolline corona of 5 pairs of fleshy ridges extending from the base of the corolla

tube to either side of the corolla lobe sinus. Staminal corona lobes minute, spreading or occasionally erect and slightly incurved, deltoid-subulate, to about half the height of the staminal column. Stylar head appendage exserted 0.3–0.4 mm from the staminal column, broadly obconic and shortly bilobed, the lobes truncate or occasionally rounded. Follicles ± 75 × 7 mm, fawn, striate.

KENYA. Kwale District: Kaya Lunguma, 11 Nov. 1992, *Luke* 3344!; Kilifi District: Arabuko Forest Reserve, Oct. 1962, *Greenway* 10821! & Kaya Kivara, 16 Nov. 1992, *Harvey et al.* 53!
TANZANIA. Tanga District: Marungu [Morongo], 14 Nov. 1957, *Faulkner* 2092!; Zanzibar: Mazazini, Nov. 1963, *Faulkner* 3320!; Pemba: Ngezi Forest, Dec. 1989, *Beentje* 4360!
DISTR. **K** 7; **T** 3, 6, 8; **Z**; **P**; Mozambique
HAB. Lowland and coastal forest on sandy soils; 0–350 m

3. **Secamone punctulata** *Decne.* in DC., Prodr. 8: 502 (1844); N.E. Br. in F.T.A. 4(1): 284 (1902); T.T.C.L.: 68 (1949); U.K.W.F. ed. 2: 176 (1994); K.T.S.L.: 497 (1994); Goyder in Fl. Eth. 4(1): 113 (2003) & in Fl. Somalia 3: 144 (2006). Type: Tanzania, Pemba, *Bojer* s.n. (P, holo.; K!, iso.)

Thin wiry climber or small shrub, glabrous or sparsely rusty-puberulent on young shoots, the inflorescence sparsely to densely covered with spreading red hairs. Leaves shortly petiolate; petioles 0.2–3.5(–6.5) mm long; lamina linear, elliptic or oblong to lanceolate or narrowly ovate, (1–)1.3–5(–6.1) × 0.1–2(–2.4) cm, apex acute or slightly attenuate to obtuse or rounded, apiculate, margins occasionally undulate, base rounded or obtuse, glabrous or glabrescent, lower surface smooth or finely granulate; veins commonly indistinct. Inflorescences terminal and axillary, numerous, 5–20 × 2–25 mm, the cymes few-flowered and subumbelliform on short peduncles or larger and more diffuse, sometimes on leafless shoots, flowers ovoid in bud, sweetly scented; peduncles (1–)4–15 mm long; pedicels 0.5–2(–5) mm long, glabrous or sparsely pubescent. Calyx lobes ovate or suborbicular, 0.5–1 × 0.5–1 mm, the apex obtuse or rounded, margins ciliate. Corolla orange or yellow, 1–2(–3) mm long, lobed to ± $^2/_3$; tube 0.3–0.6(–0.8) mm long, broadly campanulate; lobes oblong, 1.2–2 × 0.6–0.8 mm, obtuse. Corolline corona of 5 pairs of fleshy ridges extending from the base of the corolla tube to either side of the corolla lobe sinus. Staminal corona lobes erect, green or white, reaching to just below the top of the staminal column or variously reduced to an upward or outward pointing peg about $^1/_2$ the height of the column. Stylar head appendage exserted 0.2–0.4 mm from the staminal column, obconic or clavate at the apex. Follicles widely divergent, olive-green, (3.5–)4.5–5.5 × 0.3–0.5 cm, slender, attenuate, striate.

UGANDA. Karamoja District: Kadam [Debasien] Mt, *Eggeling* 2824!; Mbale District: West Bugwe Local Forest Reserve, 12 Apr. 1951, *G.H.S. Wood* 30!; Mengo District: Kajansi Forest Reserve, May 1935, *Chandler* 1212!
KENYA. Trans-Nzoia District: Kitale, Aug. 1971, *Tweedie* 4106!; Nairobi District: Karen, 2 Sep. 1975, *van Someren* in EA 15954!; Teita District: Tsavo National Park, Galana River, Lugard Falls, 2 Dec. 1966, *Greenway & Kanuri* 12645!
TANZANIA. Tanga District: Bomandani, Boma Peninsula, 3 Aug. 1953, *Drummond & Hemsley* 3617!; Morogoro District: Uluguru Mts, Kitundu, 22 Nov. 1935, *E.M. Bruce* 195!; Zanzibar: M'kokolone, 15 Aug. 1961, *Faulkner* 2887!; Pemba, *Vaughan* 696!
DISTR. **U** 1–4; **K** 1, 3, 4, 5, 7; **T** 1–4, 6, 7; **Z**; **P**; Ivory Coast, Ghana, Congo-Kinshasa, Sudan, Ethiopia, Somalia, Angola and Namibia
HAB. Thickets, riverine and coastal forests; 0–2400 m

SYN. *Secamone micrandra* K. Schum. in E.J. 17: 142 (1893); N.E. Br. in F.T.A. 4(1): 281 (1902). Type: Angola, Golungo Alto, *Welwitsch* 5942 (?B†, holo.; K!, lecto., designated here; COI!, LISU!, isolecto.)
    S. stenophylla K. Schum. in P.O.A. C: 325 (1895). Type: Tanzania, Lushoto District: Mashewa [Mascheua], *Holst* 3510 (B†, holo.; K!, W!, iso.)

S. sansibariensis K. Schum. in P.O.A. C: 325 (1895); N.E. Br. in F.T.A. 4(1): 282 (1902); K.T.S.L.: 497 (1994). Types: Zanzibar, *Stuhlmann* I: 533 (K!, lecto., designated here; B†, isolecto.); Zanzibar, *Stuhlmann* 490 (B†, paralecto.); 572 (B†; paralecto.) & 772 (B†; paralecto.)

S. punctulata Decne. var. *stenophylla* (K. Schum.) N.E. Br. in F.T.A. 4(1): 284 (1902); T.T.C.L.: 68 (1949)

S. sp. A of U.K.W.F. ed. 1: 373 (1974) & K.T.S.L.: 498 (1994) (*Archer* 213)

NOTE. Considerable variation occurs in leaf shape and flower size with a trend for reduction in dimensions west of the Flora area. Narrow leaved forms may be associated with drier or more open vegetation types.

4. **Secamone attenuifolia** *Goyder* in K.B. 47: 446 (1992); U.K.W.F. ed. 2: 176 (1994). Type: Tanzania, Masai District, Mt Longido, *Richards* 23643 (K!, holo.)

Slender woody climber, glabrous or with fine adpressed hairs on young stems and inflorescences. Leaves with petioles 1–4 mm long; lamina lanceolate, 2–5.2 × 0.1–1.1 cm, with an attenuate apex, base obtuse to subcuneate, margins commonly inrolled in dried material, thin, glabrous, lower surface minutely but densely tuberculate-papillate; veins distinct. Inflorescences terminal and axillary, numerous, 5–10 × 5–10 mm, the cymes few-flowered and subumbellate; peduncles 1–2 mm long with fine adpressed hairs; pedicels 1–3 mm long, finely pubescent in 2 lines. Calyx lobes broadly ovate, ± 0.5 × 0.5 mm, obtuse or rounded, finely ciliate. Corolla ± 1.5 mm long, ± globose in bud, yellow, lobed to ± ²/₃; tube broadly campanulate, ± 0.5 mm long; lobes broadly ovate, ± 1 × 1 mm, obtuse. Corolline corona of 5 pairs of fleshy ridges extending from the base of the corolla tube to either side of the corolla lobe sinus, the top of each ridge formed into an auriculate flange. Staminal corona lobes minute, laterally flattened, entirely adnate to the staminal column, triangular, widest at the bottom and reaching ¹/₃–¹/₂ the height of the column. Stylar head appendage scarcely exserted from the column, truncate, not dilated at the apex. Follicles closely adpressed, 3.5–5.5 × 0.3–0.5 cm, attenuate.

KENYA. West Suk, May 1969, *Tweedie* 3638!; Machakos District: Kibwezi, 1922, *Dummer* 5069!; Masai District: Karibani [Garabani] Hill, 16 Mar. 1940, *V.G.L. van Someren* 141!
TANZANIA. Mbulu District: Lake Manyara National Park, 26 May 1965, *Greenway & Kanuri* 12081!; Pare District: Kisiwani, 2 Feb. 1936, *Greenway* 4569!; Iringa District: Kitonga Gorge, 16 Dec. 1961, *Richards* 15700!
DISTR. **K** 2, 4, 6; **T** 2, 3, 5, 7; not known elsewhere
HAB. Ravines and rocky outcrops; 800–1500 m

5. **Secamone parvifolia** (*Oliv.*) *Bullock* in K.B. 9: 368 (1954); U.K.W.F. ed. 2: 176 (1994); K.T.S.L.: 497 (1994); Goyder in Fl. Eth. 4(1): 113 (2003). Type: Tanzania, Kilimanjaro, *Johnson* s.n. (K!, holo.)

Woody climber, old stems with two spiralling, corky wings alternating at the nodes; young shoots puberulent, the hairs white or red, ± adpressed. Leaves with petiole 1–4(–6) mm long; lamina usually thin, broadly ovate or suborbicular, occasionally lanceolate, (0.8–)1.1–5.6(–6.4) × 0.4–3(–4.6) cm, apex acute to acuminate or obtuse, sometimes apiculate, the base rounded or rarely subcordate, lower surface smooth or minutely but densely tuberculate-papillate; veins prominent, the margins smooth or crispate. Flowers 1 to many in terminal or axillary, simple or branched cymose inflorescences, scented, commonly on short lateral shoots, lax or condensed, 5–30 × 10–30 mm; peduncles pubescent, 1–8(–15) mm long; pedicels 1–6 mm long, glabrous or with a line of hairs. Calyx lobes oblong, broadly ovate or suborbicular, 0.5–1 × 0.5–1 mm, obtuse or rounded, ciliate. Corolla white, cream or yellow, 2–2.5 mm long, divided for ¹/₅–³/₄ of its length; tube ± 0.5 mm long, campanulate;

FIG. 54. *SECAMONE PARVIFOLIA* — **1**, habit, × ²/₃; **2**, lower stem with corky wings × ¹/₂; **3**, flowering shoot, × 2; **4**, flower, × 8; **5**, gynostegium with staminal corona lobes and emergent stylar head appendage, × 22; **6**, pollinaria, × 34; **7**, seeds, with coma × 1; **8**, without coma × 2. 1 & 7 from *Gilbert & Phillips* 8296; 2 from *Gilbert et al.* 7821; 3-6 from *Gilbert & Thulin* 171. Reproduced with permission from Fl. Eth., drawn by E. Papadopoulos.

lobes ovate or oblong, 1.5–2 × ± 1 mm, obtuse or rounded at the apex. Corolline corona of 5 pairs of fleshy ridges extending from the base of the corolla tube to either side of the corolla lobe sinus. Staminal corona lobes from ½ to as long as the staminal column, laterally compressed, falcate. Anthers sometimes with white fimbriate appendages surrounding the stylar head appendage. Stylar head appendage barely exserted to exserted for ± 1 mm from the column, entire to deeply bifid, but not spreading or strongly dilated. Follicles widely divergent, 4.5–10 × 0.5–0.7 cm, the upper face flattened, tapering gradually to a long point. Seeds reddish brown, ± 5–11 × 1–1.5 mm, channelled down one face. Fig. 54, p. 172.

UGANDA. Karamoja District: Cholol, Oct. 1956, *J. Wilson* 300!; Bunyoro District: Butiaba Flats, May 1941, *Eggeling* 4352!
KENYA. Northern Frontier District: Faiyu [Faio], 18 May 1952, *Gillett* 13225!; Kitui District: near Mutomo, 22 Nov. 1979, *Gatheri et al.* 79/127!; Tana River District: Kuwara, 6 Oct. 1961, *Polhill & Paulo* 613!
TANZANIA. Mwanza District: Massanza Is., Nyambiti, 12 Mar. 1953, *Tanner* 1277!; Uzaramo District: Dar es Salaam, 21 Nov. 1968, *Harris* 2592!; Kilwa District: Selous Game Reserve, Kingupira, 5 Jan. 1977, *Vollesen* in MRC 4282!; Zanzibar, near Chukwani, 12 Dec. 1959, *Faulkner* 2429!
DISTR. U 1, 2; K 1, 2, 4, 7; T 1–8; Z; South Sudan, Ethiopia, Malawi, Mozambique, Zimbabwe, Botswana and South Africa
HAB. Thickets, coastal forest and *Acacia-Commiphora* bushland; 0–1500 m

SYN. *Gymnema parvifolium* Oliv. in Trans. Linn. Soc., Bot., Bot. 2: 342 (1887)
    Secamone schweinfurthii K. Schum. in E.J. 17: 143 (1893); N.E. Br. in F.T.A. 4(1): 284 (1902); T.T.C.L.: 68 (1949). Type: Sudan, Kulongo, Bongo, *Schweinfurth* 2232 (B†, holo.; K!, iso.)
    S. emetica (Retz.) Schult. var. *glabra* K. Schum. in P.O.A. C: 324 (1895). Type: Tanzania, Lushoto District: Mashewa [Maschaua], *Holst* 3555 (B†, holo.; K!, iso.)
    S. kirkii N.E. Br. in K.B. 1895: 248 (1895) & in F.T.A. 4(1): 285 (1902). Type: Tanzania, Zanzibar, *Kirk* s.n. (K!, holo.)
    S. zambeziaca Schltr. in J.B. 33: 303 (1895); N.E. Br. in F.T.A. 4(1): 285 (1902). Type: Malawi, Chiromo, Shire River, *Scott Elliot* 2803 (BM!, holo.; K!, iso.)
    S. usambarica N.E. Br. in F.T.A. 4(1): 281 (1902); T.T.C.L.: 68 (1949). Type as for *S. emetica* var. *glabra*
    S. mombasica N.E. Br. in F.T.A. 4(1): 284 (1902). Type: Kenya, near Mombasa, *Hildebrandt* 1979 (K!, holo.; B†, W, iso.)
    S. zambeziaca Schltr. var. *parvifolia* N.E. Br. in Fl. Cap. 4(1): 544 (1907). Type: Mozambique, Maputo [Lourenço Marques], *Schlechter* 11669 (K!, lecto., designated by Goyder in K.B. 47: 451 (1992), as 'holotype'; B†, BOL!, iso.)

NOTE. *Newbould & Jefford* 2699, from the Mahali Mountains of T 4, resembles *S. parvifolia* in most characters, but the style apex is strongly dilated as in *S. punctulata*. There is a single comparable collection from Northern Zambia.

6. **Secamone alpini** *Schult.* in Roem. & Schultes, Syst. veg. 6: 125 (1820); N.E. Br. in F.T.A. 4(1): 279 (1902) & in Fl. Cap. 4(1): 544 (1907); T.T.C.L.: 67 (1949); U.K.W.F. ed. 2: 176 (1994); Burrows in Pl. Nyika Plateau 73 (2005). Type: '*Periploca secamone*' (LINN 307.2!, lecto., designated by Goyder & Singh in Taxon 40: 630 (1991))

Robust liane or shrub, minutely rusty puberulent on young shoots and inflorescence. Leaves with petiole 1–8.5(–11.5) mm long; lamina obovate, oblong, lanceolate- or ovate-oblong or occasionally elliptic, (1.6–)2.5–7.4(–9.4) × 0.6–2.3(–3.4) cm, apex acute, obtuse or rounded, occasionally retuse, often apiculate, base rounded or subcuneate, young leaves with scattered hairs on both surfaces, mature leaves coriaceous, glabrescent, lower surface smooth or minutely but densely tuberculate-papillate. Cymes arranged in axillary or terminal, pyramidal or corymbose panicles, rusty-puberulent except for the corolla, the inflorescence 10–55 × 10–45 mm; flowers scented with a faint musty odour; peduncles 4–12 mm to first inflorescence branch, subsequent internodes of similar length; pedicels 1–4 mm long. Calyx lobes broadly ovate, 0.5–1 × 0.5–1 mm, obtuse or subacute. Corolla greenish-yellow or white, rotate, 1.5–3 mm long, lobed almost to the base, the tube 0.5–1 mm long, lobes ovate, convex, 1–2 × 0.8–1 mm, glabrous on the outer face, pubescent with spreading white hairs inside except for the glabrous margin. Corolline corona absent. Staminal corona lobes equalling or usually exceeding the staminal column, subulate, erect or incurved over the tips of the anthers. Stylar head appendage shorter than the anthers or exserted for up to ± 0.2 mm, the apical part truncate, angled but not 2-lobed. Follicles widely divergent, dark greenish brown, 4.5–8(–10) × 0.3–0.8 cm, tapering to a long point, striate. Seeds dark reddish brown, linear-lanceolate, (4–)8–12 × 1–2 mm, channelled down one face, somewhat flattened, one face concave, the other convex, glabrous.

UGANDA. Kigezi District: Bukimbiri, Bufumbiri, Oct. 1947, *Purseglove* 2498!; Mengo District: Kampala, Makerere College, Oct. 1935, *Chandler* 56!
KENYA. Nakuru District: Eburu Forest Reserve, 17 July 2002, *Luke et al.* 8923!; Kiambu District: Limuru, June 1918, *Snowden* 645!; Teita District: E side of Mbololo Forest, 19 July 2000, *Wakanene & Mwangangi* 665!
TANZANIA. Lushoto District: Sungwi-Shumi, Dec. 1971, *Lema* 43!; Mpanda District: Kungwi-Mahali Peninsula, Selimgweni Peak, Sep. 1959, *Harley* 9585!; Songea District: Lupembe Forest Reserve, Matengos, Nov. 1956, *Semsei* 2580!
DISTR. U 2, 4; **K** 3, 4, 7; **T** 3, 4, 6–8; Zambia, Malawi, Mozambique, Zimbabwe and South Africa
HAB. Montane forest; 1300–2000 m

SYN. *Periploca secamone* L., Mant.: 216 (1771), excl. Alp. Aegypt.: t. 134; Thunb., Prodr.: 47 (1794). Type as for *Secamone alpini*
   Secamone thunbergii E. Mey., Comm Pl. Afr. Austr.: 224 (1837). Type: South Africa, Cape, 'in sylvis Houtniquas', *Thunberg* (UPS (Hb. Thunb. 6222), lecto. (designated by Goyder & Singh, Taxon 40: 630 (1991)); K!, microfiche)
   S. thunbergii E. Mey. var. *retusa* E. Mey., Comm Pl. Afr. Austr.: 224 (1837). Type: South Africa, near Glenfilling, *Drège* 3474 (B†, holo., K!, iso.)

7. **Secamone stuhlmannii** *K. Schum.* in P.O.A. C: 325 (1895); N.E. Br. in F.T.A. 4(1): 283 (1902); T.T.C.L.: 68 (1949); Malaisse in Fl. Rwanda 3: 114 (1985); U.K.W.F. ed. 2: 176 (1994); K.T.S.L.: 498 (1994). Type: Tanzania, Bukoba District: Karagwe, Kafuro, *Stuhlmann* 1894 (B†, holo.; K!, iso.)

Woody climber, the whole plant pubescent with spreading hairs which are particularly dense on the inflorescence, young stems, petioles and the underside of the leaves. Leaves with petioles 0.5–4.5 mm long; lamina oblong or ovate to lanceolate-elliptic, (1–)1.3–4(–6.5) × (0.2–)0.5–2(–2.4) cm, apex acute or occasionally rounded, base rounded, obtuse or acute. Cymes terminal and axillary, up to 6-flowered; flowers sweetly scented, ovoid or subglobose in bud; peduncles 1–12 mm long; pedicels 1–3(–8) mm long. Calyx lobes narrowly to broadly ovate, 0.5–1.5 × ± 0.5 mm, acute or obtuse, pubescent, the margins ciliate, paler than the rest of the calyx. Corolla 1.5–3 mm long, lobed almost to the base, tube 0.2–0.4 mm long, salveriform; lobes rotate, ovate-oblong, 1.5–2.5 × ± 1 mm, obtuse, yellow when fresh, drying reddish brown with a paler margin, glabrous or with scattered white hairs. Corolline corona ridges minute, forming a small pocket at the corolla lobe sinus. Staminal corona lobes attached near the base of the staminal column, $\frac{1}{3}$ to as long as the column, laterally compressed and broadly triangular or falcate with an incurved tip. Stylar head appendage exserted for ± 1 mm from the top of the staminal column, obconic or clavate, deeply 2-lobed or subentire. Follicles silvery brown or olive green, 6.5–9.5 × 0.7–0.9 cm, tapering gradually to a drawn-out point, striate, puberulent. Seeds reddish brown, 7–10 × 1–1.5 mm, channelled down one face.

subsp. **stuhlmannii**

Underside of leaves with minute but dense tuberculate papillae in addition to the reddish indumentum.

UGANDA. Ankole District: Kigarama Hill, Ruampara, Oct. 1932, *Eggeling* 658!
KENYA. Masai District: Mara Plains, Keekorok [Egalok], Oct. 1958, *Verdcourt & Frazer-Darling* 2287!
TANZANIA. Musoma District: Manchera [Manjira], Ikoma, Nov. 1959, *Tanner* 4451!; Iringa District: Tosamaganga, Dec. 1965, *Harris* 10271!; Lindi District: Rondo Plateau near Mchinjiri, Dec. 1955, *Milne-Redhead & Taylor* 7597!
DISTR. U 2; K 6; T 1, 2, 5, 7, 8; Rwanda, Burundi and South Sudan
HAB. Thickets; 800–1700 m

SYN. *Secamone floribunda* N.E. Br. in F.T.A. 4(1): 282 (1902); T.T.C.L.: 67 (1949). Type: Tanzania, Mwanza District: Usmawo, Kayenzi [Kageyi], *Fischer* 396 (K!, holo.; B†, iso.)
   S. phillyreoides S. Moore in J.L.S. 37: 182 (1905). Type: Uganda, Ankole District: Mulema, *Bagshawe* 283 (BM!, holo.)
   S. rariflora S. Moore in J.L.S. 37: 183 (1905). Type: Uganda, Mengo District: Buvuma Is., *Bagshawe* 646 (BM!, holo.)

NOTE. The second subspecies, *S. stuhlmannii* subsp. *whytei* (N.E. Br.) Goyder & T. Harris, differs from the type subspecies in that it lacks the papillae on the leaf surface – it is known only from Mt Malosa in southern Malawi.

8. **Secamone clavistyla** T. *Harris & Goyder* in K.B. 62: 281 (2007). Type: Tanzania, Lindi District, Rondo Forest Reserve, 6 km N of Forest Station, *Bidgood, Abdallah & Vollesen* 1518 (K!, holo.; NHT, iso.)

Slender woody twiner with white latex; young shoots pubescent with spreading reddish hairs; older stems glabrous. Leaves with petiole 0.2–0.4 mm long, densely reddish-pubescent; lamina narrowly ovate to elliptic, rarely linear or ovate, 0.7–1.8 × 0.2–0.6 cm, apex apiculate, base rounded, lower surface sparsely pubescent, the surface ± smooth, the stomata not obscured by minute papillae, upper surface smooth and glabrous; red hairs restricted to upper and lower midrib. Cymes mostly terminal, single-flowered; peduncle 0–1 mm long, glabrous; pedicels 2–3.5 mm long, glabrous. Calyx lobes ovate to broadly ovate, 1 × 0.6 mm, apex rounded, margins ciliate. Corolla 3–4 mm long, yellowish-green, divided for ¹/₂–²/₃ of its length; tube 0.5–1 mm long; lobes ovate-oblong, 1.9–3 mm long, apex rounded. Corolline corona of 5 fleshy ridges from the base of the corolla tube and bifurcating around the corolla lobe sinuses. Staminal corona lobes ± 0.5 × 0.1–0.2 mm, somewhat fleshy, adnate to the column from its base to ²/₃ of its height, tapering gently into the rounded apex which is free from the column for ± ¹/₆ of its length. Anther appendages fimbriate apically; anther wings linear, with ± parallel margins narrowing slightly from apex to base, drying white. Stylar head appendage exserted for ± 1 mm from the column, clavate. Follicles and seeds not seen.

TANZANIA. Lindi District: Rondo Plateau, Rondo Forest Reserve, 6 km N of Forest Station, 12 Feb. 1991, *Bidgood, Abdallah & Vollesen* 1518!
DISTR. T 8; only known from the type
HAB. Coastal forest and thicket on sand; 800 m

9. **Secamone africana** (*Oliv.*) *Bullock* in K.B. 8: 362 (1953); F.P.U.: 118, fig. 67 (1962); Bullock in F.W.T.A. ed. 2, 2: 88 (1963); Malaisse in Fl. Rwanda 3: 114 (1985); U.K.W.F. ed. 2: 176 (1994); K.T.S.L.: 497 (1994). Type: Uganda, Bunyoro [Unyoro], *Speke & Grant* s.n. (K!, holo.)

Woody climber, glabrous or rusty-puberulent on young shoots, the inflorescence covered with spreading reddish hairs. Leaves with petiole 1.7–7(–9.5) mm long; lamina ovate or ovate-oblong to lanceolate, (2.1–)3–7.2(–7.8) × 0.8–3 cm, acute or attenuate, base rounded or obtuse, glabrous, lower surface smooth; veins at 60–80° from midrib. Cymes terminal and axillary on lateral shoots, the subtending leaf often detached at time of flowering, the cymes arranged in crowded or occasionally lax, spreading panicles; inflorescence 10–40 × 5–40 mm, flowers sweetly scented; peduncle 4–35 mm long; pedicels 1–3 mm long, glabrous or rusty puberulent. Calyx lobes ovate to suborbicular, 1–1.5(–2) × 1–1.5 mm, the margins minutely ciliate. Corolla cream, (2.5–)3–4.5 mm long, entirely glabrous or with a patch of hairs near the base of the corolla lobes, divided for $^3/_4$ of its length; tube ± 0.5 mm long, campanulate; lobes oblong, 2–4 × ± 1 mm, obtuse with 2 strong lateral veins, one each side of the sinus. Corolline corona of 5 pairs of ridges running from the base of the corolla tube to about halfway along the corolla lobes. Staminal corona lobes ± as long as the staminal column, laterally compressed and somewhat incurved at the tip. Anther appendages laciniate, or with a tuft of hairs. Stylar head appendage exserted up to ± 1 mm from the staminal column, the apical part ovoid-clavate, bifid, the lobes erect. Follicles widely divergent,, olive-green to brown 5.5–10 × 0.4–0.6 cm, narrowing gradually to an attenuate point, striate. Seeds oblong, ± 8 × 2 mm, somewhat channelled down one face.

UGANDA. Toro District: Butogo, Baramba, Oct. 1840, *Eggeling* 4063!; Mbale District; West Bugwe Forest Reserve, Busia–Bugiri road, Sep. 1953, *Drummond & Hemsley* 4493!; Mengo District: Naguru Hill, May 1969, *Rwaburindore* 27!
KENYA. Trans-Nzoia District: Kitale, Sep. 1964, *Tweedie* 2905!; Kericho District: Sotik, Kibajet Estate, Oct. 1948, *Bally* 6475!
TANZANIA. Mwanza District: Geita, Nungwe Bay Road, Jan 1953, *Procter* 128!
DISTR. U 2–4; K 3, 5; T 1; Nigeria, Cameroon, Central African Republic, Congo-Kinshasa, Burundi, South Sudan and Angola
HAB. Thicket and forest edge; 1200–2000 m

SYN. *Toxocarpus africanus* Oliv. in Trans. Linn. Soc., Bot. 29: 109 t. 118A (1875)
    Secamone platystigma K. Schum. in E.J. 17: 143 (1893); N.E. Br. in F.T.A. 4(1): 280 (1902). Types: Angola, Golungo Alto, *Welwitsch* 5944 (K!, lecto., designated here; BM, C!, COI!, isolecto.); Angola, Pungo Andongo, *von Mechow* 92 (B†, Z!, paralecto.); *Welwitsch* 5935 (BM, COI!, K!, paralecto.), 5945 (COI!, K!, paralecto.) & 5947 (COI!, K!, paralecto.)

10. **Secamone racemosa** (*Benth.*) *Klack.* in Adansonia 23: 322 (2001). Type: Cameroon, Mt Cameroon, *Mann* 1273 (K!, holo.; GH, P, iso.)

Woody climber, glabrescent with youngest branches covered with short reddish hairs. Leaves with petiole 8–20 mm long with appressed reddish hairs; lamina ovate to obovate, but usually elliptic, 0.5–1.2 × 0.2–0.5 cm, the apex acuminate, the base cuneate to rarely almost truncate, surface smooth (not tuberculate/papillate), glabrous on both sides to rarely with sparse reddish appressed hairs below. Cymes axillary, up to 9-flowered; flowers narrowly ovoid in bud with obtuse apex; peduncles 14–24 mm to first inflorescence branch, subsequent internodes as long, to as short as 5 mm long; pedicels 3–10 mm long. Calyx with one to several surrounding calycoid bracts; lobes 3–4 × 1.5–2.2(–2.7) mm, ovate to usually oblong, obtuse to rounded at the apex, glabrous but with ciliate margin. Corolla 5–7 mm long, lobed for $^1/_2$–$^2/_3$ of its length, not twisted, glabrous, white or yellowish to green, sometimes pinkish; tube 1.5–2.5 mm long; lobes oblong to sometimes obovate, 4.3–5.4 × 2–3.4 mm, truncate to rounded at the apex. Corolline corona of five fleshy horizontal bars between the corolla lobes at the sinuses and forming small pockets. Staminal corona lobes about as long as to longer than the staminal column, dorsiventrally flattened, oblong, the apex crenate-truncate to bifid. Stylar head appendage projecting at least twice as long as the staminal column, 3.3–5.3 mm long, cylindrical, entire to slightly bifid at apex. Follicles straight to slightly curved, linear in outline, 18–29 × 0.4–0.5 cm, thin-walled, finely pubesent. Seeds 5–8 mm long; coma 3–4.5 cm long.

UGANDA. Toro District: Buckeragi ridge, Nyamugasani, *Ross* 913; Kigezi District: Bwindi
Impenetrable Forest, Sep. 1936, *Eggeling* 3297! & 15 Aug. 1998, *Eilu* 310!
DISTR. U 2; Cameroon, Equatorial Guinea, Congo-Kinshasa, Rwanda and Burundi
HAB. Primary montane forest on marshy ground or along rivers, also in clearings with
secondary regrowth; 1000–3000 m

SYN. *Rhynchostigma racemosum* Benth. in Hooker's Ic. Pl. 12: 77, pl. 1189 (1876); Bullock in
F.W.T.A. ed. 2, 2: 88 (1963); Malaisse in Fl. Rwanda 3: 114 (1985)
Toxocarpus racemosus (Benth.) N.E. Br. in F.T.A. 4(1): 287 (1904)

Subfam. ASCLEPIADOIDEAE

This is the largest subfamily in the expanded Apocynaceae and is of Old World
origin, but with one additional major radiation in New World tropics. The centre
of diversity is tropical and subtropical Africa. The subfamily consists of four tribes:
Fockeeae, Marsdenieae, Ceropegieae and Asclepiadeae, all of which are
represented in the Flora area. The arrangement of the account presented here
reflects the evolutionary relationships elucidated in major molecular surveys
across the subfamily in recent years as summarised by Goyder in Ghazanfar &
Beentje (eds.), Taxonomy and ecology of African plants, their conservation and
sustainable use: 205–214 (2006). This in turn was based on the global overviews
of Rapini, Chase, Goyder & Griffiths in Taxon 52: 33–50 (2003) & Liede-
Schumann, Rapini, Goyder & Chase in Syst. Bot. 30: 184–195 (2005), and the
many supporting studies of individual tribes and subtribes cited in these works, or
published since.

Approximately 172 genera and 3000 species in tropical and subtropical regions of the world.
41 genera and 275 species occur within the Flora region.

A few non-native asclepiads have been recorded in cultivation in the Flora area in
Jex-Blake's Gardening in East Africa, 4th ed. (1957), from records in regional herbaria,
and from information supplied by Ann Robertson, to whom we are most grateful.

Araujia sericifera Brot.
An attractive vine with white flowers and heart-shaped leaves. The fleshy-walled
follicle splits to release many plumose seeds. Native to Argentina, but widely
naturalised and of some concern in a number of regions with Mediterranean-type
climates. Not likely, however, to become a problem in East Africa.

Asclepias curassavica L. – see entry under *Asclepias*

Ceropegia sandersonii Hook.f.
A typical twining *Ceropegia* with large, green or whiteish, lantern flowers 4–7 cm
long – the top of the lantern somewhat umbrella-like. Native to southern
Mozambique, Swaziland, and neighbouring regions of South Africa (Mpumalanga
and KwaZulu-Natal).

C. linearis E. Mey. subsp. *woodii* (Schltr.) H. Huber
Small slender twiner from a subglobose tuber; leaves fleshy, heart-shaped, less that
2 cm in length, flowers slender, to 2.5 cm long, whitish except for the green or purple
lobes. Native of southern Africa.

Hoya carnosa (L.f.) R. Br.; U.O.P.Z.: 300 (1949)
Semisucculent vine with large leaves and heavily scented umbels of white flowers
that drip copious nectar. Native to eastern Asia.

Oxypetalum coeruleum (Sweet) Decne.
Woody herb or small straggling shrub with very characteristic blue flowers. Native to Argentina.

Stephanotis floribunda R. Br.; U.O.P.Z.: 453 (1949)
Vine with semisucculent glossy leaves and white, subcylindrical, scented flowers. Native to Madagascar.

Stapelia gigantea N.E. Br. – see under *Stapelia* in main account

Innumerable other stapeliads have been cultivated in succulent collectors' gardens, but have not generally persisted.

SPOT CHARACTERS

Stem succulence – succulent stems (frequently but not always associated with reduced leaves) are characteristic of one subtribe, and can occur in species of a handful of other genera:

1. The stapeliads – Ceropegieae subtribe Stapeliinae (genera 58–66) (*Caralluma, Desmidorchis, Echidnopsis, Edithcolea, Huernia, Monolluma, Orbea, Rhytidocaulon, Stapelia*)
2. Some species of *Ceropegia* (genus 56)
3. *Schizostephanus alatus* (genus 81) and some species of *Cynanchum* (genus 83)
4. *Fockea multiflora* (genus 47) – massive swollen stem tapering into the basal tuber

Erect herbaceous habit – two types of erect herbaceous habit have restricted distribution in this subfamily:

1. Erect ephemeral annuals – the monotypic *Conomitra linearis* (genus 53) is the only genus with this habit, behaving as an extremely short-lived annual – germinating with the rare desert rainfall, flowering and fruiting probably in weeks, then dying after setting seed. On the rare occasions on which this species has been encountered, it has been found in large numbers.
2. Perennial herbs with annually produced erect shoots from a tuberous or woody underground rootstock:
   - Characteristic of many genera in the *Asclepias* radiation (genera 72–79) (*Asclepias, Aspidoglossum, Glossostelma, Gomphocarpus, Margaretta, Pachycarpus, Stathmostelma, Xysmalobium*)
   - This growth form also occurs sporadically in *Ceropegia* & *Brachystelma* (genera 56, 57) and *Tylophora caffra* (genus 87) is a pyrophyte with erect or sprawling stems
   - Check also *Raphionacme* in subfam. Periplociodeae (genus 40)

Latex colour – although it is often incorrectly assumed that all Apocynaceae have white or milky latex, certain taxonomic groups within the Asclepiadoideae (in particular) are exceptional in having clear, or in the more succulent genera, slightly cloudy, latex. Very often, if the collection data on the label omits to mention latex, it is because white exudate was not observed. This may be an indication that the latex was clear, and that it was not recognised as latex by the collector. If your plant has clear (or cloudy) latex, it might be:

1. Tribe Ceropegieae (genera 52–66) – *Ceropegia* and allies (*Brachystelma, Ceropegia, Conomitra, Leptadenia, Neoschumannia, Riocreuxia*); and the succulent stapeliads (*Caralluma, Desmidorchis, Echidnopsis, Edithcolea, Huernia, Monolluma, Orbea, Rhytidocaulon, Stapelia*)

2. *Pleurostelma, Schizostephanus* (genera 80 & 81) and the group of genera around *Tylophora* (genera 84–87: *Blyttia, Diplostigma, Pentatropis, Tylophora*)
3. *Marsdenia* – some species apparently may have clear latex.

All other genera have white or milky latex, frequently in copious amounts!

Pollinarium structure, and orientation/attachment of pollinia to the corpusculum – although this requires careful observation under a stereo dissecting microscope, pollinarium structure provides the most reliable way of subdividing the subfamily into major tribal groups. Because these are not characters readily observable in the field, I have used them sparingly in constructing the artifical key to genera which follows, but in the herbarium, they give you a powerful diagnostic tool to ascertain the affinities of the plant under investigation. The Fockeeae (genera 47, 48: *Cibirhiza* and *Fockea*) have peculiar pollinaria that do not fit comfortably into the broad patterns outlined below – the pollen masses are attached directly to a little-differentiated translator. All other groups within the subfamily have a well-structured corpusculum with a longitudinal ventral groove, and arms (in various configurations) linking the corpusculum to the two pollinia, Fig. 55, p. 179. Look for the following:

1. Translator arm attached at the base of the pollinium so that the pollen masses are held erect in the anther cells – this takes you to tribes Marsdenieae (genera 49–51) and Ceropegieae (genera 52–66)
    *   If there is a well defined translucent germination zone on the inner margin of the pollinia, this indicates Ceropegieae (genera 52–66)
    *   If not, this indicates Marsdenieae (genera 49–51)

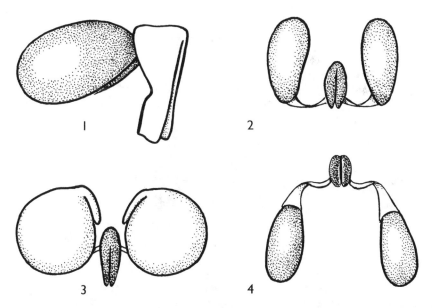

FIG. 55. Basic pollinarium types found in the ASCLEPIADOIDEAE. **1**, FOCKEEAE: two pollinia attached directly to corpusculum; **2**, MARSDENIEAE: two erect (basifixed) pollinia attached indirectly to corpusculum; **3**, CEROPEGIEAE: two erect (basi- or medifixed) pollinia with translucent germination zones on inner margins, attached indirectly to corpusculum; **4**, ASCLEPIADEAE: two pendant (apically attached) pollinia attached indirectly to corpusculum. Drawn by M. Tebbs.

2. Translator arm attached to the top of the pollinium, so the pollen masses are pendant within the anther cells – this takes you to tribe Asclepiadeae (genera 67–87) [But note that one genus in the Flora region (*Tylophora*) has such small flowers with the vertical scale so reduced that the anther sacs may be displaced laterally, the pollinia are minute, and the point of attachment is variable in position]

   • A translucent germination zone on the pollinia in this group within the Flora area is indicative of *Aspidoglossum* (genus 72)

KEY TO GENERA

1. Leaves absent or reduced to scales; stems succulent 2
   Leaves well developed; stems succulent or not 12
2. Corolla tube well-developed, much longer than wide, usually with a distinct basal swelling; corolla lobes usually shorter than the tube, remaining attached at the tips to form a lantern or cage-like structure . . . .   **56. Ceropegia** (p. 220)
   Corolla not as above; if tube longer than lobes, then stems with prominent angles or rows of tubercles 3
3. Stems smooth or striate, lacking prominent angles or rows of tubercles, frequently scrambling or twining; latex white . . . .   **83. Cynanchum** (p. 474)
   Stems with prominent angles or rows of tubercles, prostrate to erect, but never scrambling or twining; latex clear or at most cloudy 4
4. Stems heteromorphic, with a vegetative, strongly succulent basal part, and an apical flower-bearing part that tapers to a thin synflorescence stalk . . . .   **58. Caralluma** (p. 310)
   Stems not differentiated apically into a tapering, flower-bearing part 5
5. Plants robust and massive, stems erect, sharply 4-angled; flowers very large (15–40 cm diameter), or much smaller but then in terminal globoid inflorescences of 4–12 cm diameter 6
   Plants not massive (rarely robust), stems creeping to erect, obtusely 4–5(–14)-angled; flowers and inflorescences not as above 7
6. Stems smooth, glabrous; inflorescences terminal, many-flowered, globose . . . .   **59. Desmidorchis** (p. 317)
   Stems pubescent; inflorescences basal, 1–4-flowered . . . .   **66. Stapelia** (p. 356)
7. Gynostegium with a conspicuous cylindrical filament tube, separating the interstaminal corona (which is appressed to corolla tube) from the terminal staminal corona lobes . .   **64. Huernia** (p. 336)
   Gynostegium without a conspicuous cylindrical filament tube, interstaminal and staminal corona series not separated but partly or totally fused to each other 8

8. Leaf rudiments with a broadened base and so hardly separable from tubercles (stem projections/cushions carrying the leaves), triangular-deltate to conical-subulate, caducuous .... 65. **Orbea** (p. 343)

Leaf rudiments without broadened base and well separable from tubercles (occasionally absent), deltoid to lanceolate, caducuous (occasionally persistent and thorny: *Edithcolea*) 9

9. Stems (5–)6–11(–20)-angled .... 62. **Echidnopsis** (p. 325)

Stems rounded 4–5-angled 10

10. Plants sparsely branched from a main stem, stem surface papillose or tessellate; leaf rudiments strongly ascending-erect; stipular glands present, globoid; flowers small, delicate .... 63. **Rhytidocaulon** (p. 335)

Plants much branched, stem surface smooth; leaf rudiments spreading, stipular glands absent; flowers large, fleshy 11

11. Stems erect, leaf rudiments inconspicuous, caducuous; flowers 2–3 cm in diameter .... 60. **Monolluma** (p. 321)

Stems creeping, leaf rudiments a sharp and hard, conical thorn, 1 mm long; flowers 8–12 cm in diameter................... 61. **Edithcolea** (p. 323)

12(1). Slender, erect, ephemeral herb with linear leaves; plant of desert environment .... 53. **Conomitra** (p. 212)

Perennial herbs, shrub or vines 13

13. Corolla tube well developed, generally much longer than wide, usually with a distinct basal swelling; corolla lobes mostly shorter than the tube and remaining attached at the tips to form a lantern or cage-like structure (very rarely free at the tips and rigidly spreading); latex clear 14

Corolla not as above; latex white or clear 15

14. Inflorescences generally paniculate, lax; corolla mostly (in Flora region) orange or yellowish; corolla and corona glabrous throughout; staminal corona lobes often rudimentary and exceeded by the anthers; rootstock woody, with fusiform roots ...... 55. **Riocreuxia** (p. 216)

Inflorescences generally umbelliform, more compact; corolla of various colours, but rarely yellow or orange; corolla and corona glabrous or pubescent; staminal corona lobes generally exceeding the anthers; rootstock generally not woody, roots fibrous, fusiform, tuberous or rhizomatous .... 56. **Ceropegia** (p. 220)

15. Shrubs to 5 m tall, larger plants arborescent with corky stems; leaves glaucous, semisucculent, mostly obovate, to 26 × 15 cm; corona lobes laterally compressed, with a short, upturned basal spur .... 70. **Calotropis** (p. 367)

Erect perennials, twiners or scramblers, if shrubs then no more than 3 m tall; leaves much smaller than above; corona lobes lacking an upturned basal spur 16

25. Leaves with an indumentum of stiff hairs, feeling scabrid to the touch; lamina generally broad with prominent secondary veins and a truncate to cordate base; follicles generally ornamented with longitudinal wings or soft spine-like processes (but smooth and inflated in *Pachycarpus lineolatus* and *P. bisacculatus*) 26

Leaves glabrous or pubescent, if stiff scabrid hairs present, then confined to the margins; lamina narrow to broad, secondary veins often obscure, base frequently cuneate, occasionally truncate or cordate; follicles generally smooth, rarely with weak ornamentation of short filiform processes 27

26. Corona lobes solid, fleshy; corolla usually bearded within towards the apex; follicles covered with soft filiform processes .... **78. Xysmalobium undulatum** (p. 450)

Corona lobes generally dorsiventrally flattened, commonly with lobes or fleshy wings arising near the base of the ventral face; corolla not conspicuously bearded; follicles winged or smooth, lacking filiform processes .... **77. Pachycarpus** (p. 436)

27. Corona lobes solid, fleshy, occasionally with teeth or small processes apically (but laminar in *Glossostelma carsonii*) 28

Corona lobes laminar or cucullate, with or without teeth or other ornamentation within the cavity of the lobe, sometimes somewhat reduced 29

28. Rootstock with a short vertical axis from which fusiform lateral tuberous roots arise; leaves semisucculent, venation obscure .... **79. Glossostelma** (p. 458)

Rootstock a vertical napiform tuber; leaves membranous, venation mostly easily observed **78. Xysmalobium** (p. 450)

29. Corona lobes petaloid, larger and more showy than the corolla .... **74. Margaretta** (p. 404)

Corona lobes not petaloid 30

30. Pollinaria with differentially winged and contorted translator arms: proximal part broad and membranous, distal part filiform and pendulous; corolla and corona generally brightly coloured (red, orange, yellow), occasionally white with purple markings; flowers held erect .... **75. Stathmostelma** (p. 409)

Pollinaria not as above; corolla and corona dull, if brightly coloured then corona with prominent tooth arising from the cavity, or flowers nodding or held laterally .... **73. Asclepias** (p. 382)

31(16). Gynostegial corona in three series –
downward pointing staminal lobes partially
fused to form a skirt which obscures the
gynostegial stipe; a whorl of spreading or erect
interstaminal lobes; and a further whorl of 5
erect dorsi-ventrally flattened staminal lobes
with swollen bases; latex clear; robust twiner,
basally becoming woody, with a relict wet
forest distribution . . . .   **54. Neoschumannia** (p. 214)
Gynostegial corona not in three series, or if so,
not in the confuguration described above,
and plant with white latex 32
32. Twining herbs or slender to woody lianas arising
from large subterranean or partially exposed
tubers; gynostegial coronas in several series,
one of which is tubular; pollinia attached
directly to the corpusculum 33
Plants not arising from large subterranean or
partially exposed tubers; gynostegial coronas,
if present, in one or at most two series;
pollinia attached indirectly to corpusculum 34
33. Corolla lobes linear, spreading; fused part of
corolla not extending beyond corona;
tubular part of the corona longer than the
gynostegium . . . .   **47. Fockea** (p. 186)
Corolla lobes broadly ovate or triangular;
fused part of corolla extending as a flat limb
beyond the corona; tubular part of corona
shorter than the gynostegium . . . .   **48. Cibirhiza** (p. 189)
34. Corolline corona present in corolla lobe
sinuses; gynostegial corona minute or absent 35
Corolline corona absent, or forming an annulus
around the base of the gynostegium;
gynostegial corona mostly well developed (but
reduced to scales in *Pleurostelma*) 36
35. Corolla rotate, lobes lanate; stylar head enclosed
by anthers; pollinia with translucent marginal
germination zone; latex clear . . . .   **52. Leptadenia** (p. 210)
Corolla campanulate to urceolate, lobes
glabrous; apex of stylar head exserted from
staminal tube; pollinia without translucent
germination zone; latex white . . . .   **49. Marsdenia sylvestris** (p. 194)
36. Pollinia erect in the anther cells, translator arms
attached to the base of the pollinia; corolla
mostly with well-developed tube, at least at the
base (but rotate to broadly campanulate in
Marsdenia magniflora and *M. exellii*) 37
Pollinia pendent in anther cells, translator arms
attached to the apex of the pollinia (but note
this character unreliable in some *Tylophora*);
corolla rotate or broadly campanulate, united
part short or not clearly tubular (except in
Pergularia) 39
37. Staminal corona lobes with a ligule on the
inner face . . . .   **51. Telosma** (p. 208)
Staminal corona lobes lacking a ligule on the
inner face 38

38. Follicles single-seeded, with an irregular knobby ridge at the base of the beak (at least in FTEA area); much-branched shrubs with leafy short shoots; stylar head appendage long-rostrate, well exserted from anthers ... 50. **Stigmatorhynchus** (p. 206)
Follicles many-seeded, smooth or variously ornamented, but lacking irregular knobby ridge; slender to woody vines, or shrubs; stylar head exserted or not from anthers .. 49. **Marsdenia** (p. 190)
39. Corolla united for about half its length, shallowly campanulate, 1.5–2.5 cm diameter; latex white; riverine vegetation in dry country .... 68. **Oxystelma** (p. 361)
Corolla lobed almost to base; latex white or clear 40
40. Staminal corona lobes semisagittate, with a free subulate apex arched over the head of the column and a basal projection; corolla lobes with bearded margins; rampant scramblers with copious white latex .... 69. **Pergularia** (p. 364)
Staminal corona lobes not as above; corolla lobes without bearded margins 41
41. Stems thick and fleshy; latex clear or yellowish, not milky; corona united into a fluted tube surrounding and obscuring the gynostegium, tube topped with 10 free erect lobes .... 81. **Schizostephanus** (p. 466)
Plant not as above, stems not succulent; if corona tubular, surrounding the gynostegium, then latex white 42
42. Stylar head appendage long-rostrate, well-exserted from staminal column; gynostegial corona vestigial; latex clear .... 80. **Pleurostelma** (p. 464)
Stylar head appendage not strongly exserted from staminal column, or if so, then corona lobes well-developed 43
43. Latex white, generally copious 44
Latex clear, often sparse 45
44. Staminal corona lobes fused at the base only, or if more highly fused, then with clearly visible "sutures" along the lines of fusion ... 82. **Pentarrhinum** (p. 468)
Staminal corona lobes fused partially or for much of their length, without visible "sutures" along the lines of fusion .... 83. **Cynanchum** (p. 474)
45. Corona in two series, a spreading outer corona ring arising from the base of the gynostegium and appressed onto the corolla, and conspicuous rectangular staminal lobes radiating from the backs of the anthers .... 85. **Diplostigma** (p. 492)
Corona in a single series of 5 staminal lobes 46
46. Richly branched woody shrublets, young shoots frequently twining; flowers minute, less than 2 mm long .... 84. **Blyttia** (p. 491)
Plants scrambling, twining or decumbent, sometimes woody towards the base, but not forming richly branched shrublets; if flowers minute, then plants are slender vines 47

47. Guide rails (sclerified anther wing margins)
    radiating from the column and extended
    along the base of the corona lobes;
    inflorescence few-flowered .... **86. Pentatropis** (p. 494)
    Guide rails not extended along corona lobes;
    inflorescences few to many-flowered,
    frequently with clusters of flowers at the
    angles of the extended zig-zag axes ...... **87. Tylophora** (p. 497)

47. FOCKEA

Endl. in Endlicher et Fenzl, Nov. Stirp. Dec.: 17 (1839); Court in Asklepios 40: 67–74 (1987); Bruyns & Klak in Ann. Missouri Bot. Gard. 93: 535–564 (2006)

Massive lianas, or erect to twining herbs arising from subterranean or partially exposed tuber; latex white. Leaves opposite. Inflorescences extra-axillary, umbelliform. Corolla rotate, fused part not extending beyond corona; lobes linear, spreading. Corolline corona absent. Gynostegial corona forming a tube obscuring the gynostegium with several series of erect or spreading lobes at the mouth of the tube, and an inner series of 5 tall erect lobes arising from the inner face of the corona tube. Gynostegium with connivent, erect, inflated and somewhat translucent anther appendages filling the mouth of the corona tube and completely obscuring the much shorter stylar head. Pollinarium with two erect, flattened, clear or translucent pollinia attached directly to the pale brown corpusculum; corpusculum long and narrow, narrower than the pollinia; pollinia consisting of tetrads which are not enclosed by a wall. Follicles pendulous, single by abortion, fusiform and narrowing into a slender beak; seeds flattened, ovate with a narrow marginal rim, with coma at one end only (extending around margin in one species outside the Flora region).

6 species restricted to southern or south tropical Africa, but with two more widely distributed species with disjunct distributions extending into Kenya and Tanzania.

Stems slender but sometimes woody at base, arising from a basal
    tuber, at most to 2 m long; leaves less than 2.5 cm wide,
    glabrous or minutely pubescent beneath; inflorescences
    sessile, few-flowered; corolla twisted in bud, narrowly conical   1. *F. angustifolia*
Stems massive and swollen towards base, but not clearly
    differentiated into stem and basal tuber, often scrambling to
    15 m; leaves generally more than 2.5 cm wide, tomentose
    beneath especially when young; inflorescences pedunculate
    with few to many flowers; corolla not twisted in bud, ovoid .   2. *F. multiflora*

    1. **Fockea angustifolia** *K. Schum.* in E.J. 17: 146 (1893). Types: South Africa, Cape, Griqualand West, *Marloth* 1008 (B†, holo.; M!, iso., see note below)

Slender woody twiner to 1.5(–2) m from a large napiform tuber, latex white; stems often reddish, minutely pubescent. Leaves opposite, petiole 1–3 mm long, minutely pubescent; lamina narrowly linear to oblong, elliptic or ovate, 2–8 × 0.1–2.5 cm, apex rounded to subacute or acuminate, base cuneate, glabrous or minutely pubescent beneath. Inflorescences extra-axillary, sessile but sometimes on short lateral shoots, 1–6 flowers open at one time; pedicels 0–1(–2) mm long, minutely but densely pubescent. Sepals triangular, ± 1 mm long, minutely pubescent. Corolla narrowly conical and strongly contorted in bud, divided ± to the base, lobes linear,

3–6 × 0.5–1 mm, green and densely white-pubescent abaxially, adaxially green to orange or brown, sparsely pubescent or papillose. Corona white, tubular part 2–3(–6) mm long and 1–1.5 mm in diameter, urceolate or cylindrical, apical lobes 0.5–1 mm long; lobes from inner whorl of corona filiform, projecting from mouth of tube for up to 1.5 mm. Follicles grey-green often with purplish bands, 7–20 × 0.8–1.2 cm, smooth; seeds yellow-brown, 8–10 × 4–6 mm.

KENYA. Kwale District: between Samburu and Mackinnon Road, 30 Aug. 1953, *Drummond & Hemsley* 4045!; Tana River District: Tana River, near Dalu, 8 Aug. 1988, *Luke & Robertson* 1288!
TANZANIA. Pare District: Mkomazi Game Reserve, Ibaya Hill, 3 June 1996, *Vollesen* 96/22!; Singida District: Lake Kitangiri, near Chechem R.C. mission, 3 Nov. 1960, *Richards* 13501!; Dodoma District: Chenene public land, 15 May 1976, *Magogo* 719!
DISTR. **K** 7; **T** 2, 3, 5; also recorded from Angola, Zambia, Malawi, Zimbabwe, Botswana, Namibia and South Africa
HAB. *Acacia-Commiphora* bushland or coastal thicket; 0–1100 m

SYN. *Fockea sessiliflora* Schltr. in E.J. 20, Beibl. 51: 44 (1895); N.E. Br. in F.T.A. 4(1): 429 (1903). Type: South Africa, Limpopo Province, near Klippdam, *Schlechter* 4493 (B†, holo.; drawing at W, lecto., designated by Bruyns & Klak in Ann. Missouri Bot. Gard. 93: 548 (2006))
     F. lugardii N.E. Br. in F.T.A. 4(1): 429 (1903). Type: Botswana, Ngamiland, Kwebe Hills, *Lugard* 299 (K!, holo.)
     F. dammarana Schltr. in E.J. 38: 56 (1905). Type: Namibia, Damaraland, 1879, *Een* s.n. (BM!, holo.)
     F. tugelensis N.E. Br. in Fl. Cap. 4(1): 778 (1908). Type: South Africa, KwaZulu-Natal, Tugela, *Gerrard & McKen* 1310 (K!, holo.; BOL, TCD!, iso.)
     F. mildbraedii Schltr. in Mildbr., Z.A.E.: 545 (1913). Types: Tanzania, Pare District: Lembeni, *Winkler* 3803 (K!, lecto., designated here; B†, WRSL, isolecto.); Kenya, Teita District: Voi, *Mildbraed* 8 (B† paralecto.)
     F. monroi S. Moore in J.B. 52: 149 (1914). Types: Zimbabwe, Chimanimani District, Victoria, *Monro* 828 (BM!, lecto., designated here (see note below); BOL, SRGH, iso.) & *Monro* 837 (BM!, paralecto.)
     Cynanchum omissum Bullock in K.B. 10: 623 (1956); K.T.S.L.: 491 (1994). Type: Kenya, Kwale District, between Samburu and Mackinnon road, *Drummond & Hemsley* 4045 (K!, holo.; EA!, iso.)

NOTE. A disjunct distribution with East African populations separated from populations in dry tropical and subtropical regions of southern Africa by the more humid *Brachystegia* belt of southern Tanzania, Zambia and Malawi.
     The discovery of an isotype of *Fockea angustifolia* (*Marloth* 1008) in Munich makes the neotypification of this name by Bruyns & Klak in Ann. Missouri Bot. Gard. 93: 548 (2006) redundant. In the same work Bruyns & Klak designated the duplicate of *Monro* 828 at BOL as lectotype of *Fockea monroi* S. Moore. However, Spencer Moore clearly states on p. 89 of the introduction to the paper describing this species that 'types of species described are in the National Herbarium' – as the paper was published in Journal of Botany, the 'National Herbarium' is unquestionably BM, and the lectotype must be selected from material deposited in that institution.

2. **Fockea multiflora** *K. Schum.* in E.J. 17: 145 (1893); N.E. Br. in F.T.A. 4(1): 428 (1903); T.T.C.L.: 65 (1949). Type: Tanzania, Mwanza District: French mission, Usambiro, *Stuhlmann* 848 (B†, holo.; K!, iso. (fragment))

Robust twiner to 15 m, latex white; stems massively swollen and up to ± 30 cm across towards the base (distinction between stem and basal tuber rapidly obscured with age), semi-succulent with shiny grey or reddish bark, young shoots densely pubescent to tomentose. Leaves opposite, petiole 0.5–4 cm, densely pubescent; lamina obovate to elliptic or suborbicular, to ± 15 × 8 cm, apex rounded to subacute, base cuneate or rounded, tomentose beneath, at least in younger leaves, adaxial face pubescent. Inflorescences extra-axillary, forming simple or branched umbelliform clusters of up to ± 30 flowers; peduncles 5–15(–30) mm long, tomentose; pedicels 6–13 mm, tomentose or densely pubescent. Sepals triangular, ± 1 mm long,

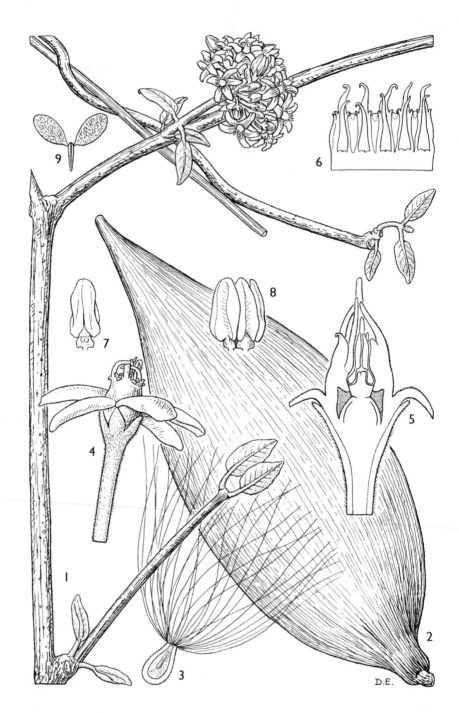

FIG. 56. *FOCKEA MULTIFLORA* — **1**, habit, × 1; **2**, follicle, × 1; **3**, seed with coma of hairs, × 1; **4**, flower, × 4; **5**, half-flower, × 8; **6**, tubular gynostegial corona, opened out and viewed from within, × 4²⁄₃; **7–8**, anthers, showing the large inflated anther appendages, × 8; **9**, pollinarium × 1. 1 from *Greenway* 9056; 2 & 3 from *Bainbridge* 139/55; 4-9 from *Bullock* 1186. Drawn by D. Erasmus.

tomentose or densely pubescent. Corolla not twisted in bud, ovoid; tube ± 1 mm long; lobes bright green or brown, oblong, 6–10 × ± 2 mm, glabrous to sparsely pubescent. Corona white, tubular part 2–3 mm long and 1.5–2.5 mm in diameter, urceolate or cylindrical, apical lobes 0.5–2 mm long; lobes from inner whorl of corona filiform, projecting from mouth of tube for ± 1 mm. Follicles purplish, 10–22 × 1.5–3 cm, broadly fusiform and tapering to a long beak, smooth; seeds 8–10 × 7–8 mm. Fig. 56, p. 188.

TANZANIA. Mwanza District: Badugu, Nassa, 20 Oct. 1952, *Tanner* 1085!; Masai District: E of Kakessio, 24 Nov. 1956, *Greenway* 9056!; Singida District: Lake Kitangiri, 3 Nov. 1960, *Richards* 13490!
DISTR. **T** 1, 2, 5, 7; also recorded from Mozambique, Zambia and Zimbabwe along the Zambezi valley, Angola and northern Namibia
HAB. Dry *Acacia-Commiphora* bushland, open woodland or semi-desert, often among rocks; 1000–1500 m

SYN. *Fockea schinzii* N.E. Br. in K.B. 1895: 259 (1895) & in F.T.A. 4(1): 428 (1903); T.T.C.L.: 65 (1949). Types: Angola, *Welwitsch* 4194 (K!, lecto., designated by Bruyns & Klak in Ann. Missouri Bot. Gard. 93: 557 (2006); P, iso.); Namibia, Ombandja, Amboland, *Schinz* 5 (K!, paralecto.)

48. **CIBIRHIZA**

Bruyns in Notes Roy. Bot. Gard. Edin. 45: 51 (1988)

Slender woody twiners arising from a large tuberous rootstock; latex white. Leaves opposite. Inflorescences extra-axillary, umbelliform. Corolla rotate, fused part extending as a flat limb beyond the corona; lobes broadly ovate or triangular. Corolline corona absent. Gynostegial corona in several series; tubular element shorter than the gynostegium; inner elements strongly lobed or divided. Anther appendages ± obsolete. Pollinarium with two pollinia attached directly to the pale brown corpusculum; corpusculum broad, both in relation to its length and in relation to the pollinia; pollinia consisting of single pollen grains enclosed by a wall. Follicles single by abortion.

3 species, one from Dhofar in western Oman, one from the Ogaden in eastern Ethiopia, and the third from Tanzania and Zambia.

Cibirhiza albersiana *H. Kunze, Meve & Liede* in Taxon 43: 368 (1994). Type: Zambia, Mazabuka District, 30 km N of Pemba near Milimo village, *White* 6969 (K!, holo.; FHO, iso.)

Slender woody twiner to ± 5 m from a large napiform tuber, latex white; stems minutely pubescent. Leaves opposite, petiole 0.5–5 cm long, minutely pubescent; lamina broadly ovate to oblong or elliptic, sometimes variable even on the same shoot, 2.5–13 × 1–9 cm, apex subacute to acuminate, base rounded to cuneate, minutely pubescent at least on the veins. Inflorescences extra-axillary, subsessile, forming a dense sub-umbelliform cluster of flowers, minutely pubescent; pedicels 3–5 mm long. Sepals ovate, ± 1 mm long, minutely pubescent. Corolla yellow or green with brownish spots, rotate, ± 10 mm in diameter and lobed for half its length; lobes broadly ovate or triangular, 2–3 mm long, apex rounded, glabrous. Gynostegial corona in several series (see note below); a low outer collar, and free erect bi- or tri-partite staminal lobes, the innermost terminating in a long drawn-out filiform projection 1 mm or more in length. Anther wings ± 0.3 mm long. Pollinaria with corpusculum ± 0.2 mm long; translator arms absent; pollinia ± 0.2 mm long, narrowly oblong. Stylar head flat, not projecting beyond anthers. Follicles and seeds unknown. Fig. 57, p. 190.

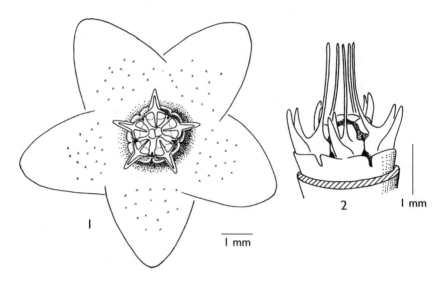

FIG. 57. *CIBIRHIZA ALBERSIANA* — **1**, flower; **2**, gynostegium and corona. Reproduced with permisson from *Kunze et al.* in Taxon 43: 370, fig. 2 & 4.

TANZANIA. Pare District: Mkomazi Game Reserve, SE of Ndea hill, 27 Apr. 1995, *Abdallah & Vollesen* 95/23!; Kondoa District: Farkwa, 18 Feb. 1966, *Newman* 173!
DISTR. **T** 3, 5; a few localities in Zambia
HAB. *Acacia-Commiphora* bushland, perhaps seasonally inundated; 800–1300 m

NOTES. Zambian material of this species clearly has a low annular corona surrounding the gynostegium, and three elements that make up the inner, staminal lobes. Material from Tanzania is limited, and difficult to observe in detail, but I have been unable to make out the annular corona and the outermost coronal element convincingly. The corona also appears more delicate than in the Zambian material, where the outer staminal lobe is brown, robust, and sheaths the inner elements. More material with well-preserved flowers is required to assess the status of the Tanzanian collections.

    Specks 248 from Mpwapwa District was cited in the protologue by Kunze *et al.* (1994). However, according to Verhoeven *et al.* in Grana 42: 75 (2003), this collection is *Fockea multiflora.*

49. **MARSDENIA**

R. Br., Prodr.: 460 (1810), *nom. cons.*; Omlor, Gen. Rev. Marsdenieae: 73–82 (1998); Forster, Australian Syst. Bot. 8: 703–933 (1995)

Gymnema R. Br., Prodromus: 461 (1810)
Dregea E. Mey., Comm Pl. Afr. Austr.: 199 (1838), *non* Eckl. & Zeyh. (1837)
Pterygocarpus Hochst. in Flora 26: 78 (1843)
Dregea sect. *Pterygocarpus* (Hochst.) K. Schum. in E. & P. Pf. 4(2): 293 (1895)
Traunia K. Schum. in N.B.G.B. 1: 23 (1895)
Dregea subgen. *Traunia* (K. Schum.) Bullock in K.B. 11: 516 (1957)
Gongronema sensu Bullock in K.B. 15: 197 (1961) pro parte, *non* (Endl.) Decne. (1844)

Woody or wiry twiners, occasionally scandent shrubs; latex white or clear. Leaves opposite, petiolate, glabrous or more usually pubescent. Inflorescences extra-axillary or axillary; sessile or pedunculate, umbelliform or irregularly branched. Corolla somewhat fleshy, mostly tubular but occasionally rotate; lobes frequently imbricate and contorted, but sometimes subvalvate. Corolline corona generally absent, if present then either forming longitudinal ridges down the corolla tube (*M. sylvestris*), or forming an annulus around the gynostegium (*M. exellii*). Gynostegial corona of 5 ovoid or flattened fleshy staminal lobes, attached to anthers basally, shorter or longer than anthers, apex free; rarely vestigial, or absent. Pollinia erect, club-shaped or rounded, basally attached to translator arms; corpusculum narrowly ovate to subcylindrical. Apex of stylar head flat, domed or rostrate, exserted beyond the anther appendages or concealed by them. Follicles single or paired, thick and often somewhat woody, smooth, weakly ribbed or with conspicuous longitudinal wings or surface entirely obscured by the strongly contorted, cristate wings; seeds flattened, with a broad or narrow marginal wing and with a coma of hairs at one end.

The delimitation of *Marsdenia* adopted here largely follows that of Omlor's generic revision of the Marsdenieae (1998), but with the addition of *Gymnema*, formerly separated on the position of its corona, which is on the corolla rather than on the gynostegium. The loss (or gain) of staminal coronas may occur more frequently than earlier thought, and has been demonstrated in sister species in the distantly related Andean genus *Philibertia* Kunth (Goyder in K.B. 59: 415–451 (2004)). In *Philibertia*, taxa lacking a staminal corona frequently compensate by the possession of swellings on the corolla which perform the same function of restricting pollinator access to five regions of the tube and gynostegium. The situation in African *Marsdenia* parallels this exactly, with two morphologically similar species distinguished florally by the position of the corona – the predominantly tropical species, *M. sylvestris*, formerly *Gymnema sylvestre*, has a corona derived from the corolla, while in the South African species *M. dregea*, the corona is gynostegial. A similar generic concept was adopted by Forster (1995) for Asian species. Comparable parallel patterns of variation in *Philibertia* and *Marsdenia* include sessile versus stipitate gynostegia, and the presence or absence of prominent stylar head appendages. Fruit ornamentation, while striking in some African *Marsdenias*, appears to behave in a similar way, with some species possessing prominent longitudinal wings, while apparently closely related species lack them.

SPOT CHARACTERS

Fruiting material [fruit not known in *M. exellii* and *M. taylori*] Fig. 58, p. 192

Follicles with 4 well-developed, longitudinal wings – *M. faulkneri*, *M. rubicunda*, *M. stelostigma*

Follicle surface ± completely obscured by the many cristate wings, or many longitudinal wings only slightly cristate – *M. abyssinica*

Follicles lacking wings
surface smooth – *M. cynanchoides*, *M. latifolia*, *M. magniflora*, *M. sylvestris*
surface with longitudinal wrinkles at least when dry – *M. crinita*, *M. macrantha*, *M. schimperi*

1. Apex of stylar head extending into a rostrate
    appendage exserted from both staminal
    column and the mouth of the corolla tube 2
    Apex of stylar head flat or domed, never long-
    rostrate 4
2. Inflorescences sessile and umbelliform; corolla
    tube only partially enclosing the gynostegium   8. *M. stelostigma* (p. 201)
    Inflorescences pedunculate and irregularly
    branched; corolla tube fully enclosing the
    gynostegium with the exception of the stylar
    head appendage 3

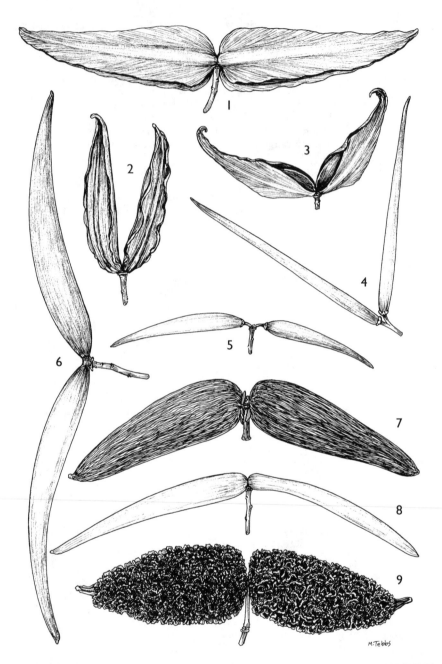

Fig. 58. *MARSDENIA* fruits — **1**, *MARSDENIA RUBICUNDA*, paired follicles; **2**, *MARSDENIA FAULKNERAE*, paired follicles; **3**, *MARSDENIA STELOSTIGMA*, paired follicles; **4**. *MARSDENIA ANGOLENSIS*, only one follicle developing in each flower; **5**, *MARSDENIA SYLVESTRIS*, only one follicle developing in each flower; **6**, *MARSDENIA SCHIMPERI*, paired follicles; **7**, *MARSDENIA MACRANTHA*, paired follicles; **8**, *MARSDENIA LATIFOLIA*, paired follicles; **9**, *MARSDENIA ABYSSINICA*, paired follicles. 1 from *Goyder & Masinde* 3952; 2 from *Faulkner* 1019; 3 from *Waterman & Agnes* 1113; 4 from *Chandler* 1045; 5 from *Greenway & Kanuri* s.n., April 1970; 6 from *Greenway* 7580; 8 from *Dhetchuvi* 873; 9 from *Bullock* 3050. Drawn by M. Tebbs.

3. Staminal corona lobes reaching beyond the anther appendages; anther wings ± 1.2 mm long; follicles with 4 well-developed longitudinal wings running the entire length of the follicle .... 10. *M. faulkneri* (p. 202)

   Staminal corona lobes shorter than the anther appendages; anther wings ± 0.8 mm long; follicles unwinged .... 9. *M. schimperi* (p. 201)

4. Gynostegium fully exposed on the base of the corolla, or only partially obscured by a short corolla tube 5

   Gynostegium (with the exception of the domed stylar head of some species) obscured by the corolla tube 9

5. Inflorescences divaricately branched with flowers scattered along the axes; flowers small – corolla lobes 1.5–2 mm long; pedicels 2–5 mm long .... 3. *M. latifolia* (p. 196)

   Inflorescences not as above, mostly umbelliform; flowers larger–corolla lobes 3–9 mm long; pedicels 5–20 mm long 6

6. Corolla lobes pubescent adaxially 7

   Corolla lobes glabrous adaxially 8

7. Sepals foliaceous; corolla lobes 7–9 mm long, ± uniformly pubescent adaxially; surface of follicles not obscured by wings or other ornamentation, finely longitudinally wrinkled when dry .... 6. *M. macrantha* (p. 198)

   Sepals not foliaceous; corolla lobes 4–5 mm long, indumentum mostly towards apex and margins adaxially, centre of the lobe glabrous; surface of follicles entirely or mostly obscured by highly cristate convoluted wings .... 12. *M. abyssinica* (p. 204)

8. Fused part of corolla 10–15 mm across; lobes 5–6 mm long; staminal corona lobes subglobose; anther wings ± 1 mm long ...... 13. *M. magniflora* (p. 205)

   Fused part of corolla 5–7 mm across; lobes 3–5 mm long; staminal corona lobes ovoid and drawn out into a weak point facing towards the column; anther wings 0.5–0.7 mm long ...... 14. *M. exellii* (p. 205)

9. Young shoots densely shaggy-pubescent with long, spreading, tawny hairs; corolla tube at least 5 mm long; lobes 10–12 mm long ...... 11. *M. crinita* (p. 203)

   Young shoots glabrous or pubescent, but lacking shaggy tawny hairs; corolla tube less than 3 mm long; lobes less than 5 mm long 10

10. Corolla lobes 3–4 mm long, densely pubescent adaxially; leaves generally with a fine mealy indumentum beneath; follicles with four strongly developed, slightly undulate, longitudinal wings .... 7. *M. rubicunda* (p. 199)

    Corolla lobes 0.5–2 mm long, glabrous adaxially or with hairs only in the throat or ciliate; leaves glabrous or pubescent, but lacking a mealy indumentum; follicles (where known) lacking wings 11

11. Corona corolline, of five fleshy, hirsute ridges
    running from the corolla lobe sinus down the
    length of the corolla tube, frequently bifid and
    projecting somewhat at the mouth of the
    corolla tube . . . .   1. *M. sylvestris* (p. 194)
    Corona gynostegial, of 5 lobes adnate to the back
    of the stamens 12
12. Leaves deeply cordate at the base; pubescent on
    both upper and lower surfaces; infloresences
    pedunculate . . . .   2. *M. angolensis* (p. 195)
    Leaves cuneate to rounded at the base; glabrous
    except for a few short hairs along the midrib
    above; inflorescences sessile or subsessile 13
13. Anther wings ± 0.4–0.5 mm long, about as broad
    as long; gynostegium stipitate; coastal Kenya to
    the coast of central Tanzania . . . .   4. *M. taylori* (p. 196)
    Anther wings ± 0.7–0.8 mm long, longer than
    broad; gynostegium subsessile; coastal
    southern Tanzania and Mozambique . . . .   5. *M. cynanchoides* (p. 197)

1. **Marsdenia sylvestris** (*Retz.*) *P.I. Forst.* in Austr. Syst. Bot. 8: 694 (1995). Type: India, *Koenig* 6733 (LD, lectotype, designated by Forster in Austrobaileya 3: 126 (1989))

Scandent shrub or woody twiner with white latex; stems to ± 5 m long, young shoots shortly villous with white or tawny hairs, older stems lenticellate, to ± 2.5 cm in diameter at base, arising from a long taproot. Leaves opposite, petiole 0.5–1.5(–2) cm long; lamina broadly ovate, obovate, elliptic or suborbicular, 3–7 × 1.5–5 cm, apex acute, shortly attenuate or narrowing abruptly into a short acumen, base rounded or very weakly cordate, pubescent on both surfaces. Inflorescences extra-axillary, single or paired, much shorter than the adjacent leaves, forming dense subumbelliform clusters of flowers; peduncles (0.3–)0.5–1.5 cm long, densely pubescent; pedicels 2–5 mm long, pubescent. Sepals lanceolate to broadly ovate or orbicular, 1–2 × 0.5–1 mm, apex rounded, pubescent with ciliate margins. Corolla yellow, orange or cream, glabrous except on the corona; tube 1–1.5 mm long, ± completely enclosing the gynostegium; lobes spreading, oblong or triangular with a rounded or somewhat truncate apex, ± 1.5 × 1 mm, margins ciliate. Corolline corona forming five fleshy ridges running from the corolla lobe sinus down the length of the corolla tube, pubescent along the lateral faces, frequently bifid and projecting somewhat at the mouth of the corolla tube. Staminal corona absent. Anther wings ± 0.8 mm long. Pollinaria minute, corpusculum ovoid, 0.1 mm long; translator arms slender, ± 0.1 mm long; pollinia ovoid, ± 0.1 mm long. Stylar head projecting for ± 1 mm beyond anthers, ovoid-conical. Follicles single, fusiform, 5–9 × ± 0.7 cm, slender, tapering gently into a long attenuate tip, smooth, glabrous. Seeds flattened, ± 9 × 5 mm, oblong with a narrow margin; coma ± 3.5 cm long. Fig. 58/5, p. 192.

UGANDA. Busoga District: Lake Victoria, Lolui Island, 19 May 1964, *G. Jackson* 4118!; Mengo District: Mpanga Forest, 11 July 1953, *Byabainazi* 37! & Kijassi Forest, May–June 1935, *Chandler* 1186!
KENYA. Meru District: Ngaia Forest Reserve, 20 Jan. 2001, *Luke et al.* 7265!; Nyanza District: Sakwa centre, Nyanza, 30 Sep. 1956, *Padwa* 450!; Kilifi District: Arabuko, *M.D. Graham* 1970!
TANZANIA. Dodoma District: Great North Road, 25 Jan. 1962, *Polhill & Paulo* 1261!; Uzaramo District: Pugu Forest Reserve, June 1954, *Semsei* 1726!; Njombe District: Lower Ndumbi Valley, 30 Mar. 1987, *Lovett & Congdon* 1886!; Zanzibar, 4 Dec. 1930, *Greenway* 2658!
DISTR. U 1–4; K 4, 5, 7; T 1, 3–7; Z; widely distributed across Old World tropics - in Africa from the northeastern provinces of South Africa northwards to Senegal and Mauritania in the west and Ethiopia in the east, then via the Arabian Peninsula to the Indian subcontinent.

HAB. Riverine forest margins, thicket and dry bushland, frequently on sand; 700–1500 m

SYN. *Periploca sylvestris* Retz., Obs. Bot. 2: 15 (1791)
   Gymnema sylvestre (Retz.) Schultes, Syst. Veg. 6: 57 (1820); N.E. Br. in F.T.A. 4(1): 413
      (1903); T.T.C.L.: 65 (1949); F.P.S. 2: 408 (1952); Bullock in F.W.T.A. ed. 2, 2: 95 (1963);
      Malaisse in Fl. Rwanda 3: 104 (1985); K.T.S.L.: 493 (1994); U.K.W.F. ed. 2: 181 (1994);
      Albers & Güldenberg in Fl. Eth. 4(1): 153 (2003)

2. **Marsdenia angolensis** N.E. Br. in K.B. 1895: 258 (1895) & F.T.A. 4(1): 423
(1903). Type: Angola, *Welwitsch* 4245 (K!, lecto., designated by Bullock in K.B. 9:
367 (1954))

Herbaceous or woody scrambler with white latex, frequently 3–4 m high, but in
suitable habitats reaching the forest canopy; stems densely pubescent with spreading
ivory or yellowish hairs to 1 mm long. Leaves with petiole 2–6 cm long, densely
pubescent; lamina broadly ovate, 5–12 × 3–7 cm, apex acuminate, base deeply
cordate, the auricles separated by a narrow sinus, both surfaces pubescent with
spreading hairs, particularly on the veins. Inflorescences extra-axillary, ± as long as
the petioles of the adjacent leaf pair, with 2–3 principal branches terminated by
subumbelliform clusters of flowers, pubescent; bracts linear, to ± 6 mm long, foliose,
densely pubescent; pedicels 4–10 mm long. Sepals mostly ovate to suborbicular with
a rounded apex, but occasionally lanceolate and acute, 1.5–2 mm long, densely
pubescent. Flower buds ovoid, the corolla lobes overlapping to the right. Corolla
cream or yellowish green, united into a broadly cylindrical tube 1.5–2.5 mm long, for
about half its length ± enclosing the gynostegium, pubescent outside, inner face with
pubescent patches towards the base, glabrous or minutely papillose above; lobes
spreading, oblong with a rounded apex, 1.5–2 mm long, 1.5 mm wide, pubescent
outside, glabrous within. Corolline corona absent. Staminal corona lobes fleshy, ±
2 mm long, quadrate to triangular in section below in lower half, adnate to the back
of the anthers for 1 mm, the upper half of the lobe free, tapering into a dorsiventrally
flattened tongue arched over the stylar head. Anther wings ± 1 mm long, curved
gently away from the column towards the base and almost completely obscured by
the corona lobes; corpusculum rhomboid, ± 0.2 × 0.2 mm; translator arms ± 0.1 mm
long; pollinia oblong-ovate in outline, ± 0.2 × 0.1 mm, somewhat flattened. Stylar
head extending above the anthers for ± 1 mm, domed. Follicles occurring singly or
paired at an angle of ± 60°, narrowly cylindrical, 8–12 × 0.5 cm, tapering to a slightly
displaced tip, densely pubescent with short spreading hairs, at least when young.
Seeds oblong, ± 7 × 2 mm, flattened with a narrow yellowish margin; coma ± 2 cm
long. Fig. 58/4, p. 192.

UGANDA. Ankole District: near Mwizi, Ruampara, Jan. 1940, *Eggeling* 3842!; Teso District: Kyere
   Rock, Dec. 1932, *Chandler* 1045!; Mengo District: Mabira Forest, 6 km SE of Mulange Hill, 28
   Aug. 1950, *Dawkins* 624!
KENYA. Trans-Nzoia District: Kitale, Aug. 1962, *Tweedie* 2403!; North Kavirondo District:
   Elgon Nyanza, Aug. 1959, *Tweedie* 1886!; Kericho District: Sotik, Kibajet Estate, 12 Sep.
   1949, *Bally* 7472!
TANZANIA. Kigoma District: Gombe Stream Reserve, Mkenke Valley, 29 Apr. 1992, *Mbago &
   Mpongo* 1013!; Iringa District: Mufindi, Lugoda Tea Estate, 17 June 1978, *Thulin & Mhoro*
   3259! & Mufindi, Lake Ngwazi, 27 Jan 1991, *Bidgood et al.* 1285!
DISTR. U 1–4; K 3, 5; T 1, 4, 7–8; Z; widespread in wetter regions of tropical Africa from West
   Africa to Ethiopia, and Angola to Zimbabwe
HAB. Scrambling over margins of evergreen forest or bushland; 700–1900 m

SYN. *Marsdenia gondarensis* Chiov. in Ann. Bot. Roma 9: 80 (1911). Type: Ethiopia, Gondar,
      Dembia, *Chiovenda* 1741 (?FT holo.)
   Gongronema angolense (N.E. Br.) Bullock in K.B. 15: 199 (1961) & in F.W.T.A. ed. 2, 2: 98
      (1963); Malaisse in Fl. Rwanda 3: 102 (1985); K.T.S.L.: 493 (1994); U.K.W.F. ed. 2: 181
      (1994); Albers & Güldenberg in Fl. Eth. 4(1): 153 (2003)

3. **Marsdenia latifolia** (*Benth.*) *K. Schum.* in Just's Bot. Jahresb. 26(1): 372 (1900). Type: Equatorial Guinea, São Tomé [St. Thomas], *Don* s.n. (K!, holo.)

Slender to robust woody scrambler to ± 4 m high with white latex; stems sparsely to densely pubescent with spreading white or ivory hairs ± 0.5 mm long. Leaves with petiole 1–6 cm long, sparsely to densely pubescent; lamina ovate to ovate-oblong, 6–12 × 3–7 cm, apex acute to stongly acuminate, subtruncate to shallowly cordate at the base, the auricles separated by a broad sinus, both surfaces pubescent with spreading hairs, particularly on the veins. Inflorescences extra-axillary, initially ± as long as the petioles of the adjacent leaf pair and subumbelliform, but branches developing rapidly into long axes with flowers scattered along their lengths, pubescent; bracts linear, 1–3 mm long, densely pubescent; pedicels 2–5 mm long. Sepals ovate to suborbicular, 1(–2) mm long, apex rounded, densely pubescent. Flower buds ovoid, the corolla lobes overlapping to the right. Corolla cream or yellowish green, united into a 0.5–1 mm tube for no more than a third of its length and only partially enclosing the gynostegium, glabrous outside, inner face entirely glabrous to minutely papillose, with or without tufts of long inward-pointing hairs in the throat; lobes spreading, oblong, 2–2.5 × 1–1.5 m, apex rounded, glabrous outside, minutely papillose within but with entirely glabrous margins. Corolline corona absent. Staminal corona very variable, lobes mostly well-developed, fleshy, ± 1–1.5 mm long, quadrate to triangular in section and adnate to the back of the anthers in the lower half, the upper half free and arching over the stylar head, but sometimes reduced to a vestigial peg. Anther wings ± 0.5–0.7 mm long, oblong, not obscured by the corona lobes; corpusculum subcylindrical, 0.2–0.25 × 0.05 mm; translator arms 0.1–0.2 mm long; pollinia ± 0.2 × 0.15 mm, ovate to suborbicular in outline, somewhat flattened. Stylar head extending above the anthers for 0.5–1 mm, conical or domed. Follicles occurring singly or in pairs, lanceolate-subcylindrical, 4–8 × 1–1.5 cm, tapering to a rounded apex, glabrescent. Seeds not seen. Fig. 58/8, p. 192.

UGANDA. Mengo District: Mabira, Mulange, Jan.-Feb. 1920, *Dummer* 4390! & Kitubulu Forest, May 1935, *Chandler* 1229!
TANZANIA. Ufipa District: Sumbawanga, Kara River Gorge, 30 Dec. 1956, *Richards* 7405! & Kasanga Escarpment, 24 Nov. 1959, *Richards* 11823!
DISTR. U 4; T 4; Sierra Leone to Uganda, Angola and Zambia
HAB. Scrambling over margins of evergreen forest or bushland; 700–1500 m

SYN. *Gongronema latifolium* Benth. in Hook., Niger Fl.: 456 (1849); Bullock in F.W.T.A. ed. 2, 2: 98 (1963)
    Marsdenia leonensis Benth. in Hook., Niger Fl.: 455 (1849); N.E. Br. in F.T.A. 4(1): 424 (1903). Type: Sierra Leone, *Vogel* s.n. (K!, holo.)
    M. glabriflora Benth. in Hook., Niger Fl.: 455 (1849); N.E. Br. in F.T.A. 4(1): 424 (1903). Type: Sierra Leone, *Vogel* s.n. (K!, holo.)
    M. racemosa K. Schum. in E.J. 17: 147 (1893). Type: Angola, Lunda, Lulua plain, *Pogge* 1249 (B† holo., K!, iso.)
    M. profusa N.E. Br. in K.B. 1895: 258 (1895) & in F.T.A. 4(1): 425 (1903). Type: Niger, Brass, *Barter* 16 (K!, holo.)
    M. glabriflora Benth. var. *orbicularis* N.E. Br. in F.T.A. 4(1): 424 (1903). Type: Nigeria, Bonny River, Oct. 1860, *Mann* s.n. (K!, holo.)

NOTE. Material from SW Tanzania is much more slender than Ugandan collections. The inflorescence appears less elongated, but this could be young material; the corolla tube is slightly longer than elsewhere, and the tufts of hairs in the throat of the corolla are very well-developed. Four Zambian collections, *Fanshawe* 4906 & 5972, *Mitchell* 15/99 and *Richards* 13704, are similar.

4. **Marsdenia taylori** *Schltr. & Rendle* in J.B. 34: 100 (1896); N.E. Br. in F.T.A. 4(1): 422 (1903); T.T.C.L.: 67 (1949). Type: Kenya, Kilifi District: Rabai Hills, 1885, *W.E. Taylor* s.n. (BM!, holo.; K! (fragment), iso.)

Slender woody scrambler to ± 5 m high, latex colour not recorded; stems glabrous. Leaves with petiole 1–2 cm long, ± glabrous; lamina somewhat succulent, oblong to obovate or elliptic, 7–12 × 4–6 cm, apex shortly acuminate, base rounded to cuneate, glabrous except for a few short hairs along the midrib above. Inflorescences extra-axillary, subsessile, appearing umbelliform but flowers on two or more short branches, inflorescence shorter than the petioles of the adjacent leaf pair, pubescent with ± appressed hairs; bracts ovate or broadly triangular, 0.5–1 mm long, glabrous but sometimes with ciliate margins; pedicels 1–4 mm long. Sepals ovate to suborbicular, 1–2 mm long, apex rounded, glabrous or sparsely pubescent, margins ciliate. Flower buds broadly cylindrical with a subglobose head, the corolla lobes overlapping to the right. Corolla cream or yellowish green, united into a 2–2.5 mm tube for ± two-thirds of its length and completely enclosing the gynostegium, glabrous outside, inner face glabrous except for 5 patches of downward-pointing hairs near the base of the tube, and a dense band of white hairs ± 1 mm long in the throat; lobes spreading or erect, 0.5–1.5 × ± 1 mm, oblong with a rounded apex, glabrous outside, minutely papillose within beyond the tuft of long hairs in the throat. Corolline corona absent. Staminal corona lobes fleshy, 1–1.5 mm long, quadrate to triangular in section and adnate to the staminal column for half their length, the upper half free and forming a slender tongue reaching the top of the stylar head. Anther wings 0.4–0.5 mm long, ± as broad as long; at top of gynostegial stipe ± 1 mm long; corpusculum subcylindrical, ± 0.15 mm long; translator arms ± 0.3 mm long; pollinia oblanceolate in outline, ± 0.4 × 0.2 mm, somewhat flattened. Stylar head not exserted beyond anther appendages. Follicles and seeds not seen.

KENYA. Kilifi District: Kaya Jibana, 12 July 1997, *Luke* 4695! & Rabai Hills near Mombasa, 1885, *W.E. Taylor* s.n.!
TANZANIA. Lushoto District: Lunguza (Lg 8), 13 Dec. 1916, *Zimmermann* G7778!; Morogoro District: 3 km from Kimboza Forest camp site, Kubangire village, 21 March 2006, *Festo, Luke & Kayombo* 2256!; Ulanga District: Mahenge, Kwiro Forest Reserve on E flank of ridge above Isongo, 20 Jan. 1979, *Cribb et al.* 11094!; Zanzibar, Haitajwe, 25 May 1935, *Vaughan* 2235!
DISTR. **K** 7; **T** 3, 6; **Z**; not known elsewhere
HAB. Dry forest, sometimes on limestone; 200–1400 m

SYN. *Gongronema taylori* (Schltr. & Rendle) Bullock in K.B. 15: 201 (1961)

NOTE. Replaced by the very closely allied *M. cynanchoides* further south (from the Rondo Plateau to central Mozambique and the Sabi valley of SE Zimbabwe), from which it can be distinguished by its much shorter anther wings and the stipitate rather than ± sessile gynostegium.

5. **Marsdenia cynanchoides** *Schltr.* in E.J. 38: 53 (1905). Type: Mozambique, Beira, Dondo [25 Miles Station], *Schlechter* 12243 (B†, holo.; K!, iso.)

Slender woody scrambler to ± 3 m high, latex colour not recorded, arising from a cluster of fleshy fusiform roots; stems glabrous or pubescent, corky below. Leaves with petiole 0.5–1.5(–2.5) cm long, pubescent on upper face; lamina oblong to elliptic, 3–12 × 1.5–5.5 cm, apex rounded to acute, usually shortly and abruptly acuminate, base cuneate or occasionally rounded, glabrous except for a few short hairs along the midrib above. Inflorescences extra-axillary, subsessile, appearing umbelliform but flowers on two or more short branches, inflorescence generally shorter than the petioles of the adjacent leaf pair, glabrous or glabrescent below; bracts 0.5–1 mm long, ovate to lanceolate, glabrous but with ciliate margins; pedicels 3–5 mm long. Sepals lanceolate to suborbicular, 1–2 mm long, apex rounded, glabrous, margins ciliate. Flower buds broadly cylindrical to subglobose, the corolla lobes overlapping to the right. Corolla cream or green, united into a tube for half to two-thirds of its length and completely enclosing the gynostegium; tube 1–2 mm long, glabrous outside, inner face pubescent, usually also with 5 patches of downward-pointing hairs near the base of the tube and a band of white hairs to 1 mm long in the throat; lobes spreading, oblong with a rounded apex, 1–1.5 × ± 1 mm, glabrous outside, glabrous or minutely papillose within beyond the tuft of long hairs in the throat. Corolline corona absent. Staminal corona lobes 1–2 mm long, fleshy, quadrate to triangular in section below and adnate to the staminal column for half their length, the upper half free and forming a slender tongue arching over the top of the stylar head. Anther wings 0.7–0.8 mm long, much longer than broad; gynostegium ± sessile; corpusculum subcylindrical, ± 0.2 mm long; translator arms ± 0.1 mm long; pollinia broadly elliptic in outline, ± 0.2 × 0.2 mm, somewhat flattened. Stylar head not exserted beyond anther appendages. Follicles poorly preserved, ± 7 cm long, 2 cm wide, smooth, glabrous. Mature seeds not seen.

TANZANIA. Kilwa District: Malemba Thicket, 24 Jan. 1977, *Vollesen* in MRC 4379!; Lindi District: Lutamba Lake, 30 Oct. 1934, *Schlieben* 5564! & Rondo Plateau, Rondo Forest Reserve, 9 Feb. 1991, *Bidgood, Abdallah & Vollesen* 1435!
DISTR. **T** 8; Also recorded from northern and central Mozambique, and the lower Sabi Valley of south-eastern Zimbabwe
HAB. Mixed deciduous woodland or thicket on sandy or stony ground; 200–700 m

SYN. *Gongronema taylori* sensu Bullock in K.B. 15: 201 (1961), pro parte excl. type

NOTE. Perhaps growing in more fire-prone habitats that the preceding species.

6. **Marsdenia macrantha** (*Klotzsch*) *Schltr.* in E.J. 51: 143 (1913); T.T.C.L.: 66 (1949). Type: Mozambique, Rios de Sena, *Peters* s.n. (B†, syn.) & surroundings of Tete, *Peters* s.n. (B†, syn.)

Woody scrambler to ± 10 m, but often much less; latex white; stems frequently corky and winged below, lenticellate above; very young shoots with short indumentum, rapidly becoming glabrescent. Leaves with petiole 1–3 cm long, densely rusty-puberulent; lamina lanceolate to elliptic or suborbicular, 3–7 × 1–6 cm (but up to ± 16 × 12 cm in older leaves), apex usually acute, but sometimes rounded or obtuse and abruptly acuminate, base truncate to rounded or very shallowly cordate, younger leaves sometimes somewhat cuneate, both surfaces puberulent when young but rapidly becoming glabrescent. Inflorescences generally crowded on short lateral shoots, extra-axillary, ± umbelliform but with flowers opening successively, densely puberulent; peduncles up to 2 cm long; bracts foliaceous, lanceolate to elliptic, 3–6 mm long; pedicels 5–20 mm long. Sepals foliaceous, lanceolate to elliptic, 5–7 × 2–3 mm, sparsely puberulent. Flowers unpleasantly scented; buds ovoid, the corolla lobes overlapping to the right. Corolla rotate, greenish cream, united into a glabrous tube for 2–3 mm, partially enclosing the gynostegium; lobes spreading, slightly contorted, oblong, 7–9 × 2–4.5 mm, apex rounded or truncate, glabrous outside, with a sparse to dense pubescence of white hairs on the adaxial face. Corolline corona absent. Staminal corona lobes ± 3 mm long, dorsiventrally flattened, adnate to the staminal column for just under half their length, the free part narrowing gradually, spreading away from the column then arching over it as a dorsally flattened falcate tongue. Anther wings ± 2 mm long, gynostegium ± sessile; corpusculum subcylindrical, ± 0.6 mm long; translator arms ± 0.2 mm long; pollinia ovate in outline, ± 0.4 × 0.2 mm, somewhat flattened. Stylar head ± flat, not exserted beyond anther appendages. Follicles occuring singly or in opposite pairs, narrowly ovoid to fusiform, 7–10 × 1–3 cm, tapering gently to a rounded or acute apex, tough and horny in texture, surface yellowish and with fine longitudinal wrinkles when dry; seeds obovate to suborbicular, ± 12 × 7–10 mm, flattened, smooth, with a shiny marginal rim; coma to ± 3 cm long. Fig. 58/7, p. 192.

KENYA. Masai District: Ol Lorgosailic [Ololkisalie], 28 Dec. 1963, *Greenway* 11181! & Ol Lorgosailic Plains, 5 Aug. 1943, *Bally* 2669! [flowering material only – fruit are of *M. rubicunda*]; Lamu District: Witu Forest at N point of reserve, 17 Nov. 1988, *Robertson & Luke* 5536!

Tanzania. Mpwapwa District: Mpwapwa, 18 Dec. 1969, *Mapunda & Raya* in DSM 1077!; Kilwa District: Kingupira, Selous Game Reserve, 27 Dec. 1975, *Vollesen* in MRC 3155! & valley above Hot Springs, Selous Game Reserve, 10 Dec. 1998, *Luke & Luke* 5584!
Distr. **K** 6, 7; **T** 1, 3, 5–8; dry zone from our area to N Namibia and N South Africa
Hab. Dry deciduous thicket or woodland, occasionally on margins of forest; 0–1300 m

Syn. *Dregea macrantha* Klotzsch in Peters, Reise Mossamb. Bot.: 272 (1861); K.T.S.L.: 492 (1994); U.K.W.F. ed. 2: 182 (1994)
Periploca petersiana Vatke in Oestr. Bot. Zeitschr. 26: 147 (1876). Types: Mozambique, near Tete, 26 Dec. 1844, *Peters* s.n. (B†, syn.) & 11 Jan. 1845 *Peters* s.n. (B†, syn.); Rios de Sena, 1846, *Peters* s.n. (B†, syn.)
Marsdenia zambesiaca Schltr. in J.B. 33: 338 (1895); N.E. Br. in F.T.A. 4(1): 420 (1903). Type: Malawi, Shire, Zambesi at Chiromo, *Scott Elliot* 3791 (BM!, holo.; K!, iso.)

Note. Some material from **T** 3 has slightly smaller flowers than elsewhere, and in the absence of fruit can be difficult to distinguish from *M. rubicunda.*

7. **Marsdenia rubicunda** (*K. Schum.*) *N.E. Br.* in F.T.A. 4(1): 421 (1903); T.T.C.L.: 66 (1949); F.P.S. 2: 411 (1952); Goyder in Fl. Somalia 3: 165, fig. 115 (2006). Type: Kenya, Mombasa Island, *Hildebrandt* 2024 (K!, lecto., designated by Bullock in K.B. 11: 515 (1957); B†, isolecto.)

Woody scrambler to ± 10 m, but often much less; latex white; stems corky and winged below, lenticellate above, very young shoots with a dense cream or reddish indumentum. Leaves with petiole 1–4 cm long, cream or rusty-puberulent; lamina broadly ovate to elliptic or suborbicular, 4–10 × 3–7 cm, apex rounded to acute, sometimes abruptly acuminate, base truncate to rounded or very shallowly cordate, both surfaces with a dense cream or reddish indumentum when young, glabrescent above but usually retaining a fine mealy indumentum beneath. Inflorescences frequently crowded on short lateral shoots, extra-axillary, ± umbelliform but with flowers opening successively, densely puberulent; peduncles up to 2 cm long; bracts ovate to triangular, 1–2 mm long; pedicels 6–10 mm long. Sepals ovate to suborbicular, 1–2.5 mm long, largely obscured by the mealy indumentum. Flowers unpleasantly scented; buds ovoid, the corolla lobes overlapping to the right; corolla rotate, greenish cream, united into a glabrous tube for ± 2 mm, ± enclosing the gynostegium; lobes spreading or reflexed, slightly contorted, oblong, 3–4 × 1–2 mm, apex rounded or truncate, glabrous outside, with a dense pubescence of white hairs on the adaxial face. Corolline corona absent. Staminal corona lobes 2–3 mm long, dorsiventrally flattened, adnate to the staminal column for under half their length, the free part narrowing gradually, exserted from the mouth of the corolla tube for at least 1 mm as a slender, subulate appendage. Anther wings 1–1.5 mm long, gynostegium sessile; corpusculum subcylindrical, 0.3–0.4 mm long; translator arms ± 0.1 mm long; pollinia ovate in outline, ± 0.3 × 0.2 mm, somewhat flattened. Stylar head ± flat, not exserted beyond anther appendages. Follicles occuring singly or in opposite pairs, narrowly ovoid to fusiform, 6–10 × 2–4 cm, with 4 strongly developed, slightly undulate, longitudinal wings running the entire length of the follicle, tough and woody and with fine longitudinal wrinkles when dry, minutely puberulent or glabrescent; seeds ± 12 × 6–8 mm, flattened, obovate to suborbicular, smooth, with a shiny marginal rim; coma to ± 3 cm long. Fig. 59, 58/1, pp. 200 & 192.

Uganda. Karamoja District: Apule River, 28 Oct. 1939, *A.S. Thomas* 3091! & 50–80 km N of Kacheliba, 9 May 1953, *Padwa* 100!
Kenya. Northern Frontier District: Moyale, at foot of escarpment, 22 Aug. 1952, *Gillett* 13746!; Elgeyo District: Kerio Valley, foot of Elgeyo Escarpment ± 5 km N of Biretwo, 4 Jan. 1995, *Goyder & Masinde* 3952!; Kwale District: Mwachi near Mombasa, Apr. 1930, *R.M. Graham* 2335!
Tanzania. Tanga District: Sawa, 25 Apr. 1966, *Faulkner* 3768! & Pongwe, 13 Jan. 1937, *Greenway* 4835!; Mpwapwa District: Kikombo, E of Dodoma, 12 July 1950, *Bally* 7923!
Distr. **U** 1, 2; **K** 1–3, 6, 7; **T** 3, 5, 6; drier zones from Sudan to W & S Ethiopia and Somalia, extending southwards to our area

FIG. 59. *MARSDENIA RUBICUNDA* — **1**, habit, × ²/₃; **2**, paired follicles, × ²/₃; **3**, inflorescence with single flower open, × 3; **4**, flower showing calyx and outside of corolla, × 3; **5**, flower from above, × 2; **6–8**, progressive dissection of the flower, × 6, with the corolla removed to show calyx and gynostegium (**6**); with gynostegial corona pulled away to reveal stamens (**7**); and with the staminal tub removed to reveal the gynoecium of paired ovaries and expanded stylar head (**8**); **9**, half-flower, × 4; **10**, pollinarium, × 20. 1 from *Jeffery* 388; 2 from *Thomas* 3091; 3-10 from *Faulkner* 735. Drawn by D. Erasmus.

HAB. In dry deciduous bushland or riverine vegetation in dry country; 0–1300 m

SYN. *Dregea rubicunda* K. Schum. in E.J. 17: 147 (1893); K.T.S.L.: 492 (1994); U.K.W.F. ed. 2: 182 (1994); Albers & Güldenberg in Fl. Eth. 4(1): 151 (2003)

NOTE. Very similar to the previous species, but with smaller flowers, more slender corona lobes, and conspicuously 4-winged follicles. The somewhat mealy undersurface to the leaves is also distinctive.

8. **Marsdenia stelostigma** *K. Schum.* in E.J. 33: 330 (1903); Goyder in Fl. Somalia 3: 165 (2006). Types: Ethiopia, Sidamo, Tarro Gumbi, Boran, *Ellenbeck* 2080 & 2086 (B†, syn.)

Woody and somewhat shrubby scrambler to ± 3 m with clear latex; stems with well-developed corky wings on older wood, young shoots pubescent. Leaves with petiole 0.5–2.5 cm long, densely pubescent; lamina broadly ovate to elliptic or suborbicular, 2–6 × 2–5 cm, apex mostly obtuse or rounded, rarely subacute, base mostly truncate, but occasionally cuneate or weakly cordate, softly pubescent on both surfaces. Inflorescences sessile, extra-axillary, ± umbelliform but with flowers opening successively, densely pubescent; bracts lanceolate, ± 1 mm long; pedicels 1–2 mm long. Sepals ovate to lanceolate, 1–2 mm long, densely pubescent. Flowers scented; buds ± conical, twisted, the corolla lobes overlapping to the right. Corolla cream or white, united into a tube for ± 1 mm partially enclosing the gynostegium; lobes slightly contorted, oblong, 2.5–3 × 1 mm, apex rounded or truncate, glabrous outside, bearded along the centre of the adaxial face and in the mouth of the corolla tube. Corolline corona absent. Staminal corona lobes somewhat fleshy, oblong, ± 0.8 mm long, dorsiventrally flattened, apex truncate, adnate to the staminal column and reaching to the top of the anthers. Anther wings ± 0.7 mm long, gynostegium sessile; corpusculum ovoid, ± 0.2 mm long; translator arms ± 0.15 mm long; pollinia ovate in outline, ± 0.2 × 0.1 mm, somewhat flattened. Stylar head rostrate, ± 2 mm long, exserted beyond anther appendages and the mouth of the corolla tube. Follicles occuring singly or in opposite pairs, narrowly ovoid to fusiform, 5–8 × 1.5–2(–4) cm, with 4 strongly developed, slightly undulate, longitudinal wings running the entire length of the follicle, tough and woody and with fine longitudinal wrinkles when dry, pubescent; seeds obovate, ± 8 × 4 mm, flattened, smooth, with a shiny marginal rim; coma ± 2 cm long. Fig. 58/3, p. 192.

KENYA. Northern Frontier District: 66 km E of Isiolo near the N tip of Nyambeni Range, 8 Dec. 1971, *Bally & Smith* 14463!; Machakos District: near Kyulu on Mombasa–Nairobi road, 30 Aug. 1959, *Verdcourt* 2395!; Tana River District: Kora National Reserve, 30 km from Kampi ya Simba along road to Tana River, 12 May 1983, *Mungai et al.* 93!
DISTR. **K** 1, 4, 7; S Ethiopia and Somalia
HAB. Dry *Acacia-Commiphora* bushland; 0–900 m

SYN. *Stigmatorhynchus stelostigma* (K. Schum.) Schltr. in E.J. 51: 141 (1913)
   Marsdenia stefaninii Chiov., Result. Sci. Miss. Stef.-Paol, Coll. Bot. 1: 116 (1916). Type: Somalia, *Stefanini* 1216 (FT, holo.)
   Dregea stelostigma (K. Schum.) Bullock in K.B. 11: 516 (1957); K.T.S.L.: 493 (1994); U.K.W.F. ed. 2: 182 (1994); Albers & Güldenberg in Fl. Eth. 4(1): 152 (2003)

9. **Marsdenia schimperi** *Decne.* in DC., Prodr. 8: 616 (1844); N.E. Br. in F.T.A. 4(1): 419 (1903); T.T.C.L.: 66 (1949); Goyder in Fl. Somalia 3: 165 (2006). Type: Ethiopia, Gennia, Mensach region, Adua, *Schimper* I: 260 (P, holo.; K!, iso.)

Woody scrambler to ± 15 m with clear latex; young shoots pubescent, older wood lenticellate. Leaves with petiole 1.5–5 cm long, densely pubescent; lamina broadly ovate, obovate or suborbicular, 5–11 × 3–8 cm, apex acute to strongly acuminate, base mostly truncate, but occasionally cuneate or weakly cordate, softly pubescent on both surfaces. Inflorescences lax, pedunculate, extra-axillary, irregularly-branched, densely pubescent; peduncles mostly 1.5–3 cm long; bracts ovate, 1–2 mm long; pedicels 6–15 mm long. Sepals ovate to lanceolate, (2–)3–5 × 1–2 mm, densely pubescent. Flowers sweetly scented; buds narrowly conical, twisted, the corolla lobes overlapping to the right. Corolla greenish-white, cream or yellow, united into a tube for 2–3 mm ± enclosing the gynostegium, the tube sparsely pubescent within with long hairs; lobes contorted, oblong, 4–6 × 1 mm, apex rounded or truncate, glabrous or sparsely pubescent outside, sparsely pubescent along the centre of the adaxial face, margins ciliate. Corolline corona absent. Staminal corona lobes ± 1.5 mm long, dorsiventrally flattened, tapering gently to an acute or occasionally rounded apex, adnate to the staminal column below and reaching to half way along the anther appendages. Anther wings ± 0.8 mm long, gynostegium sessile; corpusculum ovoid, ± 0.3 mm long; translator arms ± 0.25 mm long; pollinia oblong-elliptic in outline, ± 0.5 × 0.2 mm, somewhat flattened. Stylar head rostrate, to ± 3 mm long, exserted beyond anther appendages and the mouth of the corolla tube. Follicles occuring singly or in opposite pairs, woody, ovoid, 6–8 × 2–3 cm, lacking longitudinal wings, but with fine longitudinal wrinkles when dry, pubescent; seeds obovate, ± 10 × 6 mm, flattened, smooth, with a shiny marginal rim; coma ± 3 cm long. Fig. 58/6, p. 192.

UGANDA. Karamoja District: Lotim Forest Reserve, June 1955, *Philip* 677!
KENYA. Northern Frontier District: Mathews Range, Mantachien [Mandasion], 7 Dec. 1960, *Kerfoot* 2591!; Turkana District: Murua Nysigar [Moruassigar], 15 Feb. 1965, *Newbould* 7125!; Machakos/Masai Districts, saddle of Chyulu Hills, 18 Jan. 1997, *Luke & Luke* 4595!
TANZANIA. Masai District: Ketumbeine Forest Reserve, 5 Apr. 2000, *Festo et al.* 638! & Endulen, Ngorongoro, 16 Nov. 1965, *Herlocker* 207!; Kondoa District: Bereku Ridge at Salanga Hill, 13 Jan. 1928, *B. D. Burtt* 1066!
DISTR. U 1; K 1–6; T 2, 5; from Nigeria to Somalia and the Arabian peninsula, and S to Angola
HAB. Forest margins and secondary vegetation; 1500–2400 m

SYN. *Cynanchum schimperi* Hochst. in sched. (1840), *nom. nud.*
    Gymnema macrocarpum A. Rich., Tent. Fl. Abyss. 2: 43 (1851). Types: Ethiopia, Tigray, Beless, *Quartin Dillon* (P, syn.); Shoa, *Petit* (P, syn.)
    Traunia albiflora K. Schum. in N.B.G.B. 1: 23 (1895) & in E. & P. Pf. 4(2): 287 (1895). Type: Tanzania, Kilimanjaro, Marungu, *Volkens* 2110 (B†, holo.; K!, iso.)
    Dregea schimperi (Decne.) Bullock in K.B. 11: 518 (1957) & in F.W.T.A. ed. 2, 2: 97 (1963); K.T.S.L.: 493 (1994); U.K.W.F. ed. 2: 182 (1994); Albers & Güldenberg in Fl. Eth. 4(1): 151 (2003)

10. **Marsdenia faulkneri** (*Bullock*) *Omlor*, Gen. Revis. Marsdenieae: 79 (1998). Type: Tanzania, Tanga District: Korogwe, Magunga, *Faulkner* 1189 (K!, holo.; EA!, B iso.)

Woody scrambler to ± 5 m, latex colour not recorded; young shoots densely pubescent, older wood with thick corky wings. Leaves with petiole 1.5–6 cm long, densely pubescent; lamina broadly ovate, 6–13 × 4–10 cm, apex acute to strongly acuminate, base weakly to strongly cordate, softly pubescent on both surfaces. Inflorescences lax, pedunculate, extra-axillary, irregularly-branched, densely pubescent; peduncles 1–3 cm long; bracts lanceolate, to ± 5 mm long; pedicels 1–1.5 cm long. Sepals lanceolate, 4–5 × 1–1.5 mm, densely pubescent at least on mid-line. Flowers with foetid sweet odour; buds ovoid-conical, the corolla lobes overlapping to the right. Corolla greenish-cream, united into a tube for ± 3 mm ± enclosing the gynostegium, the throat and tube sparsely pubescent within with long hairs; lobes oblong, 6–7 × 1.5–2.5 mm, apex rounded or truncate, weakly contorted, sparsely pubescent towards the mouth of the tube but otherwise glabrous on both faces. Corolline corona absent. Staminal corona lobes ± 3 mm long, fleshy and quadrate in section below, adnate to the staminal column for ± 1 mm, the upper $^2/_3$ free, dorsiventrally flattened, tapering gently to an acute apex, and reaching well beyond the anther appendages. Anther wings ± 1.2 mm long, gynostegium sessile; corpusculum ovoid-subcylindrical, ± 0.4 mm long; translator arms ± 0.1 mm long; pollinia elliptic in outline, 0.3–0.4 × 0.2 mm, somewhat flattened. Stylar head rostrate, to ± 3 mm long, exserted beyond the mouth of the corolla tube and somewhat contorted apically. Follicles occurring singly or in pairs held at an acute angle, woody, narrowly ovoid, 6–8 × 1.5–2.5 cm, apex attenuate, with 4 well-developed longitudinal wings running the entire length of the follicle, and fine longitudinal wrinkles over the whole surface, subglabrous; seeds not seen. Fig. 58/2, p. 192

KENYA. Kilifi District: Kaya Ribe, Mleji River, 16 Sep. 1997, *Luke & Luke* 4749!
TANZANIA. Lushoto/Tanga Districts: Magunga Estate, 28 June 1952 (fl.) & 16 Sep. 1952 (fr.), *Faulkner* 1019! & Magunga, 30 Apr. 1953, *Faulkner* 1189!
DISTR. **K** 7; **T** 3; known only from a handful of collections from central Mozambique and our area, but probably occurring throughout the intervening coastal regions
HAB. Forest margins; 0–400 m

SYN. *Dregea faulkneri* Bullock in K.B. 11: 520 (1957)

11. **Marsdenia crinita** *Oliv.* in Hooker's Ic. Pl. 20: t. 1993 (1891); N.E. Br. in F.T.A. 4(1): 418 (1903). Type: Nigeria, expedition to the interor of Yoruba, Oyo, 2 May 1890, *Millson* s.n. (K!, lecto., designated by Bullock in K.B. 9: 367 (1954))

Woody scrambler to ± 8 m with clear latex; young shoots densely shaggy-pubescent with long, spreading, tawny hairs; older wood lenticellate. Leaves with petiole 1–1.5 cm long, densely pubescent; lamina ovate or somewhat obovate, 5–11 × 3–6 cm, apex acute or attenuate, base ± cuneate or truncate, softly pubescent on both surfaces. Inflorescences lax or congested, pedunculate, extra-axillary, irregularly-branched, densely pubescent; peduncles 1–3 cm long; bracts linear, to ± 1 cm long; pedicels 1.5–2 cm long. Sepals lanceolate, ± 10 × 1.5–3 mm, densely pubescent at least on mid-line. Flowers sweetly scented, buds with an arrow-head profile formed by the projecting corolla lobe sinuses, apical part ± conical, corolla lobes overlapping to the right. Corolla white, glabrous, united into a tube for 6–8 mm ± enclosing the gynostegium; lobes oblong, 10–12 × 3–4 mm, apex rounded or truncate, weakly contorted. Corolline corona absent. Staminal corona lobes 5–6 mm long, dorsiventrally flattened, with lateral wings meeting adjacent corona lobes and forming a chamber around the margins of the anther wings, corona adnate to the staminal column for ± half its length, the upper half free, tapering abruptly into an erect tongue reaching well beyond the anther appendages and about as long as the stylar head, apex acute. Anther wings 2–3 mm long, gynostegium subsessile; corpusculum ovoid-subcylindrical, ± 0.5 mm long; translator arms ± 0.2 mm long; pollinia elliptic in outline, ± 0.6 × 0.2 mm, somewhat flattened. Stylar head broad, fleshy and truncate, to ± 2 mm long. Follicles occurring singly or in pairs, woody, mottled grey-green, narrowly ovoid, to 13 × 3.5 cm, apex blunt, with longitudinal wrinkles over the whole surface, subglabrous and somewhat shiny; seeds not seen.

KENYA. Kwale District: Shimba Hills, Mkongani N, 4 May 1992, *Luke* 3123! & Shimba Hills Nature Reserve, Mwele Forest, 29 Nov. 1996, *Luke* 4545!
DISTR. **K** 7; widespread in West Africa and S to Angola
HAB. Dry forest; to ± 250 m

SYN. *Dregea crinita* (Oliv.) Bullock in K.B. 11: 519 (1957) & in F.W.T.A. ed. 2, 2: 97 (1963)

NOTE. These two collections from the Shimba Hills are the only ones from the FTEA area. The flowers are larger than in West African material, and might perhaps justify subspecific recognition.

12. **Marsdenia abyssinica** (*Hochst.*) *Schltr.* in E.J. 51: 143 (1913); T.T.C.L.: 66 (1949); F.P.S. 2: 411 (1952). Types: Ethiopia, Gondar, near Mt Sabra, 19 July 1838 (fl.), *Schimper* II: 1366 (B†, syn.; K!, isosyn.) & 9 Mar. 1840 (fr.), *Schimper* II: 1294 (B†, syn.; K!, isosyn.), see note below

Woody scrambler to several metres with white latex; older wood lenticellate, soft and corky below, young shoots minutely puberulent. Leaves with petiole 1–2.5 cm long, minutely puberulent; lamina ovate to broadly elliptic, 5–10 × 3–6 cm, apex acute to shortly attenuate or rounded and abruptly acuminate, base broadly cuneate, rounded or truncate, glabrous on both surfaces. Inflorescences extra-axillary, pedunculate, minutely puberulent, with congested, irregularly branched, subumbelliform clusters of flowers; peduncles 1–1.5(–2.5) cm long; bracts ovate, to ± 3 mm long, scarious; pedicels 0.7–1.4 cm long. Sepals broadly ovate to orbicular, 1.5–3 × 1.5–3 mm, densely puberulent at least towards the base. Flowers sweetly scented, buds subglobose, corolla lobes overlapping to the right. Corolla white, cream or green, united for 1–2 mm and not obscuring the gynostegium; lobes ovate or oblong, 4–5 × 2–3 mm, apex rounded or truncate, glabrous or sparsely pubescent abaxially, densely pubescent or even bearded towards the apex, margins and throat adaxially, the central part of the lobe glabrous or minutely papillose. Corolline corona absent. Staminal corona lobes solid, fleshy, ± 2 × 1 mm, about as tall as the anthers, adnate to the staminal column for ± half their length. Anther wings ± 1 mm long, gynostegium sessile; corpusculum somewhat rhomboid, ± 0.3 mm long; translator arms ± 0.2 mm long; pollinia elliptic in outline, ± 0.6 × 0.2 mm, somewhat flattened. Stylar head obscured by the anther appendages. Follicles occurring singly or in opposite pairs, ovoid, 6–7(–10) × 2–3 cm, tapering to an acute apex, but the outline generally obscured by the many well-developed and extremely convoluted longitudinal wings which cover the surface, woody, subglabrous or minutely pubescent; seeds obovate, ± 12 × 6 mm, flattened, smooth, with a shiny marginal rim; coma ± 4 cm long. Fig. 58/9, p. 192.

UGANDA. Karamoja District: Moroto Township, May 1956, *J. Wilson* 241!; Teso District: Serere, Jan. 1933, *Chandler* 1060! Mengo District: Mabira Forest edge, Nagojji, Mar. 1916, *Dummer* 2778!
KENYA. Northern Frontier District: Ndoto Mts, track leading up valley E of Ngurunit Mission Station, 9 June 1979, *Gilbert et al.* 5571! & 5572!; Nairobi District: Nairobi, quarry at back of Thika Road House, 17 Feb. 1951, *Verdcourt* 438!
TANZANIA. Bukoba District: Minziro Forest, Sep. 1958, *Procter* 997!; Masai District: Endulen, Ngorongoro, 12 Dec. 1966, *Herlocker* 532!; Iringa District: roadside bank below Iringa town, 8 Dec. 1994, *Goyder et al.* 3918!
DISTR. U 1–4; K 1, 3–6; T 1–7; widespread in drier regions of tropical Africa, and the south-western parts of the Arabian Peninsula
HAB. Forest margins; 400–1500 m

SYN. *Pterygocarpus abyssinicus* Hochst. in Flora 26: 78 (1843)
   Hoya africana Decne. in DC., Prodr. 8: 639 (1844). Type as for *Pterygocarpus abyssinicus*
   Dregea africana (Decne.) Martelli, Fl. Bogos.: 55 (1886)
   Dregea abyssinica (Hochst.) K. Schum. in P.O.A. C: 326 (1895) & in E. & P. Pf. 4(2): 293 (1895); Bullock in F.W.T.A. ed. 2, 2: 97 (1963); K.T.S.L.: 492 (1994); U.K.W.F. ed. 2: 182 (1994); Albers & Güldenberg in Fl. Eth. 4(1): 151 (2003)
   Marsdenia spissa S. Moore in J.B. 39: 260 (1901); N.E. Br. in F.T.A. 4(1): 420 (1903). Type: Kenya, Northern Frontier District: near Lake Marsabit, 1898, *Delamere* s.n. (BM!, holo.; K!, iso.)
   Marsdenia abyssinica (Hochst.) Schltr. forma *complicata* Bullock, K.B. 7: 423 (1952). Type: Tanzania, Ufipa District: Kate, 21 Oct. 1949, *Silungwe* s.n. (K!, holo.)

NOTE. Material from Northern Kenya, Ethiopia and Sudan has more clearly defined and less cristate wings to the follicles than further south. Bullock described *Marsdenia abyssinica* forma *complicata* for the southern material, but there are no sharp discontinuities in the degree of convolution of the wings – indeed the variation appears clinal rather than discontinuous – precluding the recognition of formal taxonomic entities.
   Bullock (in K.B. 11: 517 (1957)) designated *Schimper* 1573 as lectotype of *Pterygocarpus abyssinicus*, but this is inadmissible, as the collection was not mentioned by Hochstetter, and therefore is not a syntype.

13. **Marsdenia magniflora** P.T. Li in J. S. China Agric. Univ., 15(1): 64 (1994), as '*magriflora*'. Replacement name for *M. grandiflora* Norman. Type: Angola, Cabinda, M'Boka, Belize, Mayumbe, *Gossweiler* 7024 (BM!, holo.)

Slender wiry climber to several metres, latex colour not recorded; young shoots densely pubescent with long, spreading, tawny hairs. Leaves with petiole 0.5–1.5 cm long, densely pubescent; lamina narrowly obovate to elliptic, 7–12 × 3–7 cm, apex acute to acuminate, base rounded or very weakly cordate, sparsely pubescent on both surfaces with spreading, tawny hairs. Inflorescences extra-axillary, pedunculate, densely spreading-pubescent, subumbelliform; peduncles 0.5–2 cm long; bracts filiform, to 5 mm long; pedicels 1–4 cm long, extremely variable within a single inflorescence. Sepals lanceolate, 3–5 × 1–1.5 mm, pubescent with spreading, tawny hairs. Corolla fleshy, colour highly variable, brown, red or pinkish outside, red, white or blueish within, united for about half its length into a flat or broadly and shallowly campanulate bowl 1–1.5 cm in diameter fully exposing the gynostegium, abaxial surface becoming somewhat wrinkled when dry, glabrous, adaxial surface smooth and glabrous, but appearing somewhat velvety due to a dense covering of minute papillae; corolla lobes broadly triangular, 5–6 × 7–9 mm. Corolline corona absent. Staminal corona lobes solid, fleshy, subglobose, 2 mm diameter and about as tall as the anthers, bright red. Anther wings ± 1 mm long, gynostegium subsessile; corpusculum ± 0.4 mm long, laterally compressed; translator arms ± 0.4 mm long; pollinia elliptic in outline, ± 0.7 × 0.4 mm, somewhat flattened. Stylar head obscured by the anther appendages. Follicles narrowly fusiform, ± 10 cm long, glabrous; seeds not seen.

TANZANIA. Bukoba District: Kele Hill, Minziro, Kagera, 22 Oct. 1994, *Congdon* 372!
DISTR. T 1; almost certainly also occurs in adjacent parts of S Uganda; scattered records across West and Central Africa as far as Liberia
HAB. Forest and forest margins; ± 1250 m

SYN. *Marsdenia grandiflora* Norman in J.B. 67, suppl. 2: 97 (1929), *non* (Decne.) Choux (1923)
  M. normaniana Omlor, Gen. Revis. Marsdenieae: 81 (1998), *nom. superfl.* [Replacement name for *M. grandiflora* Norman]

14. **Marsdenia exellii** *Norman* in Exell, Cat. Vasc. Pl. S. Tome: 244 (1944). Type: Equatorial Guinea, São Tomé, Macambrará [Vanhulst], *Exell* 138 (BM!, holo.)

Slender wiry climber to several metres, latex colour not recorded; young shoots pubescent with short hairs. Leaves with petiole 0.5–1.5 cm long, sparsely to densely pubescent; lamina narrowly ovate-elliptic, 5–13 × 1.5–5 cm, apex acuminate, base rounded, cuneate or very weakly cordate, both surfaces ± glabrous except for the prominent raised veins. Inflorescences extra-axillary, umbelliform, pedunculate, densely spreading-pubescent; peduncles 0.5–3 cm long; bracts filiform, to 4 mm long; pedicels 1–2 cm long, ± equal in length within an inflorescence. Sepals linear or triangular, 2–4 × 0.5–1 mm, pubescent. Corolla colour highly variable, red, white or yellow, united basally into a shallowly campanulate bowl 0.5–0.7 cm in diameter, with a smaller pubescent cup or annulus around the base of the gynostegium, abaxial surface glabrous to sparsely pubescent, adaxial surface pubescent at least around the basal annulus, glabrous or minutely pubescent elsewhere; corolla lobes broadly oblong, 3–5 × 3–5 mm, margins ciliate. Corolline corona absent. Staminal corona lobes solid, fleshy, ovoid drawn out into a weak point facing towards the column at least when dry, ± 1.5 × 1 mm and about as tall as the anthers. Anther wings 0.5–0.7 mm long, gynostegium on stipe 1–2 mm long; corpusculum rhomboid, ± 0.3 mm long; translator arms ± 0.1 mm long; pollinia oblanceolate but slightly curved towards the base, 0.4–0.5 × 0.1 mm, flattened. Stylar head obscured by the anther appendages. Follicles not seen.

TANZANIA. Rungwe District: Kiwira [Kibila], Mulagala River, 13 Feb. 1913, *Stolz* 1878!
DISTR. **T** 7; São Tomé, E Congo-Kinshasa, Malawi (Mt Mulanje)
HAB. Forest and forest margins; ± 1500 m

NOTE. I have treated this plant as a small-flowered variant of *M. exellii*, formerly thought to be
restricted to São Tomé, where it occurs in montane forest. *Jenkins* s.n. from Mt Mulanje and
Bytebier & Luke 2899 from eastern Congo-Kinshasa are also part of this complex. Vegetatively
the specimens are almost identical, the ± glabrous leaves have prominent, arched, raised
veins on the lower surface. The corollas of East African collections are much smaller than in
São Tomé material, but have similarly distributed indumentum. Gynostegial characters differ
only slightly, with a shorter stipe and a minor reduction in anther wing length in East African
material. The differences therefore seem largely quantitative rather than qualitative. Flower
colour, however, is reported to be yellow in São Tomé, reddish brown in the Congo
collection, and red in the Malawi material. The Tanzanian collection probably had red
flowers, but this was not recorded on the label.
    The situation is parallel to that of *Tylophora anomala* N.E. Br., another forest asclepiad, that
has a widely scattered geographic distribution and locally distinctive forms. As with other
forest taxa, *M. exellii* is probably undercollected, and could perhaps be more widespread than
the four localities suggest.

50. **STIGMATORHYNCHUS**

Schltr. in E.J. 51: 141 (1913); Bullock in K.B. 9: 349–373 (1954)

Branched shrubs, sometimes with twining shoots, short shoots very leafy; latex
white. Leaves opposite. Inflorescences extra-axillary, sessile or shortly pedunculate
umbels or clusters with up to 12 flowers; pedicels very short. Corolla lobed to middle,
campanulate, outside smooth, inside densely hairy, base adnate to gynostegium.
Corolline corona absent or reduced to thickened ridges in the corolla tube.
Gynostegial corona-lobes staminal, basally adnate to the back of the anthers,
dorsiventrally flattened. Anther membranes elliptic-lanceolate, ± pointed. Stylar
head elongated and tapering. Pollinia erect, ovoid, attached basally to horizontally
oriented translator arms; corpusculum elongated. Follicles single or paired, smooth
or with an irregular knobbly ridge near the base, beaked; seeds usually one per
follicle, with a coma of hairs at one end.

Two or three disjunct species in tropical and subtropical Africa.

Stigmatorhynchus umbelliferus (*K. Schum.*) *Schltr.* in E.J. 51: 141 (1913). Type:
Tanzania, Iringa/Mbeya Districts: Uhehe, *Goetze* 478 (B†, holo.; K, iso.)

Shrub to ± 3 m, much branched; latex white; stems frequently with short lateral
shoots, minutely pubescent, older stems glabrous and somewhat lenticellate. Leaves
opposite and widely spaced on main shoots, or clustered on short lateral shoots;
petiole 0.3–0.7(–1.3) cm long, densely pubescent; lamina broadly ovate to rhomboid,
2.5–8(–11) × 1.5–3.5(–7) cm, apex subacute to attenuate, base cuneate, pubescent
on both surfaces. Inflorescences extra-axillary, forming dense umbelliform clusters
of flowers, densely pubescent; peduncles 2–5 mm long; pedicels ± 2 mm long. Sepals
broadly ovate to triangular, ± 1 mm long, pubescent. Corolla white or cream,
glabrous outside; tube 1–1.5 mm long, slightly inflated, bearded with inward-
pointing hairs in the throat; lobes erect, oblong, 1–1.5 × ± 0.7 mm, apex rounded or
truncate, pubescent adaxially, somewhat contorted. Staminal corona lobes
dorsiventrally compressed, ± 0.5 mm long, taller than the anthers, narrowly oblong.
Gynostegium concealed within corolla tube except for long-rostrate stylar head.
Anther wings ± 0.3 mm long. Pollinaria minute, corpusculum 0.1 mm long, narrowly
subcylindrical; translator arms slender, ± 0.1 mm long; pollinia ovoid, ± 0.15 mm
long. Stylar head projecting for 1–1.5 mm beyond anthers, narrowly conical to

FIG. 60. *STIGMATORHYNCHUS UMBELLIFERUS* — **1**, habit, × 1; **2**, infructescence, × 1; **3**, dehisced follicle exposing the single, plumose seed, × 1; **4**, flower, × 8; **5**, gynostegium with staminal corona lobes surrounding the base of the long-rostrate stylar head appendage, × 20; **6**, gynoecium, showing paired ovaries and base of the stylar head and its appendage, × 20; **7**, pollinarium, × 30. 1 from *Burtt* 3789; 2, from *Burtt* 3519; 3 from *Koritschoner* 1196; 4-7 from *Burtt* 999. Drawn by D. Erasmus.

subcylindrical, apex entire. Follicles generally single, if paired then held at an acute angle to each other, 4–6 × ± 0.6 cm, subcylindrical basally and forming a knobbly ridge ± 1 cm from the base, then tapering gently into a long attenuate tip, smooth, pubescent. Seeds oblong, ± 1.3 × 0.2 mm, longitudinally canaliculate, smooth; coma ± 2 cm long. Fig. 60, p. 207.

TANZANIA. Shinyanga District: Shinyanga hill, 21 Feb. 1932, *B.D. Burtt* 3519!; Kilosa District: Mwega valley, 6–8 km E of Malolo village, 31 Jan. 1988, *Pócs & Persson* 88007!; Mbeya/Njombe Districts: foot of Chimala escarpment on road to Matamba, 21 Nov. 1986, *Brummitt et al.* 18074!
DISTR. T 1–2, 4–7; not known elsewhere
HAB. Dry *Acacia-Commiphora* or *Brachystegia* woodland; 600–1500 m

SYN. *Marsdenia umbellifera* K. Schum. in E.J. 28: 460 (1900); N.E. Br. in F.T.A. 4(1): 422 (1903); T.T.C.L.: 67 (1949)

51. **TELOSMA**

Coville in Contr. U.S. Natl. Herb. 9: 384 (1905)

Slender herbaceous or somewhat woody twiners; latex white or clear. Leaves opposite, petiolate. Inflorescences extra-axillary; sessile or pedunculate, umbelliform. Corolla urceolate with an inflated base and contorted lobes. Corolline corona absent. Gynostegial corona of 5 flattened fleshy staminal lobes attached to anthers basally, and with an adaxial ligule on their inner faces. Pollinia erect, basally attached to translator arms. Apex of stylar head not exserted beyond the anther appendages. Follicles generally single, smooth. Seeds flattened, with a hairy coma.

5 species, mostly in SE Asia, with 1 in tropical Africa.

Telosma africana (*N.E. Br.*) *N.E. Br.* in Fl. Cap. 4(1): 776 (1908); T.T.C.L.: 68 (1949); Bullock in F.W.T.A. ed. 2, 2: 97 (1963). Type: Nigeria, Nupe, *Barter* s.n. (K!, lecto., designated by Goyder in K.B. 59: 651 (2004))

Slender woody twiner to ± 5 m with clear or white latex; older stems somewhat lenticellate, glabrous. Leaves with petiole 1.5–5 cm long, minutely pubescent; lamina broadly ovate to oblong, 4–10 × 1.5–6 cm, apex attenuate to acuminate, base mostly rounded or truncate, rarely cuneate, glabrous or subglabrous. Inflorescences extra-axillary, forming a dense sub-umbelliform cluster of flowers, subsessile or clearly pedunculate, minutely pubescent; peduncles to 10 mm long; pedicels 3–6 mm long. Sepals broadly ovate to triangular, 2–4 × 1–1.5 mm, glabrous or minutely pubescent. Corolla yellow or green, frequently flushed reddish purple outside, strongly twisted in bud; tube urceolate, 4–6 mm long, glabrous or sparsely pubescent outside, bearded with inward-pointing hairs in the throat and the base of the lobes; lobes spreading and somewhat contorted, 6–10 × 0.5–1 mm, narrowly oblong with a rounded or truncate apex, densely pubescent adaxially, glabrous or at most sparsely pubescent abaxially. Gynostegium concealed within corolla tube. Staminal corona lobes broadly ovate, 2–2.5 mm long, dorsiventrally compressed, with a linear or narrowly triangular ligule on the inner face extending for a further 1 mm and overtopping the gynostegium. Anther wings ± 1 mm long. Pollinaria with corpusculum ± 0.2 mm long, ovoid; translator arms ± 0.1 mm long; pollinia narrowly oblong, ± 0.6 mm long. Stylar head conical, not projecting beyond anthers. Follicles generally single, subcylindrical, ± 6 cm long, smooth. Seeds not seen. Fig. 61, p. 209.

UGANDA. Teso District: Serere, Mar. 1932, *Chandler* 551!; Mengo District: Kawanda Hill near Kampala, Oct. 1937, *Chandler* 1974! & Lake shore, Entebbe, Mar. 1923, *Maitland* 604!
KENYA. Teita District: Taita Hills, Kighombo, 12 June 2000, *Mwachala et al.* in EW 3338!

FIG. 61. *TELOSMA AFRICANA* — **1**, habit, × 1; **2**, flower, × 3; **3**, flower bud, × 3; **4**, gynostegium with staminal corona lobes, × 8; **5**, gynostegium with corona lobes removed, × 8; **6**, staminal corona lobe showing tooth on inner face, × 8; **7**, stamen viewed from within, × 8; **8**, gynoecium showing paired ovaries and stylar head with expanded stylar head appendage, × 8; **9**, pollinarium, × 36. 1 from *Chandler* 551; 2–9 from *Maitland* 604. Drawn by D. Erasmus.

TANZANIA. Bukoba District: Kikuru Forest Reserve, E of Kagera river, 14 Oct. 2000, *Festo & Bayona* 800!; Lushoto District: E Usambaras, 15 Mar. 1943, *Greenway* 6672!; Rungwe District: Kyimbila, 10 Dec. 1912, *Stolz* 1753!

DISTR. U 2–4; **K** 7; **T** 1, 3, 4, 7; widespread in West Africa, but with only a few scattered records from the rest of the continent. Probably undercollected and more widespread than these records suggest

HAB. Margins of wet forest; 900–1300 m

SYN. *Pergularia africana* N.E. Br. in K.B. 1895: 259 (1895) & in F.T.A. 4(1): 426 (1903)
    Telosma unyorensis S. Moore in J.B. 46: 307 (1908). Type: Uganda, Bunyoro District: near Mruli, Victoria Nile, *Bagshawe* 1558 (BM!, holo.)
    Pergularia tacazzeana Chiov. in Ann. Bot., Roma 9: 80 (1911). Type: Ethiopia, Shire, along Tacazzè R. below Timchet, *Chiovenda* 617 (FT!, holo.)
    "*Telosma africanum* (N.E. Br.) Coville", comb. ined., in F.W.T.A. ed. 2, 2: 97 (1963)

52. **LEPTADENIA**

R. Br., On the Asclepiadeae: 34 (1810); Bullock in K.B. 10: 265–292 (1955)

Leafless shrubs (but not in FTEA area) or woody twiners with well developed leaves; latex clear and watery. Leaves opposite. Inflorescences with many flowers in pedunculate, extra-axillary umbels. Corolla rotate with fleshy, pubescent or lanate lobes and a short campanulate tube; corolline corona lobes present in the corolla lobe sinuses. Gynostegium with or without an annular staminal corona at the base; anthers incumbent over the stylar head, lacking apical appendages. Pollinaria subhorizontal or suberect; corpusculum minute, reddish brown; translator arms translucent, flattened and obtriangular; pollinia somewhat flattened and with a translucent apical germination zone. Follicles developing singly, fusiform, slender or stout; seeds smooth, flattened, with a silky coma.

Four species bordering the Sahara and extending eastwards to Arabia and the Indian subcontinent, with a fifth on Madagascar.

Leptadenia hastata (*Pers.*) *Decne.* in DC., Prodr. 8: 551 (1844); Bullock in F.W.T.A. ed. 2, 2: 98 (1963); K.T.S.L.: 494 (1994); U.K.W.F. ed. 2: 182 (1994); Goyder in Fl. Eth. 4(1): 156 (2003). Type: "Africa" (L, holo.)

Twining shrub; branches green, minutely pubescent. Leaves with petiole 1–2 cm long; lamina extremely variable in shape, elliptic to broadly ovate or rarely hastate, 2.5–8(–12) × 1.5–4.5 cm, apex subacute to attenuate, base cuneate to truncate, minutely pubescent. Inflorescences extra-axillary with ± 20 flowers in a crowded umbel; peduncles 5–12 mm long, minutely pubescent; pedicels 2–5 mm long, minutely pubescent. Sepals oblong-lanceolate, 1.5–2.5 mm long, subacute, densely pubescent. Corolla rotate, white to cream or orange, becoming darker with age; tube ± 1 mm long, glabrous within; lobes fleshy, oblong, 4–5 × 1 mm, acute, margins slightly revolute, upper surface lanate with white hairs, underside pubescent. Staminal corona forming a minute fleshy annulus around the base of the anthers; gynostegium ± 1.5 mm long, ± obscured by the corolline corona lobes. Follicles occuring singly, lanceolate in outline, 8–11 × 1–1.5 cm, with a long-attenuate beak, stout, glabrous. Fig. 62, p. 211.

a. subsp. **hastata**

Corolla lobe sinuses with minute, glabrous swellings, and a fleshy sublanate corona lobe ± 1.5 mm long, erect or incumbent over head of the gynostegium.

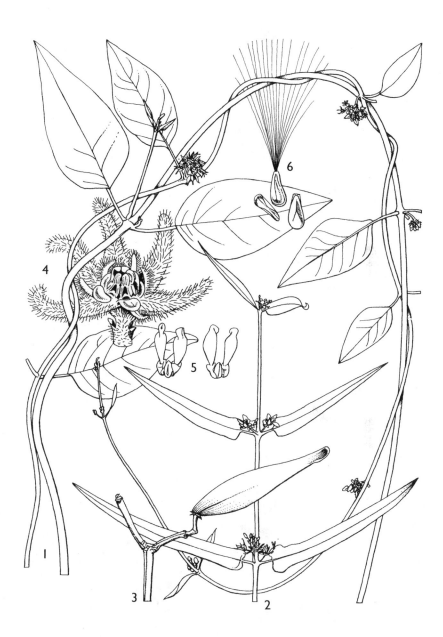

Fig. 62. *LEPTADENIA HASTATA* subsp. *HASTATA* — **1 & 2**, habit showing polymorphic leaves, × ²/₃; **3**, follicle, × ²/₃; **4**, flower, × 6; **5**, pollinaria × 26; **6**, seeds, × 2. 1, 2, 4 & 5 from *Tweedie* 2114; 3 from *Mwangangi & Gwynne* 1148; 6 from *Newbould* 6935. Reproduced with permission, from the Flora of Ethiopia and Eritrea 4, 1; drawn by E. Papadopoulos.

UGANDA. West Nile District: Leya and Ayo [Aiyu] River junction, 25 Mar. 1945, *Greenway &*
Eggeling 7246!; Karamoja District: 5–10 km N of Lokapel, 16 Nov. 1968, *Lye* 446!; Bunyoro
District: Kibiro [Kibero], near Salt Spring, Jan. 1941, *Purseglove* 1090!
KENYA. Northern Frontier District: banks of River Kerio near Lokori, 13 Aug. 1968, *Mwangangi*
& Gwynne 1148!; Turkana District: Lorengipe [Lorengeppe] Camp, 9 June 1965, *Bally*
12808!; Elgeyo District: Kerio valley, Tot, 21 Mar. 1961, *Bally* 12360!
DISTR. U 1, 2; K 1–3; across the semi-arid belt S of the Sahara from Mauritania to Ethiopia
HAB. Scrambling over bushes in hot dry regions, sometimes in dry river beds; 400–1400 m

SYN. *Cynanchum hastatum* Pers., Syn. Pl. 1: 273 (1805)
    C. lanceolatum Poir., Encycl. Suppl. 2: 430 (1811). Type: "in Africa" (P, holo.)
    C. lancifolium Schumach. & Thonn., Beskr. Guin. Pl. 150 (1827). Type: Ghana [Guinea
        coast], *Thonning* 251 (C syn. 6 sheets (IDC microfiche 29: I. 5–7; II. 1–7; III. 1–2, 4–5);
        S!, syn.)
    C. scabrum Schumach. & Thonn., Beskr. Guin. Pl. 152 (1827). Type: Ghana [Guinea],
        Thonning s.n. (C syn. 2 sheets (IDC microfiche 29: III. 6–7; 30: I. 1–2); S!, syn.)
    Tylophora incanum Brunner in Flora 23(2), Beibl. 26 (1840). Type not traced
    Leptadenia lancifolia (Schumach. & Thonn.) Decne. in Ann. Sc. Nat. sér. 2, 9: 269 (1838);
        N.E. Br. in F.T.A. 4(1): 430 (1903)
    L. lancifolia (Schumach. & Thonn.) Decne. var. *scabra* Decne. in DC., Prodr. 8: 628
        (1844). Type: South Sudan, Fazokl, Sennar, *Kotschy* 559 (K! holo. (Hb. Benth.); K! iso.
        (Hb. Hooker))

b. subsp. **meridionalis** Goyder, **subsp. nov.,** subsp. hastatae similis sed lobis coronis sublanatis
e sinibus loborum corollae carentibus differt. Typus: Tanzania, Mpanda District, 27 km along
the Inyonga–Tabora road, *Bidgood, Leliyo & Vollesen* 7852 (K! holo.; DSM!, EA!, MO!, NHT! iso.)

Corolla lobe sinuses with minute, glabrous swellings at the mouth of the corolla tube; well-
developed sublanate corona lobes absent.

TANZANIA. Mpanda District: 27 km along the Inyonga–Tabora road, 19 February 2009, *Bidgood,*
    Leliyo & Vollesen 7852!
DISTR. T 4; not known elsewhere
HAB. *Brachystegia* woodland on grey sandy soil; ± 1100 m
NOTE. This collection represents a very considerable range extension for the genus, as the
    remaining taxa in mainland Africa occur around the margins of the Sahara Desert and within
    the Flora area are restricted to northern Kenya and Uganda.

53. **CONOMITRA**

Fenzl in Endl., Nov. Stirp. Mus. Dec.: 65 (1839); Bullock in K.B. 10: 612–613
(1956); Field in K.B. 37: 341–347 (1982)

Ephemeral herb with solitary or paired extra-axillary flowers; latex colour not
recorded, but almost certainly clear. Leaves opposite. Calyx lobed almost to the base.
Corolla with a short tube and slender lobes. Corolline corona of 5 lobes arising from
the top of the corolla-tube at the corolla-lobe sinuses; gynostegial corona absent.
Anthers terminated by membranous anther-appendages. Pollinia erect in the anther
sacs and with a pellucid upper margin, attached to the minute corpusculum by short,
broad, translator-arms. Apex of stylar head erect, conical, exserted well beyond the
anthers. Follicles terete, smooth, glabrous, erect.

A single species.

Conomitra linearis *Fenzl* in Endl., Nov. Stirp. Dec.: 66 (1839); Bullock in K.B. 10:
613 (1956); Field in K.B. 37: 346 (1982); Goyder in Fl. Eth. 4(1): 157, t. 140.35 (2003)
& in Fl. Somalia 3: 168, fig. 117 (2006). Type: Sudan, Kordofan, *Kotschy* 35 (W! holo.;
K! (2 sheets), BM! iso.)

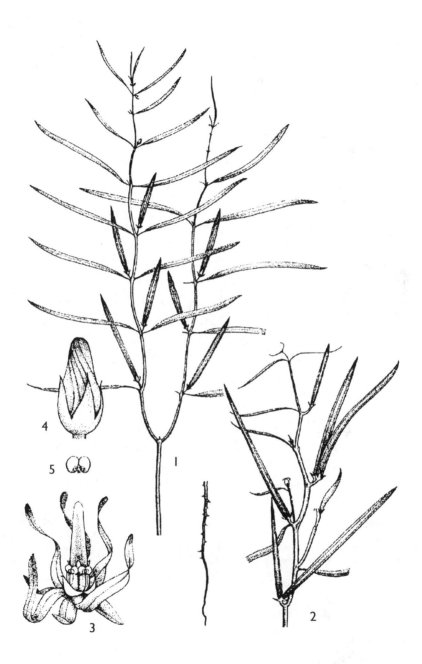

FIG. 63. *CONOMITRA LINEARIS* — **1 & 2**, habit, × ¹/₂; **3**, open flower, × 10; **4**, flower bud, × 10; **5**, pollinarium, × 20. 1 & 4 from *Gilbert & Thulin* 1260A; 2, 3 & 5 from *Stannard & Gilbert* 974. Reproduced with permission, from the Flora of Ethiopia and Eritrea 4, 1; drawn by E. Papadopoulos.

Ephemeral herb, slender, erect; stems 6–30 cm tall, unbranched or sparsely branched, verruculose or minutely pubescent with extremely short, spreading, white hairs. Leaves spreading or suberect, petiole 1–2 mm long; leaf-blade of first pair of leaves linear or linear-oblong, 10–18 × 1–3 mm, the apex rounded, tapering at the base into the petiole; other leaves linear, 25–70 × 2–3 mm, narrowing slightly towards the rounded or subacute apex, tapering at the base into the petiole; surface verruculose above, particularly near the revolute margins. Flowers solitary or paired, extra-axillary; bracts filiform, glabrous, ± 0.5 mm long; pedicels glabrous, 0.5–1 mm long, spreading or reflexed. Calyx ± 1 mm, lobes narrowly triangular, acute, minutely pubescent. Corolla dirty white, lobes twisted in bud; tube campanulate, ± 0.5 mm long; lobes 2 mm long, narrowing abruptly to a long, acuminate tip. Corona-lobes subulate, erect, 0.5 mm long, arising from the corolla-lobe sinus. Gynostegium ± 0.75 mm tall, anther-appendages minute, subulate, resting on the base of the stylar head appendage. Pollinia erect, almost semicircular in outline, 0.2 mm long, with a pellucid upper margin; translator-arms broad, translucent; corpusculum reddish, minute. Stylar head appendage stout, conical, projecting ± 1.5 mm beyond the anthers, 0.3 mm wide at the base. Follicles erect, usually single, fusiform, 30–75 mm long, 3 mm wide, tapering gradually to an acute apex, smooth, glabrous. Seeds oblong, 4–6 × 1–1.5 mm, flattened; coma ± 1 cm long. Fig. 63, p. 213.

Kenya. Northern Frontier District: 2 km N of El Wak, Apr. 1978, *Gilbert & Thulin* 1260! & N of Ndotos on South Horr–Laisamis road, Nov. 1977, *Carter & Stannard* 579! & 55 km S of Modo Gash, Dec. 1971, *Stannard & Gilbert* 974!

Distr. **K** 1; Niger, Sudan, S Ethiopia, Somalia but probably more widespread across the southern margins of the Sahara

Hab. Very rarely collected, but occurring in large ephemeral populations when conditions are suitable, on sand or silty soil in very open *Acacia-Commiphora* bushland; 250–900 m

Syn. *Glossonema lineare* (Fenzl) Decne. in DC., Prodr. 8: 555 (1844); N.E. Br. in F.T.A. 4(1): 291 (1902); F.P.S. 2: 407 (1952)

Odontanthera linearis (Fenzl) Mabberley in Manilal, Bot. Hist. Hortus Malabaricus: 89 (1980)

54. NEOSCHUMANNIA

Schltr. in E.J. 38: 38 (1905); Bullock in F.W.T.A. ed. 2, 2: 95 (1963); Meve in Pl. Syst. Evol. 197: 233–242 (1995) & in E.J. 119: 427–435 (1997); Harris & Goyder in K.B. 52: 733–735 (1997)

Neoschumannia Schltr. in Westafr. Kautschuk-Exped.: 310 (1900), *nomen nudum*
Swynnertonia S. Moore in J.B. 46: 308 (1908)

Robust twiners, becoming woody towards the base; roots fibrous; latex clear, but turning into a white powder on drying. Leaves opposite, membranous. Inflorescences extra-axillary, pedunculate, with flowers clustered towards the tip of a long-lived racemose axis; bracts persistent. Calyx fused at the base. Corolla lobes fused only at their extreme base; corolline corona absent. Gynostegium stipitate; gynostegial corona of three series – downward pointing staminal lobes partially fused to form a skirt; a whorl of spreading or erect interstaminal lobes; and a further whorl of 5 erect dorsi-ventrally flattened staminal lobes with swollen bases. Pollinia erect in the anther sacs, with pellucid inner margins, attached to the minute corpusculum by very short, broad, translator-arms. Apex of stylar head flat or depressed, not exserted from the staminal column. Follicles paired, long, slender, fusiform. Seeds ovoid-oblong with a narrow wing and a long coma.

A genus of three species with relict distributions – the W African *N. kamerunensis*, known from isolated forests in Ivory Coast, Cameroon and the Central African Republic; the East African *N. cardinea*, known from just two localities in Tanzania and one in Zimbabwe; and a third, currently undescribed species, discovered very recently in the Gishwati Forest of Rwanda – as this locality is only 50 km from the Impenetrable Forest of Western Uganda, it is likely that it will also be found there at some time.

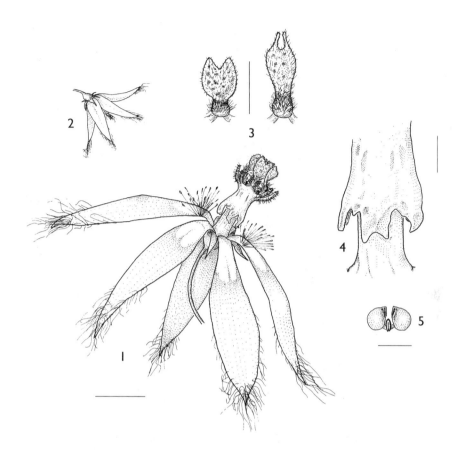

FIG. 64. *NEOSCHUMANNIA CARDINEA* — **1**, flower in lateral view; **2**, natural orientation of flower with petals fallen forwards; **3**, staminal corona lobes; **4**, base of staminal column with coronal skirt; **5**, pollinarium. Scale bars: 1 = 5 mm; 3 = 2 mm; 4 = 1 mm; 5 = 0.5 mm. 1–4 from *Swynnerton* 1080; 5 & 3 (corona lobe to the right) from *Rodgers & Hall* 1346. Adapted from Pl. Syst. Evol. 197: 236 (1995), drawn by U. Meve, mostly after Swynnerton's sketches on the holotype.

Neoschumannia cardinea (*S. Moore*) *Meve* in Pl. Syst. Evol. 197: 235 (1995). Type: Zimbabwe, Chirinda Forest, *Swynnerton* 1080 (BM! holo.; K! iso. (2 sheets))

Large woody climber; glabrous in all parts except the flower; stems slender, to ± 4 mm diameter in flowering parts. Leaves with petioles 1.5–2.5 cm long; lamina elliptic to broadly ovate, 5–15 × 3–8 cm, apex acuminate, sometimes very shortly so, base rounded to truncate, glabrous. Peduncles 1.5–6 cm long; bracts triangular, ± 2 mm long, acute, sparsely pubescent; pedicels filiform, 1–4 cm long. Calyx lobes very narrowly triangular, 4–5 × 1 mm, glabrous or sparsely pubescent. Corolla fused only at extreme base; lobes narrowly lanceolate-elliptic, 12–20 × 3–4 mm, with an acute apex, abaxial face greenish, glabrous, adaxial face purple with two white patches at the base of each lobe, glabrous except for the margins which have a line of slender crisped hairs 2–3 mm long, the tips of the corolla lobes which have a dense tuft of similar hairs, and the pale basal patches to the corolla lobes which have a tuft of slightly more robust clavate hairs ± 2 mm long. Gynostegial stipe ± 2.5 mm long, mostly concealed by the pendant white skirt of fused staminal corona lobes arising below the anthers; interstaminal lobes ± oblong, ± 1.5 × 0.5 mm, flattened dorsiventrally, with a somewhat bifid emarginate-truncate apex, spreading from the column initially, then turning erect, maroon, pubescent; innermost whorl of staminal corona lobes held erect, spatulate, 2–3 mm long, with a truncate/emarginate apex from which two acute teeth may project a further 0.5 mm, maroon, pubescent. Anther wings ± 0.4 mm long. Follicles and seeds not seen. Fig. 64, p. 215.

TANZANIA. Lushoto District: Amani Nature Reserve, Sigi Trail, 9 March 1999, *Liede & Meve* 3359; Iringa District: Sanje Forest Reserve, Mwanihana Forest Reserve, Nov. 1981, *Rodgers & Hall* 1346!
DISTR. **T** 3, 7; only known from these two localities and Chirinda Forest in E Zimbabwe
HAB. Moist submontane tropical forest; ± 1200 m

SYN. *Swynnertonia cardinea* S. Moore in J.B. 46: 309 (1908)

55. **RIOCREUXIA**

Decne. in DC., Prodr. 8: 640 (1844); Harv., Gen. S. Afr. Pl. ed. 2: 240 (1868); G.P. 2(2): 780 (1876); Schlechter in E.J. 18, Beibl. 45: 13 (1894); Schumann in E. & P. Pf 4(2): 273 (1895); Dyer, *Ceropegia, Brachystelma & Riocreuxia* in S. Afr.: 227 (1983); Retief & Herman, Pl. N. Prov. S. Afr., Strelitzia 6: 276 (1997); Bruyns in Goldblatt & Manning, Cape Plants, Strelitzia 9: 288 (2000); Masinde in K.B. 60: 410 (2005)

Ceropegia sect. *Riocreuxia* sensu H. Huber in Mem. Soc. Brot. 12: 167 (1958), partly

Perennial herbs with somewhat woody rootstock, producing a tuft of slightly fleshy or sub-fusiform roots; latex clear. Stems twining or sometimes suberect and tufted, sometimes becoming woody at base, mostly unifariously (rarely bifariously) puberulous, rarely puberulous all round. Leaves opposite, petiolate, base conspicuously cordate; interpetiolar stipules of tufts of whitish or brownish-translucent hairs. Flowers in laxly branched cymes or in fascicles at 1, 2 or 3 nodes of peduncle or flowering axis. Corolla mostly tubular, straight or campanulate, ± inflated at base; lobes 5, linear-attenuate, erect, shorter or longer than tube, mostly connate at tips to form a subglobose cage-like structure; corolline corona absent. Gynostegial corona (1–)2 seriate, variable, subsessile, stipitate. Anthers erect, dorsiventrally compressed, subquadrate, incumbent on the ± globose or ± truncate stylar head, longer or shorter than staminal lobes; pollinia erect, oblong or elliptic with an acute pellucid margin at apex resembling a small beak, attached in pairs from near base by short caudicles to a small dark-brown corpusculum. Follicles paired, terete-fusiform, acuminate, usually ± beaded from being constricted between the seeds, glabrous; seed dorsiventrally compressed, linear-elliptic or oblong, curling length-wise on drying to become convex-concave, with a very narrow marginal wing along perimeter, apex with a tuft of white fluffy hairs.

8 species restricted to Africa south of the Sahara.

1. Corolla campanulate; above 2100 m	3. *R. chrysochroma*	
Corolla not campanulate; below 1850 m 2		
2. Stems, peduncles and pedicels with hairs in 1–2 rows, leaves sparsely pubescent	1. *R. polyantha*	
Stems, peduncles and pedicels densely villous all round; leaves usually velvety pubescent	2. *R. splendida*	

1. **Riocreuxia polyantha** *Schltr.* in J.B. 33: 272 (Sep. 1895); Brown in Fl. Cap. 4(1): 801 (1908); Dyer in Fl. Pl. S. Afr. 29: t. 1124 (1953); Masinde in K.B. 60: 412, fig. 1 (2005). Type: South Africa, Transkei, near Bashee River, *Schlechter* 6291 (B†, holo.; K! (stamped 17 Jan. 1905), lecto. selected by Masinde in K.B. 60: 412 (2005); K!, PRE, isolecto.)

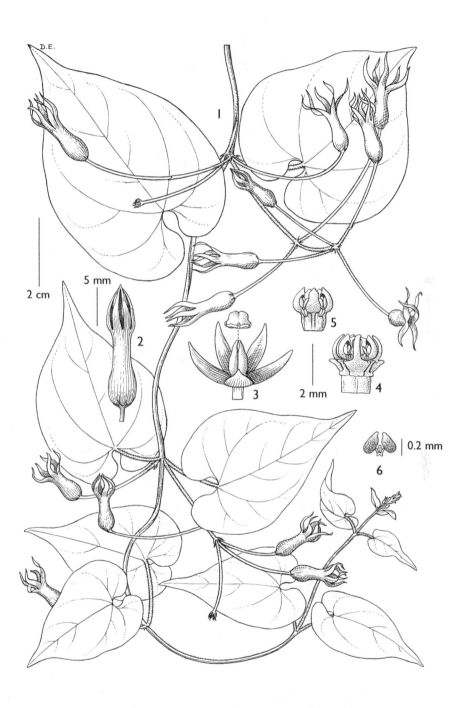

FIG. 65. *RIOCREUXIA POLYANTHA* — **1**, flowering stem; **2**, flower; **3**, calyx and gynoecium showing glabrous carpels; **4**, gynostegium; **5**, gynostegium with outer corona removed; **6**, pollinarium. All from *Milne-Redhead & Taylor* 9781. Drawn by D. Erasmus and reproduced from KB 60: 413, fig. 1.

Leafy climber, 1–3 m or more high, sparsely hairy; stem annual, twining, freely branching, stout in older parts; nodes with a distinct ring of simple, ± 0.6 mm long whitish-translucent hairs; mostly unifariously puberulous, sometimes bifarious, with minute reflexed hairs. Leaves herbaceous, spreading; petiole (10–)30–60(–90) mm long, pubescent; lamina ovate or elliptic-ovate, 3–9(–13.5) × 2–8(–18) cm, base deeply cordate, apex acuminate, occasionally acute; with a broad subtruncate sinus between the semiorbicular incurved or overlapping basal lobes; sparsely and rather minutely pubescent on both sides or rarely glabrous above, veins especially on lower surface more densely pubescent, margin ciliolate. Inflorescence of long, slender, branched, paniculate cymes up to 170 mm long; peduncle 1–2 mm diameter, unifariously puberulous, branching into up to 4 branches, each branch in turn mostly dichotomously or trichotomously branching once or twice; pseudoumbels 10–30+-flowered; bracts linear-subulate, ± 2 mm long, glabrous; flower-buds cylindric, attenuate-acute, often slightly twisted clockwise in apical half; many flowers opening simultaneously; pedicels (15–)20–35 mm, glabrous. Sepals linear-subulate or lanceolate, 2.5(–4) × 0.5(–0.7) mm, acute, often apically reflexed, glabrous. Corolla (13–)16(–28) mm long, straight; tube subcylindric or slightly inflated towards base, 10–14 mm long × 3–4(–5) mm diameter; exterior and interior whitish/yellowish/cream/pale green occasionally with faint greenish striations, glabrous throughout; lobes erect, linear-lanceolate, 4–8(–13) × 1–1.8(–2.5) mm, attenuate-acute, connate at apex to form an ellipsoid or subglobose cage-like structure 5(–8) mm diameter, occasionally breaking free, sometimes orange-red/yellow or pale yellow which is usually more intense adaxially, glabrous. Corona apparently cream or yellowish, ± subglobose, ± 1.5 × 1.5 mm, subsessile or shortly stipitate, biseriate, glabrous throughout; outer lobes ± forming 5 small pockets, emarginate or shortly bifid, with ± 0.5 mm long subhorizontally orientated teeth or lobules somewhat appearing face-to-face; inner lobes linear-obtuse, 0.5–1.2 × ± 0.3 mm, slightly longer or slightly shorter than anthers, closely applied to their backs and connivent over them; guide rails ± 300 mm long. Pollinia elliptic-oblong, 0.28–0.4 × 0.15–0.2 mm; corpusculum elliptic, 0.12–0.19 × 0.07–0.08 mm. Carpels glabrous. Follicles paired, acutely divergent, cylindrical, (90–)150–170(–210) × 2–4 mm diameter at centre, apically tapering; no distinct constrictions between seeds. Seed dark brown or blackish, oblong-elliptic, 8–10 × ± 2.5 mm, dorsiventrally compressed, with a ± 0.5 mm wide paler margin; coma ± 20 mm long. Fig. 65, p. 217.

TANZANIA. Ufipa District: 3 km on Namanyere–Chala road, 2 Mar., 1994, *Bidgood, Mbago & Vollesen* 2569!; Morogoro District: Ulugurus, 16 Apr. 1935, *E.M. Bruce* 1049!; Songea District: Litenga Hill, 19 Apr. 1956, *Milne-Redhead & Taylor* 9781!
DISTR. T 4, 6, 7, 8; Rwanda, Burundi, Angola, Zambia, Malawi, Zimbabwe, Swaziland, Lesotho, South Africa
HAB. Forest margins (including montane and riverine forest), wooded grassland and thickets; 350–1200 m

SYN. *R. torulosa* sensu Schlechter in E. J. 18, Beibl. 45: 24 (1894) and sensu Schumann in P.O.A., C: 327 (1895), *non* Decne. (1844)
   R. burchellii K. Schum. in E & P. Pf. 4(2): 273 (Oct. 1895); R.A. Dyer in F.S.A. 27, 4: 85 (1980) & in Bothalia 13: 3 & 4: 436 (1981) & *Ceropegia, Brachystelma & Riocreuxia* S. Afr.: 233 (1983). Type: South Africa, Cape, Albany Division, between Blaauw Krantz and Kowie Poort, *Burchell* 2668 (?B†, holo.; K!, M, iso.)
   R. profusa N.E. Br. in K.B. 1895: 260 (Oct. 1895) & in F.T.A. 4(1) 465 (1903); Malaisse in Fl. Rwanda 3: 112 (1985). Type: Malawi, Shire Highlands, *Buchanan* 205 & 455 (K!, syn.)
   Ceropegia burchellii (K. Schum.) H. Huber subsp. *burchellii*; H. Huber in Mem. Soc. Brot. 12: 171 (1958)
   C. burchellii (K. Schum.) H. Huber subsp. *profusa* (N.E. Br.) H. Huber in Mem. Soc. Brot. 12: 172 (1958)

Riocreuxia flanaganii Schltr. subsp. *segregata* R.A. Dyer in Bothalia 12(4): 632 (1979) & in F.S.A. 27, 4: 87 (1980). Type: South Africa, Kastrol Nek near Wakkerstroom, *FitzSimons & van Dam* s.n. in *TRV* 25981 (PRE, holo.)

NOTE. Very similar to *R. splendida* in vegetative, corolla and corona characters, and the two appear to be closely related; occasionally cultivated.

2. **Riocreuxia splendida** *K. Schum.* in E.J. 28: 460 (1900); N.E. Br. in F.T.A. 4(1): 466 (1903); Masinde in K.B. 60: 412, fig. 2 (2005). Type: Tanzania, Iringa District: Uhehe near Makombe, *Goetze* 681 (B†, holo.); Tanzania, Mbeya District: N. slopes of Poroto Mts, *St. Clair-Thompson* 722 (K!, 2 sh., neo., selected by Huber (1958: 173))

Climber, hairy, leafy to 2 m or more high; stem twining, branched, stout in older parts, villous all round, dry stems hollow at centre; nodes with a distinct ring of simple, ± 1 mm long brownish-translucent hairs. Leaves herbaceous, spreading; petiole 18–50 mm long, villous all around; lamina broadly ovate, 60–130 × 30–100 mm, base deeply cordate, apex acuminate, with a broad subtruncate sinus between the semiorbicular incurved or overlapping basal lobes, densely pubescent on both surfaces becoming velvety especially underneath and in young leaves, margin ciliolate. Inflorescence of long, branched, pedunculate cymes up to 140 mm long; peduncle densely pubescent, branching into up to 3 branches, each branch in turn mostly dichotomously or trichotomously branching once or twice; pseudoumbels up to 30+-flowered, somewhat congested, many flowers opening simultaneously; bracts linear-subulate, ± 2 × 0.3 mm at base, ± glabrous to sparsely pubescent; pedicels 10–24 mm, glabrous or pubescent. Sepals linear-subulate, 2–3 × ± 0.5 mm, ± glabrous or sparsely pubescent. Corolla (10–)17 mm long, straight; tube subcylindric, slightly inflated towards base, 7–10 mm long × 3–4 mm diameter, exterior whitish-green or orange-yellow, glabrous, interior apparently somewhat orange-red/yellow; lobes erect, orange-red/yellow, linear, 6–10 × 0.5–1 mm, attenuate, connate at apex to form an oblong-elliptic cage-like structure ± 4 mm diameter, glabrous throughout. Corona ± subglobose, ± 1.5 × 1.5 mm, subsessile, biseriate, glabrous throughout; outer lobes ± forming 5 small pockets, emarginate or shortly bifid, with ± 0.2 mm long subhorizontally orientated teeth; inner lobes linear-obtuse, 0.6–0.9 × ± 0.2 mm, slightly longer than anthers, closely applied to their backs and connivent over them; guide rails ± 0.25 mm long. Pollinia elliptic-oblong, 0.35 × 0.2 mm; corpusculum elliptic, ± 0.15 × 0.08 mm. Carpels glabrous. Follicles and seed unknown.

TANZANIA. Mpwapwa District: N of Mpwapwa on mountain top, 5 Mar. 1933, *Mr & Mrs Hornby* 447! & Kiboriani Mts, 26 Feb. 1933, *B.D. Burtt* 4547!; Mbeya District: N slopes of Poroto Mts, 3 Mar. 1932, *St. Clair-Thompson* 722!;
DISTR. **T** 5, 7; not known elsewhere
HAB. Montane forest margins; 1750–1850 m

SYN. *Ceropegia splendida* (K. Schum.) H. Huber in Mem. Soc. Brot. 12: 172 (1958)

NOTE. Further field work may reveal that *R. splendida* is within the range of variation of *R. polyantha*.

3. **Riocreuxia chrysochroma** (*H. Huber*) *Radcl.-Sm.* in K.B. 21: 298 (1967); Masinde in K.B. 60: 427, fig. 7 (2005). Type: Tanzania, Njombe District: Kitulo [Elton] Plateau, Nyarere River, *Richards* 7594 (K!, holo.)

Climber 1–3 m high, hairy, leafy; stem twining, branched, stout in older parts, pubescent all round, dry stems hollow; nodes with indistinct ring of ± 0.5 mm long brownish-translucent hairs. Leaves herbaceous, spreading; petiole (10–)20–50 mm long, terete, pubescent all around but more so in adaxial channel; lamina ovate or elliptic-ovate, (18–)40–75 × (13–)35–50 mm, base deeply or shallowly cordate, apex acuminate or acute; deeply cordate leaves with a broad subtruncate sinus between the semi-orbicular incurved or overlapping basal lobes; upper surface sparsely pubescent to ± glabrous, lower surface minutely velvety pubescent especially along veins; margin ciliolate. Inflorescence of short, pedunculate cymes up to 80 mm long; peduncle ± 0.5 mm diameter, pubescent all round, branched into 2 or 3 and then producing the umbellate cymes or dichotomously branching once before producing flowers; pseudoumbels up to 10-flowered, many flowers opening simultaneously; bracts subulate, ± 1.5 mm long, glabrous; pedicels 5–10 mm, densely pubescent all round. Sepals linear-subulate/lanceolate, ± 1.5–2 × 0.6 mm, almost glabrous except for some pubescence at the bases. Corolla straight, bell-shaped, 7–10 mm long, glabrous throughout; tube bell-shaped, 3–5.5 mm long × ± 4 mm diameter in middle, mostly slightly longer than lobes; exterior and interior cream/whitish-yellow, glabrous throughout; lobes spreading, deep orange-red adaxially, deltoid-linear, 2.5–4 × 1–1.5 mm, acute or obtuse, apically free. Corona elongate or ± subglobose, 1.8–2 × 1.8 mm, shortly stipitate, biseriate, glabrous throughout; outer lobes forming 5 small pockets, deltoid-oblong, emarginate, teeth erect, ± 0.2 mm long; inner lobes erect-connivent, linear-obtuse, ± 1 × 0.2 mm, slightly or much longer than anthers, closely applied to their backs and connivent over them; guide rails ± 0.25 mm long. Pollinia elliptic-oblong, ± 0.35 × 0.15 mm; corpusculum elliptic, ± 0.15 × 0.07 mm. Carpels glabrous. Follicles and seed unknown.

TANZANIA. Rungwe District: Kiwira River, Rungwe Range, 7 Feb. 1961, *Richards* 14256!; Njombe District: Kitulo [Elton] Plateau, Nyarere River, 8 Jan. 1957, *Richards* 7594! & Kipengere Mts, along tributary of Rumakala River, 14 Jan. 1957, *Richards* 7761!
DISTR. **T** 7; Malawi
HAB. Montane forest margins and bushland; 2100–2450 m

SYN. *Ceropegia chrysochroma* H. Huber in Mem. Soc. Brot. 12: 203 (1958)

NOTE. Rare; probably closely related to *R. splendida*.

56. **CEROPEGIA**

Ceropegia L. in Sp. Pl. 1: 211 (1753); Schumann in E & P. Pf. 4(2): 270 (1895); Werderman in B.J.B.B. 15: 222 (1938) & in E.J. 70: 189 (1939); Bally in J. East Afr. & Uganda Nat. Hist. Soc. 17: 234 (1943); Huber in Mem. Soc. Brot. 12: 1–203 (1957, publ. 1958) excl. *Riocreuxia* Decne.; Dyer, *Brachystelma, Ceropegia* & *Riocreuxia* in S. Afr. 12 (1983); Boele et al., Checklist *Brachystelma, Ceropegia, Riocreuxia* & Stapelieae (1987) & Suppl. (1990); Archer, Kenya *Ceropegia* Scrapbook (1992); L.E. Newton in Ballya 1(4): 79 (1994); Meve in Illustr. Handbook Succ. Pl. *Asclepiadaceae*: 63 (2002); de Kock in Checklist *Brachystelma, Ceropegia* & genera of Stapeliads: 84 (2007)

Perennial herbs, stems twining or less often erect, sometimes succulent and persistent; latex usually clear, occasionally cloudy or milky in the more succulent species; rootstock a fascicle of fusiform roots or a subglobose to compressed tuber or, in the case of the more succulent species, plant usually with fibrous roots, occasionally plants with a pyriform tuber or a rhizomatous root system. Leaves herbaceous, succulent or, less often, rudimentary and sometimes reduced to scales, rarely ± lobed, glabrous or pubescent; petiole often adaxially channeled. Inflorescence extra-axillary sessile or pedunculate, mostly contracted into a pseudoumbel, less often elongated and raceme-like or paniculate. Pedicel always present. Sepals small, mostly lanceolate or subulate. Corolla very variable and often complex in form, indumentum and colouring, occasionally zygomorphic; corolla tube well developed, usually with a distinct basal swelling around the corona, sometimes with an internal annulus, often with ciliate hairs on inside; corolla lobes mostly shorter than corolla tube, mostly remaining attached at tips to form a cage-like structure over the mouth of the corolla tube, less often spreading to reflexed, usually hairy. Corolline corona absent. Gynostegial corona biseriate; with 5 staminal (inner) and 5 interstaminal (outer) parts; outer lobes mostly 2-toothed, sometimes teeth very short; inner corona lobes usually dorsiventrally compressed, ± linear, mostly much longer than anthers and connivent into a column over stylar head or free, rarely rudimentary and shorter than anthers. Pollinarium of two shiny yellowish or occasionally shiny orange pollinia laterally attached to a central hard, shiny, brown corpusculum by very short simple terete whitish-translucent caudicles; pollen with prominent apical translucent margins (germination mouths). Follicles paired, narrowly or widely divergent, usually slender and smooth, occasionally thick and smooth or verrucose. Seeds dorsiventrally flattened, ovate to oblong, with a smooth generally blackish to brown central part enclosing the embryo surrounded by a narrow, thinner, paler margin; coma of white silky hairs at micropylar end.

A large genus with about 180 to 200 species, most numerous within Africa south of the Sahara but extending to the Canary Islands and through tropical Asia including the Arabian Peninsula, India, Bhutan and Nepal to China and Australia. There is considerable uncertainty as to the number of species as many are apparently extremely rare whilst others seem to have very disjunct distributions. 51 species recognised for FTEA.

Many species have flowers with somewhat cryptic coloration and members of this genus are difficult to spot in the field, even when the flowers are quite large. The basal parts are very characteristic and should always be recorded. Swollen roots of many species are recorded as edible, often as thirst quenchers.

1. Leaves rudimentary, much reduced, subulate, largest ones up to 5 mm long, sessile 2
   Leaves well developed, conspicuous, more than 7 mm long, petiolar or subsessile 6
2. Roots fibrous; stems ± 10 mm thick, smooth; leaves triangular, thickly firm, scale-like and spiny, borne on conspicuous tubercles ...... 40. *C. variegata* (p. 277)
   Roots fusiform; stems ± 3 mm thick, with fine longitudinal lines often giving a rough feel to the touch; leaves not borne on tubercles 3
3. Corolla lobes narrowed evenly to apex, with a narrow rounded keel, completely glabrous; corolla exterior white or greenish-white ..... 48. *C. ampliata* (p. 287)
   Corolla lobes usually with a broad sharp keel, with clavate, vibratile, purple hairs along margins; corolla grey or creamy with fine blackish spots in parts veined from base to apex 4
4. Corolla lobes united apically into an umbrella-like canopy; margins of canopy with long pendulous, purple clavate hairs .... 51. *C. galeata* (p. 290)
   Corolla lobes erect and narrowing towards apex, not forming an umbrella-like canopy 5
5. Corolla lobes linear from narrowly deltoid bases, not plicate or only hardly so; adaxially at bases with a dense ring of purplish-maroon 2.5–3 mm long thick clavate hairs; rest of lobes glabrous 50. *C. laikipiensis* (p. 290)
   Corolla lobes with broad conspicuous keels; basal margins with fine, short, whitish and brownish simple hairs; purplish-maroon clavate hairs arising variously from midway to apex ...... 49. *C. arabica* (p. 288)

6. Stems succulent or subsucculent, often covered
    with a waxy whitish bloom or glaucous or shiny,
    mostly > 2 mm thick 7
    Stems herbaceous, not covered with a waxy
    whitish bloom, not glaucous, occasionally
    shiny, mostly < 2 mm thick 24
7. Corolla tube with a double inflation basally
    marked by a constriction; or second inflation
    indistinctly visible externally, but internally
    divided into two chambers 8
    Corolla tube with a single inflation basally; no
    division of inflation internally 9
8. Stems cylindrical; corolla lobes narrowly
    rectangular from deltoid bases, forming a
    broad keel with a gibbous (hump-like)
    projection towards base; corolla lobes with
    clavate hairs along apical margins .... 47. *C. denticulata* (p. 285)
    Stems cylindrical, indistinctly ridged or
    quadrangular; corolla lobes triangular,
    forming a broad keel; gibbous projection
    towards base when present, gentle; corolla
    lobe faces and margins densely covered with
    simple hairs but no clavate hairs along margins 46. *C. nilotica* (p. 283)
9. Corolla lobe faces and margins glabrous; leaves
    sessile or subsessile, linear or linear-lanceolate;
    stems weakly ridged especially near nodes; root
    system fusiform (*C. yampwapwa* root system
    unknown but probably fusiform) 10
    Corolla lobe faces and/or margins hairy; leaves
    petiolate, ovate, lanceolate or oblong or a
    combination of these shapes, e.g., ovate-oblong,
    etc. (except in *C. maiuscula* and *C. richardsii* and
    sometimes *C. konasita*); stems cylindrical,
    smooth or pimpled (strongly ridged in *konasita*);
    root system fibrous (*C. maiuscula* and *C. richardsii*
    root systems unknown; *C. konasita* with chains
    of tubers) 12
10. Stems erect or scrambling, up to 15(–30) cm high;
    inflorescence with distinct peduncles (5–10 mm
    long); corolla pale green to cream; corolla lobes
    oblanceolate to lanceolate; inner corona lobes
    rudimentary, shorter than anthers and outer
    corona .... 43. *C. crassifolia* (p. 280)
    Stems erect or climbing, over 20 cm high;
    inflorescence sessile, subsessile or shortly
    pedunculate (1–4 mm long); corolla pale
    green to cream; corolla lobes narrowly
    triangular or linear; inner corona lobes well
    developed, over 5 times as high as anthers and
    outer lobes 11

11. Corolla white with pale green inflation and purplish at apex; corolla tube not or hardly inflated in throat; corolla lobes linear to filiform, only slightly curved along margins, basally with auriculate downward pointing projections .... 45. *C. stenantha* (p. 282)

    Corolla whitish to mauve with pale brown inflation; corolla tube widely inflated in throat to about twice as wide as basal inflation; corolla lobes acutely triangular, completely folded back for most part, basally without auriculate projections .... 44. *C. yampwapwa* (p. 281)

12. Leaves subsessile or sessile, linear; corolla lobes narrowly linear from short triangular bases 13

    Leaves petiolate, ovate, lanceolate or oblong, or a combination of these shapes e.g. ovate-oblong, etc.; corolla lobes not narrowly linear 15

13. Corolla interior pilose from upper level of inflation to tube apex; inner corona lobes glabrous .... 42. *C. richardsiae* (p. 279)

    Corolla interior glabrous throughout 14

14. Inner corona lobes glabrous .... 17. *C. konasita* (p. 247)

    Inner corona lobes densely minutely hairy from base to apex .... 41. *C. maiuscula* (p. 279)

15. Stems subsucculent; corolla white to greenish; corolla lobes linear to filiform, erect, appearing to arise from within tube; forming an ellipsoid cage or tightly close together, dark red, with a cover of fine spreading purple ± 0.6 mm long hairs from base to apex ...... 37. *C. speciosa* (p. 273)

    Stem and corolla parts not fulfilling all the above character combination 16

16. Stems succulent, glaucous; corolla tube slightly inflated at base, inflation > or ± $^1/_2$ as wide as widest part of throat, inflation constricted below middle and immediately above constriction with 5 indentations opposite each sepal; part below constriction creamy; corona laterally adnate to corolla at the constriction 17

    Stem, corolla and corona parts not fulfilling all the above character combination 20

17. Corolla tube ± straight, interior glabrous; corolla lobes ± lanceolate-linear, erect, as long as or slightly shorter than tube, in basal $^1/_2$–$^1/_3$ deltoid, bright greenish-yellow with distinctive purplish-brown net venation, apically ± linear, dark maroon or purple; keels well away from each other; margins ciliate with whitish or maroonish hairs ± from base to apex .... 33. *C. lugardiae* (p. 269)

    Corolla tube strongly curved through over 90°, interior hairy in basal inflation; corolla lobes not lanceolate-linear, much shorter than tube or longer than tube but in which case forming a double cage; keels close together in parts or all along their length; margins glabrous or with short simple hairs basally and long ones at apex 18

18. Leaves linear, sometimes stiffly succulent, very shortly petiolate; roots fusiform; stems not succulent, spotted and streaked; inflorescence perennial, herbaceous, raceme-like, branching with branches elongating to 40 mm or more, articulated and usually with bracts subtending the articulations; corolla silvery-grey with purple-maroon longitudinal lines, more distinct in throat; corolla lobes crescent-shaped, forming a globose cage with keels well apart; corona distinctly constricted just below outer lobes (keys out only some forms of *C. affinis*) . . . . 29. *C. affinis* (p. 260)

Leaves not as above; roots fibrous; stems succulent without spots or streaks; inflorescence not as above; corolla when silvery-grey then with large purple-maroon splodges; corolla lobes elongated, forming an elongated cage or a double cage, keels mostly closer together; corona not distinctly constricted below outer lobes 19

19. Corolla externally densely puberulent except at the base of inflation, most parts pale green to yellowish with irregular darker green splodges; corolla lobes extended above the apex of fusion into a slender, long or short column terminated by an angled head or cage; each keel in lower cage with a prominent or rudimentary deep maroon-purple knob (horn) . . . . 31. *C. somalensis* (p. 264)

Corolla externally glabrous, most parts cream to grey-green with irregular purple-maroon splodges; lobes abruptly narrowed apically to 0.5–1 mm and remaining so wide to apex or slightly expanded and spatulate at apex, margins apically with long whitish hairs or corolla lobes extended above the apex of fusion into a slender, long or short column terminated by an angled head or cage in which case with glabrous margins; without knobs on any parts of keel . . . . 32. *C. distincta* (p. 266)

20. Stems herbaceous; corolla tube thinly puberulent; corolla lobes ± lanceolate, apically blackish to cohering tips, margin of blackish part lined with few long, vibratile, whitish hairs up to tips; corona sessile, laterally adnate to corolla . . . . . 36. *C. johnsonii* (p. 272)

Stem (sub-)succulent; corolla not fulfilling all the above character combination 21

21. Stems subsucculent; leaves ovate to elliptic, herbaceous; corolla tube exterior pale white-green in basal $^1\!/_3$, passing into brown-maroon at throat, basal inflation globose and ± compressed; corolla lobe faces brown-maroon at apex and base separated by cream-yellow band, with few purple-maroon hairs at apex; outer corona lobes broadly bifid with few hairs on top margins . . . . 35. *C. sankuruensis* (p. 271)
Stems succulent; leaves and corolla colour not as above; basal inflation not appearing compressed; corolla lobes not coloured as above; outer corona lobes when broadly bifid then with a dense cover of needle-like hairs in all marginal parts 22

22. Stems succulent, usually 2–3 mm thick; leaves herbaceous; peduncles thinner than stems and not stout; flowers under 40 mm long; outer corona lobes bifid with short triangular teeth, with a dense cover of needle-like hairs in all marginal parts . . . . 34. *C. aristolochioides* (p. 270)
Stems thickly succulent, usually ± 7 mm thick; leaves usually subsucculent to succulent; peduncles thick and stout, often as thick as stems; flowers over 50 mm long; outer corona lobes deeply bifid with linear teeth, inner corona lobes erect-connivent to apex 23

23. Stems scandent, often stoloniferous and rooting at nodes when in contact with the ground; fresh leaves dark green, stiff, with distinct whitish veins on upper side; corolla lobes not apically twisted into a helix; sepals clasping basal inflation, margins glabrous . . . . 39. *C. albisepta* (p. 275)
Stems climbing, not rooting at nodes; fresh leaves shiny green, flexible, without distinct whitish veins on upper side; corolla lobes apically twisted into a helix; sepals spreading, margins ciliate . . . . 38. *C. ballyana* (p. 274)

24. Root system fibrous or fusiform, with a cluster of thick swollen roots usually with fibrous roots arising from ends, occasionally rootstock slightly swollen but not tuberous 25
Root system with a distinct swollen tuber or chains of tubers connected by rhizomes, usually with fibrous roots arising from basal parts or occasionally with swollen to fusiform roots arising from the basal part 35

25. Flowers arising from terminal part of stem,
relatively large; corolla 50–90 mm long with
basal inflation 8–15 mm diameter; stems erect;
root system a cluster of fusiform roots 26
Flowers arising laterally, relatively small; corolla <
50 mm long (when corolla relatively large then
inflation < 8 mm diameter and stems with one
or two rows of hairs or some internodes
puberulous all over – *C. gilgiana*) 27

26. Corolla tube inflated in basal $\frac{1}{2}$–$\frac{1}{3}$, urceolate
inflation large, 10–15 mm diameter; tube
exterior white or pale green with greener
longitudinal lines, passing into whitish-brown or
greenish-brown respectively in throat just below
lobes; interior of tube glabrous throughout;
corolla lobes linear, puberulous on interior . . .   28. *C. umbraticola* (p. 258)
Corolla tube inflated in basal $\frac{1}{5}$–$\frac{1}{4}$, ovoid
inflation 8–10 mm diameter; tube exterior
dark reddish-violet or pale yellow, sparsely
puberulous; interior sparsely pilose at apex of
inflation glabrous; corolla lobes elliptic-
spatulate or obovate-spatulate, plicate with a
broad keel in parts, forming a cage which is
apically cone-shaped with an apiculate tip
or canopy-shaped top . . . .   27. *C. filipendula* (p. 257)

27. Inflorescence perennial, herbaceous, raceme-like,
purple-maroon, mostly branching with branches
elongating to 40 mm or more, articulated and
usually with bracts subtending the articulations
remaining intact; corolla tube inflation almost
obsolete, slight inflation < or ± $\frac{1}{2}$ as wide as
throat, constricted below middle and
immediately above constriction with 5
indentations opposite each sepal; part below
constriction creamy; corona laterally adnate to
corolla at the constriction 28
Inflorescence and corolla tube not fulfilling all
the above character combination 29

28. Corolla lobe cage appearing ± compressed,
hence wider than high and broadly rounded;
outer corona lobes shallowly emarginate with
short erect teeth . . . .   29. *C. affinis* (p. 260)
Corolla lobe cage not as above, cage mostly
elongated and usually narrowing apically;
outer corona lobes deeply lobed ± to base with
adjacent teeth curved towards each other so
that the acute apices ± touch . . . .   30. *C. racemosa* (p. 261)

29. Leaves sessile or subsessile, mostly linear,
occasionally linear-lanceolate; stems erect 30
Leaves petiolate, mostly ovate, lanceolate or
elliptic, stems climbing 31

30. Flowers usually many, corolla > 25 mm long, externally puberulent, yellowish or pale green throughout; corolla lobes internally glabrous, externally minutely pubescent; stems with 1 or 2 rows of hairs or some internodes puberulous all over; plants growing on edges of ponds or other waterlogged habitats .... 23. *C. gilgiana* (p. 254)

Flowers arising singly, corolla < 20 mm long, externally glabrous, purplish with greenish in parts; corolla lobes internally minutely pubescent, sometimes with longer hairs basally; stems glabrous; plants not growing in waterlogged habitats .... 25. *C. keniensis* (p. 256)

31. Inflorescence sessile or subsessile, umbellate, mostly with more than one flower open at a time; corolla externally pale green to purplish with longitudinal ridges internally which are also well marked externally; corolla lobes not fully folded back especially in basal $\frac{1}{3}$, where sinuses between corolla lobes are flat, hence cage openings very narrow in basal $\frac{1}{3}$; faces and margins glabrous; roots fusiform ...... 13. *C. cufodontis* (p. 243)

Inflorescence distinctly pedunculate; corolla without distinct longitudinal ridges internally throughout; corolla lobes fully folded back for nearly their whole length, margins and usually adaxial faces distinctly hairy 32

32. Roots rhizomatous, corolla tube inflated in basal $\frac{1}{2}$; corolla lobes 3–6 mm long, bases of adjacent lobes united into channelled, spreading projection ± 5–6 mm long, margins of projections finely ciliate; inner corona lobes deeply bifid in apical $\frac{2}{3}$ to form linear obtuse lobes .... 19. *C. furcata* (p. 249)

Root system forming a cluster of thickly swollen fusiform roots; corolla and corona not as above 33

33. Corolla tube straight; outer corona lobes 2 × as long as inner lobes, lobed to half way down, teeth linear-filiform .... 3. *C. filicorona* (p. 234)

Corolla tube distinctly curved above basal inflation; outer corona lobes shorter than inner lobes 34

34. Leaves whorled, 3 or more at a node; stems stoloniferous; corolla lobe margins basally not distinctly recurved; corolla < 18 mm long ... 2. *C. verticillata* (p. 233)

Leaves 2 at a node; stems not stoloniferous; corolla lobe margins basally strongly recurved, mostly broadly incurved over the broadly funnel-shaped mouth of tube; corolla mostly > 20 mm long .... 1. *C. meyeri-johannis* (p. 231)

35. Stems usually erect even during flowering, occasionally ± twining in apical parts when stems quite elongated 36

Stems twining, usually not flowering when stems still erect 40

36. Stems always erect; pedicels, sepals and usually
    stems densely covered with long spreading
    hairs; sepals ± $\frac{1}{2}$–$\frac{3}{4}$ as long as corolla tube;
    inner corona lobes erect and free, apices
    strongly incurved at ± 90° and ± meeting above
    stylar head ....   20. *C. abyssinica* (p. 250)
    Stems erect or occasionally ± twining in apical
    parts; pedicels, sepals and stems minutely
    pubescent without long spreading hairs; sepals
    < $\frac{1}{2}$ as long as corolla tube; inner corona lobes
    not as above 37
37. Inflorescence distinctly pedunculate; corolla
    lobe margins with distinct vibratile hairs ....   15. *C. imbricata* (p. 245)
    Inflorescence sessile; corolla lobe margins
    without distinct vibratile hairs 38
38. Flowers appearing in terminal parts of stem,
    flowers large, ± 90 mm long ....   26. *C. campanulata* (p. 256)
    Flowers lateral, not appearing in terminal parts
    of stem, flowers < 40 mm long 39
39. Stems dwarf 70–180 mm long, with 2 rows of
    minute hairs; corolla externally glabrous ...   22. *C. kituloensis* (p. 252)
    Stems >180 mm long, without rows of hairs but
    pubescent all over; corolla externally sparsely
    to densely puberulent ....   21. *C. achtenii* (p. 251)
40. Pedicels, sepals and exterior of corolla including
    corolla lobes densely covered with bristly,
    pointed, reflexed, whitish-translucent hairs;
    corolla lobes free and widely spreading
    (tentacle-like); inner corona lobes erect and
    free, apices strongly incurved at ± 90° and ±
    meeting above stylar head ....   10. *C. ringoetii* (p. 239)
    Pedicels, sepals and exterior of corolla including
    corolla lobes when free, not widely spreading;
    inner corona lobes not as above 41
41. Corolla lobes united at apex into an umbrella-
    like canopy ± 6 mm diameter supported by
    slender to broad completely folded back basal
    parts of lobes; corolla tube and lobes separated
    by a deep constriction in bud ....   16. *C. rendallii* (p. 246)
    Corolla lobes united at apex but not forming
    an umbrella-like canopy or free [except in *C.
    cordiloba*]; corolla tube in bud not as above 42
42. Inflorescence distinctly pedunculate, usually
    > 3 mm long 43
    Inflorescence sessile or subsessile 46
43. Basal sinuses of adjacent corolla lobes auriculate
    with distinct downward pointing horn-like
    projections; inner corona lobes usually
    laterally compressed and falcate, occasionally ±
    dorsiventrally compressed ....   18. *C. purpurascens* (p. 248)
    Basal sinuses of adjacent corolla lobes not
    auriculate; inner corona lobes dorsiventrally
    compressed or cylindrical 44

44. Leaves usually linear, sometimes linear-lanceolate, mostly subsessile, herbaceous; outer corona lobes rectangular, suberectly radiating out to form five pouches .... 24. *C. bulbosa* (p. 255)

Leaves ovate, ovate-lanceolate, ± oblong-ovate, ± elliptic, rarely orbicular, or further variations of these shapes, mostly petiolate, herbaceous or stiffly succulent; outer corona lobes not as above 45

45. Corolla lobes not folded back, hence without keel, ± linear, margins glabrous; corolla cage usually progressively widening towards apex, sometimes slightly constricted in middle .... 14. *C. inornata* (p. 244)

Corolla lobes folded back, with a distinct keel, margins and most of faces with distinct vibratile hairs; corolla cage usually ± oblong or narrowing towards apex, not constricted in middle .... 15. *C. imbricata* (p. 245)

46. Corolla lobes slightly curved back, ± linear, apically opening out and curved at ± 90° before uniting, hence forming broad truncate apex; corolla cage slightly constricted in middle; outer corona lobes united to form a deep cup, with a very shallowly 5-lobed, serrulate margin .... 5. *C. cordiloba* (p. 235)

Corolla lobes mostly completely folded back in most parts (straight in *C. papillata*); cage and outer corona lobes not as above 47

47. Corolla lobes completely folded back from base to apex into ± thick cylindrical columns with gently rounded keel (not sharp edged), usually leaving very narrow cage openings and with prominent basal sinuses projecting below mouth of tube 48

Corolla lobes folded back but not forming thick cylindrical columns and without prominent basal sinuses projecting below mouth of tube; cage openings quite wide 50

48. Plants delicate, twiner to 30 cm high; stems with 1 or 2 rows of hairs; corolla cage broadly globose, cage openings very narrow and opening downwards so that inside of cage is not visible at all; inner corona lobes erect and free, apically strongly incurved at ± 90° and connate above stylar head .... 9. *C. sobolifera* (p. 238)

Plants not delicate, twining to 1 m high and above; stems without rows of hairs, minutely pubescent or glabrous; corolla cage not broadly globose; cage openings not so narrow as above, hence inside of cage visible from a lateral perspective; inner corona apically erect or recurved, not incurved at ± 90° 49

49. Leaves ovate to tri-lobed with margins often deeply toothed; corolla lobes distinctly shorter than tube in length – $\frac{1}{3}$ length of tube or shorter; basal inflation of tube globose; inner corona lobes not spatulate, connivent-erect or free, apically tapering and recurved; corolla cage oblong-ovoid; **T** 8 . . . .   8. *C. paricyma* (p. 238)

Leaves ovate, margins entire; corolla lobes ± equal to tube in length; basal inflation of tube urceolate; inner corona lobes spatulate, erect, ± apically connivent to free; corolla cage ± elliptical; **K** 1 . . . .   12. *C. manderensis* (p. 243)

50. Stems pubescent all round or with 1 or 2 rows of hairs; corolla lobes linear to filiform, $>\frac{1}{2}$ length of corolla tube, sometimes up to 3 × length of tube; apices sometimes very loosely joined and separating very easily hence rarely remaining attached in herbarium material in forms with distinctly long corolla lobes; corolla sometimes distinctly zygomorphic with curved tube and oblique mouth; inner corona lobes dorsiventrally compressed to cylindrical, linear, erect-connivent or free but always slanting towards each other . . . .   11. *C. stenoloba* (p. 240)

Stems pubescent all round; corolla lobes not linear or filiform, $<\frac{1}{2}$ length of corolla tube, usually distinctly much shorter; apices not as above; corolla radially symmetrical; inner corona lobes dorsiventrally compressed to cylindrical, linear to ± spatulate, erect-connivent apically remaining erect or recurved 51

51. Flowers in dense clusters, usually 10–20 at each node; corolla tube inflation pear-shaped; corolla cage often tapering towards apex; corolla lobes with coloured bands – blackish-green with a whitish central band . . . .   4. *C. papillata* (p. 234)

Flowers not in dense clusters, usually 2–3 rarely up to 8 at a node; corolla tube inflation globose; corolla cage globose to oblong, without coloured bands 52

52. Corolla tube inflation much constricted at very base before suddenly expanding; outer corona lobes ± $\frac{1}{2}$ as high as inner lobes, rectangular to triangular in outline and distinctly bifid . . . . .   7. *C. namuliensis* (p. 237)

Corolla tube globose and not constricted basally; outer corona lobes below level of bases of inner lobes, usually ± truncate . . . .   6. *C. claviloba* (p. 236)

1. **Ceropegia meyeri-johannis** *Engl.* in Hochgebirgsfl. Trop. Afr.: 343 (1892); Schumann in E & P. Pf. 4(2): 272, fig. C (1895); N.E. Br. in F.T.A. 4(1): 449 (1903); Werderm. in E.J. 70: 222 (1939); Bullock in K.B. 9: 590 (1954, publ. 1955); Huber in Mem. Soc. Brot. 12: 154 (1958); Bally in F.P.A. 35: t. 1371 (1962); Archer in U.K.W.F.: 391 (1974) & Kenya *Ceropegia* Scrapb.: 53, XI (1992); U.K.W.F. ed. 2: 183 (1994); Masinde in Cact. Succ. J. (US) 6: 107–114 (1998 publ. 1999). Type: Tanzania, Kilimanjaro, steppes between Samburi, Moshi and Marangu, *Meyer* 196 (B†, holo.). Neotype: Tanzania, Kilimanjaro, 6000' [1829 m], Oct. 1884, *Johnston* s.n. (BM!, neo., designated by Huber in Mem. Soc. Brot. 12: 155 (1958))

Herbaceous twiner to 2(–4) m high, leafy, pubescent; latex clear; rootstock a cluster of fusiform roots up to 100 × 5–10 mm; stems sparsely branched, 1–2 mm diameter, pubescent with short retrorse hairs. Leaves herbaceous, spreading; petiole 15–35 mm; lamina ovate or occasionally ovate-oblong, 30–50(–80) × 15–45(–70) mm, base cordate, apex acuminate, minutely pubescent on both surfaces, margin finely ciliate. Inflorescence umbellate, up to 8-flowered, flowers opening consecutively, 1–3 flowers open at a time, mostly without detectable scent, occasionally scented; peduncle 3–40 mm long, pubescent; bracts subulate, ± 1 × 0.3 mm, abaxially pubescent; pedicels 5–15 mm long. Sepals yellowish-green, 3–7 × 0.5–1 mm, acute with recurved tips, abaxially hispidulous-pubescent. Corolla 19–30(–35) mm long, curved; tube 12–21 mm long; in basal $\frac{1}{4}$–$\frac{1}{3}$ with a globose inflation 5–8 × 4–8 mm, with 5 slight indentations at base in between sepals, curved and narrowing above inflation into a cylindrical part ± 2 mm diameter, dilated in throat to 10 mm diameter, glabrous or pilose outside with whitish-translucent hairs especially in throat; exterior greenish-white to yellow-green with maroon spots and splodges on inflation, passing into silvery-grey to yellowish with maroon dots and streaks around throat, occasionally tube whitish-cream throughout; interior: from apex of inflation deep reddish-maroon merging into pale yellow or white in throat, sometimes with green or dark maroon vertical lines at edge of throat; lobes linear from a broadly deltoid base, 5–10 mm long, plicate, incurved and connate at their tips, with needle-like hairs on adaxial surface, adaxially with various shades of green or blackish or purplish, basal part whitish or yellowish with or without reticulated veins. Corona distinctly stipitate, basally cupular, translucent white or cream, 2–2.7(–3.5) × 2–2.8(–3.5) mm, glabrous or pubescent in parts; outer lobes triangular-rectangular in outline, apically deeply or shallowly bifid, teeth linear to triangular, acute or obtuse, suberect, splashed purplish, often finely sparsely ciliate in marginal areas; inner lobes erect, spatulate, linear or linear-spatulate, ± 1.5 × 0.3 mm, apex rounded to ± truncate, well above or overtopping outer lobes, with purplish splash in basal areas. Anthers subquadrate, overtopping the flattened to rounded stylar head; pollinia elliptic, 0.2–0.22 × 0.12–0.14 mm, corpusculum dark brown, linear, ± 0.19 × 0.04 mm. Carpels glabrous. Follicles paired, divergent at an acute angle, 100–130 mm long, 3–4 mm diameter at centre, glabrous. Seed 8–10 × 1.5–2.5 mm, with a brownish or blackish margin; coma 10–25 mm long. Fig. 66, p. 232.

UGANDA. Acholi District: Lamwo County, 4 km SE of Lomwaka, 18 July 1974, *Katende* K 2155!; Kigezi District: Rubaya, 4 July 1945, *A.S. Thomas* 4258!; Masaka District: Kabula County, 1–2 km E of Lyantonde, 16 May 1972, *Lye* 6914!
KENYA. Northern Frontier District: 3 km N of Maralal, 25 Oct. 1978, *Gilbert* 5131!; Kericho District: 5 km from Lumbwa along Idi–Molo road, July 1933, *Rogers* 515!; Masai District: Ngong near Nairobi, 28 Mar. 1961, *Munro* 6741!
TANZANIA. Lushoto District: Mtai escarpment, W Usambaras, 25 May 1953, *Drummond & Hemsley* 2753!; Morogoro District: Nguru Mts, Maskati Mission, 11 Dec. 1966, *S.A. Robertson* 344! & 9 June 1978, *Thulin & Mhoro* 3091!
DISTR. **U** 1–4; **K** 1, 3–7; **T** 1–3, 6; Congo-Kinshasa, Zambia, Malawi, Mozambique, South Africa
HAB. Forest edges and clearings, and thickets especially on hills and lower mountain slopes; 900–2500 m

SYN. *C. verdickii* De Wild. in Ann. Mus. Congo sér. 4, 3: 109 (1903). Type: Congo-Kinshasa, Lukafu, *Verdick* 389 (BR, holo.)

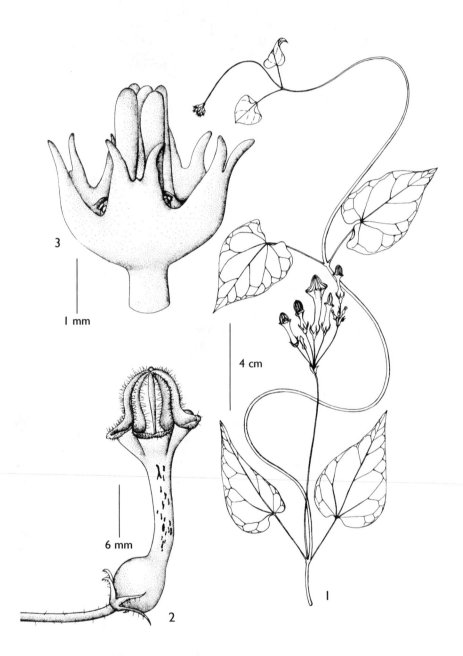

FIG. 66. *CEROPEGIA MEYERI-JOHANNIS* — **1**, habit; **2**, flower; **3**, gynostegium and corona. All from *Goyder et al.* 4027. Drawn by N. Muema based on pencil drawings by S. Masinde.

C. calcarata N.E. Br. in F.T.A. 4(1): 453 (1903). Type: Malawi, Zomba, cult. Kew flowered 30 July 1900, *Mahon* s.n. (K!, holo.)

C. angiensis De Wild. in Pl. Bequaert 4: 358 (1928). Type: Congo-Kinshasa, Angi, *Bequaert* 5781 (BR, holo.)

C. meyeri-johannis Engl. var. *verdickii* (De Wild.) Werderm. in B.J.B.B. 15: 228 (1938) & by Huber in Mem. Soc. Brot. 12: 155 (1958)

C. criniticaulis Werderm. in B.J.B.B. 15: 232 (1938). Type: Congo-Kinshasa, Virunga, Nyamlagira, *Lebrun* 1873 (BR, holo.)

C. meyeri-johannis Engl. var. *angiensis* (De Wild.) H. Huber in Mem. Soc. Brot. 12: 155 (1958)

NOTE. Related to *C. verticillata*, *C. filicorona* and probably also distantly related to *C. papillata*. A widely variable species in its flower and leaf size, corolla and corona morphology and coloration. Fresh leaves of specimens from Mt Kasigau emit a distinctive strong geranium-like scent.

2. **Ceropegia verticillata** *Masinde* in Cact. Succ. J. (US) 72: 155 (2000). Type: Kenya, Taita District, Mbololo Forest, *Bytebier* 1160 (EA!, holo.; BR!, K iso., incl. spirit mat.)

Herbaceous twiner to 2 m high, pubescent; root system a cluster of fusiform roots; stems twining or rambling, sparsely branched, wiry, cylindrical, often rooting at nodes on contact with ground, green to purplish with curved whitish-translucent hairs. Leaves whorled, mostly in 3s; petiole 0.9–2 cm long, channel rims ciliate; blade narrowly ovate, 1.2–4 × 0.6–1.3 cm, base rounded or shallowly cordate, apex narrowly acuminate, both surfaces puberulous, sometimes strongly discolorous, margins entire and finely ciliate. Inflorescence umbellately cymose, 1–2-flowered, flowers developing successively; green to purplish; peduncles 2–6 mm long, ± 1 mm diameter, sparsely puberulous; pedicels 3–10 mm long. Sepals subulate, 2–2.5 × 0.6 mm, abaxially purplish and pilose. Corolla 11–18 mm long; tube with a globose basal inflation ± 3 × 3 mm, curved and narrowed above inflation to ± 1.5 mm diameter, ascending cylindrical part expanded in throat to ± 4 mm diameter; exterior with very pale green to cream background, but paler on inflation, then above inflation with diffuse, purplish-maroon flecks which disappear in throat, glabrous; interior glabrous including basal inflation; lobes plicate, ± linear with deltoid bases, 3–4 × 1 mm in folded state, apically connate to form a lantern-like cage ± 4 × 4 mm, apical half greenish- or maroonish-black, adaxially including keels and margins densely covered with fine blackish needle-like hairs. Corona subsessile to shortly stipitate, basally cupular, ± 2.3 × 3 mm; inner lobes erect, free or apically connivent, dorsiventrally compressed, spatulate, ± 2.8 × 0.4 mm, glabrous; outer lobes sub-erect, forming 5 out-spreading open pockets, each ± 1.4 × 1.2 mm, each lobe bifid in apical half, teeth acute, divergent, adaxially sparsely, finely ciliate along parts of margins. Gynostegium with subquadrate anthers overtopping the stylar head; guide rails ± 0.2 mm long, well exposed; corpusculum linear-obovate, 0.12 × 0.04 mm; pollinia ovoid-elliptic, 0.2 × 0.12 mm. Follicles paired, acutely divergent, cylindrical, 8.5–10 cm long, 3 mm diameter at centre, apically tapering, straw coloured with fine dark dots when dry. Seed dorsiventrally compressed, elliptic, ± 12 × 2 mm, black including marginal wing; coma of silky white hairs ± 2.4 cm long.

KENYA. Teita District: Mbololo Hill, Mraru Ridge, 28 May 1972, *Faden & Faden* 72/277! & Mbololo plains, Sep. 1938, *Joanna* in EA 8988! & Wandanyi hill [Wundanyi], Oct. 1938, *Joanna* in EA 8790!
DISTR. **K** 7; only known from Mbololo area
HAB. Evergreen forest; 1600–1750 m

NOTE. *C. verticillata* is easily identifiable even from vegetative material alone, on the basis of the whorled leaves which are unusual for the genus. Most nodes have three leaves (ternate) and others have four. The leaves are considerably narrowed towards the apex resulting in an extended narrowly acuminate apex. The cluster of spindle-shaped roots is similar to that of *C. meyeri-johannis* to which it is closely related. A specimen from Chyulu Hills (**K** 4/6), end of forest, 13 Dec. 1991, *Luke* 2982 (EA, K) has somewhat intermediate characteristics between *C. verticillata* and *C. meyeri-johannis*.

3. **Ceropegia filicorona** *Masinde* **sp. nov.** *C. meyeri-johannis* similis sed tubo corollae recto, lobis interstaminalibus quam staminalibus ± duplo longioribus atque profunde bifidis lobulisque filiformibus ornatis differt. Typus: Kenya, Trans-Nzoia District, Suam, *Irwin* 339 (EA!, holo.; K!, iso.)

Herbaceous leafy twiner to 1 m high, pubescent; latex unknown; root system a cluster of fusiform roots to 120 mm long, 7 mm diameter; stem slender, sparsely branched, densely pilose. Leaves herbaceous, spreading; petiole 8–30 mm long, densely pilose all round; lamina ovate, 12–35 × 10–23 mm, base cordate, apex acuminate, both surfaces densely velvety hairy, margin ciliate. Inflorescence an umbellate cyme, up to 5-flowered, flowers developing successively; peduncle 14–35 mm, densely pubescent; bracts subulate, ± 1 × 0.5 mm, abaxially densely pilose; pedicels 5–8 mm long. Sepals lanceolate-subulate, 3–3.5 × 0.6 mm, abaxially densely pilose. Corolla 23–25 mm long, straight; tube 18–20 mm long, straight, inflated in ± basal $^1/_3$ or more to form a globose inflation, 9–10 × 5–7 mm diameter, then narrowing gradually to ± 2 mm diameter into an ascending cylindrical part, dilated shortly in throat to 5–7 mm diameter; exterior whitish-purple and glabrous from base to lower throat area, then passing into black-purple, pilose part in throat to base of lobes, inflation with weak longitudinal veins; interior probably similar in colour to outside, with conspicuous longitudinal veining from base to apex, glabrous throughout; lobes linear, arising from triangular bases, ± 5 × 0.4 mm, completely plicate, basal margins rolled back below mouth of tube, connate at apices to form a ± semi-globose cage, adaxial and abaxial faces green passing into black-purple at bases, lobes densely hairy. Corona shortly stipitate, ± 4 × 2.6 mm, uniformly purple and glabrous in all parts; outer lobes suberect, rectangular in outline, ± 3 × 0.5 mm, bifid in apical half, teeth linear, ± 1.7 mm long, obtuse, very base of unlobed part forming very shallow pouches; inner lobes cylindrical, erect, apically connivent, ± 1.8 mm high × 0.15 mm diameter, apices obtuse. Anthers subquadrate, overtopping stylar head; gynostegium, pollinia and guide rails fairly well exposed, guide rail ± 0.25 mm long. Pollinia bright yellowish, elliptic, ± 0.2 × 0.1 mm; corpusculum brownish, ± linear, 0.07 × 0.03 mm. Follicles paired, acutely divergent, pale brown, 60–70 mm long. Seed ± 6 × 2 mm, blackish with a paler thinner margin; coma ± 13 mm long.

KENYA. Trans-Nzoia District: Suam, 19 May 1957, *Irwin* 339!
DISTR. **K** 3; not known elsewhere
HAB. Bushland on hill in stony ground, in full sun; ± 1900 m

NOTE. The fusiform root system, long peduncles, dense hairs in most vegetative and floral parts including the outside of the corolla, are characters shared by forms of *C. meyeri-johannis*, to which this appears to be closely related. The corona is very characteristic by the outer lobes which are very deeply bifid and about twice as high as the inner lobes. This corona form is similar to that of the typical *C. paricyma* except that in *C. paricyma*, the outer lobes are about half as long as the inner lobes.

4. **Ceropegia papillata** *N.E. Br.* in K.B. 1898: 308 (1898) & in F.T.A. 4(1): 452 (1903); Werderm. in E.J. 70: 219 (1939); Huber in Mem. Soc. Brot. 12: 152 (1958); Cribb & Leedal, Mt Fl. S. Tanz.: 101, t. 23a (1982). Type: Malawi, Zomba Mts, ± 5500 ft, 1897, *Whyte* s.n. (K!, holo.)

Herbaceous leafy climber to 1(–3) m high; latex clear; root system a small depressed subglobose tuber ± 30 × 15 mm, with fibrous roots from sides and base; stem usually wiry, sparsely branched, pubescent. Leaves simple, occasionally ± trilobed to hastate; petiole 12–28(–50) mm long; lamina ovate, (15–)25–75(–120) × (8–)13–30(–70) mm, base cordate, apex obtusely pointed or shortly acuminate; basal lobes rounded, in young plants separated by a narrow sinus, in older plants occasionally overlapping and enclosing a small rhomboid sinus; thinly to densely pubescent on both surfaces, margin ciliate. Inflorescence sessile or subsessile, stems with many nodes carrying dense 10–20-flowered pseudoumbels, usually many flowers open simultaneously, without detectable scent; peduncle 0–3 mm long; bracts linear-subulate, 4–8 mm long, pubescent; pedicels 8–12 mm. Sepals linear-subulate, 3–5 × 0.5 mm, acute and recurved at apex, abaxially sparsely or densely pubescent, sepals sometimes reaching beyond apex of inflation. Corolla curved, 16–25 mm long; tube 11–20 mm long, in basal $^1\!/_3$ with a pear-shaped, ± depressed inflation ± 4 × 5 mm, narrowed and slightly curved above inflation into cylindrical part ± 1.5 mm diameter, dilated in throat to 2.5–4 mm diameter; exterior translucent white at base of inflation to pale green or pale purple to dark brownish in throat, glabrous; interior in throat with dark green or blackish veining on a whitish background, thinly pilose in throat; lobes linear from a deltoid base, plicate, straight, 4–7 × ± 2 mm, keeled down the upper part within, connate at the tips to form a cage which narrows towards apex, apical half and margins of basal half blackish-green, whitish with a dark green central line on the lower part of the face, adaxially with hairs. Corona stipitate, cupular, whitish-translucent, ± 2.5 × 2.5 mm; outer lobes erect or slightly incurved, ± triangular-rectangular, 1–1.8 × ± 0.5 mm, apically obtuse or bifid (3-toothed), teeth purplish; inner lobes connivent-erect, linear-spatulate, ± 1.5 × 0.2 mm, dorsiventrally compressed, apically recurved or erect, minutely to distinctly papillose and purplish. Anthers subquadrate, overtopping the ± rounded stylar head; gynostegium, pollinia and guide rails ± hidden by corona lobes in lateral view, guide rails ± 0.1 mm long, nectaries dark or blackish-purple; pollinia bright yellowish, 0.25–0.28 × 0.12–0.16 mm; corpusculum dark brown, linear, 0.12–0.16 × 0.04 mm. Follicles paired, acutely divergent, 75(–120) mm long × 2–3 mm diameter at centre. Seed 5–6 × 2–2.5 mm, dark brown or blackish with a paler margin; coma 10–15 mm long.

TANZANIA. Ufipa District: Tatanda Mission, 24 Feb. 1994, *Bidgood, Mbago & Vollesen* 2441!; Iringa District: near Kibengu, 29 km S of Dabaga, 14 Feb. 1962, *Polhill & Paulo* 1474!; Songea District: 5 km SW of Kitai by R. Mkaku, 9 Mar. 1956, *Milne-Redhead* 8677A!
DISTR. **T** 4, 6–8; Congo-Kinshasa, Zambia, Malawi
HAB. Thickets and forest patches, *Brachystegia* woodland and riverine forests; 400–2200 m

SYN. [*C. stolzii* Schltr. *in sched., nomen nudum,* based on Tanzania, Rungwe District: Kymbila, *Stolz* 707 (HBG, photo! K!, M, photo! WAG, photo!)]

NOTE. *C. papillata* is related to the group including among others *C. claviloba, C. paricyma,* and probably also *C. meyeri-johannis.* Most specimens hitherto incorrectly identified as *C. meyeri-johannis* var. *verdickii* should correctly be referred to *C. papillata* mainly on the basis of the tuberous roots and the dense sessile inflorescences.

5. **Ceropegia cordiloba** *Werderm.* in E.J. 70: 209 (1939). Type: Tanzania, Songea District: Matengo Hills at Litembo, *Zerny* 438 (B†, holo.; P, iso., photo!)

Herbaceous climber to 1.5 m high, wiry; latex not recorded but probably clear; root system a compressed globose tuber, 50 × 40 mm × 3 mm deep; stem twining, slender, tinged purplish, laxly hirsute. Leaves herbaceous, membranous, spreading; petiole 15–38 mm long; lamina ovate, 60–70 × 30–55 mm, base cordate, apex subacute or acuminate, minutely pubescent above and beneath, margins entire, rarely slightly sinuate, minutely ciliate. Inflorescence sessile or subsessile, in dense fascicles, umbellately 2–20-flowered, flowers opening in pairs or in succession, usually many flowers open at same time; bracts linear-subulate, abaxially pubescent; pedicels 6–20 mm. Sepals pale yellow-green, linear-subulate, ± 2 × 0.5 mm, acute and recurved at apex, abaxially with a few minute hairs. Corolla 2–2.2 cm long; tube 15–17 mm long, in basal $^1\!/_3$ or more with a prominent globose inflation 6 × 5–6 mm, above inflation very slightly curved and abruptly narrowing into a cylindrical part 6–8 mm long, 1.5–2 mm diameter; exterior with greenish background, delicately shaded from whitish-green on inflated base and lower part of the tube, with purplish highlights (due to the inner dark purple lining showing through) to pure pale green in throat, glabrous; interior yellowish green in throat; lobes 6–8 × ± 3 mm, hardly plicate, keeled down the upper part within, slightly constricted at $^2\!/_3$ and then broadly cordiform with obtuse apices, curved inwards above constriction and connate for some length to apices hence forming an umbrella-like canopy above the ± oblong corolla cage, leaving only narrow slit-like openings sideways; abaxially greenish-yellow, glabrous; adaxially with sparse white hairs along keel and basal parts, otherwise glabrous, shaded progressively dark green towards apex and margins; auriculate below sinuses of adjacent lobes. Corona very shortly stipitate, ± 2.8 × 2.3 mm, deeply cupular; outer lobes united laterally to form a translucent-whitish deep cup, with a very shallowly 5-lobed serrulate purplish-tinged margin; inner lobes connivent-erect, purplish-red, linear, subequalling the outer cup formed by outer lobes. Anthers subquadrate, overtopping the flattened to rounded stylar head, pollinia and guide rails completely hidden when corona is viewed laterally, guide rails 0.3 mm; pollinia 0.36 × 0.16 mm, yellowish; corpusculum dark brown, linear-elliptic, 0.12 × 0.05 mm. Follicles and seed unknown.

Tanzania. Njombe District: Ludewa near Madunda Mission, Livingstone Mts, 13 Feb. 1991, *Gereau & Kayombo* 4002!; Songea District: 9.5 km NW of Miyau, 29 Jan. 1956, *Milne-Redhead & Taylor* 8898A! & Mbinga, Kiteza Forest Reserve, 1 Mar. 1987, *Congdon* 156!
Distr. **T** 6–8; not known elsewhere
Hab. Bushland with secondary forest and post-burn vegetation; 15–2200 m

Syn. *C. papillata* N.E. Br. var. *cordiloba* (Werderm.) H. Huber in Mem. Soc. Brot. 12: 152 (1958); Bally in Fl. Pl. Afr. 43(3 & 4): t. 1716 (1976)

Note. I have reinstated the specific status of *C. cordiloba* on the basis of the corolla and corona differences between *C. papillata* and *C. cordiloba* which Huber (1958) downplayed in relegating the taxon to varietal status. Bally's (1976) colour drawing clearly shows these differences which he also mentions in his description. It is to be noted that the units for the pedicel length in Bally's description should be 6–20 mm and not cm. My investigations on *C. cordiloba* agree more or less entirely with Werdermann's original description, except for the unifarious hairs (in a single row) which have not been seen as yet. *C. papillata* is more widespread and appears to have a wide range in its morphological variation as opposed to *C. cordiloba* which is only known from southern Tanzania and is much less variable. The group comprising *C. papillata, C. cordiloba, C. meyeri-johannis, C. namuliensis* and *C. claviloba*, among others, is taxonomically confusing as was evidenced by the many undetermined and misidentified specimens seen at EA, K, and BM.

6. **Ceropegia claviloba** *Werderm.* in E.J. 70: 221 (1939); Huber in Mem. Soc. Brot. 12: 153 (1958). Type: Tanzania, Singida/Dodoma District: Turu, *Peter* 33836 (B†, holo.). Neotype: Zimbabwe, Umtali, Engwa, *Exell, Mendonça & Wild* 147 (BM!, neo., designated by Huber in Mem. Soc. Brot. 12: 154)

Herbaceous twiner to 1.5 m high; latex unknown; root system a small subglobose tuber 12–32 mm diameter × 10–24 mm deep, with fibrous roots from base; stem twining, sparsely branched, slender, thinly covered with a minute pubescence. Leaves herbaceous or rarely succulent-herbaceous; petiole 30–40 mm long, pubescent; lamina cordate or subsagittate to 3- or more-lobed, 30–70 × 20–45 mm, basal sinuses often deep, apex acuminate, occasionally retuse, ciliate, basal lobes shortly acute, both surfaces glabrous to thinly pubescent, margins entire or occasionally subsinuate to crenate especially in basal parts, finely ciliate. Inflorescence cymose, flowers umbellately 4–8-fascicled, opening successively, scent unknown; sessile to subsessile or to 3.5 mm long, glabrous to laxily pilose; bracts linear-subulate, 2–3 mm long, abaxially sparsely pubescent; pedicels 4–8 mm long. Sepals linear-subulate, ± 2 × 0.5 mm, acute and recurved at apex, abaxially with a few minute hairs. Corolla 13–17 mm long; tube ± straight, 9–11 mm long, slightly inflated ± elliptical base 3–4 mm diameter, hardly curved above inflation, narrowing into a cylindrical tube to 1.5–2 mm in diameter; exterior mostly pale at very base of inflation then passing into uniform pale or dark green ('bottle-green') up to throat, rarely with purple spots on basal inflation; lobes linear-spatulate from a deltoid base, 4–7 × ± 0.7 mm, with tapering acute apex, plicate for most of apical half except for near apex, erect, connate at the tips, keel marked by a dark band which runs down into throat area where it fades out gradually, adaxially dark purple or very dark green; in the apical half and at basal rolled sinuses where adjacent lobes intersect, with conspicuous, rather long, dark purple, fine, vibratile hairs. Corona stipitate, ± 2.5 × 2.5 mm, shallowly cupular, whitish-translucent; outer lobes forming 5 separate, cup-shaped spreading pouches ± 0.8 × 0.5 mm, adjacent lobes separated by a high wall which may appear as a lobe; margin entire, glabrous or very sparsely finely hairy; inner lobes erect, linear-spatulate, ± dorsiventrally compressed, ± 1.5 × 0.2 mm, apices rounded and connivent, glabrous. Anthers subquadrate, overtopping stylar head, gynostegium, pollinia and guide rails ± well exposed, guide rails 0.2 mm long; pollinia yellowish, elliptic with acute apex, 0.22–0.25 × 0.14 mm, corpusculum light orange, linear-obovate, 0.11 × 0.05 mm. Follicles paired, divergent at an acute angle, green with purple spots, 55–120 mm long, 2–3 mm in diameter at centre; seed 6–7(–9) × 3 mm, blackish with a brownish margin; coma ± 12 mm long.

Tanzania. Kigoma District: Tubira Forest, 1 Apr. 1994, *Bidgood & Vollesen* 3024!; Rungwe District: Kyimbila District: 8 Jan. 1914, *Stolz* 2426! & Isalala River near Bumbigi, Livingstone Mts, 5 Mar. 1991, *Gereau & Kayombo* 4224!
Distr. **T** 4, 5, 7; Congo-Kinshasa, Zambia, Malawi, Zimbabwe
Hab. Riverine open forest; 1000–2100 m

7. **Ceropegia namuliensis** *Bruyns* in Aloe 41(4): 76 (2004). Type: Mozambique, Zambezia, northern slope of Namuli, *Bruyns* 9725 (BOL, holo., photo!; K, iso.)

Herbaceous and erect or wiry twiner, 1–50 cm high; latex not recorded; roundish to compressed-globose tuber 10–15 mm deep and 10–30 mm in diameter, with fibrous roots from base; stem slender, green to purple, internodes in dwarf plants very condensed, thinly pubescent. Leaves herbaceous, simple to ± trilobate; petiole 2–40 mm long; lamina narrowly to broadly ovate or ovate-triangular, (10–)14–65 × (5–)10–35 mm, base cordate, subsagittate or hastate, apex acuminate, obtuse to retuse or rounded with an apiculate tip, veins creamy beneath, both surfaces ± glabrous to densely finely pubescent, margins entire or irregularly dentate especially in ± trilobate, subsagittate or hastate leaves, ciliate; when ± trilobate, the two basal lobes further shallowly to deeply dentate resulting in variously shaped lobes which may further be toothed. Inflorescence umbellately 2–4(–10)-flowered, flowers developing successively, sometimes with more than one flower open at a time, without detectable scent, (sub-)sessile, glabrous or densely hairy; bracts subulate, ± 2 × 0.5 mm; pedicels 5–10 mm long. Sepals linear-subulate, 1.5–3 × 0.5–1 mm, acute and recurved at apex, abaxially minutely hairy. Corolla slightly curved, 11–22 mm long; tube ± straight, 7–18 mm long, expanded into a ± pear-shaped inflation 4–5 mm in diameter, gradually narrowing into an ascending cylindrical tube 1.5–2 mm diameter; exterior translucent cream at base, (pale) purple at inflation and rest of tube, mostly glabrous, occasionally pilose; interior pale purple in throat; lobes gently folded back, ± linear from deltoid bases, 3–5 × 1 mm, connate at apices to form a globose cage; adaxially shiny green in apical areas, passing into purple at the bases, covered with fine purple hairs, keel with longer downward pointing hairs in upper part, basal margins strongly rolled back. Corona stipitate, whitish-translucent, ± 3 × 2.8 mm; outer lobes forming a deeply cupular structure, apically bifid, adaxially purplish, ciliate; inner lobes curved inwards at base before arching above stylar head, then erect and apically connivent, ± dorsiventrally compressed, ± 2 × 0.3 mm, deep purple at base, paler towards apex. Anthers subquadrate, overtopping the rounded to flattened stylar head; gynostegium, pollinia and guide rails somewhat hidden by corona lobes, guide rail 0.4 mm long, nectaries bright purple; pollinia pale orange, 0.21 × 0.16 mm; corpusculum dark orange, linear, ± 0.16 × 0.05 mm. Follicles paired, divergent at an acute angle, (40–)50–70 mm long × 2–3 mm diameter at centre, green with fine longitudinal purplish streaks, drying brown-cream with the fine streaks paler; seed 4–4.5 × 2 mm, shiny brown with a paler thinner curved margin; coma ± 12 mm long.

TANZANIA. Ufipa District: Muse–Kisungu, 23 Feb. 1950, *Bullock* 2526!; Mpanda District: 12 km on Mpanda–Uvinza road, 7 Mar. 1994, *Bidgood et al.* 2685!; Songea District: Masuguru, 1996, *Specks* 756!

HAB. *Brachystegia* woodland and thickets in grassland; 300–1000 m

DISTR. **T** 4, 7, 8; Mozambique, Zambia, Malawi

SYN. *Ceropegia tanzamalawiense* Masinde *in sched.*, *nomen nudum* based on Tanzania, Mpanda District: 12 km on Mpanda–Uvinza road, *Bidgood et al.* 2685 (K!)

8. **Ceropegia paricyma** *N.E. Br.* in K.B. 1898: 309 (1898) & in F.T.A. 4(1): 457 (1903); Werderm. in E.J. 70: 218 (1939); Huber in Mem. Soc. Brot. 12: 152 (1958). Type: Malawi, Lake Nyasa, no date, *Simons* s.n. (K!, holo.)

Herbaceous, wiry climber to 1 m high; latex not recorded but probably clear; root system a small compressed globose tuber, ± 10 mm deep × 15 mm diameter; stem twining, slender, glabrous or slightly minutely pubescent. Leaves membranous, rarely somewhat succulent; petiole 8–16(–35) mm long; lamina ovate to tri-lobed, 25–60(–90) × 13–40(–50) mm, base cordate, or the lowest leaves cuneate, apex acuminate or acute, rarely cuspidate, glabrous or sparsely pubescent on both sides, margins entire or sometimes deeply dentate or crenulate, dentate lobes cuspidate in outline and often dentate. Inflorescence cymose, flowers solitary or few, 2(–3)-fascicled, opening successively, sessile to subsessile; bracts subulate, ± 1 mm long; pedicels 4–10 mm long. Sepals linear-subulate, ± 2.5 × 0.6 mm, acute and recurved at apex, abaxially subglabrous. Corolla ± straight, 14–20 mm long; tube 9–13 mm long, in basal $^1/_3$ with prominent ± depressed globose inflation, ± 5 × 6–7 mm, gradually narrowing to into an ascending cylindrical part ± 2 mm diameter; exterior translucent white at base of inflation passing into dark purple to blackish and then whitish or whitish-green in the narrowed part up to throat, glabrous; interior whitish near throat, glabrous throughout; lobes erect, blackish to blackish-green or dark dull green, 5–7 × 0.6–1 mm, linear-spatulate from broad deltoid bases, plicate throughout, connate at the tips, adaxial faces with fine hairs. Corona shortly stipitate, whitish-translucent with deep purple in basal parts, ± 4 × 3–4 mm, shallowly cupular with suberect to spreading outer lobes, apical margins truncate or bifid or trifid, lobes ± twice as long as the unlobed part, glabrous throughout or ciliate in basal unlobed parts; inner lobes connivent-erect or free, cylindrical, ± 3 × 0.2 mm, apices tapering and recurved, glabrous. Anthers subquadrate, overtopping stylar head; gynostegium, pollinia and guide rails well exposed, guide rails ± 0.2 mm long, nectaries with circular openings; pollinia bright yellowish, 0.23–0.27 × 0.18–0.23 mm; corpusculum brownish, linear-obovate, 0.13–0.15 × 0.06 mm. Follicles paired, acutely divergent, ± 80 mm long × 2–3 mm diameter at centre; seed not recorded.

TANZANIA. Lindi District: Rondo Plateau, Rondo Forest Reserve, 8 km NW of forest station, 17 Feb. 1991, *Bidgood et al.* 1608!; Masasi District: near Masasi, 9 Mar. 1991, *Bidgood et al.* 1862! & 8 km NE of Masasi, Masasi Hill, 18 Mar. 1991, *Bidgood et al.* 2061!

DISTR. **T** 8; Zambia, Malawi, Mozambique, Zimbabwe, Namibia

HAB. Open *Milletia-Cussonia* forests and thickets on hills and in gullies; 450–900 m

SYN. *C. dentata* N.E. Br. in K.B. 1909: 327 (1909). Type: Mozambique, Macome, Madanda, *Johnson* 100 (K!, holo.)
      C. mutabilis Werderm. in E.J. 70: 218 (1939). Type: Tanganyika, Lindi District: ± 65 km W of Lindi, *Schlieben* 6152 (B†, holo.)

9. **Ceropegia sobolifera** *N.E. Br.*

var. **nephroloba** H. Huber in Mem. Soc. Brot. 12: 201 (1958). Type: Tanzania, Songea District: Matagoro Hills, *Milne-Redhead & Taylor* 8863 (K!, holo.; EA!, G, iso.)

Herbaceous twiner to 30 cm high, delicate; latex clear; root system tuberous, tubers globose, sometimes < 10 mm diameter, sometimes larger and compressed, 14–25 × 12–20 × 10–15 mm, rootstock sometimes extending below ground for 2–3 cm in which case thicker and rhizomatous with fibrous roots just below ground and then rootstock continues for ± 20 mm deep without side roots before a tuber forms; stems herbaceous, sparsely branched, wiry, slender, with 1 or 2 rows of hairs, green with a purplish tinge. Leaves herbaceous; petiole 5–10 mm long, rims of channel ciliate; lamina ovate, elliptic-lanceolate or rounded, 10–40 × 5–15 mm, base rounded to cuneate or occasionally hastate, apex acute, both surfaces with few pale hairs, margins minutely ciliate. Inflorescence sessile, pseudoumbellate, 1–2-flowered, flowers developing successively, without discernible scent; bracts subulate, ± 1 mm long, glabrous; pedicels 3–5 mm long. Sepals lanceolate-subulate, ± 2.5 × 0.7 mm, clasping inflation, pale, abaxially puberulent. Corolla straight, 15–20 mm long; tube straight, 7–12 mm long, in basal half with a globose-urceolate inflation 5–6 × 5–6 mm, narrowing gradually to ± 2 mm diameter and not dilated in throat; exterior yellowish-green with purplish tinge, interior purplish-maroon, glabrous throughout; lobes 5–7 × ± 2 mm, gently but completely folded back so that each lobe is thick and kidney-shaped, basally auriculate, connate at apex to form a ± elliptic cage ± 5 × 7 mm, openings to the cage small and oriented towards posterior so that the inside of the cage is hardly visible; adaxially purple-maroon and glabrescent, abaxially yellowish-green and with sparse whitish and purplish hairs. Corona shortly stipitate, creamy with purplish in parts, ± 3.5 × 3.5 mm, hardly cupular; outer lobes rectangular and ± horizontally spreading, apical half with two teeth arising from side margins; unlobed half shallowly cupular, pilose along margins; teeth suberect and cylindrical-acute, ± 0.8 mm long, with minute hairs, purple-maroon internally including the nectary area; inner lobes erect and meeting above the centre of the stylar-head and connate at the apices, purplish-maroon in basal half, 3 × 0.15 mm, glabrous. Anthers subquadrate, overtopping the ± flattened stylar head; gynostegium, pollinia and guide rails very well exposed; guide rails 0.4 mm, nectaries with circular openings; pollinia yellowish, 0.27 × 0.22 mm; corpusculum orange, ± linear, 0.07 × 0.05 mm. Follicles and seed unknown.

TANZANIA. Songea District: Matagoro Hills, 22 Feb. 1956, *Milne-Redhead & Taylor* 8863! & 8863A! & 55 km from Songea in direction to Mhukuru, 1996, *Specks* 765! & 761!
DISTR. **T** 8; Zimbabwe
HAB. Rocky ground, in shallow soil and crevices; 1300 m

NOTE. Closely related to *C. paricyma* and *C. cordiloba*. It differs from the typical variety which occurs in the Ethiopian highlands by the tiny roundish tuber in contrast to the rhizomatous root system in the typical variety. It is also more delicate than the typical variety. The typical variety has for many decades erroneously been thought to occur in northern Kenya due to the wrong determinations of two specimens, *Gillett* 12832 (K, incl. spirit mat.) & 12895 (EA, K) which are in fact *C. aristolochioides*. Although there is a large distribution gap between the typical variety in Ethiopia and variety *nephroloba* in Tanzania and Zimbabwe, both populations are florally and vegetatively very similar. The differences in the root systems are not so significant as to warrant the raising of the two varieties to specific rank.

10. **Ceropegia ringoetii** *De Wild.* in B.J.B.B. 4: 394 (1914); Werderm. in B.J.B.B. 15: 225 (1938), Huber in Mem. Soc. Brot. 12: 166 (1958). Type: Congo-Kinshasa, Shinsenda, *Ringoet* in *Homblé* 553 (BR, holo.) [cited incorrectly in protologue as 153]

Herbaceous twiner to 1.5 m high; root system a small globose or compressed tuber; stem sparsely branched, slender, pale green to brownish, densely covered with bristly, pointed, reflexed, whitish-translucent hairs. Leaves herbaceous; petiole 8–20 mm long, brownish pubescent; lamina narrowly ovate, 4–11 × 1.3–4 cm, base cordate or rounded, apex obtusely pointed or shortly acuminate, dark green above, midrib yellow-green, lighter green below, both surfaces pubescent, abaxially with prominent veins more densely hairy than other areas of lamina, margin ciliate. Inflorescence terminal or extra-axillary, umbellately cymose subsessile, flowers 3–12-fascicled, developing successively; bracts linear-subulate, 3–4 mm long, abaxially pubescent; pedicels 1–15 mm. Sepals linear-subulate, 5–6 × 0.8 mm at base, acute and recurved at apex, abaxially densely pubescent. Corolla straight, 20–30(–45) mm long; tube 6–8 mm long, basal inflation ovoid, ± 4 × 3 mm, abruptly constricted into a funnel-shaped tube ± 2 mm in diameter, ± 3.5 mm in diameter in throat; exterior black, brown, or yellowish-white with purplish red longitudinal bands or dots, densely pubescent on the cylindrical tube part and less so on inflation; interior with prominent longitudinal veins (probably purplish), finely pilose in throat then glabrous up to inflation; lobes linear-filiform from a triangular abaxially pale yellow or blackish base, adaxially emerald green with dark red median part dividing into a dark brown line between two pale yellow lines as it enters the throat, linear part pale yellow on both sides or abaxially blackish, (4–)12–20 mm long, hardly plicate, incurved or recurved, free at the apices, spreading out like tentacles, densely bristly outside. Corona prominently stipitate, ± 2 × 1.3 mm, basally cupular, whitish-translucent; outer lobes form five pouches ± 0.5 mm wide, ciliate in positions adjoining inner lobes and apical margins; inner lobes connivent-erect, cylindrical, ± 1.5 × 0.1 mm, apically incurved at ± 90° towards centre so that the apices ± touch at apex, ciliate in apical part and in abaxial part of bases. Anthers subquadrate and overtopping stylar head; pollinia yellow, elliptic, ± 0.27 × 0.13 mm, corpusculum ± linear, dark yellow, ± 0.14 × 0.03 mm. Follicles and seed unknown.

Tanzania. Morogoro District: Uluguru Mts, 12 Apr. 1933, *Schlieben* 3762!; Rungwe District: Kyimbila, 1915, *Stolz* K 31!; Songea District: Ruhuma River, by the river E of Kitai, 17 Apr. 1956, *Milne-Redhead & Taylor* 9326D! & 9326E!
Distr. **T** 5–8; Congo-Kinshasa, Zambia
Hab. *Brachystegia-Uapaca* woodland; 850–1400 m

Syn. *C. schlechteriana* Werderm. in E.J. 70: 196 (1939). Type: Tanzania, Morogoro District: Uluguru Hills, *Schlieben* 3762 (B†, holo.; BR, iso.)
[*C. wilmsiana* Schltr. *in sched.*, *nomen nudum*, based on Tanzania, Rungwe District: Kyimbila, *Stolz* 709 (K!, L, photo!, M, photo!, WAG, photo!, UPS!); includes *Stolz* 709B in M, but excl. *Stolz* 709 in HBG which is *C. abyssinica* Decne.]

Note. *C. ringoetii* belongs to a group of closely allied taxa comprising *C. melanops, C. ringens* (NE Africa) and *C. nigra* (W Africa). The dense multicellular hairs on all external surfaces of the flower parts and the free, spreading, tentacle-like corolla lobes characterise this species.

11. **Ceropegia stenoloba** *Chiov.* in Ann. Bot. Roma 10: 395 (1912); Werderm. in E.J. 70: 216 (1939); R.A. Dyer in F.S.A. (1983); Huber in Mem. Soc. Brot. 12: 164 (1958) & Prod. Fl. SW-Afr. 114: 25 (1967); Archer in U.K.W.F.: 391 (1974) & Kenya *Ceropegia* Scrapb. 39, VIII (1992); U.K.W.F. ed. 2: 183 (1994); M.G. Gilbert in Fl. Eth. 4(1): 160 (2003). Type: Ethiopia, Djeledjeranne [Dscheladscheranne] N. of Takazze River, *Schimper* 2048 (FT, holo.; BM!, P photo!, S, W, iso.)

Perennial wiry twiner to 2 m high, occasionally erect dwarf ± 10 cm high; latex clear; rootstock a compressed globose tuber 30–50 mm diameter, with fibrous to fusiform roots from base; stems tufted, slender, herbaceous, trailing or climbing, with fine curled whitish hairs all round or in 1 or 2 rows even on same plant, green with a maroon tinge or yellow-green with maroon spots in older parts. Leaves herbaceous, variable even on same plant; petiole 3–45 mm long; lamina narrowly to broadly ovate, lanceolate, or occasionally linear or linear-lanceolate, occasionally ± trilobate, (13–)25–75 × (3–)5–60 mm, base cuneate to rounded or cordate, apex acuminate or acute or rounded and apiculate in broadly ovate leaves, puberulous, margins entire or denticulate to ± sinuate, ciliate. Inflorescence close to leaf axils, umbellate, 2–21-flowered, sessile or subsessile, up to 10 flowers open simultaneously; bracts subulate, 1.5–2.5 × 0.5 mm, abaxially pubescent; pedicels 3–10(–22) mm long. Sepals subulate, recurved at apex, 2–3 × 0.5 mm. Corolla 10–24 mm long, curved and zygomorphic in which case sometimes S-shaped or ± straight and radially symmetrical; tube straight or distinctly curved, 3–9 mm long, hardly to distinctly inflated in basal $^1/_3$–$^1/_2$, inflation globose or urceolate, 2.5–4 × 3–5 mm, narrowing abruptly to curved 2–3 mm diameter ascending cylindrical part, hardly inflated in throat; exterior silvery-green, pale green or cream with occasional faint to distinct purple or maroon streaks, glabrous or occasionally pilose to scabrid; interior red-maroon or yellow-maroon in throat, glabrous; lobes 2–17 × 0.3–0.5 mm, bases narrowly triangular, yellowish-green to creamy passing into purplish-maroon to blackish, margins often strongly rolled back and ± auriculate so that adjacent lobes are separated by sinuses, margins glabrous or with dense needle-like purple-maroon hairs; then narrowly linear, sometimes appearing wiry and twisted, yellowish-green to cream; apices free or connate and then forming variously shaped cages much wider than tube. Corona prominently stipitate, probably whitish-translucent, 2.5–3 × 1.5–2 mm, shallowly cupular, mostly glabrous; outer lobes suberect, spreading or subhorizontal, ± 0.5 mm long forming 5 shallow pouches, apical margins ± truncate to concave; inner lobes erect-connivent or free, linear, 1.5–2 × 0.1–0.2 mm, apices rounded, dorsiventrally compressed, glabrous or pilose, bases yellowish to purplish. Anthers subquadrate and incumbent on and overtopping the rounded to flattened stylar head; gynostegium, pollinia and guide rails exposed, guide rail 0.15–0.3 mm long; pollinia bright yellow, 0.23–0.33 × 0.19–0.25 mm; corpusculum ± linear to elliptic, pale orange, 0.11–0.17 × 0.05–0.06 mm. Follicles paired, acutely divergent (40°–60°), suberect, yellow-green with grey-maroon streaks, 45–85 mm long × 1.5–2 mm in diameter at centre, apically tapering, slightly curved inwards, glabrous; seed 4.5–6 × 1.5–2 mm, dark brown with a paler margin; coma 5–20 mm long.

NOTE. *C. stenoloba* is a complex variable species that is widespread in tropical Africa. It displays a variety of vegetative and floral morphological forms across its range. A broad species concept is adopted here but an attempt is made to recognise formally the morphological variants at infraspecific rank. A fertile specimen in BM mounted on the same sheet as *Schimper* 2052, an isotype of *Ceropegia hochstetteri* (= *C. racemosa*), should be regarded as an isotype for *C. stenoloba*. It has the same collection data as *Schimper* 2048. The root tubers of *C. stenoloba* are eaten by local inhabitants as thirst quenchers. *C. stenoloba* is closely related to *C. manderensis* and appears to be distantly related to *C. abyssinica* and *C. achtenii*.

1. Stems with 2 rows of hairs, some internodes appearing to have only 1 row of hairs . . . . c. var. **schliebenii**
   Stems glabrous or wholly minutely hairy 2
2. Flowers pendulous, often with long (30 mm long or more) pedicels; corolla lobes > 2 × as long as tube, easily separating at apices; inflation mostly urceolate . . . . b. var. **moyalensis**
   Flowers not pendulous, mostly with short (3–10 mm long) pedicels; corolla lobes 2 × as long as tube, not easily separating at apices; inflation mostly distinctly globose a. var. **stenoloba**

a. var. **stenoloba**

Stems uniformly minutely pubescent; corolla sometimes asymmetrical; tube with a globose inflation.

UGANDA. Masaka District: Mawogola County, 17.5 km SE of Ntsusi, 19 Oct. 1969, *Lye & Rwaburindore* 4466!
KENYA. West Suk District: Kongoli (Kongelai) road, July 1961, *Lucas* 192!; Nakuru District: Lanet Lodge Hill, 1927, *Blunt* 127; Machakos District: 1.6 km W of Kimutwa, 10 June 1962, *Archer* 384!
TANZANIA. Ufipa District: Muse–Kisungu, 23 Feb. 1950, *Bullock* 2528!; Kondoa District: Kolo, 24 km N of Kondoa on Great North Road, 12 Jan. 1962, *Polhill & Paulo* 1147!; Njombe District: ± 20 km E of Njombe, 1996, *Specks* 338!

DISTR. **U** 4; **K** 1–4, 6; **T** 4, 5, 7; Rwanda, Burundi, Ethiopia, Zambia, Namibia, South Africa
HAB. Wooded grassland and open bushland; 300–1900 m

SYN. *C. aberrans* Schltr. in E.J. 51: 151 (1913), as *C. aberrantis*. Type: Namibia, Damaraland, Aukas, *Dinter* 843 (B†, holo.)
   C. stenoloba Chiov. var. *australis* H. Huber in Mem. Soc. Brot. 12: 164 (1957). Type: Namibia, Nama Pan, cult. Pretoria, *Story* 5309 (PRE, holo., photo! M, iso., photo!)

b. var. **moyalensis** *H. Huber* in Mem. Soc. Brot. 12: 165 (1958). Type: Kenya, Northern Frontier District: Moyale, *Gillett* 14061 (K!, holo.; G, iso.)

Stems puberulous. Leaves ovate or ± trilobate, base cordate, apex acute, thinly pubescent above and beneath. Inflorescence densely umbellate, 10–22-flowered; pedicels ± 3.5 mm long. Corolla ± 25 mm long, tube ± 5 mm long, 3–4 mm diameter, with an indistinct globose-urceolate inflation, greenish with reddish longitudinal veins or uniformly chocolate brown; lobes 7–19 mm long, ± 6 × as long as tube, filiform, brownish-red, easily snapping at connate tips and hence becoming free, mostly glabrous, sometimes very minutely puberulent basally on adaxial faces.

KENYA. Northern Frontier District: Moyale, 29 Apr. 1952, *Gillett* 12972!; Machakos District: 10 km S of Kiboko along track past KARI Rangeland Research Station, 5 Dec. 1992, *Harvey & Vollesen* 63!
TANZANIA. Masai District: 64 km on Arusha–Dodoma road, 22 Mar. 1968, *Greenway & Kanuri* in EA 13995!
DISTR. **K** 1, 4; **T** 2; Ethiopia
HAB. *Acacia-Commiphora* bushland on red volcanic soils; 1050–1200 m

NOTE. Similar to *C. stenoloba* var. *stenoloba* except for the corolla lobes which are linear-revolute with triangular base, lobes 7–17 mm, sometimes remaining attached at tips to form ± globose cage wider than tube, uniformly dark brown, minutely puberulent. The pendulous flowers and their octopus-like appearance make the taxon easily distinguishable. On herbarium sheets, the corolla lobes are almost always free and spreading. The long corolla lobes with only a very short fine indumentum are different from those of the other varieties. Black and white photos of the type specimen showing the habit, flowers and inside of corolla including the corona are at K.

c. var. **schliebenii** (*Markgr.*) *Masinde* **stat. nov.** Lectotype: Tanzania, Iringa District: Lupembe, near Ruhudje, *Schlieben* 143 (B†, holo.; M, lecto., selected by Huber in Mem. Soc. Brot. 12: 166 (1958), photo!; BM!, K!, BR, G, P, S, Z)

Stems with 1 or 2 rows of hairs. Leaves small, ovate or lanceolate. Inflorescences 2–7-flowered. Corolla tube 7–9 mm long, in basal ⅛–⅔ with an urceolate inflation up to 3.5 mm diameter, gradually narrowing to 1–2 mm diameter and not dilated in throat; lobes linear-filiform, 4–12 mm long, apices connate or free.

TANZANIA. Ufipa District: Sumbawanga, Nsanga Forest area, 31 July 1961, *Robinson* 4842!; Njombe District: Bhimale River, 6 Jan. 1957, *Richards* 7489! & Lupembe, Feb. 1931, *Schlieben* 143!
DISTR. **T** 4, 6, 7; Rwanda, Burundi
HAB. Forest edge and open bushland on hills; ± 2000 m

SYN. *C. schliebenii* Markgr. in N.B.G.B. 11: 404 (1932)
   C. chortophylla Werderm. in E.J. 70: 214 (1939); Huber in Mem. Soc. Brot. 12: 165 (1958). Type: Tanzania, Morogoro District: Uluguru Mts, Logongo, *Peter* 39161 (B†, holo.; P, iso., photo!)

NOTE. The stems with 1 or 2 rows of hairs distinguish this taxon from the other varieties. The corolla in some forms sometimes looks like that of variety *moyalensis* with the urceolate inflation and the free corolla lobes. In most specimens including the type, the corolla lobes are connate and do not appear to easily snap and become free. Variety *schliebenii* is rarely collected, making it difficult to assess this group. *C. chortophylla* is included here based on the assessment of the protologue and the image of the type specimen.

12. **Ceropegia manderensis** *Masinde* in K.B. 59: 242 (2004). Type: Kenya, Northern Frontier District, Ramu–Banissa road, 68 km from the turning to Banissa, *Gilbert & Thulin* 1443 (EA! holo.; K!, UPS!, iso.)

Herbaceous, leafy, wiry twiner to 1.5 m high; latex unknown; rootstock a depressed globose tuber to 25 mm diameter; stems slender, herbaceous, minutely pubescent. Leaves membranous, herbaceous; petiole 5–20 mm long; lamina narrowly to broadly ovate, 30–(50–70) × 15–(35–40) mm, base cordate, rounded or ± cuneate, apex acuminate, upper surface densely puberulous, lower surface less pubescent, markedly paler, margin ciliate. Inflorescence umbellate cymes close to leaf axils, sessile, up to 5-flowered, up to 3 flowers open simultaneously; bracts subulate, ± 1 mm long, abaxially pubescent; pedicels 5–15 mm long. Sepals subulate, ± 2.5 × 0.7 mm, acute and recurved at apex, abaxially minutely pubescent to scabrid. Corolla straight, 11–15 mm long; tube straight, ± 6 mm long, in basal $\frac{1}{2}$ with an urceolate inflation ± 3.5 mm diameter, slightly narrowing into a short ascending cylindrical part to ± 2 mm diameter, hardly dilated in throat; exterior whitish, glabrous; interior coloration not recorded but probably similar to external colour, completely glabrous including basal inflation; lobes erect, chocolate brown to almost black, linear, 4–6 mm long, ± equal to tube in length, completely gently folded back so that there is no sharp keel but the lobes appear ± thickly cylindrical, bases ± auriculate because they are reflexed below mouth of tube, adjacent lobes separated by deep sinuses, connate at apices to form an elliptical cage ± 6 mm in diameter; adaxially and abaxially densely minutely puberulous. Corona prominently stipitate, ± 2.5 × 1.5–2 mm, basally very shallowly cupular, probably whitish-translucent, glabrous; outer lobes spreading and forming 5 shallow saucer-like pouches, ± 0.3 × 0.7 mm, apical margins ± truncate; inner lobes erect and connivent at apices or free, spatulate, ± 1.8 × 0.2 mm, dorsiventrally compressed. Anthers subquadrate and incumbent on and overtopping the rounded stylar head; gynostegium, pollinia and guide rails well exposed, guide rails ± 0.3 mm long; pollinia bright yellowish, ovoid, ± 0.2 × 0.16 mm; corpusculum ± linear, pale orange, ± 0.1 × 0.03 mm. Follicles and seed unknown.

KENYA. Northern Frontier District: Ramu–Banissa road, 68 km from the turning to Banissa, 4 May 1975, *Gilbert & Thulin* 1443!
DISTR. **K** 1; only known from the type
HAB. *Acacia-Commiphora* woodland; ± 810 m

NOTE. *C. manderensis* has a corolla which is similar to that of *C. sobolifera* but the corona is similar to that of *C. stenoloba* and *C. melanops* and it appears closely related to this group of species.

13. **Ceropegia cufodontis** *Chiov.*, Miss. Biol. Borana Racc. Bot. 167 (1939); Bally in F.P.A. 35: t. 1370 (1962); Archer, Kenya *Ceropegia* Scrapb.: 57, XII (1992); U.K.W.F. ed. 2: 183 (1994), as *cufodontii*; M.G. Gilbert in Fl. Eth. 4(1): 166 (2003); Masinde in Cact. Succ. J. (US) 6: 107 (1999). Type: Ethiopia, Arero, Meta Gafersa, *Cufodontis* 337 (FT, holo., photo!)

Perennial, wiry climber to 2 m high; latex clear; root system a cluster of fusiform roots up to 150 × 10 mm; stem slender, twining, sparsely branched, thinly pubescent. Leaves herbaceous; petiole 5–35 mm long; lamina ovate to lanceolate, 15–50(–110) × 7–40(–55) mm, base rounded or cordate, apex acute, thinly pubescent and with distinct creamy to purplish net venation on both surfaces, margin ciliate. Inflorescence umbellate, sessile to subsessile with peduncles to 2 mm long, 5–10-flowered, flowers opening successively, up to 2 flowers open at a time, scent reminiscent of raw mushrooms; bracts linear-subulate, pubescent; pedicels 4–10 mm. Sepals linear-subulate, 2.5–5 × 0.5–1 mm, clasping inflation, apices slightly recurved, abaxially pubescent, greenish-white with maroon tinge. Corolla 23–30(–35) mm long; tube 18–25 mm long, in basal half with ovoid whitish-green inflation 9–13 × 6–10 mm, slightly curved and narrowed to ascending cylindrical pale yellowish-green part 2.5–6 mm diameter, with longitudinal purplish-maroon veins, dilated in throat to 5–12 mm, glabrous; interior with ridges becoming purple in throat; lobes ± oblong, 9–13 mm long, with sinuses between them ± flat, only slightly folded back in apical $\frac{1}{2}$–$\frac{2}{3}$, adaxially greenish to green-yellow to dull brownish-maroon with white, needle-like, downward-pointing hairs along keel; abaxially pale yellowish-green with a few maroonish longitudinal lines. Corona shortly stipitate, ± 3.5 × 4 mm, whitish-translucent with purplish-maroon spots in parts; outer lobes forming a saucer-like base with two obtuse teeth ± 0.7 mm long, appearing to arise as shoulders from bases of inner lobes, and deeply lobed almost to base hence exposing pollinia, margins purplish-maroon and with long inward-pointing whitish-translucent hairs; inner lobes connivent-erect, linear, ± 2.5 × 0.3 mm, ± dorsiventrally compressed, apices obtuse, glabrous except very base, basal half purplish-maroon, apically creamy with faint purple. Anthers subquadrate and overtopping the rounded stylar-head, guide rails ± 0.2 mm long, nectary areas deep purple-maroon; pollinia yellowish, ± 0.3 × 0.2 mm; corpusculum shiny brown, obovate, ± 0.25 × 0.05–0.15 mm. Follicles paired, divergent at acute angle, 4.5–9 cm long × 2.5–4 mm diameter at centre, apices curved inwards and bulging, at maturity yellowish-green with maroon streaks; seed ± 6.5–7.5 × 1.5–3 mm, dark brown with a lighter brown margin; coma 20–23 mm long.

UGANDA. Mbale District: Sipi, Bugishu, 31 Aug. 1932, *A.S. Thomas* 419! & Sebei Kare River, 27 Aug. 1955, *Norman* 287!
KENYA. Northern Frontier District: 3 km N of Maralal, 25 Oct. 1978, *Gilbert et al.* 5130!; Baringo District: Kapsoo off Sacho road near Kabarnet town, 4 Jan. 1995, *Masinde* 804!; Masai District: Entasekera River, 11 July 1961, *Glover et al.* 2069!
DISTR. U 3; K 1–6; Ethiopia
HAB. Open mostly evergreen wooded bushland and forest edges; 1450–2350 m

SYN. [*C. striata* P.R.O. Bally *in sched., nomen nudum,* based on Kenya, Nyeri District: Sagana fishing camp, 1942, *Copley* 5113 (K!, G)]

NOTE. *C. cufodontis* commonly occurs in montane areas. The roots are eaten by Maasai children in Aitong - Entasekera area of Narok District, Kenya, as a thirst quencher. The juice is watery, not sweet. A generally less variable species over its entire range and hence fairly easy to identify. The herbaceous, wiry, twining growth form and the cluster of fusiform roots are hardly distinguishable from species in the *C. papillata* group e.g. *C. meyeri-johannis*. The corona is somewhat similar to that of *C. claviloba*. The large ovoid pollinia without acute apical translucent margins and the large obovate corpusculum with broad basal translucent membranes are characters of the *C. distincta* group.

14. **Ceropegia inornata** *Masinde* in K.B. 53: 949 (1998); Archer in Kenya *Ceropegia* Scrapb.: 31, VI (1992); Masinde in Cact. Succ. J. (US) 6: 107 (1999); M.G. Gilbert in Fl. Eth. 4(1): 161 (2003). Type: Kenya, Taita District, Mt Kasigau, *Masinde, Goyder, Meve & Whitehouse* 838 (EA!, holo.; K!, MSUN!, iso.; incl. spirit mat.)

Herbaceous twiner to 2 m high; latex clear; root system a subglobose tuber, whitish-cream when young, brownish when mature, 18–25(–80) × 7–18(–30) mm deep, often slightly sunken at top, fibrous roots emerging from sides and base, sometimes connected by rhizomes to other smaller tubers; stems mostly sparsely branched, wiry, glabrous or hairy. Leaves herbaceous to succulent, succulent forms stiff; petioles 2–6 mm long; lamina heart-shaped, ovate to narrowly so to lanceolate, sometimes quite variable even on same plant, 19–56(–60) × 60–23(–40) mm, base rounded to broadly cuneate, occasionally cordate, apex acuminate, apiculate, glabrous or whitish pubescent on both sides. Inflorescences extra-axillary, umbellate cymes, mostly close to leaf axils, 2–6-flowered, flowers mostly opening consecutively or up to 4 flowers open at a time, without detectable scent; peduncle (1.8–)3.5–9(–35) mm long; bracts subulate ± 1.5 × 0.5 mm, glabrous; pedicels 2.5–8(–15) mm long. Sepals greenish or faint purplish, subulate-lanceolate, (0.5–)2–2.5 × 0.25–0.5 mm, glabrous, ciliate or abaxially pubescent. Corolla curved, 1.5–2.7(–3.2) cm long; tube 15–20 mm long, in basal $^1/_3$–$^1/_4$ with a globose inflation 4–9 × 3–5(–8) mm, then curved through 30°–90° and narrowing abruptly to 1–2.5 mm diameter, widening gradually to 2–5 mm diameter in throat; exterior greenish cream or greenish-yellow/purple, glabrous; interior: inside with sparse hairs in throat; lobes ± linear, 10–13 × 1–2 mm, narrowing in middle third of cage to ± 0.5 mm wide and margins slightly recurved up to apex, broadening again to 1–1.5 mm near apex and connate at apices thus forming a ± oblong cage, similar in colour to throat region or brownish-purple near apex. Corona subsessile to shortly stipitate, ± 4 × 3 mm, basally cupular, translucent white; outer lobes radiating out to form five cupular spaces ± 1 mm deep × 1 mm broad, with needle-like hairs in basal parts; inner lobes strap-shaped, ± 2 × 0.6 mm, incumbent upon anthers then erect, connivent midway up or free, apically recurved, apex sometimes tending towards maroonish, glabrous. Anthers yellow, apices triangular, ± subhorizontally orientated on the rounded stylar head; pollinia brownish-orange, ± 0.27 × 0.2 mm; corpusculum dark brown, linear, ± 0.17 × 0.07 mm. Follicles paired, divergent at an angle of 20–90°, 70–160 mm × 2–3 mm thick at centre, light green with maroon streaks and spots; seed brown, ± 6 × 1.5 mm, margin light brown; coma 25(–40) mm long, white.

KENYA. Baringo District: Radad, 31 km S of Marigat, 9 Oct. 1961, *Archer* 390!; Teita District: 3 km E of Bura Railway Station, 17 Jan. 1972, *Gillett* 19565!; Kilifi District: Arabuko Sokoke Forest, 19 Nov. 1992, *S.A. Robertson et al.* 6742!
TANZANIA. Pare District: near Lembeni off main road to Same, no date, *Specks* 1152!
DISTR. **K** 3–4, 7; **T** 3; Ethiopia
HAB. Dry coastal forest margins and rocky deciduous bushland in well drained soil; 50–1900 m

NOTE. *C. inornata* is related to *C. imbricata* and the *C. linearis* E. Mey. complex. It is however quite distinct due to the non-plicate corolla lobes which lack marginal hairs.

15. **Ceropegia imbricata** *E.A. Bruce & P.R.O. Bally* in K.B. 1950: 372 (1950); Archer, Kenya *Ceropegia* Scrapb.: 15, II (1992); Masinde in Cact. Succ. J. (US) 6: 107 (1999). Type: Kenya, Nairobi, *Copley* in *Bally* S48 (K!, holo.; EA!, iso.!)

Twining herb to 2 m high; latex clear; rootstock a subglobose or globose tuber 10–45 × 7–45 mm deep, in larger tubers ± sunken at the top, occasionally with rhizomes connecting one or more smaller tubers, with fibrous roots from equator and below, smooth or occasionally warty; stems sparsely branched, glabrous, occasionally thinly pubescent. Leaves herbaceous to succulent and ± stiff, shape variable even on same plant; petiole 2–13 mm; lamina broadly or narrowly ovate or lanceolate, elliptic, linear, rarely orbicular, (10–)17–50(–115) × (3–)12–20(–35) mm, in succulent forms 1.5–2 mm thick and sometimes concave on upper side, base cuneate or rounded, apex acute or rounded and mucronate or ± cuspidate, glabrous or occasionally minutely pubescent, dark green above with whitish veins, pale green beneath, margins occasionally slightly revolute and/or wavy, ciliate. Inflorescences pseudoumbellate, 1–4(–6)-flowered, flowers developing successively, scent in some forms very mild, not detectable in most forms; peduncle pendent, (2–)5–30 mm long, glabrous or thinly pubescent; bracts tinged maroon, lanceolate-subulate, 1–2 × ± 0.5 mm, glabrous or occasionally pubescent; pedicels 3–30 mm long. Sepals tinged purplish or brownish, linear-lanceolate, 2–3 × 0.5–0.7 mm, clasping inflation, apically recurved, occasionally pubescent. Corolla 17–42 mm long; tube 10–30 mm long, in basal $^1/_5$–$^1/_4$ with a globose inflation 3–10 × 3–7 mm, then bent through ± 90° and abruptly narrowing into an ascending cylindrical part 1–2 mm diameter, dilated in throat to 4–6 mm diameter; exterior greenish-silvery/white or purplish/pink-green near base, sometimes with brownish or purplish venation in throat, glabrous, rarely sparsely pilose; interior greenish-cream with purple spots in throat area; lobes 4–10 × 0.5–1 mm, plicate except for apical part, linear-erect from deltoid bases, connate at rounded apices to form a ± oblong cage; adaxially with dense purplish-maroon needle-like hairs, faces blackish-purple or maroon in basal part passing into greenish towards apices. Corona shortly to prominently stipitate, whitish translucent, 2.5–3 × 2.5–3 mm, basally cupular, glabrous or pubescent; outer lobes bifid in upper half or laterally connate to form a cupular structure 0.5–1.2 mm deep, so that the inner lobes appear to arise from inside the cup; inner lobes connivent ± in middle, ± subulate,1.8–2.5 × 0.2–0.4 mm, towards apices reflexed and tapering but apices obtuse. Anthers subquadrate, incumbent on the rounded stylar head and overtopping it, apices sometimes suberect; gynostegium, pollinia and guide rails well exposed; pollinia bright yellow, 0.27–0.29 × 0.15–0.2 mm; corpusculum dark orange, linear-elliptic, 0.08–0.1 × 0.05 mm. Follicles paired, divergent at acute angle of 10°–90°, 75–110 mm long × 2–4 mm in diameter at centre; seed 6–7 × 2–2.5 mm, dark or light brown with a pale brown margin; coma of silky white hairs 10–25 mm long.

KENYA. Nairobi District: Langata, 11 Dec. 1959, *Archer* s.n.!; Kiambu District: Karura Forest, 4 Apr. 1941, *Bally* 1458!; Masai District: Athi plains near Kiserian River, no date, *Foresti* 389!
TANZANIA. Musoma/Maswa Districts: Seronera River, Serengeti, 25 Apr. 1958, *Paulo* 382!; Lushoto District: Mkuzi, 6.5 km NE of Lushoto, 20 Apr. 1953, *Drummond & Hemsley* 2159!; Mpwapwa District: 6.5 km N of Kibakwe on Mpwapwa track, 10 Apr. 1988, *Bidgood et al.* 986!
DISTR. **K** 2–4, 6, 7; **T** 1–8; not known elsewhere
HAB. Wooded savannah grassland and forest edges; 200–1850 m

SYN. ?*C. collaricorona* Werderm. in E.J. 70: 224 (1939). Type: Tanzania, Masasi District: 160 km W of Lindi, Masasi area, *Schlieben* 6378 (B†, holo.) – see note
    C. sp. B, Archer in U.K.W.F.: 392 (1974); U.K.W.F. ed. 2: 183 (1994)
    C. intracolor L.E. Newton in Bradleya 13: 35 (1995); Archer in Kenya *Ceropegia* Scrapb.: 27, V (1992). Type: Kenya, Nairobi, Embakasi, *Newton* 4460 (K!, holo.; EA!, iso.)
    C. euryacme sensu U.K.W.F. ed. 2: 183 (1994)

NOTE. *C. imbricata* is a very variable species both florally and vegetatively and is related to the *C. linearis* complex of SE Africa. There seems to be no stability in the morphological forms. *C. imbricata* was resurrected in an earlier study (Masinde, 1998) as the correct name for this taxon. It is however now evident from the description of Werderm. (1939) and the sketch of the corona (*l.c.*) that the earliest name for this taxon is most likely *C. collaricorona*. The type material (holotype, *Schlechter* 6378, and paratype, *Schlechter* 6377, both from Tanzania, 160 km west of Lindi) for *C. collaricorona* were destroyed in 1943 in the Berlin Herbarium. Reverting to the earliest name would mean that *C. collaricorona* be based on only the sketch of a corona. This would be an unsatisfactory decision because it leaves many uncertainties. The epithet *imbricata* is taken as a better choice since it is typified by extant specimens. A study of the southern and eastern African material is needed so as to resolve the taxonomic complexities.

16. **Ceropegia rendallii** N.E. Br.

subsp. **mutongaensis** Masinde, **subsp. nov.** subspeciei typicae similis sed lobis staminalibus coronae dorsiventraliter (nec lateraliter) compressis differt. Type: Kenya, Embu District: Embu–Meru road, 3 km N of Mutonga River Bridge, *Archer* 519 (EA!, holo.)

Twining herb to 1 m high, with a mostly compressed globose tuber ± 22 mm diameter × 12 mm deep, sometimes forming chains connected by rhizomes, fibrous roots arising from middle to base; latex clear; stem sparsely branched, slender, green with longitudinally oriented maroon spots and streaks, glabrous. Leaves semi-succulent, somewhat stiff, purplish when well exposed to sun; petiole 5–6 mm long; lamina ovate to rounded, 10–20 × 10–20 mm, base rounded or broadly cuneate, apex acute or rounded with an apiculate tip ± 2 mm long, glabrous above and beneath, margin sparsely ciliate. Inflorescence pseudo-umbellate cyme, 1–3-flowered, flowers developing successively; peduncle 5–8 mm long, glabrous; bracts subulate, ± 2 × 0.3 mm, glabrous. Sepals lanceolate, 2–2.5 × ± 0.5 mm, apically recurved, maroon-green or pale green to purplish, glabrous. Corolla 14–25 mm long; tube whitish-cream outside with faint purplish-maroon in lower part, 8–15 mm long, in basal ± ¼ with a globose inflation (sometimes greenish with maroon spots) 4–5 × 3–4.5 mm, then curved through 45°–100° and abruptly narrowing into an ascending cylindrical part 1–1.5 mm diameter, dilated to 3–5 mm diameter in throat with faint grey-green or maroon veins outside; lobes 3–5 mm long, completely plicate in basal $^3/_4$–$^1/_2$ or only for a short length in the middle and the rest of the parts not completely folded back, the slightly curved back apical part much broader, ± elliptic-spatulate, apically connate forming a cage with a broad (4–5 mm diameter) umbrella-like canopy with narrow arching keels; adaxially greenish-black with maroon tips, basally bluish-black especially on keel with the colour extending into throat, keels and most adaxial surfaces sparsely covered with vibratile needle-like purple hairs. Corona shortly stipitate, translucent white, ± 2 × 2.3 mm, basally cupular; outer lobes cupular, with 5 deep pockets, ± 0.8 × 0.8 mm, glabrous or ciliate, margins apically ± truncate to slightly raised in middle; inner lobes arising from within the cup formed by outer corona lobes, linear, 1–1.5 × 0.1–0.2 mm, apices obtuse, incumbent on anthers then erect-connivent, strongly recurved in apical half or only slightly at apices. Anthers subquadrate, incumbent on the rounded stylar head and overtopping it; gynostegium, pollinia and guide rails well exposed, guide rails basally hang in the hollowed, purple-maroon nectaries; pollinia yellowish, ± 0.18 × 0.09 mm; corpusculum orange, linear-elliptic, ± 0.1 × 0.03 mm. Follicles paired, divergent at acute angle, ± 60°, ± 90 × 3.5 mm at centre, pointing upright, curved inwards slightly and apically tapering; seed 8–9 × 2 mm, dark brown with a light brown margin; coma 20–30 mm long.

KENYA. Meru/Embu Districts: 3 km N of Mutonga River Bridge along Embu–Meru road, 22 Jan. 1966, *Archer* 519!; Nairobi, ex hort. Specks, 1998, *Specks* 3142!
DISTR. **K** 4; not known elsewhere
HAB. Wooded grassland and bushland; 650–700 m

NOTE. See Archer in Kenya *Ceropegia* Scrapb.: 19, III (1992); Masinde in Cact. Succ. J. (US) 6: 107 (1999).
    Subspecies *mutongaensis* has dorsiventrally compressed inner corona lobes as opposed to the laterally compressed ones of subsp. *rendallii*. *C. rendallii* belongs to the *C. linearis* complex. Members of the group are characterised by tuberous root systems, twining mostly glabrous stems, and a cupular corolla whose outer corona lobes are not lobed. However, *C. rendallii* is the only species of the *C. linearis* complex with a corolla cage forming a canopy/umbrella-shaped top. Subspecies *mutongaensis* is East African as opposed to the typical variety that occurs in southern Africa.

17. **Ceropegia konasita** Masinde in Cact. Succ. J. (US) 7: 146 (1999); Masinde in Cact. Succ. J. (US) 6: 107 (1999). Type: Kenya, Coast Province, Taita District, Bura, *Masinde* 819 (EA!, holo.; K!, MSUN!, iso.; incl. spirit mat.)

Climber to 2 m high; latex clear; rootstock often forming chains of ovoid to globose tubers, (20–)30(–40) mm diameter, 20–30 mm deep, with fibrous roots emerging; stem succulent, sparsely branched, mostly leafless in old growth, dark green, 1.5–2 mm thick, 6-angled, ridges often twisted in older growth looking like threads of a screw, glabrous, sometimes trailing on ground and rooting at nodes to form more tubers. Leaves sessile and small, occasionally shortly petiolate and large, sometimes variable even on same plant; petiole 0–8 mm long; lamina herbaceous or stiffly sub-succulent, linear or lanceolate, 10–25(–45) × 1–2(–15) mm, margins entire, apex acute, glabrous to very sparsely ciliate. Inflorescence sub-umbellate, close to leaf axils, up to 6-flowered, flowers opening consecutively, scent ± intense, pungent, short-lived; peduncles shiny purplish-green, (4–)10–23 mm long, 0.5–10 mm diameter, glabrous; bracts subulate, ± 1 × 0.4 mm; pedicels (2–)5–9 mm long. Sepals greenish-purple, subulate-lanceolate, 1–2 × 0.5–0.7 mm, adpressed on inflation, glabrous. Corolla 20–27(–32) mm long; tube 11–17(–19) mm long, inflated in basal $^1/_5$ to form ± conical bulb (4–)7–9 × (3–)6–7 mm, then curved through ± 50° and narrowing to 1.5–2.5 mm diameter, widening to 3.5–6 mm diameter in throat; exterior whitish translucent at base of inflation then whitish-cream/greenish with purple-maroon tint and blotches and streaks, glabrous; interior pale greenish-yellow in throat, glabrous throughout; lobes 6.5–10 × 2–2.5 mm, connate at apex, folded back in middle $^1/_3$, but less so towards base and apex, pale greenish-yellow basally, adaxially blackish-purple in upper $^2/_3$–$^3/_4$, adaxial faces and margins with ± 1 mm long maroon hairs in upper $^2/_3$. Corona distinctly stipitate, whitish translucent, ± 2.7 × 3 mm, glabrous; outer lobes ± rectangular with upward curved margins, ± 1 × 1 mm, connate radially to form ± horizontally spreading out saucer-shaped structure; inner lobes linear, ± 2 × 0.25 mm, incumbent on anthers then erect-connivent. Gynostegium with distinct stipe; guide rails prominent, completely visible, ± 0.2 mm long, slightly widened at their mouths; guide rail mouths hang in bright purple maroon hollow leading to nectary opening; pollinia orange, ± 0.26 × 0.15 mm; corpusculum dark orange, linear-obovate, ± 0.2 × 0.08 mm. Follicles paired, divergent at 85°–140°, 60–100 × 1–3 mm diameter at centre, apically tapering, green with pale purplish streaks, drying straw coloured, glabrous; seed dark brown, dorsiventrally flattened, 6–8 × 2.5–3 mm, margin ± 0.7 mm wide, curved, straw coloured; coma of white silky hairs (15–)30(–40) mm long.

KENYA. Teita District: Bura, 18 Oct. 1961, *Archer* 428! & 1 km towards Voi from Bura, 25 Apr. 1974, *Faden & Faden* 74/492!; Kilifi District: Kaya Mudzimuvya, no date, *Mbinda & Pakia* 218!
DISTR. **K** 7; not known elsewhere
HAB. Bushland and thickets; 60–1250 m

SYN. *C. sp.* Archer in Kenya *Ceropegia* Scrapb.: 35, VII (1992)

NOTE. Related to the *C. imbricata* – *bulbosa* – *linearis* complex. Readily distinguished by the succulent, angled stems and reduced mostly sessile, linear leaves.

18. **Ceropegia purpurascens** *K. Schum.* in E.J. 17: 152 (1893); N.E. Br. in F.T.A. 4(1): 450 (1903); H. Huber in Mem. Soc. Brot. 12: 110 (1958); Malaisse in B.J.B.B. 54: 222, figs. 4, 7, B-D (1984); Malaisse & Schaijes in Asklepios 58: 28, fig. 4, 15 (1993). Type: Angola, Cuanza Norte, Pungo Andongo, *von Mechow* 122 (Z, holo.)

Twiner to 2 m high, leafy; latex clear; rootstock a roundish tuber 20–60 mm diameter; stems sparsely branched, slender, dying back during dry seasons, glabrous. Leaves herbaceous; petiole 5–10 mm; lamina ovate, ovate-oblong or rarely broadly elliptic, 17–40(–70) × 9–25 mm, base cordate or rounded, apex acute, acuminate, rarely mucronulate or truncate, glabrous or occasionally sparsely pilose, margin minutely ciliate. Inflorescences umbellate, up to 10-flowered, flowers developing successively, without detectable scent; peduncle 10–35 mm long, pendent, glabrous, rarely thinly pubescent; bracts subulate, ± 1 × 0.5 mm; pedicels 4–8 mm long. Sepals purplish, linear-subulate, 1–2 × 0.5 mm, glabrous. Corolla 19–25 mm long; tube purplish-white to greenish, 12–16 mm long, in basal $^1/_3$ with an inflation 4–5 × 3–4 mm with purple flecks, with 5 indentations at base opposite sepals, then curved through 90° and narrowing into a long ascending cylindrical part ± 2 mm diameter, dilated to 6–7 mm in throat; glabrous or rarely pilose; interior pale maroon to creamy/greenish, sparsely purplish pilose in throat; lobes 12–16 × ± 0.5 mm, ± subequal to tube length, linear from broad deltoid bases and with conspicuous basal, horn-like auricles below sinuses of adjacent lobes, plicate for most of the their length, connate at their tips to form an elongated ± oblong or elliptic cage, adaxially from bases yellowish-green passing into some bluish in middle $^1/_3$; in apical $^2/_3$–$^3/_4$ ciliate with whitish to purplish vibratile hairs on faces and margins, keels with inward-pointing hairs more dense in basal areas; abaxially visible parts yellowish-green and glabrous. Corona distinctly stipitate, whitish translucent, 3–4 × 2.8 mm; outer lobes forming 5 separate cupular spaces ± 1 mm deep, apically broadly bifid, teeth triangular, diverging, appearing to arise as shoulders from the bases of inner lobes, adaxially purplish-maroon and sparsely whitish-ciliate in marginal areas; inner lobes laterally compressed and falcate or rarely linear-subulate, 2–2.5 × 0.2–0.6 mm, apically recurved, mostly free from bases to apex, wholly glabrous or pilose in parts and pilose-crenulate at apices. Anthers subquadrate, overtopping stylar head; gynostegium, pollinia and guide rails well exposed; pollinia bright orange, elliptic, ± 0.25 × 0.2 mm; corpusculum dark orange, ± linear-obovate, ± 0.15 × 0.05 mm. Follicles paired, obtusely divergent, ± 110 mm long × 2–3 mm diameter at centre; seed not recorded.

TANZANIA. Buha District: Kakombe Valley, 6 Feb. 1964, *Pirozynski* P 348!; Morogoro District: Uluguru, 24 Apr. 1933, *Schlieben* 3818!; Mbeya District: Utengele, Usangu, Feb. 1979, *Leedal* 5388!
DISTR. **T** 4, 6–7; Congo-Kinshasa, Angola, Zimbabwe, Botswana, Namibia
HAB. Thickets and forest edges; 700–800 m

SYN. *C. kwebensis* N.E. Br. in F.T.A. 4(1): 456 (1903). Type: Botswana, Kwebe Hills, *Lugard* 116 (K!, holo.)
  C. kaessneri S. Moore in J.B. 48: 256 (1910). Type: Congo-Kinshasa, Kitimbo, *Kassner* 2349 (K!, holo.; P, E, iso.)
  C. thysanotos Werderman in E.J. 70: 225 (1939). Type: Tanzania, Morogoro District: Uluguru Mts, *Schlieben* 3818 (B†, holo.; M, lecto., photo!; BM!, BR, G, P photo!, Z, isolect.)
  C. purpurascens K. Schum. subsp. *thysanotos* (Werderm.) H. Huber in Mem. Soc. Brot. 12: 112 (1958)

NOTE. A variable species; variation in material hitherto known as subspecies *thysanotos* also occurs in the typical subspecies, therefore no infraspecific taxa are recognised here. *C. purpurascens* is distantly related to members of the *C. linearis* - *C. imbricata* complex. The corolla and corona are quite distinct.

19. **Ceropegia furcata** *Werderm.* in E.J. 70: 200 (1939); Huber in Mem. Soc. Brot. 12: 175 (1958). Type: Tanzania, Rondo [Muera] Plateau, *Schlieben* 6126 (B†, holo.; P, iso., photo!)

Twining herb; latex unknown; rootstock a rhizome with unknown form of root system; stem herbaceous, glabrous. Leaves herbaceous; petiole up to 22 mm long; lamina membranous, ovate, ± 60 × 55 mm, base cordate, apex acute, upper surface glabrous, lower surface pilose on the thickly prominent veins, margins finely pilose. Inflorescences extra-axillary, ± umbellate or shortly racemose to helicoid, 3–8-flowered, flowers developing successively, scent unknown; peduncle up to 20 mm long, glabrous; bracts linear-lanceolate to filiform, 3–5 mm long, glabrous; pedicels 10–15 mm long. Sepals linear to filiform, 10–12 mm long, glabrous. Corolla ± straight, 20–25 mm long; tube 17–19 mm long, in basal $\frac{1}{2}$ with a globose inflation ± 11 mm diameter, narrowing abruptly above inflation to 3–4 mm diameter into an ascending cylindrical part which is hardly dilated in throat but abruptly opens into a wide mouth of ± 20 mm diameter, externally whitish-green, glabrous; internally coloration unknown, with a ring of pilose hairs at entrance to inflation, rest of parts glabrous; lobes 3–6 mm long, bases of adjacent lobes united into channelled spreading projection 5–6 mm long, margins of projections finely ciliate, then narrowed in middle part, apically plicate or ± so and broadened into a subdiscoid or explicate cordiform part 1–1.3 mm wide, apices shortly connate, margins of broadened part with 1.5–2 mm long violet-purple clavate hairs; abaxial sides of lobes glabrous. Corona stipitate, ± 5 × 4 mm, basally very shallowly cupular; outer lobes bifid to base to form linear-filiform lobes ± 2 mm long, densely villose at base; inner lobes 4–4.5 mm long, deeply bifid in apical $\frac{2}{3}$ to form linear obtuse lobes, glabrous. Anthers unknown; pollinarium unknown. Follicles and seed unknown.

TANZANIA. Lindi District: Plateau, 17 May 1935, *Schlieben* 6126, photo!
DISTR. **T** 8; only known from the type
HAB. Woodland/bushland, climbing in shrubs; ± 500 m

Note. May be related to *C. speciosa*. A good species that should be maintained despite the scanty material. The description here is based on the translation of the original Latin description. This species seems to have a unique corolla and corona form. No other species in its alliance has inner corona lobes divided. Based on the interpretation of the protologue, the corolla lobe shape and the resulting cage in addition to the hardly inflated throat of the corolla tube may be similar to that in some forms of *C. variegata*. However, it does not appear to be allied to *C. variegata* in all other characters. Werderm. (loc. cit.) cites *C. denticulata* as the closest relative. However since *C. furcata* has no succulence in its aerial parts and considering that vegetative parts and floral parts are quite different from *C. denticulata*, the two taxa are not likely to be related. *C. furcata* is probably related to the group of species with herbaceous, leafy, twining stems, and fusiform or tuberous roots, comprising among others *C. meyeri-johannis*, *C. papillata*, and *C. ringoetii* which all occur in southern Tanzania. These views are based on the protologue only since the type material was not investigated.

20. **Ceropegia abyssinica** *Decne.* in DC., Prodr. 8: 644 (1844); A. Rich., Tent. Fl. Abyss. 2: 46 (1851) pro parte; N.E. Br. in F.T.A. 4(1): 462 (1903); Werderm. in E.J. 70: 206 (1939); Bullock in K.B. 3: 424 (1952) excl. syn. *C. achtenii*; Archer in U.K.W.F.: 391 (1974) & Kenya *Ceropegia* Scrapb.: 11, I (1992); U.K.W.F. ed. 2: 183 (1994); M.G. Gilbert in Fl. Eth. 4(1): 160 (2003). Type: Ethiopia, Tigre near Gafla, *Schimper* 1416 (G, holo., photo!; K!, FI, S, W, iso.)

Erect herb, very rarely twining (Kwale, K7), usually with a solitary stem to 35 cm high; root system a subglobose to globose tuber 10–50 mm in diameter × 10–20 mm deep, sunken at top, with many fibrous roots arising from above equator; stem light green, with a dense or occasionally minute cover of brownish hairs. Leaves herbaceous, subsessile or petiolate, petiole (0–)2–5 mm long; lamina variable even on the same plant, linear, oblong-linear, elliptic, ovate or obovate, 30–90 × 2–30 mm, base cuneate to rounded, apex acute to acuminate or rounded; broadest leaves at the base of stem, upper leaves sometimes ± V-shaped; both surfaces pubescent, margins entire or occasionally wavy, ciliate. Inflorescence a subsessile umbellate cyme, 1–5-flowered, flowers usually in uppermost nodes to apex of stem, without detectable scent; bracts subulate, 2–5 mm long, abaxially hairy; pedicels 6–10 mm long, with long spreading hairs. Sepals linear-subulate, 6–9 × 0.5 mm, with long spreading hairs. Corolla 14–20(–25) mm long; tube 4–13 mm long, with an indistinct basal ± urceolate inflation ¹/₂–²/₃ length of tube, 3–5 mm diameter, then narrowing gradually to ± 2 mm in diameter, not dilated in throat; exterior whitish/pale green on inflation with pale maroon inside markings faintly showing through, narrowed part pale grey-white with or without grey-maroon streaks, glabrous or puberulous in upper part; internally white to pale yellow with black longitudinal lines in throat, glabrous; lobes linear, plicate in apical part, 8–10 × ± 1 mm, lower ²/₃ not folded right back, connate at tips to form ± spherical cage, with protruding sinuses at bases, abaxially whitish, glabrous or sparsely puberulous, adaxially black, puberulent in basal part. Corona with a long basal stipe, whitish-translucent, ± 2 × 1.6 mm, basally very shallowly cupular; outer lobes form 5 shallow pouches ± 0.7 × 0.3 mm, ciliate adjoining inner lobes, apex truncate, entire, black at margin; inner lobes cylindrical, ± 1.6 × 0.15 mm, apically incurved at ± 90° towards centre so that the apices ± touch, ciliate, black or black with white apices. Anthers subquadrate and overtopping stylar head; pollinia bright yellow, ± 0.24 × 0.14 mm, corpusculum ± linear, orange, ± 0.1 × 0.03 mm. Follicles paired, divergent at an acute angle, 20–30°, 100–115 × 2.5–3 mm in diameter at centre, curved outwards towards apex, pale yellow-green with a few brown-maroon markings; seed 5–7 × ± 2 mm, brownish; coma 20–30 mm long.

a. var. **abyssinica**

Plant stiffly erect and densely bristly hairy. Leaves oblong-lanceolate, ovate or obovate. Corolla lobes shorter than tube.

KENYA. Trans-Nzoia District: NE Elgon, Aug. 1949, *Tweedie* 767!; South Nyeri District: Kirinyaga District: Murundiko area, 1.5 km S of natural bridge, 9 Jan. 1972, *Robertson* 1693!; Machakos/Masai Districts: 1.6 km WSW of main peak of Chyulu Hill, 3 June 1967, *Archer* 542!
TANZANIA. Moshi, June 1928, *Haarer* 1493!; Iringa District: Kasanga, West Mufindi, 24 Mar. 1962, *Polhill & Paulo* 1867!; Rungwe District: Kyimbila, 1915, *Stolz* K 31!
DISTR. **K** 3–7; **T** 1–2, 5, 7–8; Central African Republic, Congo-Kinshasa, Eritrea, Ethiopia, Angola, Zambia, Zimbabwe
HAB. Grassland and open areas in miombo woodland; 80–2450 m

SYN. *C. hirsuta* Decne. in DC., Prodr. 8: 644 (1844), *nom. inval.* in synonymy
  C. steudneri Vatke in Linnaea 40: 217 (1876); N.E. Br. in F.T.A. 4(1): 463 (1903). Type: Eritrea, Keren, *Steudner* 765 (B†, holo.)
  C. leucotiana K. Schum. in E.J. 17:1 51 (1893) & in E. & P. Pf. 4, 2: 272 fig. B (1895); N.E. Br. in F.T.A. 4,1: 451 (1903); Norman in J.B. 67, suppl 1,2: 99 (1929); Werderm. in B.J.B.B. 15: 227 (1938) & in E.J. 70: 217 (1939). Type: Angola, Malange, *von Mechow* 417 (B†, holo., K!, lecto., designated here)
  C. steudneriana K. Schum. in E. & P. Pf. 4,2: 272 (1895). Type: as for *C. steudneri* Vatke
  C. gilletii De Wild. & Durand in B.S.B.B. 38: 95 (1899); N.E. Br. in F.T.A. 4(1): 452 (1903). Type: Congo-Kinshasa, lower Congo, Ndembo (Dembo), *Gillet* s.n. (BR, holo.)
  C. hispidipes S. Moore in J.B. 46: 309 (1908); Werderm. in E.J. 70: 217 (1939). Type: Zimbabwe, near Chirinda, *Swynnerton* 1137 (BM!, holo., K!, iso.)
  C. bequaerti De Wild. in Rev. Zool. Afr. 8, fasc. I, suppl. Bot. 1 (1920); Werderm. in B.J.B.B. 15: 231 (1938). Type: Congo-Kinshasa, Bogoro-Mboga, *Bequaert* 4975 (BR, holo.)
  C. filicalyx Bullock in K.B. 1933: 145 (1933); Werderm. in E.J. 70: 206 (1939). Type: Tanzania, Kondoa District: Kikori, *Burtt* 2755 (K!, holo.)

NOTE. *C. abyssinica* is related to *C. achtenii* and probably also distantly related to *C. filipendula*, *C. umbraticola* and *C. stenoloba*. Variety *abyssinica* occupies the northern distribution range extending up to Ethiopia. The isotype in Kew is numbered as *Schimper* 368. This specimen was the type for the now invalid name, *C. hirsuta*.

b. var. **songeensis** *H. Huber* in Mem. Soc. Brot. 12: 202 (1958). Type: Tanzania, Songea District, Chandamara Hill, *Milne-Redhead & Taylor* 8807 (K!, holo., inc. spirit material; EA, BR!, iso.)

Tuber ± globose, about 2–3 cm deep; stems erect but apically sometimes twining, purplish red. Leaves dull green, very variable in shape; abaxially base of beneath reddish; veins sunken above and raised beneath, not very conspicuous. Inflorescence 1–few-flowered; pedicels reddish; sepals pale yellow-green. Corolla tube whitish with 10 distinct and 10 less distinct dull reddish-purple veins; lower half of lobes whitish with very deep dark purplish brown veins, the upper half uniform deep mauve-purple or mahogany, coloured hairs at sinuses deep purple.

TANZANIA. Songea District: Chandamara Hill, 16 Feb. 1957, *Milne-Redhead & Taylor* 8807!
DISTR. **T** 8; not known elsewhere
HAB. *Brachystegia-Uapaca* woodland near rocky outcrops; ± 1150 m

NOTE. A taxon which is still only known from the type. Variety *songeensis* is a smaller form which occurs locally in southern Tanzania. According to Huber (1958), the main key character distinguishing var. *songeensis* from the typical variety is "Hairs on the pedicels shorter or equal to the diameter of the pedicel". Pubescence length is a variable character that cannot be used in distinguishing *Ceropegia* taxa. The stable characters which distinguish this variety from var. *abyssinica* are the smaller flowers, narrowly linear leaves and corolla lobes which are longer than the tube.

21. **Ceropegia achtenii** *De Wild.* in Pl. Bequaert. 4: 356 (1928); Werderm. in B.J.B.B. 15: 231 (1938); Malaisse & Schaijes in Asklepios 58: 27 (1993). Type: Congo-Kinshasa, Katende, *Achten* 589 (BR, holo.; EA, photo!)

Erect herb to 1 m high; latex unknown but probably clear; root system a small discoid tuber 10–23 × 10–14 mm, with fibrous roots from base, sometimes with a rootstock with fibrous roots ± 10 mm below ground and at base with a tiny tuber which also has fibrous roots at base; stem herbaceous, greenish to reddish, mostly solitary, pubescent with whitish-translucent, brown or purplish hairs. Leaves herbaceous; petiole 0–8 mm long; lamina lanceolate, linear-lanceolate or ovate, 20–70 × 3–15 mm, base cuneate, apex acute, both surfaces densely minutely pubescent, margins entire, finely ciliate. Inflorescence sessile, pseudoumbellate, 3–8-flowered, flowers developing successively, up to 2 flowers may be open at same time, scent unknown; bracts subulate, ± 1 mm long, minutely pubescent; pedicels 5–12 mm long. Sepals lanceolate or subulate, 3–7 × 0.5–1.5 mm, abaxially densely minutely pubescent. Corolla 18–26 mm long; tube 13–26 mm long, in ± basal ¹/₃ with ovoid inflation 4–5 × 3–5 mm, then narrowing into a long ascending cylindrical part 1–2 mm diameter, dilated in throat to 1.5–2.5 mm diameter; exterior greenish-white, speckled and striped greyish pink or dark purple in throat, sparsely to densely covered with whitish-translucent fine hairs; interior glabrous including inflation, colouration unknown; lobes erect, linear, 7–13 mm long, from narrowly triangular bases, connate at apices to form an oblong cage, strongly recurved along midrib but not completely folded back, 0.2–0.5 mm broad, basal margins strongly recurved; adaxially glabrous including keel, dark greenish-purple in bases passing into dark green towards apices; abaxially greenish-white and similarly pubescent as tube. Corona distinctly stipitate, 2–3 × 2–2.5 mm, basally cupular, whitish-translucent but colour in life unknown, glabrous; outer lobes erect, ± rectangular, basally adnate to bases of inner lobes and forming 5 pouches 0.6–1 × 0.5 mm, apically shallowly and broadly emarginate to obtusely bidentate; inner lobes erect, spatulate, ± 1.2–1.4 × 0.2, at apices free or ± connivent. Anthers subquadrate and overtopping the ± flattened stylar head, gynostegial parts including guide rails well exposed; pollinia bright yellow, ± 0.22 × 0.13 mm; corpusculum narrowly elliptic-obovate, ± 0.1 × 0.03 mm. Follicles 80–100 × 2.5 mm at centre; seed unknown. Fig. 67, p. 253.

TANZANIA. Ufipa District: Tatanda Mission, 25 Feb. 1994, *Bidgood et al.* 3315!; Rungwe District: Kyimbila, 11 Apr. 1911, *Stolz* 1215!; Songea District: Unangwa Hill, 22 Mar. 1956, *Milne-Redhead & Taylor* 9322!
DISTR. **T** 4, 7–8; Togo, Congo-Kinshasa, Angola, Zimbabwe
HAB. *Brachystegia-Uapaca-Pterocarpus* wooded grassland; 1000–1900 m

SYN. *C. adolfi* Werderm. in E.J. 70: 214 (1939). Type: Tanzania, Rungwe District: Kyimbila, *Stolz* 1215 (B†, holo.; K!, WAG!, iso.; M photo!, S photo!, P photo!)
    C. adolfi Werderm. var. *gracillima* Werderm. in B.J.B.B. 15: 231 (1938). Type: Congo-Kinshasa, upper Katanga, Kisamba, Ferme Selemo, *Quarré* 2404 (BR, holo.)
    C. achtenii De Wild. subsp. *adolfii* (Werderm.) H. Huber in Mem. Soc. Brot. 12: 158 (1958)

NOTE. *C. achtenii* is related to *C. kituloensis* and possibly *C. abyssinica.* The following specimens from southern and western Tanzania exhibit intermediate characters between *C. abyssinica* and *C. achtenii*: Ufipa District: Muse-Kisungu, 23 Feb. 1950, *Bullock* 2531 & Songea District: top of Chandamara Hill, 23 Mar. 1956, *Milne-Redhead & Taylor* 9332.

22. **Ceropegia kituloensis** *Masinde & F. Albers* in E.J. 122: 161 (2000). Type: Tanzania, Mbeya/Njombe District, Ishinga Mt on edge of Kitulo Plateau, *Leedal* 5293 (K!, holo. (2 sheets), incl. spirit mat.)

Dwarf, erect herb 7–18 cm high; latex colour unknown; rootstock roundish tuber, ± 10 × 8 mm deep, smooth; stems herbaceous, leafy, rarely branching, ± 1.5 mm diameter, with 2 rows of whitish-translucent pubescence. Leaves herbaceous; petiole 1–2 mm long; stipules interpetiolar, of minute tufts of hairs; lamina narrowly ovate-lanceolate or narrowly-elliptic, (0.8–)12–33 × 3–9 mm, base acuminate, apex acute, green, pubescent above, ± glabrous or sparsely pubescent beneath, margins finely ciliate. Inflorescence extra-axillary, 1-flowered, sessile; single accompanying bract lanceolate-subulate, 2–3 × 1 mm, acute; pedicel 5–11 mm long. Sepals narrowly-lanceolate, 2–6 × 5 mm, apically recurved, abaxially pubescent. Corolla 26–38 mm long; tube 18–21 mm long, in basal ± ¹/₄–¹/₃ with a globose inflation ± 6 × 7 mm,

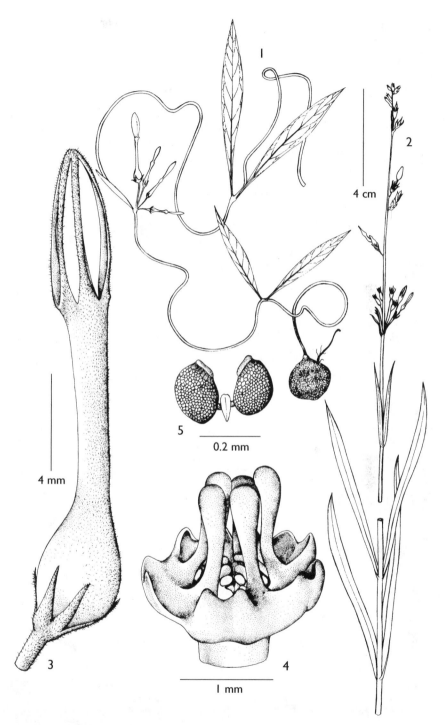

FIG. 67. *CEROPEGIA ACHTENII* — **1**, habit, twining plants; **2**, habit, erect plants; **3**, flower; **4**, gynostegium and corona; **5**, pollinarium. 1, from *Milne-Redhead & Taylor* 9322; 2, from *Milne-Redhead & Taylor* 9332; 3–5, from *Milne-Redhead & Taylor* 9172. Drawn by N. Muema based on pencil drawings by S. Masinde.

narrowing into an ascending cylindrical part ± 2.5 mm diameter, expanding gradually to ± 3.5 mm diameter in throat; exterior greenish, glabrous; interior colour unknown, completely glabrous including basal inflation; lobes erect, linear, 10–22 × ± 1 mm, apically connate, margins recurved, apically brownish or blackish, entirely glabrous adaxially and abaxially including margins. Corona subsessile, ± 3 × 3 mm, basally cupular; inner lobes connivent-erect or free, spatulate, ± 2 × 0.3 mm, glabrous; outer lobes suberectly spreading, forming 5 open pockets, each ± 1.4 × 1.4 mm, margins entire, glabrous. Gynostegium subsessile. Anthers incumbent on and overtopping stylar head, guide rails 0.3–0.6 mm long, ± straight, well exposed, not widened at base; pollinia 0.25–0.31 × 0.17–0.24 mm; corpusculum obovate, 0.14–0.19 × 0.05–0.06 mm. Follicles and seed unknown.

TANZANIA. Mbeya District: N of Ishinga Mt, 9 Feb. 1979, *Cribb et al.* 11355!; Njombe District: Kitulo [Elton] Plateau, above Matamba on Kitulo road, 13 Jan 1989, *Gereau et al.* 2837! & Matamba Pass, 20 Jan. 1978, *Leedal* 4852!
DISTR. **T** 7; not known elsewhere
HAB. On ± treeless crests of Kitulo Plateau, in rocky ground; 2400–2750 m

NOTE. *C. kituloensis* is related to *C. achtenii* and probably also to *C. gilgiana* and *C. abyssinica*. The specimen, *Richards* 7560 (K), Kitulo [Elton] Plateau, 2400 m, 7 Jan. 1957, may be a hybrid between *C. achtenii* De Wild. and *C. kituloensis* and not *C. achtenii* × *C. abyssinica* Decne. as previously supposed by H. Huber in his determination of 1957. It shows a much closer resemblance to *C. kituloensis* in its flowers and bifarious-hairy stems but it has a climbing habit. Another collection showing intermediate characteristics between *C. achtenii* and *C. kituloensis* is *Bidgood et al.* 2319 from Chunya District: 8 km on Makongolosi-Mbeya road, 17 Feb. 1994.

23. **Ceropegia gilgiana** *Werderm.* in E.J. 70: 205 (1939); Huber in Mem. Soc. Brot. 12: 150 (1958). Type: Tanzania, Tabora District: Unyamwezi, km 991.6 E of Kombe, *Peter* 35763 (B†, holo.); neotype: Tanzania, Ufipa District: Sumbawanga, *Bullock* 2363 (K!, neo., designated by Huber, l.c.)

Herbaceous leafy herb, with a solitary erect stem to 30 cm high; latex clear; root system a cluster of whitish fusiform roots ± 110 × 5 mm; stem herbaceous, mostly with 1 or 2 rows of hairs, sometimes minutely uniformly pubescent. Leaves herbaceous, subsessile or petiole to 2 mm long; lamina lanceolate to linear-lanceolate, 10–60 × 3–9 mm, base cuneate, apex acute, both surfaces densely minutely pubescent with whitish-translucent hairs, margins finely ciliate. Inflorescence a sessile umbellate cyme, stems richly flowered with flowers usually arising from middle to uppermost nodes, nodes 1–2-flowered, flowers opening consecutively, scent unknown; bracts subulate, 2–3 × 0.5 mm, minutely pubescent beneath; pedicels 3–10 mm long. Sepals lanceolate or subulate, 4–5 × 1 mm, abaxially minutely pubescent or subglabrous. Corolla 25–75 mm long; tube 20–40 mm long, in basal ⅕ with a small ovoid inflation ± 8 × 5 mm, narrowing into a long ascending cylindrical part 1–1.5 mm diameter, dilated in throat to 3–4 mm diameter; exterior green or yellowish, minutely pilose; interior probably greenish to yellowish, glabrous throughout including inflation; lobes filiform from narrowly deltoid bases, 20–40 × ± 0.5 mm, often ± equal to tube in length, adaxially greenish or yellowish, glabrous; abaxially minutely puberulent. Corona shortly stipitate, ± 3.5 × 2 mm, basally shallowly cupular, whitish-translucent; outer lobes form 5 shallow pouches ± 1 × 0.5 mm, in ± apical half bifid, teeth ± 0.4 mm long with rounded apices and appearing to arise as shoulders from the bases of inner lobes, glabrous except for the margins which are vaguely papillate; inner lobes connivent-erect, dorsiventrally compressed, ± spatulate, ± 2.5 × 0.4 mm. Anthers subquadrate and overtopping the ± flattened stylar head; pollinia bright yellow, ± 0.29 × 0.22 mm; corpusculum ± obovate, ± 0.14 × 0.05 mm. Follicles (immature) paired, divergent at an acute angle, relatively thick and stout; seed unknown.

TANZANIA. Ufipa District: Tatanda Mission, 23 Feb. 1994, *Bidgood et al.* 2407!; Dodoma District: Kazikazi Railway Station, 14 Jan. 1933, *B.D. Burtt* 4579!; Njombe District: Ilembula, 09 Jan. 1975, *Brummitt & Polhill* 13664!

DISTR. **T** 4–5, 7; Zambia

HAB. Wooded grassland and miombo woodland in and around seasonal pools; 750–1900 m

NOTE. This is among the few *Ceropegia* species which are adapted to wetland habitats. It is frequently found growing in water at edges of seasonal ponds in the savannahs. The species seems to survive the frequent fires in the grasslands as an underground rootstock with fleshy roots and puts on new growth in the wet season when there is plenty of water and less disturbance from fire or grazers. It is related to *C. filipendula, C. umbraticola* and probably also to *C. achtenii.*

24. **Ceropegia bulbosa** *Roxb.,* Pl. Coast Coromandel. 1: 11, t. 7 (1795). Lectotype: tab. 7 of Roxb. (1795), designated by Huber in Mem. Soc. Brot. 12: 61 (1958))

Climber to 60 cm high; root system a discoid or depressed globose tuber, ± 40–60 mm diameter × 30 mm deep, with many fibrous roots arising from sides and base, smooth or with a rough, grey peeling bark; latex clear; stems sparsely branched, wiry, mostly glabrous. Leaves herbaceous, subsessile to shortly petiolate; lamina linear, 30–70 × 3–5 mm, base cuneate, apex acute and sometimes apiculate, both surfaces glabrous and sometimes tinged purplish, margin sparsely ciliate. Inflorescence a pseudoumbellate cyme, 5–6-flowered, flowers developing successively, scent very faint musky; peduncle 6–15 mm long, glabrous; bracts subulate, ± 1 × 0.5 mm, glabrous; pedicel 3–6 mm long. Sepals 1–2 × 0.5 mm, apices recurved, greenish or faintly purplish, glabrous. Corolla 17–28 mm long; tube 13–23 mm, in basal $^1/_4$–$^1/_3$ with a globose inflation ± 5 × 5 mm, narrowing abruptly and bent through ± 90° into an ascending cylindrical part 2–3 mm diameter, dilated in throat to ± 5 mm diameter; exterior greenish-white with faint maroon spots in inflation and lower part of tube, passing into dark-maroon in throat to bases of lobes; lobes linear to narrowly triangular from deltoid bases, 5–6 × ± 2 mm, tightly folded back for most of length but slightly parting at apices, apices subacute and connate to form a cage which is wider at base than at apex, cage apex truncate to obtuse; lobes adaxially with downward-pointing needle-like purple hairs along basal half of keel, apical half of keel glabrous, apical $^1/_2$–$^2/_3$ of margins with fine purple vibratile hairs, deep maroon in bases of faces and becoming paler and passing into green with minute purple dots in apical $^2/_3$. Corona shortly stipitate, translucent white, ± 2 × 2.5 mm, basally shallowly cupular, glabrous; outer lobes rectangular-spatulate, ± 0.7 × 0.6 mm, apical margin slightly pointed in middle and flanked by two very short horns; inner lobes erect-connivent, linear to linear-spatulate, dorsiventrally compressed, ± 1.5 × 0.3 mm, apically recurved. Anthers subquadrate, with rounded apices, overtopping and ± subhorizontally oriented on the rounded stylar head; pollinia yellowish, ± 0.28 × 0.19 mm; corpusculum dark orange, linear-elliptic, ± 0.1 × 0.04 mm. Follicles paired, acutely diverging, 74 × 3 mm diameter at centre, apically tapering, straw coloured on drying, without spots or streaks; seed unknown.

KENYA. Kisumu District: Onjiko, Nov. 1939, *Opiko* B697!

DISTR. **K** 5; Cameroon, Ethiopia, Somalia, North Yemen, Oman, Saudi Arabia, India

HAB. Wooded grassland; ± 1100 m

SYN. *C. vignaldiana* A. Rich. in Tent. Fl. Abyss. 2: 48 (1851); Schumann in E. & P. Pf. 4(2): 271, fig E (1895); N.E. Br. in F.T.A. 4(1): 463 (1903); Werderm. in E.J. 70: 203 (1939), M.G. Gilbert in Fl. Eth. 4(1): 161 (2003) & in Fl. Somalia 3: 173 (2006). Type: Ethiopia: Wojerat [Ouodgerate], Sensata, *Quartin-Dillon & Petit* s.n. (P, holo.; K, photo!)

    C. humilis N.E. Br. in F.T.A. 4(1): 464 (1903). Type: Ethiopia, Tigre, Lake Amba, *Schimper* 763 (?B†, holo.)

    C. brosima E.A. Bruce & P.R.O. Bally in K.B. 1950: 368 (1951); Archer in U.K.W.F.: 392 (1974); U.K.W.F. ed. 2: 183 (1994). Type: Kenya, Kisumu District: Onjiko, *Opiko* s.n. in *Bally* S 83 (K!, holo. (in spirit); EA! iso.)

NOTE *C. bulbosa* is better known in Africa as *C. vignaldiana* but a global investigation shows that they are conspecific. It is florally and vegetatively very variable. In East Africa and also elsewhere, the tubers are eaten raw by the local people as thirst quenchers. It appears to be closely related to *C. imbricata* and and is probably also distantly related to *C. keniensis*.

25. **Ceropegia keniensis** *Masinde* in K.B. 54: 477 (1999). Type: Kenya, Uasin Gishu District, Eldoret, *C. Harvey* 1374 (EA!, holo.; K!, iso.)

Erect herb to 40 cm high; root system a cluster of fusiform roots; latex clear; stems herbaceous, sparsely to much branched, cylindrical, glabrous or with fine whitish-translucent hairs. Leaves opposite, sessile; lamina very narrowly elliptic-linear, 31–33 × 15–3 mm, apex acute, glabrous above, with whitish-translucent hairs on midrib underneath and margins. Inflorescence a sessile umbellate cyme, up to 2-flowered, flowers opening consecutively with mostly one flower open at a time; bracts subulate ± 1 mm long, pilose; pedicels 8–11 mm long. Sepals greenish or faintly purplish, subulate, ± 2.2 × 0.4 mm, pilose. Corolla 12–17 cm long; tube 8–12 mm long, in basal $\frac{1}{3}$ with a moderate subglobose inflation ± 4 × 4 mm, hardly kinked, narrowing gently to ± 1.5 mm diameter and gradually broadening to 2.5 mm diameter in throat; exterior greenish-purple, glabrous; interior glabrous including basal inflation; lobes ± linear with triangular bases, ± 6 × 4 mm, not plicate, externally creamy to very pale green, glabrous; internally densely whitish-ciliate or with longer hairs in middle (in *Harvey* 1374) but basally glabrous, apices obtuse and connate to form an oblong-elliptic cage ± 3.5 mm diameter. Corona shortly stipitate, basally cupular, translucent white, biseriate, ± 2 × 3 mm; outer lobes rectangular, suberectly radiating out to form five cupular spaces ± 1 × 0.5 mm, finely ciliate along margins; inner lobes erect, spatulate, ± 2 × 0.4 mm, apically connivent. Anthers yellowish with rounded apices, ± subhorizontally oriented on the conical stylar head; pollinia yellowish, ± 0.23 × 0.18 mm; corpusculum dark brown, ± linear, ± 0.1 × 0.05 mm. Follicles paired, diverging at acute angle, ± 60°, 84–87 mm long, ± 3 mm diameter at centre, gradually tapering towards apices, straw coloured with faint purplish-brown spots in some parts; seed dark brown, ovate, 6–7 × ± 3 mm, dorsiventrally flattened, acute, with ± 1 mm wide light brown or straw-coloured margin; coma a tuft of white hairs, ± 7 mm long.

KENYA. Elgeyo District: Elgeyo Escarpment, 1926, *Harger* s.n.!; Uasin Gishu District: Moiben, Sergoit Rock, 10 Mar. 1964, *Polhill* 2407!; Masai District: near Ntululei, off Mai Mahiu–Narok road, 28 Nov. 1996, *Foresti & Indiaka* 707!
DISTR. **K** 3, 6; not known elsewhere
HAB. Wooded grassland; 2000–2450 m

NOTE. *C. keniensis* is related to *C. bulbosa* and is easily distinguished by the cluster of fusiform roots; the erect often branching stems with sessile very narrowly elliptic-linear leaves; and greenish-purple sessile flowers with non-plicate corolla lobes lacking vibratile hairs along margins except for the fine pubescence internally. Apparently well known by the pastoralist Maasai people of Narok, Kenya, who eat the fleshy roots for their juice as a thirst quencher.

26. **Ceropegia campanulata** *G. Don*, Gen. Hist. 4: 112 (1837); Bullock in K.B. 9: 592 (1955). Type: locality unclear, *Don* s.n. (BM, holo.)

Perennial erect herb to 20 cm high; latex unknown; root system a small rounded tuber; stem annual, herbaceous, solitary, pubescent with whitish-translucent hairs. Leaves sessile, herbaceous; lamina linear-lanceolate, ± 60 × 5 mm, base cuneate, apex acute, both surfaces densely minutely pubescent, margins finely ciliate. Inflorescence sessile, pseudoumbellate, terminal or in apical part of stem, 1–2-flowered, flowers large, developing successively, scent unknown; bracts subulate ± 1 mm long, pubescent beneath; pedicels ± 5 mm long. Sepals lanceolate, ± 3 × 0.7 mm, abaxially minutely pubescent. Corolla ± 90 mm long; tube 50–60 mm long, in basal $\frac{1}{3}$ with an ovoid inflation ± 12 mm long × 9 mm diameter, narrowing gradually into a long ascending cylindrical part ± 3 mm diameter, shortly dilated in throat to ± 8 mm diameter; exterior flushed with numerous dark red spots especially in throat area, glabrous; interior colour unknown; lobes erect, linear from broadly deltoid bases, ± 2.5 mm long, plicate and in folded state ± 0.5 mm broad, apices connate, apical part of cage twisted, basal margins strongly rolled back on tube mouth, basal part of keel covered with whitish-translucent hairs; adaxially with dark-red spots, margins glabrous. Corona sessile, ± 4 × 4 mm, basally cupular, whitish-translucent but colour in life unknown, glabrous; outer lobes ovate and basally forming 5 shallow pouches ± 1.5 × 1 mm, apically deeply bifid, teeth parallel, triangular, obtuse, ± 0.4 mm long; inner lobes cylindrical with rounded apices, dorsiventrally compressed, ± spatulate, ± 2.7 mm high × 0.4 mm broad, erect and arching above stylar head then connivent ± midway and finally towards apices recurved. Anthers subquadrate and overtopping the rounded stylar head; pollinia bright yellow, ± 0.4 × 0.25 mm, corpusculum orange-brown, obovate, ± 0.33 × 0.21 × 0.1 mm. Follicles and seed unknown.

var. **pulchella** *H. Huber* in Mem. Soc. Brot. 12: 130 (1958). Type: Uganda, West Madi District: Leya River watershed, 1500 *Greenway & Eggeling* 7247 (EA!, holo.)

UGANDA. West Nile District: Leya River watershed, 25 Mar. 1945, *Greenway & Eggeling* 7247!
DISTR. U 1; only known from the type
HAB. Stony hill slope in *Terminalia-Crossopteryx-Combretum* scattered tree grassland; ± 1500 m

NOTE. Huber (1958) treated the only Ugandan collection, which is the only easternmost relict, as a new variety. It is however very close to *C. insignis* R.A. Dyer of southern Africa and the typical *C. campanulata* of West Africa. The Ugandan collection differs from typical *C. campanulata* by the corona form and glabrous margins of the corolla lobes. The characters given by Huber (1958) in the key to distinguish *C. campanulata* and *C. insignis* do not hold. *C. campanulata - C. insignis* is however a very variable group of rare taxa which are only known from very few collections. Based on the few specimens seen and the literature, it is likely that a full study with more material representing its entire range may show the group to represent a singe species complex with fewer infraspecific taxa than those currently recognised.

27. **Ceropegia filipendula** K. Schum. in E.J. 17: 150 (1893) & in E. & P. Pf. 4(2): 272, fig. D (1895); N.E. Br. in F.T.A. 4(1): 462 (1903); H. Huber in Mem. Soc. Brot. 12: 150 (1958); Stopp in E.J. 83: 123 (1964); Malaisse in B.J.B.B. 54: 218, fig. 3 (1984); Malaisse & Schaijes in Asklepios 58: 27 (1993). Type: Angola, Cuanza Norte, Cissacola, River Coanga, *von Mechow* 553B (B†, holo.). Neotype: Angola, Ganda, *Damann* 1228 (K!, neo., designated by Stopp, l.c.)

Erect herb to 50 cm high; latex unknown; root system a cluster of white fusiform roots, ± 6 mm diameter; stem solitary, green, puberulous or sparsely scabrid. Leaves herbaceous, green but often turning yellow on drying; petiole 1–10 mm long; lamina ovate, ovate-lanceolate/oblong, 17–57 × 6–34 mm, base rounded, subcordate or cuneate, apex subacute, paler beneath, shortly and softly pilose on both sides, rather more dense along the margins. Inflorescence sessile, racemose, mostly terminal, 1(–2)-flowered, flowers large and mostly appearing singly, opening consecutively, scent unknown; bracts subulate to lanceolate, minutely pubescent; pedicel 4–10 mm long. Sepals pale green with reddish tinge, lanceolate-subulate, 5–7 × 1 mm, abaxially pubescent. Corolla (40–)60–70 mm long; tube 35–55 mm long, dilated in basal ¹/₅–¹/₄ into a small ovoid inflation 8–12 mm long × 8–10 mm diameter, narrowing gradually into a long ascending cylindrical part ± 3 mm diameter, dilated in throat to 10–12 mm diameter; exterior whitish at base of tube then rest of tube dark reddish-violet or pale yellow, sparsely puberulous; interior yellow, throat glabrous; lobes 20–25 mm long, completely folded back in middle ¹/₃, connate for ¹/₄–¹/₃ of their length at the elliptic-spatulate or obovate-spatulate slightly plicate apical part into a short broad apiculate cone or umbrella-like top, with a broad wing-like keel down their inner face; adaxially yellowish throughout or yellow-green in basal half and in apical half glossy reddish/purplish-brown, glabrous; abaxially similar in colour to adaxial side, puberulous or glabrous. Corona sessile or subsessile, whitish translucent, ± 3.5 × 4.5 mm, basally cupular, glabrous or margins of outer lobes hairy in '*C. peteri*', outer lobes suberect, forming 5 pouches in basal half, ± 1.5 × 1 mm, in apical half bifid, teeth linear, 2–2.5 mm long; inner lobes cylindrical to somewhat compressed, linear-subulate, 2–2.5 × ± 0.2 mm, arching above stylar-head then erect, ± as high as outer teeth or longer. Anthers subquadrate and overtopping the ± rounded stylar head; pollinia bright yellow, ± 0.53 × 0.36 mm, corpusculum orange, obovate, ± 0.43 × 0.18 × 0.09 mm. Follicles paired, erect, acutely divergent, stout, ± 100 × 4 mm diameter at centre, shortly tapering towards apex, at maturity dull reddish speckled pale green; seed ± 6 × 3 mm with a coma of white silky hairs.

TANZANIA. Njombe District: Ukango Hill, Apr., *Goetze* 839!; Rungwe District: Kyimbila, 4 Mar. 1914, *Stolz* 2578!; Songea District: ± 6.5 km W of Songea, 30 Mar. 1956, *Milne-Redhead* 9380!
DISTR. **T** 7–8; Congo-Kinshasa, Angola, Zambia, Malawi, Mozambique
HAB. Regenerating *Brachystegia–Uapaca* woodland; 900–1000 m

SYN. *C. medoensis* N.E. Br. in K.B. 1895: 262 (1895) & in F.T.A. 4(1): 460 (1903). Huber in Mem. Soc. Brot. 12: 149 (1958). Type: Mozambique, Medo country between Lugenda River and Ibo, Feb. 1888, *Last* s.n. (K!, holo.)
      C. dichroantha K. Schum. in E.J. 30: 385 (1901). Type: Tanzania, Njombe District: Ukangu Mt near Lumbila [Langenburg], Tanganyika, *Goetze* 839 (B†, holo.; K, fragment, lecto., designated here)
      C. peteri Werderm. in E.J. 70: 204 (1939); Huber in Mem. Soc. Brot. 12: 150 (1958); Stopp in E.J. 83: 123 (1964). Type: Tanzania, Tabora District: near Kombe, *Peter* 35383 (B†, holo.). Neotype: Angola, Sandu-Tjigaka (Quingege), *Damann* 1225 (K!, neo., designated by Stopp, l.c.)
      C. mirabilis H. Huber in Mem. Soc. Brot. 12: 149 (1958); Malaisse in B.J.B.B. 54: 220, fig. 7A (1984). Type: Malawi, 40 km W of Karonga, *Williamson* 200 (BM!, holo.)
      C. renzii Stopp in E.J. 83: 124 (1964). Type: Angola, Huilla and Moçâmedes, 6 km SE of Nova Lisboa, *Stopp* BO106 (K!, holo.)

NOTE. *C. filipendula* is closely related to *C. umbraticola*. It is a variable species over its entire range of distribution, especially in its corolla. The corolla may be externally puberulous or glabrous. It may have one cage as in all known East African specimens or a double cage. The double cage forms occur in Malawi, SE Congo-Kinshasa and Angola. Although *C. peteri* is included here, it is likely that it may represent an infraspecific taxon on the basis of the different corona form with inner lobes longer than outer lobes, and outer lobes densely hairy in marginal areas. The corolla exterior of *C. peteri* is described as more or less glabrous (Werdermann, 1939).

28. **Ceropegia umbraticola** *K. Schum.* in E.J. 17: 153 (1893) & in E. & P. Pf. 4(2): 272, fig. F (1895); N.E. Br. in F.T.A. 4(1): 461 (1903); Norman in J.B. 67, suppl. 1 & 2: 99 (1929); Huber in Mem. Soc. Brot. 12: 148 (1958); Stopp in E.J. 83: 117 (1964); Malaisse in B.J.B.B. 54: 215, fig. 1 (1984); Malaisse & Schaijes in Asklepios 58: 27, fig. 3, 8 (1993). Type: Angola, Malange, *von Mechow* 370 (B†, holo.). Neotype: Congo-Kinshasa, Station de Keyberg, 8 km SE of Lubumbashi [Elisabethville], *Schmitz* 1055 (BR, neo., designated by Stopp, l.c.)

Perennial herb to 30 cm high; latex unknown; root system a cluster of whitish fusiform roots up to 18 cm long; stem solitary, erect, herbaceous, mostly unbranched, dying back at end of growing season, with a relatively thick fibrous bark, densely pubescent, glabrescent near base. Leaves herbaceous, shortly petiolate; petiole 5(–18) mm long; lamina linear-lanceolate, broadly ovate to rounded, 26–48 × 25–38 mm, base rounded to cordate, apex acute, both surfaces densely minutely pubescent, margins finely ciliate. Inflorescence sessile, terminal, flowers relatively large, solitary and developing successively, scent unknown; bracts subulate, ± 1 mm long, pubescent; pedicels 8–20 mm long. Sepals lanceolate, 5–6 × ± 1 mm,

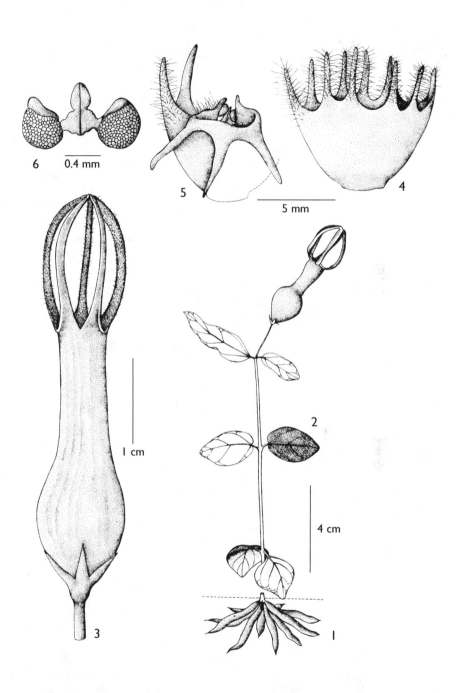

FIG. 68. *CEROPEGIA UMBRATICOLA* — **1**, root system; **2**, habit; **3**, flower; **4**, corona, lateral view; **5**, corona, to show interior parts arrangement; **6**, pollinarium. 1, 2–4, from *Gossweiler 2286;* 3, from *Davies* 678. Drawn by N. Muema based on pencil drawings by S. Masinde.

puberulous to pubescent. Corolla 50–86 mm long; tube 30–60 mm long, in $^1/_3$–$^1/_2$ with an urceolate inflation 10–15 mm diameter and narrowing gradually to 6–10 mm diameter in throat, not dilated below lobes; exterior white or pale green with greener lines, passing into whitish-brown or greenish-brown below lobes, glabrous; interior glabrous throughout; lobes linear-erect from deltoid bases, 15–40 × 1–2 mm, gently folded back with non-connate basal margins distinctly rolled back, connate at tips to form a ± elliptical cage, purplish-brown to orange in basal $^1/_5$ on both adaxial and abaxial sides, rest of lobes whitish to yellowish, adaxially minutely pilose, abaxially very sparsely pilose or glabrous and often with a tinge of brown. Corona sessile, probably whitish-translucent, 5–8.5 × 6–7 mm; outer lobes basally connate and forming a deep cup-shaped part 3–5 mm deep, bifid in apical $^1/_3$ to form 10 erect linear teeth 2–3 × 0.2 mm, covered on the inner face and margins with 1 mm long fine hairs; inner lobes vestigial, 0.6–1 × 0.2 mm, incumbent on and ± as long as anthers, glabrous. Anthers subquadrate, ± erect and overtopping the ± flattened stylar head, guide rails 3–5 mm long; all gynostegial parts and inner lobes hidden by the cup of the outer lobes when corona is viewed laterally; pollinia ± 0.65 × 0.52 mm; corpusculum brownish, obovate, ± 0.63 × 0.26 × 0.12 mm. Carpels glabrous. Follicles paired, divergent at a very acute angle, relatively thick and stout, 200–400 × 4–6 mm in diameter at centre, shortly tapering at apex; seed brownish, ± 8 × 4 mm; coma ± 30 mm long. Fig. 68, p. 260.

TANZANIA. Mbeya District: Mbosi-Zambi, 19 Nov. 1932, *Davies* 678! & Mbosi circular road, 12 Jan. 1961, *Richards* 13885 & Tunduma path, 27 Nov. 1963, *Richards* 18425!
DISTR. **T** 7; Congo-Kinshasa, Angola, Zambia, Malawi
HAB. Miombo woodland; 1500–1600 m

SYN. *C. wellmanii* N.E. Br. in K.B. 1908: 408 (1908); Huber in Mem. Soc. Brot. 12: 148 (1958). Type: Angola, Benguela, Ulondo Mts, *Wellman* 1781 (K!, holo.)
     C. rostrata E.A. Bruce in K.B. 3: 464 (1948). Type: Zambia, Mwinilunga District, just SE of Dobeka Bridge, *Milne-Redhead* 3702 (K!, holo.)
     C. chipiaensis Stopp in E.J. 83: 124 (1964). Type: Angola, Chipia, *Stopp* BO105 (K!, holo.)

NOTE. Characterised by very large flowers by *Ceropegia* standards and the stiffly erect stems.

29. **Ceropegia affinis** *Vatke* in Linnaea 40: 218 (1876); N.E. Br. in F.T.A. 4(1): 445 (1903); Werderm. in E.J. 70: 212 (1939); Huber in Mem. Soc. Brot. 12: 97 (1958); Archer in Kenya *Ceropegia* Scrapb.: 47, X, A & B (1992); Masinde in Cact. Succ. J. (US) 6: 107 (1999); M.G. Gilbert in Fl. Eth. 4(1): 166 (2003) & in Fl. Somalia 3: 172 (2006). Type: Ethiopia, Chilguana in Sabra, *Schimper* 301 (B†, holo.; K, lecto.!, designated by Huber, l.c., G, P, S, photo!)

Perennial twiner to 3 m high; latex clear; with a cluster of fusiform roots up to 180 mm long, 5–10 mm diameter, with fibrous roots to 160 mm arising from ends, or with a swollen rootstock to 15 mm diameter from which fusiform roots arise, or with a rhizome swollen at the end and with a further cluster of fusiform roots; stems herbaceous and green with maroon spots and streaks, or semi-succulent and glaucous tinged maroon with faint white spots in lines, sparsely branched, 1–3 mm diameter, glabrous or rarely hairy. Leaves herbaceous or older ones stiffly succulent, spreading; petiole 2–8 mm long; lamina ovate, linear or narrowly lanceolate to oblong-ovate, 20–60 × 5–20, base rounded or cuneate, apex acute to acuminate, subtruncate and apiculate, glabrous or thinly puberulous on both sides, dark green to purplish, margin ciliate, sometimes narrowly revolute. Inflorescence perennial, pseudoumbellate at first and after several flowering seasons elongating into a long articulated zigzag ± racemose inflorescence with old bracts mostly intact, greyish-maroon to greenish-purple in most parts, up to 9-flowered, flowers opening successively, with intense scent reminiscent of raw mushrooms, sometimes combined with production of clear liquid droplets in apical part of lobes; peduncle 10–45 mm long, mostly glabrous; bracts subulate, ± 1 × 0.5 mm; pedicel 5–15 mm long. Sepals purplish to maroonish-yellow, lanceolate or subulate, 1–3 × 0.5 mm, basally clasping inflation, sometimes apices recurved, glabrous. Corolla 21–29 mm; tube 16–21 mm long, with five indentations at base, in basal $^1\!/_4$–$^1\!/_3$ with a slight inflation 5–6 × 4–5 mm with a gentle constriction in middle, then narrowing and gently curved into a long ascending cylindrical part to 2.5–3 mm diameter, dilated in throat to 4–7 mm diameter; exterior whitish-translucent to creamy at very base, passing into silvery-grey to whitish-cream with purple-maroon longitudinal lines or short purple-maroon longitudinal streaks and spots continuing up to the lobes, glabrous; lobes 5–8 × 1.5–2 mm, basally gently curved back with margins strongly rolled back, auriculate hence producing sinuses, in apical half completely folded back; apices connate to form a ± globose cage 6–9 mm diameter; adaxially green-yellow and greener at apices with red-maroon veins or reddish-maroon throughout, keels with long needle-like reddish-maroon downward-pointing hairs, faces and margins with dense reddish-maroon outward-pointing hairs. Corona (sub-)sessile, ± 4.5 × 4 mm, constricted in basal $^1\!/_4$, basally cupular, whitish to cream with purple-maroon in parts; outer lobes ovate to rounded, ± 1.5 × 1.5 mm, emarginate, teeth ± 0.3 mm long, margins purple-maroon and covered with whitish and purple-maroon hairs; inner lobes linear, erect-connivent, 2.5–3 × 0.4 mm, apically recurved, rising well above outer lobes; bases purplish-maroon, dorsally with whitish and purplish needle-like hairs. Anthers subquadrate, incumbent on the ± rounded to flattened stylar head and overtopping it, guide rails ± 700 μm long; pollinia ± 0.3 × 0.17–0.22 mm; corpusculum ± T-shaped, 0.22–0.24 × 0.15–0.18 × 0.05–0.08 mm. Carpels glabrous. Follicles paired, erect, acutely divergent 70–100 × 3 mm diameter at centre, at maturity grey-green with black-maroon streaks; seed reddish-brown, ± 5 × 1.5 mm, with a pale yellowish-brown margin; coma ± 20 mm long.

UGANDA. Karamoja District: Magosi Hills, 1964, *J. Wilson* 1489!
KENYA. Turkana District: Lokitanyala, 8 Sep. 1968, *Archer* 589!; Teita District: 6.5 km W of Voi, 18 Oct. 1961, *Archer* 397! & Voi, near Voi River, 1963, *Archer* C110!
DISTR. U 1; K 2, 7; Eritrea, Ethiopia, Somalia
HAB. *Acacia-Commiphora* woodland; 450–1700 m

SYN. *C. biddumana* K. Schum. in Ann. Ist. Bot. Roma 7: 41 (1897). Type: Ethiopia, Bidduma, *Ruspoli & Riva* 1202 (FT, holo., photo!)
     C. ruspoliana K. Schum. in Ann. Ist. Bot. Roma 7: 42(1897). Type: Ethiopia, Monte Coromma, *Ruspoli & Riva* 1373 (FT, holo., photo!)

NOTE. Closely related to *C. racemosa* and *C. carnosa*.

30. **Ceropegia racemosa** *N.E. Br.* in K.B. 1895: 262 (1895) & F.T.A. 4(1): 456 (1903); Werderm. in E.J. 70: 207 (1939); Bullock in K.B. 9: 591 (1955); Huber in Mem. Soc. Brot. 12: 93 (1958); Bullock in F.W.T.A. ed. 2, 2: 102 (1963); Archer in Kenya *Ceropegia* Scrapb.: 43, IX (1992); U.K.W.F. ed. 2: 184 (1994); Masinde in Cact. Succ. J. (US) 6: 107–114 (1999); M.G. Gilbert in Fl. Eth. 4(1): 166 (2003) & in Fl. Somalia 3: 173 (2006). Type: South Sudan, Bahr el Ghazal, Seriba Jur Ghattas, *Schweinfurth* 2105 (K!, holo.)

Leafy twiner to 3 m high; latex clear; cluster of fusiform roots up to 100 mm long, 5–10 mm diameter, sometimes with fibrous roots arising from ends, sometimes with a swollen rhizome and with a further cluster of fusiform roots; stems herbaceous, glabrous or pubescent, tinged purple-maroon all over or with faint spots and streaks in lines, young stems blue-green, shiny green or purplish. Leaves herbaceous or sometimes older ones stiffly succulent, variable even on same plant; petiole 2–15 mm long; lamina ovate, obovate, elliptic or linear, 15–80 × 5–45 mm, base rounded or cuneate, apex acute, acuminate, or rounded to notched with apiculate tip, glabrous or puberulous, margins ciliate, sometimes narrowly revolute or wavy. Inflorescence racemose and sometimes branching, elongating into an up to 30 cm articulated zigzag inflorescence with old bracts mostly intact, greyish-maroon to greenish-purple in most parts, (1–)6-flowered, flowers opening simultaneously or successively, scent of raw mushrooms or fermentation intense; peduncle 10–80 mm long, mostly glabrous; bracts subulate, 1–2 × 0.5 mm long; pedicel 4–15 mm long. Sepals purplish to maroonish-yellow, lanceolate or subulate, 2–4 × 0.5–0.8 mm, basally clasping inflation, sometimes apices recurved. Corolla 19–31 mm; tube 16–22 mm long, base whitish-translucent to creamy sometimes with few purple spots, in basal ¼ with a slight inflation 5–6 × 4–6 mm, then narrowing and gently curved into a long ascending cylindrical part to 3–4 mm diameter, yellow or yellow-green with diffuse orange-maroon and streaks, dilated in throat to 7–12 mm diameter, mostly glabrous; lobes plicate, triangular, ovate or oblong-ovate, 4–10 × 2–3 mm, bases deltoid, margins not rolled back; apices connate to form a rounded cage to 10 mm diameter, is narrower towards apex; adaxially greenish-yellow to whitish with dark maroon or dark red in apical ¼, keel with whitish or purplish downward-pointing hairs, margins lined with similar vibratile hairs; abaxially yellow-green with diffuse orange-maroon and streaks. Corona sessile, 4 × 3.5–3.7 mm, basally shallowly cupular, whitish to cream with purple-maroon parts; outer lobes deeply bifid almost to base, teeth ± 1 mm long, curved towards each other, leaving a large opening below them which reveals nectaries and bases of guide rails, marginal areas purple-maroon and with dense whitish and purple-maroon needle-like hairs; inner lobes linear, erect-connivent, 2.5–3 × 0.3 mm, apically recurved, rising well above outer lobes; bases purplish maroon and dorsally with whitish and purplish needle-like hairs. Anthers subquadrate with tapering apices, incumbent on the ± rounded to flattened stylar head and overtopping it, guide rails ± 0.7 mm long; pollinia 0.3–0.55 × 0.2–0.29 mm; corpusculum ± T-shaped, 0.14–0.24 × 0.15–0.25 × 0.04–0.08 mm. Follicles paired, erect, mostly obtusely divergent to 180°, sometimes acutely divergent, 70–130 × 2.5–3.5 mm diameter at centre, at maturity grey-green with dark blackish-maroon streaks; seed dark brown, 6–10 × 2.5–3 mm, with a pale brown margin; coma 10–40 mm long.

1. Corolla lobes hardly plicate with corolla cage openings
     of very narrow longitudinal slits . . . .   c. var. **voiensis**
   Corolla lobes plicate with corolla cage openings wider
     and not narrow slit like 2
2. Plant twining to over 1 m high, inflorescences produced
     from the upper nodes of stem; corolla tube mostly
     yellow or yellow-green sometimes with diffuse orange-
     maroon and streaks to base of corolla lobes . . . .   a. var. **racemosa**
   Plant mostly erect to 50 cm high or occasionally
     longer; inflorescences produced from basal nodes
     near the ground and then continuously ± at each
     node to apex of stem; corolla tube wine-red purple-
     maroon to base of corolla lobes . . . .   b. var. **tanganyikensis**

a. var. **racemosa**

Twining herb; stems mostly thinner than in var. *tanganyikensis*, mostly glabrous. Leaves sometimes subsucculent, lanceolate, elliptic to ± oblanceolate, densely puberulent on both surfaces. Inflorescence produced from the upper nodes of stem, racemose, sometimes branching and elongating to 30 cm long; peduncle, pedicel, bracts and sepals mostly glabrous sometimes puberulent. Corolla 19–31 mm long, slightly curved above the slight inflation; tube exterior below constriction whitish-cream passing into yellow or yellow-green with diffuse orange-maroon and streaks to base of corolla lobes, glabrous or occasionally puberulous.

UGANDA. Karamoja District: Lodoketeminit, N Moroto, 15 July 1959, *Kerfoot* 1284! & Pian
     County, near Moroto, 15 July 1959, *Kerfoot* 1289! & Kailekong, Sep. 1958, *J. Wilson* 497!
KENYA. Laikipia District: Colchochio, 17 km N of Kisima Ranch, 6 Feb. 1995, *Masinde* 832!;
     North Nyeri District: Naro Moru River Lodge, 6 Apr. 1977, *Gillett* 21041!; Fort Hall District:
     Thika behind Blue Posts Hotel, 7 Dec. 1969, *Faden* 69/2067!

TANZANIA. Ufipa District: Chala Mt, 26 Jan. 1981, *Leedal* 6246; Morogoro District: Morogoro University Campus, cemetery grove, 6 May 1974, *Wingfield* 2660!; Iringa District: Ihene, 40 km S of Iringa on Great North Road, 24 Feb. 1962, *Polhill & Paulo* 1598A!

DISTR. U 1; K 3–4, 6–7; T 2, 4, 6–8; Guinea, Burkina Faso, Ivory Coast, Ghana, Nigeria, Cameroon, Central African Republic, Congo-Kinshasa, South Sudan, Ethiopia, Somalia, Angola, Zambia, Malawi, Swaziland, Namibia, South Africa; Madagascar

HAB. Bushland and thickets; 100–1900 m

SYN. *C. angusta* N.E. Br. in K.B. 1895: 262 (1895) & in F.T.A. 4(1): 458 (1903). Type: Angola, Pungo Andongo, Cuanza Norte, by streams of Pedra de Cazella, *Welwitsch* 4275 (K!, holo.)

C. hochstetteri Chiov. in Ann. Bot. Roma 10: 396 (1912). Type: Ethiopia, near Dscheladscheranne, *Schimper* 2025 (BM!, holo., spec. sterile)

C. kamerunensis Schltr. in E.J. 51: 154 (1913). Type: Cameroon, between Ngerik & Limbameni, *Ledermann* 4289 (?B†, holo.)

C. atacorensis A. Chev. in Bull. Soc. Bot. France 61, Mem. 8: 273 (1917). Type: Burkina Faso, Ouagadougou, *Chevalier* 24689 (P!, holo., photo!, K!, iso.)

C. gourmacea A. Chev. in Bull. Soc. Bot. France 61, Mem. 8: 274 (1917). Type: Burkina Faso, Ouagadougou to Ouahigouya, *Chevalier* 24732 (P! holo., photo!, K!, iso.)

C. glabripedicellata De Wild. in Rev. Zool. Afr. 8 Suppl. Bot.: 2 (1920). Type: Congo-Kinshasa, Kasonero, Semliki Valley, *Bequaert* 5044 (BR, holo.)

C. pedunculata Turrill in K.B. 1921: 389 (1921); F.W.T.A. 2: 62 (1931); Type: Nigeria, Bomu road in Kilba country, *Dalziel* 96 (K!, holo.)

C. butaguensis De Wild., Pl. Bequart. 4: 360 (1928); Werderm. in B.J.B.B. 15: 234 (1938). Type: Congo-Kinshasa, Ruwenzori, Butagu Valley, *Bequaert* 3694 (?BR, holo.)

C. setifera Schltr. in E.J. 20, Beibl. 51: 48 (1895); N.E. Br. in F.T.A. 4(1): 457 (1903) & in Dyer, Fl. Cap. 4(1): 821 (1908) & in Fl. Pl. S. Afr. 13, t. 519 (1933). Type: South Africa, Transvaal, *Schlechter* 4515 (B†, holo.; Z, lecto.)

C. setifera Schltr. var. *natalensis* N.E. Br. in Dyer, Fl. Cap. 4,1: 821 (1908). Type: South Africa, Natal, *Wood* 8261 (K!, holo.; NH. iso.)

C. secamonoides S. Moore in J.B. 50: 364 (1912). Type, Angola, Laussingua, *Gossweiler* 2552 (BM!, holo.; K!, iso.)

C. racemosa N.E. Br. subsp. *setifera* (Schltr.) H. Huber in Mem. Soc. Brot. 12: 96 (1958)

C. racemosa N.E. Br. subsp. *secamonoides* (S. Moore) H. Huber in Mem. Soc. Brot. 12: 97 (1958) & in Fl. SW. Afr. 114: 25 (1967)

C. cynanchoides Schltr. in E.J. 51: 153 (1913). Type: Namibia, Damaraland, Gaub, *Dinter* 2410 (SAM!, holo. photo!)

C. sp. D. in U.K.W.F.: 392 (1974)

C. bajana Schltr., *nom. nud.*

NOTE. A very widely distributed species in tropical Africa and also very variable especially in its floral morphology. *C. racemosa* is closely related to *C. affinis* and *C. carnosa*. Huber (1958) recognised 3 subspecies: subsp. *glabra* Huber, with glabrous leaves from Madagascar; subsp. *secamonoides* (Schltr.) Huber, with a puberulous corolla exterior from Angola, Zimbabwe and Namibia; and subsp. *setifera* (Schltr.) Huber from South Africa, but this classification is not satisfactory. The characters used to separate the taxa are quite variable and are neither specific to any particular species nor to any particular geographical area as suggested by Huber. The isolated, disjunct Madagascan subsp. *glabra* might be a good taxon since specimens are tetraploid. However, even these are still morphologically difficult to distinguish from the taxa on mainland Africa. The SE of Kenya to N Tanzania has rather distinct populations with the corolla resembling *C. affinis* in colouration and corolla cage shape, but the corona resembles that for *C. racemosa*; hence implicating possible hybridisation between *C. affinis* and *C. racemosa*. These intermediates include the following collections from K 7 and T 2: Kenya, Teita District: Mt Kasigau, 4 km S of Rukanga, 2 May 1981, *Gilbert & Gilbert* 6120; & S of Mt Kasigau, climb from Makwasinyi village along pipeline path, 13 Feb. 1996, *Masinde* 844; & Msau River Valley, 18 May 1985, *Kabuye et al.* 587; Kwale District: between Mariakani and Kinango, 27 Mar. 1964, *Archer* 466 = 479; Tanzania, near Lake Chala, Feb. 1999, *Meve & Liede* 3355.

b. var. **tanganyikensis** *Masinde* **var. nov.** varietati typicae similis sed caulibus plerumque brevibus et erectis, nodis omnibus flores ferentibus, floribus plerumque omnino intense purpureo-marroninis differt. Typus: Tanzania, Iringa District, Penny Penn's farm, 20 km W of Mafinga on Madibira road, by Ndembera River, *Brummitt et al.* 18161 (K!, holo.)

Plant erect to 50 cm high or occasionally scrambling to 120 cm; stems mostly thicker than in var. *racemosa*, densely pubescent. Leaves sometimes subsucculent, often folded along midrib into a V-shape, lanceolate, elliptic to ± oblanceolate, densely puberulent on both surfaces. Inflorescence produced from the basal nodes near the ground and then continuously ± at each node to the apex of stem, racemose, sometimes branching and elongating to 8 cm long; peduncle, pedicel, bracts and sepals mostly purple and densely puberulent. Corolla 19–26 mm long, slightly curved above the slight inflation; tube externally below constriction whitish-cream passing into wine-red purple-maroon to bases of lobes, glabrous, rarely puberulous (in Malawi); lobes 4–9 mm long, adaxial faces greenish-yellow/cream with dark red/ purple net venation; apex dark red/maroon with similar coloured hairs on outside edge, rest of lobes glabrous except for the longer inward and downward-pointing hairs along keel. Follicles and seed not known.

TANZANIA. Iringa District: Penny Penn's farm, 20 km W of Mafinga, 25 Nov. 1986, *Brummitt* 18161!; Songea District: off Njombe-Songea road, 22 Mar. 1991, *Bidgood, Abdallah & Vollesen* 2101!
DISTR. **T** 7–8; Zambia
HAB. *Brachystegia-Uapaca* woodland on rock outcrops and near river banks; 1000–1700 m

NOTE. This seems to be a high altitude variant. It differs from typical variety by mostly short erect stems with flowers produced from basal nodes to apex; and flowers mostly intensely purple-maroon throughout.

c. var. **voiensis** *Masinde* **var. nov.** varietati typicae similis sed lobis corollae leniter tantum retroflexis atque ad apicem connatis structuram tecto similem formantibus, spatiis longitudinalibus inter lobos corollae angustissimis differt. Typus: Kenya, Kwale District, Shimba Hills, *Archer* 402 (EA!, holo.)

Twining herb to 2 m high. Leaves stiffly semi-succulent, ovate to lanceolate, 28–58 × 11–27(–40) mm, acute. Inflorescence an up to 3-flowered pseudoumbel; peduncle 20–30 mm long; pedicel 4–5 mm long. Corolla tube 16–22 mm long, not significantly expanded in the throat, silvery-grey exterior due to whitish-translucent hairs; lobes linear, slightly plicate, 4–6 × ± 3.5 mm, fused throughout the top to form a canopy-like structure with a central pentagonal bulla where the five lobe apices meet and five peripheral ridges where adjacent pairs of lobes meet, with lobe cage openings looking like longitudinal slits on the sides of a corolla which has been slightly compressed from the top, with sparse needle-like downward-pointing purplish-maroon hairs only on the keels. Follicles divergent at ± 90°, 13 cm long, 3 mm diameter in fresh state greenish with grey maroon spots; seed ± 10 × 3 mm; coma of white hairs, 3.5–4 cm long.

KENYA. Teita District: Voi, *Specks* s.n.!; Kwale District: Shimba Hills, 3 km NW of Kwale, 22 Sep. 1962, *Archer* 402!
DISTR. **K** 7; only known from the type (see note)
HAB. Bushland and thicket; 50–650 m

NOTE. See Archer, Kenya *Ceropegia* Scrapb.: 69, XV (1992); Masinde in Cact. Succ. J. (US) 6: 107–114 (1999). I have seen one collection each from Socotra and Zimbabwe with flowers very similar to this variety, but the overall character combinations of each of these collections differ from the East African collections. The Socotra collection is poor while the Zimbabwe collection has small flowers on a dwarf plant.

31. **Ceropegia somalensis** *Chiov.* in Result. Sci. Miss. Stefanini-Paoli Somalia Ital. 1: 116 (1916); Werderm. in E.J. 70: 200 (1939); Huber in Mem. Soc. Brot. 12: 90 (1958); Brandham in Asclepiadaceae 15: 7 (1978); Bruyns in Notes Roy. Bot. Gard. Edin. (1989); Archer, Kenya *Ceropegia* Scrapb.: 125, XXV (1992); U.K.W.F. ed. 2: 183 (1994); M.G. Gilbert in Fl. Eth. 4(1): 164 (2003) & in Fl. Somalia 3: 169 (2006). Type: Somalia, El Bar to Al Ellan, *Paoli* 889 (FT, holo., photo!)

Often leafless climber to 2 m high, with many branched fibrous roots; latex clear; stem succulent, 3–5 mm diameter, sparsely branched, glaucous-green, pruinose near apex, at base developing a thin smooth grey peeling papery bark. Leaves herbaceous, mainly only on new stems, soon deciduous; petiole 5–15 m long; lamina ovate to broadly lanceolate, 10–25(–52) × 5–10(–25) mm, base rounded-cordate, apex

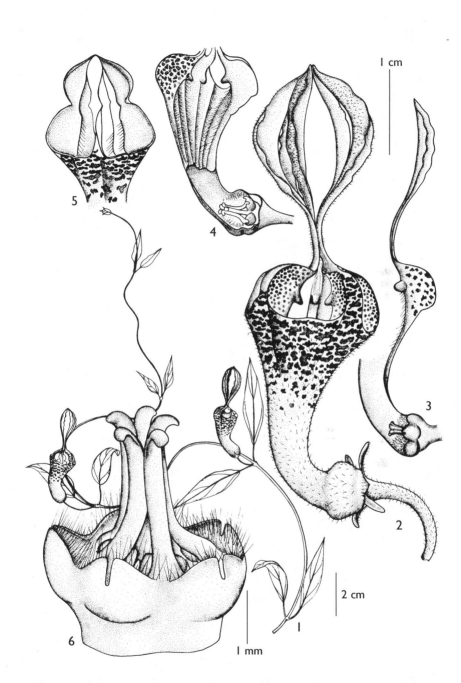

FIG. 69. *CEROPEGIA SOMALENSIS* — **1**, habit; **2**, flower; **3**, longitudinal section of flower; **4**, longitudinal section of corolla tube showing ornamentation; **5**, corolla cage of plants without 'horn' corolla lobe keels; **6**, gynostegium and corona. 1–4, 6, from *Masinde* 868; 5, from *Masinde* 835. Drawn by N. Muema based on pencil drawings by S. Masinde.

acuminate or rounded with apiculate tip, glabrous but for minutely ciliate margins. Inflorescence of shortly pedunculate, umbellate cymes, 1–(2)-flowered, flowers developing successively, in southern Kenya with intense citrus scent; peduncle succulent, descending, (2–)5–15 mm long, glabrous; bracts subulate, ± 2 × 0.5 mm; pedicel 5–13 mm long. Sepals greenish, linear-subulate, 3–5 × ± 0.5 mm, mostly strongly recurved, pubescent. Corolla 25–45 mm long; tube 15–25 mm long, base creamy-white with 5 indentations opposite each sepal, in basal and horizontal $^1/_5$ with a slight inflation 4–5(–7) × 5–7 mm, then narrowing and curved through ± 90° to an ascending part 3–4 mm diameter, widening gradually to 12–16 mm diameter in throat; exterior creamy-white at very base, then yellowish-green, pale yellowish-green near apex with faint to darker green spots, entirely finely velvety-pilose; lobes semi-ovate, 11–30 × 3–5 mm, initially erect, then connivent to form a stalk-like part or keels well apart for their whole length, then diverging again; finally connate at apices hence forming a secondary upper cage; adaxially with prominent or rudimentary maroon-purple knob on keel in lower cage; glabrous except for downward-pointing pale hairs along keel, generally yellowish-green sometimes with darker green spots on faces of lower cage or forms with a rudimentary knob with a broad maroon-purple area around it; abaxially pale green and finely pilose as in rest of corolla. Corona sessile and basally adnate to corolla, most parts translucent white/yellow, 4–5 × 3.5–4 mm, basally cupular; outer lobes deltoid-ovate, 0.5–1 × 1.4–1.8 mm, very shortly bifid, marginally purplish-maroon with dense pale and maroon-purple needle-like inward-pointing hairs, sometimes with a maroon-purple blotch below base; inner lobes 3–3.5 × ± 0.4 mm, incumbent upon anthers then connivent-erect, apically recurved, with pale and purplish-maroon hairs dorsally at bases, background colour of bases purplish. Anthers subquadrate and overtopping the ± flattened styar-head; guide rails ± 0.6 mm long; pollinia 0.39–0.4 × 0.2 mm; corpusculum obovate, ± 0.3 × 0.1 mm. Follicles paired, acutely divergent, 12–18.5 cm long × 3.5–4 mm at centre, apically tapering and curved outwards, maroon-purple and somewhat bristly, drying straw-colour with brownish-maroon streaks; seed dark brown, ± 7 × 2 mm, with light brown margin; coma 25–40 mm long. Fig. 69, p. 265.

KENYA. Turkana District: 10 km from Loiya on Lodwar road, 8 Nov. 1977, *Carter & Stannard* 282!; Masai District: 20 km from Namanga to Amboseli, 30 Oct. 1960, *Archer* C28! & *Archer* 126!; Teita/Kwale Districts: Mackinnon Road on Mombasa–Nairobi road, 21 Jan. 1975, *Adams* 145!
DISTR. **K** 1–3, 6–7; Ethiopia, Somalia, North Yemen, Saudi Arabia
HAB. Open bushland; 300–1650 m

SYN. *C. somalensis* Chiov. forma *erostrata* H. Huber in Mem. Soc. Brot. 12: 90 (1958). Type: Somalia, boundary pillar 93, 8° 37'N, 45° 9'E, *Gillett* 4192 (K!, holo., 2 sheets)

NOTE. Distinguished from its closest ally, *C. distincta*, by the densely finely pilose exterior of the corolla. Populations occurring around the Equator have a prominent dark maroon-purple horn on each keel, which becomes less prominent in southern Kenya. Flowers of populations in southern Kenya have flowers with a short secondary (upper) corolla cage which is not separated from the lower one by a stalk-like structure. They also emit a strong citrus scent. Old stems become woody and have a grey papery bark which often peels off.

32. **Ceropegia distincta** *N.E. Br.* in K.B. 1895: 262 (1895) & in F.T.A. 4(1): 443 (1903); Werderm. in E.J. 70: 207 (1939); Huber in Mem. Soc. Brot. 12: 88 (1958); D.V. Field in K.B. 36: 441 (1981); Archer, Kenya *Ceropegia* Scrapb.: 131, XXVI (1992); Masinde in Cact. Succ. J. (US) 6: 107 (1999). Type: Tanzania: Zanzibar, *Kirk* 28 (K!, holo.)

Glabrous twiner to 3–6 m high, sometimes flowering when leafless, with fibrous roots; latex clear; stems twining, sparsely branched, succulent, cylindrical, 3–6 mm diameter, finely verrucose, greyish-green due to whitish bloom, sometimes showing a maroon mottling on older stems; young stems pruinose, glaucous-green. Leaves herbaceous; petioles 4–20 mm long; lamina elliptic, narrowly ovate-oblong or ±

orbicular, (10–)20–76(–100) × (5–)35–55(–78) mm, base cuneate to subcordate to cordate, apex acuminate or rounded with apiculate tip or rarely notched, upper surface often glaucous, margins sparsely ciliate. Inflorescence pseudoumbellate, 1–3(–5)-flowered, flowers opening successively or in pairs or occasionally several together, without scent; peduncle horizontal or slightly descending, (10–)30–50 mm long, succulent; bracts linear or subulate, 1–2 × 0.5 mm, acute; pedicel 4–20 mm long. Sepals subulate to linear, 3–5(–13) × 0.5–2(–3) mm, attenuate-acuminate, apically reflexed, abaxially with purplish-maroon spots. Corolla (20–)25–38 mm long; tube 14–25 mm long, with 5 indentations opposite each sepal, in basal ¼ with inflation 5 × 5–6 mm, then curved through ± 90° and narrowing into an ascending cylindrical part 2–4 mm diameter, gradually dilated to 8–17 mm diameter in throat; exterior whitish to creamy with purple-maroon spots at the constriction, with progressively larger spots to throat; lobes (5–)8–12(–25) × (2–)3–14 mm, broadly deltoid-ovate basally, with parallel plate-like keels meeting over centre of tube or well apart, lower margin with sparse downward-pointing pale hairs, adaxially yellowish with distinct purplish-maroon/green reticulate venation; with two main forms of lobes and the resulting cage: (i) lobes narrowed and forming a beak-like apex or slightly expanded and spatulate at apex, only joined at extreme tips, finely ciliate with vibratile whitish hairs along margins of purplish-brown or brownish-green apical part; (ii) lobes narrowed, opening out and marginally connate to form a linear stalk-like part up to 7 × 0.8 mm, finally spreading apart into slightly recurved linear parts, connate at the extreme tips and form a second smaller, elliptical cage ± 4 × 2 mm acute at apex, lobes glabrous except for the hairs on bases of keels. Corona sessile and basally adnate to corolla, ± 3.5 × 4–4.5 mm, basally cupular, most parts purplish-maroon/green; outer lobes deltoid-ovate, 0.5–1 × ± 1.4 mm, shortly bifid, marginally purplish-maroon with dense whitish-translucent, needle-like inward-pointing hairs; inner lobes strap-shaped, ± 2.5 × 0.5 mm, incumbent upon anthers, then connivent-erect, with purplish-maroon hairs dorsally from bases to midway up, otherwise glabrous. Anthers subquadrate and overtopping the rounded to flattened stylar head; guide rails ± 0.7 mm long; pollinia 0.39–0.45 × 0.2–0.28 mm; corpusculum obovate, 0.3–0.36 × 0.16 (apex) × 0.09 (base) mm. Follicles paired, sharply divergent, 160–240 mm long × 3–4 mm, apically tapering and sometimes curved, dull grey-green to grey-maroon, drying straw-colour with brownish-maroon streaks; seed dark brown, 10–12(–18) × 2–3(–5) mm, with light brown margin; coma 20–30 mm long.

1. Apex of corolla lobes joined into a filiform column, then
     dilated at top of column to form a smaller secondary
     cage; margins glabrous .... c. var. **rostrata**
   Apex of corolla lobes not as above, without a secondary
     cage; apices attenuate or expanded, margins lined with
     fine, spreading, whitish, vibratile hairs 2
2. Keels of plicate corolla lobes produced at base, meeting
     over centre of tube; vertical edges of keels lying parallel
     and in contact; corolla lobe apices usually attenuate to
     extreme tips, close together and beak-shaped .... b. var. **brevirostris**
   Keels of plicate corolla lobes not produced at base, not
     meeting over centre of tube; vertical edges of keels
     lying parallel and usually not in contact; corolla lobes
     mostly attenuate in middle then narrowly to broadly
     expanded at apices, usually not beak-shaped .... a. var. **distincta**

a. var. **distincta**

Keels of plicate corolla lobes not produced at base, not meeting over centre of tube; vertical edges of keels parallel and usually not in contact; corolla lobes mostly attenuate in middle then expanded at apices, usually not beak-shaped.

KENYA. Kilifi District: Kambe Kaya, 27 Mar. 1981, *Hawthorne* 217! & Bamba, 15 Feb. 1996, *Masinde* 860!; Kwale District: km 48 along Mombasa–Tanga road, 1960, *Bally* 12191!
TANZANIA. Uzaramo District: Pugu Hills Forest Reserve, 24 July 1971, *Hansen* 370!; Iringa District: 6.5 km E of Iringa on Morogoro road, 6 Feb. 1962, *Polhill & Paulo* 1364!; Lindi District: 45 km W of Lindi, 17 Mar. 1935, *Schlieben* 6127! Zanzibar: Chukwani bush, 11 Nov. 1930, *Vaughan* 1509!;
DISTR. **K** 7; **T** 3, 7–8, **Z**; southern Africa
HAB. Coastal bush, sometimes on low coral cliffs, miombo woodland, dry bushland, riverine; 0–1600 m

SYN. *C. cyrtoidea* Werderm. in E.J. 70: 209 (1939). Type: Tanzania, Lindi District: 45 km west of Lindi, *Schlieben* 6127 (B†, holo.; BM, BR, K! (photo of BR dup.), P, iso., photo!)

NOTE. *C. distincta* is closely related to *C. lugardiae* and is a distant relative of *C. somalensis*. In southern Africa, *C. distincta* is represented by subspecies *haygarthii* (Schltr.) R.A. Dyer and subsp. *verruculosa* R.A. Dyer. Specimens from Zanzibar which represent the type material differ in their relatively long (9–13 mm) and broad (2–3 mm) sepals so that their apices reach halfway up the corolla tube with their basal margins often overlapping. They also have relatively large leaves 80–100 × 70–78 mm.

b. var. **brevirostris** (*P.R.O. Bally & D.V. Field*) *Masinde* **stat. nov.** Type: Tanzania, Lushoto District: W Usambara Mts, Soni, 9.5 km on Lushoto–Mombo road, *Drummond & Hemsley* 2527 (K!, holo., EA!, iso.)

Leaves usually larger than in other varieties, ovate-oblong, ± 80 × 50 mm, deeply cordate, apex rounded with apiculate tip. Corolla tube widely expanded in throat to 17 mm wide or more. Corolla lobes broadly deltoid, 10–12 × 13–14 mm, plicate and with plate-like keels produced and meeting over centre of tube, abruptly narrowed at top, lying in close proximity but only joined at extreme tips forming an erect beak-like part.

TANZANIA. Pare District: Kisiwani, 20 Apr. 1950, *Mohamed & Verdcourt* 163!; Lushoto District: Mgambo [Mgombo], 26 Mar. 1940, *Greenway* 5927!; Morogoro, Nov. 1951, *Eggeling* 6367!
DISTR. **T** 2–3, 6; not known elsewhere
HAB. Bushland or forest edges and clearings; 40–1150 m

SYN. *C. brevirostris* P.R.O. Bally & D.V. Field in K.B. 36: 448 (1981)

NOTE. There are many intermediates with variety *distincta* in sympatric areas. The taxon is therefore a local variant which shows morphological extremities in allopatric areas. This variety is similar in flower form to subsp. *haygarthii* of southern Africa. Subspecies *haygarthii* differs from this variety by: keels of folded back corolla lobes produced at base and the upper secondary cage usually larger with cilate margins and separated from the lower cage by a prominent stalk.

c. var. **rostrata** *Masinde* **var. nov.** varietati typicae similis sed lobis corollae glabris ad apicem filiformibus atque structuram carcero similem formantibus differt. Typus: Kenya, Teita District, 1 km towards Voi from Bura Station, *R.B. & A.J. Faden* 74/494 (EA!, holo.; K!, iso.)

Corolla lobes abruptly very much narrowed above the broad, folded back, deltoid-ovate part, then opening out and marginally connate to form a long stalk-like part up to 7 mm long × ± 0.8 mm diameter, and finally spreading apart into non-plicate but slightly recurved ± linear parts ± 0.5 mm wide, connate at the extreme tips and form a second smaller, elliptical cage ± 4 × 2 mm which may be acutely pointed at apex, lobes completely glabrous in this case except for the hairs on bases of keels.

KENYA. Teita District: 1 km towards Voi from Bura, 25 Apr. 1974, *Faden & Faden* 74/494!
DISTR. **K** 7; not known elsewhere
HAB. Forest edges and clearings and near streams or rivers; ± 1000 m

NOTE. The long corolla lobes ± equalling the tube in length and the double cage easily distinguishes this variety from the others. The corolla form and especially the corolla lobes and the double cage are similar to *C. distincta* subsp. *verruculosa*.

33. **Ceropegia lugardiae** *N.E. Br.* in Gard. Chron. 30: 302 (1901) & in Dyer, F.T.A. 4(1): 455 (1903); Werderm. in E.J. 70: 20 (1939); R.A. Dyer in F.S.A. 27(4): 57 (1980); Archer, Kenya *Ceropegia* Scrapb.: 137, XXVII (1992); Masinde in Cact. Succ. J. (US) 6: 107 (1999). Type: Botswana, Kwebe Hills, *Lugard* 262 (K!, holo.)

Vigorous twiner to 2 m high or more, generally glabrous, with fibrous roots; latex clear; stems sparsely branched, succulent, terete, 3–5 mm diameter, smooth or finely verrucose, greyish-green due to whitish bloom, young stems pruinose, glaucous. Leaves herbaceous, occasionally semi-succulent; petioles 4–20(–40) mm long; lamina ovate, ovate-oblong to broadly elliptic or ± round, 18–85 × 6–55 mm, base cordate to rounded, sometimes with basal lobes overlapping, apex rounded with an apiculate tip to 3 mm long, margins occasionally wavy, finely ciliate. Inflorescence pseudoumbellate, flowering stems are often short branches up to 160 mm long, up to 9-flowered, flowers opening successively, up to 2 flowers open at a time, scent not detectable or in some plants very mild; peduncle horizontal or slightly ascending, 6–20(–55) mm long; bracts subulate to lanceolate, 1–2.5 × ± 0.3 mm; pedicel (6–)10–20 mm long. Sepals linear-lanceolate, 3–5 × 0.5–1 mm, apically reflexed, abaxially yellowish-green with purplish-maroon spots. Corolla 40–50 mm long; tube 17–27 mm long, with an indentation opposite each sepal, in basal ¹⁄₅ with pale greenish-yellow inflation 4–7 × 4–7 mm, then cream and ± straight or slightly curved and gradually narrowing into an ascending cylindrical part 3–4 mm diameter, gradually widening to 10–18 mm diameter in sulphur-yellow throat, with dark maroon/purple or reddish spots progressively becoming larger and more distinct in throat; occasionally tube generally uniformly greenish to greenish-yellow throughout; glabrous or finely sparsely pubescent; lobes completely folded back, 15–26 × 4–5 mm, apical ¹⁄₂–²⁄₃ linear-lanceolate from broader bases, connate at apices to form an oblong cage with acute apex, keels rarely touching basally, basal parts bright yellow-green with purplish-brown or brownish-maroon venation, upper part dark-maroon/purple, margins ciliate with pale or maroon vibratile hairs, keels in basal part with whitish downward-pointing hairs. Corona sessile, 4–5 × 4(–5) mm, basally laterally adnate to corolla and cupular, creamy or yellowish with purple-maroon; outer lobes ovate, forming 5 pouches, ± 1 × 1.5 mm, shallowly bifid (emarginate) with apices ± touching or ± erect, lobes with dense 0.5 mm needle-like inward-pointing purple-maroon hairs; inner lobes linear to ± spatulate, 3–4 × 0.5–0.8 mm, incumbent on anthers, then erect-connivent and apically recurved, bases purple-maroon with few needle-like hairs dorsally, passing into yellow-white or creamy and then purplish-maroon at apices, glabrous for the most part. Anthers subquadrate, yellowish, overtopping the rounded to flattened stylar head; pollinia 0.4–0.45 × 0.2–0.25 mm; corpusculum obovate, 0.3 × 0.16 (apex) × 0.08 mm (centre). Follicles divergent at ± 45°, 90–145 × 4–5 mm, slightly curved, yellow-green with many dark maroon stripes and streaks, drying brownish with dark-brown/blackish streaks running lengthwise; seed 9–10 × 3 mm long, light brown with paler wings encrusted and wrinkled; coma 15–35 mm.

KENYA. Machakos District: Yatta Plateau, *Archer* 186! & base of Lukenya bluff, 6 Apr. 1960, *Archer* 63! & Kabaa, near Athi River, 23 July 1967, *Archer* 585!
TANZANIA. Kondoa District: Kolo, 24 km N of Kondoa on Great North Road, 13 Jan. 1962, *Polhill & Paulo* 1155!; Mpwapwa District: 6.5 km N of Kibakwe on Mpwapwa track, 10 Apr. 1988, *Bidgood et al.* 985! & Mpwapwa, 3 Apr. 1937, *Mr & Mrs Hornby* 757!
DISTR. **K** 4, 6; **T** 2, 3, 5–7; Zimbabwe, Namibia, Botswana
HAB. *Acacia-Commiphora* bushland and thickets; 550–2150 m

SYN. *C. distincta* N.E. Br. subsp. *lugardiae* (N.E. Br.) H. Huber in Mem. Soc. Brot. 12: 88 (1958)
 [*C. archeri* Bally *in sched., nomen nudum* based on Kenya, *Bally* B12256 (K!)]
 Ceropegia sp. C, Archer in U.K.W.F.: 392 (1974)

NOTE. *C. lugardiae* is a close relative of the *C. distincta* complex and *C. aristolochioides*.

34. **Ceropegia aristolochioides** *Decne.* in Ann. Sci. Nat., Bot. sér. 2, 9: 263 (1838). Type: Senegal, near Lebar, *Heudelot* 477 (P, holo., photo!; G, iso; K, photo. & pencil ill.!)

Mostly glabrous climber to 3(–5) m high, with fibrous roots, occasionally stoloniferous; latex clear; stem sparsely branched, succulent, 2–3(–7) mm diameter, glaucous-green, pruinose near apex. Leaves herbaceous or semi-succulent and sometimes stiff; petiole 5–15 m long; lamina broadly to ovate-lanceolate, ovate-oblong or elliptic, 10–40(–60) × 5–30(–40) mm, base rounded, cordate or broadly cuneate, apex acuminate or rounded-apiculate, margins minutely ciliate. Inflorescence pseudoumbellate, 3–4(–12)-flowered, flowers opening consecutively, scent reminiscent of red wine or fermenting sugars; peduncle 10–45(–70) mm long, succulent, ± 2–3 mm thick; bracts subulate, 1–2 × 0.3 mm; pedicel 5–10 mm long. Sepals linear-subulate, 1–2(–5) × 0.5–1 mm, greenish-cream with maroon spots. Corolla 18–30(–40) mm long; tube 12–28 mm long, in basal $^1/_5$–$^1/_3$ with ovoid inflation 6–9(–11) × 4–8(–10) mm, curved and narrowing into a long ascending cylindrical part 2–4 mm diameter then gradually expanded to (6–)8–13 mm diameter in throat; exterior cream at base passing into pale greenish/yellowish with reddish-maroon spots and streaks becoming darker towards throat, or red-maroon throughout, rarely yellowish-cream throughout ; rarely minutely pubescent; lobes mostly ovate, ovate-linear, crescent-shaped or ± lanceolate, 7–13(–16) mm long, connate at apex to form a globose or ellipsoid cage; visible adaxial faces yellowish or greenish-yellow with reddish maroon reticulated veins and marked brownish/reddish-maroon in apical part, sometimes with few white or maroon needle-like hairs (often vibratile) along margins of apices and/or along basal $^1/_3$–$^1/_2$ part of keel; abaxially similar in colour and indumentum to tube. Corona shortly stipitate or subsessile, 4–5 × 4–5 mm, basally cupular, creamy or yellowish with purple-maroon in parts; outer lobes deltoid-ovate to rectangular, 1.4–2.5 × 1.5–2 mm, bifid, margins dense purple-maroon inward-pointing hairs, with dense tuft at teeth apices; inner lobes ± linear, 1.7–3 × 0.2 mm, incumbent upon anthers then erect-connivent, apices recurved and free, maroon in basal parts with few maroon needle-like hairs then apically passing into creamy to yellowish, glabrous part or occasionally apically sparsely pilose or papillate. Anthers subquadrate, overtopping or much shorter than stylar-head; guide rails 0.6–0.8 mm long; pollinia 0.39–0.41 × 0.24–0.25 mm; corpusculum 0.25–0.3 × 0.15 (apex) × 0.05–0.06 mm. Follicles paired, acutely divergent, 12.5–24 cm long, 3.5–5 mm thick, greenish-brownish/maroon with many white spots and streaks, drying straw-coloured with blackish streaks; seed dark brown, 7–10 × 2–3 mm with light brown marginal wing; coma 30–50 mm long.

subsp. **aristolochioides** *Decne.* in Ann. Sci. Nat. ser. 9: 263 (1838) & in DC., Prodr. 8: 642 (1844); N.E. Br. in F.T.A. 4(1): 464 (1903); Bullock in F.W.T.A. ed. 2, 2: 102 (1963); Engl. in Hochgebirgsfl. Trop. Afr. 343 (1892); Werderm. in E.J. 70: 211 (1939); Archer, Kenya *Ceropegia* Scrapb.: 113, XXIII & 121, XXIV (1992); Masinde in Ghazanfar & Beentje, Taxon. Ecol. Afr. Pl. 583 (2006)

UGANDA. Acholi District: Agoro, Chua, 14 June 1944, *Forbes* B 4092!; Bunyoro District: near Kisansya on Butiaba Flats, 16 June 1969, *Lye et al.* 4002!; Toro District: 8 km W of Muhokya, on Kasese–Kitunguru road, 5 Nov. 1968, *Lock & Ferreira* 68/267!

KENYA. Northern Frontier District: Moyale, 22 Apr. 1952, *Gillett* 12895!; Nairobi/Machakos Districts: Athi near Athi River, 26 Nov. 1960, *Polhill & Lucas* 317!; Kwale District: Mackinnon Road, 14 Aug. 1994, *Luke* 4052!.

TANZANIA. Arusha District: Kimosonu on Belam Malang Farm, at base of Mt Meru, 16 Dec. 1970, *Richards & Arasululu* 26539!; Lushoto District: Mazinde, 1 May 1953, *Drummond & Hemsley* 2325!; Tanga District: Moa District: Mtotohovu near Tanga, 8 Dec. 1935, *Greenway* 4238!

DISTR. **U** 1–2; **K** 1, 3–7; **T** 1–3; Senegal, Burkina Faso, Cameroon, Central African Republic, Congo-Kinshasa, South Sudan, Ethiopia

HAB. Bushland, thickets and dry forest; 15–2200 m

SYN. *C. beccariana* Martelli, Fl. Bogos 56 (1886); N.E. Br. in F.T.A. 4(1): 446 (1903); Werderm. in
  E.J. 70: 222 (1939). Type: Eritrea, near Keren, *Beccari* 306 (FT, holo., photo!)
C. perrottetii N.E. Br. in K.B. 1898: 308 (1898) & in F.T.A. 4(1): 448 (1903). Type: Senegal,
  Perrottet 791 (BM, holo.)
C. albertina S. Moore in J.B. 45: 51 (1907); Werderm. in E.J. 70: 222 (1939). Type: Uganda,
  Bunyoro District: Butiaba Plain, E shore of Lake Albert, *Bagshawe* 848 (BM!, holo.)
C. crassula Schltr. in E.J. 51: 152 (1913). Type: Cameroon, Garua, *Ledermann* 4608a (B†, holo.)
C. aristolochioides Decne. var. *wittei* Staner in Werderm. in B.J.B.B. 15: 237 (1938). Type:
  Congo-Kinshasa, Kivu, Mayi ya Moto, *De Witte* 2028 (BR, holo.)
C. seticorona E.A. Bruce in Cact. Succ. J. (US) 13(11): 181 (1941); P.R.O. Bally in Candollea
  20: 13 (1965) & in Fl. Pl. Afr. 41: t. 1616A (1970); Archer in Kenya *Ceropegia* Scrapbook:
  XXII (1992); U.K.W.F. ed. 2: 184 (1994); Masinde in Cact. Succ. J. (US) 6: 107 (1999).
  Type: Kenya, Masai District: near Namanga, *Bally* 7319 (K!, holo., EA!, iso.)
C. seticorona E.A. Bruce var. *dilatiloba* P.R.O. Bally in J. East Afr. Nat. Hist. Soc. 17: 234–243
  (1943) in part & in Candollea 20: 24 (1965) & in Fl. Pl. Afr. 41: t. 1616B (1970); Archer
  in U.K.W.F.: 392 (1974); U.K.W.F. ed. 2: 184 (1994); Masinde in Cact. Succ. J. (US) 6: 107
  (1999). Type: Kenya, 3 km S of Kajiado, *Milne-Redhead & Taylor* 7151 (K!, holo., EA!, iso.)
C. aristolochioides Decne. subsp. *albertina* (S. Moore) H. Huber in Mem. Soc. Brot. 12: 92 (1958)
C. volubilis N.E. Br. var. *crassicaulis* H. Huber in Mem. Soc. Brot. 12: 200 (1958), *nom. superf.*
  Type as for *C. seticorona* E.A. Bruce var. *dilatiloba* P.R.O. Bally
C. maasaiorum Halda & Prokes, Cactaceae etc. (Bratislava) 10(2): 43–45 (2000) & in
  Repert. Pl. Succ. 51: 8 (2000, publ. 2001). Type: no data, *Halda & Prokes* 20000125 (herb.
  Halda, holo.)
C. erergotana M.G. Gilbert in Nord. J. Bot. 22(2): 205 (2003) & in Fl. Eth. Erit. 4(1): 164
  (2003). Type: Ethiopia, Harerge region, 11 km from Dire Dawa towards Erer Gota, *de
  Wilde & Gilbert* 446 (K! holo., WAG, iso.)
C. burgeri M.G. Gilbert, Nord. J. Bot. 22(2): 205 (2003) & in Fl. Eth. Erit. 4(1): 165 (2003):
  Type: Ethiopia, Uadenadao Plateau, 36 km ESE Harar, then 20 km S, *Burger* 2071 (K!,
  holo.; ETH!, iso.)
  [*C. baringii* P.R.O. Bally *in sched., nomen nudum* based on Kenya, Lukenya, *Bally* 9891 (K!, G)]

NOTE. A vary variable taxon in the corolla morphology and colouration but much less variable
in vegetative form. The variants with externally pubescent corollas occur in Kenya and
Ethiopia. *C. aristolochioides* complex is related to *C. sankuruensis* and probably also distantly to
C. rupicola Defl. of the Arabian Peninsula.

35. **Ceropegia sankuruensis** *Schltr.* in E.J. 51: 155 (1913), as *sankurnensis*
(corrected by De Wild. in B.J.B.B. 7: 29 (1920)); Werderm. in B.J.B.B. 15: 233 (1938);
Huber in Mem. Soc Brot. 12: 79 (1958); Bullock in F.W.T.A. ed. 2, 2: 102 (1963);
U.K.W.F. ed. 2: 183 (1994); Archer, Kenya *Ceropegia* Scrapb.: 155, XXX (1992);
Masinde in Cact. Succ. J. (US) 6: 107 (1999). Type: Congo-Kinshasa, between
Kondue and Sankuru, *Ledermann* 59 (B†, holo.); lectotype: Congo-Kinshasa,
Yangambi, 470 m, May 1906, *Luja* s.n. (BR, lecto., designated by Huber, l.c., photo!)

Leafy glabrous climber to 2(–4) m high; root system fibrous; latex clear; stem
sparsely branched, subsucculent to succulent, somewhat stiff, ± 5 mm diameter,
shiny green to maroonish. Leaves herbaceous; petiole (10–)35–40 mm long; lamina
slightly coriaceous, ovate, ovate-oblong to elliptic, 20–90(–150) × 12–35(–80) mm,
base cuneate-rounded, apex acuminate, shiny dark green, glabrous or very sparsely
pilose. Inflorescence cymose, umbellately 3–6-flowered, flowers opening
successively, with up to 2 flowers open at a time, scent reminiscent of cacao;
peduncles 6–20 mm long; bracts subulate ± 1.5 × 0.3 mm, abaxially pubescent;
pedicels 10–25 mm long. Sepals subulate, 2–4 × 0.5–1 mm, clasping base of tube
inflation, green, abaxially pubescent. Corolla 25–30 long; tube 15–18 mm long, in
basal $^{1}/_{3}$ with a prominent globose or obovoid inflation 9–10 × 9–12 mm, then
curved through ± 90° or more and abruptly narrowing into an ascending cylindrical
part 2–3 mm diameter, dilated gradually in throat to 10–15 mm diameter; exterior
creamy-white with a few pale grey longitudinal lines and maroon spots to top part
of inflation, passing into cream-white with brown-maroon blotches in throat,

glabrous; lobes ± lanceolate, 8–12 × 3 mm, plicate for most of apical part, visible adaxial faces brown-maroon base and apex separated by a cream-yellow band and with distinct reticulated veins in basal part, margins of apical brown-maroon part with few maroon, vibratile hairs, ± 2 mm long, exposed abaxial parts brown-maroon. Corona shortly stipitate, ± 3.5 × 3.5 mm, whitish for most part, basally ± cupular; outer lobes rectangular in outline, apically broadly bifid, teeth parallel and apically obtuse, margins of teeth with few needle-like hairs, inside of lobes yellowish; inner lobes cylindrical, erect-connivent, ± 2.5 × 0.2 mm, apex rounded, bases greenish, glabrous. Pollinia ovoid, ± 0.38 × 0.22 mm; corpusculum 0.22 × 0.16 mm at centre. Follicles and seed unknown.

UGANDA. Kigezi District: Rubimbwa, Kinkizi, Feb. 1945, *Purseglove* P2706!; Masaka District: Bugambo, Bukasa Sese, 3 July 1935, *A.S. Thomas* 1374!
KENYA. Kericho District: 3 km S of Sotik township on Sisei River bank, 26 Aug. 1961, *Archer* 433!
TANZANIA. Pare District: Ngulu, Dao, May 1928, *Haarer* 65; Rungwe District: Kyimbila District: Ukinga, 13 Nov. 1913, *Stolz* 2277!
DISTR. U 2, 4; K 5; T 3, 7; Sierra Leone, Liberia, Nigeria, Cameroon, Congo-Kinshasa
HAB. Forest edges and clearings; 800–1800 m

SYN. *C. anceps* S. Moore in Cat. Talbot's Nigerian Pl.: 66 (1913); F.W.T.A. 2: 63 (1931). Type: Nigeria, Oban, *Talbot* 174 (BM!, holo., K!, iso.)
   C. degemensis S. Moore in J.B. 57: 214 (1919). Type: Nigeria, Degema District, *Talbot* 3652 (BM!, holo.)
   C. batesii S. Moore in J.B. 64: 41 (1926). Type: Cameroon, Bitye, River Dja [Ja], *Bates* 1435 (BM!, holo.)
   C. aristolochioides sensu F.W.T.A. 2: 63 (1931) in part

NOTE. Rare in East Africa and only known from few collections. Related to *C. aristolochioides*, but more of an equatorial forest species whereas *C. aristolochioides* occupies drier habitats.

36. **Ceropegia johnsonii** *N.E. Br.* in F.T.A. 4(3): 451 (1903); Bullock in K.B. 9: 592 (1954) & in F.W.T.A. ed. 2, 2: 102 (1963). Type: Ghana, Aburi Gardens, *Johnson* 768 (K!, holo., GC, iso. photo!)

Herbaceous twiner to 3 m high; root system fibrous; latex colour unknown; stem sparsely branched, herbaceous, twining, 2–3 mm diameter. Leaves herbaceous; petiole 4–40 mm long; lamina elliptic or ovate-oblong, 38–140 × 20–70, base cuneate, apex varying from obtuse to acuminate but in obtuse forms always terminating with an apiculate linear-filiform tip 3–5 mm long, very thinly pubescent or almost glabrous above, densely pubescent with spreading hairs on the midrib and principal veins beneath, and thinly so between them, margin ciliate. Inflorescence cymes pedunculate, umbellately 3–6-flowered; peduncle 10–60 mm long; bracts linear, ± 2 mm long, acute; pedicel 6–13 mm long. Sepals linear, ± 4 × 0.7 mm, acute, clasping base of corolla, spreading at the tips, abaxially shortly hairy. Corolla 44–46 mm long; tube 32–36 mm long, in basal $^1/_6$ with a slight ovoid inflation 4–5 mm diameter, narrowing gradually above to 3–4 mm diameter then gradually expanded to 7–10 mm diameter at the throat; exterior whitish-cream with purple-maroon spots which are paler and more diffuse on inflation at base and more distinct in upper $^3/_4$ of tube, thinly pubescent; interlobes erect, ± lanceolate, 8 × 5–6 mm in plicate state, cohering at the incurved tips, visible adaxial faces whitish, with purple veins on the upper part and with green ones on the basal part, faces and keel glabrous, apically blackish to cohering tips, margin of blackish part lined with few long vibratile whitish hairs up to tips. Corona sessile, basally adnate to corolla, ± 4.5 × 3 mm, basally cupular, whitish-translucent in most parts; outer lobes obtusely deltoid-ovate, 0.7–1.4 × 0.5–1 mm, very shortly emarginate with parallel teeth; teeth marginally with long, needle-like inward-pointing hairs; inner lobes, linear, 2–2.6 × 0.3 mm, incumbent upon anthers then connivent-erect, bases pubescent on dorsal sides, apex sometimes shortly recurved and somewhat papillate. Anthers subquadrate, overtopping stylar head; pollinia ± 0.35 × 0.25 mm; corpusculum obovate, ± 0.4 × 0.18 at apex × 0.06 mm in middle. Follicles linear-terete, (21)–25–28(–32) cm long, 3–(4–5) mm thick at centre, beaked at apex, glabrous; dark brown, seed linear-oblong, 12–13 × 2.5–3.2 mm, tapering into a short beak, with broad thickened light brown margin; coma of white hairs ± 3 mm long.

UGANDA. Masaka District: Sese Islands, 1902, *Mahon* 9!; Mengo District: Kitububa Forest near Entebbe, May 1935, *Chandler* 1205! & Mengo, May 1914, *Dummer* 796!
TANZANIA. Kigoma District: Kasekela Valley, Gombe Stream Reserve, 10 Feb. 1964, *Pirozynski* P381!
DISTR. **U** 4; **T** 4; Ghana, Congo-Kinshasa
HAB. Forest edges and clearings; 1000–1300 m

SYN. *C. lujai* De Wild. in B.J.B.B. 7: 28 (1920); Werderm. in B.J.B.B. 15: 235 (1938). Type: Congo-Kinshasa, Kasai District: Sankuru Forest, 1906, *Luja* s.n. (BR, holo.)

NOTE. A rare species in East Africa, only known from about three collections. Closely related to *C. fusiformis* of West Africa. Previously not recorded for East Africa due to incorrect identifications of specimens, e.g. *Dummer* 796 (BM) was cited by Huber (1958) under *C. sankuruensis*. East African material match the type for *C. johnsonii* quite well.

37. **Ceropegia speciosa** *H. Huber* in Mem. Soc. Brot. 12: 144 (1958). Type: Tanzania, Mpwapwa District: Kiboriani Hills, *B.D. Burtt* 4632 (K!, holo., EA!, iso.)

Twiner to 2 m high, glabrous; root system unknown; latex unknown; stem sparsely branched, subsucculent to succulent, 3–5 mm diameter, glaucous-green. Leaves herbaceous; petiole 10–25 m long; lamina ovate to ovate-oblong, 40–80 × 12–35 mm, base cuneate, apex acuminate, margins minutely ciliate. Inflorescence of pseudoumbellate cymes, 1–(3)-flowered, flowers developing successively, scent unknown; peduncle 20–35 mm long; bracts subulate or lanceolate, ± 3 × 0.5 mm wide at base; pedicel 20–30 mm long. Sepals linear-lanceolate, 7–11 × ± 1 mm, greenish, apices strongly reflexed. Corolla 44–60 mm long; tube 24–40 mm long, in basal ¹⁄₃–¹⁄₂ with a subglobose inflation 15–33 × 10–14 mm, then curved and narrowing abruptly into an ascending cylindrical part ± 3 mm diameter, gradually dilating to 13–15 mm diameter in throat; exterior white to greenish, glabrous; lobes 20–30 mm long, plicate, linear-filiform, connate at apices and very close to each and forming an ellipsoid cage; in dried flowers appearing connivent so that there is hardly any cage; when viewed laterally, corolla lobes appear as if arising from within tube; lobes adaxially dark red, with a dense cover of spreading, vibratile purple hairs 6–10 mm long from base to apex. Corona stipitate, ± 5 × 4 mm, translucent white in dried material, yellow in life; outer lobes suberect, oblong-ovate, ± 3 × 1 mm, basally connate and cupular, apically very shortly bifid, externally densely hairy and internally only hairy towards margins and in apical area including teeth; inner lobes spatulate, ± 3 × 0.5 mm, incumbent upon anthers then erect, apically somewhat connivent, densely pilose in apical ³⁄₄ with whitish-translucent hairs. Anthers subquadrate, incumbent on and overtopping the rounded styar head; gynostegial parts not readily visible when corona is viewed laterally; pollinia ovoid, ± 0.48 × 0.28 µm; corpusculum obovate, orange, ± 0.24 × 0.17 × 0.06 mm at centre. Follicles and seed unknown.

TANZANIA. Mpwapwa District: Kiboriani Hills, 25 Jan. 1933, *B.D. Burtt* 4632!; Iringa District: N of Gologolo Mts, 13 Sep. 1970, *Thulin & Mhoro* 945!
DISTR. **T** 5, 7; Malawi
HAB. *Brachystegia* woodland; 1500–1600 m

NOTE. A rare species, only known from about three collections, with very showy large attractive flowers. It is probably distantly related to *C. distincta* and *C. somalensis*.

38. **Ceropegia ballyana** *Bullock* in K.B. 4: 625 (1955); Bally in Fl. Pl. Afr. 41 (1 & 2): t. 1614 (1970); Archer in U.K.W.F.: 392 (1974) & Kenya *Ceropegia* Scrapb.: 149, XXIX (1992); Blundell, Wild Fl. E. Afr.: 145, t. 476 (1987); U.K.W.F. ed. 2: 184 (1994); Masinde in Cact. Succ. J. (US) 6: 107 (1999); Masinde & Albers in Kakt. u. a. Sukk. 50(12): 303 (1999). Type: Kenya, Northern Frontier District, Mathew's Range, *Adamson* in *Bally* S153 (K! holo.; EA!, iso.)

Climber to 5 m high, with fibrous roots; base of stem occasionally thickened to 20 mm diameter; latex clear; stem succulent, terete, 4–10 mm diameter, often covered by a greyish sheen, glaucous (blue-green) especially in young growth, nodes with prominently protruding leaf bases, glabrous. Leaves mostly semi-succulent but supple; petiole 10–20 mm long; lamina ovate, elliptic, obovate or oblong, 70–90 × 50–60 mm, base rounded to cuneate, margins revolute, apex rounded and apiculate or acuminate, glabrous, upper surface with faint creamy veins, margins ciliate. Inflorescence pseudoumbellate, 1–3-flowered, mostly 1 flower open at a time, without detectable scent; peduncle succulent, 30–100 mm long, often as thick as stem, glaucous, glabrous; bracts prominent, leafy, elliptic, 10–35 × 5–15 mm, margins ciliate; pedicel 10–30(–50) mm long. Sepals spreading or reflexed, linear-subulate to lanceolate, 10–25(–30) × 2–5(–7) mm, margin ciliate. Corolla basally curved, 60–120 mm long; tube 40–50 mm long, in basal $\frac{1}{3}$ with an ovoid inflation 10–20(–25) × 10–17 mm, then narrowing abruptly into a long ascending cylindrical part 2–4 mm diameter, dilating to 13–25 mm diameter in throat; tube exterior glabrous, with background of pale green, greenish-yellow or cream, spotted with purple-maroon on the whole tube or only in throat or occasionally without spots; lobes 35–75 × 8–12 mm, plicate so that only the adaxial faces are visible, with long narrow often purplish-maroon to brown apical parts which are tightly or loosely spirally twisted into a helix-like sceptre, connate at apex; adaxially finely and densely pilose with purplish to whitish hairs, background paler than tube and often with distinct purplish net venation in bases, or with fine purple-maroon spots at base of spiral. Corona distinctly stipitate, 4–5.5 × 3–4 mm, yellowish-cream or green, sometimes with purplish-maroon/brown spots in basal area, glabrous or with sparse pale hairs; outer lobes deeply bifid, 2–3.5 × ± 0.2 mm; inner lobes glabrous, linear, connivent-erect, 3–4.5 × 0.3–0.7 mm. Anthers subquadrate and mostly with club-shaped apex, overtopping the rounded stylar head; pollinia 0.37–0.4 × 0.25 mm; corpusculum obovate, dark-orange ± 0.15 × 0.1 mm. Follicles paired, divergent at acute angle, (80–)100–200 mm long, 7–10 mm diameter at centre, stout, shortly tapering at apices, with a greyish/whitish sheen when still green, drying brownish; seed 11–15 × ± 4 mm, brown to dark-brown with a thinner wrinkled paler margin; coma ± 40 mm long.

KENYA. Northern Frontier/Meru Districts: Isiolo District: on hill behind Veterinary Primary School, 12 Apr. 1971, *Kimani* 266!; West Suk District: Morun River, Sebit, 27 July 1994, *Masinde* 683!; Teita District: Taita Hills, Chawia, Wusi, May 1931, *Napier* 1145!
TANZANIA. Masai/Mbulu Districts: Manyara escarpment, 14 Mar. 1960, *Bally* 12132! & Chem Chem River Gorge, Lake Manyara National Park, 21 Nov. 1963, *Greenway & Kirrika* 11065! & Manyara National Park, Endabash, 21 Nov. 1969, *Richards* 24719!
DISTR. **K** 1, 3–4, 7; **T** 2; not known elsewhere
HAB. Forest patches in wooded bushland and riverine areas; 600–1550 m

SYN. *C. helicoides* E.A. Bruce & P.R.O. Bally in K.B. 5: 371 (1950), *nom. illeg.*, *non C. helicoidea* Choux in Bull. Mus. Hist. Nat. Paris 31: 400 (1925). Type as for *C. ballyana*

NOTE. A stout large plant which is the largest *Ceropegia* in eastern Africa, and only rivalled in size by *C. albisepta* var. *robynsiana* to which it is closely related. The corolla lobes which are twisted into a helix-like structure distinguish *C. ballyana* from *C. albisepta* var. *robynsiana*. Up to about three main variants which are easily distinguishable by their colouration are known in Kenya.

39. **Ceropegia albisepta** *Jum. & H. Perrier* in Ann. Inst. Bot.-Géol. Colon. Marseille, sér. 2 6: 227 (1908). Type: Madagascar, Andranomandavo, near Andranomavo, *Perrier* s.n. (P, holo.)

Climber to 2 m high, with fibrous roots; latex clear; stem succulent, sometimes sparsely branched, often trailing on ground to 6 m long, forming loops and rooting at nodes, terete, 4–7 mm diameter, glaucous (blue-green) especially in young growth, glabrous, nodes purplish. Leaves semi-succulent, stiff; petiole 8–15 mm long; lamina ovate or oblong-elliptic, occasionally obovate, (40–)50–80 × (20–)30–60 mm, base cuneate to ± rounded, apex sub-acute, glabrous, veins above whitish, margins sometimes revolute, sometimes ciliate. Inflorescence pseudoumbellate, 1–3-flowered, up to 3 flowers open at a time, without detectable scent or with very mild sweetish scent; peduncle succulent, 40–150 mm long, up to 5 mm diameter, glaucous, glabrous; bracts linear-lanceolate, 4–9 × 1 mm; pedicel 10–25 mm long. Sepals linear-subulate, 5–10 × ± 1.5 mm, apex often recurved, greenish, usually clasping inflation, glabrous. Corolla 35–50(–70) mm long; tube 12–40 mm long, in basal ¹/₃ with obovoid (ovoid) inflation 11–16 × 7–15 mm, then narrowing into a long ascending cylindrical part 3–4 mm diameter, dilating gradually to 11–22 mm diameter in throat; tube glabrous, green-white to greenish-yellow with purple-maroon spots; lobes 12–31 × 5–7 mm, connate at apices, plicate, acutely triangular or linear-lanceolate with long linear apical parts; keels connivent or well apart in apical ¹/₃; adaxially variously coloured: (**i**) whitish to creamy-yellow in basal ²/₃ area, apical ¹/₃ bright green, paler green, maroon or green with maroon along margin, basal ¹/₃ densely covered with short whitish hairs, margins purple with a dense cover of purple hairs; (**ii**) green faces with purple hairs in bases, apices orange, margins maroon covered with purple hairs; (**iii**) cream with fine purplish spots with purple bristle-like hairs arising from the spots, apices maroon, margin covered with purple hairs. Corona stipitate, 4–4.5 × 3–4 mm, yellowish-cream to whitish-green, sometimes with purplish-maroon spot in the depression at the base of outer lobes; sparsely or densely pilose with whitish-translucent hairs; outer lobes deeply bifid, lobules ± 2 × 0.3 mm, pilose with fine translucent white hairs, internally with purple-maroon spots, marginally pilose; inner lobes linear, connivent-erect, 3–3.5 × 0.4–0.7, glabrous or pilose. Anthers subquadrate, overtopping the ± flattened stylar head; pollinia ± 0.36 × 0.25 mm; corpusculum obovate, ± 0.25 × 0.15 at apex × 0.07 mm basally. Follicles paired, divergent at acute angle, 125–230 mm long, 4–8 mm diameter at centre, stout, apices curved inwards and shortly tapering, when immature glaucous-green, drying brownish; seed dark brown to blackish, 10–15 × 2–4 mm, with a thinner wrinkled paler margin; coma 20–55 mm long. Fig. 70, p. 276.

var. **robynsiana** (*Werderm.*) *H. Huber* in Mem. Soc. Brot. 12: 199 (1958); Malaisse & Schaijes in Asklepios 58: 28, fig. 4, 16 (1993); Masinde in Cact. Succ. J. (US) 6: 107 (1999); Masinde & Albers in Kakt. u. a. Sukk. 50(12): 303 (1999). Type: Congo-Kinshasa, Matadi, Grotte de Guadi, *Dacrémont* 399 (BR, holo., sketch of type in K!)

UGANDA. Toro District: Fort Portal, Kilcloony Crater Lake, 21 Jan. 1945, *Forbes* 65!
KENYA. Nairobi, near Karura River 14 Oct. 1967, *Mwangangi & Abdallah* 243!; Masai District: Ngong Forest, Feb. 1934, *van Someren* 5869!; Teita Hills, Mwatate River Valley, 19 Oct. 1961, *Archer* 468!
DISTR. **U** 2; **K** 1, 3–4, 6–7; Congo-Kinshasa
HAB. Upland forest patches and riverine areas; 900–1900 m

SYN. *C. robynsiana* Werderm. in B.J.B.B. 15: 238 (1938)
    C. succulenta E.A. Bruce in Cact. Succ. J. (US) 13: 181 (1941); Archer in U.K.W.F.: 392 (1974) & Kenya *Ceropegia* Scrapb.: 141, XXVIII (1992); Blundell, Wild Fl. E. Afr.: 145, fig. 479 (1987); U.K.W.F. ed. 2: 184 (1994). Type: Kenya, near Nairobi, 6 Jan. 1914, *Barber* s.n. (K!, holo.)

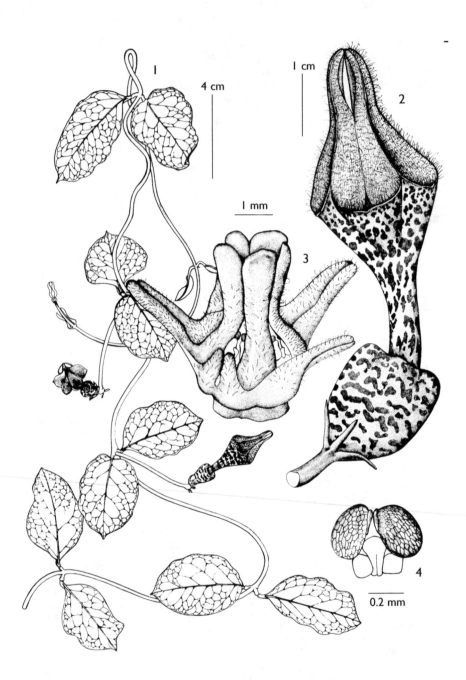

FIG. 70. *CEROPEGIA ALBISEPTA* var. *ROBYNSIANA* — **1**, habit; **2**, flower; **3**, gynostegium and corona; **4**, pollinarium. All from *Masinde* 878. Drawn by N. Muema based on pencil drawings by S. Masinde.

C. evelynae E.A. Bruce & P.R.O. Bally in K.B. 5: 370 (1950). Type: Uganda, Toro District: Fort Portal, Kilcloony Crater Lake, *Forbes* 65 (K!, holo., EA!, iso.)

C. albisepta Jum. & H. Perrier var. *bruceana* H. Huber in Mem. Soc. Brot. 12: 199 (1958). Type as for var. *robynsiana* (Werderm.) H. Huber

NOTE. Closely related to *C. ballyana*. The dark green, stiff, leaves with whitish veins above and the mostly stoloniferous growth easily distinguish *C. albisepta* var. *robynsiana* from *C. ballyana*.

40. **Ceropegia variegata** *Decne.* in Ann. Sci. Nat. ser. 2, 9: 262 (1838) & in Prod. 8: 642 (1844); O. Schwartz in Fl. Trop. Arab.: 191 (1939); Archer in Kenya *Ceropegia* Scrapb.: 159, XXXI (1992); Masinde in Cact. Succ. J. (US) 6: 107 (1999); M. Gilbert in Fl. Somalia 3: 170, t. 2a (2006). Type: North Yemen, Jebel Ra's, 1837, *Botta* s.n. (P, holo. not seen)

Climber to ± 2 m high, mostly leafless, glabrous, with fibrous roots sometimes from a slightly swollen (to 23 mm) rootstock; latex clear; stem sparsely branched, thickly succulent, terete, (4–)10–12 mm diameter, glaucous green, sometimes with darker flecks. Leaves sessile, arising from prominent tubercles, rudimentary, triangular, thickly firm and spiny, 3–4 × 1.5–3 mm wide at base, with a pair of thorny stipules, caducous. Inflorescence up to 2–5-flowered, flowers developing successively; peduncle thickly succulent, stem-like, short then elongating, 10–50 mm long, 4–12 mm diameter, glaucous green; bract triangular-lanceolate, 2–3 × ± 1 mm, apex thorny; pedicel 3–8 mm long. Sepals subulate-lanceolate, 3–4 mm long, apices spreading and recurved. Corolla 30–35 mm long; tube pale greyish-white with fine greyish-purple spots/blotches all over, (15–)25–35 mm long, with 5 indentations at base, in basal $^1/_3$–$^1/_2$ with double inflation, lower chamber ovoid, 6–13 × 6–8 mm, narrowing to ± 4 mm then dilating to upper chamber, 4–5 × 7–8 mm, more densely spotted, upper part dilated to 5–6 mm diameter in throat; lobes 5–30 mm long, variable with two main forms: (**i**) erect lobes with short (up to 5 mm long) spreading canaliculate part; (**ii**) lobes with a very short basal erect part and long (up to 30 mm long) spreading or downward pointing canaliculate apical part; basal part erect, connivent near apex, erect part first white or pale pink-green at apex, often grey-brown along margin, then with horizontal purple-black band, followed by a yellowish to greenish band and finally purple-brown to grey-green before uniting into a linear projection 1–5 mm long, apical bases of adjacent lobes united into channelled, spreading to slightly ascending projection 5–40 mm long, margin rolled outwards, inner surface of channel white to pale green covered with purple-white hairs; abaxially glabrous and similar to tube in colour. Corona distinctly stipitate, yellow-white, 3.5–5 × 3.5–4 mm, glabrous throughout; outer lobes basally cupular, ascending, deeply bifid, 1–2 × 0.3 mm; inner lobes linear-cylindrical, 0.7–3 × ± 0.2 mm, incumbent on anthers, sometimes rudimentary and not exceeding them in length, erect above stylar head and connivent to apices. Anthers subquadrate, overtopping the rounded to flattened stylar head; pollinia 0.45–0.5 × 0.25–0.3 mm, corpusculum obovate ± 0.25 × 0.08 µm. Follicles paired, divergent at acute angle, 85–90 × 10 mm diameter at centre, apically tapering, apex rounded, pale greenish with greyish-purple speckles/blotches all over hence ± similar in colour and thickness to corolla tube and stems, smooth or verrucose. Seed ± 5 × 2 mm, with a paler margin; coma ± 50 mm long. Fig. 71, p. 278.

UGANDA. Karamoja District: Upe County, Amudat, Jan. 1958, *J. Wilson* 548!
KENYA. Northern Frontier District: 2 km S of El Wak township, 11 Dec. 1971, *Bally* 14546!; Baringo District: 6.5 km N of Murgurin, 20 Oct. 1969, *Archer* 621!; Masai District: Kedong Valley, ± 12 km SE of Suswa, 13 Jan. 1995, *Foresti* 422!
DISTR. **U** 1; **K** 1, 3, 6–7; ?**T**; Somalia, Ethiopia, Saudi Arabia, Yemen
HAB. *Acacia-Commiphora* bushland; 300–1500 m

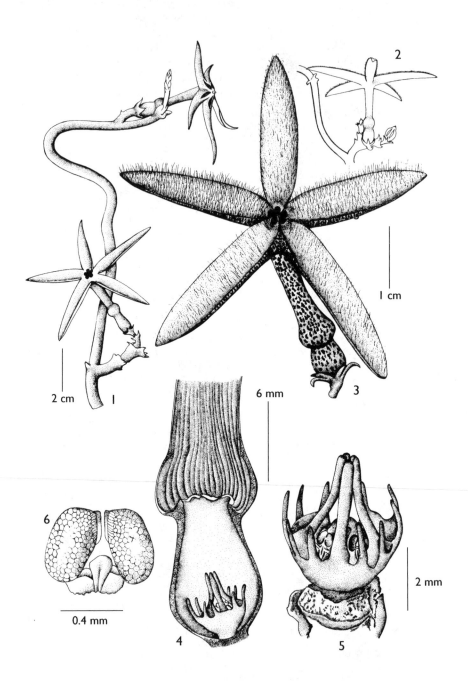

FIG. 71. *CEROPEGIA VARIEGATA* — **1 & 2**, habit; **3**, flower; **4**, longitudinal section of double inflation area of corolla tube; **5**, gynostegium and corona; **6**, pollinarium. All from *L. Newton* in *Masinde* 833. Drawn by N. Muema based on pencil drawings by S. Masinde.

SYN. *Stapelia variegata* Forssk., Fl. Aegypt.-Arab.: 51 (1776), *nom. illeg.*, non *S. variegata* L. Type:
North Yemen, Wadi Surdud, *Forsskal* s.n. (no material preserved according to Hepper &
Friis (1994: 79) & Bruyns in Notes Roy. Bot. Gard. Edin. 45: 287 (1989))

Ceropegia tubulifera Deflers in Bull. Soc. Bot. France 43: 111 (1896) & in Mem. Inst. Egypt
3: 2, t. 1 (1896). Type: South Yemen, near Serrya in a gorge in the Jebel al `Urays, *Deflers*
799 (not located)

C. devecchii Chiov., Fl. Somala 2: 301 (1932); Bally in F.P.A. 35: t. 1368 (1962); U.K.W.F. ed.
2: 183 (1994). Type: Somalia, Cisgiuba, Buracaba, *Senni* 459 (FT, holo., photo!)

C. variegata Decne. var. *cornigera* H. Huber in Mem. Soc. Brot. 12: 141 (1957). Type as for
C. devecchii Chiov.

C. devechii Chiov. var. *adelaidae* P.R.O. Bally in Candollea 17: 79 (1959, description) & ibid.
29: 390 (1974, validation). Type: Kenya, Kwale District, Maji ya Chumvi, *A. Bally* in *Bally*
10540 (G, holo., photo!; K! iso.)

NOTE. *C. variegata* is closely related to the group comprising *C. cimiciodora* and *C. stapeliiformis* of
southern Africa. It has a remarkable corolla which is widely variable especially in the shape of
the corolla lobes. The forms with long spreading sinuses are reminiscent of a helicopter's
propeller blades. In the Arabian Peninsula, all the different forms of the corolla lobes occur,
but in East Africa, there are apparently two main forms of corolla lobes. For instance, *Lavranos*
s.n. from Baringo has short corolla lobes matching those of the coloured plate by Deflers
(1896: t. 1, as *C. tubulifera*), drawn from a North Yemeni plant, but the inner corona lobes are
rudimentary. *Newton* s.n. has long corolla lobes and well developed inner corona lobes.

41. **Ceropegia maiuscula** *H. Huber* in Mem. Soc. Brot. 12: 116 (1958). Type:
Tanzania, Kondoa District: along Kondoa–Dodoma road, *Vaughan* 3123 (BM!, holo.;
EA!, iso.)

Climber to 3 m, mostly leafless, glabrous; root system unknown. Stem twining,
sparsely branched, succulent, 2–3 mm diameter, covered with whitish-grey wax. Leaves
subsessile to sessile; lamina thickly succulent, linear, 50–100 × 1–5 mm, ± 1 mm thick,
± V-shaped. Inflorescence pseudo-umbellate, few flowered, flowers developing
successively, flowering branches at each consecutive node; peduncle 15–25 mm long,
1–1.5 mm diameter; bracts subulate, ± 2 × 0.5 mm; pedicel 5–8 mm long. Sepals
subulate, 2–3 × 0.8 mm. Corolla 35–37 mm long; tube 20–24 mm long, in basal $^1/_3$ with
an ovoid inflation 8 × 5 mm in diameter, then gradually narrowing into a cylindrical
part ± 2 mm in diameter, gradually expanded in throat to ± 5 mm in diameter;
exterior green, glabrous; interior glabrous throughout including basal inflation; lobes
linear, plicate, 12–16 × 0.5 mm in plicate state, connate at apices to form a ± elliptic
cage, ± 6 mm in diameter, greenish, visible adaxial faces finely ciliate with vibratile
hairs, abaxially glabrous. Corona shortly stipitate or subsessile, ± 3.5 × 2.5 mm, basally
very shallowly cupular, whitish-translucent; outer lobes form 5 shallow pouches ± 0.7
× 0.4 mm, margins entire, glabrous; inner lobes cylindrical, filiform, erect, ± 0.3 ×
0.1 mm, apically recurved, pilose throughout. Pollinia elliptic ± 0.35 × 0.26 mm,
corpusculum obovate-linear, ± 0.16 × 0.05 mm. Follicles and seed unknown.

TANZANIA. Kondoa District: between Kondoa and Dodoma, 23 Apr. 1940, *Vaughan* 3123!
DISTR. **T** 5; only known from the type
HAB. Bushed grassland and bushland; ± 1550 m

NOTE. A distinct species but only known from the type material. It has been grossly
misinterpreted by earlier workers (e.g. Huber, 1958) with regard to its relationships. The
supposed allies according to these workers are most unlikely candidates. The stems, corona
and pollinaria point to a relationship with *C. subaphylla* K. Schum.

42. **Ceropegia richardsiae** *Masinde* **sp. nov.** *C. subaphyllae* similis sed foliis multo
evolutis atque linearibus, lobis corollae pilos longos vibratorios ferentibus,
marginibus loborum ad apicem integris (nec bifidis) differt. Type: Tanzania, Moshi
District: Sanya Juu–Engare Nanyuki road, *Richards* 20127 (K!, holo.)

Glabrous twiner; root system unknown; stem twining, branched, succulent, 1–2 mm diameter. Leaves sessile to subsessile, stipules subulate, 1.5 × 0.5 mm, finely pubescent; lamina linear, 20–62 × 10–15 mm, base cuneate, apex acute, margin sparsely ciliate. Inflorescence pseudoumbellate, 1–3-flowered, flowers developing successively with one flower open at a time; peduncle 13–32 mm long; bracts subulate; pedicel 6–10 mm long. Sepals subulate-lanceolate, ± 2–3 × 0.5 mm. Corolla 25–28 mm long; tube ± 19 mm long, in basal ¼ with an ovoid inflation ± 4 × 3 mm, slightly curved above inflation and narrowing into a long ascending cylindrical part ± 2 mm diameter, expanded in throat to ± 4 mm diameter; exterior dull green, glabrous; interior probably also pale greenish, finely pilose from throat down the tube up to entrance to inflation but inside inflation glabrous; lobes linear, plicate, 9–13 × ± 0.5 mm wide, apices connate to form a ± oblong cage, adaxially finely ciliate on faces and keels, abaxially glabrous. Corona shortly stipitate, ± 4 × 2.5 mm, basally deeply cupular, colouration unknown but in dried state translucent white, all parts glabrous; outer lobes forming five shallow, suberect to ± spreading, apically ± truncate pouches ± 0.7 mm wide × 0.5 mm deep, margins entire, glabrous; inner lobes cylindrical, linear, ± 3 × 0.2 mm, incumbent on anthers then erect-connivent or free, apically recurved. Anthers overtopping the rounded stylar head; pollinia 0.36 × 0.26 mm; corpusculum linear-obovate, 0.15 × 0.05 mm. Follicles and seed unknown.

Tanzania. Moshi/Masai Districts: Sanya Juu–Engare Nanyuki road, 9 Apr. 1965, *Richards* 20127!
Distr. **T** 2; only known from the type
Hab. In open bushland, in a lugga among lava stones; altitude, ± 1350 m

Note. A distinct taxon which is only known from one collection. It is probably allied to *C. subaphylla*. It differs in the following characters: leaves well developed and linear, corolla lobes with conspicuous long vibratile hairs, outer corona lobe apical margins entire, not bifid.

43. **Ceropegia crassifolia** *Schltr.* in J. Bot. Lond. 33: 273 (1895); N.E. Br. in Fl.Cap. 4,1: 818 (1908); U.K.W.F. ed. 2: 183 (1994). Type: South Africa, Cape, King William's Town, *Sim* 312 (B†, holo.; BOL, iso.).

Herb, erect or occasionally twining in apical parts, to 15(–30) cm high; latex clear; rootstock a cluster of whitish fusiform roots; stem sparsely branched, succulent, cylindrical, 2.5–5 mm diameter, somewhat flattened below node on opposite sides, weakly ridged in some internodes, shiny grey-green with grey-maroon streaks. Leaves suberect, succulent, sessile or subsessile; lamina linear or linear-lanceolate, 40–90 × 3–5 mm, V-shaped above, greyish, glabrous above and beneath, midrib prominent beneath and whitish. Inflorescence pseudoumbellate, 3–5-flowered, flowers developing successively, scent strong like cow dung or putrid; peduncle suberect, succulent, 5–10 mm long, 1.5–2.5 mm diameter, glabrous; bracts subulate, ± 2 × 0.3 mm at base, glabrous; pedicels 3–6 mm long. Sepals lanceolate-subulate, ± 2 × 1 mm, apex acuminate, greenish with maroonish tinge, glabrous, tightly clasping corolla inflation. Corolla 13–25 mm long; tube 9–20 mm long, dilated in basal ⅓ into an obovate inflation 4–7 × 4–5 mm, then moderately curved and narrowed gradually above inflation to 2–2.5 mm diameter to form an ascending cylindrical part which dilates gradually in throat to ± 5 mm diameter, glabrous, uniformly white-green or yellow-green or inflation sometimes with a pinkish blotch in between sepals; lobes 3.5–6 mm long, not folded fully back, in semi-plicate state ± oblanceolate, ± 5 mm broad in apical area, apices connate to form a cage that is about twice as wide as inflation; both sides cream-green or yellow-green, keel with long whitish downward-pointing needle-like hairs, adaxial faces at very base with inward-pointing to suberect needle-like hairs. Corona shortly stipitate, ± 1 × 2.5 mm, cupular, greenish-cream; outer lobes ± 1 mm high, forming 5 cupular pouches, very shallowly emarginate or entire, margin; inner lobes rudimentary, incumbent on dorsal sides of anthers and less than half as long as anthers, ± 3 × 2 mm, bases brownish. Anthers subquadrate, incumbent on the rounded stylar head and overtopping it; pollinia ± 0.26 × 0.24 mm;

corpusculum linear, ± 0.15 × 0.04 mm. Follicles paired, divergent at ± 60°, cylindrical and 75–90 mm long, ± 3.5 mm diameter at centre, shortly tapering at apices, at maturity pale green with grey-maroon streaks and spots; seed 6–7 × 2–3 mm, dark brown with a paler margin; coma 20–25 mm long.

var. **copleyae** (*E.A. Bruce & P.R.O. Bally*) *H. Huber* in Mem. Soc. Brot. 12: 102 (1958); Archer in Kenya *Ceropegia* Scrapb.: 65, XIV (1992); Masinde in Cact. Succ. J. (US) 6: 107 (1999). Type: Kenya, Kiambu, 24 km E of Nairobi, *Armstrong* s.n. in *Bally* S44 (EA!, holo.; K!, iso., only spirit mat.)

KENYA. Kiambu District: 24 km E of Nairobi, Mar. 1941, *Armstrong* S44!; Nairobi District: Embakasi, near junction of entrance to Jomo Kenyatta Airport, Nairobi, 5 July 1994, *Newton* 4461! & Embakasi, between JKIA Airport and Mombasa road, 1 Mar. 1996, *Masinde* 874!
TANZANIA. Lushoto District: Mkomazi Game Reserve, Mbula, 8 Jan. 1996, *Abdallah & Mboya* in NHT 3948
DISTR. **K** 4; **T** 3; not known elsewhere
HAB. Wooded grassland with *Acacia drepanolobium*, in half shade on mostly black cotton soil; 1600–1850 m

SYN. *C. copleyae* E.A. Bruce & P.R.O. Bally in K.B. 5: 369 (1950)

NOTE *Ceropegia crassifolia* var. *copleyae* is probably distantly related to *C. stenantha* and *C. nilotica*. The typical variety occurs in southern Africa: Botswana, Namibia, South Africa, Swaziland and Zimbabwe. In Kenya, it is often found beneath scattered low shrubs of *Acacia drepanolobium* in grassland where there is much grazing by domestic and wild animals, especially around JKIA Airport, Nairobi and Kapiti plains. It is kept low and dwarf most of the time because of the grazing pressure. However when the stems are not grazed, they grow taller and start scrambling in grass and may even adopt a twining habit in the apical parts. The unpleasant scent from the flowers may be a an adaptation to attract pollinators and at the same time make the plants unattractive to grazers especially during anthesis. The generally monochromatic colour of the flowers and the unique margins of the outer corona lobes as well as the vestigial inner corona lobes make the flowers of variety *copleyae* quite distinct.

44. **Ceropegia yampwapwa** *Masinde* **sp. nov.** *C. stenanthae* similis sed fauce corollae inflata 11 mm diameter, parte inflata omnino glabra, lobis corollae acute triangularibus non auriculatis atque ad apicem usque ²/₃ longitudinem loborum plicatis, antheris ad apicem ± clavatis, pollinariis caudiculas comparate longas ferentibus differt. Type: Tanzania, Mpwapwa District, 13 km S of Mpwapwa on Gulwe track, *Bidgood, Mwasumbi & Vollesen* 961 (K!, holo.)

Herbaceous, leafy twiner to 50 cm high; root system unknown; stem sparsely branched, succulent, 2–3 mm diameter glabrous. Stipules subulate, persistent, 1–3 × 0.5 mm, abaxially pilose. Leaves sessile or subsessile, subsucculent; lamina linear to linear-lanceolate, 30–140 × 2–8 mm, base cuneate, apex acute, glabrous except for midrib beneath, margin ciliolate. Inflorescence sessile, extra-axillary, umbellate, 1–3-flowered, flower buds apically sharply beaked, flowers developing successively, one flower open at a time; bracts subulate, ± 1 × 0.3 mm, pale green with purplish spots, glabrous; pedicels 5–6 mm long. Sepals subulate-lanceolate, ± 4 × 0.5 mm, glabrous, pale green with purplish-mauve specks. Corolla ± 34 mm long; tube ± 26 mm long, basal ¹/₃ with an ovoid inflation ± 8 × 5 mm diameter, then narrowing gradually into a long ascending cylindrical part ± 3 mm diameter widely dilated in throat to 11 mm diameter; exterior pale-brown in inflation then passing into whitish or mauve in middle part to base of lobes, glabrous; interior probably similar in colour to outside, inflation glabrous then at narrowed apically part with a dense ring of downward-pointing whitish hairs which gradually decrease and eventually disappear halfway up the tube, then rest of tube to throat glabrous; lobes ± 9 mm long, plicate in apical ²/₃, acutely triangular, with distance from keel to margin of ± 2 mm in middle, erect, connate at apex to form a ± conical cage ± 20 mm diameter at widest, with an apical beak, margins glabrous; adaxially pale brownish, very sparsely finely ciliate in basal part of keel to ± glabrous; abaxially mauve, glabrous. Corona distinctly stipitate, ± 3 × 2.5 mm diameter, translucent white with purplish blotches and specks in parts, glabrous in all parts; outer lobes apically much reduced to below bases of pollinaria, forming five shallow pouches ± 0.4 × 0.7 mm wide, margins entire; inner lobes cylindrical, ± 2 × 0.2 mm, erect, free, apically slightly recurved, apex obtuse. Anthers yellowish, prominently visible with thick apices, subhorizontally orientated above the conical to flattened stylar head; pollinia ovoid, ± 0.33 × 0.24 mm; corpusculum obovate, ± 0.11 × 0.07 mmm; caudicles unusually long, ± 0.1 mm long. Follicle and seed unknown.

TANZANIA. Mpwapwa District: 13 km S of Mpwapwa on Gulwe track, 9 Apr. 1988, *Bidgood et al.* 961!
DISTR. **T** 5; only known from the type
HAB. *Acacia-Delonix* wooded bushland on white sandy soil; ± 800 m

NOTE. *C. yampwapwa* is related to *C. stenantha* and only known from a single collection. It is vegetatively very similar to *C. stenantha* in its leaves and stems but florally very distinct by the relatively larger flowers with a widely dilated tube. The characters distinguishing *C. yampwapwa* from *C. stenantha* are: *C. yampwapwa*: corolla much dilated in throat to 11 mm in diameter, corolla lobes acutely triangular, plicate in apical ²/₃, not auriculate; tube inflation completely glabrous throughout, anther apices ± club-shaped, pollinaria with relatively long caudicles. *C. stenantha*: corolla not or very slightly dilated in throat to 3 mm in diameter, corolla lobes linear, only slightly or not plicate, basally auriculate; tube inflation pilose in ± basal ¹/₂, anther apices not club-shaped, pollinaria with shorter caudicles.
  The caudicles of *Ceropegia* species are usually quite short so that when the pollinaria are separated from the gynostegium, the pollinia tend to move towards the corpusculum and may even conceal them making them only observable with difficulty. The conspicuous caudicles are reminiscent of those prevalent in such groups like the *Cynanchinae* and *Marsdenieae*. The collection *Fries* 1383, from Zambia (as *Rhodesia borealis in sched.*), Northern Province, Kalambo near Tanganyika, 28 Nov. 1921, probably belongs to this species. It is vegetatively similar to *Bidgood et al.* 961 and has a cluster of fusiform roots. It has only very young tiny flower buds, making it difficult to identify the specimen positively.

45. **Ceropegia stenantha** *K. Schum.* in E.J. 17: 152 (1893); N.E. Br. in F.T.A. 4(1): 459 (1903); Werderm. in E.J. 70: 223 (1939); Bullock in K.B. 7: 425 (1953); Huber in Mem. Soc. Brot. 12: 123 (1958); Archer in Kenya *Ceropegia* Scrapb.: 61, XIII (1992); U.K.W.F. ed. 2: 183 (1994); Masinde in Cact. Succ. J. (US) 6: 107 (1999). Type: South Sudan, Bahr el Ghazal, Djur Ghattas [Jeuba Ghattas], *Schweinfurth* 2104 (B†, holo.; K!, lecto.!, 2 sheets)

Glabrous twiner to 2 m high; latex milky; root system a cluster of whitish fusiform roots, up to 12 cm long × ± 5 mm thick, arising laterally for ± 5 cm off the main vertical rootstock; stem mostly sparsely branched, occasionally much branched from near base to form a dense bush to 30 cm wide, vaguely angular, succulent, 2–5 mm diameter, greenish, often with some greyish, or grey-maroon streaks. Leaves subsucculent to succulent; petiole 0–4 mm long; lamina narrowly or broadly linear or linear-elliptic, (5–)20–60(–160) × 2–10 mm, base cuneate, apex acute, glabrous on both sides, margin ciliolate. Inflorescence extra-axillary, umbellate, usually dense at or near almost every successive node, sublateral at nodes, 1–10-flowered, flowers developing successively with up to several flowers open at a time, without detectable scent, subsessile to shortly pedunculate; peduncle 0–4 mm long, ± 1.5 mm diameter; bracts linear-subulate, ± 1–2 × 0.3 mm, maroon or pale green with purplish speckling; pedicels 3–10 mm long. Sepals subulate, 2–3 × 0.5–1 mm wide at base, apex recurved, pale green with maroonish apex or spotted purplish-maroon. Corolla 20–30(–35) mm long; tube 10–20 mm long, in basal ¹/₄–¹/₃ or more with elongated ovoid inflation 4–8 × 2–3 mm, narrowing gradually into a long ascending cylindrical part ± 2 mm diameter, not dilated in throat; exterior creamy-white, greenish-white/yellow or yellowish, glabrous; lobes linear, (6–)15–17 × 0.5–1 mm, not completely folded back, margins gently recurved, with conspicuous basal horn-like auricles below sinuses of adjacent lobes, connate at apices to form an elliptical cage; glabrous, pale yellow-green to creamy-white, sometimes purplish or maroon at connate apices. Corona shortly stipitate, 2.5 × 1.3–1.6 mm diameter, mostly translucent white; outer lobes apically much reduced, forming five shallow pouches ± 0.5 mm deep, apex truncate, entire; inner lobes 1.5 × 0.2 mm, cylindrical, erect, free, apically recurved, apex rounded. Pollinia 0.3–0.36 × 0.2–0.27 mm; corpusculum linear-obovate, 0.11–0.14 × 0.05 mm at centre, brownish-orange. Follicles paired, acutely divergent (45–)60–100 mm long × ± 3 mm at centre, at maturity green to yellow green with maroon streaks; seed dark brown, 6–7 × 1.5–2 mm, with a marginal wing; coma 20–30 mm long.

UGANDA. Ankole District: Gayaza, 25 Apr. 1946, *A.S. Thomas* 4442! & Ruizi River, 27 Sep. 1950, *Jarrett* 271!; Masaka District: Maogola County, 18 km S of Ntsusi, 19 Oct. 1969, *Lye & Rwaburindore* 4468!

KENYA. Fort Hall District: near Makuyu, 2 Dec. 1966, *Agnew et al.* 8825! & 5 km S of Tana River on Thika–Wamumu road, 8 Jan. 1972, *Gillett* 19434!; Masai District: 1.6 km W of Olenarau River along Kajiado–Selengai road, 31 Jan. 1965, *Archer* 500!

TANZANIA. Shinyanga District: Shinyanga Block 4B, Apr. 1935, *B.D. Burtt* 5158!; Ufipa District: Muse–Kisungu, 23 Feb. 1950, *Bullock* 2529!; Iringa District: Msembi–Causeway track, km 6.2, 24 Mar. 1970, *Greenway & Kanuri* 14190!

DISTR. U 2, 4; K 4, 6; T 1–8; Congo-Kinshasa, Rwanda, South Sudan, Zambia, Malawi, Mozambique, Zimbabwe, Namibia, South Africa

HAB. Thickets in grassland and open woodland; 500–1300(–1900) m

SYN. *Riocreuxia longiflora* K. Schum. in E.J. 28: 459 (1901). Type: Tanzania, Iringa District: between Ukutu [Khutu] and Uhehe near River Mloa, *Goetze* 495 (B†, holo.)
  Ceropegia stenantha K. Schum. var. *parviflora* N.E. Br. in F.T.A. 4(1): 459 (1903). Type: Mozambique, Coast District: *Hannington* s.n. (K, holo.)
  C. infausta N.E. Br. in F.T.A. 4(1): 459 (1903). Type as for *Riocreuxia longiflora* K. Schum.
  C. angustiloba De Wild. in Ann. Mus. Congo sér. 4, fasc. 3: 109 (1903); N.E. Br. in F.T.A. 4(1): 621 (1904); Werderm. in B.J.B.B. 15: 229 (1939) & in E.J. 70: 216 (1939). Type: Congo-Kinshasa, Katanga, Lukafu, *Verdick* 367 (BR, holo.; K, iso., photo!)
  C. mazoensis S. Moore in J.B. 46: 309 (1908); Werderm. in E.J. 70: 215 (1939). Type: Zimbabwe, Mazoe, *Eyles* 518 (BM, holo., photo!)
  C. quarrei De Wild. in Contr. Fl. Katanga, suppl. 1: 71 (1927); Werderm. in B.J.B.B. 15: 229 (1938). Type: Congo-Kinshasa, Katanga, Katuba, *Quarré* 80 (BR, holo.)
  C. tenuissima S. Moore in J.L.S. 37: 185 (1905); Werderm. in E.J. 70: 213 (1939). Type: Uganda, Ankole District: 'Wazinga' near Mulema, *Bagshawe* 254 (BM!, holo.)

NOTE. *C. stenantha* is related to *C. yampwapwa* and probably distantly related to *C. nilotica*. The specimen *Polhill & Paulo* 1353 from Tanzania is unusual by its very short corolla lobes in relation to the tube and the fine longitudinal ridges inside the inflation.

46. **Ceropegia nilotica** *Kotschy* in Sitz. Akad. Wien. Math.-Nat. Abh. 1, 51: 356 (1865); N.E. Br. in F.T.A. 4(1): 447 (1903); Werderm. in E.J. 70: 202 (1939); Bullock in K.B. 8: 58 (1954); Lisowski & Malaisse in B.J.B.B. 44: 417, figs. 7 & 8 F (1974); H. Huber in Mem. Soc. Brot. 12: 103 (1958) in part; Dyer in F.S.A. 4: 59 (1980); Archer in Kenya *Ceropegia* Scrapb.: 93, XX (1992); Malaisse & Schaijes in Asklepios 58: 24, figs 1–1 & 3–1 (1993); Masinde in Cact. Succ. J. (US) 6: 107 (1999); M.G. Gilbert in Fl. Eth. 4(1): 165 (2003). Type: South Sudan, Gondokoro, *Knoblecher* 35 (W, holo.; EA! & K! photo)

Twiner to 4 m high, occasionally trailing or decumbent, with a cluster of whitish slightly swollen roots 40–130 × 2.5–4 mm, often rhizomatous; latex milky, copious; stem sparsely to much-branched, many-angled or distinctly quadrangular, succulent, 3–4 mm diameter, in dry habitats mostly leafless except for young growth, glabrous. Leaves herbaceous, slightly fleshy to thickly and stiffly succulent; petiole 1–15 mm long; lamina narrowly to broadly ovate or -lanceolate, 8–40(–100) × 3–15(–40) mm, base broadly cuneate, apex acuminate, acute or mucronate, glabrous, margins indistinctly undulate or crenate-serrulate, sparsely ciliate. Inflorescence pseudoumbellate, 1–2-flowered, flowers developing successively, scent intense and sweet or in group (iii) variants (see note) pungent and nauseating; peduncle succulent, 2–60 mm long, 2–3 mm thick, long ones descending, short ones suberect, vaguely angled or distinctly quadrangular, glabrous; bracts subulate or lanceolate, 3–10 × 0.5–1.5 mm, glabrous; pedicel succulent, 4–15 mm long. Sepals subulate or lanceolate, 0.7–12 × 1–5 mm, apically reflexed, yellowish-green or apically tinged pinkish to purplish, glabrous. Corolla 25–45(–50) mm long; tube 13–30(–40) mm long, in basal $^1/_3$ with ovoid inflation 11–17 × 5–9 mm, with a constriction in apical $^1/_3$ dividing it into two chambers, then narrowing gradually into an ascending cylindrical part 3–4 mm diameter, gradually expanded to 13–37 mm diameter in throat, glabrous; coloration quite variable: yellowish, greenish-cream or white throughout, or white to yellowish to greenish-cream to silvery-grey with purple or purplish-maroon spots in throat; lobes triangular, 10–24 × 6–10 mm, plicate to form a broad keel with gibbous projection towards base, apices connate to form a conical to triangular cage, but usually narrowing apically, sometimes apex ± truncate with cage openings oriented to the top; adaxially densely pilose including margins with hairs taking the colour of background or sometimes uniformly purple; sometimes glabrous at apex towards keel, with 3–4 variously coloured transverse bands, dull olive green, reddish or pale-brown then black, then white, and finally blackish-green or reddish. Corona shortly stipitate, 4–5 × 3–4.5 mm, basally shallowly cupular, cream to yellowish with pinkish-purple to dark maroon usually in hidden parts, glabrous or rarely sparsely pilose; outer lobes forming 5 concave to truncate, shallow pouches 1–1.5 × 0.8 mm; inner lobes linear, 2.5–3.5 × 0.3–0.5 mm, incumbent upon anthers then connivent-erect, usually glabrous. Anthers subquadrate and overtopping the flattened to rounded styar-head; guide rails ± 0.6 mm long, area around nectaries dark purple to maroon, gynostegial parts well exposed; pollinia 0.35–0.45 × 0.25–0.3 mm; corpusculum obovate, 0.2–0.26 × ± 0.15 at apex × 0.06–0.09 mm medially. Follicles paired, pointing sideways, obtusely divergent at mostly over 180°, 100–220 mm long, 2.5–4 mm at centre, apically tapering, dull yellow-maroon/brown; seed reddish-brown to blackish, 8–10 × 2.5 mm, with a paler margin; coma 20–40 mm long.

KENYA. Kilifi District: 16 km W of Ganze, 18 June 1964, *Archer* 477! & Arabuko Sokoke Forest, 3 km N of 'Lake Mastangoni' seasonal pools, 9 Jan. 1995, *Masinde* 816!; Teita District: 20 km from Maungu Station on Maungu–Rukanga road, 4 Apr. 1969, *Faden et al.* 69/405!.

TANZANIA. Tanga District: Ngole, 9 June 1937, *Greenway* 4945!; Morogoro District: Turiani, 22 Nov. 1955, *Milne-Redhead & Taylor* 7352!; Uzaramo District: Dar es Salaam, University Campus, 18 Mar. 1972, *Gardiner* 6201!

DISTR. **K** 6, 7; **T** 3, 4, 6; tropical and subtropical Africa including Ghana, Togo, Congo-Kinshasa, South Sudan, Angola, Zambia, Mozambique, Zimbabwe, Namibia, South Africa

HAB. *Acacia-Commiphora* bushland and forest edges; 0–1300 m

SYN. *C. constricta* N.E. Br. in K.B. 1895: 260 (1895). Type: Congo-Kinshasa, Kawal Islands in Lake Tanganyika, May 1888, *Carson* 35 (K!, holo., mixed collection)
  C. mozambicensis Schltr. in J.B. 33: 273 (1895); N.E. Br. in F.T.A. 4(1): 447 (1903); Werderm. in E.J. 70: 202 (1939); Bruce in Hook. Icon. Pl.: t. 3441 (1943); Bullock in K.B. 8: 59 (1953 publ. 1954). Type: Mozambique, mouth of Pungwe R., S of Zambezi, *Schlechter* 7106 (B†, holo.)
  C. gemmifera K. Schum. in E.J. 33: 328 (1903); N.E. Br in F.T.A. 4(1): 620 (1904); Bullock in F.W.T.A. ed. 2, 2: 102 (1936). Type: Togo, near Lome, *Warnecke* 242 (B†, holo.; BM!, iso.)
  C. boussingaultifolia Dinter in Neue Pfl. SW-Afr.: 21 (1914). Type: Namibia, Otjiwarongo, cult. in Okahandja, *Dinter* 2780 (SAM, holo., photo!)
  C. gossweileri S. Moore in J.B. 67 Suppl. 1 & 2: 99 (1929). Type: Angola, Banks of Caringa R. near to where it enters river Mumbeje – N'Dalatando – Cazengo, *Gossweiler s.n.* (BM!, holo.)
  C. plicata E.A. Bruce in Fl. Pl. S. Afr. t. 675 (1937). Type: South Africa, Muden Valley, near Greytown, *Cronwright* 16 (PRE!, holo., photo!)

C. mozambicensis Schltr. var. *ulugurensis* Werderm. in E.J. 70: 202 (1939). Type: Tanzania, Morogoro District: Uluguru Mts, *Schlieben* 3822 (B†, holo.; BM!, P photo!, M photo!, PRE photo!, iso.)

C. nilotica Kotschy var. *plicata* (E.A. Bruce) H. Huber in Mem. Soc. Brot. 12: 105 (1958)

C. grandis E.A. Bruce in Fl. Pl. S. Afr. t. 1113 (1951). Type: South Africa, Natal, Zululand, Ngoma, cult. by Phillips in Pretoria, *Carnegie* s.n. (PRE, holo., photo!; EA, iso.)

C. decumbens P.R.O. Bally in Les Succulentes 53: 133 (1957); U.K.W.F. ed. 2: 184 (1994). Type: Kenya, Machakos District: Chyulu, E slope, *Bally* 7933 (EA!, holo.; K!, iso.)

C. nilotica Kotschy var. *simplex* H. Huber in Mem. Soc. Brot. 12: 104 (1958) in part, excl. *C. denticulata* K. Schum. (1895) & *C. brownii* Ledger (1909). Type: as for *C. gemmifera* K. Schum.

[*C. archeri* Masinde, *in sched.*, *nomen nudum* based on Teita District: Mt Kasigau, S side, *Masinde* 840 (EA!, K!)]

NOTE. *C. nilotica* is closely related to *C. denticulata*. It is a variable tropical-African species complex with two cytotypes and many morphological and colour forms that have in the past been recognised at species or subspecific level, thus creating a proliferation of synonyms when comprehensive re-examinations are done. The main morphological forms that are found in East Africa but which also sometimes occur elsewhere and seemingly without any clear geographical patterns are as follows: (i) The typical common forms where plants have terete or weakly many angled stems, sepals small and short, not overlapping, leaves with short or long petioles mostly spreading and subsucculent; widespread throughout East Africa. (ii) Plants with distinctly 4-angled stems; leaves subsessile or shortly petiolate, minute, stiffly succulent; sepals narrow and short, not overlapping. Poor climbers often decumbent and short, stems segmented hence nodes well articulated and easily break at nodes to produce new propagules as in the west African *C. nilotica* forms (formally known as *C. gemmifera* K. Schum.), occurring mostly in highlands of **K** 7 area, previously known as *C. decumbens* (e.g. Taita-Taveta District: Mt Kasigau, S side along pipeline path from Makwasinyi Village, 1100 m, 13 Feb. 1996, *Masinde* 840 & 841). (iii) Plants with terete or weakly many angled stems; sepals broad and long, overlapping especially in flower buds; peduncles quite long and descending; corolla lobe cage very broad and mostly ± apically truncate with the cage openings oriented to the top. Flowers mostly creamy throughout or with reddish-maroon spots in throat, occurring mostly in lowlands of **K** 7 area and able to survive in waterlogged black cotton soil (e.g. Kwale District: between Mariakani and Kinango, 27 Mar. 1964, *Archer* 481).

47. **Ceropegia denticulata** *K. Schum.* in P.O.A. C: 327, t. 40 F. (1895); K. Schum. in E. & P. Pf. 4(2): 272, fig. A (1895); N.E. Br. in F.T.A. 4 (1): 448 (1903); Werderm. in E.J. 70: 201 (1939); Bullock in K.B. 8: 58 (1954); Archer in U.K.W.F.: 391 (1974) & Kenya *Ceropegia* Scrapb.: 103, XXI (1992); Blundell, Wild Fl. E. Afr.: 145 (1987); U.K.W.F. ed. 2: 183 (1994); Masinde in Cact. Succ. J. (US) 6: 107 (1999). Type: Tanzania, Lushoto District: Eastern Usambaras, Mashewa [Massaua], Silai, *Holst* 3583 (B†, holo.; K!, lecto., designated here)

Twiner to 4 m high, in dry habitats mostly leafless except for the young growth, with a cluster of whitish slightly swollen roots 40–200 × 5 mm with occasional fibrous rootlets especially at ends; latex cloudy or milky, copious; stem often branched, succulent, terete or weakly angled, 3–5(–7) mm diameter, shiny green with whitish-grey longitudinal lines and streaks, sometimes with two flat opposite sides extending for some distance below the nodes; glabrous. Leaves herbaceous or thickly and stiffly succulent; petiole 2–15 mm long; lamina narrowly to broadly ovate or elliptic, 20–85 × 10–45 mm, base cuneate, apex acuminate to acute, glabrous, margins serrulate to denticulate, sparsely ciliate. Inflorescence pseudoumbellate, 1–2-flowered, flowers developing successively, scent sweet; peduncle succulent, 3–22 mm long, 2–2.5 mm thick, suberect, weakly angled or quadrangular, glabrous; bracts subulate, 2–5 × 0.5–1 mm, glabrous; pedicel 4–13 mm. Sepals subulate or lanceolate, 2.5–6 × 0.5–1 mm, apically mostly recurved, greenish to brownish-purple, glabrous. Corolla (28–)40–50(–70) mm long; tube 19–38 mm long, in basal ¹/₃ with ovoid inflation 11–24 × 6–10 mm (sometimes with constriction in apical ¹/₃, dividing it into two), then narrowing gradually into an ascending cylindrical part 3–4 mm diameter, gradually expanding to 13–20 mm; tube exterior glabrous, from base yellowish to whitish-green, sometimes with faint purplish-maroon lines on upper chamber, above inflation green or whitish-green or yellowish flushed brown-maroon, sometimes with faint purplish-maroon longitudinal lines; throat usually with many purplish-maroon or brownish spots; lobes narrowly rectangular from deltoid base, 10–24 × 3–6 mm, plicate except for apices, forming a broad keel with a gibbous projection towards base, apices connate to form a cage with side openings; margins strongly rolled back; adaxially with dense bristle-like hairs except for apical $^1\!/_3$ part where face glabrous and margin with (2–)9 vibratile purple (rarely white) clavate hairs 3–6 mm long; adaxial faces with transverse bands, commonly starting from top, greenish, pink-green or pink-yellow then blackish, then creamy-white or yellowish, then brown-maroon or greenish with brighter green towards keel. Corona shortly stipitate, 3.5–4 × 3 mm, basally shallowly cupular, yellowish-cream to yellowish-green with purple-maroon in parts, glabrous; outer lobes forming 5 shallow pouches 0.5–0.8 × 1–1.5 mm; inner lobes linear to spatulate, 2.5–3 × ± 0.4 mm, incumbent upon anthers then erect and apically connivent, glabrous, purple-maroon in bases. Anthers subquadrate and overtopping the flattened to rounded stylar-head; guide rails ± 0.6 mm long, area around nectaries dark purple to maroon; gynostegial parts well exposed; pollinia 0.35–0.4 × 0.25–0.3 mm; corpusculum obovate, 0.2 × 0.15 apex × 0.06 mm medially. Follicles paired, pointing sideways or drooping, obtusely divergent at mostly >180°, 100–210 mm long, 3 mm diameter at centre, apically tapering, at maturity dull yellow to maroonish. Seed dark brown or blackish, 6–7 × 2–2.5 mm, with a paler margin; coma ± 25 mm long.

a. var. **denticulata**

Basal inflation not or weakly constricted to form two chambers.

Kenya. Nairobi, Kileleshwa, Nov. 1960, *Archer* s.n. in EA 12113!; Machakos District: NE slopes of Chyulu Hills near Mombasa–Nairobi road on way to microwave repeater station, 1 June 1981, *Gilbert* 6260!; Teita District: Bura Market, 11 Jan. 1995, *Masinde* 820!
Tanzania. Pare District: North Pare Hills, Dec. 1953, *Bally* S215!; Usambara, 4 Aug. 1993, *Holst* 3583!
Distr. **K** 1, 3–7; **T** 3; not known elsewhere
Hab. Bushland and forest edges; 500–1900 m

Syn. *C. nilotica* Kotschy var. *simplex* H. Huber in Mem. Soc. Brot. 12: 104 (1958) in part, excl. type

Note. *C. denticulata* is closely related to *C. nilotica* but differs by the clavate hairs on corolla lobes and the elongated, somewhat rectangular-linear corolla lobes.

b. var. **brownii** (*Ledger*) *P.R.O. Bally* in Candollea 20: 13 (1965). Type: Uganda, Mengo District: Mabira Forest, Jinja, Dec. 1908, cult. UK, Wimbledon, by Ledger, *E. Brown* 466 (K!, holo., EA!, iso.)

Basal inflation distinctly constricted to form two chambers.

Uganda. Toro District: Mweya, Queen Elizabeth National Park, 22 Dec. 1967, *Lock* 67/178! and 12 Nov. 1968, *Lock* 68/289!; Teso, Apr. 1932, *Chandler* 692!; Masaka District: 1 km E Lugalama, 16 May 1971, *Lye* 6092
Kenya. Nairobi, Karura Forest, 4 Mar. 1941, *Bally* B1457!; Kisumu-Londiani District: Muhoroni, 2 May 1964, *Archer* 475!; Masai District: Masai Mara Game Reserve, 11 July 1979, *Gilbert* 5746!
Tanzania. Lindi District: Rondo Plateau, Rondo Forest Reserve, 10 Feb. 1991, *Bidgood, Abdallah & Vollesen* 1466! & 14 Feb. 1991, *Bidgood et al.* 1567! & 3 km NW of Rondo Forest Reserve station, 15 Feb. 1991, *Bidgood et al.* 1588!
Distr. **U** 2–4; **K** 3–7; **T** 8; not known elsewhere
Hab. Bushland and forest edges; 700–2100 m

Syn. *C. brownii* Ledger in K.B. 1909: 326 (1909)

Note. Variety *brownii* is mainly distributed to the west of the range of var. *denticulata*. The corolla lobes are commonly narrowly triangular.

48. **Ceropegia ampliata** *E. Mey.*, Comm Pl. Afr. Austr. 194 (1837); Decne. in DC., Prodr. 8: 645 (1844); Huber in Mem. Soc. Brot. 12: 143 (1958); Archer in Kenya *Ceropegia* Scrapb.: 89, XIX (1992); Masinde in Cact. Succ. J. (US) 6: 107 (1999). Type: South Africa, Fish River Valley, hills near Trumpeter's Drift, *Drège* 4949 (P, holo., photo!; K!, W, MEL, iso.)

Climber to 2 m high, with fibrous roots; latex somewhat milky; stem often scrambling and hanging, sparsely branched, succulent, (3–)4–5 mm diameter, with very fine longitudinal ridges giving a rough feel to the touch, grey-green, glabrous, new stems grey-green/maroon and somewhat rigid, leafless except at young tips of stem. Leaves rudimentary, succulent, sessile to subsessile, mainly in terminal, younger growth, soon caducous; petiole 0–1 mm long; lamina lanceolate, 4–8(–40) × 1.5(–7) mm, V-shaped above, green with a purplish-maroon tinge, glabrous. Inflorescence sessile to subsessile, up to 4-flowered, flowers developing successively, without discernible scent; peduncle 0–1 mm long; bracts subulate 0.3–1 mm long, glabrous; pedicel 6–15 mm long. Sepals lanceolate, glabrous, 2–3 × ± 0.5 mm, resting on ridges in between the 5 shallow indentations at base of inflation. Corolla 20–35 mm long; tube 11–24 mm long, in basal ± $\frac{1}{3}$ with subglobose inflation 8–18 × 5–12 mm, then slightly curved and slightly or distinctly narrowing into an ascending cylindrical part 4–8 mm diameter, gradually dilated in throat to 9–11 mm diameter; exterior white to greenish-white or cream with numerous pale green longitudinal veins running from base to throat, glabrous; lobes 5–12 mm long, curved back, narrowly lanceolate from broad deltoid bases, claw-shaped, connate at apices to form a conical to semi-spherical cage; adaxially and abaxially bright olive-green with shiny yellow in adaxial bases and with an emerald green splodge on the strongly rolled back margin at the basal sinuses of adjacent lobes, completely glabrous throughout. Corona shortly stipitate, ± 5 × 2.5 mm, basally shallowly cupular, greenish-yellow in most parts; outer lobes with a little maroon on edge at centre, forming 5 shallow pouches ± 0.5 × 0.8 mm, apically broadly bifid, with a tuft of 0.5 mm long hairs; inner lobes filiform, tall and delicate, ± 3.5 mm long, basally shortly incumbent on anthers then erect and free or connivent and sometimes apically slightly recurved, glabrous, translucent white with greenish at base. Pollinia ± 0.5 × 0.29 mm; corpusculum ± 0.27 × 0.11 mm at apex × 0.07 mm in middle. Follicles paired, divergent at acute angle, 100–110 mm long, 4–6 mm diameter, pink-grey with maroon spots. Seed 5–6 × 2–2.5 mm, brown with light brown margins; coma 25–30 mm long.

Kenya. Kwale District: Shimba Hills, 22 Nov. 1962, *Archer* 401!; Teita District: Mulemwa Hill, W of Kasigau Hill, 22 June 1968, *Archer* 558! & Rukanga–Maungu, 1967, *Archer* s.n.!
Tanzania. Uzaramo District: Msasani, N of Dar es Salam, 24 May 1939, *Vaughan* 2806!
Distr. **K** 7; **T** 6; South Africa, Madagascar
Hab. Coastal forest edges and bushland in rocky ground; 100–1000 m

Syn. *C. ampliata* E. Mey. var. *oxyloba* H. Huber in Mem. Soc. Brot. 12: 143 (1958). Type: Tanganyika, Uzaramo District: Msasani, N of Dar es Salaam, *Vaughan* 2806 (BM!, holo.; EA!, iso.)

Note. A rare taxon in East Africa. Vegetatively, *C. ampliata*, *C. arabica*, and *C. laikipiensis* are hardly distinguishable. They all have clusters of fusiform roots; readily rooting succulent stems with fine longitudinal ridges which are rough to the touch; rudimentary, sessile or subsessile, lanceolate, deciduous leaves and sessile or subsessile inflorescences. *C. ampliata* may however be distantly related to the rest of the species since it has flowers which are unique in their morphology (corolla and corona) and colouration.

49. **Ceropegia arabica** *H. Huber* in Mem. Soc. Brot. 12: 138 (1957). Type: 'Arabia', *Ogilvie-Grant* 759 (E, holo.)

Twiner to 3 m high, mostly leafless, glabrous, with a cluster of spreading creamy-white fusiform roots ± 10 × 6 mm; latex clear but turning cloudy on exposure; stem climbing but sometimes stiff and unable to twine, sparsely branched, succulent, 2–6 mm in diameter, grey-green or purplish-green in older growth, minutely ridged with very fine longitudinal lines and sometimes minute scabrid along ridges, glabrous. Leaves stiffly succulent, suberect or almost clasping stem, rapidly caducous, sessile; lamina 4–5 × 1.5–2 mm, acute, glabrous. Inflorescence of umbellate cymes, 1–3(–5)-flowered; flowers developing successively, one flower open at a time, scent sweetish or citrus but in some plants not detectable; peduncle 0–4(–7) mm long, ± 1 mm thick; bracts subulate, ± 1 × 0.3 mm; pedicels 5–10 mm long. Sepals subulate, 2–5 × 0.7–1.5 mm, pink-cream/green-brown with faint spots. Corolla 28–47(–54) mm long; tube 20–42 mm long, with ovoid or subglobose inflation 8–15 × (4–)6–13 mm, then slightly curved and narrowing into cylindrical part 2–3(–4) mm in diameter, gradually expanding to 6–15 mm in throat; silvery green-cream with greyish-black speckles/spots especially on inflation and apical part of tube, somewhat weakly ridged on inflation, or greyish-purple with weak longitudinal veins but not spotted, glabrous; lobes triangular or linear-acute, 8–24 × 1.5–4 mm, sharply plicate with plate-like keels but well apart, lower edge occasionally with sparse hairs, longitudinal edges close but only touching at apex where lobes unite, adaxially in apical $^1/_3$–$^2/_3$ pale green, grey-green, grey-yellow, greenish-purple or rarely red merging into a small irregular blackish band, then a grey-black band which may have purple-maroon on edge, at base greenish-cream with purplish or blackish speckles and sometimes sparse fine whitish hairs; margin in basal $^1/_3$ velvety black/purple and mostly strongly rolled back and glabrous, then along the coloured part above the small bright band, wavy with 2–8 mm long clavate vibratile greyish-grey/purple/black hairs arising from fine grooves along margins. Corona distinctly stipitate, 2–4 × ± 3 mm, basally cupular, greenish/yellowish-cream with purplish flecks or deep maroon in most parts, glabrous; outer lobes triangular, apically bifid, basally forming pouches 0.7–1 × 1 mm; inner lobes cylindrical, connivent-erect, 0.5–3 × 0.1–0.2 mm, vestigial in which case they are narrower and shorter than anthers or connivent-erect and mostly longer than outer lobes or ± equalling outer lobes, apex rounded. Pollinia 0.34–0.4 × 0.23–0.27 mm; corpusculum obovate, orange, 0.2–0.28 × 0.05–0.08 mm at centre. Follicles paired, acutely divergent 7.5–13 cm long × 3–7 mm at centre, pale grey-green with pale grey or pink-maroon lines and spots. Seed dark brown, 5–6 × 2–3 mm, with a pale marginal wing; coma 18–30 mm long.

var. **powysii** (*D.V. Field*) *U. Meve & R.M. Mangelsdorff* in J.L.S. 137: 105 (2001). Type: Kenya, Laikipia District, Lower Narok Farm, 35 km N of Rumuruti, on track to Uaso Narok River, *Field & Powys* 115 (K!, holo.). Fig. 72, p. 289.

KENYA. Masai District: Ol Esakut Hill, 11 km SW of Ngong Hills, May 1960, *Archer* 469; Laikipia District: Lower Narok Farm, 35 km N of Rumuruti on track to Uaso Narok River, 22 May 1977, *Field & Powys* 115!; Teita District: Mt Maktau, 14 Feb. 1996, *Masinde* 850!
DISTR. **K** 3, 6, 7; Ethiopia
HAB. Bushland and thickets in rocky places; 1200–1950 m

SYN. *Ceropegia* sp. A, Archer in U.K.W.F.: 391 (1974)
    C. powysii D.V. Field in K.B. 37: 308 (1982); Archer in Kenya *Ceropegia* Scrapb.: 77, 83, XVII & XVIII (1992); U.K.W.F. ed. 2: 183 (1994); Masinde in Cact. Succ. J. (US) 6: 107 (1999)
    C. barbigera Bruyns in K.B. 44: 271 (1989); Archer in Kenya *Ceropegia* Scrapb.: 83, XVIII in part (1992); Masinde in Cact. Succ. J. (US) 6: 107 (1999); M. Gilbert in Fl. Somalia 3: 169 (2006). Type: Ethiopia, Harar Region, *Lavranos & Gilbert* 9241 (K!, holo.)
    [*C. powysii* D.V. Field var. *gigantica* Masinde, *in sched.*, *nomen nudum* based on Kenya, Taveta District, base of Mt Maktau, *Masinde* 851 (EA!)]

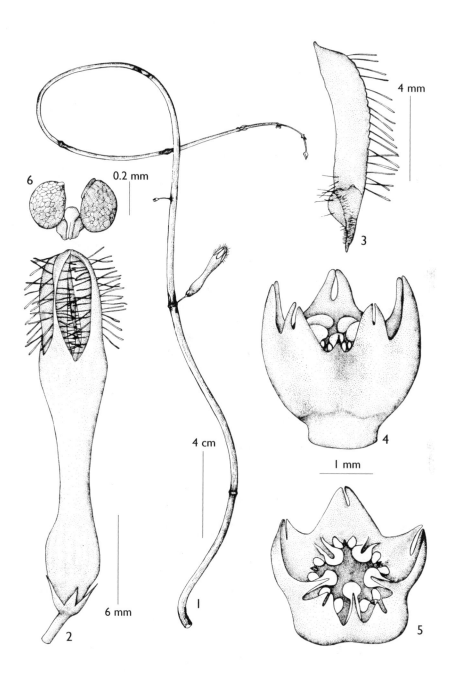

FIG. 72. *CEROPEGIA ARABICA* var. *POWYSII* — **1**, habit; **2**, flower; **3**, corolla lobe, lateral view; **4**, gynostegium and corona, lateral view; **5**, gynostegium and corona, top view; **6**, pollinarium. All from *Masinde* 828. Drawn by N. Muema based on pencil drawings by S. Masinde.

NOTE. Closely related to *C. laikipiensis*. A very variable species in its flower size and morphology but vegetatively fairly uniform. Plants with flowers that have larger flowers with distinctly dilated basal inflation are what was known as *C. barbigera*. Those with smaller flowers with a slightly dilated basal inflation are what was known as *C. powysii*. In between these extremes are many intermediate morphological forms that do not seem to lend themselves to a clear classification.

50. **Ceropegia laikipiensis** *Masinde* in K.B. 59: 241 (2004). Type: Kenya, Longopito, *Powys* 20 (K!, holo., incl. spirit mat.)

Twiner to 1 m high, mostly leafless, glabrous; latex unknown; root system unknown; stem climbing, sparsely branched, succulent, ± 2.5 mm in diameter, grey-green, minutely ridged with very fine longitudinal lines. Leaves sessile, minute, succulent, caducous, only found near growing apex; lamina subulate, 2 × 0.5–1 mm, acute. Inflorescence umbellate cymes, 1–3-flowered, flowers developing successively, one flower open at a time, scent unknown; peduncle 0–4 mm long, thickening to 1.5–2 mm in diameter; bracts subulate, ± 1 × 0.4 mm; pedicels up to 4 mm long. Sepals subulate, 3–4 × 0.5–0.7 mm, clasping corolla. Corolla 28–38 mm long; tube 14–22 mm long, in basal $^1/_4$–$^1/_3$ with ovoid inflation 5–7 × 4–5 mm, then abruptly narrowing into a long ascending cylindrical part ± 2 mm in diameter, curved through ± 120° in the middle and gradually dilating to ± 5 mm in diameter and suddenly narrowing to ± 4 mm in diameter just at bases of corolla lobes; exterior with a light background, unspotted or spotted purplish-maroon especially on inflation and in the apical half of tube, inflation with weak longitudinal veins, glabrous throughout; lobes linear from narrowly deltoid bases, 14–17 × ± 2 mm and narrowing to ± 0.7 mm for most of the length, not plicate or only hardly so, connate at apices to form a globose cage 2–3 times as wide as the tube; adaxially at bases with a dense cover of 2.5–3 mm thick inward-pointing purplish-maroon vibratile clavate hairs; rest of lobe glabrous including abaxial side and margins, light-coloured for most part. Corona distinctly stipitate, ± 3.5 × 2.5 mm, basally cupular, yellowish-cream; outer lobes triangular, apically deeply bifid basally forming pouches ± 1 × 0.7 mm; inner lobes connivent-erect or ± free, ± 2.6 × 0.15 mm. Pollinia ± 0.38 × 0.28 mm; corpusculum ± 0.24 × 0.1 mm at centre. Follicles paired, acutely divergent, 70–100 mm long × ± 4 mm at centre, tapering towards bases and apices, drying straw coloured. Seed ± 4 × 2 mm, brownish with a ± 0.5 mm wide, paler, corky marginal wing; coma of silky white hairs ± 20 mm long.

KENYA. Northern Frontier District: Barsalinga drift on S bank of Uaso Nyiro River, no date, *Powys* s.n.!; Laikipia District: Longopito, 1975, *Powys* 20! & Longopito, no date, *Powys* s.n!
DISTR. **K** 1, 3; not known elsewhere
HAB. Semi-arid bushland/woodland; ± 1700 m

SYN. *Ceropegia* sp., Archer in Asklepios 62: 8, b/w photo only (1994)

NOTE. The stems and general appearance are very close to *C. powysii* but the distribution of hairs in the interior of the corolla tube and lobes is similar to that in the poorly known *C. subaphylla*. The curvature in the middle of the tube is reminiscent of *C. botrys* K. Schum. or *C. tihamana* Defl.

51. **Ceropegia galeata** *H. Huber* in Mem. Soc. Brot. 12: 200 (1958); Archer in Kenya *Ceropegia* Scrapb.: 73, XVI (1992); Masinde in Cact. Succ. J. (US) 6: 107 (1999). Type: Kenya, Kwale District: Kinango, cultivated in Likoni - Mombasa, *Bayliss* 31 (K!, holo., spirit mat.)

Twiner to 2 m high, mostly leafless, glabrous, with a cluster of fibrous rhizomatous roots, 20–60 mm long, from which fusiform roots up to 160 × 5–10 mm arise; latex clear to milky; stem twining, sparsely branched, succulent and fairly stiff, 3–4.5 mm in diameter, grey-green or green, minutely ridged with very fine longitudinal lines, scabridulous. Leaves rudimentary, minute, sessile, succulent, rapidly caducous; lamina subulate, 3–5 mm long, glabrous. Inflorescence umbellate cymes, 1–5-flowered, flowers developing successively, up to 3 flowers open at a time, scent reminiscent of geranium; peduncle 0–2 mm long; bracts subulate, 1.5–2 × 0.5 mm, greyish or maroonish; pedicel 10–18 mm long. Sepals subulate, 4–6 × 1–1.5 mm, often apically recurved, abaxially pinkish-yellow with a few maroon spots and streaks. Corolla 35–55 mm long; tube 24–30(–40) mm long, straight or slightly curved above inflation, in basal $\frac{1}{3}$ with prominent ovoid inflation 18–24 × 14–16 mm, then narrowing abruptly to ± 4 mm diameter and then dilated to 12–16 mm diameter in throat; pale beige mottled with pale olive green, paler in throat, inflation with longitudinal contours of pale maroon spots, apical part of tube just below lobes spotted pale grey-maroon, glabrous; lobes 18–25 × 7 mm, arising from broadly deltoid bases, hardly plicate, $\frac{1}{4}$ way up narrowing to ± 2 mm, completely folded back and constricted to give the cage a ± 9 mm diameter, then opening out rapidly, to ± 13 mm wide and curved inwards at a level $\frac{3}{4}$ way up, eventually narrowing and connate towards apices to form a cage with an umbrella-like canopy ± 22 mm in diameter at its widest; adaxially in basal $\frac{1}{4}$ with two colour bands, white on inside with fine whitish hairs and black on outside with fine blackish hairs, with an olive-green spot at the constricted part, umbrella-like part yellowish or creamy-green and glabrous with recurved margins lined with ± 3 mm long, readily vibratile, pale grey-maroon clavate hairs; abaxially creamy mottled in pale olive green or pale grey-maroon, glabrous. Corona shortly stipitate, 3.5–4 × 3(–5) mm, basally cupular, with creamy maroon spots, glabrous; outer lobes rectangular, 1.2 mm high, bifid in apical $\frac{1}{3}$, basally forming pouches ± 0.7 × 1 mm; inner lobes cylindrical, 2.5–4 × 0.2 mm, apices connivent. Anthers subquadrate, overtopping and resting on the rounded stylar head; pollinia ± 0.4 × 0.25 mm; corpusculum ± 0.25 × 0.15 at apex × 0.07 mm at centre. Follicles paired, acutely to obtusely divergent, straight but sometimes curved outwards, 85–190 mm long × 6–9 mm at centre, rather flat on inward facing lateral surfaces, pale grey-green with grey-maroon streaks and blotches which may be slightly raised. Seed 7–9 × 4–4.5 mm, dark brown with pale brown marginal wing; coma of white silky hairs 25–47 mm long.

KENYA. Kwale District: km 58 on Mombasa–Tanga road, 1960, *Bally* 12184! & 5 km S of Kinango, 1 Nov. 1957, *Bayliss* 3!; Kwale/Kilifi Districts: Maji ya Chumvi, W of Mombasa, Oct. 1955, *Bally* 10432A = *Bally* S220!
DISTR. **K** 7; not known elsewhere
HAB. Semi-arid bushland; 50–250 m

NOTE. A rare species only known from old collections. *C. galeata* is allied to the *C. arabica* complex due to the close similarities in vegetative form: root system, stem and rudimentary leaves; and floral form: corolla tube and clavate hairs. The major difference is in the canopy- or umbrella-shaped top of the corolla cage that is reminiscent of *C. sandersonii* Hook. f. of Mozambique, South Africa and Swaziland.

EXCLUDED SPECIES

Ceropegia tihamana Chaudhary & Lavranos in Notes Roy. Bot Gard. Edin. 42: 316 (1985). Type: Saudi Arabia, Abu Arish, near Jizan, *Chaudhary* 901A (RIY, E)

Bruyns in his account of *Ceropegia* in Arabia (Notes Roy. Bot. Gard. Edin. 45(2): 306 (1989)) records *C. tihamana* as occuring in Kenya. His report was based on plants collected by Patricia and Gilfrid Powys. No voucher specimens were cited and there is no indication as to where the specimens are kept. Therefore no specimens referred to this species were located during the present investigation. The available data are insufficient to allow for the inclusion of this species.

57. BRACHYSTELMA

Sims in Bot. Mag. 49, t. 2343 (1822), *nom. cons.*, excl. syn.; Bentham in G.P. 2: 781 (1876); Schlechter in E.J. 18, Beibl. 45: 52 (1895) & Beibl. 51(2): 144 (1913) & in J.B. 35: 292 (1897); Schumann in E. & P. Pf. 4(2): 263 (1897); N.E. Br. in F.T.A. 4(1): 466 (1903); Dyer, *Brachystelma, Ceropegia* & *Riocreuxia* in S. Afr. 12 (1983); Walker in Asklepios 25: 92–106 (1982); Boele et al, Checklist *Brachystelma, Ceropegia, Riocreuxia* & Stapelieae (1987) & Suppl. (1990); Meve & Porembski in E.J. 115: 315 (1993); L.E. Newton in Ballya 1(4): 78 (1994); Meve in Illustr. Handbook Succ. Pl. *Asclepiadaceae:* 20 (2002) & in Checklist *Brachystelma, Ceropegia* & genera of Stapeliads: 116 (2007); Masinde in K.B. 62: 47 (2007)

Tenaris E. Mey., Comm Pl. Afr. Austr.: 198 (1838); Decne. in DC., Prodr.: 606 (1844); Bentham in G.P. 2: 775 (1876); Victor & Nicholas in S. Afr. Journ. Bot. 64: 205 (1998)

Geophytic perennial herbs, usually dwarf, with a single tuber or a cluster of fleshy, fusiform roots; latex clear; stems prostrate or procumbent to erect, never twining, single or variously branched. Leaves opposite, herbaceous, never rudimentary, sessile or with a short petiole, without glands at base. Inflorescence lateral or terminal, pedicellate, 1–many-flowered umbellate cyme, rarely a raceme-like pedunculate panicle (e.g. *B. rubellum*). Sepals without basal glands within. Corolla with united base, sometimes ± tubular, mainly campanulate to shallowly bowl-shaped, tube rarely longer than lobes; lobes free at tips and rotately spreading, rarely connate at tips (e.g. *B. floribundum* var. *mlimakito* and *B. gracile*), flat or replicate, broadest at base, rarely broader at apex (e.g. *B. rubellum*). Corolline corona absent. Gynostegial corona arising from staminal column, sessile or subsessile, 2-seriate or sometimes appearing 1-seriate; interstaminal (outer) parts cupular (with the inner corona lobes arising from within), disciform, or variously lobed or toothed or pocket-like and confluent with lateral margins of inner lobes; staminal (inner) parts incumbent on backs of anthers and occasionally elongated above staminal column, sometimes reduced to cushion-like swellings at bases of anthers. Anthers 2-locular, oblong or subquadrate, without appendages, incumbent or inflexed on the conical-convex to flattened stylar head at centre of staminal column. Pollinia subglobose, oblong or suborbicular, often D-shaped; subhorizontal or ascending, solitary in each anther theca, pellucid on inner margin to varying degrees, attached in pairs by short delicate caudicles to a hard, often winged, corpusculum. Follicles fusiform, linear-fusiform or ellipsoid, smooth, green, sometimes mottled. Seeds oblong, convex on one side, flat or concave on other, with narrow marginal wing, crowned with a coma of tufted long hairs at the chalazal end.

100–120 species distributed entirely in the Old World, predominantly in southern Africa with ± 70 species; also in India, SE Asia and one species in Australia.

Widespread, but most taxa are rare and poorly represented in herbaria. Tubers of *B. gracile*, *B. johnstonii*, *B. plocamoides*, *B. simplex* and *B. rubellum* are consumed by humans. Competition for food by humans and wild animals contributes to the rarity of the species.

1. Stems erect, long and slender; vegetative parts
   glabrous; mature internodes usually 100 mm
   long and over; leaves sessile, very narrowly
   linear-acute and stem-like, usually not more
   than 2 mm wide; corolla rotate, tube ± obsolete 2

   Stems erect, prostrate or decumbent but not long
   and slender; vegetative parts mostly pubescent;
   mature internodes to 50 mm long at most but
   mostly less than 20 mm; leaves sessile or
   petiolate and broad but if linear-acute then
   mostly not elongate and stem-like: corolla not
   rotate, tube at least distinct 3

2. Root a globose tuber, inflorescence terminal,
   paniculate; corolla lobes mostly spatulate, pale
   mauve to pink . . . .   13. *B. rubellum* (p. 306)

   Root a fusiform cluster, inflorescence lateral,
   umbellate; corolla lobes filiform, dark purple
   to greenish . . . .   14. *B. arachnoideum* (p. 308)

3. Plants stout; leaves oblong, oblanceolate, elliptic
   or combinations of these but not linear, mostly
   > 30 mm long; stems mostly 3–4 mm in diameter 4

   Plants not stout; leaves linear, or if non-linear
   then leaves elliptic, ovate or obovate and mostly
   < 30 mm long (to 75 mm in *B. lineare*), stems
   mostly < 3 mm in diameter 6

4. Inflorescence lateral, up to 8-flowered at most,
   flowers opening successively, not forming a
   head, lobules of outer corona lobes densely
   covered with needle-like hairs except for the
   very tips . . . .   8. *B. floribundum* (p. 300)

   Inflorescence a terminal umbellate head to
   60 mm diameter with 20–30 flowers opening
   simultaneously; outer corona lobes not as above 5

5. Plant dwarf to 10 cm high; inflorescence produced
   when leaves absent; corolla lobes connate at
   apex to form a globose cage, corolla interior
   mostly liver-red or dark-brown . . . .   8. *B. floribundum* (p. 300)

   Plants stout and tall to 30 cm high; inflorescence
   produced when leaves present; corolla
   campanulate, lobes free and spreading, interior
   with yellow and liver-red concentric zonation . . .   7. *B. buchananii* (p. 300)

6. Stems decumbent, prostrate or procumbent 7
   Stems erect 8

7. Leaves obovate or oblong; sepals conspicuous,
   much exceeding corolla lobe bases; corolla
   lobes adaxially uniform green in long narrowly
   linear posterior part, glabrous, often united at
   apex to form a cage . . . .   1. *B. megasepalum* (p. 295)

   Leaves ovate to circular; sepals tiny, not exceeding
   corolla lobe bases; corolla lobes adaxially spotted,
   posterior part linear-oblong, lobes spreading,
   free and never uniting to form a cage . . . .   2. *B. coddii* (p. 296)

8. Corolla lobes adaxially densely covered with long
   vibratile hairs 9
   Corolla lobes adaxially glabrous or puberulent but
   without vibratile hairs 11
9. Leaves broadly linear or lanceolate-elliptic; flowers
   not congested, extra-axillary in lower nodes;
   apical part of corolla lobes or whole length with
   hanging, readily vibratile hairs which are beaded
   towards apex; outer corona saucer-shaped .....   4. *B. tavalla* (p. 297)
   Leaves mostly very narrowly linear; flowers
   congested and appearing verticillate in the axils
   of the terminal or subterminal pairs of leaves;
   outer corona not saucer-shaped 10
10. Corolla 60–100 mm in diameter; corolla lobes
    tapering from deltoid bases into very long linear
    or linear-filiform tentacle-like tails; adaxial faces
    densely covered with white woolly hairs ...... 10. *B. johnstonii* (p. 302)
    Corolla 20–30 mm in diameter; corolla lobes
    triangular, not tapering into long tails; adaxial
    surface with a felt of purple hairs .... 9. *B. festucifolium* (p. 301)
11. Corolla lobes plicate for most of length into linear-
    filiform processes, united at apex to form an
    oblong cage ....   16. *B. gracile* (p. 310)
    Corolla lobes not plicate, free at apex and mostly
    spreading 12
12. Corolla lobes never appearing to be fully open,
    thick and fleshy, slightly to strongly folded back
    along midrib, outer corona saucer-shaped .....   5. *B. plocamoides* (p. 298)
    Corolla lobes always fully open, not thick and
    fleshy or folded back 13
13. Plants with few broad leaves, leaves obovate or
    elliptic, densely pilose on both surfaces 14
    Plants with narrowly linear leaves, usually glabrous
    on both surfaces (except *B. simplex*, both
    surfaces pubescent) 15
14. Corolla stellate, ± 70 mm diameter in total, exterior
    yellowish-green, papillose-hispid; tube dish-
    shaped, ± 3 mm diameter at mouth, 1.5(–3) mm
    deep, not completely enclosing the corona;
    exterior red, white puberulent; interior light red
    with few whitish spots ....   3. *B. lancasteri* (p. 296)
    Corolla not stellate, ± 8 mm diameter in total; tube
    cup-shaped, 5–7 mm in diameter at mouth,
    5–6 mm deep, enclosing the corona; exterior
    purple, glabrous; interior yellow marked with
    many concentric dark purple spots ....   6. *B. maritae* (p. 299)
15. Stems, leaves and sepals densely pubescent; corolla
    ± 9 mm diameter in total, greenish-yellow; outer
    corona lobes forming a saucer-shaped structure
    with margins entire or undulate ....   15. *B. simplex* (p. 309)
    Stems, leaves and sepals not as above; corolla over
    9 mm diameter, outer corona lobes forming a
    cup-shaped structure 16

16. Several stems usually < 50 mm tall; leaves usually
    elliptic; inflorescence 1–3-flowered; flowers
    with long wiry pedicels well away from nodes;
    corolla ± 10 mm, glabrous throughout, white or
    greenish-yellow with purple flushes . . . . 12. *B. keniense* (p. 305)
    Single or few stems usually > 50 mm tall; leaves
    usually linear; inflorescence usually of two
    opposite sessile umbels of up to 8 flowers each,
    appearing verticillate; flowers with short non-
    wiry pedicels close to nodes; corolla ± 50 mm
    diameter, puberulent on interior, variably
    coloured with purple, greenish-yellow, cream
    with brown or black spots . . . . 11. *B. lineare* (p. 303)

1. **Brachystelma megasepalum** *Peckover* in Kakteen u. a. Sukk.: 249 (1996); Meve in Illustr. Handbook Succ. Pl. *Asclepiadaceae*. 35, t. III, d (2002); Masinde in K.B. 62: 50, fig 1 (2007). Type: Tanzania, District unclear, Ruvuma S of Mpepo near Mozambique border, originally collected in 1992 and cultivated in various countries, type specimen probably prepared in 1996, *M. & E. Specks* 385 (PRE, holo.)

Herb to 20 cm high; root a discoid tuber 50–60 mm in diameter; stem one to several, with few procumbent branches, purplish-green, 1–1.5 mm in diameter, ± hispid all round. Leaves herbaceous; petiole 2–3(– 15) mm long; lamina obovate or oblong, 10–30 × 10–18 mm, base obtuse to ± rounded, apex obtuse or rounded, apiculate; both surfaces minutely pubescent especially on the veins and midrib beneath, margin undulate or not, ciliolate. Inflorescence extra-axillary, sessile, arising from apical nodes, up to 4-flowered, flowers almost all opening simultaneously, emitting a putrid smell reminiscent of dung; bracts subulate, ± 2 × 0.3 mm, pubescent; pedicels 14–30 mm, shortly pubescent. Sepals spreading, subulate, 8–10 × 2–3 mm, densely pubescent at least on the midvein and towards apex. Corolla in total ± 70 mm diameter, lobed ± to the base, exterior pubescent, interior glabrous throughout; tube campanulate, shallow dish-shaped, 2–3 mm deep, 6–12 mm in diameter at mouth, partially enclosing corona, interior brown-violet with yellow-green spots, glabrous, exterior similar in colour to interior but paler; lobes 25–38 × 2.5–3 mm, deltoid at the base and narrowing abruptly into conspicuously long linear-oblong processes with acute apex, apically united to form a globose cage, or free and spreading, adaxial surface yellow-green, margins revolute. Corona yellowish with red-brown mottling, globular, ± 3.5 × 2 mm, sessile, biseriate, glabrous throughout; outer lobes cupular, margin deeply incised at centre into U-shaped sinuses, 10-toothed, teeth divergent, erect, deltoid-subulate, ± 0.6 mm long; inner lobes linear-oblong, 1–1.5 × ± 0.4 mm, incumbent on backs of anthers and equalling or surpassing them; guide rails ± 0.31 mm long, well exposed. Pollinia ± 0.4 × 0.3 mm; corpusculum obovate, ± 0.32 × 0.07 mm. Follicles paired, thickly fusiform, ± 50 × 15 mm, glabrous; seed dorsiventrally compressed, ± 10 × 7 mm.

TANZANIA. Songea District: Kitumbaloma, no date, *M. & E. Specks* 748!
DISTR. **T** 8; Zambia
HAB. Grassland and bushland on rocky hills with shallow alkaline soil; 1500–1700 m

NOTE. *B. megasepalum* is probably related to *B. coddii* subsp. *kituloense* and other southern
  African relatives of *B. coddii* such as *B. pulchellum* (Harv.) Schltr. and *B. bruceae* R.A. Dyer. *B.
  megasepalum* has relatively large elegant flowers and makes an attractive plant in cultivation.

2. **Brachystelma coddii** *R.A. Dyer* in Fl. Pl. Africa 30: t. 1181 (1955). Type: Swaziland, Piggs Peak, *Codd* 7826 (PRE, holo.)

Dwarf herb 3.5–12 cm high; root a discoid tuber 20–40(– 200) mm in diameter; stem procumbent, branching near the base, branches renewed annually, 1–2 mm in diameter, pubescent all round, nodes slightly swollen. Leaves herbaceous; petiole 1–3(–6) mm; lamina broadly ovate to orbicular, (6–)9–20 × 4–16 mm, base obtuse to ± rounded, apex obtuse or rounded, minutely pubescent especially on the veins and midrib beneath, margin undulate or not, ciliolate. Inflorescence axillary, sessile, 1–2(–3)-flowered; bracts subulate, ± 1.5 mm long, abaxially pubescent; pedicels 3–13(–16) mm, shortly pubescent. Sepals subulate, 2.5–4 × 0.5–1 mm, minutely pubescent at least on the midvein and towards apex. Flower buds at maturity ± 10 mm long, basal half distinctly globular and abruptly narrowing into a broadly triangular beak-like apex. Corolla 10–11 mm long, 8–15 mm in diameter, exterior glabrous throughout; tube campanulate, dish-shaped, 3–4 mm deep, 4–7 mm in diameter, enclosing the corona; exterior probably yellowish, glabrous, interior cream with red, brown, or black spots, glabrous; lobes 7–8 × 2.5–3 mm, deltoid basal part greenish-yellow and narrowing abruptly into linear-oblong processes with acute apex, adaxial surface becoming purple-red or brown towards the apex, covered with spreading white hairs, abaxially yellow. Corona discoid, ± 3.5 × 2.5 mm, sessile, biseriate but appearing as if uniseriate, dark brown, pubescent in parts; outer lobes separated by deep V-shaped sinuses, ± 1.2 mm high, fused for ± 0.5 mm at the base, ± square with paired tooth-like auricles, the inner surface of these auricles with stiff white- or purple tinged hairs; inner lobes linear-oblong, 1–1.5 mm long, incumbent on backs of anthers and slightly exceeding them; guide rails ± 0.5 mm long; carpels glabrous. Pollinia ± 0.38 × 0.28 mm; corpusculum elliptic ± 0.28 × 0.12 mm, with broad, basal lateral membranes. Follicles paired, acutely divergent, fusiform, tapering at both ends, ± 60 × 4 mm in diameter at centre; seed not known.

subsp. **kituloense** (*Goyder*) *Masinde* in K.B. 62: 50, fig. 2 (2007). Type: Tanzania, Njombe District: Kitulo Plateau above Matamba, *Brummitt & Goldblatt* 18092 (K! holo., EA!, DSM, MO photo!, PRE photo! iso.)

TANZANIA. Mbeya/Njombe Districts: Kitulo [Elton] Plateau, Ndumbi River, *Richards* 7693! & Kitulo Plateau, *Richards* 18501! & Kitulo Plateau, top of ridge above Matamba, 9°01'S 33°58'E, 22 Nov. 1986, *Brummitt & Goldblatt* 18092!
DISTR. **T** 7; not known elsewhere
HAB. Short montane grassland on gritty shallow soils in granite outcrops; 2200–2800 m

SYN. *Brachystelma kituloense* Goyder in K.B. 45: 729 (1990), as '*kituloensis*'

NOTE. In the typical subspecies *coddii*, the plants are more delicate with thinner stems (± 1 mm diameter when dry), and smaller leaves and flowers. The corolla lobes and corona are mostly completely glabrous throughout; the basal deltoid part of lobes has a pattern of pale circular white or cream markings or none at all in which case the corolla is in different shades of red or maroon throughout; the narrower distal part of lobes are narrowly triangular or linear-subulate and ± equal in length to the basal deltoid part. *B. coddii* belongs to the southern African *B. pulchellum* complex comprising *B. bruceae*, *B. tuberosum* R. Br., *B. foetidum* Schltr., *B. modestum* R.A. Dyer and *B. ngomense* R.A. Dyer among others. On *Brummitt & Goldblatt* 18092 it is noted that the tuber is about 20 cm in diameter. However, all the tubers on the sheet are only up to 4 cm in diameter at the maximum.

3. **Brachystelma lancasteri** *Boele* in Excelsa 16: 30 (1993); L.E. Newton in Bradleya 14: 96 (1996); Meve in Illustr. Handbook Succ. Pl. *Asclepiadaceae:* 33 (2002); Masinde in K.B. 62: 52, fig. 3 (2007). Type: Zimbabwe, Bulawayo, near Bulawayo Station, *Rogers* 5444 (BOL, holo. photo!; SRGH, iso.)

Dwarf, erect, reddish-white, pubescent herb, (5–)11–16 cm high; root a depressed-discoid tuber 30–70 mm in diameter. Stem sparsely branched near base, 1–1.5 mm in diameter, pubescent all round. Leaves herbaceous; petiole 0.2–5 mm; lamina broadly elliptic, 5–35 × 3–21 mm, base obtuse, apex acute, densely white-puberulent especially on the veins and the prominent midrib beneath or sometimes upper surface glabrous; margin ciliolate. Inflorescence extra-axillary in apical parts of stem, sessile, up to 6-flowered, 1–2 flowers opening at a time; bracts subulate, ± 0.6 mm long, abaxially pubescent; pedicels 1–4 mm, puberulent. Sepals subulate or triangular, 2–3 × ± 1 mm, abaxially white-puberulent. Corolla in total ± 70 mm in diameter, exterior yellowish-green, papillose-hispid; tube campanulate, dish-shaped, 1.5–3 mm deep, ± 3 mm in diameter at mouth, not completely enclosing the corona, exterior red, white-puberulent; interior light red with few whitish spots; lobes ± 8 × 2 mm, deltoid in basal quarter and narrowing gradually into linear-oblong semi-erect or rotate spreading or reflexed processes, strongly folded back along midrib, adaxial surface light red with few whitish spots on broad basal part, dark liver-red or yellowish-green in most of the apical part, glabrous or thinly puberulous, abaxially red, white-puberulent. Corona globular, ± 1.8 × 1 mm, subsessile, biseriate, yellowish and tinged ± purple or with purple margins; outer lobes cupular at base, ± 1.5 mm long, margin deeply incised at centre ± to the base to form elliptic-shaped sinuses, 10-toothed, teeth deltoid, convergent, erect; inner lobes ± 0.75 mm long, subulate, incumbent on backs of anthers and not exceeding them; guide rails ± 0.38 mm long. Pollinia 0.27 × 0.2 mm; corpusculum ± obovate-elliptic, ± 0.17 × 0.06 mm. Follicles paired, acutely divergent, fusiform, ± 60 × 4 mm diameter at centre, tapering at both ends; seed unknown.

TANZANIA. Songea District: S of Songea near Mhukuru, 21 Dec. 1993, *M. & E. Specks* 456! & *Specks* 762!
DISTR. **T** 8; Zimbabwe
HAB. Grassy Miombo woodland; ± 850 m

NOTE. *B. lancasteri* appears to be closely related to *B. richardsii*, and *B. punctatum* Boele from Zimbabwe as well as to *B. tavalla*.

4. **Brachystelma tavalla** *K. Schum.* in E.J. 28: 459 (1900); N.E. Br in F.T.A. 4 (1): 470 (1903); Percy-Lancaster in Excelsa 13: 71 (1989); Boele in Excelsa 13: 47 (1989a) & in Excelsa 14: 48 (1989b); Meve in Illustr. Handbook Succ. Pl. *Asclepiadaceae*: 44 (2002); Masinde in K.B. 62: 55, fig. 4 (2007). Type: Tanzania, Iringa District: Uhehe, near Rugaro, *Goetze* 541 (B†, holo.). Neotype: Zimbabwe, Harare, Hatfield, *Whellan* 34949 (SRGH, neo., selected by Boele (1989b: 48))

Dwarf erect herb, 23–30 cm high, pubescent, dirty olive-green in overall appearance; root a depressed-discoid tuber 10–30 mm in diameter; stem sparsely branched near base, branches spreading, 1–1.5 mm in diameter, densely pubescent all round. Leaves herbaceous, sessile or subsessile; lamina linear-lanceolate or lanceolate-elliptic, 10–30 × 3–5 mm, base cuneate, apex acuminate; both surfaces densely tomentose, margin revolute. Inflorescence extra-axillary in apical parts of stem, sessile, 1–(4)-flowered, flowers foetid or not; bracts subulate, ± 0.6 mm long, abaxially pubescent; pedicels 4–6 mm, puberulent, strongly decurved holding bud facing downwards, at anthesis becoming more horizontal. Sepals linear-lanceolate, 2–3 × ± 1.5 mm, abaxially tomentose. Corolla rotate, in total 18–20 mm in diameter, exterior violet-green; tube campanulate, dish-shaped, ± 1 mm deep, ± 2 mm in diameter at mouth, not completely enclosing the corona, exterior and interior violet-green, glabrous; lobes lanceolate, ± 8 × 2 mm, flat, rotate-spreading; adaxial surface violet, with 6–10 mm long vibratile hairs in the apical half or completely covered with purple-violet bristles; abaxially violet-green, glabrous. Corona dish-shaped, ± 3.5 × 1.6 mm, sessile, biseriate; outer lobes cupular at base, ± 1.5 mm long, margin deeply incised at centre to form elliptic-shaped sinuses, 10-toothed, teeth deltoid, ± convergent, suberect; inner lobes rudimentary and forming short bulges at bases of backs of anthers or longer to ± 0.9 mm but not exceeding anthers in length; guide rails ± 0.4 mm long. Pollinia ± 0.3 × 0.28 mm; corpusculum elliptic-linear, ± 0.25 × 0.06 mm. Follicles and seed unknown.

TANZANIA. Iringa District: Uhehe, near Rugaro, *Goetze* 541
DISTR. **T** 7; Zimbabwe
HAB. Wooded grassland; no altitude given

NOTE. The description given here is based on the protologue as well as observations on Zimbabwean material. *B. tavalla* is in the group of related species comprising *B. lancasteri, B. richardsii* and *B. punctatum*. It should be noted that the collection *Whellan* 34949 (SRGH) was selected by Boele (1989b: 48) as the neotype for *B. tavalla* but there is no indication in that paper that a duplicate is housed at K. The only specimen of *B. tavalla* that Boele (*loc. cit.*) cited as being at K is *Whellan* 34818 (Fig. 4G), for which I found herbarium and spirit specimens. There is need to reconfirm the identity of the original neotype specimen, *Whellan* 34949 at SRGH in order to set the record straight about the K and SRGH specimens as well as the typification.

5. **Brachystelma plocamoides** *Oliv.* in Trans. Linn. Soc., Bot. 29: 112, t. 77, fig. 1 (1875); Schumann in P.O.A. C: 327 (1895) & in E. & P. Pf. 4(2): 264, 268, fig. 77/H (1895); N.E. Br in F.T.A. 4(1): 470 (1903); Bullock in K.B. 17: 190 (1963); Percy-Lancaster in Excelsa 13: 71 (1988); Boele in Excelsa 14: 48 (1989); Meve & Porembski in E.J. 115: 318 (1993); Meve in Illustr. Handbook Succ. Pl. *Asclepiadaceae:* 39 (2002); Masinde in K.B. 62: 57, fig. 5 (2007). Type: Tanzania, near Tabora/Dodoma District Boundary, Uyansi, Ngunda-Nukali, Jiwa la Mkoa, 1 Jan. 1861, *Speke & Grant* s.n. (K!, holo.)

Herb 10–40 cm high, glabrous; root a globose tuber, 5–15 mm in diameter; stem erect to ascending, very leafy, repeatedly trichomotously or dichotomously branched from the base; 1–2 mm in diameter, glabrous; nodes laterally compressed, slightly swollen. Leaves herbaceous, suberect; petiole 0–3 mm long; lamina narrowly or broadly linear, 35–100 × 1–10 mm, base narrowly cuneate or attenuate, apex shortly and abruptly uncinate-mucronate or acute. Inflorescence sessile in the axils of the subterminal pair of leaves, 1–3-flowered, flowers opening consecutively, usually only one open at a time, flower buds and open flowers usually pendulous due to the downward curved pedicels; bracts linear-subulate, ± 0.5 mm long, ± glabrous; pedicels 12–14 mm long, accasionally bearing a linear bract at about their middle. Sepals linear-lanceolate, 3–4 × ± 0.7 mm, glabrous. Corolla up to 4 mm in diameter, stellate-rotate, but more often lobes not spreading hence clustered together; exterior probably purplish; tube campanulate and ± obsolete to ± 0.5 mm long, ± 3 mm in diameter at the mouth; exterior glabrous; interior dark purple, glabrous; lobes thick and fleshy, linear-lanceolate from deltoid bases, 13–20 × 1–2(– 3) mm, often slightly to strongly folded back along their length, adaxially dark purple, glabrous, abaxially glabrous. Corona cupular, ± 4 × 2 mm, sessile, biseriate, entirely glabrous; outer lobes forming a saucer-like structure ± 1.5 mm high, 10-toothed; teeth in 5 contiguous pairs, ± 0.6 mm long, rounded, strongly pleated at centre and at adjoining points behind the inner lobes where they spread outwards; inner lobes rudimentary and only at bases of anthers to 0.25 mm long, obtuse; carpels glabrous. Pollinia ± 0.4 × 0.25 mm; corpusculum linear, ± 0.2 × 0.08 mm. Follicles paired, acutely divergent, short ellipsoid, fusiform, 40–60 × 10–15 mm diameter, with with red stripes, glabrous. Seed ovate, ± 10 × 6 mm, dorsiventrally compressed, brown with a narrow pale margin; coma ± 15 mm long.

TANZANIA. Ufipa District: about half-way on Mbala [Abercorn]–Sumbawanga road, 21 Nov. 1958, *Napper* 985! & 13 km on Sumbawanga–Tunduma road, 20 Feb. 1994, *Bidgood, Mbago & Vollesen* 2349!; Iringa, no date, *Specks* 338!

DISTR. **T** 4, 4/5, 7; Congo-Kinshasa, Zambia, Malawi, Zimbabwe

HAB. Frequently burnt *Brachystegia* woodland; 1100–1900 m

SYN. *B. linearifolium* Turrill in K.B. 1914: 248 (1914): Type: Zimbabwe, without locality, cult. at Kew, *Hislop* 81 (K!, holo.)

NOTE. *B. plocamoides* is related to *B. mortonii* and probably also to *B. tavalla*. On *Speke & Grant* s.n. (K, *in sched.*), it is noted that the liquorice tasting tuber is peeled like a turnip and consumed. Peckover in Cact. Succ. J. (Los Angeles) 68: 3–5 (1996) reports that *B. plocamoides* occurs in the area around Namchwea Hill near Lake Malawi, Ruvuma Prov., near the type locality of *B. maritae*. I have not yet seen specimens or photographs of *B. plocamoides* from the area to substantiate this report. From a photograph of *B. plocamoides* in habitat, it is evident that the pedicels curve downwards and become distinctly pendant so that the buds and open flowers face the ground. The curvature of the pedicel may be a geotropic response and the disposition may be important in the pollination processes.

6. **Brachystelma maritae** *Peckover* in Cact. Succ. J. (Los Angeles) 68: 3 (1996); Meve in Illustr. Handbook Succ. Pl. *Asclepiadaceae:* 35 (2002); Masinde in K.B. 62: 57, fig. 6 (2007). Type: Tanzania, Songea District: Namchwea Hill, cultivated in South Africa, Sunnyside, *M & E. Specks* 419 (PRE, holo.)

Herb to 30 cm high; root a discoid tuber to 50–100 mm in diameter and up to 30 mm deep. Stem single, erect, 2–3 mm diameter, nodes slightly swollen, finely pubescent all round. Leaves ± subsessile, herbaceous; petiole ± 1 × 2 mm; lamina obovate, 20–45 × 15–20 mm, base acute, apex obtuse or rounded, finely pubescent especially on the veins and midrib beneath, margin ciliolate. Inflorescence in the uppermost leaf axils, usually two clusters on stem, sessile, up to 10-flowered, flowers strongly scented with a pungent odour; bracts subulate, ± 2 × 0.2 mm, abaxially pubescent; pedicels 3–4 mm long, shortly pubescent. Sepals linear-lanceolate, ± 5 × 0.5 mm, abaxially minutely pubescent. Flower buds ovoid in basal fifth that is purple-red then constricted into a green linear beak with purple-maroon splodges in parts, apex obtuse; at maturity the lobes dehisce at one fissure only for 4 to 5 days whilst the others remain fused into a tube. Corolla ± 8 mm in diameter; tube campanulate, cup-shaped, 5–6 mm deep, 5–7 mm in diameter, enclosing the corona; exterior purple, glabrous; interior yellow marked with many concentric dark purple spots, glabrous; lobes erect, deltoid in basal half, 25–30 × ± 3 mm, narrowing gradually to 1 mm into long linear processes with obtuse apex, adaxial surface with concentric dark purple spots in the broad deltoid bases, the spots becoming smaller and disappearing completely in the greenish-yellow, narrowed linear part, glabrous, abaxial surface greenish-yellow with dark purple spots in parts. Corona discoid, ± 3 × 2 mm, sessile, biseriate but appearing as if uniseriate, reddish in most parts; outer lobes separated by deep V-shaped sinuses, ± 0.3 mm apart, cupular at the base, 10-toothed, teeth deltoid-obtuse suberect, furnished with a tuft of minute hairs at the apex on the inner face; inner lobes linear-oblong, ± 1.2 × 0.4 mm, obtuse, incumbent on backs of anthers and slightly exceeding them, glabrous; guide rails ± 0.25 mm long; carpels glabrous. Follicles paired, acutely divergent, upright, fusiform, ± 100 × 4 mm, glabrous, containing 20–24 seeds. Seed elliptic-oblong, ± 9 × 3 mm, dorsiventrally compressed, brownish-black with a light brown margin; coma ± 20 mm long.

TANZANIA. Songea District: Namchwea Hill, Jan. 1993, flowering in hort. Peckover at South Africa, Sunnyside, *M & E Specks* 419

DISTR. **T** 8; known only from the type

HAB. Grassland and bushland on rocky hills, in pockets of thin soil; ± 1700 m

NOTE. The nearest relative to *B. maritae* is probably *B. barbarae*.

7. **Brachystelma buchananii** *N.E. Br.* in K.B. 1895: 263 (1895) & in F.T.A. 4(1): 467 (1903); Bullock in K.B. 17: 188 (1963); Boele in Excelsa 14: 46 (1989); Lauchs in Kakteen Sukk. 53: 236 (2002) & in Asklepios 86: 20 (2002); Meve in Illustr. Handbook Succ. Pl. *Asclepiadaceae:* 23 (2002); Masinde in K.B. 62: 60, fig. 7 (2007). Type: Malawi, Shire Highlands, *Buchanan* 116 (K!, holo., fragment: one leaf & one flower in envelope)

Stout, erect herb to 30 cm high; root a discoid tuber to 5 cm in diameter; stem erect, stout, 3–4 mm in diameter, nodes slightly swollen, pubescent all round. Leaves herbaceous; petiole 0–4 mm long; lamina broadly elliptic, elliptic-obovate or oblanceolate-oblong, 38–127 × 19–70 mm, base cuneate-acute, apex acute, subobtuse or acuminate, shortly pubescent especially on the veins and midrib beneath, margin ciliolate. Inflorescence terminal, sessile; umbels 20–30-flowered, up to 80 mm in diameter, flowers opening simultaneously; bracts subulate, ± 1 × 0.2 mm, pubescent; pedicels 5–38 mm long, shortly pubescent. Sepals linear-lanceolate or lanceolate-attenuate, 4–6 × ± 1 mm, abaxially minutely pubescent. Corolla in total 15–26 mm in diameter, rotate or broadly saucer-shaped, glabrous throughout; exterior dark mauve or blackish purple; interior concentrically zoned with yellowish and blackish purple or dark purple-brown; tube cup- or bowl-shaped, 6–16 mm deep, 12–16 mm in diameter at mouth; lobes triangular-acute, 3–6 × 2–5 mm, lobed to half-way down or reduced to deltoid teeth. Corona globular, ± 2.5 × 4.5 mm, subsessile, biseriate, red-brown; outer lobes cupular at the base, margin deeply incised at centre into ± U-shaped sinuses, 10-toothed; teeth 1–2 mm long, deltoid-subulate, ascending, densely whitish hairy in upper half; inner lobes linear or oblong, 1–1.5 × ± 0.4 mm, obtuse, incumbent on backs of anthers and equalling or slightly exceeding them, glabrous; guide rails ± 0.4 mm long. Pollinia 0.25–0.5 × 0.17–0.4 mm, corpusculum obovate ± 0.15 × 0.05 mm. Follicles thickly fusiform, ± 68 × 9 mm, smooth; seed not known.

TANZANIA. Tabora District: Kakoma, on termite mounds in swamp grassland, 25 Jan. 1936, *Lloyd* 31!; Mpanda District: Sibwesa, 6°30'S 30°44'E, 1969, *Kielland* 48!; Songea District: Ruvuma, fl. in cult. Münster, Germany, *Specks* 378!
DISTR. T 4, ?6, 8; Congo-Kinshasa, Zambia, Malawi, Zimbabwe
HAB. Seasonally wet grassland; 1150–1600 m

SYN. *B. magicum* N.E. Br. in K.B. 1895: 263 (1895) & in F.T.A. 4(1): 467 (1903). Type: Tanganyika, ?Ulanga District: near Ujiji, 1884, *Belgian Consul at Zanzibar* s.n. (K!, holo.)
  B. shirense Schltr. in J.B. 33: 339 (1895). Type: Malawi, Mt Sochi, *Scott Elliot* 8666 (B†, holo.; K!, lecto., selected by Bullock in K.B. 17: 188 (1963))
  B. nauseosum De Wild. in Ann. Mus. Congo Belge sér. 5: 191 (1904). Type: Congo-Kinshasa, Ufuru R. Valley, 23 Oct. 1901, *Cabra & Michel* s.n. (BR, holo. photo!)

NOTE. In terms of the overall size of stems, leaves and height, this species is the largest *Brachystelma* in East Africa. It is closely related to *B. omissum* Bullock and *B. togoense* Schltr. which occur more frequently from central to West Africa, although *B. omissum* reaches Zambia.

8. **Brachystelma floribundum** *Turrill* in K.B. 1922: 197(1922), *non* Dyer in Fl. Pl. Africa t. 1224 (1956); Bullock in K.B. 17: 188 (1963); Percy-Lancaster in Excelsa 13: 67 (1988); Boele in Excelsa 14: 47 (1989); Meve in Illustr. Handbook Succ. Pl. *Asclepiadaceae:* 29 (2002); Masinde in K.B. 62: 62, fig. 8 (2007). Type: Zimbabwe, without locality, 1922, *Hislop* s.n. (K!, holo.)

Dwarf herb, 7–10(–20) cm high; root a discoid tuber to 150 mm in diameter; stem erect, simple or sparingly branched near base, leafy, subsucculent, 3–4 mm diameter, glabrous; nodes laterally compressed, slightly swollen. Leaves herbaceous; petiole 0–3 mm long; lamina linear-lanceolate or narrowly elliptic, 25–60 × 1–9 mm, base narrowly cuneate or attenuate, apex acute, adaxially glabrous, abaxially puberulous, midrib prominent beneath, margin, ciliolate. Inflorescence of sessile quite floriferous umbels, up to 8-flowered with the flowers congested and verticillate in the axils of the terminal or subterminal pair of leaves, flowers opening successively or simultaneously; bracts subulate, 1–2 mm long, abaxially puberulent; pedicels 7–8 mm long, thinly puberulous. Sepals linear-lanceolate, 4–5 × ± 0.3 mm, abaxially minutely pubescent. Corolla ± 25 mm in diameter, rotate with long prominent tentacle-like lobes; exterior with purple/brown spots on a light greenish background, interior liver-coloured, exterior and interior glabrous throughout; tube campanulate 3–5 mm long, 4–6.5 mm in diameter at mouth, urceolate, interior with thick concentric wrinkles starting at mouth; lobes deltoid, 10–38 × 2–4 mm, gradually tapering to 1 mm into spreading or reflexed linear-lanceolate tails, margins slightly recurved, often more deeply coloured purple/brown than the tube. Corona globose but more elongate, 3.2–5 × 3.2–3.5 mm, sessile or subsessile, biseriate, yellowish with purple spots; outer lobes cupular but deeply divided at centre ± to the base into a U- or V-shaped outline, with a well-developed pair of divergent cylindrical teeth 1–1.5 × 0.2–0.5 mm rising up to 1 mm above the inner lobes and stylar head, mostly covered with fine whitish hairs except for a small glabrous area at the very apex towards the adaxial side; inner lobes rudimentary with traces only at bases of anthers (e.g. *Napper* 865) or well-developed, compressed linear-oblong, ± 2 × 0.7 mm, obtuse, incumbent on backs of anthers and slightly surpassing them with apices touching and overlapping each other at centre of stylar head (e.g. *Drummond & Hemsley* 1785), glabrous; guide rails 0.17–0.4 mm long. Pollinia 0.42–0.5 × 0.25–0.4 mm; corpusculum obovate, 0.3–0.31 × 0.1–0.15 mm. Follicles and seed unknown.

NOTE. The subsucculent stem and prominent cylindrical and densely finely pubescent lobules of the interstaminal lobes which are divergent distinguish *B. floribundum* from all other species in East Africa. It is closely related to *B. maritae* and probably more distantly to *B. barberae*.

a. var. **floribundum**

Plant leafless when flowering. Flowers mostly solitary. Corolla ± 30 mm long, mostly solitary; outer corona lobules densely covered with needle-like conspicuous hairs.

TANZANIA. Morogoro District: Morogoro-Dakawa road, 15 km N of Morogoro, 25 Mar. 1953, *Drummond & Hemsley* 1785!; Iringa District: ± 48 km S of Iringa on Mbeya road, 14 Nov. 1958, *Napper* 865!
DISTR. **T** 6, 7; Mozambique, Zimbabwe
HAB. Miombo woodland; 600–1800 m

b. var. **mlimakito** *Masinde* in K.B. 62: 62, fig. 9 (2007). Type: Tanzania: Ufipa District: foot of Mt Kito, *Richards* 10224 (K!, holo.)

Plant leafy when flowering. Flowers umbellate. Corolla 15–20 mm long; outer corona lobules sparsely pilose or glabrous.

TANZANIA. Ufipa District: foot of Mt Kito, 21 Nov. 1958, *Richards* 10224!
DISTR. **T** 4; known only from the type
HAB. Open wooded grassland; 1500 m

9. **Brachystelma festucifolium** *E.A. Bruce* in Hooker's Icon. Pl. 34(3): t. 3369 (1938); Meve in Illustr. Handbook Succ. Pl. *Asclepiadaceae:* 29 (2002); Masinde in K.B. 62: 64, fig. 10 (2007). Type: Tanzania, Tabora District: Kakoma, *Lloyd* 68 (K!, holo.)

Suberect herb to 15 cm high; root not recorded but probably a tuber; stem sparsely branching above the base, 2–3 mm in diameter, pale brown, glabrous, nodes slightly swollen. Leaves herbaceous, ascending, sessile; lamina linear, 35–80 × 1–1.5 mm, base narrowly cuneate, apex acute; glabrous, midrib prominent and bulging on abaxial side, margin incurved, glabrous. Inflorescence terminal, sessile, umbels up to 8-

flowered; bracts subulate, ± 1.5 × 0.3 mm, glabrous or abaxially minutely pubescent; pedicels 20–25 mm long, sparsely pubescent. Sepals ovate-lanceolate, 4–5 × 0.5–1 mm, acuminate, abaxially sparsely pubescent, apex often recurved. Corolla subrotate, in total 20–30 mm in diameter, generally dark-mauve, exterior glabrous throughout, interior variously pubescent; tube campanulate, bowl-shaped, ± 3 mm deep, 4–5 mm in diameter at mouth, only partially enclosing the corona, interior pale yellow with purple spots, glabrous; interior cream with red, brown, or black spots; lobes triangular, 7–12 × 5–8 mm, apex acute, adaxial surface dark mauve with an adpressed felt of outwardly directed bristle-like hairs, becoming shorter and less dense towards the centre; margins densely bearded with 2–3 mm long pale purple vibratile hairs; abaxial surface dark mauve, glabrous. Corona globular, ± 3 × 2 mm, sessile, biseriate, purple, glabrous throughout; outer lobes cupular at base, incised into elliptic-shaped sinuses, 10-toothed, teeth deltoid, up to 0.5 mm long, erect and incurved, tending to converge; inner lobes linear-oblong, ± 1.5 mm long, apex obtuse, conspicuously spotted, incumbent on backs of anthers and surpassing them; anthers subquadrate, incumbent on the ± flattened stylar head, guide rails straight, ± 0.25 mm long, well exposed. Pollinia ± 0.4 × 0.3 mm; corpusculum obovate, ± 0.27 × 0.15 mm. Follicles and seed unknown.

TANZANIA. Tabora District: Kakoma, 11 Jan. 1936, *Lloyd* 68!
DISTR. **T** 4; known only from the type
HAB. Miombo woodland; ± 1170 m

NOTE. In the protologue, the leaves subtending the inflorescence were misinterpreted to be bracts. The bracts are tiny subulate structures.

10. **Brachystelma johnstonii** *N.E. Br.* in F.T.A. 4(1): 468 (1903) & in Hooker's Icon. Pl. 28: t. 2754 (1903); Bullock in K.B. 17: 189 (1963); U.K.W.F.: 392 (1974) & ed. 2: 184 (1994); Meve in Illustr. Handbook Succ. Pl. *Asclepiadaceae:* 32 (2002); Masinde in K.B. 62: 66, fig. 11 (2007). Type: Kenya, Kisumu–Londiani District: Lumbwa [Fort Ternan], *Johnston* s.n. (K!, holo.)

Herb, 15–20 cm high; root a discoid tuber to 5 cm in diameter; stem erect, leafy, slightly swollen, 2.5–3 mm in diameter, pubescent all round, nodes laterally compressed. Leaves herbaceous; petiole 0–1 mm long; lamina narrowly linear, occasionally elliptic, 10–90 × 2–6 mm, base narrowly cuneate or attenuate, apex acute, usually longitudinally folded along midrib or with incurved undulate or non-undulate margins, adaxially glabrous, abaxially minutely pubescent, midrib very prominent beneath, margin entire or undulate, ciliolate. Inflorescence terminal or subterminal, sessile, umbels 4–7-flowered with flowers congested and appearing verticillate in the axils of the terminal or subterminal pair of leaves; foetid; bracts subulate, ± 1.5 × 0.3 mm, abaxially puberulent; pedicels 2–4 mm long, minutely pubescent. Sepals linear-lanceolate or lanceolate-attenuate, 6–7 × ± 1.5 mm, sometimes adaxially shallowly channelled in the apical part, abaxially minutely pubescent. Corolla 60–100 mm in diameter, rotate with long prominent tentacle-like lobes; exterior purplish- or dull greenish-brown or maroon with yellow stripes, exterior glabrous throughout; tube campanulate urceolate, 6–10 mm long, 5–8 mm in diameter at the mouth, slightly constricted at mouth before distinctly spreading out for 2–3 mm to the bases of the lobes; the constriction is marked on the interior by wrinkles of up to 6 concentric rings; exterior dark red; interior dark purple-brown, marked with a few narrow whitish or yellowish concentric zones at the mouth, densely pubescent; lobes deltoid, 55–85 × 4–9 mm, gradually tapering into very long linear or linear-filiform tails, ascending to spreading, adaxially sometimes covered with 1.5–2.5 mm long white hairs on a dark purple-brown or dark-green background, abaxially glabrous. Corona globose, ± 4 × 2.5 mm, sessile, biseriate, blackish-purple; outer lobes cupular at the base, ± 1 mm long, 10-toothed, teeth deltoid-oblong ± 0.5 mm long and broad, obtuse, furnished with a tuft of minute deflexed hairs at the apex on the adaxial face; inner lobes linear-oblong, ± 1 × 0.4 mm, obtuse, incumbent on backs of anthers and just surpassed by them, glabrous; carpels glabrous. Pollinia ± 0.53 × 0.4 mm; corpusculum obovate, ± 0.4 × 0.18 mm. Follicles thickly fusiform, ± 30–70 × 8–13 mm; seed not recorded.

UGANDA. Karamoja District: Lochoi, 24 May 1940, *A.S. Thomas* 3529! & 3530!; Toro District: Kitakwenda, *Bagshawe* 1223! Queen Elizabeth National Park, Ishasha Northern Circuit, 17 Mar. 1968, *Lock* 68/41!
KENYA. Turkana District: Kacheliba, N of Suam River, 6 Aug. 1969, *Forbes-Watson* in EAH 14188!; Trans-Nzoia District: Endebess, 2 May 1953, *Irwin* 157! & 4 June 1955, *Irwin* in EAH 98/55! & NE Elgon, June 1960, *Tweedie* 2017!
DISTR. U 1, 2, 4; **K** 2, 3, 5; Senegal, Mali, Ghana, Nigeria, Central African Republic
HAB. Seasonally wet grassland and bushland in shallow sandy soil; plants often sprouting from underground tubers after burning; 900–2100 m

SYN. *B. bagshawei* S. Moore in J.B. 45: 330 (1907). Type: Uganda, Toro District: Kitakwenda, *Bagshawe* 1223 (BM!, holo.)
    B. lanceolatum Turrill in K.B. 1922: 197 (1922). Type: Uganda, Mengo District: Entebbe, 6 Jun. 1919, no collector's name, *Hort. Kew* s.n. (K!, single fl., holo.)
    B. constrictum J.B. Hall in K.B. 20: 251 (1966); L.E. Newton in Natl Cact. Succ. J. 33: 15 (1978). Type: Ghana, ± 3 km NE of Kwahu Tafo, *Hall* 3019 (K!, holo.; CCG, GC photo!, iso.)
    B. medusanthemum J.-P. Lebrun & Stork in Adansonia 1984(4): 491 (1985). Type: Mali, 25 km S of Sikasso, *Demange* 2938 (ALF, holo.)

NOTE. A species variable in vegetative and floral characters over its distribution range, especially in leaf and corolla lobe shape, length and size of indumentum, but there are no distinct forms to warrant recognition of species or infra-specific taxa especially in Central to West Africa.

11. **Brachystelma lineare** *A. Rich.*, Tent. Fl. Abyss. 2: 49 (1851) & Atlas t. 72 & in Walp. Ann. 3: 68 (1852); Schumann in E. & P. Pf. 4(2): 268 (1895); N.E. Br. in F.T.A 4(1): 470 (1903); Bullock in K.B. 17: 190 (1963); U.K.W.F.: 393 (1974); Jenkins in Asklepios 25: 114 (1982); U.K.W.F. ed. 2: 184, t. 72 (1994); Meve in Illustr. Handbook Succ. Pl. *Asclepiadaceae*: 34, t. III, b (2002); M.G. Gilbert in Fl. Eth. 4(1): 169 (2003); Masinde in K.B. 62: 69, fig. 12 (2007). Type: Ethiopia, Tacazze River, *Quartin Dillon* s.n. (P, holo.)

Erect herb, 2.5–10(– 45) cm high; root a discoid tuber up to 75 mm in diameter, orange-brown, sitting ± 7 cm deep in the soil, with many fleshy roots from the base; stem branching at the base, 1.5–2.5 mm in diameter, pubescent or puberulent all round, nodes slightly swollen. Leaves herbaceous, ascending, sessile or subsessile; lamina linear, 25–57(– 75) × 2–3(– 6) mm, plicate along midrib to form a V-shape, base narrowly cuneate, apex acute; upper surface glabrous, lower surface sparsely and minutely pubescent especially on midrib and towards base, margin glabrous or ciliolate. Inflorescence extra-axillary, terminal or subterminal, usually in two opposite sessile umbels of up to 8 flowers in each, appearing verticillate due to congestion, many of them opening simultaneously, emitting a penetrating foetid smell; bracts subulate, 1.5(–4) × 0.3(–1) mm, glabrous or abaxially minutely pubescent; pedicels 2–3 mm long, sparsely pubescent. Sepals linear-lanceolate to subulate, 4–5 × 0.5–1 mm, sparsely pubescent beneath at least on the midvein. Flower buds ± 10 mm long at maturity, basal half distinctly globular and rather abruptly narrowing into a broadly triangular beak-shaped apex. Corolla in total 10–11 mm long, ± 50 mm in diameter, exterior glabrous throughout, interior minutely pubescent in tube and to ± halfway up the lobes; tube campanulate, bowl-shaped, 4–5 mm deep, 5–7 mm in diameter at mouth, enclosing the corona, interior cream with red, brown, or black spots, interior and exterior glabrous; lobes deltoid, 22–25 × 3–4 mm, narrowing gradually to ± 1 mm wide into linear-oblong tails, recurved along margins, sometimes folded back along mid-rib into a V-shape (e.g. in *Newton* 4230), apex acute or obtuse, adaxial surface becoming purple-red or brown towards

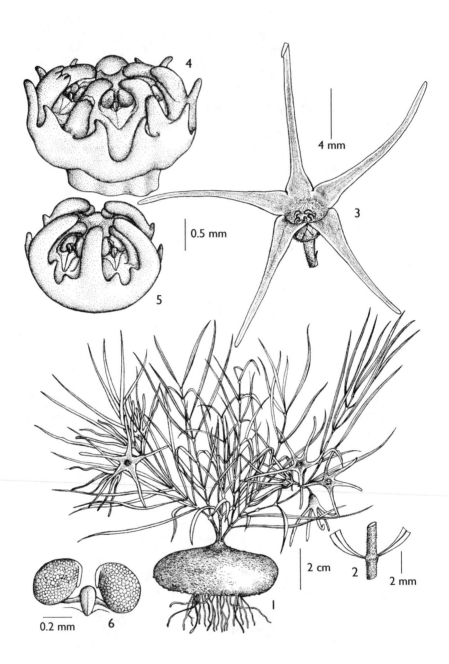

FIG. 73. *BRACHYSTELMA LINEARE* — **1**, whole plant with discoid tuber; **2**, enlarged stem to show minute dense indumentum; **3**, flower; **4 & 5**, gynostegium and corona of two flowers to show variation; **6**, pollinarium. 1 & 2 from Archer 196; 3, 5 & 6 from *Powys* in *Newton* 4230; 4 from *Tweedie* 2821. Drawn by N. Muema based on pencil drawings by S. Masinde. Reproduced from KB 62: 71, fig. 12.

the apex, covered with spreading white hairs that gradually thin out and disappear halfway along the lobes, abaxially yellow. Corona globular, 2.5–3 × 1.5–1.7 mm, subsessile or shortly stipitate, biseriate, glabrous throughout; outer lobes cupular at base, incised into V-shaped sinuses, 10-toothed, teeth deltoid, divergent, erect, 0.2–0.5 mm long; inner lobes linear-oblong, 0.7–1 × 0.2–0.4 mm, apex obtuse and sometimes uneven (e.g. in *Tweedie* 2821), incumbent on backs of anthers and equalling them or reaching halfway up; guide rails 0.2–0.38 mm long. Pollinia 0.37–0.4 × 0.3–0.39 mm; corpusculum obovate to elliptic, 0.22–0.24 × 0.1–0.11 mm, with broad, basal lateral membranes; carpels glabrous. Follicles paired, acutely divergent, narrowly cylindrical, ± 90 × 4 mm diameter at centre, tapering at both ends; seed not recorded. Fig. 73, p. 304.

KENYA. Laikipia District: Laikipia Plateau, 45 km NE of Rumuruti, 12 Jan. 1993, *P. Powys in Newton* 4230!; Nairobi District: Langata, May 1950, *Archer in Bally* 7777! & Golf Range, between Wilson Airport and Army Barracks, just outside Nairobi National Park, 1977, *Gilbert et al.* 4940!
DISTR. **K** 3, 4; South Sudan, Eritrea, Ethiopia
HAB. Seasonally wet shallow soil in wooded grassland; 1600–1800 m

SYN. *B. ellipticum* A. Rich., Tent. Fl. Abyss. 2: 49 (1851) & Atlas t. 72 & in Walp. Ann. 3: 68 (1852); Schumann in E. & P. Pf. 4(2): 268 (1895); N.E. Br in F.T.A 4(1): 472 (1903). Type: Ethiopia, Taccaze River, *Quartin Dillon* s.n. (P, holo.)
B. phyteumoides K. Schum. in E.J. 17: 149 (1893); N.E. Br in F.T.A 4(1): 469 (1903). Type: South Sudan, near Jur Ghattas, *Schweinfurth* ser. III. 37. (B†, K!, lecto., fragment, fl and stem, selected by Masinde in K.B. 62: 69 (2007))
B. asmarense Chiov. in Ann. Bot. (Roma) 10: 390 (1912). Syntypes: Eritrea, Asmara, *Baldrati* D80 & D92 (FT, syn., photo!)
B. pellacibellum L.E. Newton in Bradleya 14: 97 (1996). Type: Kenya, Laikipia District: Laikipia Plateau, 45 km NE of Rumuruti, *P. Powys* in *Newton* 4230 (K!, holo., EA!, iso.)

NOTE. A broad species concept is followed here. A variable species in vegetative and floral characters over its entire range but there is need for more material in order to interpret the variation fully.

12. **Brachystelma keniense** *Schweinf.* in Reise zum Rudolf-See u. Stephanie-See, Append. 8: 8 (1892) & in Abhandl. Preuss. Akad. Wiss. 2: 344 (1892); Engler in Koen. Akad. Wiss: 344 (1892); Schumann in P.O.A. C: 327 (1895) & in E. & P. Pf. 4(2): 268 (1895); N.E. Br in F.T.A 4(1): 471 (1903); Bullock in K.B. 17: 189 (1963); U.K.W.F.: 393 (1974); Jenkins in Asklepios 25: 114 (1982); U.K.W.F. ed. 2: 184 (1994); Meve in Illustr. Handbook Succ. Pl. *Asclepiadaceae:* 32 (2002); Masinde in K.B. 62: 72, fig. 13 (2007). Type: Kenya, Kikuyu, at the foot of Mt Kenya, *von Hohnel* 29 (B†, holo.). Neotype: Kenya: Aberdare Range, Kinangop, cult. Kew, *Chandler* 2323 (K!, neo., incl. alc., designated by Masinde in K.B. 62: 72 (2007))

Dwarf, delicate, erect herb 10(– 80) mm high; root a discoid tuber, 50–80 mm in diameter, buried about 50 mm deep in the soil; stem erect, sometimes sparsely branching at the base, ± 1 mm in diameter, puberulent all round, nodes slightly swollen. Leaves herbaceous, ascending; petiole 0–1 mm long; lamina elliptic-lanceolate or narrowly linear-lanceolate, 13–25 × (1.5–)3–7 mm, base narrowly cuneate, apex acute; upper surface glabrous, lower surface rather minutely pubescent especially on midrib and towards base, margin ciliolate. Inflorescence subterminal in apical leaf axils, sessile, 1–3-flowered, flowers well spaced away from the node due to long wiry pedicels, opening successively, one at a time; bracts subulate, ± 0.7 mm long, abaxially puberulous; pedicels 4–18 mm long, puberulous. Sepals linear-lanceolate, 1–1.5 × ± 0.4 mm, abaxially sparsely pubescent. Flower buds at maturity linear-ovate, ± 3 mm long, apex obtuse, gradually narrowed from base to apex. Corolla in total ± 10 mm in diameter, glabrous throughout, tube whitish or a delicate shade of pale mauve dotted with purple spots and lobe apices a bright apple green or greenish yellow; tube campanulate, bowl-shaped, 2–3 mm deep, 3–5 mm in diameter at mouth, slightly enclosing the corona; lobes suberectly spreading, deltoid, 5–6 × 2.5–3 mm, narrowing gradually to ± 1 mm wide into subulate tails, recurved along margins. Corona globular to discoid, 1.6–2 × 2–3.5, subsessile or shortly stipitate, biseriate, purple, glabrous in most parts; outer lobes cupular, shallowly incised into broad sinuses, 10-toothed, teeth deltoid, divergent, erect, 0.25 mm long, adaxially glabrous or white-barbate; inner lobes linear-oblong to triangular, 0.6 × 0.25 mm, apex obtuse, incumbent on backs of anthers and equalling them or surpassed by them; guide rails ± 0.25 mm long; carpels glabrous. Pollinia 0.27–0.3 × 0.22–0.27 mm; corpusculum elliptic, ± 0.22 × 0.1 mm. Follicles and seed unknown.

KENYA. Laikipia District: Powys [Kisima or Pinguone] Ranch, 40 km N of Rumuruti by [Powys'] house [next to air strip], *McLeod* s.n.!; Naivasha District: between Kinangop and Naivasha, 4 Apr. 1922, *Fries* 2767!; South Nyeri District: 16 km from Kiganjo at the foot of Mt Kenya, 27 Nov. 1959, *Moore* 18!
DISTR. K 3, 4; not known elsewhere
HAB. Dry grassland with loamy soil in upland plateaus; 1600–2700 m

NOTE. *B. keniense* is related to *B. lineare* because of the similarities in corolla and corona form and to some extent the habit.

13. **Brachystelma rubellum** (*E. Mey.*) *Peckover* in Aloe 33: 43 (1996); Harold in Asklepios 46: 91 (1989); Meve in Illustr. Handbook Succ. Pl. Asclepiadaceae: 41, t. 3g (2002); Masinde in K.B. 62: 74, fig. 14 (2007). Type: South Africa, Eastern Cape, Uitenhage, Addo, *Drège* 2227 (K!, holo.)

Glabrous herb, 20–50 cm high; root system a discoid tuber 30–50(–90) mm in diameter, brownish grey skinned; stem mostly solitary, occasionally branched above the base in upper parts, erect with few leaves far apart due to the long internodes, 1–1.5(– 2) mm in diameter, nodes hardly laterally compressed or swollen, glabrous. Leaves spreading, sessile; lamina linear-lanceolate, 20–70 × 1–2 mm, base cuneate, apex acute, completely glabrous. Inflorescence terminal, pedunculate and open-branched or a narrow lax raceme-like panicle to 160 mm long, subtended by bract-like leaves, 1–3-flowered, up to 2 flowers opening simultaneously, often pendent due to thin pedicels; bracts subulate, ± 1.5 mm long, glabrous; pedicels long and wiry, 5–12 mm long, glabrous. Sepals linear-lanceolate or lanceolate-attenuate, 1–2 × ± 0.3 mm, abaxially pubescent or glabrous. Corolla in total 8–30 mm in diameter, lobed almost to the base, appearing somewhat rotate, pale mauve or whitish cream on interior and exterior, occasionally maroon; tube campanulate, shallowly bowl-shaped, 0.7–1 mm long, 2–3 mm in diameter at the mouth; lobes mostly linear-spatulate, sometimes linear-oblong, 8–16(–20) × 1.5–3 mm, apex apiculate or gently notched, suberectly spreading, slightly folded back along the midrib, basally with purple-red papillae, sometimes with fine hairs as well, margins sometimes revolute. Corona cupular, ± 1.8 × 2 mm, shortly stipitate, pink-orange, glabrous throughout; outer lobes ascending, cupular concave-ovate or deltoid-ovate, ± 0.7 × 0.5 mm, apically bifid, teeth deltoid-oblong, parallel, ± 0.1 mm long, glabrous; inner lobes linear-oblong, 0.7–1 × 0.15 mm, obtuse, incumbent on backs of anthers and slightly longer than them or erect and about twice as long, glabrous; guide rails ± 0.4 mm long, not well exposed; carpels glabrous. Pollinia ovoid ± 0.32 × 0.18 mm; corpusculum obovate, ± 0.17 × 0.07 mm. Follicles paired, very narrowly elongate cylindrical, 85–115 mm long by 2–3 mm broad at centre, tapering at both ends, glabrous; seed elliptic 5–7 × 1.5–2 mm, with pale brown margin; coma 10–20 mm long. Fig. 74, p. 307.

UGANDA. West Nile District: Logiri, 17 Mar. 1945, *Greenway & Eggeling* 7218! & Apr. 1939, *Hazel* 726!; Mengo District: Buvuma Island, 8 Dec. 1942, *A.S. Thomas* 4108!
KENYA. Naivasha District: Lake Naivasha area, base of Mau Escarpment, 13 May 1973, *Magius* in Mrs S.F. Polhill 396!; Machakos/Masai Districts: Chyulu Hills, Outside End Forest, 13 Dec. 1991, *Luke* 2979!; Masai District: Garabani Hill, 16 Mar. 1940, *van Someren* 152!

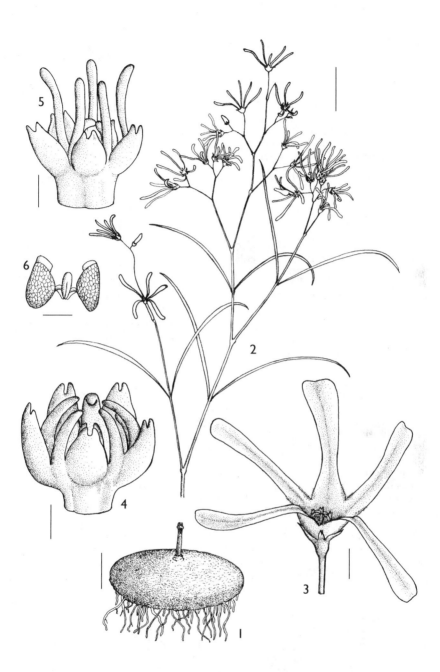

FIG. 74. *BRACHYSTELMA RUBELLUM* — **1**, discoid tuber; **2**, flowering stem; **3**, flower; **4** & **5**, gynostegium and corona of different flowers to show variation; **6**, pollinarium. 1 & 2 from *Greenway & Turner* 12003; 3, 4 & 6 from *Hopwell* 8626 in *Harold* Y104; 5 from *Thomas* 4443. Scale bars: 1 = 1 cm, 2 = 2 cm, 3 = 2 mm, 4–5 = 0.5 mm, 6 = 0.2 mm. Drawn by N. Muema based on pencil drawings by S. Masinde. Reproduced from KB 62: 75, fig. 14.

TANZANIA. Bukoba District: Thangiro, Aug. 1931, *Haarer* 2103!; Musoma District: Wogakuria Hill, 30 Dec. 1964, *Greenway & Turner* 12003!; Pare District: Kamuri–Kisungu road, Oct. 1927, *Haarer* 841!

DISTR. U 1, 2, 4; **K** 3, 4, 6; **T** 1–3, 5; Swaziland, South Africa

HAB. Seasonally burnt wooded grassland; 1700–2300 m

SYN. *Tenaris rubella* E. Mey., Comm Pl. Afr. Austr. 1(2): 198 (1838); Decne. in DC., Prod.: 606 (1844); Schumann in P.O.A. C: 327 (1895); N.E. Br in F.T.A 4(1): 795 (1908); Bullock in K.B. 8: 357 (1953); Malaisse in Fl. Rwanda: 117 (1985); Blundell, Wild Fl. E. Afr.: t. 759 (1987); Victor & Nicholas in S. Afr. Journ. Bot. 64: 205 (1998)

   T. rostrata N.E. Br. in Gard. Chron. ser. 2, 24: 39 (1885) & in F.T.A. 4(1): 473 (1903); Schumann in P.O.A. C: 327 (1895); U.K.W.F.: 389 (1974) & ed. 2: 182 (1994). Type: Tanzania, Mpwapwa District: Sagara [Usagara], cult. Kew, *Last* s.n. (K!, holo.)

   T. volkensii K. Schum. in P.O.A. C: 327 (1895) & in E. & P. Pf. 4(2): 295 (1895). Type: Tanzania, Kilimanjaro, Himo River below Marangu, *Volkens* 2183 (B†, holo.). Neotype: Fig. 91 in E & P. Pf. 4(2): 295 (1895) selected by Masinde in K.B. 62: 74 (2007)

   T. simulans N. E. Br in F.T.A 4(1): 796 (1908); Schlechter in E.J. 20, Beibl. 51: 45 (1895) & in J.B. 35: 291 (1897). Type: South Africa, Transvaal, Elands Spruit Mountains, *Schlechter* 3858 (K!, holo.)

NOTE. *Brachystelma rubellum* is related to the former members of the genus *Tenaris* which are all found in South Africa, namely *B. christianeae* Peckover and *B. chloranthum* (Schltr.) Peckover. It is the only East African *Brachystelma* species with a pedunculate inflorescence. Specimens from West Nile and Acholi Districts of N Uganda appear to be morphologically different from the rest of the East African specimens. They have white or cream flowers with much narrower corolla lobes that are less distinctly spatulate. Due to their small stems, leaves and flowers, they look on the outset much more like *B. chloranthum*.

14. **Brachystelma arachnoideum** *Masinde* in K.B. 62: 76, fig. 15 (2007). Type: Tanzania, Sumbawanga District: Tatanda Mission, *Bidgood, Mbago & Vollesen* 2474 (K!, holo.; DSM, NHT, iso.)

Erect, glabrous herb, 60–100 cm high; root a fascicle of fusiform roots, 3–4 mm in diameter; stem solitary, erect, mostly leafless or with 1–2 pairs of leaves when in flower, with leaves far apart due to the long internodes, 1–1.5(– 2) mm in diameter, glabrous, nodes hardly laterally compressed or swollen. Leaves sessile or subsessile; lamina linear-lanceolate, 40–120 × 1–5 mm, base rounded, apex acute, completely glabrous. Inflorescence extra-axillary, in the apical nodes to the terminal ones, sessile, umbels 1–3-flowered, flowers often opening simultaneously; apical flowering nodes with tiny leaves looking like bracts or just a bit longer; flowers pendulous; bracts subulate, ± 0.8 × 0.3 mm, abaxially glabrous or occasionally pubescent; pedicels recurved downwards at maturity, 12–22 mm long, long and wiry, purplish, glabrous. Sepals linear-lanceolate or lanceolate-attenuate, 2–3 × 0.3–0.4 mm, abaxially pubescent or ± glabrous. Corolla ± 40 mm in diameter, rotate with long filiform, feathery, tentacle-like lobes; tube ± obsolete, campanulate, 0.7–1 mm long, 1–1.5 mm in diameter at the mouth, only slightly enclosing the corona, interior dark purple, exterior similar in colour to interior but paler, with a dense cover of purple/maroon hairs; lobes deltoid, 20–27 × 1.5–2 mm, green, gradually tapering into long filiform spreading tails, adaxially covered with 0.5–1 mm long purple hairs on a green or dark-purple background, margins recurved, abaxially gabrous. Corona globular, subsessile, 1–1.5 × 1–2 mm, purple, brown or pale yellow, glabrous throughout; outer lobes cupular, only slightly raised from bases of inner lobes thus quite low, only very shallowly emarginate to form a wave or tiny V-shaped sinus, teeth absent or deltoid and to ± 0.12 mm long, glabrous; inner lobes linear-oblong, 0.2–0.4 × 0.2 mm, obtuse, incumbent on backs of anthers and reaching to halfway up them, glabrous; anthers subquadrate, incumbent on the flattened stylar head, guide rails straight, ± 0.3 mm long, well exposed. Pollinia ovoid or ± D-shaped, 0.27–0.3 × 0.2–0.21 mm; corpusculum orange-brown, linear-elliptic, 0.2–0.25 × 0.05–0.08 mm. Follicles and seed unknown.

TANZANIA. Ufipa District: Tatanda Mission, 25 Feb. 1994, *Bidgood et al.* 2474!

DISTR. **T** 4; Zambia, Malawi, Mozambique
HAB. *Brachystegia* woodland in rocky soil; 1000–1800 m

NOTE. Unique by its fusiform root system, probably related to the W. African *B. exile* Bullock and the Zimbabwean *B. bikitaense* Peckover. The specimen *Lloyd* 60 from Tanzania, Kakoma, S of Tabora, in miombo woodland, 1150 m, 21 Feb. 1936, which is 46 cm high but without roots or flowers, although recorded to possess dark brown flowers, is most probably *B. arachnoideum*.

15. **Brachystelma simplex** *Schltr.* in E.J. 38: 40, fig. 5 (1905); Meve in Illustr. Handbook Succ. Pl. Asclepiadaceae: 43, t. III, f (2002); Masinde in K.B. 62: 78, fig. 16 (2007). Type: Mozambique, *Schlechter* 12121 (B†, holo.). Lectotype: Fig. 5 in E.J. 38(1): 41(1905), designated by Masinde l.c.

Herb to 30 cm high with all vegetative parts densely and minutely pubescent; root a discoid tuber 45–80 mm in diameter, up to 70 mm deep, basal part often with several thinly fleshy fusiform roots to 5 mm in diameter; stem erect, branching from near the base to apical regions, 1.5–2 mm diameter, brownish pilose all round. Leaves ascending or spreading, sessile; lamina narrowly linear, 10–55 × 2–3 mm, weakly plicate along midrib to form a V-shape, base narrowly cuneate, apex acute; both surfaces densely and minutely brownish pubescent, margin ciliolate. Inflorescence extra-axillary, arising almost from every other node in the flowering branches, small and very close to the stem thus appearing as if arising at the bases of leaves, congested, 1–4-flowered, opening consecutively, scent not recorded; bracts subulate, ± 1.5 mm long, abaxially pubescent; pedicels ± 2 mm long, densely pubescent. Sepals subulate, ± 2 × 0.3 mm, pubescent beneath. Flower buds globose, congregated in leaf axils. Corolla in total ± 5 mm in diameter, rotate, glabrous throughout, greenish-yellow; tube campanulate, bowl-shaped, ± 1 mm deep, ± 2 mm in diameter at mouth, only slightly enclosing the corona, mouth greenish-yellow and distally brownish-yellow with small purple spots at centre; lobes deltoid-acuminate, ± 3 × 1.5 mm, apex acute, greenish-yellow. Corona discoid, saucer-shaped, ± 1.5 × 0.7 mm, subsessile or shortly stipitate, appearing uniseriate, glabrous throughout; outer lobes cupular, ascending, ± 0.5 × 0.5 mm, margin entire or gently undulate; inner lobes rudimentary, forming studs at the bases of anthers; anthers subquadrate, yellow, incumbent on the flattened green stylar head; guide rails ± 0.3 mm long, well exposed. Pollinia ± 0.33 × 0.25 mm; corpusculum obovate-elliptic, ± 0.12 × 0.08 mm, carpels glabrous. Follicles paired, narrowly fusiform, 75 × 6 mm at centre, tapering at both ends; seed not recorded.

KENYA. South Nyeri/Embu Districts: Mbeere District: Mwea area, Riakanau Village, 23 Dec. 2004, *Musili & Muasya* 18!; Nairobi: Langata, 8 Nov. 1961, *Archer* 287!; Machakos District: base of Lukenya [Hills] on western side, 32 km SE of Nairobi, 14 May 1967, *Archer* 536!
DISTR. **K** 4, 6; Burkina Faso, Ivory Coast, Benin, Nigeria, Zambia, Mozambique
HAB. Grassland in shallow well drained soils on rocky outcrops; 1650–1800 m

SYN. *Brachystelma* sp. A, U.K.W.F.: 393 (1974) & ed. 2: 184 (1994)
  B. simplex Schltr. subsp. *banforae* J.-P. Lebrun & Stork in Adansonia 11(1): 71 (1989); Meve & Porembski in E.J. 115: 322 (1993); Type: Nigeria, Oyo Prov, Igbetti, *Stanfield* 196 (K!, holo.)

NOTE. *Brachystelma simplex* is probably closely related to the tropical African *B. dinteri* Schltr. There are similarities in vegetative form and in the dense pubescence, as well as in the flower form and corona. In both species, flowers are borne very close to leaf axils and are small and comparable in size. The corona in both species has rudimentary staminal lobes and the interstaminal lobes spread out to form a shallow saucer-shaped structure. *Brachystelma simplex* has a vegetative form similar to *B. lineare* and *B. gracile* but it does not appear to be related to these species. It differs from *B. lineare* in much smaller flowers, the broadly triangular corolla lobes and in the corona form. *B. simplex* is a common plant in Riakanau area of Mbeere District where it is known as 'ithinza' in the local Kamba language (M. Muasya pers. comm.). In Riakanau, it grows in sandy well drained soils in *Combretum - Terminalia* bushland. Although Meve (2002) did not give reasons for not recognizing subspecies *banforae*, I have followed his treatment because the characters used to distinguish subspecies *banforae* such as the outer corona shape are variable and not restricted to any particular geographical area.

16. **Brachystelma gracile** *E.A. Bruce* in Fl. Pl. Afr. 27: t. 1077 (1949); Percy-Lancaster in Excelsa 13: 67 (1989); Boele in Excelsa 14: 46 (1989); Meve in Illustr. Handbook Succ. Pl. *Asclepiadaceae:* 31 (2002); Masinde in K.B. 62: 80, fig. 17 (2007). Holotype: Zimbabwe, Plumtree, *Porter* s.n. in *PRE* 27227 (PRE, holo. photo!)

Erect herb, ± 60 cm high; root a discoid tuber ± 9 cm in diameter; stem solitary, leafy, dichotomously branching, ± 1.5 mm in diameter, densely finely pubescent. Leaves sessile or subsessile; lamina narrowly linear to linear-elliptic, 20–120 × 1.5–3 mm, cuneate, apex acute, both surfaces finely pubescent but more so on abaxial side, margin recurved, ciliolate. Inflorescence extra-axillary, sessile; flowers solitary, pendulous, cage-like, appearing in apical to terminal nodes; flower buds with a long beak; bracts subulate, ± 0.8 × 0.3 mm, glabrous or occasionally abaxially pubescent; pedicels recurved downwards at maturity, 2–4 mm long, purplish, glabrous. Sepals linear-lanceolate, ± 1.5 mm long, abaxially pubescent. Corolla ± 15 mm long, ± 6 mm in diameter; tube ± obsolete; lobes ± 13 mm long, arising from deltoid bases and narrowed in apical ²/₃ into very narrowly linear or filiform parts ± 1.5 mm broad, somewhat folded back along the midrib, apically united to form an oblong cage that is constricted ¹/₃ from the base, finely ciliate; lobe exterior dark purple to blackish in the broad deltoid bases merging into brown in the narrow filiform parts, glabrous. Corona not seen. Follicles and seed not seen.

TANZANIA. Photographs of plants in flower taken in habitat – no specimens seen
DISTR. **T** 8; Zimbabwe, Botswana, South Africa
HAB. Seasonally wet grassland; ± 1700 m

NOTE. The inclusion of *B. gracile* in this account is based on colour photographs that were sent to me in 2006 by Leonard E. Newton for identification. The photographs were taken in southern Tanzania by Ernest Specks who supposedly also collected specimens for cultivation.

58. CARALLUMA

R. Br. in Asclepiadeae: 14 (1810) & Mem. Wern. Nat. Hist. Soc. 1: 25 (1811); Meve & Liede in Pl. Syst. Evol. 234: 195 (2002)

Spathulopetalum Chiov. in Ann. Bot. (Rome) 10(3): 392 (1912)

Erect, stem-succulent perennials; latex clear or at most cloudy; stems heteromorphic, with basal sections cylindrical, roundly or sharply 4-angled, green, blue-green or light brown, and apical sections tapering, often much elongated. Leaves reduced to scales, opposite and decussate, sessile, caducuous; stipules reduced to a few trichomes. Inflorescences along the apical tapering sections (synflorescence stalks), 1–4-flowered, bostrychoid, sessile. Flowers subsessile or petiolate, inodorous or foetid. Corolla rotate, apopetalous or corolla lobes basally fused, adaxially uniformly coloured or maculate, banded or reticulate, glabrous, papillose or with verrucose trichomes, lobes mostly strongly replicate along midrib, margins glabrous or ciliate. Corolline corona absent. Gynostegium sessile or seated on a stalk. Gynostegial corona in 2 series: "outer" corona of 5 staminal and 5 interstaminal parts fused basally or up to half of length, cyathiform or rotate; interstaminal corona lobes subulate to deltoid, often bilobed filiform; "inner" corona of ovoid or trianguloid staminal lobes, erect or inflexed. Pollinia ovoid or transversely rectangular; caudicles rectangular. Mericarps two, narrowly oblong. Seeds ovate, broadly winged, brownish.

23 species distributed in Africa, Arabia and Asia, with East Africa and India as centres of distribution. The genus reaches its southern limits of distribution in Tanzania.

1. Corolla lobes at least 25 mm long .... 7. *C. longiflora*
   Corolla lobes no longer than 15 mm 2
2. Flowers with spider-like appearance due to spreading to
   recurved, long and small lanceolate corolla lobes 3
   Flowers rotate or with drooping, not lanceolate corolla lobes 5
3. Gynostegial corona atop a stalk 2–3 mm long, interstaminal
   lobes basally pouch-like, apically blunt .... 2. *C. gracilipes*
   Gynostegial corona atop a stalk 0.5–1 mm long, interstaminal
   corona lobes bifid into two subulate, laterally spreading
   teeth 4
4. Corolla lobes brownish to purple, often spotted darker
   brownish or purplish; interstaminal corona lobes deeply
   bifid into two subulate, curved, spreading teeth .... 1. *C. arachnoidea*
   Corolla lobes pale yellow-green; interstaminal corona lobes
   bifid into two small triangular, straight to slightly curved,
   hardly spreading teeth .... 4. *C. flavovirens*
5. Corolla rotate with spreading, (narrowly) triangular corolla
   lobes; gynostegial corona flattened discoid, staminal
   corona lobes incumbent on anthers .... 8. *C. priogonium*
   Corolla lobes drooping, oblanceolate, folded back along
   midrib; gynostegial corona conical 6
6. Drooping corolla ovoid in outline; staminal corona lobes
   shorter than anthers .... 6. *C. peckii*
   Drooping corolla ± linear in outline with a slight constriction
   at the position of the corona; staminal corona lobes much
   longer then anthers, ascending 7
7. Interstaminal corona lobes shortly rectangular, bifid into
   bluntly deltoid lobules; staminal corona lobes 3 mm long 3. *C. dicapuae*
   Interstaminal corona lobes rectangular or square, deeply
   bifid into erect or spreading, subulate teeth; staminal
   corona lobes 1–2 mm long .... 5. *C. turneri*

1. **Caralluma arachnoidea** (*P.R.O. Bally*) *M.G. Gilbert* in Natl. Cact. & Succ. J. 32(2): 26 (1977) & in Bradleya 8: 12 (1990); U.K.W.F. ed. 2: 185, t. 74 (1994); Meve & Liede in Pl. Syst. Evol. 234: 196 (2002); Gilbert in Fl. Eth. 4(1): 171 (2003). Type: Uganda, North Karamoja, *Eggeling* 5692 sub *Bally* 6294 (G, holo.; K, iso.)

Plants tufted, often spreading irregularly; stems 5–25 cm long, strictly 4-angled, serrate due to acute, straight tubercles orientated upwards; flower-bearing apical region terete to 4-angled, often spreading nearly horizontally. Inflorescences usually 2-flowered, shortly bracteate; pedicel ± 1 cm long, directed downwards. Corolla deeply incised, brownish to purple, often spotted darker brownish or purplish; tube strongly reduced; lobes 10–15 mm long, stiff, horizontally spreading, almost completely reflexed along the midrib, upper face pubescent to pilose, basally mostly with clavate cilia. Gynostegial corona bowl-shaped, on a stalk 0.5–1 mm long, yellowish, reddish or dark purple, occasionally spotted; interstaminal corona lobes spreading to ascending-erect, deeply bifid into horn-like, subulate teeth, laterally spreading, acute; staminal corona lobes ribbon-like to deltoid, connivent above style-head.

a. var. **arachnoidea**

Flower-bearing apical parts 4-angled; corona 3 mm diameter; interstaminal teeth < 0.8 mm long, staminal corona lobes ± 2 mm long, narrowly deltoid to linear, connivent-erect into cone-shaped column above style-head, much longer than the anthers.

UGANDA. Karamoja District: Loyoru, 9 June 1940, *A.S. Thomas* 3732 & 48 km SE of Moroto township, 1961, *Kerfoot* 4728 & 8 km from Moroto on Soroti road, May 1948, *Eggeling* 5781
KENYA. West Suk District: between Kabernet and Kapenguria, 1989, *Hartmann & Newton* 28406!; Teita District: Bungule village, foot of Mt Kasigau, 13 Feb. 1996, *Goyder et al.* 4026 (≡*Meve et al.* 934!)
TANZANIA. Mbulu District: Lake Manyara National Park, Msasa River Valley slopes, 9 Jan. 1965, *Greenway & Kanuri* 12036; Pare District: Kisiwani, 25 June 1942, *Greenway* 6492; Lushoto District: 12 km from Mkomasi to Kisiaroni, 22 Mar. 1999, *Liede & Meve* 3384!
DISTR. U 1; **K** 1, 2, 4, 6, 7; **T** 1–3; Ethiopia
HAB. Grass- or succulent-dominated communities, in the open or within small shrubs, on (red) sand or in sand pockets on rocks; 600–2000 m

SYN. *Caralluma gracilipes* K. Schum. subsp. *arachnoidea* P.R.O. Bally in Candollea 24: 10 (1969)
    Spathulopetalum arachnoideum (P.R.O. Bally) Plowes in Haseltonia 3: 56 (1995)

NOTE. The most frequent *Caralluma* in East Africa, and one of the most frequently found stapeliads of Kenya and Tanzania; with various colour forms and patterns from brown to purple. *C. flavovirens*, morphologically most similar, differs in colouration of corolla and corona, and shorter but wider, less spreading interstaminal corona lobes.

b. var. **breviloba** (*P.R.O. Bally*) *M.G. Gilbert* in Natl. Cact. Succ. J. 32(2): 29 (1977) & in Bradleya 8:12 (1990); Meve & Liede in Pl. Syst. Evol. 234: 196 (2002); Gilbert in Fl. Eth. 4(1): 172 (2003). Type: Kenya, Masai District: Olorgesailie, *Bally* 12267 (≡*Bally* S163) (K, holo.; ZSS, iso.)

Flower-bearing apical parts of stems terete to rounded–4-angled; corona 2.5 mm diameter; interstaminal corona teeth ≥ 1 mm long; staminal corona lobes narrowly deltoid to linear, shorter than 1 mm, not rising above the style-head, acute, partly touching apically, as long as or shorter than the anthers.

KENYA. Northern Frontier District: Maralal–Baragoi, 10 km N of Martii, 3 June 1979, *Gilbert, Kanuri & Mungai* 5472 & Lerogi Forest area, 55 km N of Maralal, 22 Feb. 1974, *Bally & Carter* 16541; Masai District: Olorgesailie, *Bally* 12267 (≡*Bally* S163)
DISTR. **K** 1, 4, 6; Ethiopia
HAB. Bare rocky hills and on diatomaceous lake bed soils; 1100–1700 m

SYN. *Caralluma gracilipes* K. Schum. subsp. *breviloba* P.R.O. Bally in Candollea 24: 14 (1969)

2. **Caralluma gracilipes** *K. Schum.* in P.O.A. C: 328 (1895); White & Sloane, Stapelieae ed. 2, 1: 189 (1937); F.P.U.: 121 (1962); U.K.W.F. ed. 2: 185 (1994). Type: Kenya, Kitui District: *Hildebrandt* 2700 (B†, holo.; K, photo.!)

Flowering plants up 30 cm high, sparsely branched; stems 8–15 mm diameter, rather sharply 4-angled, distantly serrate due to acute, straight tubercles oriented upwards; leaf rudiments erect, narrowly lanceolate, up to 5 mm long, acute; flower-bearing apical region 2–4 times longer than stems, slender, terete. Inflorescences 2-flowered, flowers spreading; pedicel 7–8 mm long. Corolla ± 2.5 cm diameter, nearly divided to base; tube small, little inflated; lobes horizontally spreading, whitish or yellow, spotted with purple or maroon, stiff, 10 × 1 mm, pubescent, margins revolute, basally with clavate, apically with simple hairs and cilia. Gynostegial corona on a distinct, narrowly cylindrical stalk, 2–3 mm long, abruptly broadening at bases of corona lobes; interstaminal lobes basally pouch-like, apically bluntly deltoid, occasionally shortly bifid; staminal coronal lobes oblong-spatulate, 1.5 mm long, apically rounded and touching each other, erect, surpassing the style-head. Mericarps two, to 10 cm long, narrowly oblong.

KENYA. Northern Frontier District: Furroli, 20 Sep. 1952, *Gillett* 13964 & Huri Hills, 25 Feb. 1963, *Bally* 12528; Kitui District: Ukamba, *Hildebrandt* 2700
TANZANIA. Pare District: Pare Mts, no date, *Specks* 637!
DISTR. **K** 1, 4; **T** 3; not known elsewhere
HAB. On granite rock, and on lava flows; 800–1400 m

SYN. *Spathulopetalum gracilipes* (K. Schum.) Plowes in Haseltonia 3: 56 (1995)

NOTE. *C. gracilipes* can be confused with *C. arachnoidea*, but the latter is more robust.

3. **Caralluma dicapuae** (*Chiov.*) *Chiov.* in White & Sloane, Stapelieae ed. 2, 1: 187 (1937); U.K.W.F. ed. 2: 185, t. 74 (1994); Gilbert in Fl. Eth. 4(1): 173 (2003); Lavranos in Fl. Somalia 3: 177 (2006). Type: Eritrea, between Chelamet and Oazata, *Terracciano & Pappi* 498 (FT, holo.)

Dwarf perennials to 60 cm high in flowering state; stems numerous, up to 25 cm long, weakly branching, grey-green, robust, 4-angled with obtuse edges, with conical tubercles 5–8 mm high; leaf rudiments short-lived, 2 mm long; flower-bearing apical region ± 20 cm long, terete. Inflorescences 2-flowered; flower buds basally clavate-cylindrical, apically abruptly articulated, 8–10 × 2–3 mm; pedicels pointed downwards, 1–1.6 cm, slender. Corolla nearly divided to base, tube 0.5 mm in diameter; lobes at very first spreading then drooping, linear to narrowly spoon-shaped, 11–14 × 3 mm, mucronate, lamina only very basally spreading, otherwise strictly folded back along midrib with the margins in close contact, abaxially glabrous, dark purple, often spotted yellowish, adaxially light green or yellowish, dotted or banded chestnut-brown to purple, densely pubescent, basally with clavate cilia. Gynostegial corona 2.5 × 3 mm; interstaminal corona lobes rectangular, upright-spreading, shortly bifid into bluntly deltoid lobules, outside occasionally with 0.5 mm long stiff hairs; staminal corona lobes erect to connivent-erect, narrowly deltoid to linear, 3 mm long, overtopping the style-head, dotted white on chestnut-brown. Mericarps two, 10–13 cm long, 4–5 mm wide, very narrowly oblong.

KENYA. Northern Frontier District: Archers Post, by Agricultural Dept. quarters, 14 June 1979, *Gilbert et al.* 5653; Turkana District: 122 km N of Lodwar on road to Lokichokio, 27 Dec. 1987, *Hartmann & Newton* 28438; Laikipia District: 6 km S of Suguta Marmar on Rumuruti–Maralal road, 25 Oct. 1978, *Gilbert, Gachathi & Gatheri* 5100
DISTR. **K** 1–3; Eritrea, Ethiopia and Somalia
HAB. In dry bushland or open succulent communities (*Sansevieria robusta*), also on laterite; 600–1700 m

SYN. *Spathulopetalum dicapuae* Chiov. in Ann. Bot. 10: 392 (1912)
    Caralluma dicapuae (Chiov.) Chiov. subsp. *seticorona* P.R.O. Bally in Candollea 24: 17 (1969). Type: Somalia N, SE of Odweina, *Bally* 10429 (≡*Bally* S248) (K, holo.; G, iso.)

4. **C. flavovirens** *L.E. Newton* in Asklepios 74: 25 (1998); Bruyns, Al Farsi & Hedderson in Taxon 50: 1031–1043 (2010). Type: Kenya, Teita District, Kizima Hill, *Newton* 5589 (K, holo., EA, iso.)

Plants tufted; stems erect to spreading, laxly and irregularly branched, green, 5–12 cm long, 0.7 cm wide, strictly and acutely 4-angled, tubercles inconspicuous; leaf rudiments ascending, 4 mm long, persistent; flower-bearing apical region terete to 4-angled, up to 26 cm long, often spreading nearly horizontally. Inflorescences usually 1-flowered, shortly bracteate; pedicel directed downwards, ± 1 cm long. Corolla pale yellow-green, basally spotted with red, stellate; corolla lobes fused basally into a tube 0.5 mm long; lobes ± 8 × 1 mm, basal 1 mm erect, with vibratile purple clavate cilia 1 mm long, apical 7 mm horizontally spreading, linear-lanceolate, almost completely reflexed along the midrib, upper face pubescent to pilose, cilate with simple hairs 1 mm long. Gynostegial corona atop a stalk 1 mm long, pale yellow; interstaminal corona lobes spreading horizontally, 0.75 mm long, dark red, bifid into triangular, fairly straight, hardly spreading lobules; staminal corona lobes linear, ± 2 mm long, apically obtuse, connivent above style-head, basally mottled pale red to pink, apically dark red.

KENYA. Teita District: Taita Ranch, Kizima Hill, 15 Oct. 1996, *Newton* 5589 & same loc., Dec. 1998, *Luke et al.* 5574
DISTR. **K** 7; only known from this hill
HAB. Isolated rocky, gneissic hill; 450–600 m

NOTE. This Teita endemic appears like a yellow-green flower variant of *Caralluma arachnoidea* with less subulate and less spreading interstaminal lobules. However, molecular studies (Bruyns *et al.* in Taxon 50: 1031–1043, 2010) placed it in sister relationship to *C. turneri*. With the latter (and with *C. arachnoidea*) it shares a highly similar corona structure, but not the mobile and drooping corolla lobes.

5. **Caralluma turneri** *E.A. Bruce* in Hook. Ic. Pl. 34: t. 3339 (1937); Blundell, Wild Fl. E. Afr.: t. 550 (1987); U.K.W.F. ed. 2: 185 (1994); Meve & Liede in Pl. Syst. Evol. 234: 197 (2002); Gilbert in Fl. Eth. 4(1): 172 (2003). Type: Kenya, South Kavirondo, Kanam, foot of Homa Mt, *A. Turner* s.n. in *CM* 3629 (K, holo.)

Plants shrubby, up to 50 cm high; stems distinctly heteromorphic, grey-green to brown, apically tapering; vegetative basal regions of stems 4–20 cm long, 1–2 cm thick, leaf rudiments lanceolate or subulate, fleshy; flower-bearing apical region 20–30 cm long, 4-angled, erect. Inflorescences 1–4-flowered; flowers freely pendant; pedicel spreading, 3–6 mm long; sepals ± 2 mm long, long-acuminate. Corolla 8–15 mm long, divided almost to base; lobes oblanceolate to spatulate, 8–12 × 2–3 mm, mobile, at very first spreading then drooping, lamina ± completely folded back along the midrib with the margins touching each other, base claw-like, ciliate, apex rounded with minute mucro, adaxially glabrous or scattered pubescent, cream-coloured striped with blackish- to brownish-purple, rarely entirely dark, bases marginally with simple or clavate partially vibratile purplish hairs. Gynostegial corona 0.7–1.6 mm long, on a short stalk, purple (occasionally with white); interstaminal corona lobes rectangular or square, 0.7 mm long, deeply bifid into erect or spreading subulate teeth; staminal corona lobes ribbon-shaped, 1–2 mm long, apically occasionally dentate, tips connivent high above over the style head. Mericarps two, 10–12 cm long, narrowly oblong, bright brown, striped purplish, seeds 5–6 × 3.5 mm, ovate, broadly winged, bright brown. Fig. 75, p. 315.

NOTE. The drooping and versatile corolla lobes make *C. turneri* an exquisite species, only confusable with *C. dicapuae* and *C. peckii*.

a. subsp. **turneri**

Plants robust, stems sharply 4-angled, edges distinctly serrate by acute tubercles. Leaf rudiments 2 mm long, ascending. Corolla lobes adaxially glabrous or scattered hairy. Stalk carrying corona cylindrical, staminal corona lobes forming a column.

KENYA. South Kavirondo District: Kanam, Kavirondo gulf, Dec. 1934, *A. Turner* s.n. in *CM* 6622; Kajiado District: near Kenya Marble Quarry, SSW of Kajiado, 27 Nov. 1977, *Gilbert* 4936; Teita District: Mt Maktau, lower slopes, 14 Feb. 1996, *Goyder et al.* 4033 (≡*Meve et al.* 945!)
DISTR. **K** 3, 5–7; S Ethiopia and Somalia
HAB. Sparse grassland or (succulent) bushland, on sandy to loamy soil; 900–1600 m

SYN. *Caralluma dicapuae* (Chiov.) Chiov. subsp. *turneri* (E.A. Bruce) P.R.O. Bally in Candollea 24: 17 (1969)
Spathulopetalum turneri (E.A. Bruce) Plowes in Haseltonia 3: 56 (1995)

b. subsp. **ukambensis** (*P.R.O. Bally*) *L.E. Newton* in Asklepios 72: 9 (1997). Type: Kenya, Kitui District: Ithumbi Hill, *MacArthur* s.n. sub *Bally* S135 (ZSS, holo.)

Plants rather delicate and more slender in all respects than subsp. *turneri*; stems almost terete with edges rounded, not serrate; leaf rudiments ± 1.5 mm long, spreading to ascending. Corolla lobes adaxially hairy allover; stalk carrying corona conical, staminal corona lobes forming a cone.

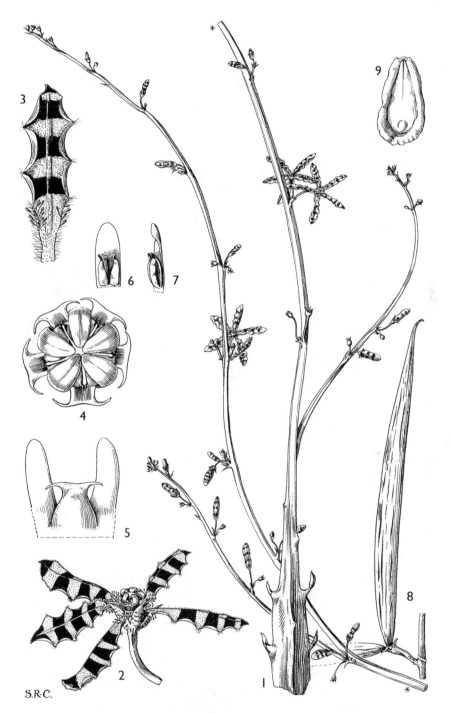

FIG. 75. *CARALLUMA TURNERI* — **1**, habit showing transition to fertile part of flowering stems, ×
1; **2**, flower, × 4; **3**, corolla lobe, upper surface, × 6; **4**, inner and outer corona from above, ×
16; **5**, inner and outer corona from outside, × 16; **6**, anther and inner corona lobe from within,
× 16; **7**, lateral view of **6**, × 16; **8**, paired follicles, × 1; **9**, seed (without coma), × 4. From *Turner*
3629. Reproduced with permission from Hook. Ic. Pl. t. 3339, drawn by Stella Ross-Craig.

KENYA. Kitui District: Waita Hill, 21 June 1992, *Newton* 4063! & Yambyu Dam, 31 May 1997, *Pearce* 1333 & Mwingi, 5 km towards Garissa, 31 May 1997, *Pearce* 1344!
DISTR. **K** 4; not known elsewhere
HAB. In soil pockets on gneissic inselbergs and other rocky outcrops; 900–1200 m

SYN. *Caralluma dicapuae* (Chiov.) Chiov. subsp. *ukambensis* P.R.O. Bally in Candollea 24: 19 (1969)

6. **Caralluma peckii** *P.R.O. Bally* in Candollea 18: 14 (1962) & in Fl. Pl. Afr. 35: t. 1394 (1962); U.K.W.F. ed. 2: 185 (1994); Meve & Liede in Pl. Syst. Evol. 234: 197 (2002); Gilbert in Fl. Eth. 4(1): 173 (2003). Type: Kenya, Northern Frontier District: Archers Post near Isiolo, *Peck* s.n. sub *Bally* S61 (K, holo.)

Plants 8–20 cm high; stems somewhat shrubbily branching, heteromorphic, green to grey-green, ± 8 cm long, 0.8 cm in diameter, obtusely 4-angled, apically lighter green, finely spotted with brown-purple, emitting an unpleasant odour; tubercles rather distant from each other, conical-subulate; leaf rudiments lanceolate, curved upwards, 1.5–2 mm long, acute; flower-bearing apical region 10–15 cm long, much tapering, acutely 4-angled. Inflorescences 1–2-flowered; flowers freely pendant; bracts lanceolate, ± 3 mm long; pedicels bent downwards, ± 10 mm long. Sepals 3 mm long, receptacle slightly swollen. Corolla ovoid in outline, 1–1.5 cm long, divided nearly to base; lobes drooping, oblanceolate, 7.5–9 × 2.5 mm, folded completely outwards with the margins being in contact so that they appear flatly double-layered, yellowish with fine red dots forming horizontal ribbon-like markings, both sides glabrous or nearly so, corolla lobe bases tapering, not folded, connivent, whitish, with purplish, clavate cilia. Gynostegial corona clubbed, 1.3–2 mm long, tapering into a conical stalk; interstaminal corona lobes reduced to two lateral appendages, spreading, scale-like, dark purple; staminal corona lobes erect, ribbon-shaped to deltoid, less than 1 mm long, only the tips bend inward but not completely covering the white anthers.

KENYA. Northern Frontier District: junction of Wamba and Isiolo–Maralal roads, 23 Oct. 1994, *Newton & Powys* 4516!; Elgeyo District, Kerio Valley 34 km S of Tot, 21 Mar. 1961, *Bally* 12359
DISTR. **K** 1, 3, 7; Ethiopia
HAB. Dry bushland; 800–1200 m

SYN. *Spathulopetalum peckii* (P.R.O. Bally) Plowes in Haseltonia 3: 56 (1995)

NOTE. A tiny *Caralluma* with delicate flowers.

7. **Caralluma longiflora** *M.G. Gilbert* in Bradleya 8: 13 (1990); Meve & Liede in Pl. Syst. Evol. 234: 197 (2002); Gilbert in Fl. Eth. 4(1): 173 (2003); Lavranos in Fl. Somalia 3: 176 (2006). Type: Somalia, 53 km on Maas–Bulo road, *Thulin & Warfa* 4604 (K, holo.)

Stems loosely branched in irregular clusters, green, distinctly heteromorphic; vegetative basal region up to 35 cm long, 0.6–0.7 cm in diameter, 4-angled but edges weakly rounded, slightly paler; tubercles indistinct, peg-like; leaf rudiments narrowly triangular, 2 mm long, acute, margins with large, unicellular hairs; flower-bearing apical region up to 40 cm long, ± terete, curved with tip nearly horizontal, sometimes branched. Inflorescences 1 to 4-flowered; tips of buds contorted clockwise; pedicels pointing downwards at anthesis, 0.8–1 cm long. Flowers 5.5–7.5 cm diameter, corolla very deeply divided, adaxially ochre dotted pale purple; lobes linear-lanceolate, 25–35 × 1.5–1.8 mm, curved outwards at an angle of 45°, margins revolute, finely ciliate, corolla tube flat, ± 3.5 mm diameter. Gynostegial corona forming a flat bowl ± 3.5 mm diameter, on a narrow stalk, completely excerted from corolla tube; interstaminal corona lobes spreading horizontally, concave, truncate, apically slightly incised, yellow with darker tips; staminal corona lobes ± 5 mm long, basally ribbon-like, apically broadly lanceolate, acuminate, partly twisted around each other, pale yellow, connivent-erect to form a column.

KENYA. Northern Frontier District: Ramu to Malka Mari road, 6 May 1978, *Gilbert & Thulin* 1535!
DISTR. **K** 1; Ethiopia, Somalia
HAB. *Acacia-Commiphora* bushland on limestone; ± 350 m

NOTE. The extremely long and connivent staminal corona lobes of this species are unique in the genus.

8. **Caralluma priogonium** *K. Schum.* in E.J. 34: 327 (1905); White & Sloane, Stapelieae, ed. 2, 1: 190 (1937); U.K.W.F. ed. 2: 185 (1994); Liede & Meve in Pl. Syst. Evol. 234: 197 (2002); Gilbert in Fl. Eth. 4(1): 171 (2003); Lavranos in Fl. Somalia 3: 176 (2006). Type: Tanzania, Pare District: between Gonja and Kihurio, *Engler* 1521/a (B†, holo.; K, iso. (flower) & photo.)

Plants shrubby, up to 50 cm high; stems erect or ascending, branching freely, distinctly heteromorphic; vegetative region 10–25 cm long, 2.5–4 cm in diameter, acutely 4-angled, edges serrate by acute tubercles; leaf rudiments oblong-ovate, 3.5–4 mm long, acuminate; flower-bearing apical region 1–3 mm in diameter, tapering, 4-angled. Inflorescences 1–3-flowered; flowers ascending; pedicels 3–6 mm long, slender. Sepals 2 mm long, acute, abaxially weakly papillate. Corolla flattened star-shaped, 2–3.5 cm diameter, deeply divided with corolla tube to 2 mm diameter, abaxially silvery-green, dotted purple; lobes horizontally spreading, stiff, lanceolate, 1–1.5 × 3 mm, apically mucronate, inside with scattered hairs, margins along lower half with purple vibratile cilia, tips with a tuft of swollen setae, lobes adaxially dark purple, sometimes tinged olive, with basal third whitish, dotted purple. Gynostegial corona flattened, dark purple (occasionally cream-coloured), usually glabrous; interstaminal corona lobes erect, ovate-rectangular, 0.5 mm long, notched or bilobed into spreading subulate teeths, staminal corona lobes spatulate, apically blunt, incumbent on anthers, tips occasionally hairy. Mericarps flattened in cross-section.

KENYA. Northern Frontier District: Sololo, 4 Sep. 1952, *Gillett* 13782 & 15 km from Moyale on Wajir road, 6 Nov. 1952, *Gillett* 14149; Baringo District: Radad, ± 25 km S of Marigat, 5 Jan. 1995, *Goyder & Masinde* 3954
TANZANIA. Pare District: Mkomazi, slopes of Passa Mt, 30 Nov. 1935, *B.D.Burtt* 5320A & B & Mkomazi–Kihurio, 9 Sep. 1935, *Greenway* 4066 & near Kirya, SE of Lembeni near Pangani River, 8 Dec. 2000, *Bruyns* 8677
DISTR. **K** 1, 2, 6, 7; **T** 2, 3; Ethiopia and Somalia
HAB. In dry bushland or open succulent or grassland communities; often on laterite soil; 700–1300 m

SYN. *Caralluma mogadoxensis* Chiov. in Fl. Somala 2: 299 (1932). Type: Somalia, Mogadishu, *Senni* 199 (FT, holo.)
   C. elata Chiov. in Miss. Biol. Borana, Racc. Bot. 4: 169 (1939). Type: Ethiopia, Sidamo, between Negele and Ganale river, *Cufodontis* 35 (FT, holo.; W iso.)
   Spathulopetalum priogonium (K. Schum.) Plowes in Haseltonia 3: 56 (1995)
   S. mogadoxense (Chiov.) Plowes in Haseltonia 3: 56 (1995)

59. **DESMIDORCHIS**

Ehrenb. in Linnaea 4: 94 (1829) & in Abh. Königl. Akad. Wiss. Berlin 1829: 31 et 39 (1832, publ. 1831); Plowes in Haseltonia 3: 58–59 (1995) & in Excelsa 17: 69–78 (1996); Meve & Liede in Pl. Syst. Evol. 234: 201–203 (2002); Mottram in Taxon 58(2): 648–649 (2009); Jonkers in Bradleya 28: 67–78 (2010)

Sarcocodon N.E. Br. in J.L.S. 17: 169 (1878)

Erect, massive stem-succulents; latex clear or at most cloudy; stems rounded to sharply 4-angled, smooth. Leaf rudiments spreading to slightly ascending, sessile, ovate, basally cordate or rounded, apically obtuse or acuminate, fleshy and caducuous or constituting thorns; stipules reduced to a few hairs. Inflorescences in terminal pseudo-umbels, 10–80(–200)-flowered, sub-sessile. Flowers open simultaneously, usually with dung-like odour. Corolla rotate or campanulate, fleshy, adaxial surfaces often warty or rugose, glabrous, papillose or with trichomes, green, yellow, brown or purple or patterned; corolla lobes triangular, ovate to lanceolate, occasionally ciliate; corolla tube flat, funnel-shaped, cup-shaped or campanulate. Corolline corona absent. Gynostegium sub-sessile, concealed in corolla. Gynostegial corona in 2 series: "outer" corona of 5 staminal and 5 interstaminal parts fused to half of length, cyathiform, interstaminal corona subulate, bifid, spreading, occasionally with trichomes; "inner" staminal corona lobes triangular to subulate, inflexed. Pollinia globose, ovoid or D-shaped; caudicles rectangular. Mericarps two 10–15 cm long, narrowly fusiform or narrowly oblong, occasionally winged; seeds ovate, broadly winged, brown or light brown.

14 species, distributed in Africa north of and around the equator, and on the Arabian Peninsula. The genus reaches its southern limits of distribution in Tanzania with *D. speciosa*.

1. Inflorescences many-flowered, mostly with more than 100
   flowers open at time; 10–12 cm in diameter . . . . 2. *D. acutangula*
   Inflorescences with 15–40 flowers open at time; 3–7 cm in
   diameter 2
2. Corolla rotate-campanulate, 1.3–1.8 cm in diameter, yellowish
   green or cream, spotted purplish . . . . 1. *D. foetida*
   Corolla campanulate 2.5–5 cm in diameter, corolla lobes
   black-purple, corolla tube yellow . . . . 3. *D. speciosa*

1. **Desmidorchis foetida** (*E.A. Bruce*) *Plowes* in Haseltonia 3: 58 (1995); Meve & Liede in Pl. Syst. Evol. 234: 202 (2002). Type: Uganda, Karamoja District: Moroto River, *Eggeling* 2955 (K, holo.)

Massive perennials to 25 cm high; stems numerous, growing in (large) clumps, grey-green, 15–25 cm long, 2–3.5 cm diameter, stout, 4-angled, edges with light brown corky band when aged; tubercles blunt to acutely triangular. Leaf rudiments short, acute, thorny when aged. Inflorescences 30–40-flowered, up to 4 cm diameter. Flowers with foetid odour; pedicels 6–12 mm long, 2–3 mm thick; sepals ± recurved, ± 4 × 1.5–2.5 mm, adaxially slightly tuberculate. Corolla rotate-campanulate, 1.3–1.8 cm diameter, fused for half of its length; tube 5–7 mm long, lobes spreading horizontally, ovoid-triangular, ± 5 × 4 mm, outside with distinct venation, inside tuberculate-papillate, adaxially and abaxially yellowish-green (cream), spotted purple, margins densely ciliate with 3 mm long clavate purple vibratile hairs. Gynostegial corona cream, 0.8–1 cm diameter; interstaminal lobes 2–3 mm, deeply bifid, appendages triangular, blunt, often overlapping with neighbouring lobes, curved outwards (or upwards); staminal corona lobes 1.2–1.5 × 0.8–1 mm, apically rounded, shorter than anthers. Pollinia 0.5 × 0.4 mm, roundish, translator wings short. Mericarps two, erect, parallel to spreading, ± 12 cm long, 0.8 cm wide. Fig. 76, p. 319.

UGANDA. Karamoja District: Moroto River, [no date], *Eggeling* 2955 & May 1948, *Eggeling* 5793 & Mt Moroto, Jan. 1959, *Tweedie* 1784
KENYA. Northern Frontier District: Ngorinit [Nguronit] Mission Station near Ndoto Mts, 31 Oct. 1978, *Gilbert & Gachathi* 5261 & Archer's Post, by Ewaso Nyiro; 14 June 1979, *Gilbert, Kanuri & Mungai* 5655; Baringo District: 5 km from Marigat on road to Lake Bogoria, 3 Jan. 1995, *Goyder & Masinde* 3950
DISTR. U 1; K 1, 3, 4; not known elsewhere
HAB. *Acacia* bushland or wooded grassland; 700–1700 m

SYN. *Caralluma foetida* E.A. Bruce in Hook. Icon. Pl. 34: t. 3371 (1938); U.K.W.F. ed. 2: 185, t. 73 (1994)

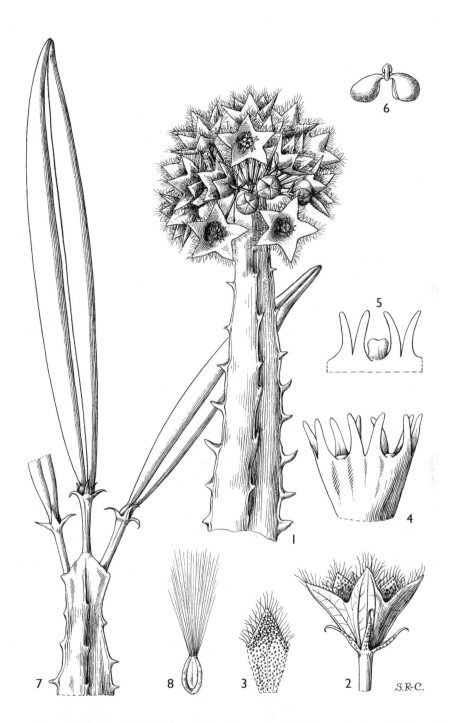

FIG. 76. *DESMIDORCHIS FOETIDA* — **1**, flowering stem, × 1; **2**, flower, × 2; **3**, corolla lobe and part of tube from within, × 2; **4**, corona from outside, × 6; **5**, outer and inner corona lobes, × 6; **6**, pollinarium (inverted), × 20; **7**, upper part of stem in fruit, × 1; **8**, seed with coma of hairs, × 1. From *Eggeling* 2955. From Hook. Ic. Pl. t. 3371, drawn by Stella Ross-Craig.

2. **Desmidorchis acutangula** *Decne.* in Ann. Sc. Nat. ser. 2, 9 : 265 (1838); Meve & Liede in Pl. Syst. Evol. 234: 201 (2002) as *acutangulus*. Type: Senegal, *Perrottet* s.n. (holo., in spiritu, not traced); Mali, Niger R. banks, Gao towards Berra, *de Wailly* 4872 (P, neo., designated by Gilbert & Raynal in Adansonia ser. 2, 19: 323 (1980))

Plants robust, much and densely branched, to 75 cm diameter; stems robust, (light) olive-green or whitish due to thick wax cover, 40–80(–150) cm long, 2–6 cm thick, 4-angled, side faces concave; tubercles pointing downwards, broadly triangular, acute. Leaf rudiments ± 1 × 1 mm, rather persistent. Inflorescences many-flowered with more than 100 open at time, outline globose, 10–12 cm diameter; pedicel 1–5 cm long. Sepals 2 mm long, occasionally with some hairs. Corolla black-purple, rotate, funnel-shaped or flat, 1.5–2 cm diameter; tube short, saucer-shaped; lobes broadly ovate-triangular, 6–8 × 6–8 mm, acute, often ciliate with purple vibratile clavate hairs, adaxial surface rugose to tuberculate, glabrous, papillate or pubescent. Gynostegial corona purplish, ± 5 mm diameter, weakly pubescent. Interstaminal corona lobes deeply bifid, teeth falcately recurved, medium interstaminal region broadened, cilate, occasionally with a short central appendage. Staminal corona lobes deltoid, blunt, touching but not surpassing the anthers. Pollinia inversely pear-shaped. Mericarps two, 12–18 cm long, narrowly fusiform, diverging at 0–45 degrees; seeds 8–9 × 4–5 mm, ovate, winged.

UGANDA. Uganda/Kenya border, Karamoja/Turkana Districts: Lorengkipi, Feb. 1958, *J. Wilson* 697
KENYA. Northern Frontier District: Nachori, 30 July 1968, *Mwangangi & Gwynne* 1056 & Lokori, 11 km S of Kangetet, 15 May 1970, *Mathew* 6216; Tana River District: valley road running NW from Garissa on E side of Tana River, 13 Jan. 1972, *Gillett* 19504
DISTR. U 1; K 1–3, 7; Sahel region from Mauritania to Ethiopia and Somalia; Saudi Arabia, N Yemen
HAB. Semi-desert, in open, overgrazed situations; 0–1500 m

SYN. *Desmidorchis retrospiciens* Ehrenb. in Abh. Königl. Akad. Wiss. Berlin 1829: 30, t. 2, fig. 8 (1831), *nomen nudum*; Lavranos in Fl. Somalia 3: 180 (2006)
Boucerosia acutangula (Decne.) Decne. in DC., Prodr. 8: 648 (1844)
B. russeliana Brongn. in Bull. Soc. Bot. Fr. 7: 900 (1863). Type: Ethiopia, "inter Sero et Mequedel", ?*Courbon* s.n. (P, holo.)
Caralluma retrospiciens N.E. Br. in Gard. Chron. ser. 3, 12: 370 (1892) & in F.T.A. 4(1): 480 (1904). Type as for *Boucerosia russeliana* Brongn
C. acutangula (Decne.) N.E. Br. in Gard. Chron. 12: 369 (1892); U.K.W.F. ed. 2: 185, t. 73 (1994)
C. hirtiflora N.E. Br. in K.B. 1895: 264 (1895). Type: Yemen, Hanish Island in the Red Sea, *Slade* 20 (K, holo.)
Boucerosia tombuctuensis A. Chev. in Actes Congr. Int. Bot. Paris: 271 (1900). Types: Mali, Timbouctou, Arnassay, Goundam, Bankorré Hills, *Chevalier* s.n. (P, syn.)
Caralluma tombuctuensis (A. Chev.) N.E. Br. in F.T.A. 4(1): 622 (1904)
C. retrospiciens N.E. Br. var. *glabra* N.E. Br. in F.T.A. 4(1): 481 (1904); F.P.U.: 121 (1962). Type: Kenya, Lamu District: Witu, *A.S. Thomas* s.n. (K, holo.)
C. retrospiciens N.E. Br. var. *hirtiflora* (N.E. Br.) Berger, Stap. Klein.: 71 (1910)
C. retrospiciens N.E. Br. subsp. *tombuctuensis* (A. Chev.) A. Chev. in Rev. Bot. Appl.: 266 (1934)
C. retrospiciens N.E. Br. subsp. *tombuctuensis* (A. Chev.) A. Chev. var. *acutangula* (Decne.) A. Chev. in Rev. Bot. Appl.: 266 (1934)
C. retrospiciens N.E. Br. var. *acutangula* (Decne.) White & Sloane in Stap., ed. 2, 1: 242 (1937)
C. retrospiciens N.E. Br. var. *tombuctuensis* (A. Chev.) White & Sloane in Stap., ed. 2, 1: 242 (1937)
C. retrospiciens N.E. Br. var. *laxiflora* Maire, in Bull. Soc. Hist. Nat. Afr. Nord 30: 357 (1939). Type: Mauritania, Adrar, Oued Amogjar, *Murat* 2387 [cited in error as 2384] (MPU, holo.)
C. russeliana (Brongn.) Cufod. in E.P.A.: 30 (1969); Blundell, Wild Fl. E. Afr.: t. 570 (1987), as *russelliana*

NOTE. Despite Plowes' attempt to validate *D. retrospeciens* Ehrenberg in Excelsa 17: 69–78 (1996), this name is a *nomen nudum* and the earliest available name is *D. acutangula*.

3. **Desmidorchis speciosa** (*N.E. Br.*) *Plowes* in Haseltonia 3: 59 (1995); Meve & Liede in Pl. Syst. Evol. 234: 202 (2002); Lavranos in Fl. Somalia 3: 179 (2006). Type: Somalia, "near Brava Magadoxo," *Kirk* s.n. (K, holo.)

Massive perennials to 30 cm height; stems erect, light green to grey-green, rather soft, 10–30 cm long up to 5 cm thick, but often tapering towards apex, 4-angled, side faces concave, edges with small, brown, corky band when aged; tubercles conical, acute. Leaf rudiments short, spreading, thorn-like. Inflorescences multi-flowered, up to 7 cm diameter, globose; flowers foetid; pedicels 4.5–6.5 mm long, 2.5–3.5 mm thick. Sepals ± 3 × 1.5 mm, acuminate, abaxially with scattered thickish hairs. Corolla campanulate, abaxially black-purple, 2.5–5 cm diameter, smooth; tube deeply cupular, ± 2 cm long, 2 cm wide, adaxially canary-yellow; lobes horizontally spreading, adaxially blackish-purple, broadly ovate-triangular, 1.2–1.5 × ± 1.3 cm, acuminate, margins slightly revolute, densely ciliate with clavate purple hairs. Gynostegial corona cyathiform, 4–4.5 mm diameter, flattened; interstaminal corona lobes bright yellow, narrowly triangular or lanceolate, 4–5 × ± 1 mm, bifid in upper third to half, appendages parallel, pressed against corolla tube; staminal corona lobes 1.5–2 × ± 1 mm, obtuse or retuse, shorter than anthers. Mericarps two, 8–12(–18) cm long, up to 1 cm wide, spreading somewhat irregularly.

UGANDA. Karamoja District: Manimani River near Kangole, 11 Sep. 1950, *Dawkins* 648 & 8 km from Moroto on Soroti road, May 1948, *Eggeling* 5779 & near Moroto, 28 Oct. 1939, *A.S. Thomas* 3081
KENYA. Northern Frontier District: 60 km along the Wajir–El Wak road, 29 Apr. 1978, *Gilbert & Thulin* 1179 & Dandu, 24 Mar. 1952, *Gillett* 12634; Teita District: 5.7 km towards Nairobi from Maungu Station, 30 Apr. 1974, *Faden & Faden* 74/521
TANZANIA. Masai District: Olduvai Gorge, 1 May 1987, *Chuwa* 2609; Mbulu District: Lake Manyara National Park, main road to Mbulu on the W rift wall from Mto ya Mbu, 25 Mar. 1968, *Greenway & Kanuri* 13244; Pare District: 6.5 km N of Kihurio, 23 July 1960, *Leach* 10337
DISTR. U 1; K 1–2, 7; T 2–3, 5–6, 8; Sudan, South Sudan, Ethiopia, Somalia
HAB. Dry bushland and wooded grassland, in alkaline soil and between volcanic rocks; 400–1300 m

SYN. *Sarcocodon speciosus* N.E. Br. in J.L.S. 17: 170 (1878)
    Caralluma speciosa (N.E. Br.) N.E. Br. in Gard. Chron. ser. 3, 12: 370 (1892); Blundell, Wild Fl. E. Afr.: t. 548 (1987); U.K.W.F. ed. 2: 185, t. 73 (1994)
    C. codonoides K. Schum. in P.O.A. C: 328 (1895). Type: Tanzania, Lushoto District: between Gonja and Kinhiro, *Volkens* 2382 (B†, holo.)
    C. oxydonta Chiov. in Fl. Somala 2: 298 (1932). Type: Somalia, Chisimaio near Afmadù, *Gorini* 37 (FT, holo.)

60. **MONOLLUMA**

Plowes in Haseltonia 3: 64 (1995); Meve & Liede in Pl. Syst. Evol. 234: 172–209 (2002)

Sanguilluma Plowes in Haseltonia 3: 65 (1995)

Erect or clump-forming stem succulents; latex copious, colourless but usually turning white when dry; stems pale green, cylindrical, 4-angled; surface often with pungent smell. Leaf rudiments minute, persistent or caducuous, ciliate; stipules reduced to hairs or absent. Inflorescences occasionally (sub-)axillary, 1–2- or 5–15-flowered, sessile. Corolla rotate or campanulate, fused for about half of length, occasionally warty, rugose or papillose; corolla lobes triangular. Corolline corona absent. Gynostegial corona in 2 series: "outer" corona of 5 staminal and 5 interstaminal parts fused to half of length, cyathiform or tubular (barrel-shaped), yellow or purplish-red; interstaminal corona lobes erect or reflexed, bilobed, occasionally ciliate; "inner" staminal corona lobes lingulate or triangular, inflexed. Mericarps fusiform or narrowly oblong, up to 10 cm long, occasionally winged. Seeds broadly winged.

A genus of five species in Africa and Arabia.

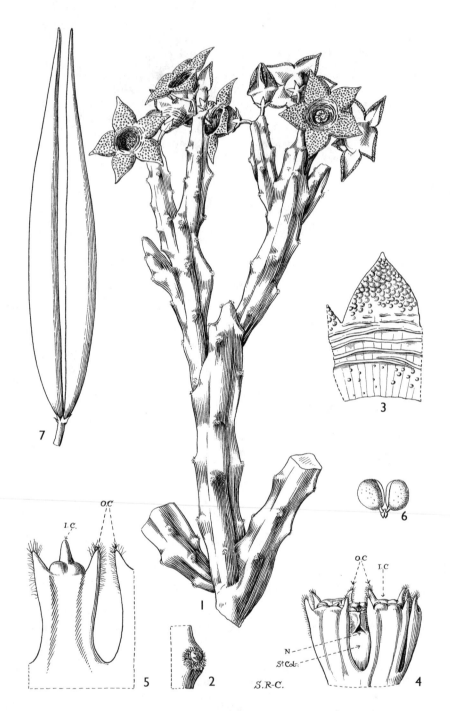

Fig. 77. *MONOLLUMA SOCOTRANA* — **1**, flowering stem, × 1; **2**, part of stem showing tooth,
× 6; **3**, corolla lobe and part of tube from within, × 3; **4**, gynostegium and corona from
the outside, × 8; **5**, outer and inner corona lobes from outside, × 12; **6**, pollinarium, × 20;
7, paired follicles, × 1. o.c. – outer corona; i.c. – inner corona; n. – nectar cavity; st.col. –
staminal column. From Hook. Ic. Pl. t. 3372, drawn by Stella Ross-Craig.

Monolluma socotrana (*Balf. f.*) *Meve & Liede* in Pl. Syst. Evol. 234: 203 (2002). Type: Yemen, Socotra, sine loc., Apr. 1880, *Balfour* s.n. (K, lecto., designated by Bruce in Hook. Icon. Pl. 34: 3372, 1938).

Clump-forming, much branching stem-succulent; stems bright (grey-)green, 15–20 mm long, 8–12 mm in diameter, slightly tapering towards tips, with pungent smell, obtusely 4-angled; outline undulate by small oblong-conical tubercles; axillary areas coloured red(-brown). Leaf rudiments deltoid, less than 0.5 mm long, cartilaginous, rapidly caducuous; stipular areas hairy. Inflorescences subterminal, subaxillary, 1-flowered, occasionally 2- or 3-flowered; pedicels 2–5 mm long. Sepals 2–2.5 mm long. Corolla shiny bright to dark red, campanulate, 2–3 cm diameter; tube broadly funnel-shaped, 5 mm long, concentrically rugose, glabrous; lobes ovate-triangular, ± 5 × 5 mm, shortly acuminate, horizontally spreading to halfway recurved, margins strongly recurved, ± sculptered by rounded and smooth emergences. Gynostegial corona purple-red, rather barrel-shaped, 3.5–4.5 mm diameter, enclosed in a central depression of tube. Interstaminal corona lobes erect, 1.5–2 × 1 mm, divided up to base, appendages narrowly triangular, less than 0.5 mm long, outer margins fused with the base of staminal corona, apical thirds ciliate. Staminal corona lobes narrowly triangular to lanceolate, ± 1 × 0.5 mm, slightly overtopping the anthers but not covering them in total. Mericarps two, fusiform, not spreading, acute, lined to inconspicuously winged. Fig. 77, p. 322.

KENYA. Masai District: Kajiado, 14 Sep. 2007, *KEFRI* 517 & S Magadi, 8 Apr. 1956, *Greenway* 8991 & Magadi Golf Course, 9 Apr. 1960, *A.J. Wood* s.n. in EA 11979
DISTR. **K** 3, 6; Ethiopia, Somalia and Yemen (Socotra)
HAB. Dry deciduous bushland; 600–1700 m

SYN. *Boucerosia socotrana* Balf. f. in Proc. Roy. Soc. Edin. 12: 79 (1884)
    Caralluma socotrana (Balf. f.) N.E. Br. in Gard. Chron., ser. 3, 12: 370 (1892); Blundell, Wild
    Fl. E. Afr.: 144 (1987); U.K.W.F. ed. 2: 185 (1994)
    Sanguilluma socotrana (Balf. f.) Plowes in Haseltonia 3: 66 (1995); Lavranos in Fl. Somalia
    3: 178 (2006)
    Caralluma rosengrenii Vierhapper in Österr. Bot. Zeitschr. 55: 91 (1905), as *Coralluma*. Type:
    Yemen, Socotra, foot of Derafonte Mts, 18 Feb. 1899, *Pauley* s.n. (WU, holo., WU-spirit
    0042712)
    C. corrugata N.E. Br. in K.B. 1912: 280 (1912). Type: Somalia, sine loc., *Drake Brockman* 477
    & 478 (K, syn.)
    C. rivae Chiov. in Fl. Somala 1: 222 (1929). Type: Ethiopia, Ogaden, 5 hrs from Lafarug, 10
    Dec. 1892, *Ruspoli & Riva* s.n. (FT?, not traced).

61. **EDITHCOLEA**

N.E. Br. in K.B. 1895: 220 (1895)

Small, creeping stem-succulents; latex clear or at most cloudy. Stems cylindrical, roundly 4–5-angled, with conical tubercles. Leaf rudimentary; stipules reduced to a few trichomes. Inflorescences near tip of stems, 1-(rarely 2–)flowered; pedicels short, thickish. Corolla rotate, fused for almost ³⁄₄ of length, fleshy; tube with small annulus encircling gynostegium and corona in total. Corolline corona absent. Gynostegial corona in 2 series: "outer" corona with 5 free staminal and 5 interstaminal parts fused basally to a ring-like structure, rotate; interstaminal corona with emarginate-dentate margin; "inner" staminal corona lobes connivent-erect. Anther wings basally forming a distinct mouth. Pollinia ovoid. Mericarps two, fusiform, winged.

Monotypic genus closely related to *Monolluma*.

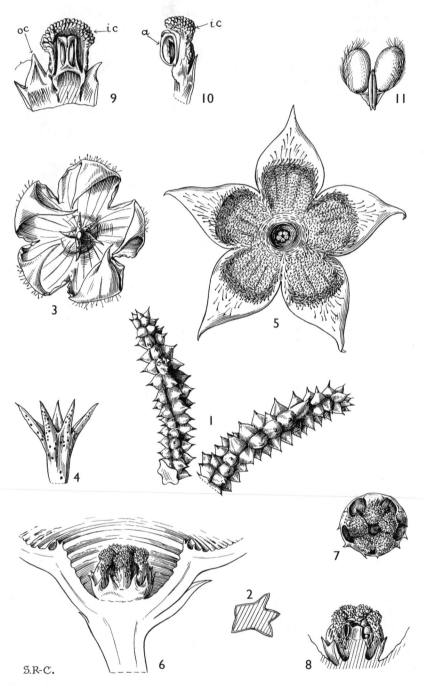

Fig. 78. *EDITHCOLEA GRANDIS* — **1**, part of stem, × 1; **2**, cross-section of stem, × 1; **3**, flower from the back, × 1; **4**, calyx, × 2; **5** flower from above, × 1; **6**, section through corolla exposing gynostegium, × 4; **7**, gynostegium from above, × 4; **8**, gynostegium with section removed, × 4; **9**, stamen and associated corona from within, × 8; **10**, anther and inner corona lobe, lateral view, × 8; **11**, pollinarium, × 24. a – anther; ic – inner corona; oc-outer corona. From *Greenway* 4928. From Hook. Ic. Pl. t. 3413, drawn by Stella Ross-Craig.

Edithcolea grandis *N.E. Br.* in K.B. 1895: 220 (1895) & in F.T.A. 4(1): 492 (1903); F.P.U.: 121 (1962); Field in Bot. Mag. 177: t. 562 (1970); Blundell, Wild Fl. E. Afr.: t. 551 (1987); U.K.W.F. ed. 2: 185 (1994); Gilbert in Fl. Eth. 4(1): 184 (2003); Lavranos in Fl. Somalia 3: 184 (2006). Type: Somalia, Henweina Valley, about 3000 ft., *Cole* s.n. (K, syn.) & *Lort Phillips* s.n. (K, syn.).

Creeping stem-succulents, 5–15 cm high, richly branching, mat-forming; latex clear or at most cloudy. Stems green or grey-green, cylindrical, 2–30 cm long, 10–18 mm wide, roundly 4–5-angled, with dark green or reddish markings around conical tubercles. Leaf rudiments flattened conical, hard, 1–2 × 1 mm, acute, light brown. Inflorescences near tip of stems, 1-(rarely 2–)flowered, sessile. Sepals lanceolate. Corolla fleshy, rotate, 8–12 cm diameter, abaxially light brownish, adaxially creamish-yellow, maculate with purplish red to red-brown, rugose, papillose; tube 5–6 mm deep and high; corolla lobes triangular, marginal and apical areas with emergences tipped by a long trichome. Gynostegial corona sessile, "outer" fleshy, ivory or brownish; "inner" staminal corona lobes connivent-erect, 2–3 mm long, clavate, profusely tuberculate. Mericarps two, 9–20 cm long, fusiform, winged; seeds 7–8 × 5.5 mm, broadly ovate, wing 1.5 mm wide, flat. Fig. 78, p. 324.

UGANDA. Karamoja District: Amudat, Feb. 1946, *Tweedie* 669
KENYA. Northern Frontier District: Dandu, 14 Apr. 1952, *Gillett* 12792; Turkana District: Loro Lodwar area, 18 Sep. 1963, *Paulo* 957; Embu District: 2 km from Ishiara town on Ena road, 28 July 2001, *Kirika et al.* in NMK 149
TANZANIA. Masai District: near Longido on Nairobi road, 29 Mar. 1970, *Richards* 25700; Lushoto District: 5 km NW of Mombo, 29 Apr. 1953, *Drummond & Hemsley* 2286; Pare District: near W end of Kihurio, 3 Dec. 1992, *Newton & Carter* 4181
DISTR. U 1; K 1–4, 7; T 2, 3; also in Ethiopia, Somalia and Yemen (Socotra)
HAB. Dry, deciduous bushland, mostly on rocky ground; 400–1350 m

SYN. *Edithcolea sordida* N.E. Br. in Forbes, Nat. Hist. Sokotra & Abd el Kuri: 486 (1903). Type: Socotra, on the Garrieh plain and on E slopes of Hughier [Hagghiher] Mountains, *Forbes* 145 (K, holo.)
  Edithcolea grandis N.E. Br. var. *baylissiana* Lavranos & D.S.Hardy in J. South Afr. Bot. 29: 21 (1963). Type: Tanzania, Pare District: 3 km N of Kihurio between Southern Pare Hills and Usambara Mts, *Bayliss* 74 (PRE, holo.)

62. **ECHIDNOPSIS**

Hook. f. in Bot. Mag.: t. 5930 (1871); Bruyns in Bradleya 6: 1–48 (1988); Plowes in Haseltonia 1: 65–85 (1993); Thiv & Meve in Pl. Syst. Evol. 265: 71–86 (2007)

Virchowia K. Schum. in Monatsschr. Kakteenk. 3: 101 (1893), *non Virchowia* A. Schenk (1852)
Pseudopectinaria Lavranos in Cact. Succ. J. (Los Angeles) 43: 9 (1971)

Small, creeping stem-succulents, occasionally rhizomatous; latex clear or at most cloudy; stems cylindrical, roundly 5–20-angled, tubercles cushion-shaped, often papillose and/or rugose. Leaf rudiments sessile, withering soon, reflexed to ascending; stipules ovate or globoid, glandular. Inflorescences usually appearing at upper regions of stems, 1–6-flowered, bostrychoid, sessile. Corolla fleshy, rotate-campanulate, globose or tubular, fused to ¹/₅ of length, glabrous, papillose or with trichomes. Gynostegial corona in 2 series: "outer" corona of 5 staminal and 5 interstaminal parts fused almost completely to form a fleshy, rotate, cyathiform, urceolate or subglobose structure, occasionally with trichomes; interstaminal parts fused as described above, occasionally spur-shaped (*E. dammanniana* Spreng.) or saccate; "inner" staminal parts subulate, lingulate or triangular. Gynostegium sessile. Pollinia minute, globoid or ovoid; caudicles trapezoidal. Mericarps two; seeds winged.

28 species from East Africa and Arabia

Echidnopsis is well characterized by its small, creeping, tessellate stems. However, *Rhytidocaulon* not only shows many similarities with *Echidnopsis*, it also shares a clade with East African species in the molecular phylogeny of Thiv & Meve in Pl. Syst. Evol. 265: 71–86 (2007). Therefore, *Echidnopsis* is paraphyletic without *Rhytidocaulon*, and, possibly, both are to be taxonomically unified in future.

1. Stem surface sculptured (rugose, papillate or
   pubescent) 2
   Stem surface not noticeably sculptured 5
2. Stem surface markedly rugose or pubescent; leaf
   rudiments spreading 3
   Stem surface papillate or faintly rugose; leaf
   rudiments erect/ascending 4
3. Stem surface rugose; leaf rudiments glabrous
   (occasionally papillate) .... 8. *E. leachii* (p. 331)
   Stem surface ± pubescent; leaf rudiments hispid 13. *E. malum* (p. 334)
4. Corolla surface with globoid papillae; corolla
   tube encircling corona .... 2. *E. virchowii* (p. 327)
   Corolla surface with conical papillae; corona
   exposed from corolla tube .... 1. *E. dammanniana* (p. 327)
5. Stems with 14–20 angles (ribs); leaf rudiments
   2.5–3.5 mm long .... 12. *E. urceolata* (p. 333)
   Stems with 6–12 angles (ribs); leaf rudiments at
   the most 2(–3) mm long 6
6. Corolla clearly campanulate to urceolate 7
   Corolla rotate, rotate-campanulate or funnel-
   shaped 10
7. Corona conspicuously cup-shaped with entire
   upper limb; staminal corona lobes short, not
   exceeding anthers (and coronal cup) .... 7. *E. archeri* (p. 330)
   Corona shallowly cup-shaped with notched
   upper limb; staminal corona lobes long,
   connivent erect, much exceeding anthers 8
8. Corolla tube ovoid to globoid, mouth of tube ±
   3 mm in diameter .... 9. *E. radians* (p. 332)
   Corolla tube inversely pear-shaped to urceolate,
   mouth of tube ± 2 mm in diameter 9
9. Corolla lobes deltoid, 1–2 mm long, erect to
   spreading .... 11. *E. ericiflora* (p. 333)
   Corolla lobes small deltoid to subulate, 6–11 mm
   long, spreading to recurved .... 10. *E. watsonii* (p. 332)
10. Leaf rudiments 2–3 mm long, lanceolate, persistent
    as white spines .... 3. *E. angustiloba* (p. 328)
    Leaf rudiments 1–2 mm long, caducous 11
11. Stems 6–8 angled; corolla rotate; corona excerted,
    stipular glands absent .... 5. *E. sharpei* (p. 329)
    Stems 8–11-angled; corolla rotate-campanulate;
    corona encircled by corolla tube; stipular
    glands present 12
12. Stems creeping; leaf rudiments lanceolate;
    corolla pale yellowish-brown .... 6. *E. repens* (p. 330)
    Stems procumbent-ascending to erect; leaf
    rudiments deltoid-cordate, thickish; corolla
    wine-red .... 4. *E. scutellata* (p. 328)

The species in this account have been grouped artificially by gross corolla morphology: species 1–7 have rotate or shallowly campanulate corollas, species 11–14 have urceolate or subglobose corollas, and the remainder are somewhat intermediate, or more variable in corolla form.

1. **Echidnopsis dammanniana** *Spreng.* in Cat. Dammann & Co.: 4, fig. 5 (1892); Bruyns in Bradleya 6: 12 (1988); Plowes in Haseltonia 1: 69 (1993); Gilbert in Fl. Eth. 4(1): 179 (2003); Lavranos in Fl. Somalia 3: 186 (2006); Thiv & Meve in Pl. Syst. Evol. 265: 73 (2007). Type: Ethiopia, Harerge Region, *Gilbert* 2374 (K, neo., designated by Bruyns l.c.; ETH, iso.)

Stems creeping to ascending, brownish- or grey-green, 5–15(–60) cm long, 1.2–2 cm diameter, 8-angled, often tinged purplish, faintly papillate and rugose; tubercles 4(–6)-angled. Leaf rudiments ascending, deltoid, ± 1 mm long, thickish, papillate. Inflorescences 1- to 3-flowered; pedicels <1 mm long. Sepals deltoid, finely papillose. Corolla recurved, rotate, 0.7–1.1 cm diameter, appressed to stem surface, abaxially greenish-brown, with scattered papillae, adaxially greenish spotted purplish (tube) or purplish (lobes), papillate and setose; lobes ovate-deltoid or -lanceolate, 2.5–3.2 × ± 2.4 mm. Gynostegial corona discoid, blackish-purple, sometimes mottled with yellow, brilliant due to nectar coat; interstaminal corona lobes deltoid to lanceolate, 0.2–0.5 mm long, spur-shaped, channelled above, spreading-ascending; staminal corona lobes truncate-deltoid, ± 0.4 mm long, covering anthers, outer dorsal-marginal zones forming a small limb together with the interstaminal parts. Mericarps two, 6–10 cm long, slender fusiform (narrowly oblong), seeds ± 3 mm long, 1.5 mm wide, ovate, winged.

KENYA. Northern Frontier District: Dandu, 24 Mar. 1952, *Gillett* 12628 & Mt Kulal, S end, Oct. 1947, *Bally* 8409 (≡S 176) & 3 km S Marsabit, 11 Mar 1987, *Hartmann & Newton* 21352!
DISTR. **K** 1; Ethiopia and Somalia
HAB. Rocky ground; 900–1500 m

SYN. *Echidnopsis somalensis* N.E. Br. in F.T.A. 4(1): 477 (1903). Type: Somalia, sin loc., *Lort Phillips* s.n. (K, holo.)

2. **Echidnopsis virchowii** *K. Schum.* in Monatsschr. Kakt. kunde 3: 98 (1893); Bruyns in Bradleya 6: 13 (1988); Plowes in Haseltonia 1: 69 (1993); Lavranos in Fl. Somalia 3: 186 (2006); Thiv & Meve in Pl. Syst. Evol 265: 75 (2007). Type: Tanzania, region near Tanga, *Hildebrandt* s.n., cult. Berlin 3 Oct. 1887 (K, holo.)

Stems procumbent, cylindric, green or (brownish-) grey, 5–20 cm long, 1.5–2 cm diameter, 6–8-angled, finely papillate; tubercles tetragonal, rather flat. Leaf rudiments ascending, thickish, (ovate-)deltoid, ± 0.7 mm long, papillate. Inforescences 1–3-flowered; pedicels < 1 mm long. Sepals (ovate-)triangular, 1 mm, acute, finely papillose. Corolla rotate, flat, appressed to stem, 7–13 mm diameter, abaxially grey-green, scattered papillate, adaxially greenish, densely spotted purple- or red-brown, densely papillate all over; tube flat; lobes reflexed, ovate to deltoid, 3–5.5 × 2–3 mm, fully expanded. Gynostegial corona purple-red, fused to form a discoid structure, 2.5–2.8 mm diameter, yellow towards nectar cavities; interstaminal corona lobes basin-shaped, with crenulate or dentate limb; staminal corona lobes deltoid, ± 1 mm long, with the broad and emarginate back confluent with interstaminal corona, decumbent on anthers, longer than anthers. Mericarps two, 6–10 cm long, slender fusiform (narrowly oblong), seeds ± 3 mm long, 1.5 mm wide, ovate, winged.

TANZANIA. Tanga District: near Tanga, cult. Berlin, 3 Oct. 1887, *Hildebrandt* s.n.
DISTR. **T** 3; N Somalia (Erigavo)
HAB. Rock crevices on bare rocky hillsides; 0–1200 m

SYN. *Echidnopsis stellata* Lavranos in Cact. Succ. J. (Los Angeles) 46: 182 (1974). Type: Somalia,
    Erigavo, escarpment below Tabah Pass, *Lavranos* F342 (FT, holo.)
    E. virchowii K. Schum. var. *stellata* (Lavranos) Plowes in Haseltonia 1: 69 (1993)

3. **Echidnopsis angustiloba** *E.A. Bruce & P.R.O. Bally* in Cact. Succ. J. (Los Angeles)
13: 180 (1941); Bruyns in Bradleya 6: 15 (1988); Plowes in Haseltonia 1: 70 (1993);
U.K.W.F. ed. 2: 184 (1994); Thiv & Meve in Pl. Syst. Evol. 265: 73 (2007). Type: Kenya,
Northern Frontier District: 8 km from Archer's Post on road to Isiolo, *Copley* s.n. in
Bally S26 (K, lecto.; ZSS, iso.)

Stems creeping to ascending, grey-green, 5–12 cm long, 1–2 cm diameter, 10–11-
angled, smooth; tubercles polygonal to rounded, conical. Leaf rudiments lanceolate,
2–3 mm long, acute, finely papillate, tapering towards apex from a broad base,
caducuous, peristent as white spines. Inflorescences 1–3-flowered; pedicels ± 1.5 mm
long. Sepals narrowly ovate-lanceolate, to 2 mm long, glabrous. Corolla yellow, rotate
to shallowly campanulate, 10–15 mm diameter, adaxially occasionally tinged purplish
or purplish, glabrous, adaxially occasionally with scattered white hairs; tube ± 1 mm
long; lobes 3–6 mm long, (narrowly) triangular, tapering into a pointed apex,
margins slightly to (apically) stronger revolute. Gynostegial corona yellow, bluntly
pentagonal, 2.3–3 mm diameter, brilliant due to nectar coat; interstaminal corona
fused to form a shallow cup, upper rim weakly to conspicuously dentate, adaxially
occasionally with few short hairs; staminal corona lobes connivent-erect, linear from
a deltoid base, ± 0.4–0.8 mm long, narrow, usually surpassing level of interstaminal
corona. Mericarps two, 6–10 cm long, slender fusiform (narrowly oblong), seeds ± 3 mm
long, 1.5 mm wide, ovate, winged.

KENYA. Northern Frontier District: 5 km N of Longobito, 22 Feb. 1996, *Goyder et al.* 4059
    (≡*Meve et al.* 959!) & 8 km from Archer's Post on road to Isiolo, 18 Dec. 1939, *Copley* s.n. in
    Bally S26; Laikipia District: Rumuruti, Dec. 1983, *Powys* 524
DISTR. **K** 1, 3; not known elsewhere
HAB. Succulent thornbush, open succulent communities (*Sansevieria robusta*), also on laterite;
    1000–1800 m

4. **Echidnopsis scutellata** *(Deflers)* A. Berger in Stap. Klein.: 26 (1910); Plowes in
Haseltonia 1: 70 (1993); Thiv & Meve in Pl. Syst. Evol. 265: 74 (2007). Type: Yemen,
Bilad Subehi, Wadi M'Adi, *Deflers* 1167 (G, holo.).

Stems procumbent-ascending to erect, profusely branching, green, occasionally
brownish-green to purplish (in full sunlight), 2–9 cm long, ± 1 cm diameter, 8–11-
angled; tubercles conical to hemispherical, 6-angled, polygonal to rounded. Leaf
rudiments spreading to reflexed, broadly deltoid to cordate, 1–2 mm long, scattered-
papillate, thickish, caducuous; stipular glands present. Inflorescences 1-to 2-
flowered, near tip of stems; pedicels 1.5 mm. Sepals lanceolate, ± 1 mm long. Corolla
rotate to shallowly campanulate, 6–12 mm diameter, abaxially pale green, adaxially
pale yellowish, occasionally suffused brown or spotted red; corolla broadly deltoid,
1.5–2 × 2.5–3 mm, apically recurved, margins slightly recurved revolute, occasionally
ciliate. Gynostegial corona yellow, yellowish-brown to purple-brown, flattened cup-
shaped, 2–2.5 mm diameter, standing free or embedded in funnel-shaped corolla
tube, circular to (sub)pentagonal, margins only very slightly emarginate; staminal
corona lobes linear-subulate, 0.5–0.8 mm long, basally decumbent-ascending, then
connivent-erect to form a column, surpassing level of coronal cup. Mericarps two,
6–10 cm long, slender fusiform (narrowly oblong), seeds ± 3 mm long, 1.5 mm wide,
ovate, winged.

SYN. *Caralluma scutellata* Deflers in Bull. Soc. Bot. Fr. 43: 114 et t. 4 (1896) subsp. **australis** *Bruyns* in Bradleya 6: 19 (1988); Thiv & Meve in Pl. Syst. Evol. 265: 74 (2007). Type: Kenya, Northern Frontier District, Gof Choba, Marsabit, *Bally* 12565 (K, holo.)

Stems 8–11-angled, tubercles conical, rounded. Leaf rudiments broadly deltoid to cordate, ± 1.2 mm long, spreading. Corolla pale (yellow-)brown, 6–9 mm diameter. Gynostegial corona embedded in funnel-shaped corolla tube, yellowish-brown, occasionally spotted red.

KENYA. Northern Frontier District: 48 km on the Ramu–El Wak road, May 1978, *Gilbert & Thulin* 1635! & Gof Choba, Marsabit, *Lavranos & Bleck* 19527 & Gof Choba, Marsabit, *Bally* 12565
DISTR. **K** 1; Ethiopia
HAB. Under bushes in *Acacia-Commiphora* bushland; 300–600 m

SYN. *Echidnopsis mariae* Lavranos in Cact. Succ. J. (Los Angeles) 54: 215 (1982). Type: Kenya, Marsabit, Gof Choba crater, *Lavranos & Bleck* 19527 (E holo.; EA, K, iso.)

5. **Echidnopsis sharpei** *A.C. White & B. Sloane* in Cact. Succ. J. (Los Angeles) 11: 67 (1939); U.K.W.F. ed. 2: 184 (1994); Thiv & Meve in Pl. Syst. Evol. 265: 82 (2007). Type: Kenya, Northern Frontier District: Baragoi, 48 km S of Lake Turkana, 1937, *Sharpe & Jex-Blake* s.n. in *Bally* S/7 (K, lecto., designated by Bruyns in Bradleya 6: 33 (1988); ZSS, iso.)

Stems creeping to ascending, greyish- or reddish-green, 5–15 cm long, 8–15 mm diameter, mostly 6-, rarely 8-angled; tubercles evenly hexagonal, slightly convex. Leaf rudiments usually strongly reflexed, deltoid, ± 1 mm long, withering rapidly. Inflorescences mostly 2-flowered; pedicels ± 1 mm long. Sepals ovate-lanceolate, 1–1.5 mm long, tips recurved. Corolla rotate, 6–10 mm diameter, abaxially greyish-green or green mottled with purple, adaxially velvety deep wine-red; tube shallowly campanulate, pentagonal, framing the corona but enclosing just its basal third; lobes spreading to reflexed, deltoid, 3–4.5 mm long, margins and tips revolute, with scattered to dense hairs, up to 1 mm long, white or red. Gynostegial corona sessile to shortly stalked, cup-shaped to suburceolate, 1–1.5 × ± 2.5 mm, circular or pentagonal, bright yellow or yellow or whitish patterned red, upper rim entire to finely dentate, abaxially occasionally with tufted hairs, staminal corona lobes narrowly deltoid, at the most as long as anthers, decumbent on anthers. Mericarps two, 6–10 cm long, slender fusiform (narrowly oblong), seeds ± 3 mm long, 1.5 mm wide, ovate, winged.

subsp. **sharpei**

Stems mostly 6-angled (though sometimes 8-angled, too), gynostegial corona sessile.

KENYA. Northern Frontier District: 95 km S of Maralal, *Lavranos & Newton* 17698!; Baringo District: Parmalok Island, Lake Baringo, 31 Oct. 1992, *Harvey, Mungai & Vollesen* 23; Tana River District: 13 km from Galole on Garsen road, 10 July 1974, *Faden & Faden* 74/1045
DISTR. **K** 1–3, 7; Kenyan endemic
HAB. Rocky slopes with *Adenium* and *Aloe*, or in dense vegetation under *Acacia, Grewia, Cadaba* and *Maerua*; 400–1800 m

NOTE. Bruyns in Bradleya 6: 33 (1988) subsumed *E. bavazzanoi* Lavranos from Ethiopia and Somalia with similar flowers under *E. sharpei*. However, this is not supported: by different ploidy levels (*E. sharpei* is diploid whereas *E. bavazzanoi* is tetraploid), by the stems that are always 8-angled in *E. bavazzanoi*, by the gynostegial corona which is sessile in *E. sharpei* but on a short stalk in *E. bavazzanoi*, and also by the phylogeny as presented by Thiv & Meve in Pl. Syst. Evol. 265: 79 (2007). Thiv & Meve therefore reinstated *E. bavazzanoi* treating it at subspecific rank as *E. sharpei* subsp. *bavazzanoi*.

6. **Echidnopsis repens** *R.A. Dyer & I. Verd.* in Cact. Succ. J. (Los Angeles) 11: 68 (1939); Plowes in Haseltonia 1: 81 (1993); Thiv & Meve in Pl. Syst. Evol. 265: 82 (2007). Type: Tanzania, Arusha District: Aldenyo, near Mt Meru, *Pole-Evans & Erens* 1020 (PRE, holo.)

Stems prostrate, rooting almost over whole length, green, 5–15 cm long, 6–10 mm diameter, 8–10-angled; tubercles elongated, quadrangular to hexagonal, only slightly convex. Leaf rudiments lanceolate, 1–1.5 mm long, acute, withering rapidly. Inflorescences 1–3-flowered; pedicels 2 mm long. Sepals deltoid-lanceolate, 2–2.5 mm long. Corolla rotate, 7–9 mm diameter, abaxially greenish, adaxially purple-red to wine-red, usually with some scattered long hairs around margins of lobes and on tube; tube campanulate, strictly framing the corona; lobes ascending to spreading, ovate-deltoid or ovate, 4.5–6 mm long, occasionally slightly reflexed, sometimes tipped yellow, margins reflexed. Gynostegial corona purplish-red, cup-shaped, ± 1 × 2.5 mm, prominently pentagonal with rounded edges; staminal corona lobes variable, nearly absent to rectangular and then covering the anthers, purplish-red. Mericarps two, 4–10 cm long, slender fusiform (narrowly oblong), seeds ± 3 mm long, 1.5 mm wide, ovate, winged.

KENYA. Masai District: 4 km S of Ilbisil, *Lavranos & Bleck* 19571!; Teita District: Mt Maktau, 14 Feb. 1996, *Goyder et al.* 4032!; Kwale District: Kilibasi, 8 Jan. 1988, *Luke* 919
TANZANIA. Arusha District: Aldenyo, near Mt Meru, 1938, *Pole-Evans & Erens* 1020; Masai District: Engare Nanyuki, 17.5 km N, 20 Mar. 1966, *Greenway & Kanuri* 12448 & Engare Nanyuki, 29 Dec. 1970, *Richards* 26594
DISTR. **K** 6–7; **T** 2; not known elsewhere
HAB. Shaded by shrubs in sandy to loamy soil (pockets); 450–1600 m

SYN. *Echidnopsis sharpei* A.C. White & B. Sloane subsp. *repens* (R.A. Dyer & I. Verd.) Bruyns in Bradleya 6: 37 (1988)

7. **Echidnopsis archeri** *P.R.O. Bally* in Cact. Succ. J. Gr. Brit. 19: 59 (1957); Bruyns in Bradleya 6: 40 (1988); Plowes in Haseltonia 1: 82 (1993); U.K.W.F. ed. 2: 184 (1994); Lavranos in Fl. Somalia 3: 186 (2006); Thiv & Meve in Pl. Syst. Evol. 265: 73 (2007). Type: Kenya, Masai District: Nguruman Escarpment, Oloibitato River near Hayton's Falls, *Archer* s.n. in *Bally* S235 (K, holo.; ZSS, iso.)

Stems decumbent, dark green, 3–6 cm long, 1–1.5 cm diameter, 8-angled, finely papillate; tubercles hexagonal. Leaf rudiments reflexed, deltoid, 2–3 mm long, papillate, withering soon, caducuous or remaining as white remnants. Inflorescences mostly close to apex, 1–2-flowered; pedicels 1 mm long. Sepals ovate-lanceolate, 1–2 mm, abaxially papillate, apex mostly recurved. Flowers spreading or facing downwards. Corolla deeply cupular to bell-shaped, 5–8 mm long, 5–6 mm diameter, abaxially purple to purplish-red (sometimes pink), smooth, adaxially dark velvety purple to crimson, smooth to slightly rugose, occasionally with some hairs; tube 3–4 mm long, 3–4 mm diameter; lobes triangular, erect, apex recurved, pointed, margins revolute, occasionally finely ciliate. Gynostegial corona pentagonal, ± 1.2 × 2–2.2 mm, light yellow, occasionally the outer margins purplish; interstaminal corona fused to form a shallow cup overtopping staminal corona, upper rim weakly to conspicuously dentate, adaxially mostly with stiff, short, white hairs at inner side; staminal corona lobes short, deltoid, glabrous, shorted than anthers, appressed to anthers, backwards fused to coronal cup. Mericarps two, 6–10 cm long, slender fusiform (narrowly oblong), seeds ± 3 mm long, 1.5 mm wide, ovate, winged.

KENYA. Masai District: Kedong Valley 24 km from junction with Ngong West on Narok road, 27 Feb. 1994, *Newton* 4451 & Nguruman Escarpment, *McCoy* s.n.! & Oloibitato River near Hayton's Falls, 11 Sep. 1952, *Archer* s.n. in *Bally* S235!
DISTR. **K** 6; Somalia
HAB. Rock pavement; 700–1800 m

SYN. *Echidnopsis similis* Plowes in Haseltonia 1: 82 (1993). Type: Somalia, north slope of Ga'an Libah, NE of Hargeisa, *Bally* 11726 (K, holo.)

8. **Echidnopsis leachii** *Lavranos* in Natl. Cact. Succ. J. 27: 69 (1972); Bruyns in Bradleya 6: 8 (1988); Plowes in Haseltonia 1: 67 (1993); Thiv & Meve in Pl. Syst. Evol. 265: 83 (2007). Type: Tanzania, Kilosa/Iringa District: 48 km W of Mikumi, Ruaha River Gorge, *Leach & Brunton* 10143 (EA, holo.; K, ZSS, iso.)

Stems procumbent-ascending to 15 cm long, glaucous-green, greyish-green or brownish, 0.8–1 cm diameter, 6-angled, rugose; tubercles hexagonal. Leaf rudiments broadly deltoid, ± 0.8 mm long, pointed, caducuous. Inflorescences 2-flowered; pedicels less than 1 mm long. Sepals lanceolate, 1–1.5 mm long, pointed. Corolla dark purple to pink, occasionally yellowish, shallowly campanulate, 3–8 mm diameter, or urceolate, 3.5–6 mm long (viz. "*E. oviflora*"), glabrous, abaxially glossy; lobes ascending to erect, deltoid, 0.8–2.5 × 1–1.8 mm. Gynostegial corona decagonal, pentagonal or ± circular, 2–2.5 mm diameter, yellow, yellow suffused with purple, or purple; interstaminal corona fused to form a shallow cup, glabrous, margins occasionally incurved, upper rim dentate; staminal corona lobes very short, apically rounded or triangular, appressed to anther, much shorter than anthers, bases of corona lobes extended and fused into coronal cup. Mericarps two, 6–10 cm long, slender fusiform (narrowly oblong), seeds ± 3 mm long, 1.5 mm wide, ovate, winged. Fig. 79, p. 331.

TANZANIA. "Morogoro Province", *Specks* 1075!; Kilosa District: Mbaga Mt, N of the track from Malolo to Idodoma, *Specks* 1264! (≡ *McCoy* 2392); Kilosa/Iringa Districts: 48 km W of Mikumi, Ruaha River Gorge, June 1960, *Leach & Brunton* 10143!

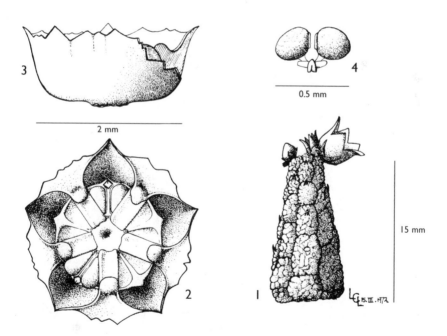

FIG. 79. *ECHIDNOPSIS LEACHII* — **1**, flowering shoot; **2**, gynostegium and surrounding corona lobes from above; **3**, lateral view of corona surrounding gynostegium; **4**, pollinarium. From *Leach & Brunton* 10143. From Natl. Cact. Succ. J. 27: 69, drawn by L.C. Leach.

DISTR. **T** 5–7; not known elsewhere
HAB. Rocky, volcanic slopes, in shrubby vegetation; 500–700 m

SYN. *Echidnopsis oviflora* T.A. McCoy in Kakt. & Sukk. 54(8): 214 (2003). Type: Tanzania, Kilosa District: Mbaga Mt, N of the track from Malolo to Idodoma, *Specks* 1264 (≡ *McCoy* 2392); (MO, holo.; UBT!, UPS, iso.)

NOTE. The morphological variability of the flower of *C. leachi* from open campanulate to urceolate is unmatched in the stapeliads.

9. **Echidnopsis radians** *Bleck* in Cact. Succ. J. (Los Angeles) 49: 263 (1977); Plowes in Haseltonia 1: 76 (1993); Thiv & Meve in Pl. Syst. Evol. 265: 74 (2007). Type: Kenya, Northern Frontier District, near Barsaloi, *Powys* s.n. in *Lavranos* 12554 (MO, holo.; E, EA, ZSS, iso.)

Stems procumbent-ascending to 20 cm long, bluish grey-green, 8–10 mm diameter, 10–12-angled; tubercles hexagonal, hemispherical. Leaf rudiments spreading, deltoid, ± 0.8 mm long, caducuous without leaving a scar. Inflorescences 1–2-flowered; pedicels ± 1 mm long. Sepals 2–3 mm long, lanceolate. Corolla abaxially purple, campanulate to suburceolate, 5–8 mm diameter in top view, glabrous; tube 2–5 mm long, mouth 3 mm wide, adaxially yellowish or whitish, with hairs; lobes (sub)erect to spreading horizontally, deltoid to lanceolate, 1.5–4 mm long, occasionally slightly recurved (apically), margins recurved, adaxially yellowish, glabrous. Gynostegial corona purple and yellow, circular or bluntly 5- or 10-angled, 2–3 mm diameter; interstaminal corona fused to form a shallow cup, upper rim recurved, weakly to conspicuously dentate; staminal corona lobes connivent-erect, narrowly linear from a thickened base, 0.3–0.8 mm long, surpassing level of interstaminal corona. Mericarps two, 6–10 cm long, slender fusiform (narrowly oblong), seeds ± 3 mm long, 1.5 mm wide, ovate, winged.

KENYA. Northern Frontier District: Barsaloi, *Powys* s.n. sub *Lavranos* 12554!; Baringo District: 10 km N Loruk, Lake Baringo, *Lavranos* 17328, cult. sub *Plowes* 6311!; Teita District: Tsavo East National Park, SW Sobo Rock, 67 km from Voi Gate, 3 Jan. 1967, *Greenway & Kanuri* 12954
DISTR. **K** 1, 3, 7; Somalia
HAB. Shallow soil pocket on rock pavement; 800–1200 m

SYN. *Echidnopsis modesta* Plowes in Haseltonia 1: 77 (1993). Type: Kenya, Tanaland, 50–65 km W of Garissa, Jan. 1940, *Ritchie* s.n. in *Bally* S47 (K, holo.)

10. **Echidnopsis watsonii** *P.R.O. Bally* in Candollea 18: 343 (1963); Plowes in Haseltonia 1: 75 (1993); U.K.W.F. ed. 2: 184 (1994); Lavranos in Fl. Somalia 3: 187 (2006); Thiv & Meve in Pl. Syst. Evol 265: 75 (2007). Type: Somalia, Northern Province, 13 km SE of Borama, *Bally & Watson* 9997 (K, holo.)

Stems creeping or prostrate, green or grey-green, 5–20 cm long, 0.6–0.8 cm diameter, 8–10-angled; tubercles tetragonal or indistinct hexagonal, flattened to convex. Leaf rudiments spreading to reflexed, lanceolate from a broadened base, 1.5–2 mm long, finely papillate, withering very soon, remaining as white spine. Inflorescences usually 1-flowered; pedicels to 1 mm long. Sepals lanceolate from an ovate base, 2–3 mm long. Corolla urceolate with long small corolla lobes, abaxially purplish-red, glabrous; tube yellow to whitish, obovoid to inverted pear-shaped, 5–11 mm long, 4–8 mm wide, with scattered hairs inside; lobes spreading to recurved, small deltoid to subulate, 6–11 mm long, margins reflexed, yellow. Gynostegial corona shallowly cup-shaped, 1.5–2.2 mm wide, circular or obtuse pentagonal, upper rim dentate; staminal corona lobes linear, ± 0.7 mm long, decumbent-ascending, then connivent-erect to form a short column, surpassing level of coronal cup. Pollinaria ovoid. Mericarps two, 6–10 cm long, slender fusiform (narrowly oblong), seeds ± 3 mm long, 1.5 mm wide, ovate, winged. Fig. 80, p. 333.

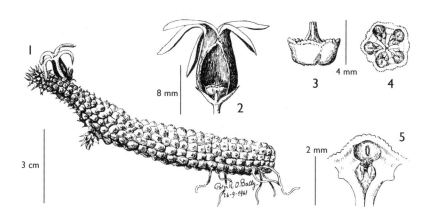

FIG. 80. *ECHIDNOPSIS WATSONII* — **1**, flowering stem; **2**, corolla partially cut away to reveal gynostegium; **3**, corona surrounding gynostegium, lateral view and **4**, from above; **5**, closer view of one section of gynostegium and corona, revealing position of pollinarium, from above. From *Bally & Watson* 9997. From Candollea 18: 344, drawn by P.R.O. Bally.

KENYA. Northern Frontier District: ± 30 km from Baragoi to Maralal, 1° 29' N, 36° 42 E, *Hartmann* & *Newton* 21332!
DISTR. **K** 1; Somalia
HAB. Dry, rocky slopes; 400–1050 m

11. **Echidnopsis ericiflora** *Lavranos* in Natl. Cact. Succ. J. 27: 70 (1972); Bruyns in Bradleya 6: 27 (1988); Plowes in Haseltonia 1: 77 (1993); Thiv & Meve in Pl. Syst. Evol. 265: 73 (2007). Type: Kenya, Teita District: 42 km ESE of Voi on Mombasa road, *Lavranos* 9305 (EA!, holo.; PRE, ZSS, iso.)

Stems creeping and rooting over their whole length, grey-green, to 20 cm long, 4–8 mm diameter, 6–8-angled, finely papillate; tubercles hexagonal. Leaf rudiments reflexed, deltoid, ovate-deltoid, 2–3 mm long, papillate, withering soon, caducuous or remaining as dry, white remnants. Inflorescences close to apex, 2-flowered; pedicels 1–1.5 mm long. Sepals ovate, 1.5 mm long, acute, tips recurved. Corolla wine-red, urceolate, 5–8 mm long, abaxially glossy; tube 4–4.5 mm diameter, at mouth ± 2 mm diameter, bottom of tube inside with scattered hairs; lobes erect to spreading when fully expanded, adaxially yellowish to bright reddish, deltoid, 1–2 mm long. Gynostegial corona yellow with purple margins, pentagonal, 2.7–3 mm diameter; interstaminal corona fused to form a shallow cup, upper rim recurved, occasionally weakly dentate, glabrous; staminal corona lobes linear from a deltoid base, ± 1 mm long, narrow, connivent-erect, surpassing level of interstaminal corona, tips finely papillate. Mericarps two, 6–10 cm long, slender fusiform (narrowly oblong), seeds ± 3 mm long, 1.5 mm wide, ovate, winged.

KENYA. Teita District: 42 km ESE of Voi on road to Mombasa, 15 Jan. 1972, *Lavranos* 9305!; Kilifi District: Lali Hills, Galana River area, Nov. 1985, *Powys & Heath* 793
DISTR. **K** 7; not known elsewhere
HAB. Dry bushland on red sandy soil; 250–450 m

12. **Echidnopsis urceolata** *P.R.O. Bally* in Candollea 18: 341 (1963); Bruyns in Bradleya 6: 29 (1988); Plowes in Haseltonia 1: 78 (1993); Lavranos in Fl. Somalia 3: 189 (2006); Thiv & Meve in Pl. Syst. Evol. 265: 83 (2007). Type: Kenya, Northern Frontier District: Malka Mari [Malka Murri], *Williams* s.n. in *Bally* 8008 (K, holo. (not traced); ZSS, iso.)

Stems procumbent-ascending to erect, dark green (purplish in full sunlight), 2–9 cm long, 1.8–2.5 cm diameter, 14–20-angled; tubercles indistinctly hexagonal, convex. Leaf rudiments spreading, lanceolate, 2.5–3.5 mm long, acute, withering soon and remaining as dry, white spine. Inflorescences 2-flowered; pedicels 1–1.5 mm long, holding flower erect. Sepals lanceolate, 2–3 mm long, tips often recurved. Corolla abaxially and adaxially pale yellow and dark purple or purple-red towards base, globoid-, obovoid- to cylindrical-urceolate, 8–18 mm long, 3.5–7 mm wide, glabrous; lobes erect to spreading, pale yellowish or greenish, deltoid, 1–3 mm long, margins slightly reflexed. Gynostegial corona purple with yellowish base, cup-shaped, 1–2 × 2–3 mm, circular to pentagonal, upper rim slightly dentate; staminal corona lobes decumbent-ascending, linear from a broadened base, ± 0.8 mm long, occasionally connivent-erect to form a short column, surpassing level of coronal cup or not. Mericarps two, 6–10 cm long, slender fusiform (narrowly oblong), seeds ± 3 mm long, 1.5 mm wide, ovate, winged.

KENYA. Northern Frontier District: Malka Mari, *Newton & Lavranos* 12210 & Mandera District: 30 km along the Ramu–Malka Mara raod, 8 May 1978, *Gilbert & Thulin* 1566
DISTR. **K** 1; Ethiopia
HAB. *Acacia-Commiphora* bushland; 500–700 m

SYN. *Echidnopsis specksii* T.A. McCoy in Kakt. & Sukk. 54: 215 (2003). Type: Ethiopia, Sidamo Province, *Specks* 787, cult. sub *McCoy* 2486 (MO, holo.; UPS, iso.)

NOTE. *E. urceolata* is extremely polymorphic with regard to corolla size and shape; but the flowers are always erect, urceolate and bicoloured. Vegetatively, the multi-angled stems and long, acute leaf rudiments are also significant.

13. **Echidnopsis malum** (*Lavranos*) *Bruyns* in Bradleya 6: 43 (1988); Thiv & Meve in Pl. Syst. Evol. 265: 81 (2007). Type: Somalia, 22 km N of Erigavo on road to Mait, *Lavranos* 6721 (FT, holo.)

Stems creeping and rooting over the whole length, grey-green or brown-green, up to 20 cm long, 7–10 mm diameter, 5–6-angled, scattered-pubescent, angles occasionally compressed; tubercles indistinctly hexagonal, protruding. Leaf rudiments spreading, broadly deltoid, 1–1.5 mm long, hispid, withering soon, caducuous. Inflorescences 1-flowered; pedicels 10–15 mm long, presenting flower erect, papillate. Sepals lanceolate, 2–3 mm, papillate. Corolla abaxially greenish with red streaks, globose-urceolate, 14–25 mm long, 18–22 mm diameter, apically with a circular opening, papillose, adaxially dark purple, rugose, with lots of stiff purple-whitish hairs at bottom an top regions, 1–1.5 mm long; lobes purple, very small, deltoid, apically connate, inverted into the tube, ± 5 mm long, folded, adaxially expanded into a broad deltoid, dark purple limb. Gynostegial corona red, globose-urceolate, 3.5–5 mm diameter, with circular opening, inner face with stiff hairs, margin recurved, dentate; staminal corona lobes deltoid, ± 1.2 mm long, developed from base of coronal urn, incumbent, touching base of anthers. Pollinia globose-ovoid. Mericarps two, 6–10 cm long, slender fusiform (narrowly oblong), seeds ± 3 mm long, 1.5 mm wide, ovate, winged.

KENYA. Northern Frontier District: 46 km W of Mandera, 8 Aug. 1975, *Newton* 12201!
DISTR. **K** 1; Ethiopia, Somalia
HAB. Deciduous bushland and woodland; 300 m in Kenya (0–1250 m in Somalia)

SYN. *Pseudopectinaria malum* Lavranos in Cact. Succ. J. (Los Angeles) 43: 10 (1971) & in Fl. Somalia 3: 190, fig. 133 (2006)

NOTE. Molecular studies have clearly shown that this species is nested in *Echidnopsis* and that it is superfluous to uphold a separate genus *Pseudopectinaria*. Morphologically also it is not so different to justify exclusion from *Echidnopsis*.

63. RHYTIDOCAULON

P.R.O. Bally in Candollea 18: 335 (1962); Bruyns in Edinburgh J. Bot. 56: 211–228 (1999); DeKock in Asklepios 101: 5–15 (2008)

Creeping or erect stem-succulents, often scrambling in bushes, sparsely branched from main stem; latex mostly slightly milky; stems cylindrical, roundly 4-angled, papillose or tessellate, thickly wax-covered. Leaf rudiments strongly ascending, caducuous; stipules glandular, globoid. Inflorescences 1–3-flowered, lateral, slightly sunken. Corolla rotate or elongated-conical, basally fused, abaxially papillose, glabrous, papillose or hairy; corolla lobes linear, lanceolate or ovate, spreading (in the one Kenyan species) or remaining connected at tips. Corolline corona absent. Gynostegial corona in 2 series: "outer" corona of 5 staminal and 5 interstaminal parts fused basally up to half of length, rotate to cyathiform, interstaminal corona subulate, deltoid or 2-lobed, erect or reflexed; "inner" staminal corona lingulate, triangular or rectangular, inflexed. Gynostegium sessile. Pollinia ovoid or pyriform. Mericarps two.

Around 13 species (depending on the species concept) in NE Africa and Arabia. A genus in close but complex and still not fully understood relationship to *Echidnopsis*.

Rhytidocaulon paradoxum *P.R.O. Bally* in Candollea 18: 339 (1963); Bruyns in Bradleya 4: 29 (1986); Jonkers in Asklepios 70: 28 (1997); Bruyns in Edinburgh J. Bot. 56: 224 (1999); Gilbert in Fl. Eth. 4(1): 178 (2003); Lavranos in Fl. Somalia 3: 183 (2006); DeKock in Asklepios 101: 6 (2008). Type: Ethiopia, Ogaden, W of Shillave, *Ellis* 405 (K, holo.; ZSS, iso.)

Creeping, rarely scrambling stem succulent, with distinct main stem and smaller lateral stems; stems greyish olive-green, 2–11 cm long, 12–15 mm diameter, obtusely 4-angled, surface wrinkled, furrowed, tessellate, ± papillate; tubercles flattened ovoid. Leaf rudiments narrowly triangular, ± 2 mm long, 1 mm wide, acute. Inflorescences 1-flowered, scattered over stem faces; pedicels less than 1 mm long. Sepals triangular-lanceolate, 1.5–2 × 1.3 mm wide, abaxially papillate. Corolla

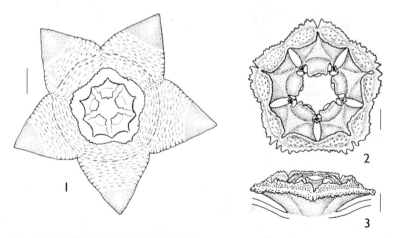

FIG. 81. *RHYTIDOCAULON PARADOXUM* — **1**, flower; **2 & 3**, gynostegium and corona from above and lateral view. Scale bars: 1 = 1 mm; 2 & 3 = 0.5 mm. From *Powys* 520. Drawn by P.V. Bruyns. Reproduced with permission from Edinburgh J. Bot. 56: 225, fig. 8.

abaxially reddish grey, rotate, 8–10 mm diameter, covered with short white hairs; tube adaxially cream-coloured, maculated reddish, ± 5 mm diameter, shallow; lobes cream with plain red(dish) tips, broadly triangular, 3–4 × 3.5 mm. Gynostegial corona greenish, fused to form a lens-shaped structure, ± 1 × 3–4 mm, brilliant due to nectar cover; interstaminal corona appressed to corolla, with nearly continuous flat but laciniate, papillate margin; staminal corona lobes broadly rectangular, ± 0.5 mm long, apically emarginate. Mericarps two, 3–4 cm long, narrowly oblong to slender fusiform, spreading. Seeds with thick, puffed margin. Fig. 81, p. 335.

Rarely collected species, and only once reported from Kenya.

KENYA. Laikipia District: Longobito, NE of Rumuruti, *Powys* 520!
DISTR. **K** 3; E Ethiopia and Somalia
HAB. In *Acacia-Commiphora* bushland, dry thornbush and open succulent communities; ± 1800 m

64. **HUERNIA**

R. Br. in Asclepiadeae: 11 (1810); Leach in Excelsa Tax. Ser. 4: 1–196 (1988)

Clump-forming or creeping stem-succulents; latex clear or at most cloudy. Shoots green or blue-green, cylindrical, roundly or sharply 4- to 6-angled; leaf rudiments spreading to ascending, sessile, caducous, sometimes spinescent; stipules absent. Inflorescences, 1–6-flowered, basal, pedunculate. Flowers often foetid. Corolla 5-lobed, but often with small intermediate bulges or lobules in the sinuses of the principal lobes, rotate, campanulate, cyathiform, elongated-conical, urceolate or globose, fused for most of its length, tube often with with central annulus (corolline corona), corolla lobes triangular, abaxially occasionally warty, adaxially often warty or rugose, glabrous, often scuptured with massive conical to cylindrical emergences, often tipped by a papilla or trichome, variously coloured and/or patterned. Gynostegial corona in 2 series: "outer" corona with 5 free staminal and 5 interstaminal parts fused to form a basal ring or disc appressed to corolla tube, free interstaminal corona lobes, if differentiated, lingulate or rectangular; "inner" corona of staminal lobes atop gynostegium, dolabriform, triangular or clavate, often with a humped back, erect, reflexed or inflexed. Gynostegium sessile, with elongated filament tube and little humps below each widened mouth of vertical anther wings. Pollinia oblong-ellipsoid; corpusculum with triangular to oblong basal projections. Mericarps two, fusiform, seeds ovate, broadly winged.

Sixty-seven species in Africa and Arabia, with a centre of distribution in southern Africa.

1. Corolla tube raised to form a fleshy and bulging annulus 2
   Corolla tube flat or concave, without annulus 3
2. Corolla lobes 12 mm long, 8 mm wide, smooth, glabrous   1. *H. tanganyikensis*
   Corolla lobes 10 mm long, 6 mm wide, medial areas with
      red-pointed emergences . . . .   2. *H. andreaeana*
3. Corolla yellowish spotted purple/maroon, surface densely
      sculptured by conical emergences 1–2.8 mm long 4
   Corolla brownish, red to purple-black, surface with or
      without emergences up to 1 mm long 6
4. Staminal corona lobes ± 2.5 mm long, cream, lined and
      spotted purplish, apically expanded into a massive
      head of 1 mm diameter . . . .   3. *H. archeri*
   Staminal corona lobes 3–3.5 mm long, whitish, sparsely
      spotted purplish, apically slightly enlarged at the most 5

5. Corolla adaxially densely sculptured by broadly conical
   emergences up to 1.5 mm long .... 4. *H. erinacea*
   Corolla adaxially densely sculptured by cylindrical
   emergences from 1 to 2.8 mm long .... 5. *H. recondita*
6. Pedicels 15–25 mm long, erect, bent below calyx to
   present the flower vertically .... 7. *H. lenewtonii*
   Pedicels 5–12 mm long, straight or bent, presenting the
   flower vertically to drooping 7
7. Staminal corona lobes inconspicuous, triangular,
   incumbent on anthers, hardly 0.5 mm long, but dorsally
   with a much larger, spreading, ovoid, obtuse hump .... 9. *H. schneideriana*
   Staminal corona lobes conspicuous, 1–2(–4) mm long,
   (narrowly) deltoid, suberect to connivent-erect, dorsal
   humps not much expanded 8
8. Stems up to 20 cm long, creeping, tubercles flat, blunt; leaf
   rudiments 1–1.5 mm long; inflorescences 1-flowered ... 8. *H. aspera*
   Stems ascending-erect (rarely creeping), 2–12 cm long,
   tubercles rather prominent, tapering; leaf rudiments
   2–4 mm long; inflorescences with few to many flowers
   in succession .... 6. *H. keniensis*

1. **Huernia tanganyikensis** (*E.A. Bruce & P.R.O. Bally*) *L.C. Leach* in Bothalia 10: 54 (1969); Meve in The genus Duvalia: 43 (1997). Type: Tanzania, Arusha District: Mt Longido, *Bally* S19 (K, holo.!; PRE, photos only)

Creeping, dense mat-forming stem-succulent; stems few-branched, bright green, 10–65(–90) cm long, 10–18 mm diameter, obtusely 5-angled, often spotted reddish; tubercles small, slightly acute. Leaf rudiments triangular-lanceolate, ± 1 mm long. Inflorescences few-flowered, pedunculate; flowers pointed upwards; pedicels ± 2 cm long. Sepals lanceolate, ± 12 mm long, protruding from the corolla lobe sinuses. Corolla rotate, 3–3.5 cm diameter, with a massive tube bulging to a wine-red annulus enclosing and encircling the gynostegium; lobes pale ochre, triangular, 12 × 8 mm, spreading, slightly channelled, acute, glabrous, margins occasionally wine-red. Gynostegial corona 3–3.5 mm diameter. Interstaminal corona lobes fused to form a thickened disc, yellow-ochre, occasionally reddish; staminal corona lobes spreading horizontally, deltoid appressed to anthers, ± 1.2 mm wide, with flattened ovoid dorsal humps, reddish or yellow-ochre. Mericarps two, 8–12 cm long, fusiform, stout. Fig. 82, p. 338.

TANZANIA. Mbulu District: Lake Manyara National Park, Chem Chem river gorge, 24 Mar. 1964, *Greenway & Kanuri* 11414; Arusha District: Olmutonyi Plain, 13 Dec. 1969, *Richards* 24997!; Lushoto District: Makayuni to Mombo, 16 Dec. 1959, *Greenway* 9676
DISTR. **T** 2, 3; not known elsewhere
HAB. In partial shade of *Acacia, Euphorbia, Dobera* etc. on shallow sandy loam overlying sandstone and gneiss; 800–1400 m

SYN. *Duvalia tanganyikensis* E.A. Bruce & P.R.O. Bally in Cact. Succ. J. (Los Angeles) 13: 179 (1941)

NOTE. Very closely related to *H. andreaeana* of southern Kenya, with which it is most likely conspecific.

2. **Huernia andreaeana** (*Rauh*) *L.C. Leach* in Bothalia 10: 54 (1969) & in J. South Afr. Bot. 40(1): 21 (1974); Plowes in Excelsa 9: 49 (1980); Leach in Excelsa Tax. Ser. 4: 154 (1988). Type: Kenya, between Mombasa and Voi, *Rauh* Ke867 (HEID, holo.; K, clono.).

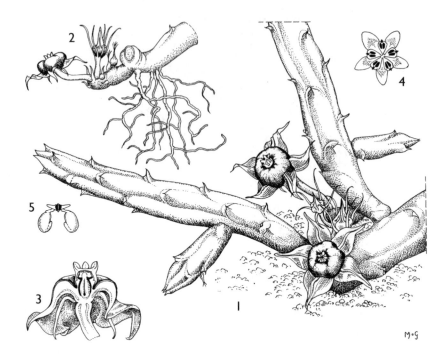

FIG. 82. *HUERNIA TANGANYIKENSIS* — **1**, habit, × 1; **2**, shoot showing roots and inflorescence, × ²/₃; **3**, half-flower, showing swollen annular corolla-encircling corona and gynostegium, × 2; **4**, gynostegium and staminal corona from above, showing position of pollinaria, × 4; **5**, pollinarium (inverted), × 16. Drawn by M. Grierson.

Creeping, mat-forming stem-succulent; stems few-branched, bright green, often spotted reddish, 10–50 cm long, 10–15 mm diameter, rooting over the whole length, 4(–5)-angled; tubercles small, obtuse. Leaf rudiments triangular-lanceolate. Inflorescences 1(–2)-flowered; flowers pointed upwards; pedicels ± 2 cm long. Sepals lanceolate, ± 10 mm long, protruding corolla lobe sinuses. Corolla rotate, 2.5–2.7 cm diameter, with a massive tube bulging to a pale wine-red annulus enclosing and encircling the gynostegium; corolla lobes spreading, pale ochre, triangular, 10 × 6 mm, acute, apical ³/₅ folded upwards to form a channel, with undulate margins, medial areas with red-pointed emergences. Interstaminal corona lobes fused to form a thickened disc, yellow-ochre; staminal corona lobes yellow-ochre, deltoid appressed to anthers, with flattened ovoid dorsal humps. Mericarps two, 8–12 cm long, fusiform, stout.

KENYA. Kwale District: km 85 from Mombasa to Voi, 9 Mar. 1960, cult. Heidelberg (no date), *Rauh* Ke 867; Teita District, Taita Ranch, Kisima Hill, 5 Dec. 1998, *Selempo* in *Luke* 5576
DISTR. **K** 7; not known elsewhere
HAB. Dry, deciduous bushland, mostly on rocky ground; 400 m

SYN. *Duvalia andreaeana* Rauh in Kakt. & Sukk. 12: 117 (1961)

NOTE. Rarely collected taxon, very closely related to *H. tanganyikensis*, with which it is most likely conspecific.

3. **Huernia archeri** *L.C. Leach* in Excelsa Tax. Ser. 4: 88 (1988); Newton in Asklepios 89: 16 (2003). Type: Kenya, near Kilifi, *Archer* s.n. in *Leach* 15561 (PRE, holo.; K, clono. sub *Bally* 13154)

Creeping, mat-forming stem-succulent; stems up to 30 cm long, 5–7 mm diameter, rounded 4-angled; tubercles flat, blunt. Leaf rudiments narrowly triangular, ± spreading, ± 2 mm long. Inflorescences 1-flowered, basal; buds with long beak; pedicels up to 18 mm long. Sepals ± 9 mm long, attenuate. Corolla adaxially (bright) yellow, mottled with maroon, rotate-campanulate, 30–40 mm diameter, with rather flat central depression with fine concentric streaks or spots, 8–9 mm diameter, abaxially finely scabrid; lobes spreading, elongate-triangular, ± 14 × 5–6 mm, caudate, adaxially densely sculptured by narrowly conical emergences up to 2 mm long, rim of tube finely verrucose. Interstaminal corona lobes spreading, blackish, subquadrate, ± 2.5 × 2 mm, apically slightly crenate; staminal corona lobes erect, cream-coloured with maroon pattern, cylindrical, ± 2.5 mm long, apically expanded into a tuberculate, subcircular head 1 mm diameter. Mericarps two, 8–12 cm long, fusiform, stout.

KENYA. Kilifi District: Dakabuko Hill, Galana Ranch, 1 Dec. 1990, *Luke & Robertson* 2546, cult.
  Malindi 26 Nov. 1994 sub *Robertson* 7061; Kwale District: Mwaluganji Forest Reserve, near
  Kaya Mtae, 14 Nov. 1989, cult. 28 Jan. 1992, *Robertson & Luke* 6014
DISTR. **K** 7; not known elsewhere
HAB. Rocky ground with sparse shrubs; 200–300 m

NOTE. A poorly known taxon, possibly just a variant of *H. erinacea*.

4. **Huernia erinacea** *P.R.O. Bally* in Fl. Pl. Afr. 31: t. 1206 (1956); Leach in Excelsa Tax. Ser. 4: 87 (1988). Type: Kenya, Northern Frontier District: Dandu, *Gillett* 12629 (K, syn.; PRE!, SRGH! syn.)

Decumbent to prostrate stem-succulent; stems 2–6 cm long, ± 1 cm diameter, cylindrical, rounded 5-angled, rarely ascending; tubercles flat. Leaf rudiments deltoid, barely 1 mm long, spreading. Flowers upright; pedicels 6–18 mm long. Sepals ± 6 × 1.5–2 mm, ovate-acuminate. Corolla dull yellow, adaxially ivory-coloured to yellowish, finely to coarsely spotted red(-purple), rotate-campanulate, 40–50 mm diameter, with a rather flat central depression, 12–15 mm diameter at the slightly constricted mouth, abaxially glabrous; tube coloured and sculptured as lobes except for glabrous basal central depression; lobes spreading, stippled purplish, triangular, 15–20 mm long, acute, adaxially densely sculptured by emergences up to 1.5 mm long, broadly conical. Gynostegial corona 8–10 mm diameter; interstaminal corona lobes transversely rectangular, usually emarginate, black-purple; staminal corona lobes erect, whitish, basally purple, cylindrical, ± 3.5 mm long, apically slightly diverging, blunt, subspinulate, only weakly spotted purplish. Mericarps two, 8–12 cm long, fusiform, stout.

KENYA. Northern Frontier District: Dandu, Mar. 1952, *Gillett* 12629 & Mt Furole, *Gillett* 13798 &
  S end of Burole Mt, 13 Mar. 1987, *Hartmann & Newton* 21380!
DISTR. **K** 1; not known elsewhere
HAB. Shallow soil on granite; 750–1100 m

5. **Huernia recondita** *M.G. Gilbert* in Cact. Succ. J. (Los Angeles) 47: 6 (1975) & in Fl. Eth. 4(1): 193 (2003). Type: Ethiopia, Gamu Gofa, near Hammer Koke, cult. Addis Abeba, *M.G. & S. Gilbert* 1729 (K, holo.; ETH, iso.)

Prostrate stem-succulent, to 50 cm long, 1–1.2 cm diameter, forming loose mats, 4(–7)-sided, with rounded angles and obtuse tubercles. Leaf rudiments erect, deciduous. Inflorescences often concealed beneath stems; pedicels 15–30 mm long. Sepals 7–10 mm long. Corolla shallowly campanulate with spreading limb and intermediate lobules, 35–45 mm wide, outside red with whitish reticulate markings, inside dull yellow with concentric red lines within tube, and dense large reddish blotches on limb and lobes; tube 8–9 mm long, 10–12 mm diameter, slightly constricted at the mouth, adaxially densely sculptured by rather cylindrical emergences 1–2.8 mm long, banded red and yellow; lobes spreading, 10–15 mm long, acuminate, with a prominent central vein, intermediate lobules 2–3 mm long, acute. Interstaminal corona lobes velvety purplish-black, oblong, 1–2 × ± 2 mm, irregularly obtusely bidentate; staminal corona lobes red, 3–4 mm long, attenuate, tips of lobes convergent then reflexed, apex minutely papillate, basally with rectangular dorsal hump. Mericarps two, 8–12 cm long, fusiform, stout; seeds ovate, 5–6 × 3.5 mm, wing occasionally with folds.

KENYA. Baringo District: Lake Baringo, *Pfennig* s.n., cult. Gaborone, 16 Jan. 1986, sub *Plowes* 4778!
DISTR. **K** 3; southern Ethiopia
HAB. *Acacia* bushland on red or brown lateritic soils; 1000 m in Kenya, 1250–1600 m in Ethiopia

6. **Huernia keniensis** *R.E. Fr.* in Acta Hort. Berg. 9: 79, t. 2 (1929); White & Sloane in Stapelieae ed. 2, 3: 830 (1937); R.A. Dyer in Fl. Pl. Afr. 37: t. 1472 (1957); Leach in Excelsa Tax. Ser. 4: 114 (1988); U.K.W.F. ed. 2: 186, t. 74 (1994). Type: Kenya, W side of Mt Kenya near Coles Mill, *R.E. & T.C.E. Fries* 1024 (cult. hort. Bergius 1926: S, holo.; K, clono., sterile)

Decumbent or ascending-erect stem-succulent; stems green, 2–12 cm long, ± 1 cm diameter, 5-angled, occasionally blue-green and spotted purplish; tubercles rather prominent, tapering. Leaf rudiments narrowly triangular to deltoid, 2–4 mm long, acute. Inflorescences with few to many flowers in succession; flowers horizontal to erect, buds usually nodding; pedicels 5–12 mm long. Sepals narrowly lanceolate to subulate, 5–8 mm long. Corolla purple or cream, adaxially purple to black-purple, campanulate to subglobose, 2–5 cm diameter, abaxially scabrid or glabrous, sculptured with short emergences of variable density and length, occasionally absent in the tube, emergences conical to subcylindrical; lobes broadly triangular, often tips revolute. Gynostegial corona purple, often black-purple; interstaminal corona lobes fused into an emarginate to crenate disc, 3.5–4.5 mm diameter; staminal corona lobes narrowly deltoid, 1.3–4 mm long, flattened but with band-like gibbosity at basal back, connivent-erect, apically pointing upwards, apices blunt, subspinulate-tuberculate. Mericarps two, 8–12 cm long, fusiform, stout.

A taxonomically difficult complex in urgent need of revision. The most variable species in Kenya, with many different forms, some of them formally recognized as varieties, some others not.

1. Flowers 2–3.5 cm in diameter 2
   Flowers ± 5 cm in diameter . . . . d. var. *grandiflora*
2. Corolla lobes 5-veined; abaxial face scabrid, adaxial
   emergences conspicuous . . . . a. var. *keniensis*
   Corolla lobes 3-veined (occasionally with 2 inconspicuous
   veins); abaxial face glabrous or nearly so, adaxial
   emergences minute 3
3. Corolla globose-urceolate, ± 2.5 cm in diameter . . . . b. var. *globosa*
   Corolla broadly funnel-shaped to bowl-shaped, ± 3.5 cm in
   diameter . . . . c. var. *nairobiensis*

a. var. **keniensis**

Corolla deeply and broadly campanulate with funnel- or bowl-shaped tube and spreading to reflexed lobes, 2–3.5 cm diameter; lobes abaxially prominently 5-veined. Staminal corona lobes 3–4 mm long.

KENYA. Northern Frontier District: 30 km N of Maralal, 23 Feb 1996, *Goyder et al.* 4065 (≡*Meve et al.* 963!); Baringo District: N of Lake Baringo, 1987, *Hartmann* & *Newton* 31321!; Masai District: on circular route W of Ngong Hills, 16 May 1971, *Robertson* 1518
DISTR. **K** 1–6; not known elsewhere
HAB. Preferring rocky soil; 1000–1350 m

SYN. *Huernia keniensis* R.E. Fries var. *quintitia* L.C. Leach in Excelsa Tax. Ser. 4: 118 (1988), *nom. inval.*

b. var. **globosa** *L.E. Newton* in Asklepios 74: 23 (1998). Type: Kenya, Masai District: 4 km N of Narok on road to Mau Narok, *Newton* 2927 (K!, holo.; EA, MSUN, iso.)

As var. *keniensis* but differing from it by the corolla, globose-urceolate, ± 2.5 cm diameter, adaxially with minute emergences uniform in size; corolla lobes short, ± 5 mm long, broadly triangular, erect to slightly spreading.

KENYA. Masai District: 4 km N of Narok on road to Mau Narok, 21 Dec. 1985, *Newton* 2927!
DISTR. **K** 6; only known from the type
HAB. Preferring rocky soil; 1950 m

c. var. **nairobiensis** *White & Sloane* in Stapelieae ed. 2: 3: 837 (1937); Leach in Excelsa Tax. Ser. 4: 117 (1988). Type: Kenya, near Nairobi, *Molony* s.n. (K!, lecto., in alc. sub no. 6754, designated by Leach (1988))

As var. *keniensis* but differing in: stems erect, robust, up to 12(–18) cm long, 14 mm wide; corolla 3.5 cm diameter, broadly funnel-shaped to bowl-shaped; corolla lobes abaxially 3-veined, scabrid, adaxially densely covered by small emergences; staminal corona lobes comparatively short, ± 1.3 mm long.

KENYA. Naivasha District: 1 km S of Naivasha, 12 Jan. 1994, *Newton* 4443; Uasin Gishu District: Moiben, 10 Mar. 1964, *Polhill* 2404A; Nairobi, *Newton* 951!
TANZANIA. Masai District: Ngare Nanyuki, 16 Dec. 1968, *Greenway, Kanuri & Field* 13197; Mpwapwa District: Mangalisa to Ikuyu, 12 Apr. 1988, *Lovett & Congdon* 3238; Mbeya District: Kimani Falls, Chimala, 16 Feb. 1992, *Congdon* 322
DISTR. **K** 3, 4; **T** 2, 5, 7; not known elsewhere
HAB. On rocks, in shade; 1100–2100 m

SYN. *Huernia keniensis* R.E. Fries var. *molonyae* White & Sloane in Stapelieae ed. 2: 3: 832 (1937). Type: Kenya, near Nairobi, *Molony* s.n. (not traced)

d. var. **grandiflora** *P.R.O. Bally* in Fl. Pl. Afr. 38: t. 1511B (1967); Leach in Excelsa Tax. Ser. 4: 117 (1988). Type: Kenya, Northern Frontier District: Baragoi, Sep. 1951, *Hennings* s.n. in *Bally* S204 (G, holo.; EA, iso.)

As var. *keniensis* but differing in the large bell-shaped corolla, ± 5 cm diameter, abaxially glabrous, whitish or purplish with pale venation, adaxially densely covered with very small emergences, dark purple; corolla lobes broadly triangular, with 3 prominent and 2 inconspicuous veins.

KENYA. Northern Frontier District: Lopet Plateau, 25 km S of Baragoi, 23 Feb. 1974, *Bally & Carter* 16544 & Baragoi, *Jex-Blake* s.n. in *Bally* S17 & *Hennings* s.n. in *Bally* S204
DISTR. **K** 1; only known from around Baragoi
HAB. Dry *Acacia* bushland; 1300 m

NOTE. This variety could also be regarded as a large-flowered, extremely bell-shaped form of var. *nairobiensis*. Bally, when describing var. *grandiflora*, says that it hybridises freely with var. *nairobiensis* in cultivation.

7. **Huernia lenewtonii** *Plowes* in Asklepios 64: 21, pl. 5–6 (1995). Type: Kenya, Northern Frontier District: E of Lake Turkana, just S of Ethiopian border at Sul-sul Mudde, *Newton & Powys* 3703 (K, holo.; EA, ZSS, iso.)

Prostrate-decumbent stem-succulent; stems green, often spotted purplish, 2–10 cm long, ± 1.5 cm diameter, 5-angled; tubercles prominent, tapering. Leaf rudiments triangular, 3–4 mm long. Inflorescences 2- to 3-flowered; pedicels erect, 15–25 mm long, bent below calyx to present the flowers vertically. Sepals narrowly triangular, 4–5 mm long, acute. Corolla greenish yellow, adaxially (bright) ruby-red, broadly funnel-shaped with horizontally spreading lobes, ± 3.5 cm diameter, abaxially subscabridulous, velvety by minute rounded or slightly pointed emergences; corolla lobes broadly triangular, ± 10 × 15 mm, acuminate, abaxially prominently 5-veined. Gynostegial corona ruby-red; interstaminal corona lobes fused into a fleshy, crenate disc, lobes oblong, ± 2 × 2 mm; staminal corona lobes narrowly deltoid, 1.5–2 mm long, flattened but with gibbosity at basal back, connivent-erect, apices blunt, subspinulate-tuberculate. Mericarps two, 8–12 cm long, fusiform, stout.

Kenya. Northern Frontier District: just S of Ethiopian border at Sul-sul Mudde, E of Lake Turkana, 19 Nov. 1990, *Newton & Powys* 3703! & Jibisa Hill, 18 Nov. 1990, *Newton & Powys* 3681
Distr. **K** 1; not known elsewhere
Hab. Creeping amongst rocks on dry rocky slopes; 400 m

Note. This taxon belongs to the *H. keniensis* complex, and would be better treated there as another variety characterized by long, erect, bent pedicels.

8. **Huernia aspera** *N.E. Br.* in Gard. Chron., ser. 3, 2: 364 (1887); J.D. Hooker in Curtis's Bot. Mag. 114: pl. 7000 (1888); Blundell, Wild Fl. E. Afr.: t. 552 (1987); Leach in Excelsa Tax. Ser. 4: 102 (1988); Newton in EANHS Bull. 22: 13 (1992); U.K.W.F. ed. 2: 186 (1994). Type: Tanzania, without precise locality, "probably ... a native of the region of Zanzibar", *Kirk* s.n. (K, holo.)

Creeping stem-succulent forming diffuse mats, occasionally pendulous from rocks; stems (bluish-)green, often mottled purplish, up to 20 cm long, 10–14 mm diameter, rounded 5- to 6-angular; tubercles flat, blunt. Leaf rudiments triangular, 1–1.5 mm long, acute, spreading to recurved. Inflorescences 1-flowered, basal, occasionally with short peduncle; flowers slightly nodding; pedicels 5–12 mm long. Sepals spreading, lanceolate, ± 7 mm long, acuminate. Corolla purple, broadly campanulate, 2–2.5 cm diameter, with hemispherical tube and minute intermediate lobules, abaxially scabrid, adaxially densely covered with erect, conical, occasionally shortly tipped emergences up to 1 mm long (lesser on the tube), dark purple; lobes spreading, occasionally recurved, triangular, ± 8 mm long, acute. Interstaminal corona lobes black-purple, transversely rectangular to broadly deltoid, ± 0.8 mm long; staminal corona lobes suberect, yellowish to reddish, deltoid, ± 1 mm long, apically blunt, subspinulate. Mericarps two, 8–12 cm long, fusiform, stout; seeds ± 6 × 4 mm, ovate, winged.

Kenya. 11 km SW of Nairobi, *McLoughlin* s.n.; Kitui/Machakos Districts: summit of Mbuinzau, 190 km SE of Nairobi, 18 Aug. 1975, *Lavranos & Newton* 12348; Teita District: Mt Maktau, 14 Feb. 1996, *Meve et al.* 943b!
Tanzania. Between Morogoro and Dodoma, 1990, *Carter, Abdallah & Newton* 2652!
Distr. **K** 4, 7, **T** 5/6; Malawi
Hab. Preferring rocky soil; 1000–1500 m

Note. Belongs to the *H. keniensis* complex but is most easily distinguished by the rounded and prostrate stems with flattened tubercles.

9. **Huernia schneideriana** *A. Berger* in Monatsschr. Kakt. 23: 177 (1913). Type: Tanzania, Rungwe District, Kyimbila, *Stolz* 1407 (B†, holo.; K, probable clono. sub *Leach* 14997 & sub *Brown* 12 Sep. 1925); Neotype: Hort. Bot. Monacensis, fl. 31 Oct. 1927 (M, neo., designated by Leach in Bothalia 10: 52 (1969) – probable clone of *Stolz* 1407)

Stems erect, creeping or pendulous, to 40 cm long, 1.2–1.4 cm diameter, (6–)7-angled; tubercles small, obtuse. Leaf rudiments spreading. Inflorescences many-flowered, near base of stems, flowers facing outwards, sometimes 2 flowering simultaneously; peduncle very strong, ± persistent; pedicel to 1 cm long. Sepals ± 5 × 2 mm. Corolla outside pale brown, 26–30 mm diameter with a hemispherical tube and abruptly horizontally spreading limb and lobes, glabrous; inside tube ± 9 mm across at the mouth, 5 mm deep, sparingly micropapillate, dark brown or black; limb and lobes brownish, densely dark brown pubescent, hairs tapering from the base, arising from minute papillae; lobes deltate, ± 8 × 8 mm, acute, with a prominent median vein, intermediate lobules small, reflexed in face view. Corona ± 2.5 × 5 mm, broadly subconical at the base, interstaminal corona lobes short, irregularly subcrenulate, brown or purple; staminal corona lobes ± 0.5 mm long, deltoid, yellowish, incumbent on anthers, dorsally with a much enlarged, spreading, ovoid, obtuse hump ± 1.5 mm long. Mericarps two, 8–12 cm long, fusiform, stout.

TANZANIA. "Rungwe District, Kyimbila," cult. *Leach* 14997; hort. N.E. Brown 12 Sep. 1925, *Brown* s.n.
DISTR. **T** 7; only known from the type
HAB. Not recorded

NOTE. Stolz's original living collection was widely disseminated via Berlin, and the few surviving herbarium collections are assumed to derive from this material. Leach suggested a possible hybrid origin for this taxon, postulating that one parent might be *H. verekeri* Stent.

65. **ORBEA**

Haw. in Syn. Pl. Succ.: 37 (1812); Leach in Excelsa Tax. Ser. 1: 1–75 (1978); Bruyns in Aloe 37: 72–76 (2001) & in Syst. Bot. Monogr. 63: 1–196 (2002)

Orbeopsis L.C. Leach in Excelsa Tax. Ser. 1: 71 (1978)
Pachycymbium L.C. Leach in Excelsa Tax. Ser. 1: 69 (1978)
Angolluma Munster in Cact. Succ. J. (Woollahra) 17: 63 (1990)

Decumbent to prostrate, often clump-forming stem-succulents; latex clear or at most cloudy; stems cylindrical, conical or club-shaped, roundly 4-angled to irregularly 4–5-angled, glabrous, occasionally rhizomatous; tubercles in 4 or 4–5 rows, abscission zone with leaves not always visible. Leaf rudiments sessile, triangular-deltate to conical-subulate, acute, ± caducous; stipules, if present, pointed, ovoid or globoid, glandular. Inflorescences lateral, with 1–40 flowers developing in succession on short peduncles. Flowers mostly with scent of faeces. Corolla rotate to campanulate, fused to half of length, occasionally with a central annulus (corolline corona), adaxially usually warty or rugose, glabrous, papillate or with hairs; corolla lobes lanceolate, ovate or triangular, spreading to reflexed. Gynostegial corona in 2 series: "outer" corona of 5 staminal and 5 interstaminal parts fused just basally to nearly completely to form a rotate or cyathiform structure; free interstaminal corona lobes deltoid-trianguloid, saccate, lingulate or bilobed; "inner" staminal corona lobes subulate, lingulate, ovoid, trianguloid or clavate, occasionally with hump. Gynostegium sub-sessile or atop a column. Guide rails normally vertical, basally widened. Mericarps two, obclavate-fusiform; seeds ovate, brownish, broadly winged.

Nearly 60 species in Africa and Arabia.

Pachycymbium and *Angolluma* in the sense of Gilbert in Bradleya 8: 1–32 (1990) and Plowes in Excelsa 16: 103–123 (1994), respectively, have been shown to be nested within *Orbea* and are therefore no longer accepted as separate genera (cf. Meve & Liede in Pl. Syst. Evol. 234: 171–209, 2002; Bruyns in Syst. Bot. Monogr. 63: 1–196, 2002).

Gilbert in Fl. Eth. 4(1): 186 (2003) states that *Pachycymbium laticoronum* (= *Orbea laticorona*) occurs in northern Kenya; this was based on a *Bally* specimen, which I have not seen.

1. **Orbea schweinfurthii** (*A. Berger*) *Bruyns* in Aloe 37: 76 (2001) & in Syst. Bot. Monogr. 63: 33 (2002); Müller, Kiel, Albers & Meve in Handb. Succ. Pl. Asclepiadaceae: 199 (2002); Bruyns in Stap. south. Afr. Madag. 1: 244 (2005). Type: Congo-Kinshasa, Rutihuru plains SW of Lake Edward, *Stuhlmann* 2208 (B†, holo.); iconotype: Berger, Stap. Klein.: 104, fig. 22, no.5 (lecto., designated by Bruyns (2001))

Small, decumbent stem-succulents, growing in loose mats; stems prostrate with ascending apex, green, patterned purplish(-brown), 2–12 cm long, 5–12 mm diameter, rounded; tubercles in 4 obtuse rows, ascending, conical, 3–18 mm long. Leaf rudiments subulate, 2–5 mm long, acute; occasionally subulate stipules present. Inflorescences near apex, 1–4-flowered; flowers with putrid-fruity odour; pedicels 2–4 mm long. Sepals deltoid-lanceolate, 2–3 mm long. Corolla rotate, 10–15 mm diameter, incised for more than half of length, abaxially green with few red spots, adaxially yellow to brown-yellow, spotted to banded wine-red or purplish, sculptured with rounded emergences, tipped by a short seta or not; tube shallow, ± 5 mm diameter; lobes spreading, triangular, 3–4 mm long. Gynostegial corona cream, finely spotted red, flattened-discoid, 5.5–7 mm diameter, interstaminally and staminal-dorsally much expanded, fused, many-toothed, brilliant due to nectar coat; staminal corona lobes deltoid, ± 1 mm long, appressed to anthers, occasionally toothed.

UGANDA. Toro District: Kasese, *Pfennig* 1292!; Ankole District: Ruizi River; 4 Feb. 1951, *Jarrett* 441 & Mbarara, Fort Portal Road, km 91, 1943, *Eggeling* 5502
TANZANIA. Dodoma District: 86 km S of Kondoa, 21 Dec. 2000, *Bruyns* 8724; Iringa District: 48 km W of Mikumi, *Leach & Brunton* 10139! & Kitonga Gorge, near Mahenge Village, 21 Apr. 1991, *Bidgood, Congdon & Vollesen* 2218
DISTR. U 2; T 5–7; Congo-Kinshasa, Rwanda, Zambia, Malawi, Zimbabwe, Namibia, Botswana
HAB. In seasonally very wet soils in mopane and *Brachystegia* woodland and *Acacia-Commiphora* bushland; 400–1700 m

SYN. *Caralluma schweinfurthii* A. Berger in Stap. Klein.: 103 (1910)
    C. piaranthoides Oberm. in Flow. Pl. S. Afr. 15: t. 599 (1935). Type: Zimbabwe, Hwange [Wankie] District, *Levy* 8444 (PRE, holo.)
    Pachycymbium schweinfurthii (A. Berger) M.G. Gilbert in Bradleya 8: 23 (1990)
    Angolluma schweinfurthii (A. Berger) Plowes in Excelsa 16: 118 (1994)

2. **Orbea sprengeri** (*Schweinf.*) *Bruyns* in Aloe 37: 76 (2001); Bruyns in Syst. Bot. Monogr. 63: 46 (2002); Müller, Kiel, Albers & Meve in Handb. Succ. Pl. Asclepiadaceae: 200 (2002). Type: Eritrea, Mitsiwawa [Massowa], *Schweinfurth* s.n. (holo., not traced); Iconotype: Berger, Stap. Klein.: 104, fig. 22, nos 1–4 (neo., designated by Bruyns (2002))

Small, decumbent stem-succulents, in loose mats; stems 5–15 cm long, ± 1 cm diameter, rounded 4-angled, occasionally rhizomatous, grey-green, patterned purplish-brown; tubercles in 4 rows, spreading, conical, 5–11 mm long. Leaf rudiments conical, 1–2 mm long, acute, without stipules. Inflorescences 1–4-flowered, close to apex; pedicels 4–10 mm long; flowers usually presented vertically, with strong foetid odour. Sepals lanceolate, 3–5 mm long. Corolla rotate, 2–4.4 cm diameter, adaxially pale green spotted purplish, adaxially, greenish or brownish, occasionally spotted purple, or plain purple-brown, smooth, rugose or beset with hemispherical emergences, occasionally tipped by a short seta; tube flat but with central depression, 4–5 mm diameter, enclosing the corona, depression often with thickened mouth (annulus); lobes spreading to slightly reflexed, ovoid-lanceolate, 9–20 × 5–8 mm. Gynostegial yellow, occasionally with reddish marking, or red, corona thickish lense-shaped in outline, circular or rounded pentagonal, 2–3 × 4–6 mm, ± coated by nectar; interstaminal corona lobes fused to a thickish ring, staminal corona lobes dorsally integrated within the coronal ring, apical parts oblong-elliptic or rectangular, 1–1.5 mm long, appressed to anthers, obtuse, occasionally toothed.

DISTR. Sudan, Eritrea, Ethiopia, Somalia, and Kenya; Saudi Arabia, Yemen

subsp. **sprengeri**

Stems rhizomatous, corolla pale brownish, completeley covered with hemispherical emergences, nearly each tipped with a short white seta; corolla lobes ± 18 mm long, spreading.

KENYA. Naivasha/Masai Districts: Kedong Valley, *Perkins* s.n.
DISTR. **K** 3/6; also known from northern Ethiopia and Eritrea
HAB. Dry bushland or open woodland; ± 1500 m

SYN. *Huernia sprengeri* Schweinf. in Cat. Dammann & Co. 1893: 46 (1893)
    Caralluma sprengeri (Schweinf.) N.E. Br. in K.B. 1895: 263 (1893)
    Pachycymbium sprengeri (Schweinf.) M.G. Gilbert in Bradleya 8: 22 (1990)
    Angolluma sprengeri (Schweinf.) Plowes in Excelsa 16: 110 (1994)

NOTE. A complex relationship is postulated by Bruyns (2002) in four subspecies meeting each other in Ethiopia, except for the Arabian subspecies *commutata*. In the FTEA region only the subsp. *sprengeri* occurs.

3. **Orbea gemugofana** (*M.G. Gilbert*) *Bruyns* in Aloe 37: 74 (2001); Bruyns in Syst. Bot. Monogr. 63: 80 (2002); Müller, Kiel, Albers & Meve in Handb. Succ. Pl. Asclepiadaceae: 192 (2002); Gilbert in Fl. Eth. 4(1): 186 (2003); Lavranos in Fl. Somalia 3: 193 (2006). Type: Ethiopia, Gemu Gofa Region E of Arba Minch, *Gilbert* 1731 (K, holo.; EA, ETH, iso.)

Small, decumbent stem-succulent, forming loose mats; stems grey-green, patterned purplish, 5–10 cm long, ± 8 mm diameter, rounded 4-angled; tubercles 8–15 mm long, conical, ascending. Leaf rudiments subulate, 2–4 mm long, attenuate, acute. Inflorescences 1–3-flowered, at tips of stems; pedicels 2–5 mm long; flowers unscented. Sepals lanceolate, 3–5 mm long, acute. Corolla rotate (to shallowly campanulate), 2–3.2 cm diameter, deeply incised, abaxially pale green patterned purplish, adaxially yellowish, greenish to pale brownish, occasionally (spotted) reddish, glabrous, smooth; tube shallowly bow-shaped, 1–3 mm long, 7–8 mm diameter; lobes spreading or slightly reflexed, ovoid-deltoid, 7–9 × 4–5 mm, margins slightly recurved. Gynostegial corona dark red, thickly discoid in outline, ± 3 × 4.5–6 mm, obtusely pentagonal; interstaminal corona lobes spreading, thickened but flattened fringes fused ring-like to staminal lobe bases; staminal corona lobes deltoid-lanceolate from a basal projection, ± 1.2 mm long, bent on anthers, pointed.

UGANDA. Mengo District: near Bukomero, Singo, Sep. 1932, *Eggeling* 940 (≡*Bally* S21)
KENYA. Northern Frontier District: Mt Furole, 14 Mar. 1987 & cult. Nairobi Nov. 1988, *Hartmann & Newton* 21398!; Baringo District: 8 km from Marigat, Mar. 1940, *Ritchie* s.n. sub *Bally* S39
DISTR. **U** 4; **K** 1, 3; Ethiopia
HAB. *Acacia* bushland on loamy to stony soil, also in barren, overgrazed spots; 900–1400 m

SYN. *Caralluma gemugofana* M.G. Gilbert in Cact. Succ. J. Gr. Brit. 40: 43 (1978)
    Pachycymbium gemugofanum (M.G. Gilbert) M.G. Gilbert in Bradleya 8: 23 (1990); U.K.W.F. ed. 2: 186 (1994)
    Angolluma gemugofana (M.G. Gilbert) Plowes in Excelsa 16: 118 (1994).

NOTE. Readily distinguishable from the other species of the "Ango-group" (cf. Gilbert in Cact. Succ. J. Gr. Brit. 40: 43, 1978) by its rather small flowers with pale, smooth corollas and a shallowly bowl-shaped corolla tube.

4. **Orbea laikipiensis** (*M.G. Gilbert*) *Bruyns* in Aloe 37: 75 (2001); Bruyns in Syst. Bot. Monogr. 63: 68 (2002); Müller, Kiel, Albers & Meve in Handb. Succ. Pl. Asclepiadaceae: 194 (2002). Type: Kenya, Kerio Valley, *Bally* 12376 (K, holo.)

Small, decumbent stem-succulent, clump-forming; stems ascending, green, patterned purplish, 5–8 cm long, ± 1 cm diameter, obtusely 4-angled, rhizomatous; tubercles in 4 rows, ascending, conical, ± 5 mm long. Leaf rudiments subulate, ± 3 mm long, acute. Inflorescences (1–)3(–5)-flowered, up to 5 inflorescences developed at time along stem; pedicels ascending-erect, 3–4 mm long; flowers unscented. Sepals lanceolate, 2.5–3.5 mm long, acute. Corolla abaxially pale green, adaxially wine-red, centrally often paler, rotate, 1.8–2.5 mm diameter, incised nearly down to corona, glabrous, smooth; lobes ascending, narrowly triangular-lanceolate, 10–12 × 2.5–3.5 mm, replicate, tips slightly recurved. Gynostegial corona broadly cup-shaped, ± 2.5 × 4 mm, yellow; interstaminal corona lobes pouch-like, apex truncate, spreading; staminal corona lobes ± 1 mm long, apically acute, occasionally toothed, with dorsal, ascending-erect, deltoid projection fused to upper rim of interstaminal corona.

KENYA. Laikipia District: Suguta Naibor, N of Rumuruti, 18 Oct. 1985, *Linden & Carter* 852028 & 20 km S of Suguta Marmar, *Gilbert et al.* 5340 & 43 km NE Rumuruti, *Newton* 4221!
DISTR. K 1, 3; not known elsewhere
HAB. Growing on volcanic soil, often among grass tussocks; 600–1900 m

SYN. *Pachycymbium laikipiense* M.G. Gilbert in Bradleya 8: 24 (1990); U.K.W.F. ed. 2: 186 (1994)
Angolluma laikipiensis (M.G. Gilbert) Plowes in Excelsa 16: 119 (1994)

5. **Orbea tubiformis** (*E.A. Bruce & P.R.O. Bally*) *Bruyns* in Aloe 37: 76 (2001); Müller, Kiel, Albers & Meve in Handb. Succ. Pl. Asclepiadaceae: 201 (2002). Type: Kenya, Northern Frontier District: Ewaso Ngiro, near Archer's Post, *Copley* s.n. sub *Bally* S33 (K, holo.; iso, ZSS).

Decumbent stem-succulent; stems ascending-erect, rarely branching, grey-green, spotted red-green or dark brown, 10–15 cm long, 8–12 mm diameter, rounded 4-angled; tubercles ascending, conical-triangular, 10–15 mm long, pointed. Leaf rudiments subulate, acute. Inforescences subterminal or lateral, 1-(2–)flowered; pedicels erect, 5–8 mm long. Sepals ± 5 mm long, acuminate. Corolla deeply campanulate, 18–25 mm long, 2–2.5 diameter, abaxially pale green, streaked purplish, adaxially dark flesh-coloured, brown to purple-brown, sculptured with low dome-shaped emergences tipped by a white seta, up to 3 mm long, apically swollen; tube cupular, 9–12 mm long, 8–12 mm diameter, enclosing the corona; lobes spreading to ascending, deltoid, 10–12 × 6–7 mm, margins recurved. Gynostegial corona fleshy, dark red-brown, cup-shaped, 4–6 × 6–7.5 mm, with protruding, rounded, baggy interstaminal corona lobes ± 3 mm long, apically constricted; staminal corona lobes ascending, deltoid-lanceolate, 1.5–2.5 mm long, closely apressed against anthers.

KENYA. Northern Frontier District: 3 km N Longobito, 22 Feb. 1996, *Meve, Masinde, Goyder & Newton* 958!; Turkana District: 137 km S of Lokitaung, 29 Oct. 1977, *Carter & Stannard* 101; Baringo District: Kampi ya Samaki, 13 Apr. 1974, *Renner s.n.* in *Bally* 16601
DISTR. K 1–3; not known elsewhere
HAB. In dry bushland or succulent communities (*Sansevieria robusta*), on stony soils; 600–1300 m

SYN. *Caralluma tubiformis* E.A. Bruce & P.R.O. Bally in Cact. Succ. J. (Los Angeles) 13: 167 (1941); Blundell, Wild Fl. E. Afr.: t. 475 (1987)

Pachycymbium tubiforme (E.A. Bruce & P.R.O. Bally) M.G. Gilbert in Bradleya 8: 21 (1990);
U.K.W.F. ed. 2: 186, t. 74 (1994)
Angolluma tubiformis (E.A. Bruce & P.R.O. Bally) Plowes in Excelsa 16: 105 (1994)

6. **Orbea dummeri** (*N.E. Br.*) *Bruyns* in Aloe 37: 74 (2001); Bruyns in Syst. Bot.
Monogr. 63: 80–82 (2002); Müller, Kiel, Albers & Meve in Handb. Succ. Pl.
Asclepiadaceae: 192 (2002). Type: Uganda, "Bukoba District" without precise
locality, cultivated locally, *Dummer* s.n. (K, holo.)

Stem-succulent, decumbent, forming loose clumps or mats; stems prostrate-
ascending, cylindrical or rounded 4-angled, 6–9 cm long, ± 1 cm diameter, pale
grey-green, patterned purplish; tubercles spreading horizontally, conical-subulate,
6–15 mm long. Leaf rudiments subulate, 1–2 mm long, acute. Inflorescences
1–3(–5)-flowered, at upper regions of stems; pedicels 6–12 mm long. Sepals ovate-
lanceolate, 3–5 mm long. Corolla campanulate, 2.8–4.5 cm diameter, deeply
incised, abaxially pale greenish, spotted reddish, adaxially intensively greenish-
yellow, conspicuously sculptured with up to 1.5 mm long, conical emergences
usually tipped by a long transparent seta or drop-shaped hair (at mouth of tube);
tube broadly cupular, 4–5 cm deep, 8–9 mm wide; lobes spreading, triangular, 11–16
× 4–8 mm, convex. Gynostegial corona white, occasionally pale yellow, 4–5 × ± 5 mm,
seated on a short, pentagonal stipe; interstaminal corona lobes deeply pouched in
interstaminal position, apically with rectangular, spreading, bi- or tri-lobed lobules;
staminal corona lobes rectangular, 1.5–2 mm long, bent over anthers and
overlapping, apically toothed. Fig. 83, p. 349.

UGANDA. NE Elgon, May 1953, *Tweedie* 1114; Mengo District: Kampala, Nov. 1930, *Hargreaves* 1855
KENYA. Laikipia District: Kisima Ranch, 21 Feb. 1996, *Goyder et al.* 4056!; South Kavirondo
    District: Rusinga Island, 14 Mar. 1989, *Hartmann & Newton* 28586!; Masai District: Ol
    Lorgasailic [Ol Orgasaile], 9 Apr. 1960, *Verdcourt* 2626
TANZANIA. Masai District: Olduwai [Olduvai] River crossing, on Malambo road, Mar. 1990,
    Hartmann & Newton 29288; Mbulu District: Lake Manyara National Park, Msasa River, 1 Dec.
    1963, *Greenway & Kirrika* 11103; Pare District: 9 km S of Himo Bridge on Tanga road, 27 Jan.
    1974, *Bally & Carter* 16369
DISTR. U ?1, 3, ?4; K 1–6; T ?1, 2, 3; E Congo-Kinshasa, Burundi
HAB. Predominantly in *Acacia* bushland, dry scrub or succulent communities on shallow
    limestone or volcanic soil, mostly rocky outcrops; 700–1700 m

SYN. *Stapelia dummeri* N.E. Br. in Gard. Chron., ser 3, 61: 132 (1917); F.P.U.: 121 (1962)
    Caralluma dummeri (N.E. Br.) White & Sloane in J. Cact. Succ. Soc. (Los Angeles) 12: 82
       (1940); Blundell, Wild Fl. E. Afr.: t. 16 (1987)
    Pachycymbium dummeri (N.E. Br.) M.G. Gilbert in Bradleya 8: 22 (1990); U.K.W.F. ed. 2: 186,
       t. 74 (1994)
    Caralluma dummeri (N.E. Br.) White & Sloane forma *colorata* Lodé in Fichier Encycl. Cact.,
       Ser. 8: 737 (1992). Type: not traced
    Angolluma dummeri (N.E. Br.) Plowes in Excelsa 16: 110 (1994)

NOTE. At least in Kenya this species represents one of the most frequent stapeliads.

7. **Orbea denboefii** (*Lavranos*) *Bruyns* in Aloe 37: 74 (2001); Bruyns in Syst. Bot.
Monogr. 63: 82 (2002); Müller, Kiel, Albers & Meve in Handb. Succ. Pl.
Asclepiadaceae: 191 (2002). Type: Kenya, Kajiado Distr., 6–10 km ENE
Oloitokitok, 2°54'N 37°33'E, *den Boef* sub Lavranos 21027 (EA, not traced); Kenya,
Rift Valley Prov., near Oloitokitok, *Foresti* 824 (K, neo., designated by Newton in
Asklepios 72: 13 (1997))

Decumbent stem-succulent, forming loose clumps; stems rarely branched, obtusely
pale green, spotted with dark or brownish-green, 2–20 cm long, 1–1.2(–1.5) mm
diameter, 4-angled; tubercles conical, spreading to ascending. Leaf rudiments
subulate, 2–4 mm long, acuminate. Inflorescences 1(–3)-flowered; flowers scentless,

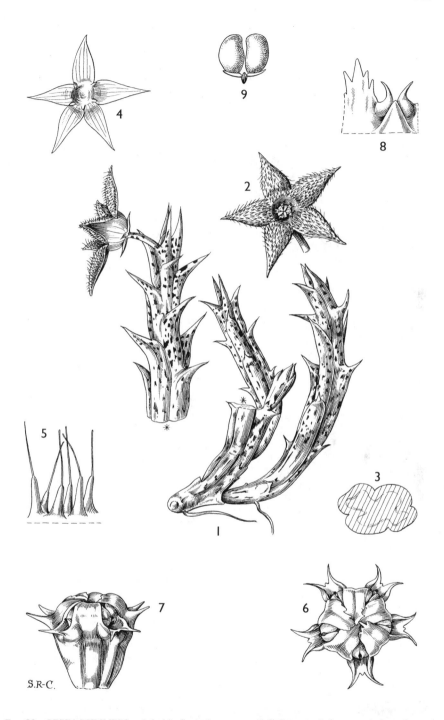

Fig. 83. *ORBEA DUMMERI* — **1**, habit, flowering stem, × 1; **2**, flower, × 1; **3**, cross-section of stem, × 2; **4**, calyx from above, × 3; **5**, papillae on upper surface of corolla lobe, × 8; **6**, gynostegium and corona from above, × 6; **7**, lateral view of gynostegium and corona, × 6; **8**, inner and outer corona from outside, × 6; **9**, pollinarium, × 16. From Hook. Ic. Pl. t. 3414, drawn by Stella Ross-Craig.

occasionally nutant; pedicels 5–12 mm long. Sepals lanceolate, ± 4 mm long. Corolla campanulate, 2.5–3.5 cm diameter, abaxially whitish- to yellowish-cream with brown longitudinal streaks, adaxially pale to golden-yellow or greenish, occasionally suffused redbrown; tube 7–12 mm long, 6–8 mm wide, cylindrical to cup-shaped, basally rounded to pentagonal, enclosing the corona; lobes weakly ascending, narrowly triangular, 8–13 × 5–7 mm, margins revolute, inside smooth, glabrous, papillate or with long whitish hairs (incl. mouth of tube). Gynostegial corona barrel-shaped in outline with exposed pouches in interstaminal position, bright or golden-yellow, 4–5 × 5–6 mm; interstaminal corona lobes forming deep pouches, margins centrally lowered, entire, occasionally emarginate; staminal corona narrowly triangular, 1–2 mm long, apically mostly overlapping. Guide rails spreading exceptionally wide for nearly whole length.

KENYA. Kitui/Kilifi Districts: Galana River, Nov 1985, *Heath & Powys* 790; Masai District: near Oloitokitok, 28 May 1997, *Foresti* 824 & Kajiado District: 7 Nov. 1993, *Newton* 4377
TANZANIA. Arusha District: 15 km S of Arusha, 9 Dec. 2000, *Bruyns* 8709a; Pare District: Njiapanda, 3 km from Himo along Dar es Salaam road, 21 Mar. 1997, *Masinde* 889! & Ngulu, near Lembeni, 9 Dec. 2000, *Bruyns* 8679
DISTR. **K** 4, 6, 7; **T** 2, 3; not known elsewhere
HAB. *Acacia-Commiphora* woodland, usually in loamy soil; 700–1500 m

SYN. *Caralluma denboefii* Lavranos in J. Cact. Succ. Soc. (Los Angeles) 55: 119 (1983)
    Pachycymbium denboefii (Lavranos) M.G. Gilbert in Bradleya 8: 22 (1990); U.K.W.F. ed. 2: 186 (1994)
    Angolluma denboefii (Lavranos) Plowes in Excelsa 16: 106 (1994)

8. **Orbea distincta** (*E.A. Bruce*) *Bruyns* in Aloe 37: 74 (2001); Bruyns in Syst. Bot. Monogr. 63: 85 (2002); Müller, Kiel, Albers & Meve in Handb. Succ. Pl. Asclepiadaceae: 192 (2002). Type: Tanzania, Tanga District: between Moa and Mwakijembi, cult. at Mtotohovu, Moa by *Boscawen* sub *Greenway* 4260 (K, lecto., designated here – see note)

Decumbent stem-succulent, often prostrate, forming loose clumps; stems rarely branched, grey-green, spotted purplish or (brownish) green, 6–10 cm long, 5–12 mm diameter, tapering basally, obtusely 4-angled; tubercles spreading to ascending, conical, much longer than leaf rudiments. Leaf rudiments subulate, ± 10 mm long, acuminate, very acute. Inflorescences 1-(2)-flowered, near apex; pedicels strictly erect, 5–12 mm long. Sepals spreading, lanceolate, ± 6 mm long, acute. Corolla tubular-campanulate, 3.5–4 mm diameter, abaxially cream-coloured or whitish with dark brownish longitudinal streaks, adaxially brownish or flesh-coloured, changing into pale yellowish towards base of lobes and within tube; tube occasionally spotted reddish inside, deeply cupular, 13–17 mm long, 8–12 mm wide, enclosing corona, basally pouching in staminal position, narrowing slightly towards mouth, with a ring of white hairs inside tube at median position; lobes ascending, ovate-lanceolate, 16–28 × 6–10 mm, margins strongly revolute, adaxially rather rugose, glabrous. Gynostegial corona (reddish) brown, variably shaped but usually with pouches in interstaminal position, 5–7 × 6–8 mm; interstaminal corona lobes forming pouches ± 2 mm long, margins centrally incised; staminal corona lobes 2–3 mm long, basally triangular and occasionally winged, apically often linear and laterally compressed to a dorsal, longitudinal, wing-like projection.

KENYA. Teita District: Taita Ranch, Galla Camp, 5 Dec. 1998, *Luke & Luke* 5571; Kilifi District: Dakawachu, 30 Nov. 1990, *Luke & Robertson* 2537A; Tana River District: Kosi, 11 km W of Tana River, Jan. 1940, *Ritchie s.n.* in *Bally* S52
TANZANIA. Pare District: W of Lake Jipe, 10 Dec. 2000, *Bruyns* 8684 & 5 km E of Lembeni, 8 Dec. 2000, *Bruyns* 8676
DISTR. **K** 7; **T** 3; not known elsewhere
HAB. In bushland on sandy to loamy soils; 100–1150 m

SYN. *Caralluma distincta* E.A. Bruce in Hook. Icon. Pl. 35: t. 3415 (1940)
   Pachycymbium distinctum (E.A. Bruce) M.G. Gilbert in Bradleya 8: 22 (1990)
   Angolluma distincta (E.A. Bruce) Plowes in Excelsa 16: 106 (1994)

NOTE. Sharing quite a few similarities with *O. denboefii* – including being a Kenya-Tanzania endemic.

   E.A. Bruce cited *Greenway* 4260 when describing this species, but also stated that the plant was cultivated by Lt. Col. Boscawen in Mtotohovu and by Mrs. Moreau in Amani (making at least three potential collection events). She also said that the drawing accompanying the description was prepared from material preserved in spirit. Two spirit bottles of *Greenway* 4260 are preserved at K, both from Boscawen's cultivated material, and we have chosen to designate the first of the two to be sent by Greenway to Kew, which bears the inscription "type specimen", as lectotype (sent 26 Jan. 1940 – the second was sent in Apr. of the same year). This is bottle number 6724.000. D.J. Goyder & U. Meve.

9. **Orbea subterranea** (*E.A. Bruce & P.R.O. Bally*) *Bruyns* in Aloe 37: 76 (2001); Bruyns in Syst. Bot. Monogr. 63: 91 (2002); Müller, Kiel, Albers & Meve in Handb. Succ. Pl., Asclepiadaceae: 201 (2002). Type: Kenya, Teita District: Sagalla, *Joanna* s.n. in *Bally* S20 (K, holo.)

Stem-succulents, growing in small clumps, rhizomatous, interconnected by thin white stolons up to 15 cm long; stems mostly erect from the rhizomes, (grey-)green, mostly spotted purplish, 2–10 cm long, 5–15 mm diameter; tubercles in 4 rows, spreading to ascending, 5–13 mm. Leaf rudiments deltoid, sublanceolate to subulate, 2–3 mm long, acute; stipules often present. Inflorescences 1–3-flowered, several developed at time, lateral at upper stems; flowers usually presented vertically, faintly scented; pedicels 0.5–2 mm long, occasionally up to 5 mm long. Sepals ovoid-lanceolate, 1–1.5 mm long, acuminate. Corolla rotate, 0.8–2 cm diameter, deeply incised, adaxially pink, reddish to black red, occasionally yellowish, covered with 0.5–1.5 mm long hairs, curved towards corolla lobe tips, rather thickish, white or reddish; tube flat to convex 3–4 mm diameter; lobes spreading, 4–7 mm long, convex. Gynostegial corona purplish, bowl-shaped, 2–3 × 2.5.–4.5 mm; interstaminal corona lobes pouch-like with spreading outer appendages, subquadrate, entire, crenate or toothed; staminal corona lobes oblong-rectangular, 1–1.5 mm long, appressed against anthers, apically occasionally ascending, dorsally often humped, ridged or pectinate.

KENYA. Northern Frontier District: Mt Kulal, 16 Nov. 1992, *Newton, Carter & Powys* 4148! & 3 km S Marsabit, 11 Mar. 1987, *Hartmann & Newton* 21355!; Naivasha District: Lake Naivasha area, Apr. 1975, *Magius* s.n. in *Hepper & Field* 5108
TANZANIA. Masai District: Ngare Nanyuki, 29 Dec. 1970, cult. 12 Mar. 1971, *Richards & Arasululu* 26597 & W of Ngare Nanyuki, 14 Dec. 2000, *Bruyns* 8699; Arusha District: Ormutonyi Plains, *Richards* 24997
DISTR. **K** 1, 3–5, 7; **T** 2; not known elsewhere, but likely to occur in N Uganda and perhaps S Ethiopia
HAB. Bare ground, thin grassland or in succulent communities, in sandy to loamy soil (pockets); 700–1600 m

SYN. *Caralluma subterranea* E.A. Bruce & P.R.O. Bally in Cact. Succ. J. (Los Angeles) 13: 165 (1941); Blundell, Wild Fl. E. Afr.: t. 549 (1987)
   Pachycymbium baldratii (White & Sloane) M.G. Gilbert subsp. *subterraneum* (E.A. Bruce & P.R.O. Bally) M.G. Gilbert in Bradleya 8: 24 (1990) & in Fl. Eth. 4(1): 187 (2003)
   Angolluma subterranea (E.A. Bruce & P.R.O. Bally) Plowes in Excelsa 16: 119 (1994)
   A. lenewtonii Lavranos in Asklepios 73: 15 (1998). Type: Kenya, Northern Frontier District, Mt Kulal, *Newton, Carter & Powys* 4148 (K, holo.)

NOTE. Aerial stems can be very stout and short.
   Note that Bruyns in Syst. Bot. Monogr. 63: 91 (2002) incorrectly cited *Bally* S4 as the type. There is also some dispute over where the type locality really is – see Newton in Asklepios 103: 7–8 (2008) – either "Sagalla Hill, Onjika near Kisumu" or Sagala Hill in Teita District.

10. **Orbea wilsonii** (*P.R.O. Bally*) *Bruyns* in Aloe 37: 76 (2001); Bruyns in Syst. Bot. Monogr. 63: 93 (2002); Müller, Kiel, Albers & Meve in Handb. Succ. Pl. Asclepiadaceae: 204–205 (2002); Dodds in Asklepios 90: 12–13 (2004). Type: Uganda, Toro District: 48 km S of Kasese, Queen Elizabeth Park, Lokitunyala near Mweya Lodge, *J. Wilson* 13 (K, holo.)

Stem-succulents, forming small clusters; stems decumbent-erect, rhizomatous, grey-green, spotted dark green to reddish-brown, 5–6 cm long, 5–12 mm diameter, rounded 4-angled; tubercles spreading to ascending, conical, 3–15 mm long, occasionally depressed wing-like. Leaf rudiments deltoid-subulate, 2–5 mm long, acute; stipules occasionally present. Inflorescences 1–6-flowered, subapical to lateral; pedicels 1–3 mm long, holding flowers vertically. Sepals lanceolate-subulate, 2–4 mm long, acute. Corolla shallowly campanulate, 12–17 mm diameter, abaxially grey-green, occasionally patterned purple-brown, adaxially lemon-yellow to red-brown, finely to coarsely spotted dark purple, covered with conical emergences tipped by a spreading purplish seta 0.2–2 mm long; tube cupular, 1–2.5 mm long, 4–6 mm diameter, enclosing the corona; lobes spreading, deltoid, 3.5–5 × 2–3 mm. Gynostegial corona red with yellow spots to dark purple, 2–3 × 3–4 mm; interstaminal corona lobes forming shallow, spreading pouches, rectangular in top view; staminal corona lobes narrowly triangular, 1.2–1.6 mm long, appressed against anthers, dorsally rugose or verrucose, apically crenate to toothed.

UGANDA. Toro District: Ruwenzori [Queen Elizabeth] National Park, Lokitunyala near Mweya Lodge, 1963, *J. Wilson* 13
KENYA. Northern Frontier District: Lake Kisima, between Maralal and Suguta Marmar, *Taylor* sub *Bally* 14004 & Mugie Ranch near Suguta Marmar, Oct. 2000, *A. & M. Dodds* 1
DISTR. U 2; **K** 1; not known elsewhere
HAB. Bushland with grasses and succulents; 1000–2000 m

SYN. *Caralluma wilsonii* P.R.O. Bally in Candollea 21: 371 (1966) & Candollea 29: 390 (1974)
  Pachycymbium wilsonii (P.R.O. Bally) M.G. Gilbert in Bradleya 8: 22 (1990); U.K.W.F. ed. 2: 186 (1994)
  Angolluma wilsonii (P.R.O. Bally) Plowes in Excelsa 16: 110 (1994)

NOTE. *O. wilsonii* has been recollected several times recently in central Kenya, and three different colour forms have been depicted by Maria Dodds in Asklepios 90: 12–13 (2004) under the name *Angolluma wilsonii*.

11. **Orbea semitubiflora** (*L.E. Newton*) *Bruyns* in Aloe 37: 76 (2001); Bruyns in Syst. Bot. Monogr. 63: 95 (2002); Müller, Kiel, Albers & Meve in Handb. Succ. Pl. Asclepiadaceae: 200 (2002). Type: Tanzania, Arusha District: Kisite crater, 1988, cult. Apr. 1990, *Newton* 3419 (K, holo.; EA, SRGH, iso.)

Decumbent stem-succulents, in dense clusters up to 50 cm diameter; stems ascending-erect, rhizomatous, pale green, patterned purplish-brown, 3–8 cm long, ± 9 mm diameter, rounded 4-angled; tubercles in 4 rows, spreading, conical, 4–10 mm long. Leaf rudiments 1–2 mm long; stipules occasionally present. Inflorescences 2–3(–8)-flowered, some buds often caducous; pedicels 5–9 mm long. Sepals 2–4 mm long, lanceolate. Corolla shallowly campanulate, 1.5–2.5 cm diameter, abaxially pale green, margins reddish, adaxially golden-yellow or purple-brown, occasionally mottled; tube 3–4 mm long, 5–7 mm diameter, enclosing the corona; lobes ascending, ovoid-lanceolate, 7–9 mm long, convex, margins revolute. Gynostegial corona golden-yellow or chestnut brown, cup-shaped, 3–4 × 4–5 mm; interstaminal corona lobes pouch-like, upper rim much lowered, curved, entire; staminal corona lobes deltoid, minute, appressed to back of anthers, dorsally humped, occasionally winged.

TANZANIA. Masai District: Engare Nanyuki, 12 Apr. 1968, *Greenway & Kanuri* 13479 & Ngorongoro Crater, Apr. 1941, *Bally* S76 & near Ol Doinyo Lengai, 2000, *Bruyns* 8705b
DISTR. **T** 2; not known elsewhere

HAB. Typically found in craters of the Rift Valley, confined to azonal stands like small lava outcrops, often together with grasses and dwarf-shrubs; 900–1700 m

SYN. *Angolluma semitubiflora* L.E. Newton in Cact. Succ. J. (Los Angeles) 65: 198 (1993)
   Pachycymbium semitubiflorum (L.E. Newton) M.G. Gilbert in Nord. J. Bot. 22: 210 (2003), as *semitubiforme*

NOTE. One of the species showing two flower colour variants – a plain yellow and a purple-brown (like *O. semota*).

12. **Orbea taitica** *Bruyns* in Syst. Bot. Monogr. 63: 97 (2002). Type: Kenya, Teita District: Taita Hills, below Maragua, *Luke et al.* 5561 (BOL, holo.; EA, K, iso.)

Decumbent stem-succulent, forming loose mats, rhizomatous or not; stems green, patterned purplish, 5–15 cm long, ± 9 mm diameter, rounded 4-angled; tubercles ascending, conical, 3–6 mm long. Leaf rudiments subulate, 1–2 mm long, acute; occasionally with stipules. Inflorescences 1–3-flowered, several at flanks or tips of stems; pedicels spreading to erect, 2–20 mm long. Sepals lanceolate, ± 2 mm long, acute. Corolla rotate to shallowly campanulate, 15–20 mm diameter, abaxially pale green, adaxially dark purple, velvety, with scattered hemispherical emergences, finely papillate throughout; tube cup-shaped, ± 2.5 mm long, 4–5 mm diameter; lobes spreading to ascending, deltoid-lanceolate, 5–8 × 4–5 mm, convex. Gynostegial corona purplish-red, ± 3.5 × 3–4.5 mm; interstaminal corona lobes pouch-like, upper rim much lowered, curved, emarginate; staminal corona lobes deltoid, ± 1 mm long, appressed to back of anthers, dorsally humped and winged, wings fused to interstaminal corona.

KENYA. Kitui District: 175 km NE of Nairobi, Ngomeni Rock, *McCoy* 2632 & near Mutomo (Ngomeni Rock?), *Rauh* Ke786 sub *Plowes* 3718!; Teita District: below Maragua, 4 Dec. 1998, *Luke et al.* 5561
DISTR. **K** 4, 7; known only from these three collections
HAB. On rocky outcrops and inselbergs; 900–1700 m

SYN. *Angolluma doddsiae* Plowes & T.A. McCoy in Asklepios 103: 14 (2008). Type: Kenya, Kitui District: 175 km NE of Nairobi, Ngomeni Rock, *McCoy* 2632 (MO, holo.; P, iso.)

NOTE. Exceptionally long pedicels, pointed buds and a larger corolla mark *Angolluma doddsiae*, here treated as synonym of *O. taitica*, as deviant from the type – probably due to long isolation of the two known populations of *O. taitica*. Treatment of *A. doddsiae* as a subspecies of *O. taitica* could be therefore also be justified.

13. **Orbea vibratilis** (*E.A. Bruce & P.R.O. Bally*) *Bruyns* in Aloe 37: 76 (2001) & in Syst. Bot. Monogr. 63: 98 (2002); Müller, Kiel, Albers & Meve in Handb. Succ. Pl. Asclepiadaceae: 204 (2002). Type: Kenya, Baringo District: Marigat near Lake Baringo, 26 Mar. 1940, *Ritchie* s.n. in *Bally* S35 (K, holo.)

Stem-succulents, spreading over larger areas; stems rhizomatous, erect from the rhizomes, solitary (rarely branching and clustering), grey(bluish)-green, spotted dark green to reddish-brown, 5–8 cm long, 5–10 mm diameter, rounded 4-angled; tubercles spreading, conical, ± 5 mm long. Leaf rudiments deltoid-spatulate, ± 1 mm long, acute; stipules present. Inflorescences 1–5-flowered, lateral; flowers strongly foetid; short peduncle present; pedicels erect or curved downwards, 5–6 mm long. Sepals lanceolate, 3–5 mm long, acute. Corolla campanulate, 0.5–1 cm long, 1–2 cm diameter, abaxially pale green, adaxially dark flesh-coloured to (mostly) black-purple, occasionally with yellow-green areas, smooth, occasionally rugose; tube cupular, 4–7 mm long, 4–7 mm diameter, enclosing corona; lobes ascending to reflexed, deltoid, 4–8 × 3–5 mm, lower half ciliate with vibratile clavate hairs. Gynostegial corona dark purple, ± 3 × 3–4.5 mm; interstaminal corona lobes flattened pouch-like, ± 2 mm long, upper rim bifid; staminal corona lobes ascending-erect, narrowly rectangular, 2–3 mm long, rising above anthers, obtuse.

UGANDA. Bunyoro District: Butiaba Flats near scarp foot, Nov. 1941, *Eggeling* 4707
KENYA. Northern Frontier District: 36 km from North Horr, Mar. 1987, *Cumming* 1443!;
  Laikipia District: Kifuku Ranch, Rumuruti, *Dodds* s.n.; Nairobi/Machakos Districts: Athi River
  Station, *MacArthur* s.n.
TANZANIA. Mbulu District: Tarangire National Park, Tarangire Camp, 1 Dec. 1969, *Richards*
  24856; sin loc., cult. at Mtotohovu, Moa by Boscawen, *Greenway* C & D
DISTR. U 2; **K** 1, 3, 4; **T** 2; S Ethiopia
HAB. *Acacia* wooded grassland, in seasonally wet areas on black cotton soil; 1000–2000 m

SYN. *Caralluma vibratilis* E.A. Bruce & P.R.O. Bally in Cact. Succ. J. (Los Angeles) 13: 179 (1941)
  Pachycymbium vibratile (E.A. Bruce & P.R.O. Bally) M.G. Gilbert in Bradleya 8: 22 (1990);
    U.K.W.F. ed. 2: 186, t. 74 (1994); Gilbert in Fl. Eth. 4(1): 190 (2003), as *vibratilis*
  Angolluma vibratilis (E.A. Bruce & P.R.O. Bally) Plowes in Excelsa 16: 106 (1994)

NOTE. The Ugandan collection was sent to Kew in spirit, but arrived in a crushed bottle, and
lacks a flower. However, there is no reason to question its identity.

14. **Orbea caudata** (*N.E. Br.*) *Bruyns* in Aloe 37: 73 (2001) & in Syst. Bot. Monogr.
63: 102 (2002); Müller, Kiel, Albers & Meve in Handb. Succ. Pl., Asclepiadaceae: 189
(2002); Bruyns in Stap. south. Afr. Madag. 1: 256 (2005). Type: Malawi, Namasi,
Cameron 25 (K, holo.)

Clump- or mat-forming stem-succulents; stems decumbent-ascending to erect,
olive-green or grey-green, spotted dark green or purplish, 5–15 cm long, 0.6–1 cm
diameter, rounded 4-angled; tubercles spreading to ascending, narrowly conical,
5–20 mm long. Leaf rudiments 5–9 mm long, subulate, acute. Inflorescences 4–7-
flowered, from lower stem regions; flowers open simultaneously, with strong, dull
excrement odour; pedicels 5–20 mm long. Sepals spreading, lanceolate, 6–7 mm
long. Corolla rotate, 4–9 cm diameter, deeply incised, abaxially pale green,
occasionally spotted reddish, adaxially slightly rugose, ± papillate (especially inside
tube), pale or dark yellow, spotted with purple or crimson (rarely completely
purple-brown, or spotted pale yellow on green); tube shallowly bowl-shaped, 10–15 mm
diameter, with cup-shaped central depression encircling the corona; lobes
spreading, 20–40 × 6–10 mm, acuminate to caudate, convex, margins of upper half
revolute, cilate with purple clavate vibratile hairs. Gynostegial corona yellowish
patterned red or purple, 3–4 × 6–8 mm, atop a short, pentagonal stipe;
interstaminal corona lobes spreading, subquadrate, 1.5–2.5 × ± 1.5 mm, truncate
or bidentate, occasionally crenate; staminal lobes narrowly to broadly ribbon-
shaped, 1–2.5 × ± 0.8 mm, obtuse or tapering, incurved on back of anthers,
occasionally ascending, occasionally tips overlapping, basally extended and fused
with interstaminal corona, dorsally occasionally with hump-like excrescence.
Mericarps two, 12–15 cm long, fusiform.

A species of wide distribution with a single collection within the Flora area of the more
northerly of the two subspecies.

subsp. **caudata**

Stems olive-green spotted dark green or purplish; tubercles plus leaf rudiments 5–10 mm
long. Corolla variously coloured or patterned but reddish typically predominates.

TANZANIA. Ufipa District: near Zambian border on Mbala [Abercorn]–Sumbawanga road, 17
  Apr. 1960, *Leach & Brunton* 10078!
DISTR. **T** 4; Zambia, Malawi and Mozambique
HAB. In shallow soil pockets on granite outcrops; 600–1700 m

SYN. *Caralluma caudata* N.E. Br. in F.T.A. 4(1): 485 (1903)
  C. caudata N.E. Br. var. *fusca* C.A. Lückhoff in White & Sloane, Stapelieae ed. 2, 1: 352 & 3:
    1144 (1937). Type: White & Sloane, Stapelieae ed. 2, 1: fig. 287, p. 252 (1937), lecto.,
    designated by Bruyns in Syst. Bot. Mongr. 63: 105 (2002)

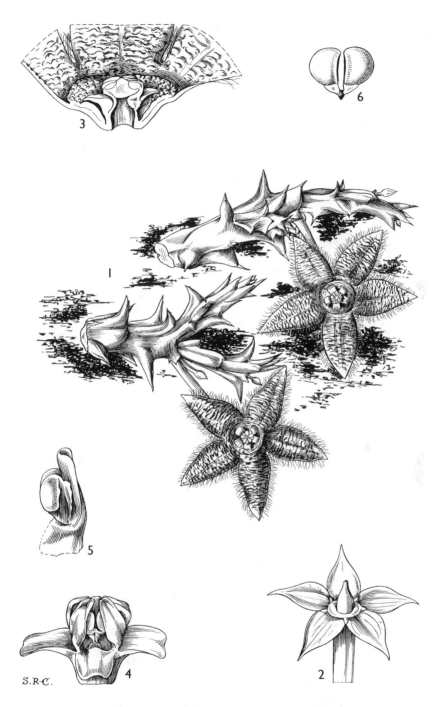

FIG. 84. *ORBEA SEMOTA* — **1**, flowering shoots, × 1; **2**, calyx and paired ovaries, × 3; **3**, longitudinal section through corolla, showing annular corona, and gynostegium showing staminal corona lobes, × 3; **4**, gynostegium, × 6; **5**, anther and inner corona lobe from side, × 12; **6**, pollinarium, × 12. From *Greenway* 4599. From Hook. Ic. Pl. t. 3416, drawn by Stella Ross-Craig.

Caralluma praegracilis Oberm. in White & Sloane, Stapelieae ed. 2, 3: 1161 (1937). Type: White & Sloane, Stapelieae ed. 2, 3: fig. 1212, p. 1160 (1937), lecto., designated by Bruyns in Syst. Bot. Mongr. 63: 105 (2002)
Orbeopsis caudata (N.E. Br.) L.C. Leach in Excelsa Tax. Ser. 1: 68 (1978)

15. **Orbea semota** (*N.E. Br.*) *L.C. Leach* in Kirkia 10: 290 (1975); Leach in Excelsa Tax. Ser. 1: 46 (1978); Meve in Kakt. & Sukk. 50(9): 222 (1999); Bruyns in Syst. Bot. Monogr. 63: 134 (2002). Type: Tanzania, Kondoa District: Kondoa-Irangi, *B.D. Burtt 1450* (K, holo.)

Clump-forming stem-succulents; stems decumbent to erect, green, mostly spotted purplish, 3–10 cm long, ± 8 mm diameter, 4-angled; tubercles ascending, conically triangular, 5–10 mm long, acuminate. Leaf rudiments 3–5 mm long; stipular glands present. Inflorescences 1–3-flowered, from basal half of stems; pedicels 15–40 mm long. Sepals lanceolate, 4–7 mm long, acuminate. Corolla rotate, 2–5.5 cm diameter, deeply incised, adaxially dark purple-brown, brown, yellow or mottled, smooth or rugose; tube shallow, most of centre raised to a circular to pentagonal fleshy annulus, 7–13 mm diameter; lobes spreading to slightly reflexed, ovoid-lanceolate, 7–27 × 6–10 mm, convex, ciliate with vibratile red or white hairs 2–3 mm long. Gynostegial corona atop a short stipe, dark brown-red, star-shaped in top view, ± 3 × 6–10 mm, presented on top of annulus, occasionally mottled yellowish; interstaminal corona lobes spreading, subquadrate, 2–3.5 mm long, staminal corona lobes spatulate, 0.8–1.7 mm long, appressed against anthers, entire or emarginate, often yellowish. Fig. 84, p. 355.

KENYA. Nairobi District: Nairobi River valley behind Governor's house, 21 May, *Napier* 6206; Machakos District: Musalala, 20 March 2005, *Newton* 5906; Masai District: Masai Mara, Olololo escarpment, Dec. 1998, *Dodds* s.n.
TANZANIA. Arusha, *Specks* 921B!; Pare District: Kisiwani, 3 Feb. 1936, *Greenway* 4592; Kondoa District: Great North Road 8 km N of Kondoa, 11 Jan. 1962, *Polhill & Paulo* 1131A
DISTR. **K** 4, 6, 7; **T** 2, 3, 5; Rwanda
HAB. Rocky places; 800–2000 m

SYN. *Stapelia semota* N.E. Br. in Cact. Succ. J. (Los Angeles) 4: 393 (1933); Blundell, Wild Fl. E. Afr.: t. 349 (1987); U.K.W.F. ed. 2: 186 (1994)
   S. discoidea Oberm. in White & Sloane, Stapelieae ed. 2, 3: 1167 (1937). Type: South Africa, sin. loc., *Transvaal Museum* 35740 (not traced)
   S. molonyae White & Sloane in Stapelieae ed. 2, 2: 435 (1937). Type: Kenya, Nairobi, *Molony* s.n. (not traced)
   S. kagerensis Lebrun & Taton in Expl. Parc Nat. Albert 1: 108 (1948). Type: Rwanda, Lugadzi, S Gabiro, *Lebrun* 9523 (BR, holo.)
   Orbea semota (N.E. Br.) L.C. Leach subsp. *orientalis* Bruyns in Syst. Bot. Monogr. 63: 135–137 (2002). Type: Kenya, Teita District: near Bura, *Luke* 5511 (BOL, holo.)

NOTE. A critical reexamination of recent collections of *O. semota*, including the living material available in private collections, revealed that the continued recognition of subsp. *orientalis* is not supported. This taxon is to be regarded as a small-flowered form of the many local forms of this widespread and variable species.

66. **STAPELIA**

L. in Sp. Pl.: 217 (1753); Leach in Excelsa Tax. Ser. 3: 1–157 (1985); Bruyns in Stap. south. Afr. Madag. 2: 418–489 (2005)

Gonostemon Haw. in Syn. Pl. Succ.: 27 (1812)

Clump-forming stem-succulents; latex clear or at most cloudy; stems erect, cylindrical, rounded or sharply 4-angled, glabrous or pubescent. Leaf rudiments sessile, ascending, caducous; stipules glandular, globoid. Inflorescences basal, 1–10-flowered, pedunculate. Flowers strongly foetid, not nectariferous. Corolla rotate, shallowly cupular or (rarely) campanulate, fused basally to half of length, fleshy, rugose or unsculptured, glabrous or with trichomes; lobes reflexed or spreading, often ciliate. Corolline corona absent. Gynostegial corona in 2 series, with 5 staminal and 5 interstaminal parts fused only at the very base; "outer" interstaminal corona lobes ovate, rectangular or deltate, spreading, "inner" staminal corona lobes subulate, often bilobed, erect, occasionally with inflexed subulate or falcate adaxial appendage shorter than staminal parts. Gynostegium (sub-)sessile or atop a column. Pollinia ovoid or reniform. Mericarps two, glabrous or pubescent.

Around forty species in southern Africa. The classical "carrion flower".

FIG. 85. *STAPELIA GIGANTEA* — Habit. Reproduced from The Gardeners' Chronicle 1888(4): 729, fig. 101.

Stapelia gigantea *N.E. Br.* in Gard. Chron., ser. nov. 7: 684 (1877); Leach in Excelsa Tax. Ser. 3: 10 (1985); Bruyns in Stap. south. Afr. Madag. 2: 472 (2005). Type: South Africa, Natal, *Gerrard* 717 (K, holo.)

Compact, dense clump-forming stem-succulents; stems 12–25(–30) cm long, 2–3 cm wide, sharply 4-angled with compressed faces; tubercles inconspicuous. Leaf rudiments deltoid, 2–3 mm long, acute. Inflorescences 1–4-flowered; flowers strongly carrion-scented; peduncle stout, up to 1 mm long; pedicels 3–6 cm long. Sepals ovate, 10–16 mm long. Corolla rotate, 15–40 cm diameter, abaxially minutely pubescent, adaxially yellowish to pale reddish, transversely rugolose, wrinkles red-brown or purple, covered by fine, whitish to reddish hairs up to 12 mm long; tube flat to slightly campanulate, 3–10 cm diameter; lobes spreading, ovate to triangular, 8–15 cm long, caudate. Gynostegial corona purple, usually atop a short, yellow stipe; interstaminal corona lobes erect-diverging, rectangular, occasionally spatulate, 5–6 × 1.5–2.5 mm, apically crenate, with or without mucro; staminal corona lobes erect, 8–12 mm long, 1–2 mm wide, with central subulate and dorsal broadly winged appendages. Mericarps erect, two, 9–13 cm long, fusiform. Fig. 85, p. 357.

Frequently cultivated in the Flora area and perhaps naturalising locally; native to C & S Zambia, southern Malawi, Mozambique, Botswana, Swaziland and South Africa

SYN. *Stapelia nobilis* N.E. Br. in Hook.f. in Bot. Mag. 127: t. 7771 (1901). Type: South Africa, Port Elizabeth, 1897, cult. Kew May 1900, *Griffiths* s.n. (K, holo.)
    S. marlothii N.E. Br. in K.B. 1908: 436 (1908). Type: Zimbabwe, Matopo Hills, *Marloth* 3414 (K, holo.; PRE, STE, iso.)
    S. gigantea N.E. Br. var. *pallida* E. Phillips in Flow. Pl. South. Afr. 5: t. 181 (1925). Type: Ex hort. PRE ex hort. *Janse* (1924) (no type material traced)
    S. youngii N.E. Br. in K.B. 1931: 43 (1931). Type: Zimbabwe, near Harare [Salisbury], *Young* in Herb. Moss 17301 (K, holo.)
    S. cylista C.A. Lückh. in South Afr. Gard. & Country Life 23: 139 (1933). Type: White & Sloane, Stapelieae ed. 2, 2: plate 15, opposite p. 524 (1937), icono.
    Gonostemon giganteus (N.E. Br.) P.V. Heath in Calyx 1: 17 (1992)
    G. giganteus (N.E. Br.) P.V. Heath var. *nobilis* (N.E. Br.) P.V. Heath in Calyx 3: 7 (1993)
    G. giganteus (N.E. Br.) P.V. Heath var. *marlothii* (N.E. Br.) P.V. Heath in Calyx 3: 7 (1993)
    G. giganteus (N.E. Br.) P.V. Heath var. *pallidus* (E. Phillips) P.V. Heath in Calyx 3: 7 (1993)
    G. giganteus (N.E. Br.) P.V. Heath var. *youngii* (N.E. Br.) P.V. Heath in Calyx 3: 7 (1993)

67. **GLOSSONEMA**

Decne. in Ann. Sci. Nat., sér. 2, 9: 335, t.12D (1838); Bullock in K.B. 10: 613–619 (1956); Field in K.B. 37: 344 (1982)

Odontanthera sensu Mabberley in Manilal, Bot. Hist. Hortus Malabaricus: 89 (1980) & in Taxon 29: 606 (1980), *non* Wight, Contr. Bot. India: 48 (1834)

Dwarf perennial herbs with white latex. Leaves opposite, subsessile or petiolate. Inflorescences lateral or subaxillary, 1–many-flowered. Calyx of 5 free sepals. Corolla-tube short; lobes spreading or suberect, often thickened below the apex. Corolline corona arising from about half-way up the corolla-tube and alternating with the corolla-lobes, adnate to the corolla-tube below, the lobes free above. Gynostegial corona absent. Anther-appendages erect or inflexed, membranous. Pollinaria with a pair of pendant or subhorizontal pollinia attached to the corpusculum by slender translator arms. Apex of stylar head level with the top of the anthers or exserted beyond. Follicles inflated, with scattered, soft, spine-like processes. Seeds compressed, ovate, with a coma of silky hairs.

5 or 6 species in sub-Saharan and NE tropical Africa, Arabia, Iran, Baluchistan and Sind.

Corolla-tube funnel-shaped, enclosing a distinct gynostegial stalk ±
    0.5 mm long; corolla yellow with red or purple centre   ......   1. *G. revoilii*
Corolla-tube open, not funnel-shaped, gynostegial stalk ± 0.25 mm
    long; corolla green tinged with pink on lobes ....   2. *G.* sp. A

1. **Glossonema revoilii** *Franch.*, Sert. Somal. in Révoil, Faune et Fl. Çomalis: 40, t.
3 (1882); Bullock in K.B. 10: 615 (1956); Field in K.B. 37: 344 (1982); Blundell, Wild
Fl. E. Afr.: t. 480 (1987); U.K.W.F. ed. 2: 176 (1994); Goyder in Fl. Eth. 4(1): 115, t.
140.10 (2003) & in Fl. Somalia 3: 162, fig. 113 (2006). Type: Djibouti, Tigiéh
[Medjourtines], *Révoil* 69 (P!, holo.)

Perennial herb to 30 cm, with branches arising from a single woody stem and
strong taproot; all parts except the interior of the flower sparsely to densely
tomentose with short, white, stiff, spreading hairs. Leaves with petiole 5–20(–30) mm
long; lamina elliptic-oblong or obovate-oblong to broadly ovate, 15–75 × 8–40 mm,
lower leaves markedly larger than the upper, apex obtuse, rounded, truncate, margin
smooth or occasionally crisped; strong pair of lateral veins arising from the base of
the lamina and up to 6 other major lateral veins at an acute angle to the midrib.
Inflorescences numerous, axillary, umbelliform, 4–7(–11)-flowered; peduncle
absent; bracts linear, 1.5–5 mm long; pedicels slender, flexuous, variable in length
within a single inflorescence – at least one pedicel ± 1 mm long, the rest to 7(–12) mm
long. Sepals linear or lanceolate, 2–3 mm. Corolla rotate, 5–7 mm in diameter; tube
red or purple, ± 1 mm deep, funnel-shaped with a slight lip at the mouth; lobes
spreading or reflexed, yellow, ovate, 2–3 mm long, obtuse with a thickened subapical
patch on the inner face, sparsely pubescent outside, glabrous or with a few scattered
hairs within. Corona lobed almost to the base, narrowly triangular to ovate below,
adnate to the corolla-tube, the free apical part subulate, 1–2 mm long, spreading, the
distal part recurved over the gynostegium to varying degrees and commonly
resembling a fish-hook. Gynostegium stipitate, the stipe ± 0.5 mm long, dilating
abruptly to a pentagonal head ± 0.5 × 1.5 mm at the mouth of the corolla-tube.
Anthers erect or divergent-erect. Apex of stylar head pentagonal, flat or slightly
domed, level with the top of the anthers. Pollinia ovoid, ± 0.25 mm long, twice as
long as the brown corpusculum. Follicles ovate in outline, 4–6.5 cm long, 1.5 cm
wide, echinate, the soft spine-like processes 5–8 mm long with straight or hooked
tips. Fig. 86, p. 360.

KENYA. Northern Frontier District: 24 km E of El Wak, Dec. 1971, *Bally & Smith* 14570! &
    Burrole Mt, Jan. 1972, *Bally & Smith* 14885A!; Teita District: Tsavo East National Park, Mzinga
    Hill, Apr. 1966, *Gillett* 17239!
TANZANIA. Musoma District: Magungu R., Banagi area, Nov. 1953, *Tanner* 1713!; Masai
    District: Lake Natron, Dec. 1944, *Bally* 4108! & Kitumbeine [Ketumbane], Jan. 1936,
    Greenway 4276!
DISTR. **K** 1, 4, 6, 7; **T** 1, 2; Ethiopia, Djibouti, Somalia and Socotra
HAB. *Acacia-Commiphora* bushland, in sandy or stony soil over limestone; 250–1200 m

SYN. *Gilgia candida* Pax in E.J. 19: 80 (1894). Type: Somalia, Ahl Mts, *Hildebrandt* 889C (B†, holo.)
    Glossonema elliotii Schltr. in J.B. 33: 304 (1895). Type: Kenya, Machakos District,
        Kikumbuliyu, *Scott Elliot* 6184 (BM!, holo.; K!, iso.)
    G. rivaei K. Schum. in E.J. 33: 323 (1903). Type: Somalia, Dolo, near Daua River, *Riva* 1129
        (?B†, holo.; FT!, K!, iso.)
    G. macrosepalum Chiov., Result. Sc. Miss. Stefan.-Paoli Coll. Bot. 1: 113 (1916). Type:
        Somalia, El Ure, *Paoli* 1082 (FT!, holo.)
    Odontanthera reniformis sensu Mabberley in Manilal, Bot. Hist. Hortus Malabaricus: 89
        (1980) & in Taxon 29: 606 (1980), *non* Wight (1838)

FIG. 86. *GLOSSONEMA REVOILII* — **1–3**, habit with fruiting and flowering shoots, × ²/₃; **4**, flower, × 5; **5**, pollinarium, × 26; **6**, seeds, with coma × 3; **7**, without coma × 4. 1,4 & 5 from *Brown* 96; 2 from *Stannard & Gilbert* 1037; 3 & 6 from *Gillett* 21293. Reproduced with permission, from the Flora of Ethiopia and Eritrea 4, 1; drawn by E. Papadopoulos.

2. **Glossonema** sp. **A**

Herb to 20 cm, unbranched; all parts except the flower sparsely to densely tomentose with short, white, stiff, spreading hairs. Leaves with petiole 3–4 mm long; lamina oblong or lanceolate-oblong, 10–40 × 3–6 mm, apex subacute, obtuse or rounded, margin smooth; a strong pair of lateral veins arising from the base of the lamina. Inflorescences numerous, axillary, umbelliform, 2–6-flowered; peduncle minute or absent; pedicels slender, flexuous, uniform in length within each inflorescence, 2–3 mm long. Sepals linear, ± 1.5 mm. Corolla rotate-campanulate, ± 5 mm in diameter, green flushed with pink towards the tips of the lobes, united for 1–1.5 mm, open, cup-like, not forming a distinct tube; lobes ovate, 1.5 mm long, obtuse with a slight subapical thickening on the inner face, glabrous or sparsely pubescent outside, glabrous within. Corona green, lobed almost to the base, the lower part adnate to the corolla-tube, ovate, the free apical part subulate, ± 1 mm long, recurved over the gynostegium. Gynostegium shortly stipitate, the stipe ± 0.25 mm long, dilating abruptly to a pentagonal head ± 0.25 × 1 mm. Anthers erect or divergent-erect. Apex of stylar head pentagonal, flat or slightly domed, level with the top of the anthers. Pollinia ovoid, ± 0.25 mm long; translator arms ± 0.2 mm long; corpusculum ± 0.15 mm long, brown. Follicles ovate in outline, with straight or hooked spine-like processes.

KENYA. Turkana District: 20 km NW of Lomoru Itae on Kaiemothia road, Nov. 1977, *Carter & Stannard* 165!; Meru District: NE end of Nyambeni Range, Lagadema Hill, Dec. 1971, *Bally & Smith* 14717!
DISTR. **K** ?1, 2, 4; known only from these collections
HAB. Sparse bushland or tussock grassland, on gravel soils; 800–1100 m

NOTE. This taxon appears close to *Glossonema boveanum* (Decne.) Decne., a species distributed around the margins of the Sahara and in the Arabian peninsula. It differs in lacking the highly contorted filiform tips to the corona lobes and the more domed stylar head appendage typical of that species. Further material is needed to resolve the status and affinities of this taxon.

68. **OXYSTELMA**

R. Br. in Mem. Wern. Soc. 1: 40 (1809); N.E. Br. in F.T.A. 4(1): 382 (1902)

Scrambling perennial herbs with white latex. Leaves opposite. Inflorescences racemose or subumbelliform. Calyx divided to the base. Corolla united to about half-way, shallowly campanulate; lobes triangular, acute, fringed with white hairs. Corolline corona annular, pubescent. Gynostegial corona of five staminal lobes, the basal parts gibbous, somewhat crumpled, adnate to the gynostegium; the apical parts free, erect, lanceolate-attenuate. Stamens with anthers erect; appendages short, membranous, inflexed. Pollinia pendulous, cylindrical or elongate-clavate, attached in pairs by the attenuate ends to the short ovoid corpusculum. Apex of stylar head pentagonal, truncate, not exceeding the anthers. Follicles solitary or paired, lanceolate or inflated and subglobose. Seeds compressed, ovate, surrounded with a narrow wing and terminating in a coma of white hairs.

A genus of two species, one endemic to tropical Africa, the other principally distributed in Asia, but also occuring in E and NE tropical Africa.

Follicles lanceolate, not inflated; seeds (including wing) less than
    1.5 mm long; leaves rarely weakly cordate at the base . . . . . .   1. *O. esculentum*
Follicles inflated, subglobose; seeds more than 2 mm long
    (including wing); leaves commonly cordate at the base . . . .   2. *O. bornouense*

1. **Oxystelma esculentum** (*L. f.*) Sm., Cyclopaedia 25: [unnumbered page 760]
(1813). Type: *Koenig* s.n. (LINN 307.7, lectotype, designated by Huber in Rev.
Handb. Fl. Ceylon Fasc. 1(1): 38 (1973))

Slender climber or scrambler; young shoots tomentose with soft white hairs,
becoming glabrous. Leaves with petiole 3–9 mm, glabrescent; lamina thin, linear-
lanceolate, 3–9 × 0.2–0.7 cm, acute, base rounded or cuneate; veins conspicuous
beneath, sparsely pubescent at least towards the base of the lower surface.
Inflorescence extra-axillary, cymes 2(–4)-flowered, glabrous or sparsely pubescent;
peduncle 1.9–3.3 cm long; bracts lanceolate, ± 1 mm, deciduous; pedicels 11–16 mm
long, slender, becoming robust in fruit. Calyx-lobes reddish, ovate-oblong to ovate-
lanceolate, acute, 3 × 1–1.5 mm, pubescent, particularly towards the apex. Corolla
white with pink or maroon markings, lobed to about half-way; tube saucer-shaped, ±
10 mm in diameter, ± 5 mm deep, glabrous except for the annular corona which
forms a shallow, minutely but densely pubescent rim around the base of the
gynostegium; lobes triangular, ± 6 × 5–7 mm with acute, reflexed apices, margins
fringed with a dense band of white hairs on the upper surface. Staminal corona lobes
swollen and wrinkled towards the base, adnate to the gynostegium for ± 2.5 mm,
narrowing abruptly to the 1.5–2 mm long, acuminate, terminal part which is inflexed
over the head of the gynostegium. Gynostegium ± 3 × 1 mm, anther-appendages
acute, membranous. Pollinia sausage-shaped, 1 mm long; translator arms short. Apex
of stylar head flat, not exceeding the anthers. Follicles lanceolate and somewhat
sickle-shaped, ± 3 cm long, 8 mm wide (immature), rounded at the base and tapering
to an obtuse or truncate apex. Seeds flattened, broadly ovate, ± 1 × 0.5 mm wide,
surrounded by a narrow wing 0.1–0.2 mm wide.

TANZANIA. Lushoto District: Korogwe, Magunga Estate, Vigai, Pangani R., Feb. 1953, *Faulkner*
1175!
DISTR. T 3; Egypt, Sudan, Ethiopia; also widely distributed in tropical and subtropical Asia
HAB. Scrambling through river or lake-shore vegetation; 300 m

SYN. *Periploca esculenta* L. f., Suppl. Pl.: 168 (1781)
    Oxystelma esculentum (L. f.) Schult., Syst. Veg. ed. 15, 6: 89 (1820), later homonym
    Oxystelma alpini Decne. in DC., Prodr. 8: 543 (1844). Type: "Egypt" (P, holo.)
    O. secamone sensu H. Karst., Deut. Fl.: 1031 (1883), non *Periploca secamone* L.
    O. esculentum (L. f.) Schult. var. *alpini* (Decne.) N.E. Br. in F.T.A. 4(1): 382 (1902)
    Sarcostemma esculentum (L. f.) Holm in Ann. Missouri Bot. Gard. 37: 482 (1950)
    S. secamone sensu Bennet, Indian Forester 95: 692 (1969), non *Periploca secamone* L.

2. **Oxystelma bornouense** *R. Br.* in Denham & Clapperton, Travels in N & Central
Africa: 239 (1826); Bullock in F.W.T.A. ed. 2, 2: 89 (1963); Blundell, Wild Fl. E. Afr.:
t. 553 (1987); U.K.W.F. ed. 2: 176 (1994); Liede in Fl. Eth. 4(1): 149, fig. 140.31
(2003); Goyder in Fl. Somalia 3: 148, fig. 102 (2006). Type: Chad, Bornou, *Oudney* 6
(BM!, holo.)

Perennial herb with slender, twining stems, very young shoots glabrescent,
otherwise glabrous throughout except for corolla margins and base of corolla-tube.
Leaves with petiole 7–14(–20) mm long; lamina linear-lanceolate to ovate-lanceolate
or lanceolate-oblong, (25–)30–70 × (5–)9–25(–30) mm, apex acute or acuminate,
base cordate to somewhat hastate in older leaves, often truncate in younger leaves.
Inflorescences numerous, (25–)50–100 mm long, arising laterally at the nodes,
(1–)2–5(–7)-flowered, racemose or subumbellate; peduncle (15–)30–50 mm long;
bracts lanceolate, deciduous, ± 1 mm; pedicels flexuous, 12–25 mm; flowers pendent;
calyx-lobes ovate, 2.4–4 × 1 mm, acute, dull reddish brown. Corolla sometimes
persistent in fruit, white outside, white or pale pink with dark crimson or maroon
centre and veins within, lobed to about half-way; tube saucer-shaped, 15–20 mm in
diameter, 5–10 mm deep, glabrous except for the basal 1 mm which forms a shallow,
minutely but densely pubescent collar around the base of the gynostegium; lobes

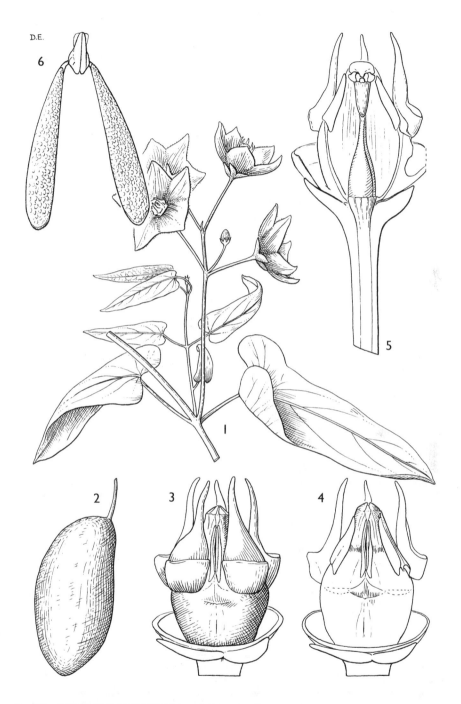

D.E.

FIG. 87. *OXYSTELMA BORNOUENSE* — **1**, flowering shoot, × ³/₄; **2**, follicle, × ³/₄; **3**, gynostegium with corona, and **4**, with 2 corona lobes removed, × 5; **5**, flower in section, × 5; **6**, pollinarium, × 30. 1 & 2 from *Schweinfurth* 982; 3 from *Keller* 103; 4 & 5 from *Chevalier* 10295. Reproduced with permission, from the Flora of Ethiopia and Eritrea 4, 1; drawn by D. Erasmus.

shallowly triangular, 5–7 × 8–10 mm, apex acute, reflexed, margins fringed with a dense band of white hairs on the upper surface. Corona-lobes ± 4 mm tall, basal part swollen and adnate to the column, upper part subulate and inflexed at the apex. Gynostegium ± 3 × 1 mm, anther-appendages acute, membranous. Pollinia ± 1 mm, tapering gradually into the short caudicle. Style apex flat, not exceeding the anthers. Mature follicles ovoid to subglobose, inflated, papery, 50–55 mm long, 33–38 mm wide, green, occasionally tinged with purple. Seeds flattened, ovate, 2 × 1 mm, surrounded by a flat or slightly undulate wing 0.25–0.5 mm wide. Fig. 87, p. 363.

KENYA. Machakos District: 200 km from Mombasa on road to Nairobi, island in Tsavo R., Aug. 1959, *Verdcourt* 2396!; Tana River District: Kora Game Reserve, Tana R., Aug. 1983, *G.R.C. van Someren* 1072! & Garissa, Bura, S bank of Tana R., Sep. 1957, *Greenway* 9245!
DISTR. **K** 4, 7; West Africa from Senegal to Cameroon, Chad, Sudan, Ethiopia and Somalia
HAB. Riverine fringe vegetation; 60–900 m

SYN. *Oxystelma senegalense* Decne. in DC., Prodr. 8: 543 (1844). Types: Senegal, *Heudelot* s.n. (P, syn.); *Leprieur* s.n. (P, syn.); *Perrottet* s.n. (P, syn.); *Richard* s.n. (P, syn.). A sheet at K labelled "*Leprieur* or *Heudelot*" bears the number 423, and probably relates to one of these syntypes

69. PERGULARIA

L. in Systema Naturae ed. 12 vol. 2: 191 (1767) & in Mantissa plantarum: 53 (1767); Goyder in K.B. 61: 245–256 (2006)

Doemia R. Br., On *Asclepiadeae* 39 (1810)

Tomentose, pubescent or occasionally glabrous twiners or twining shrubs with white latex. Leaves opposite, strongly cordate. Inflorescences extra-axillary, initially appearing umbelliform but lengthening into a raceme. Corolla with a short tube, adnate to gynostegium for whole length, and spreading lobes. Corolline corona absent. Gynostegial corona in 2 series; outer corona ring with 5 interstaminal lobes arising from the column at or just above the mouth of the corolla tube; inner corona of 5 staminal lobes adnate to the staminal column about half way up the anther wings, semi-sagittate in outline with a free subulate apex arched over the head of the column and a basal projection. Pollinaria with pendant, subsessile pollinia with translucent outer margins. Follicles generally paired, reflexed on the pedicel or occasionally spreading, fusiform to narrowly ovoid, curved slightly into a long attenuate beak, ± smooth to strongly echinate, glabrous or pubescent. Seeds flattened, ovate, margins entire, dentate or crenulate, pubescent on both sides, with a coma of silky hairs.

Two species in Africa, Arabia and eastwards through India and Sri Lanka to Bangladesh. Records from China and SE Asia refer to species now assigned to other genera.

Pergularia daemia (*Forssk.*) *Chiov.* in Result. Sci. Miss. Stef.-Paoli, Coll. Bot. 1: 115 (1916); Bullock in F.W.T.A. ed. 2, 2: 90 (1963); Blundell, Wild Fl. E. Afr.: t. 17 (1987); U.K.W.F. ed. 2: 180 (1994); Goyder in Fl. Eth. 4(1): 133, fig. 140.20 (2003) & in Fl. Somalia 3: 148, fig. 101 (2006). Type: Yemen, Zabid, 1763, *Forsskål* s.n. (not traced). Neotype: Yemen, Sana'a, *Rathjens* 37/7 (BM!, neo., designated by Goyder in K.B. 61: 249 (2006))

Scrambling herbaceous twiner, sometimes woody towards the base; stems and inflorescences glabrous or pubescent with stiff, spreading hairs. Leaves with petiole (1–)2–5(–12) cm long, minutely pubescent; lamina thin, broadly ovate to suborbicular, 2–11 × 1.5–11 cm, apex attenuate to acuminate, base strongly cordate, glabrous to pubescent with stiff hairs principally restricted to veins beneath to

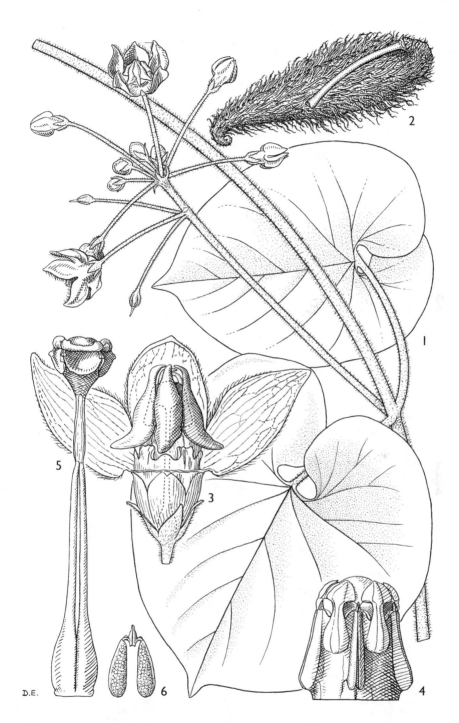

FIG. 88. *PERGULARIA DAEMIA* — **1**, habit, × 1; **2**, single follicle, × 1; **3**, flower, × 4; **4**, upper part of gynostegium with corona removed, × 8; **5**, paired ovaries and stylar head, × 8; **6**, pollinarium, × 12. From *Faulkner* 2317. Reproduced with permission, from the Flora of Ethiopia and Eritrea 4, 1; drawn by D. Erasmus.

densely pubescent on both surfaces; peduncles (2 –)4–11(– 20) cm long, glabrous or
pubescent; pedicels 1.5–4 cm long. Sepals narrowly ovate to oblong, 1.5–4 × 0.6–1 mm,
acute, glabrous or pubescent. Corolla greenish white or yellow-green, sometimes
marked with pink or brown outside; tube 2–6 mm long; lobes ovate or oblong to
elliptic, 5–10(–16) × 1.8–5 mm, subacute or acute, glabrous abaxially, upper surface
bearded towards the margins and frequently at the base of the lobes, otherwise
glabrous. United, tubular part of outer corona barely visible beyond mouth of corolla
tube or exserted for 0.5–1 mm, lobes arising from the interstaminal parts of the tube,
oblong, 0.5–1 × 0.5–1.5 mm, truncate, entire or shallowly toothed; inner corona
lobes slender to stout, 2.5–7(– 9) mm long, the apical projection reaching to the top
of the column, often projecting much further, the basal tails spreading outwards
from the column. Filament tube extending 2–4 mm beyond mouth of corolla tube
but frequently obscured by staminal corona lobes; anther wings 0.8–2 mm long;
pollinaria with laterally compressed corpusculum ± 0.3 × 0.1 mm in dorsal view, with
a keel ± 0.2 mm deep viewed laterally; translator arms ± 0.2 mm long, flattened and
curved; pollinia flattened, oblanceolate, 0.6–0.9 × 0.3 mm, the outer margin of each
pollinium translucent for most of its length. Follicles solitary or paired, narrowly
ovoid to fusiform, 5–8 × 0.8–2 cm, with a long, curved beak, subglabrous to densely
pubescent, with or without soft, pubescent processes. Seeds ovate, 4–7 × 2–4 mm,
entire or sometimes with a somewhat crenulate or dentate margin, densely
pubescent on one or both faces. Fig. 88, p. 365.

subsp. **daemia**

Stems and inflorescences pubescent. Leaf lamina 2.5–11 × 2.5–11 cm, pubescent. Corolla
tube campanulate, 2–4(–5) mm long, ± ¹/₂–²/₃ the length of the lobes; lobes ovate to elliptic,
(5–)6–10(–16) × 3–5 mm. United part of outer corona generally barely visible beyond mouth
of corolla tube but occasionally exserted for 0.5–1 mm; inner corona lobes slender to stout,
3.5–7(–9) mm long, the apical projection reaching at least 1 mm beyond top of column, often
much more. Anther wings (1.1–)1.3–2 mm long. Follicles fully or partially recurved alongside
the pedicel, pubescent, sometimes covered with soft, pubescent processes to ± 1 cm in length,
processes sometimes reduced or absent. Seeds with an entire or somewhat dentate margin.

UGANDA. Karamoja District: Moroto, Mt Karamoja, Mar. 1959, *J. Wilson* 740!; Toro District:
   Bunyangabo county, 7 km S of Kibeto, 20 Oct. 1996, *Lye & Katende* 22073!; Mengo District:
   Kyadondo, Kisaasi, 6 June 1991, *Rwabindore* 3208!
KENYA. Northern Frontier District: Dandu, 16 Mar. 1952, *Gillett* 12553!; Laikipia District: 40 km
   NNW of Nanyuki, W bank of R. Ewaso Ngiro, 2 Jan. 1995, *Goyder & Masinde* 3949!; Teita
   District: Ndara Ranch, 7 Jan. 1995, *Goyder & Masinde* 3956!
TANZANIA. Bukoba District: Kagera, Minziro Forest Reserve, W of Bulemba Hill, 20 Nov. 1999,
   Kayombo 3112!; Arusha District: Ngobok, Engari Nanyuki to Dutch Corner Road, 21 Apr.
   1968, *Greenway & Kanuri* 13489!; Dodoma District: Ikowa Dam, ± 60 km E of Dodoma, 29 July
   1970, *Thulin & Mhoro* 528!; Zanzibar: Kaskazini Unguja District, Kiwengwa Forest, 28 Nov.
   1999, *Fakih* 547!
DISTR. U 1, 2, 4; **K** 1–7; **T** 1–7; **Z**; **P**; widespread across drier tropical or subtropical regions of
   subsaharan Africa, the Arabian peninsula and the Indian subcontinent
HAB. Scrambling over shrubs in dry bushland or occurring near seasonal watercourses in more
   arid areas; 0–2000 m

SYN. *Asclepias daemia* Forssk., Fl. Aegypt.-Arab.: 51 (1775)
   A. glabra Forssk., Fl. Aegypt.-Arab.: 51 (1775). Types: Yemen, Taizz, *Forsskål* (not traced);
      Yemen, Kuhlan to Hajjah road, *Miller & Long* 3266 (K!, neo., designated by Goyder in
      K.B. 61: 249 (2006); E, isoneo.)
   Cynanchum cordifolium Retz., Obs. Bot. fasc. 2: 15 (1781). Type not traced
   C. extensum Jacq., Obs. Bot. 2: 353 (1781) & Icones pl. rar. 1: 6, t. 54 (1782)). Type: Jacquin,
      Icones pl. rar. 1: t. 54 (1782), lecto., designated by Liede in Asklepios 51: 65 (1990)
   Asclepias convolvulacea Willd., Sp. Pl. 1: 1269 (1798). Type: 'Guinea', *Herb. Willdenow* 5271
      (B-W, holo., IDC microfiche 360: I. 3!)
   A. scandens P. Beauv., Flore d'Oware et de Benin en Afrique 1: 93, t. 56 (1807). Type: Fl.
      Oware & Benin 1: t. 56

Cynanchum bicolor Andr. in Botanists Repository 9: (1809: t. 562). Type: Bot. Rep. 9, t. 562

Doemia extensa (Jacq.) W.T. Aiton, Hort. Kew. ed. 2, 2: 76 (1811), as *Daemia*

Cynanchum pendulum Poir., Encycl. Méth. Bot. Suppl. vol. 2: 429 (1812). Type: Senegal, *Dupuis* s.n. (P, holo.)

Doemia forskalii Schult. in Roemer & Schult., Syst. Veg. 6(3): 113 (1820), as *Daemia*. Type as for *Pergularia daemia*

D. glabra (Forssk.) Schult. in Roemer & Schult., Syst. Veg. 6(3): 113 (1820), as *Daemia*

Cynanchum echinatum Thunb., Observationes in *Cynanchum* (1821: 8). Type: *Herb. Thunberg* 6298 (UPS, holo., IDC microfiche 264: III. 7)

Doemia bicolor Sweet, Hortus Britannicus ed. 2: 361 (1830), as *Daemia*. Type as for *Cynanchum bicolor*

D. scandens (P. Beauv.) Loud., Hort. Britt.: 94 (1830), as *Daemia*

Asclepias echinata Roxb., Fl. Indica ed. 2, 2: 44 (1832), *non* Thunb. (1821). Type as for *Cynanchum extensum*

Doemia angolensis Decne. in Ann. Sc. Nat. sér. 2, 9: 337 (1838), as *Daemia*. Type: "Angola" (P, holo.)

D. aethiopica Decne. in DC., Prodr. 8: 544 (1844), as *Daemia*. Type: Sudan, *Kotschy* 249, as '240' (P, holo.; K!, iso. –2 sheets)

D. guineensis G. Don, Gen. Syst. 4: 156 (1837), as *Daemia*. Type as for *Asclepias scandens*

Raphistemma ciliatum Hook. f. in Curtis's Bot. Mag.: t. 5704 (1868). Type: '*Raphistemma* ?sp. nov. Hort. Kew 10/67' (K!, holo.)

Doemia cordifolia (Retz.) K. Schum. in Abhandl. Preuss. Akad. Wiss.: 17 (1894) & in P.O.A. C: 324 (1895), as *Daemia*

D. barbata Schltr. in E.J. 20, Beibl. 51: 43 (1895), as *Daemia*, *non* Klotzsch (1861). Type: South Africa near Ramakopa, *Schlechter* 4507 (B†, holo.; BOL!, iso.)

Pergularia extensa (Jacq.) N.E. Br. in Fl. Cap. 4(1): 758 (1908)

P. daemia (Forssk.) Chiov. var. *macrantha* Chiov., Result. Sci. Miss. Stef.-Paoli, Coll. Bot. 1: 115 (1916). Type: Somalia, between Mansùr and Avàile, *Paoli* 587 (FT!, syn.); Baidoa, *Paoli* 1108 (FT!, syn.)

P. glabra (Forssk.) Chiov. in Bull. Soc. Bot. Ital. 1923: 114 (1923)

NOTE. Hugely variable, with many local forms. A rather poorly preserved collection from Nondora in southern Tanzania labelled *Daemia barbata* (*Braun* 1173 (EA)) should probably be referred to subsp. *daemia*.

70. **CALOTROPIS**

R. Br. in Mem. Wern. Soc. 1: 39 (1809)

Madorius Kuntze, Rev. Gen.: 421 (1891)

Erect shrubs. Leaves large, succulent. Inflorescences terminal and extra-axillary, pedunculate, with several clusters of umbelliform cymes. Corolla lobed for ± $^2/_3$ of its length. Corolline corona absent. Gynostegial corona lobes laterally compressed with an upturned spur at the base, radiating out from the gynostegium to which they are adnate for their entire length. Pollinaria with oblong, slightly winged corpuscula and short, cylindrical translator arms; pollinia flattened, pear-shaped. Stylar head pentagonal, depressed or raised slightly above the top of the anthers. Follicles large, ovate or subglobose, smooth, usually only one of the pair developing, with spongy fibrous mesocarp and a papery endocarp. Seeds ovate, plano-convex, pubescent, with a coma of silky hairs.

Three species in Africa, Arabia and southern and SE Asia, introduced elsewhere in the tropics.

Corona lobes oblong, broadest at the top; corolla campanulate;
   follicles inflated, subglobose . . . . 1. *C. procera*
Corona lobes ovate, broadest around the middle; corolla rotate or
   slightly reflexed; follicles not inflated, ovate in outline . . . . . . 2. *C. gigantea*

1. **Calotropis procera** (*Aiton*) *W.T. Aiton*, Hort. Kew. ed. 2, 2: 78 (1811); U.O.P.Z.: 165 (1949); I.T.U.: 37 (1951); F.P.U.: 120 (1962); Bullock in F.W.T.A. ed. 2, 2: 91 (1963); Ali in Notes Roy. Bot. Gard. Edin. 38: 290 (1980); Blundell, Wild Fl. E. Afr.: fig. 569 (1987); K.T.S.L.: 490 (1994); U.K.W.F. ed. 2: 177 (1994); Goyder in Fl. Eth. 4(1): 119, fig. 140.12 (2003) & in Fl. Somalia 3: 144 (2006). Types: Jacq., Obs. Bot. Icon. 3: t. 69 (1768) as *Asclepias gigantea* (lecto., designated by Ali, 1980); Jamaica, Kingston, Palisadoes, along road E of airport, 30 March 1956, *Stearn* 625 (topo., designated by Ali, 1980)

Succulent shrub, 2–5 m tall, stout, weakly to strongly branched; young parts shortly tomentose, often appearing farinose, glabrescent. Leaves with petiole up to 5 mm long; lamina ovate, oblong-ovate, elliptic or more usually broadly obovate, 7–26 × 4–15.5 cm, apex rounded or obtuse with a short acuminate tip, base cordate, often obscuring the petiole, slightly succulent, with conspicuous midrib and arched lateral veins, glaucous. Inflorescences many-flowered, usually with at least 2 major branches each with successive subumbelliform clusters of flowers; peduncles robust, 3–8 cm long, very occasionally absent; bracts caducous, ovate-lanceolate, 7–12 mm long, acute; pedicels slender, 10–35 mm long. Calyx lobes ovate, 4–5 × 3–4 mm, acute or subacute. Corolla white with purple tips to the lobes, campanulate, united for ± ¹/₃ of its length, glabrous; lobes erect or spreading, ovate, 7–10 × 6–7 mm, acute. Corona white or purple; lobes 5–6 × ± 3 mm, oblong, obliquely truncate or rounded, cleft radially in the upper half, minutely scabrid along the margins, basal spur ± 2 mm long. Gynostegium ± 7 mm high, the head of the column forming a drum ± 2 × 5 mm; corpusculum ± 0.4 mm long, slightly winged; translator arms 0.2–0.3 mm long; pollinia 1.2–1.5 × 0.6–0.7 mm. Follicles ovoid to subglobose, 8–13 cm long, with one side somewhat flattened, obtuse or depressed at the apex. Seeds ovate, ± 7 × 5 mm, with a narrow rim, minutely pubescent. Fig. 89, p. 369.

UGANDA. Karamoja District: Kokumongola, 6 Jan. 1937, *A.S. Thomas* 2195!; Bunyoro District: Jan. 1941, *Purseglove* 1089!
KENYA. Turkana District: 80 km W of Lodwar, 11 May 1953, *Padwa* 135!; Meru District: Meru National Park, Rojwero, 16 Apr. 1972, *Ament & Magogo* 13!; Teita District: 80 km from Mombasa on Nairobi Road, 17 Aug. 1965, *Strange* 90!
TANZANIA. Mbulu District: near Lake Manyara, 3 Dec. 1958, *Napper* 1219!; Singida District: Tulia, near Lake Kitangiri, 29 Apr. 1962, *Polhill & Paulo* 2248!; Iringa District: Mwagusi sand river track from Mbagi, 19 Oct. 1970, *Greenway & Kanuri* 14587!
DISTR. U 1–2; K 1–7; T 1–7; drier parts of tropical Africa, Arabia and south Asia; naturalised elsewhere in the tropics
HAB. Close to villages on disturbed or degraded land, usually on sand, also along banks of dry river beds in the north; (0–)400–1400 m

SYN. ?*Apocynum syriacum* S.G. Gmel., Reise 2: 198 (1774), *nomen nudum*
   A. procera Aiton, Hort. Kew. 1: 305 (1789)
   Madorius procerus (Aiton) Kuntze, Rev. Gen. p. 421 (1891)
   Calotropis syriaca (S.G. Gmel.) Woodson in Ann. Missouri Bot. Gdn 17: 148 (1930); T.T.C.L.: 64 (1949), *nom. illegit.*
   C. gigantea (*L.*) *W.T. Aiton* var. *procera* (Aiton) P.T. Li in J. S. China Agric. Univ. 12(3): 39 (1991)

2. **Calotropis gigantea** (*L.*) *W.T. Aiton*, Hort. Kew. ed. 2, 2: 78 (1811); K.T.S.L.: 489 (1994). Type: Herb Hermann 2: 74 (BM, holo.)

Like *C. procera* but corolla ovoid-truncate in bud, the lobes rotate or somewhat reflexed, margins revolute. Corona lobes ovate, 8–11 mm high, narrowing gradually towards the rounded upper margin, not split apically, but with an auricle ± 1 mm long, either side of the scabrid margin; basal spur 2–4 mm long. Corpusculum 0.6–0.8 mm long; translator arms ± 0.4 mm long; pollinia ± 1.6 × 0.6 mm. Follicles ovate, 7–10 × 2.5–4 cm, acute and somewhat beaked. Seeds 5–7 × 3–4 mm.

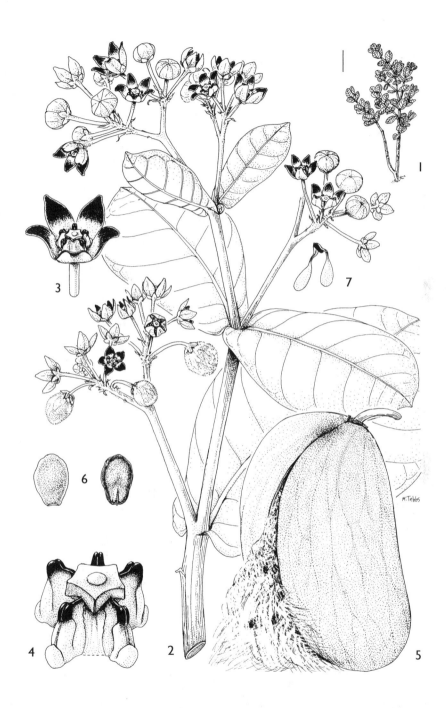

FIG. 89. *CALOTROPIS PROCERA* — **1**, habit, scale bar 0.5 m; **2**, flowering branch, × ²/₃; **3**, flower, × 2; **4**, gynostegium with corona, × 4; **5**, follicle, × ²/₃; **6**, seeds (without coma), × 4; **7**, pollinarium, × 8. 1 & 7 from *Goyder et al.* 3212; 2–5 from *Greenway & Kirrika* 10969. Reproduced with permission, from the Flora of Ethiopia and Eritrea 4, 1; drawn by M. Tebbs.

KENYA. Baringo District: 2 km E of Loruk at the northern end of Lake Baringo, 31 Oct. 1992, *Harvey et al.* 25!; Mombasa District: Shanzu, 3 Dec. 1972, *E. Polhill* 389! & Bamburi, May 1971, *Tweedie* 4032!

TANZANIA. Mwanza District: Saanane Island, 14 Nov. 1964, *Carmichael* 1145!; Tanga District: near Kange Gorge, 13 Nov. 1956, *Milne-Redhead & Taylor* 7284!; Kilosa/Morogoro Districts: 50 km W of Morogoro on road to Mikumi, 11 Dec. 1994, *Goyder et al.* 3933!; Zanzibar: Bungumandani, 6 Dec. 1999, *Fakih* 579!

DISTR. **K** 3, 7; **T** 1, 3, 6; **Z**; Mozambique. Almost certainly introduced in Africa; often associated with Indian settlements. Native to S and SE Asia from India to Indonesia; introduced and becoming naturalised in New World tropics

HAB. Bare and degraded land, often near villages, on sandy soil or coral rock; 1–100 m or ± 1100–1200 m on the shores of Lake Victoria and in the Kenyan Rift Valley

SYN. *Asclepias gigantea* L., Sp. Pl.: 214 (1753)
   Madorius giganteus (L.) Kuntze, Rev. Gen. Pl. 421 (1891)

71. KANAHIA

R. Br., On the Asclepiadeae: 39 (1810); Field *et al.* in Nordic J.B. 6: 790 (1986)

Erect, multistemmed riverine shrubs branching mainly near the base, glabrous in all parts except the corolla. Leaves opposite with stipules or colleters distributed along interpetiolar line or restricted solely to the leaf axils. Inflorescences extra-axillary; flowers arranged in an indeterminate condensed spiral. Corolla ± rotate to campanulate with a shaggy indumentum of white hairs towards the apex and margins of the lobes. Corolline corona absent. Gynostegium stipitate. Gynostegial corona lobes arising near the base of the anther wings, subglobose or somewhat compressed laterally. Anther wings vertical, frequently somewhat flared at base; anther appendages membranous, inflexed over apex of stylar head. Pollinaria with an ovoid corpusculum, straight or slightly articulated translator arms and pollinia circular or only slightly flattened in section. Stylar head flat. Follicles single or paired, inflated or not; seeds generally inflated, not differentiated into disc and rim, with a coma of silky hairs.

Two species in tropical Africa and Arabia.

Kanahia laniflora (*Forssk.*) *R. Br.* in Salt, Voy. Abyssinica, App.: 64 (1814), as *Kannahia*; N.E. Br. in F.T.A. 4(1): 296 (1902); T.T.C.L.: 66 (1949); Bullock in K.B. 7: 421 (1952); Bullock in F.W.T.A. ed. 2, 2: 91 (1963); Field *et al.* in Nordic J.B. 6: 790 (1986); Blundell, Wild Fl. E. Afr.: t. 71, 72 (1987); K.T.S.L.: 494 (1994); U.K.W.F. ed. 2: 178 (1994); Goyder in Fl. Eth. 4(1): 121, fig. 140.13 (2003). Type: Yemen, Djöbla, *Forsskål* s.n. (C, holo.)

Erect shrub to 2.5 m tall, multistemmed; stems branching mainly near the base and arising from a sandy-coloured non-tuberous rootstock. Leaves linear-lanceolate, 6–15(–20) × 0.3–1.5(–2.5) cm, tapering gradually into the attenuate apex and the channeled petiole, margins smooth, glabrous. Inflorescences extra-axillary; peduncles 1.5–9 cm long; rachis up to 1.5(–3) cm long; bracts deciduous, basal bract in each inflorescence lanceolate, to 15(–25) × 2(–4) mm, acute, other bracts smaller and more filiform; flowers spreading or erect on slender pedicels 1–2.5(–3) cm long. Sepals ovate to narrowly lanceolate, 3–10 × 1–3 mm, acute. Corolla rotate to campanulate, divided ± to the base or occasionally lobes united for about half their length, cream or white, sometimes tinged with green; lobes ovate to elliptic, 7–10(–13) × 2.5–5 mm, subacute, glabrous on outer face, inner face minutely pubescent with a shaggy indumentum of white hairs towards the apex and margins. Gynostegium with stipe 1–2.5 mm long. Corona lobes white, arising near the base of the anther wings and reaching the top of the column or to about halfway, 2–4 × 1–2 mm,

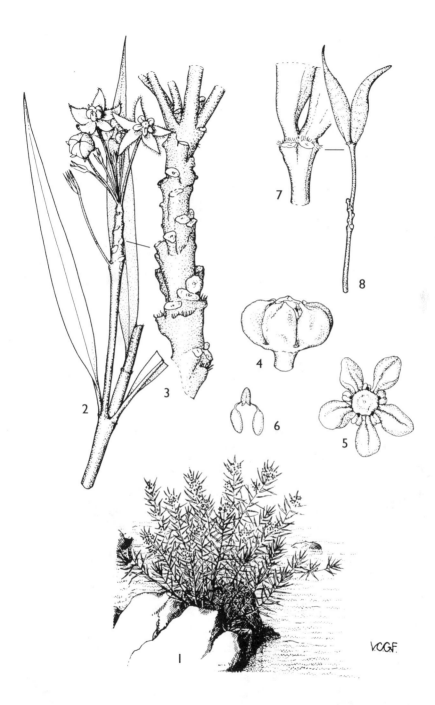

Fig. 90. *KANAHIA LANIFLORA* — **1**, habit; **2**, part of flowering branch, × ²⁄₃; **3**, axis of inflorescence; **4 & 5**, lateral and apical view of gynostegium and corona, × 5; **6**, pollinarium, × 10; **7**, base of paired follicles; **8**, infructescence, × ²⁄₃. 1 from Mike Gilbert colour slide; 2–6 from *Mooney* 8580; 7 & 8 from *Mesfin & Kagnew* 1815. From Nordic J. Bot. 6: 789, drawn by V.C. Friis.

subglobose or somewhat compressed laterally, truncate and with a deep or shallow groove distally, the margins sometimes produced into a pair of teeth arching towards or over the head of the staminal column. Anther wings 2–3 mm long, vertical, frequently somewhat flared at base; pollinaria dark brown, ovoid, with corpusculum 0.4–0.5 × 0.2 mm; translator arms 0.2–0.3 mm long, slender, straight or slightly articulated; pollinia ovoid-subcylindrical or linear-oblanceolate, 0.8–1.5 × 0.2–0.4 mm, slightly flattened in section. Stylar head flat, level with the top of the anthers. Follicles single or paired, inflated or not, ovoid, (3–)3.5–6 × (0.5–)1–2 cm, with a rounded apex to subcylindrical with a slender beak; fruiting pedicel not contorted. Seeds generally inflated, ovoid, 3–4 × 1–2 mm, not differentiated into disc and rim, tapered towards the coma, smooth; coma 6–10 mm long. Fig. 90, p. 371.

UGANDA. West Nile District: West Madi, Leya and Aiyu River junction, 25 Mar. 1945, *Greenway & Eggeling* 7254!; Karamoja District: Chosan, 12 June 1959, *Symes* 571!; Mbale District: Suam River, 1 Jan. 1937, *A.S. Thomas* 2107!
KENYA. Northern Frontier/Laikipia Districts: Uaso Nyiro River, 22 Feb. 1996, *Goyder et al.* 4057!; Machakos District: Tsavo East National Park, Athi River, 25 Dec. 1987, *S.A. Robertson* 5015!; Tana River District: 7 km W of Kora base camp, 1 August 1976, *Kibuwa* 2465!
TANZANIA. Musoma District: Nyamuma Guard Post, Mbalageti River, 25 Apr. 1961, *Greenway* 10092!; Tanga District: Muheza, 9–10 km from the junction on the Korogwe–Muheza road, 4 May 1987, *Iversen et al.* 87201!; Kilosa/Iringa Districts: Ruaha valley, 10 Dec. 1994, *Goyder et al.* 3931!
DISTR. U 1, 3; **K** 1–4, 7; **T** 1, 3, 5–7; from Ivory Coast to Somalia and S to South Africa; Yemen and Saudi Arabia
HAB. In seasonally inundated watercourses, on sand or among rocks; 0–1500 m

SYN. *Asclepias laniflora* Forssk., Fl. Aegypt.-Arab.: 51 (1775)
    Kanahia kannah Schult. in Roem. & Schult., Syst. Veg. 6: 94 (1820). Type as for *Kanahia laniflora*
    Asclepias laniflora Del., Cent. Pl. d'Afr. Voy. Caillaud: 49, t. 64 (1826), *non* Forssk. (1775). Type: Sudan, Mt Aqarô, *Caillaud* s.n. (MPD, holo.)
    Kanahia delilei Decne. in Ann. Sci. Nat., sér. 2, 9: 330 (1838) & in DC., Prodr. 8: 537 (1844). Type as for *Asclepias laniflora* Del., *non* Forssk.
    K. forsskalii Decne. in DC., Prodr. 8: 537 (1844). Type as for *Kanahia laniflora*
    Gomphocarpus glaberrimus Oliv. in Trans. Linn. Soc., Bot. 29: 110, t. 120 (1875). Type: Tanzania, Mpwapwa District: Marenge Mlali [M'Khali], *Speke & Grant* s.n. (K!, holo.)
    Asclepias glaberrima (Oliv.) Schltr. in J.B. 33: 335 (1895)
    Kanahia glaberrima (Oliv.) N.E. Br. in F.T.A. 4(1): 297 (1902); T.T.C.L.: 66 (1949)
    K. consimilis N.E. Br. in F.T.A. 4(1): 298 (1902). Type: Cameroon, Batanga, Lobe River, *Bates* 322 (K!, lecto., designated by Field *et al.* in Nordic J.B. 6: 791 (1986))
    Asclepias coarctata S. Moore in J.B. 46: 297 (1908). Types: Mozambique, lower Umswirizwi River, *Swynnerton* 248 (BM!, syn.; K!, isosyn.) & lower Buzi River, *Swynnerton* 1895 (BM!, syn.; K!, isosyn.)
    Kanahia monroi S. Moore in J.B. 49: 156 (1911). Type: Zimbabwe, Victoria, *Monro* s.n. (BM!, holo.)
    Asclepias rivalis S. Moore in J.B. 52: 337 (1914). Type: Angola, Lucalla River, *Gossweiler* 5771 (BM!, holo.)
    A. fluviatilis A. Chev. in Bull. Soc. Bot. France, 1914, 61, Mém. 8: 271 (1917). Type: Ivory Coast, Bandama River near Marabadiassa, *Fleury* in *Chevalier* 22021 (P, holo.)

72. ASPIDOGLOSSUM

E. Mey., Comm Pl. Afr. Austr.: 200 (1838); Kupicha in K.B. 38: 599–672 (1984)

Perennial herb with flowering shoots produced annually and dying back or burnt to ground level each year, all parts exuding a milky latex; stems erect, branched or not. Leaves opposite or rarely irregularly inserted, verticillate (sect. *Verticillus*), sessile or subsessile, narrowly linear or rarely broader (sect. *Verticillus*), margins revolute. Flowers in sessile or subsessile fascicles at upper nodes; pedicels pilose. Corolla campanulate, spreading or rarely reflexed, rarely tips united (*A. connatum*). Corolline corona absent. Gynostegial corona lobes alternating with the corolla lobes, mostly of two types; type A (fig. 91.1) comprising an erect, dorsally compressed, ± quadrate basal part with peaked, square or sloping shoulders, the apex produced into an attenuate tooth, ventral face with two faint vertical wings or ridges running into an appendage similar to the apical tooth of the main lobe; type B (fig. 91.2) with basal part of lobe ± as in type A but with truncate apex, apical tooth absent, ventral surface with two often well-developed vertical ridged running into an inflexed appendage. Pollinarium with subcylindrical corpusculum, rarely broader and with lateral flanges (*A. hirundo*); translator arms short, slender or flattened and ribbon like, attached subapically to the pollinia; pollinia subcylindrical or pyriform and dorsally compressed, with or without translucent germination zone near point of attachment to translator arms. Follicles single, pubescent, with or without soft appressed bristles; fruiting pedicel contorted to hold follicle erect; seeds flattened, ovate, with a coma of silky hairs.

A genus of ± 35 species from tropical and southern Africa, with somewhat uncertain generic delimitation.

In addition to the species listed below, there is one unnamed collection from southern Tanzania (*Milne-Redhead & Taylor* 8868, from Matagoro Hill near Songea) that looks as if it ought to belong here, but which Kupicha has excluded from the genus.

1. Corona lobes of type A (fig. 91.1, p. 374) . . . .   1. *A. breve*
   Corona lobes not as above 2
2. Plants up to 75 cm tall; pedicels ± 10 mm long; corolla
   glabrous or pubescent on ventral surface; corona lobes
   not as above 3
   Plants very slender, usually at least 75 cm long; pedicels
   1–8 mm long; corolla usually pubescent on ventral
   surface, the hairs sometimes very long, those at the
   tip often stouter than the rest; corona lobes of type B
   (fig. 91.2) or 'penguin'-shaped (*A. elliotii*) 4
3. Pedicels with blackish hairs; corolla glabrous or
   subglabrous on ventral face; anther wings ± 1 mm long   8. *A.* sp. A
   Pedicels without blackish hairs; corolla pubescent on
   ventral face; anther wings ± 0.5 mm long . . . .   9. *A.* sp. B
4. Corolla lobes 5.5–9 mm long, often united at their tips to
   form a lantern . . . .   4. *A. connatum*
   Corolla lobes 2–4.5 mm long, never united at tip 5
5. Corona lobes 'penguin'-shaped; corpusculum laterally
   compressed and somewhat arcuate viewed from side . .   7. *A. elliotii*
   Corona lobes of type B (fig. 91.2); corpusculum not
   curved 6
6. Corona pilose . . . .   2. *A. lanatum*
   Corona glabrous 7
7. Ventral surface of corolla lobes glabrous or shortly
   pubescent, sometimes with a few stout hairs at tip . . . .   5. *A. masaicum*
   Ventral surface of corolla lobes long-pilose, particularly
   towards apex and margins 8
8. Corolla lobes 2–3 mm long; upper margin of corona
   lobes produced into 3 teeth, the middle one much
   longer than outer pair . . . .   6. *A. interruptum*
   Corolla lobes 3–4.5 mm long; upper margin of corona
   lobes not as above . . . .   3. *A. angustissimum*

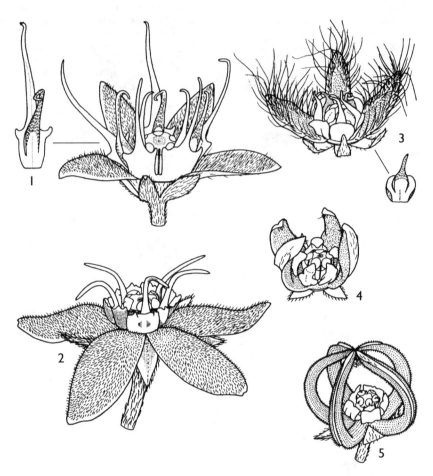

FIG. 91. *ASPIDOGLOSSUM* flowers — **1**, flower with corona of type A; **2**, flower with corona of type B; **3**, flower and corona lobe of *A. angustissimum*; **4**, flower of *A. elliotii*; **5**, flower of *A. connatum*. 1 from *Corby* 553; 2 from *Devenish* 1088; 3 from *de Wilde* 510; 4 from *Smeds* 1426; 5 from *Williamson et al.* 574A. From K.B. 38: 639, drawn by F.K. Kupicha.

1. **Aspidoglossum breve** *Kupicha* in K.B. 38: 643 (1984). Type: Malawi, Nyika Plateau, near Chelinda camp, *Hilliard & B.L. Burtt* 4399 (E, holo.; K!, iso.)

Perennial herb with slender vertical napiform tuber; stems erect, single or occasionally paired, 10–35 cm long, slender and unbranched, pubescent with spreading white hairs. Leaves opposite and paired even in inflorescence, erect, sessile, linear, (1–)2–6 × 0.1–0.2 cm, margins strongly revolute, pubescent with spreading white hairs on upper surface and on midrib below. Inflorescences sessile or with peduncle to 0.6 cm, 3–8-flowered at each of the upper nodes; pedicels 3–8 mm long, sparsely to densely pubescent with blackish hairs. Calyx lobes 1.5–2.5 × 0.5–0.8 mm, narrowly triangular to ovate, attenuate, sparsely to densely pilose in upper half of outer face, lower half with shorter reddish hairs. Corolla campanulate, lobed ± to the base, yellow-green or reddish brown outside, paler within and on margins; lobes ovate, 2.5–4 × 1.5–2 mm, acute, weakly replicate, outer surface with white hairs towards the apex, commonly with reddish ones below; inner face glabrous. Corona lobes of type A (fig. 91.1), green or yellow, sometimes tinged purple on the horns, as tall as or longer than the staminal column; basal part spreading, slightly fleshy, 0.5–0.7 × 1 mm, with square shoulders, apical tooth erect, narrowly triangular, 0.7–1 mm long; inner tooth erect or inflexed over apex of stylar head, about as long as outer tooth. Staminal column ± 1 mm long; corpusculum reddish brown, subcylindrical, 0.2–0.3 × 0.05–0.1 mm; translator arms slender, terete, becoming contorted to inver pollinia when pollinarium removed from anthers; pollinia attached subterminally to the translator arms, 0.4–0.6 × 0.2 mm, subcylindrical, not compressed and with a clearly defined apical germination zone. Follicles not seen.

TANZANIA. Mbeya District: Kitulo [Elton] Plateau, 29 Nov. 1963, *Richards* 18447! & 2 Dec. 1963, *Richards* 18524!
DISTR. **T** 7; Nyika Plateau in Malawi
HAB. Broken turf among rocks in montane grassland; 2400–2700 m

NOTE. Tanzanian material of this species has a more pronounced indumentum than collections from Malawi. The corona lobe teeth are shorter also, not overtopping the head of the staminal column. Note that fruiting material of *Richards* 18524 belongs to *Asclepias alpestris*.

2. **Aspidoglossum lanatum** (*Weim.*) *Kupicha* in K.B. 38: 652 (1984). Type: Zimbabwe, near Inyanga village, *Norlindh & Weimarck* 4183 (LD, holo.; BM!, iso.)

Perennial herb with slender vertical napiform tuber; stems single, 35–75 cm long, simple or with a single erect branch, pubescent with short curled white hairs. Leaves opposite and paired even in inflorescence, erect, sessile, 2.5–4 × 0.1–0.15 cm, minutely appressed-pubescent on upper surface and on midrib below. Inflorescence 3–9(–11)-flowered at each of the upper nodes; pedicels 2–5 mm long, pubescent. Calyx lobes ovate to triangular, 1–2.5 × 0.5–1 mm, acute, pubescent. Corolla campanulate, lobed ± to the base, yellow-green, sometimes tinged purple; lobes ovate, 2.5–3.5 × 1–1.5 mm, acute, flat or replicate, outer face pilose, inner face evenly pubescent. Corona lobes of type B (fig. 91.2); basal part 1–1.5 × ± 1 mm, ± as long as the staminal column, shoulders sloping or acute, moderately to densely pilose along upper margin; appendage erect or inflexed, narrowly triangular, ± 0.5 mm long, densely pilose. Corpusculum ± 0.25 mm long; translator arms ± 0.1 mm long; pollinium ± 0.5 mm long, subcylindrical, without clearly defined germination zone. Follicles (very immature) densely pubescent, apparently with soft linear processes.

TANZANIA. Kupicha in K.B. 38: 652 (1984) reports this species from SE Tanzania
DISTR. ?**T** 8; E Zimbabwe
HAB. Grassland and open woodland; 200–1800 m elsewhere

SYN. *Schizoglossum lanatum* Weim. in Bot. Notis. 1935: 392 (1935)

3. **Aspidoglossum angustissimum** (*K. Schum.*) *Bullock* in K.B. 7: 418 (1952); Bullock in F.W.T.A. ed. 2, 2: 92 (1963). Type: Congo-Kinshasa, Niamniamland, Gumango Hill, *Schweinfurth* 3879 (B†, holo.; K!, lecto., designated by Kupicha in K.B. 38: 653 (1984))

Perennial herb with vertical slender woody napiform tuber; stems erect, single, (30–)80–135 cm long, simple or commonly with several erect branches, pubescent with short curled white hairs. Leaves opposite and paired even in inflorescence, erect, sessile or subsessile, linear, 3–10(–17) × 0.1–0.4 cm, margins strongly revolute, appressed-pubescent on upper surface and on midrib below. Inflorescences sessile, (1–)6–15-flowered at each of the upper nodes; pedicels 1–7 mm long, pubescent. Calyx lobes ovate or trinagular, 1–2 × 0.5–0.8 mm, acute, pilose. Corolla campanulate, divided ± to the base, white, sometimes tinged brown or purple; lobes

(2–)3–4.5 × 1–2 mm, ovate but strongly replicate so appearing linear, incurved at the tip, outer face pilose, inner face pubescent with conspicuous white hairs to 2 mm long towards the apex and margins. Corona lobes of type B (fig. 91.2); white or greenish, sometimes spotted with brown or purple; basal part 0.5–1.5 mm, shorter than or ± as long as the staminal column, shoulders square or rounded; tooth linear to broadly triangular, 0.6–1 mm long, entire or the broader teeth shallowly to deeply bifid, erect or inflexed. Staminal column 0.5–1 mm long; corpusculum cylindrical, 0.2–0.3 × ± 0.08 mm; translator arms ± 0.1 mm long; pollinia attached apically to translator arms, subcylindrical, ± 0.6 × 0.15 mm, with a distinct apical germination zone. Follicles fusiform, 5–7 cm long, with a long attenuate beak, pubescent; fruiting pedicel contorted to hold follicles erect.

UGANDA. West Nile District: Zeu [Zeio], Apr. 1940, *Eggeling* 3904!; Bunyoro District: Hoima, May 1943, *Purseglove* 1561! & July 1907, *E. Brown* 411!
KENYA. Mt Elgon, Aug. 1951, *Tweedie* 918!
TANZANIA. Njombe District: Mdapo, Feb. 1954, *Semsei* 1626!; Iringa District: Iheme, 30 km S of Iringa, 24 Feb. 1962, *Polhill & Paulo* 1593!; Songea District: 3 km NE of Kigonsera, 15 Apr. 1956, *Milne-Redhead & Taylor* 9657!
DISTR. U 1–4; K 3/5; T 1, 2, 4, 6–8; from Central African Republic, Congo-Kinshasa and South Sudan to Zambia, Malawi and Zimbabwe
HAB. Seasonally waterlogged grassland; 1200–1800 m

SYN. *Schizoglossum angustissimum* K. Schum. in E.J. 17: 123 (1893); N.E. Br. in F.T.A. 4(1): 357 (1902)
    S. elatum K. Schum. in E.J. 17: 123 (1893). Type: Kenya, *Fischer* 398 (B†, holo.)
    S. whytei N.E. Br. in F.T.A. 4(1): 357 (1902). Types: Malawi, Kondowe to Karonga, *Whyte* 353 (K!, lecto., designated by Bullock in K.B. 7: 419 (1952))
    ?*Schizoglossum ledermannii* Schltr. in E.J. 51: 132 (1913). Types: Cameroon, *Ledermann* 3795 & 4491 (B†, syn.)
    ?*S. zernyi* Markgraf in N.B.G.B. 14: 116 (1938). Type: Tanzania, Songea District: Matengo Plateau, above Ugano, *Zerny* 427 (B†, holo.)
    Aspidoglossum whytei (N.E. Br.) Bullock in K.B. 7: 418 (1952)

NOTE. All East African material is of subsp. *angustissimum*. However, some specimens from the southern highlands of Tanzania are anomalous in having fewer flowers per node, less conspicuous indumentum on the corolla lobes and much reduced coronal teeth. They are undoubtedly close to *A. angustissimum*, perhaps approaching subsp. *brevilobum*, but the differences are not sufficiently constant to warrant taxonomic recognition.

4. **Aspidoglossum connatum** (*N.E. Br.*) *Bullock* in K.B. 7: 419 (1952); U.K.W.F. ed. 2: 176 (1994). Type: Zambia, Fwambo, S of Lake Tanganyika, *Carson* 17 (K!, holo.)

Perennial herb with subglobose to napiform tuber; stems single, erect, 40–90 cm long, simple or commonly with one or more erect branches, minutely pubescent with short curled white hairs. Leaves opposite and paired even in the inflorescence, linear, 3–9 × 0.05–0.3 cm, margins strongly revolute, appressed-pubescent on upper surface and on midrib below. Inflorescences sessile, 3–7-flowered at each of the upper nodes; pedicels 1–6 mm long, minutely pubescent. Calyx lobes triangular to ovate, 1.5–2.5 × 0.4–1 mm, acute, pilose. Corolla spreading to campanulate, divided ± to the base, white, cream, yellow, pink or purple; lobes linear-lanceolate, 5.5–9 × 1–2 mm, plane or more usually strongly replicate, united at the tips to form a lantern or free and spreading, glabrous or pilose on outer face, inner face glabrous to densely pubescent. Corona lobes of type B (fig. 91.2), green, cream or purple, adjacent lobes abutting to form a drum around the column; basal part 1–1.5 × 1–1.5 mm, ± as long as or longer than the staminal column, shoulders rounded, square or toothed; ventral tooth minute. Staminal column 0.5–1 mm long; corpusculum subcylindrical, ± 0.2 × 0.05 mm; translator arms ± 0.15 mm.; pollinia obovoid, ± 0.5 × 0.2 mm, attached apically to translator arms, with a conspicuous proximal germination zone. Follicle fusiform, ± 7 cm long, with a long attenuate beak, pubescent; fruiting pedicel contorted to hold follicle erect. Fig. 92, p. 377.

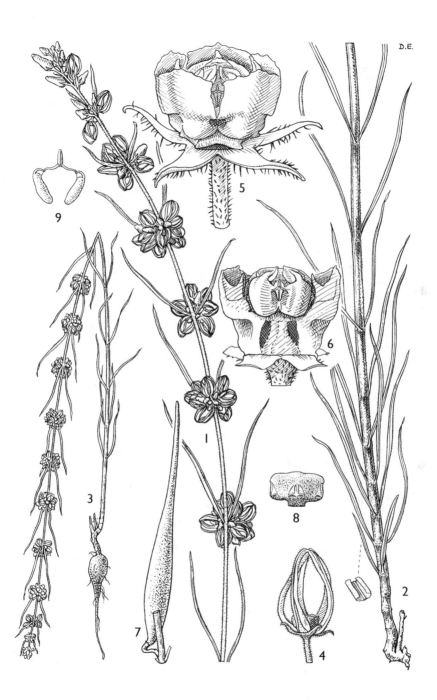

FIG. 92. *ASPIDOGLOSSUM CONNATUM* — **1 & 2**, habit, × 1; **3**, habit including tuber, × ¹/₃;
4, flower, × 3; **5**, gynostegium and corona, × 12; **6**, gynostegium with two corona lobes cut
away to expose stamens and ovaries, × 12; **7**, stylar head, × 12; **8**, follicle, × 1; **9**, pollinarium,
× 20. 1–3 from *Richards* 12012; 4–7 & 9 from *Watermeyer* 3; 8 from *Chandler* 1569. Drawn by
D. Erasmus.

UGANDA. Bunyoro District: Buruli County, 2.5 km N of Kafu River on Gulu road, 16 Feb. 1969,
  Lye 2002!; Busoga District: Bugabula, 12 km W of Kamuli, 29 Apr. 1955, *G.H.S. Wood* 725!;
  Mengo District: Bombo road, Kampala, Feb. 1936, *Chandler* 1569!
KENYA. Masai District: Lolgorien, Sep. 1933, *Napier* 5236!
TANZANIA. Ufipa District: Nkunde–Chapota, 29 Nov. 1949, *Bullock* 1961!; Iringa District: 48 km
  W of Iringa on Mbeya road, 4 Feb. 1976, *Cribb et al.* 10533!; Mufindi District: Ngwazi Swamp,
  22 Dec. 1986, *Lovett* 1139!
DISTR. U 2–4, K 6, T 4, 7; Guinea Bissau, Congo-Kinshasa, South Sudan, Zambia, Malawi
HAB. Seasonally waterlogged grassland; 900–1700 m

SYN. *Schizoglossum connatum* N.E. Br. in K.B. 1895: 69 (1895) & in F.T.A. 4(1): 356 (1902)
     S. vulcanorum Lebrun & Taton in B.J.B.B. 17: 66 (1943). Type: Congo-Kinshasa, between
       Kingi and Busogo, *Lebrun* 8647 (BR, holo.)

5. **Aspidoglossum masaicum** (*N.E. Br.*) *Kupicha* in K.B. 38: 656 (1984); U.K.W.F.
ed. 2: 176, t. 66 (1994), as *massaicum*); Goyder in Fl. Eth. 4(1): 117 (2003). Type:
Kenya, near Kilimanjaro, *Johnston* s.n. (K!, holo.; BM, iso.)

Perennial herb with globose to napiform tuber; stems single, 30–110 cm long,
simple or branched, usually extremely slender, pubescent. Leaves opposite and
paired even in inflorescence, linear, 1–6 × 0.05–0.2 cm, margins strongly revolute,
pubescent on upper surface and on midrib below. Inflorescences sessile at each of
the upper nodes, 2–10-flowered; pedicels 1–5 mm long, pilose. Calyx lobes lanceolate
or narrowly ovate, ± 1.5 × 0.2–0.5 mm, acute, pilose. Corolla spreading to
campanulate, white or greenish, sometimes flushed pink or purple, divided ± to the
base, lobes ovate, 2–4 × 1 mm, subacute, plane (but commonly replicate in East
Africa), pilose on outer surface, glabrous on inner face, often with stout white hairs
at apex. Corona lobes of type B (fig. 91.2); basal part ± 1 × 0.5 mm, with square or
rounded shoulders, tooth inflexed, triangular, to ± 0.5 mm long. Corpusculum
0.2–0.3 mm long; pollinia 0.4–0.6 mm long, with distinct proximal germination zone.
Follicles narrowly fusiform, 3–6 cm long, with a pronounced beak, pubescent;
fruiting pedicel contorted to hold follicle erect.

KENYA. Nairobi, Kilimani, 26 May 1961, *Williams* in EA 12335! & Langata, 24 May 1980, *Gilbert*
  5952!; Kwale District: Shimba Hills, 1 May 1992, *Luke* 3110!
TANZANIA. Bukoba District: Buhamila [Buhamira], Oct. 1931, *Haarer* 2322!; Tanga District:
  Machui, 10 May 1965, *Faulkner* 3513!; Songea District: near Lumecha Bridge, 3 Jan. 1956,
  Milne-Redhead & Taylor 3038!; Zanzibar: near Tungu, 2 Feb. 1929, *Greenway* 1300!
DISTR. K 3, 4, 6, 7; T 1–3, 8; Z; from Congo-Kinshasa and Ethiopia to Angola and Namibia
HAB. Seasonally waterlogged grassland; 0–2000 m

SYN. *Schizoglossum masaicum* N.E. Br. in K.B. 1895: 252 (1895) & in F.T.A. 4(1): 358 (1902)
     S. fuscopurpureum Schltr. & Rendle in J.B. 34: 98 (1896); N.E. Br. in F.T.A. 4(1): 361 (1902).
       Type: Angola, near Huilla, *Welwitsch* 4177 (BM, holo.; K!, iso.)
     S. baumii N.E. Br. in F.T.A. 4(1): 361 (1902). Type: Angola, near Kavenga on R. Kubango,
       Baum 413 (K!, holo.; BM, COI, E, Z, iso.)
     S. altum N.E. Br. in K.B. 1906: 250 (1906). Type: Malawi, Thondwe (Ntondwe), *Cameron*
       107 (K!, holo.)
     S. semlikense S. Moore in J.B. 50: 361 (1912). Type: Congo-Kinshasa, Ruwenzori District:
       Semliki Valley, *Kässner* 3282a (BM, holo.)
     Aspidoglossum kulsii Cuf. in Senck. Biol. 43: 278 (1962). Type: Ethiopia, Sidamo, 20 km SW
       from Soddu to Gofa, *Kuls* 609 (FT, holo.)

6. **Aspidoglossum interruptum** (*E. Mey.*) *Bullock* in K.B. 7: 419 (1952); Bullock in
F.W.T.A. ed. 2, 2: 92 (1963); Goyder in Fl. Eth. 4(1): 119, fig. 140.11 (2003). Type:
South Africa, Cape Province, Wittebergen, *Drège* s.n. (K!, lecto., designated by
Kupicha in K.B. 38: 658 (1984); BM, CGE, K!, MO, iso.)

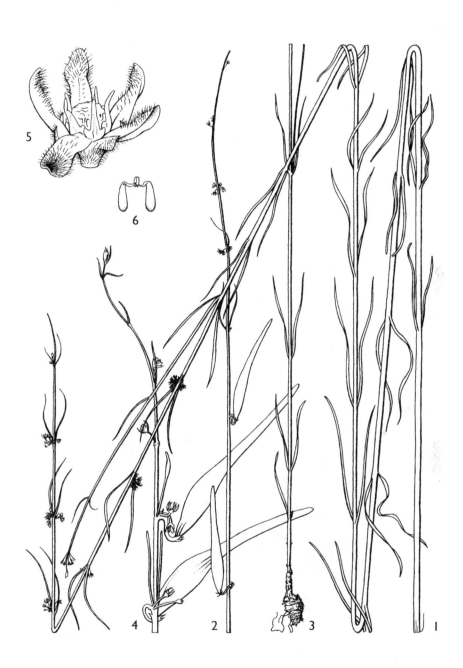

Fig. 93. *ASPIDOGLOSSUM INTERRUPTUM* — **1–3**, habit, × ²/₃; **4**, upper part of stem with fruit, × ²/₃; **5**, flower, × 10; **6**, pollinarium, × 26. 1, 5 & 6 from *Bruce* 96; 2 from *Maitland* 1925; 3 & 4 from *Wallace* 282. Reproduced with permission, from the Flora of Ethiopia and Eritrea 4, 1; drawn by E. Papadopoulos.

Perennial herb with napiform tuber; stems single, 30–110 cm long, simple or branched, extremely slender, pubescent. Leaves opposite and paired even in inflorescence, linear, 1.5–4 × ± 0.05 cm, margins strongly revolute, pubescent. Inflorescences sessile at each of the upper nodes, 1–11-flowered; pedicels 2–5 mm long, pilose. Calyx lobes lanceolate, 0.7–1.5 × 0.3–0.6 mm, acute, pilose. Corolla campanulate, yellow-green to brown, sometimes pinkish outside, divided ± to the base; lobes ovate, 2–3 × ± 1 mm, acute, plane or more usually replicate, outer surface pilose, inner surface long-pilose particularly towards the apex and margins. Corona lobes with basal part ± 0.5 × 0.8 mm, the upper margin developed into three teeth, the central one narrowly triangular, ± 0.5 mm long, much longer than the lateral pair. Gynostegium ± 0.5 mm long. Pollinaria minute; corpusculum subovoid, ± 0.1 mm long; pollinia with clear germination zone. Follicles (immature) fusiform with an attenuate beak, densely puberulent. Fig. 93, p. 379.

UGANDA. Ankole District: 1925, *Maitland* s.n.!; Teso District: Serere, Dec. 1931, *Chandler* 254!; Masaka District: Katera, Oct. 1925, *Maitland* 824!
TANZANIA. Morogoro District: Ulugurus, 3 Jan. [no year], *E.M. Bruce* 416! & Morogoro, 23 Feb. 1932, *Wallace* 282!; Kilwa District: Balenje, *W.A. Rodgers* in MRC 1221
DISTR. U 2–4; T 6, 8; widespread but local in tropical and southern Africa
HAB. Grassland; 700–1200 m

SYN. *Lagarinthus interruptus* E. Mey., Comm Pl. Afr. Austr.: 208 (1838)
   L. abyssinicus Benth. & Hook., G.P. 2: 753 (1876), *nom. inval.*
   Schizoglossum barberae Schltr. in E.J. 18, Beibl. 45: 27 (1894); N.E. Br. in Fl. Cap. 4(1): 658 (1907). Type: South Africa, Cape Province, Tsomo R., *Barber* 847 (K!, lecto., designated by Kupicha in K.B. 38: 658 (1984))
   S. abyssinicum K. Schum. in E. & P. Pf. 4, 2: 233 (1895). Type not listed: Ethiopia, ?*Schimper* 1633 (K!)
   S. altissimum Schltr. in E.J. 20, Beibl. 51: 13 (1895); N.E. Br. in Fl. Cap. 4(1): 660 (1907). Type: South Africa, Transvaal, Iuxta R., near Lydenburg, *Schlechter* 3944 (K, lecto.!, designated by Kupicha in K.B. 38: 358 (1984); BM, BR, GRA, K, NH, Z, iso.)
   S. interruptum (E. Mey.) Schltr. in J.B. 34: 450 (1896); N.E. Br. in Fl. Cap. 4(1): 660 (1907)
   ?*S. morumbenense* Schltr. in E.J. 38: 28 (1905). Type: Mozambique, Inhambane District, near Morumben, *Schlechter* 12098 (?B†, holo.)
   S. lasiopetalum Schltr. in E.J. 38: 29 (1905). Type: Mozambique, Matola near Delagoa Bay, *Schlechter* 11685 (COI, lecto., designated by Kupicha in K.B. 38: 659 (1984); BM, BR, E, GRA, K!, PRE, Z, iso.)
   ?*S. garuanum* Schltr. in E.J. 51: 131 (1913). Type: Cameroon, Garoua District: near Schuari, *Ledermann* 3541 (?B†, holo.)
   ?*S. kamerunense* Schltr. in E.J. 51: 131 (1913). Types: Cameroon, *Ledermann* 3873 (?B†, syn.) & 5825 (?B†, syn.)
   S. gracile Weim. in Bot. Notis. 1935: 384 (1935), *non Lagarinthus gracilis* E. Mey. Type: Zimbabwe, Maconi, near Maidstone village, *Norlindh & Weimarck* 4093 (LD, holo.)

7. **Aspidoglossum elliotii** (*Schltr.*) *Kupicha* in K.B. 38: 659 (1984); Goyder in Fl. Eth. 4(1): 119 (2003). Type: without locality, *Scott Elliot* s.n. (BM!, holo.)

Perennial herb with napiform tuber; stems single, 80–120 cm long, simple or branched, pubescent. Leaves opposite even in inflorescence, 2–6 × 0.05–0.1 cm, minutely pubescent. Inflorescences sessile at each of the upper nodes, 2–6-flowered; pedicels 4–6 mm long, pilose. Calyx lobes ovate, ± 1.5 × 0.7 mm, acute, pilose. Corolla campanulate, green or brown, divided ± to the base; lobes ovate, 3–4 × 1.5–2 mm, subacute, plane or weakly replicate, pilose on outer surface, long-pilose within especially towards the apex and margins. Corona lobes fleshy, ± 1.5 mm long, with lateral flanges to ± 0.5 mm from base and an inward-pointing linear or triangular tooth ± 0.5 mm long at apex (lobe described by Kupicha 1984 as 'penguin-shaped'); corpusculum ± 0.2 mm long, laterally compressed and somewhat arcuate when viewed from the side; pollinia with clear germination zone. Follicles not seen.

UGANDA. Masaka District: near Kasasa, 5 June 1971, *Lye* 6200! & Buddu, 1893–4, *Scott Elliot* 7471!

TANZANIA. Musoma District: Wogakuria Hill, 30 Dec. 1964, *Greenway & Turner* 12010!; Mbulu District: Mbulumbul, 23 June 1944, *Greenway* 6934!; Arusha District: Lake Tulusia, 30 Dec. 1970, *Vesey-Fitzgerald* 6856!

DISTR. U 4; T 1, 2; Congo-Kinshasa, Rwanda, Burundi, Ethiopia, Zambia and Zimbabwe

HAB. Grassland; 1200–1600 m

SYN. *Schizoglossum elliotii* Schltr. in J.B. 33: 305 (1895); N.E. Br. in F.T.A. 4(1): 359 (1902)
    S. debile Schltr. in J.B. 33: 305 (1895). Type: Uganda, Masaka District: Buddu, *Scott Elliot* 7471 (BM, holo.; K!, iso.)

NOTE. This species is very closely allied to *A. interruptum*, which differs in the dorsally flattened corona and the more conventional, subovoid, form of the corpusculum.

8. Aspidoglossum sp. A

Perennial herb with single stem arising presumably from an underground tuber; stem erect, 75 cm long. Leaves opposite and paired even in inflorescence, erect, sessile, linear in inflorescence, lanceolate below, 3.5–5 × 0.1–0.5 cm, apex acute, margins strongly revolute, pubescent with spreading white hairs on upper surface and on midrib below. Inflorescence sessile or shortly pedunculate, with ± 12 flowers; pedicels ± 10 mm long, slender with an indumentum of mostly blackish hairs; bracts filiform. Calyx lobes blackish purple, narrowly triangular, ± 4 × 1 mm, acute, with an indumentum of spreading white hairs. Corolla broadly campanulate, greenish yellow on inner surface, brown or purplish on outer surface, lobed ± to the base; lobes ovate-oblong, ± 4.5 × 2 mm, apex subacute, pubescent on outer face, glabrous within. Corona lobes ± 1 mm high, 3-lobed in dorsal view; intermediate lobules absent. Corpusculum ± 0.5 mm long; translator arms short, slender; pollinia ± 0.6 mm long with conspicuous apical germination zone. Follicles not known.

TANZANIA. Ufipa District: Mbisi Forest, Malonje Plateau. 17 Mar. 1959, *Richards* 12181! & *Bidgood et al.* 2523!; ?Njombe District: near Salala Falls on edge of Kitulo Plateau, 21 Feb. 1982, *Nicholson* s.n.!

DISTR. T 4, 7; not known elsewhere

HAB. Montane grassland; 2100–2400 m

SYN. *Aspidoglossum* sp. B of Kupicha in K.B. 38: 666 (1984)

NOTE. Kupicha could not place this specimen in her revision of the genus (K.B. 38: 599–672, 1984) and commented that despite its floral structure it does not have the facies of an *Aspidoglossum* species.

9. Aspidoglossum sp. B

Superficially similar to *Aspidoglossum* sp. A, but the blackish hairs on the pedicels of that species are absent, the corolla is pubescent above, the corona lobes are shorter and with a lobe on the inner face, and the anther wings are only ± 0.5 mm long.

TANZANIA. Ufipa District: Tatanda Mission, 22 Feb. 1994, *Bidgood, Mbago & Vollesen* 2378!

DISTR. T 4; only known from this collection

HAB. Rocky hill with *Brachystegia* woodland; ± 1900 m

73. **ASCLEPIAS**

L., Sp. Plantarum: 214 (1753); Goyder in K.B. 64: 369–399 (2009)

Trachycalymma (K. Schum.) Bullock in K.B. 8: 348 (1953); Goyder in K.B. 56: 129–161 (2001)

Perennial herbs with annual stems arising from a tuber or fleshy taproot; latex white; stems prostrate to erect, simple or branched. Leaves opposite. Inflorescences terminal or extra-axillary, nodding or erect, umbelliform, sessile or pedunculate. Flowers 5-merous. Corolla divided almost to the base, lobes campanulate, spreading or reflexed. Corolline corona absent. Gynostegial corona of 5 generally cucullate fleshy lobes arising from the staminal column in a staminal position; minute interstaminal lobes sometimes also present. Pollinia pendant in anther cells; translator arms slender and terete or flattened, sometimes clearly geniculate, but never with a massively expanded proximal part and slender distal part. Stylar head rarely projecting much beyond top of anthers (but long-rostrate in *A. longirostra*). Follicles mostly single by abortion, generally held erect, smooth, occasionally ribbed or with lines of soft pubescent processes. Seeds ovate, discoid, with a coma of silky hairs.

A genus with two major centres of distribution, one New World with ± 120 species mostly centred on southern parts of the North American continent (*Asclepias* sensu stricto), the other Old World, with ± 80 species, 38 of these in tropical Africa with the remainder in southern Africa.

In addition, many segregate genera have been recognised in the Old World. In the broad sense the *Asclepias* radiation comprises some 380–400 species. Molecular surveys of this group do not lend support to current generic delimitations (see Goyder *et al.* in Ann. Missouri Bot. Gard. 94: 423–434, 2007 and Fishbein *et al.* in Syst. Bot. 36: 1008–1023, 2011). However, they do not suggest a workable alternative, so the morphologically distinctive African segregate genera *Margaretta*, *Stathmostelma*, *Gomphocarpus*, *Pachycarpus*, *Xysmalobium* and *Glossostelma* have been maintained in this treatment. I can only echo N.E. Brown's view on the subject published in his openly artificial treatment of the group for the F.T.A. 4(1): 299 (1902): "Undoubtedly *Xysmalobium*, *Asclepias* and *Schizoglossum* are but artificial divisions of one natural genus, since they cannot be separated by characters that do not break down at some point..." but he then goes on to say how he has allocated species to each. We have moved on considerably since then in recognising more natural units, but the phylogenetic structure is still inadequate for a stable realignment of the group at generic level.

Note: Only generic synonyms relevant to the Old World have been cited above.

3. Inflorescences solitary and terminal 4
Inflorescences extra-axillary, sometimes initially
appearing terminal 5
4. Corona lobes about as tall as the column, less
than 3 mm long . . . . 14. *A. nuttii* (p. 397)
Corona lobes twice as long as the column, at
least 5 mm long . . . . 20. *A. grandirandii* (p. 403)
5. Corona shorter than, or about as tall as the
column 6
Corona clearly longer than the column 7
6. Outer face of corolla glabrous . . . . 17. *A. mtorwiensis* (p. 400)
Outer face of corolla pubescent . . . . 16. *A. breviantherae* (p. 399)
7. Corona lobes dorsiventally flattened, broadly
obovate . . . . 15. *A. inaequalis* (p. 398)
Corona lobes cucullate, or if dorsiventrally
flattened then broadest near the base 8
8. Leaves narrowly to broadly oblong, secondary
veins clearly visible . . . . 20. *A. grandirandii* (p. 403)
Leaves narrowly linear, only the mid-vein visible 9
9. Corona with well-developed attenuate appendage
on adaxial face; found only above 2500 m in
the Kitulo and Kipengere mountains of
southern Tanzania . . . . 18. *A. alpestris* (p. 400)
Corona lacking an adaxial appendage 10
10. Corona lobes 3–3.5 mm long, papillae present
in both cavity of lobe and on midline;
translator arms of uniform thickness . . . . . . 21. *A. pygmaea* (p. 403)
Corona lobes 4–8 mm long, papillae present
only along the midline; translator arms
broader distally . . . . 19. *A. randii* (p. 401)
11. Leaves ovate or broadly oblong, generally with
a rounded, truncate or cordate base 12
Leaves linear to lanceolate, tapering gradually
into the petiole or with a narrowly truncate
base 13
12. Corona lobes laterally compressed and with a
central cavity . . . . 9. *A. fulva* (p. 391)
Corona lobes dorsiventrally flattened, at least
distally; margins inrolled, but not forming a
distinct central cavity . . . . 11. *A. buchwaldii* (p. 394)
13. Plant generally taller than 50 cm; most leaves
more than 10 cm long 14
Plant generally shorter than 50 cm; most leaves
less than 10 cm long 17
14. Corona lobes 7–10 mm long, twice the height
of the column; corolla lobes at least 10 mm
long; sepals at least 8 mm long . . . . 2. *A. stathmostelmoides* (p. 385)
Corona lobes up to 7 mm long, 1–1.5 times
height of column; corolla lobes 6 mm or less;
sepals at most 6 mm long 15
15. Corona lobes 4–7 mm long; anther wings
2.5–3 mm long . . . . 3. *A. longissima* (p. 386)
Corona lobes 2–3.5 mm; anther wings no more
than 2 mm long 16

16. Peduncles to 1.5 cm long; anther wings with
curved margins; corona fleshy, not laterally
compressed, with a pair of acute teeth
0.5–1 mm long on the proximal margins
and a fleshy tooth on the distal margin all
pointing towards head of column . . . .   4. *A. crassicoronata* (p. 388)
Peduncles 4–13 cm long; anther wings
triangular; corona lobes laterally compressed,
with tooth half way along upper margin . . . .   5. *A. tanganyikensis* (p. 388)
17. Corona lobes laterally compressed for their
entire length, not drawn out into a tongue
distally 18
Corona lobes rounded in section or
dorsiventrally flattened, if appearing
laterally compressed near attachment to
column, then distal margins drawn out into
a tongue 20
18. Leaves densely pubescent with spreading white
hairs on both sides; corona lobes taller than
broad . . . .   8. *A. pseudoamabilis* (p. 391)
Leaves glabrous or minutely rusty pubescent;
corona lobes as broad as tall 19
19. Corona lobes variously lobed or toothed on
upper margins, outer margins fused;
growing in seasonally waterlogged grassland   6. *A. amabilis* (p. 389)
Corona lobes lacking teeth on the upper
margins, outer margins free; found in
montane grassland . . . .   7. *A. edentata* (p. 389)
20. Corona lobes subglobose; a tuft of papillae
entirely filling the cavity of the corona . . . .   10. *A. palustris* (p. 392)
Corona lobes generally longer than broad;
papillae, if present, microscopic 21
21. Gynostegium stipitate, the stipe 2–3.5 mm
long; free part of corona lobes forming an
erect dorsiventrally flattened tongue 4–5 mm
long with a truncate apex and inrolled
margins . . . .   11. *A. buchwaldii* (p. 394)
Gynostegium sessile; corona lobes not as above 22
22. Corolla white, frequently suffused with pink or
purple, but never strongly veined; anther
wings ± 1.5 mm long; proximal margins of
corona lobes reaching ± halfway up anther
wings . . . .   13. *A. foliosa* (p. 395)
Corolla white or pink with deeper veins within;
anther wings ± 2 mm long; corona lobes with
proximal margins not reaching base of
anther wings . . . .   12. *A. schumanniana* (p. 395)

T.T.C.L.: 63 (1949) cites *A. rostrata* based on *Braun* 1583 from Tanga, *non* N.E. Br. – most likely a *Gomphocarpus*, but not *G. fruticosus* subsp. *rostrata*.

1. **Asclepias curassavica** *L.*, Sp. Pl. 215 (1753); Bullock in F.W.T.A. ed. 2, 2: 92 (1963). Type: West Indies, Curaçao, Linn. Herb. 310: 19 (lecto. LINN, designated by Woodson in Ann. Missouri Bot. Gard. 41: 59 (1954); see also Jarvis, Order out of Chaos: 321 (2007))

Short-lived perennial herb with a single stem arising from fibrous rootstock; stems erect, 0.5–2 m, simple or branched, young shoots minutely pubescent, becoming glabrous with age; interpetiolar line often prominent. Leaves with petiole 5–15 mm long, glabrous or minutely puberulent; lamina narrowly oblong, 6–13 × 1–3 cm, apex attenuate, base narrowly cuneate, subglabrous except for the minutely puberulous midrib and margins. Inflorescences extra-axillary, umbelliform with (3–)5–10 flowers; peduncles (2–)3–5 cm long, glabrous or minutely puberulent; bracts filiform, 1–2 mm long, pubescent; pedicels 1–1.5 cm long, glabrous or minutely pubescent. Sepals narrowly oblong to triangular, 2–2.5 × 0.5–1 mm, acute, pubescent. Corolla strongly reflexed, bright red; lobes narrowly obovate, 6–7 × 2–3 mm, acute, glabrous or minutely pubescent. Gynostegial stipe slender, 1.5–2 mm long. Corona lobes orange or yellow, arising at the top of the gynostegial stipe, 3–4 × 1.5 mm, ± 1.5 times the height of the column, somewhat fleshy and semi-cylindrical, not strongly compressed, a falcate tooth ± 2 mm long arising from the cavity of the lobe and arching over the head of the column. Anther wings 1.8 mm long, the margin slightly convex, and with a minute notch at the base of the guide rail; corpusculum brown, ovoid, ± 0.3 × 0.15 mm; translator arms ± 0.4 mm long, slender, geniculate near attachment to pollinia; pollinia oblanceolate, ± 0.9 × 0.3 mm, strongly flattened. Stylar head ± flat. Follicle erect on erect pedicel, 6–7 × 1 cm, fusiform and tapering towards both towards the base and apex, smooth, glabrous. Seeds broadly ovate, ± 5 × 3.5 mm, flattened, with a marginal rim ± 1 mm wide, smooth or verrucose sparsely; coma ± 2.5 cm long.

UGANDA. Toro District: Ibonde, camp garden, 13 Aug. 1938, *A.S. Thomas* 2341!
KENYA. Kilifi District: Kibarani [Kiborani], 28 Sep. 1945, *Jeffery* K332!
TANZANIA. Tanga District: Usambara, *Heinsen* 129!; Lushoto District: nursery, Amani Research Station, 5 Apr. 1932, *Greenway* 2961!; Mbeya District: gardens of the Mt Livingstone Hotel, Mbeya, 29 Nov. 1994, *Goyder et al.* 3841!; Zanzibar, 1927, *Toms* 16!
DISTR. U 2; K 7; T 3, 7; Z; pantropical weed, native to the neotropics, cultivated and naturalised widely in Old World tropics
HAB. Moist forest or grassland; 0–1800 m

SYN. *Asclepias nivea* L. var. *curassavica* (L.) O. Kuntze, Rev. Gen. Pl.: 418 (1891)
　　A. bicolor Moench, Method. Pl. Hort. Marburg.: 717 (1794), *nom illegit.* Type as for *A. curassavica* L.
　　A. aurantiaca Salisb., Prodr. Stirp.: 150 (1796), *nom illegit.* Type as for *A. curassavica* L.
　　A. margaritacea Schult., Syst. Veg. 6: 86 (1820). Type: Brazil, Camete, *Hoffmannseg* s.n. (not traced)
　　A. cubensis Wender. in Bot. Zeit. 1: 830 (1843). Type: Cuba, Jan. 1841, *Pfeiffer* s.n. (?MAR† (fide Stafleu & Cowan, Tax. Lit.)
　　A. curassavica L. var. *concolor* Krug & Urb. in Urb., Symb. Antillae 1: 389 (1899). Type: Puerto Rico, *Sintennis* 3949 (lecto. MO!, designated by Goyder in K.B. 64: 374 (2009))

2. **Asclepias stathmostelmoides** *Goyder* in K.B. 64: 375 (2009). Type: Congo-Kinshasa, Katanga, *Verdick* 361 (BR!, holo.; K, photo.)

Perennial herb with a single annual stem arising from a vertical tuber; stems erect, 0.3–1.3 m long, unbranched, densely spreading-pubescent, frequently lacking leaves in upper parts. Leaves with petiole 1–8 mm long; lamina narrowly linear to lanceolate, 14–28 × 0.7–3.4 cm, apex attenuate, the base cuneate or narrowly truncate, secondary veins numerous and frequently conspicuous, ± at right angles to the midrib, pubescent with stiff white hairs on both surfaces. Inflorescences extra-axillary with 6–16 flowers in a nodding umbel; peduncles 2–4.5(–12) cm long, spreading-pubescent; bracts filiform, pubescent; pedicels (1.5–)2–5 cm long, pubescent. Sepals narrowly triangular, 8–10 × 0.5–1 mm, attenuate, densely pubescent. Corolla rotate to weakly reflexed, green outside, tinged brown or purple within, pubescent on outer surface and towards margins within, inner surface papillate elsewhere; lobes ovate, 10–15 × 5–9 mm, acute, plane or weakly replicate. Gynostegium ± sessile. Corona lobes attached ± at the base of the staminal column, 7–10 × 3–4 mm, ± twice height of column, laterally compressed, complicate above, solid below line from point of attachment to column to just below distal end of upper margin, quadrate, upper margins rounded, highest towards outer margin, pale green or white, frequently with a purple tinge or band and a white upper margin. Anther wings ± 3 mm long, triangular, flared slightly towards the base; anther appendages ± 2 mm long; corpusculum 0.5 × 0.1 mm, subcylindrical, black; translator arms ± 0.5 mm long, flattened and slightly geniculate, broadening abruptly into a short clasping overlap with the pollinium; pollinia narrowly oblong, ± 1.6 × 0.4 mm, flattened. Stylar head raised above top of anthers by ± 1.5 mm, flat. Fruiting pedicel curved to hold follicle erect, follicle (immature) fusiform, ± 6 × 1.5 cm, not inflated, smooth, densely pubescent or tomentose with short white hairs. Fig. 94/1–3, p. 387.

TANZANIA. Mpanda District: 29 km on Namanyere–Karonga road, 5 Mar. 1994, *Bidgood, Mbago & Vollesen* 2647!; Singida District: Matalele, 22 Mar. 1928, *B.D. Burtt* 1422!; Njombe/Mbeya Districts: 54 km on Chimala–Iringa road, 3 Apr. 2006, *Bidgood et al.* 5304!
DISTR. **T** 1, 4, 5, 7; Congo-Kinshasa, Burundi, Zambia, Malawi and Mozambique
HAB. Grassland and open *Brachystegia* or other deciduous woodland; 500–1500 m

SYN. *Stathmostelma verdickii* De Wild. in Ann. Mus. Congo, sér. 5, 1: 188 (1904), *non Asclepias verdickii* De Wild. in Ann. Mus. Congo, sér. 5, 1: 305 (1906)

3. **Asclepias longissima** (*K. Schum.*) *N.E. Br.* in F.T.A. 4(1): 338 (1902). Type: Tanzania, District unclear, Lake Malawi [Nyasa], *Goetze* s.n. (B†, holo.; K!, iso.)

Perennial herb with a single annual stem arising from a tuber; stems erect, 0.5–1 m long, unbranched, glabrous or minutely pubescent above, lacking leaves in upper parts. Leaves with petiole 1–5 mm long; lamina narrowly linear to linear-lanceolate, 13–27 × 0.5–1.5 cm, apex attenuate, the base cuneate or narrowly truncate, secondary veins numerous and frequently conspicuous, ± at right angles to the midrib, margins revolute, glabrous. Inflorescences extra-axillary with 2–4 flowers in a nodding umbel; peduncles 1–2.5 cm long, minutely pubescent; bracts filiform, pubescent; pedicels 1.5–2 cm long, minutely pubescent. Sepals triangular, 2–3 × 1 mm, acute, pubescent. Corolla strongly reflexed, purple outside, green tinged red within, pubescent on outer surface and towards margins within, inner surface papillate elsewhere; lobes ovate, 7–10 × 4–6 mm, acute, plane or weakly replicate. Corona lobes attached ± 1 mm above the base of the staminal column, 4.5–7 × 3–5 mm, 1–1.5 times height of column, laterally compressed, complicate above, solid in lower half, quadrate, upper margins rounded and densely papillate, highest towards inner margin, pale green, sometimes tinged with purple. Anther wings narrowly triangular, 2.5–3 mm long, flared slightly towards the base; anther appendages ± 2 mm long; corpusculum brown, narrowly ovoid, 0.4 × 0.2 mm; translator arms ± 0.5 mm long, flattened and slightly geniculate, broadening into a short clasping overlap with the pollinium; pollinia oblanceolate, ± 1 × 0.4 mm, flattened. Stylar head raised above top of anthers by ± 1.5 mm, flat. Follicles not seen. Fig. 94/4, p. 387.

TANZANIA. Ufipa District: escarpment above Kasanga, 30 Mar. 1959, *Webster* s.n. in *Richards* 11018!; Rungwe District: Kyimbila, 4 Mar. 1914, *Stolz* 2579!; Songea District: between the two crossings of the River Luhekea at foot of Mbamba Bay escarpment, 6 Apr. 1956, *Milne-Redhead & Taylor* 9484!
DISTR. **T** 4, 7, 8; Zambia (around Mbala) and Mozambique (Niassa Province)
HAB. Generally on rocky ground in open *Brachystegia* woodland or grassland; 500–1500 m

SYN. *Gomphocarpus longissimus* K. Schum. in E.J. 17: 382 (1901)

NOTE. Very close to *A. stathmostelmoides*, from which it differs principally in the papillate corona lobes which are shaped slightly differently above; and in the fewer and smaller flowers.

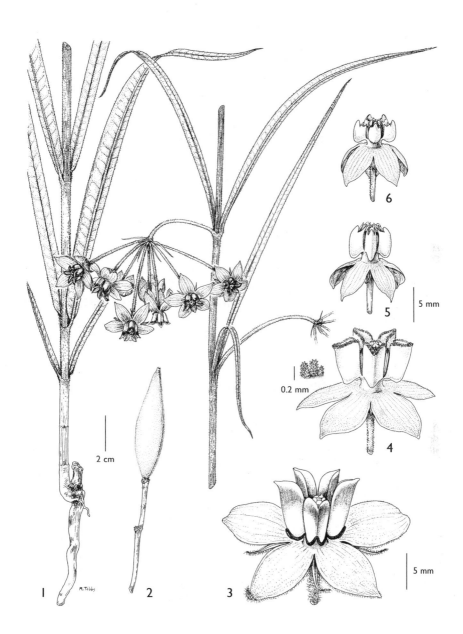

FIG. 94. *ASCLEPIAS STATHMOSTELMOIDES* — **1**, habit; **2**, follicle; **3**, flower. *ASCLEPIAS LONGISSIMA* – **4**, flower, with detail of coronal papillae. *ASCLEPIAS CRASSICORONATA* – **5**, flower. *ASCLEPIAS TANGANYIKENSIS* — **6**, flower. 1 & 3 from *Bidgood et al.* 2647; 2 from *Congdon s.n.*; 4 from *Johnston* 257; 5 from *Bidgood et al.* 2388; 6 from *Lynes* IH90. From K.B. 64: 376, drawn by M. Tebbs.

4. **Asclepias crassicoronata** *Goyder* in K.B. 64: 377 (2009). Type: Zambia, 30 km ESE of Kasama, *Robinson* 4390 (K!, holo.)

Perennial herb with a single annual stem arising from a tuber or fleshy taproot; stems erect, ± 1 m long, unbranched, lacking leaves in upper parts, glabrous or minutely pubescent above. Leaves with petiole 1–5 mm long; lamina narrowly linear, 11–20 × 0.3–0.4 cm, apex attenuate, base cuneate, secondary veins inconspicuous, margins revolute, glabrous except along margins. Inflorescences extra-axillary with 4–6 flowers in a nodding umbel; peduncles 0.5–1.5 cm long, minutely pubescent; pedicels 1.5–2 cm long, slender, minutely pubescent. Sepals triangular, 2–3 × 1 mm, acute, pubescent. Corolla strongly reflexed, green or purple, glabrous or sparsely pubescent on outer surface, minutely papillate within; lobes ovate, 4–6 × 2–3 mm, acute, plane. Corona lobes attached ± 1 mm above the base of the staminal column, fleshy, 3–3.5 × 1.5 mm, slightly taller than column, concave above, not laterally compressed, with a pair of acute teeth 0.5–1 mm long on the proximal margins and a fleshy tooth on the distal margin all pointing towards head of column, green. Anther wings 1.5 mm long, curved; anther appendages ± 1 mm long; corpusculum brown, narrowly ovoid, 0.5 × 0.2 mm; translator arms ± 0.3 mm long, flattened and slightly geniculate; pollinia obovate, ± 0.9 × 0.3 mm, flattened. Stylar head raised above top of anthers by ± 0.5 mm, flat. Follicles not seen. Fig. 94/5, p. 387.

TANZANIA. Ufipa District: Tatanda Mission, 22 Feb. 1994, *Bidgood, Mbago & Vollesen* 2388!; Iringa District: Iheme, 30 km S of Iringa, 24 Feb. 1962, *Polhill & Paulo* 1591!; Mbeya District: between Mshewe and Muvwa villages, 2 Feb. 1990, *Lovett et al.* 4046!
DISTR. **T** 4, 7; N Zambia and central and northern Malawi
HAB. *Brachystegia* or mixed deciduous woodland on rocky hills; 1000–1900 m

NOTE. The flowers of this species are smaller than the similar *Asclepias stathmostelmoides*, *A. longissima* and *A. tanganyikensis*, and the corona is not laterally compressed and cucullate, but fleshy and with an inward-pointing tooth.

5. **Asclepias tanganyikensis** *E.A. Bruce* in K.B. 1934: 303 (1934). Type: Tanzania, Iringa, Mt Luhoto [Lukota], *Lynes* Ih 90 (EA, holo. (not seen); K!, iso.)

Perennial herb with a single annual stem, rootstock not seen; stems erect, 0.6–0.8 m long, simple or branched at base, subglabrous or minutely pubescent above, upper nodes leading into inflorescence lacking leaves. Leaves sessile; lamina narrowly linear, 10–18 × 0.2–0.4 cm, apex attenuate, the base cuneate, secondary veins inconspicuous, margins revolute, glabrous. Inflorescences extra-axillary with 4–8 flowers in a nodding umbel; peduncles (measured from uppermost, leafless, node) 4–13 cm long, minutely pubescent; bracts filiform; pedicels 1.5–3 cm long, slender, minutely pubescent. Sepals triangular, 2–3 × 1 mm, acute, pubescent. Corolla strongly reflexed, green or purple (?green on adaxial surface, purple abaxially), glabrous or sparsely pubescent on outer surface, glabrous within; lobes ovate, ± 6 × 4 mm, subacute. Corona lobes attached 0.5–1 mm above the base of the staminal column, laterally compressed, ± 2.5 × 2.5 mm, ± as tall as the column, the upper margin raised distally into a pair of erect lobes or teeth ± 1 mm long. Anther wings triangular, 1 mm long; corpusculum brown, broadly cylindrical, ± 0.3 × 0.15 mm; translator arms ± 0.15 mm long, flattened and geniculate; pollinia oblanceolate, ± 0.7 × 0.3 mm, flattened. Stylar head ± level with top of anthers, flat. Follicles not seen. Fig. 94/6, p. 387.

TANZANIA. Iringa District: Mt Luhoto [Lukota], 12 Feb. 1932, *Lynes* Ih 90! & Signal Hill, 20 Feb. 1932, *St. Clair Thompson* 468! [but note that flower in packet and the drawing on the K sheet are from *Asclepias crassicoronata*]
DISTR. **T** 7; known only from these two collections
HAB. 'Frequent and gregarious in poor soil under *Brachystegia*'; 1700–1800 m; flowering in February.

SYN. *Gomphocarpus tanganyikensis* (E.A. Bruce) Bullock in K.B. 8: 340 (1953)

6. **Asclepias amabilis** *N.E. Br.* in K.B. 1895: 70 (1895). Types: Zambia, Fwambo, *Carson* 55 (K!, lecto., designated by Goyder in K.B. 64: 378 (2009)); Zambia, Fwambo, *Carson* 35 (K!, paralecto.)

Perennial herb with a single annual stem arising from a slender woody taproot or tuber; stems erect, 0.2–0.5 m long, unbranched, subglabrous to minutely rusty-pubescent above. Leaves subsessile; lamina narrowly linear, 3–10 × 0.1–0.3 cm, glabrous or minutely rusty-pubescent. Inflorescences extra-axillary with 5–9 flowers in a nodding umbel; peduncles 2–6 cm long, minutely rusty-pubescent; bracts filiform, to ± 0.6 cm long, pubescent; pedicels 1–2.5(–3) cm long, minutely rusty-pubescent. Sepals 3–4 × 1 mm, lanceolate to narrowly triangular, attenuate, pubescent, generally reddish brown. Corolla ± rotate to weakly reflexed; lobes ovate, 6–9(–11) × 3–4.5(–6) mm, subacute, adaxial surface pale green or greenish cream and papillose, occasionally only minutely so, abaxial face dull brown, purple or occasionally pink, pubescent, at least towards to apex. Corona lobes attached 1–1.5 mm above the base of the staminal column, green, cream or yellowish, 2.5–3.5(–5.5) × 2–2.5(–3.5) mm, as tall or taller than the column, laterally compressed, complicate above, quadrate, upper margin extremely variable: from almost entire with teeth or projections reduced to indistinct lobes, to strongly dissected with a pair of erect proximal teeth, the distal upper margins raised and variously undulate, linked to the proximal teeth by a lateral flap or tooth pointing back towards the column. Anther wings broadly triangular, 1.3–1.5 mm long; corpusculum brown, subcylindrical, ± 0.35 × 0.1 mm, with narrow translucent flanges down the sides; translator arms ± 0.4 mm long, flattened and weakly geniculate; pollinia oblanceolate, ± 0.8 × 0.35 mm, flattened. Stylar head ± level with top of anthers. Follicles and seed not seen. Fig. 95/3, p. 390.

TANZANIA. Ufipa District: 20 km from Kawimbe, 29 Jan. 1957, *Richards* 8029!; Chunya District: 80 km along Chunya–Itigi road, 21 Mar. 1965, *Richards* 19785!; Mbeya District: 8 km NE of Tunduma, 10 Jan. 1975, *Brummitt & Polhill* 13676!
DISTR. T 4, 7; N Malawi, SE Congo-Kinshasa and the Mbala and Mwinilunga areas of Zambia
HAB. Seasonally waterlogged grassland; 1200–2300 m

SYN. *Gomphocarpus amabilis* (N.E. Br.) Bullock in K.B. 8: 341 (1953)

NOTE. *Carson* 55 was chosen as lectotype to avoid any possible confusion over the identity of the other syntype. *Carson* 35 from Fwambo in Zambia is a syntype of this name, but a collection with the same number, from an island in Lake Tanganyika, was described as *Ceropegia constricta* N.E. Br.

7. **Asclepias edentata** *Goyder* in K.B. 64: 378 (2009). Type: Tanzania, Njombe District, Kipengere Mts, *Richards* 7774 (K!, holo., EA!, iso.)

Perennial herb with a single annual stem; rootstock not seen; stems erect, 0.2–0.4 m long, unbranched, minutely pubescent above. Leaves sessile or with petiole to ± 5 mm; lamina narrowly linear to linear-lanceolate, 7–11 × 0.3–1 cm, glabrous except on the midrib below. Inflorescences extra-axillary with 4–8 flowers in a nodding umbel; peduncles (2–)4.5–7 cm long, minutely pubescent; bracts filiform, to ± 0.8 cm long, pubescent; pedicels 1–2.5 cm long, minutely pubescent. Sepals lanceolate to narrowly triangular, ± 4–5 × 1.5 mm, attenuate, pubescent, generally reddish brown. Corolla reflexed, lobes adaxial surface green, ovate, 7–9 × 4–5 mm, subacute, minutely papillose, abaxial face dull brown or purple, sparsely pubescent, at least towards to apex. Corona lobes attached 0.5–1 mm above the base of the staminal column, green, 3–4 × 3–4 mm, ± as tall as the column, laterally compressed, complicate, oblong but somewhat rounded, upper margins entire, slightly raised distally, then falling away to the base, only uniting bear the base of the outer margin. Anther wings triangular, 1.6–1.8(–2) mm long; corpusculum brown, ovoid, ± 0.35 × 0.15 mm, with narrow translucent flanges

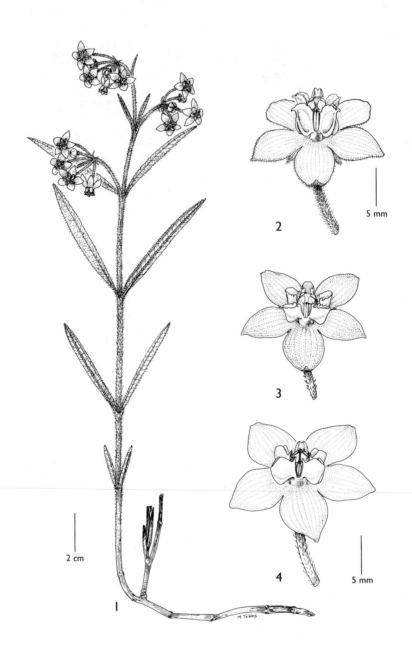

FIG. 95. *ASCLEPIAS PSEUDOAMABILIS* — **1**, habit; **2**, flower. *ASCLEPIAS AMABILIS* — **3**, flower. *ASCLEPIAS EDENTATA* — **4**, flower. 1 from *Richards* 6952 (above-ground part) and *Bullock* 1947 (rootstock); 2 from *Richards* 6952; 3 from *Richards* 8029; 4 from *Richards* 7651. From K.B. 64: 379, drawn by M. Tebbs.

down the sides; translator arms ± 0.15 mm long, flattened and strongly geniculate; pollinia oblanceolate, ± 0.9 × 0.25 mm, flattened. Stylar head ± level with top of anthers. Follicles and seed not seen.

TANZANIA. Njombe District: Kipengere Mts, 10 Jan. 1957, *Richards* 7651! & 14 Jan. 1957, *Richards* 7774!; Mbeya District: Kitulo Plateau, ridge E of Ishinga Mt summit, 9 Feb. 1979, *Cribb et al.* 11356!
DISTR. **T** 7; endemic to the Kitulo and Kipengere mountains
HAB. Rough mountain grassland; 2400–2700 m

NOTE. *Asclepias edentata* differs from *A. amabilis* in the absence of teeth on the corona, and the free outer margins of the corona lobes. Corolla and corona lobes are green in *A. edentata*, rather than pink or purplish in *A. amabilis*. *Robinson* 4174 (K) from northern Zambia is very similar to *A. edentata* but has a slightly distorted corona. It was collected at much lower altitude in woodland rather than montane grassland.

    A. edentata and the allied species *A. pseudoamabilis* are found in montane grassland, whereas *A. amabilis* is a plant of seasonally waterlogged dambos, generally at somewhat lower elevations. The altitude of 1500 m recorded on the label of *Richards* 7774 appears to be a typographic error for 2500 m, as Mary Richards' collecting notebook at K has the altitude 8000 ft. for this and neighbouring collections.

8. **Asclepias pseudoamabilis** *Goyder* in K.B. 64: 380 (2009). Type: Tanzania, Ufipa District, Mbisi, *Bullock* 1947 (K!, holo.)

Perennial herb with a single annual stem, arising from very deep-seated rootstock (not seen); stems erect, ± 0.3 m long, unbranched, densely pubescent above. Leaves sessile or subsessile; lamina linear to lanceolate, 5–9 × 0.5–1.5 cm, densely pubescent with spreading white hairs on both surfaces. Inflorescences extra-axillary with 2–6 flowers in a nodding umbel; peduncles 2–4 cm long, densely pubescent; bracts filiform, to ± 0.6 cm long, pubescent; pedicels 1–2.5 cm long, densely pubescent. Sepals lanceolate to narrowly triangular, ± 4–5 × 1.5 mm, attenuate, densely pubescent, generally reddish brown. Corolla reflexed; lobes abaxial face dull brown or purple, ovate, 7–9 × 4–6 mm, subacute, adaxial surface green, minutely papillose, pubescent towards margin, densely pubescent. Corona lobes attached 0.5–1 mm above the base of the staminal column, oblong and taller than wide, 3–5 × 2–2.5 mm, taller than the column, laterally compressed, complicate, upper margins entire, slightly raised distally, then falling away and slightly recurved, fused along the outer margin, green. Anther wings triangular, 1.9–2.1 mm long; corpusculum brown, narrowly ovoid, ± 0.4 × 0.1 mm, with narrow translucent flanges down the sides; translator arms ± 0.4 mm long, flattened and strongly geniculate; pollinia oblanceolate, ± 1.2 × 0.25 mm, flattened. Stylar head ± level with top of anthers. Follicles and seed not seen. Fig. 95/1–2, p. 390.

TANZANIA. Tabora District: Urambo, Igwisi Hill, 12 Dec. 1950, *Moors* K11!; Ufipa District: Mbisi, 27 Nov. 1949, *Bullock* 1947!; road from Muse Gap to Sumbawanga, 2 Sep. 1959, *Richards* 11410!
DISTR. **T** 4; endemic to the highlands around Sumbawanga and the volcanic Igwisi hills WNW of Tabora
HAB. Very dry seasonally burnt grassland; (1100 at Igwisi) 1700–2400 m

NOTE. Essentially a pubescent form of the previous species, with additional differences in the corona. The corona lobes are taller than broad, the outer margins fused not free. Leaves are densely pubescent on both sides rather than glabrous or subglabrous in *Asclepias edentata*.

9. **Asclepias fulva** *N.E. Br.* in K.B. 1895: 254 (Oct. 1895). Type: Uganda, [no further details], *J. Wilson* 112 (K!, holo.)

Perennial herb with annual stems arising from a vertical napiform or fusiform tuber; stems erect, 0.2–0.6 m long, usually single and unbranched, but sometimes branched below, densely puberulent with short rusty hairs. Leaves with petiole 1–6 mm long; lamina narrowly oblong to ovate-oblong, 3.5–7.5(–10) × 1.2–4 cm, apex obtuse to acute, base rounded or slightly cordate, margins scabrid, venation prominent with numerous parallel secondary veins at an acute angle to the midrib, indumentum of reddish hairs on both upper and lower surfaces. Inflorescences extra-axillary with 4–10 flowers in a nodding umbel; peduncles 1.5–7(–9) cm long, densely rusty-puberulent; bracts filiform, 5–10 mm long, puberulent; pedicels 1–3 cm long, puberulent. Calyx lobes lanceolate, 4–11 × 1.5–3 mm, acute, pubescent, green or brown. Corolla often somewhat reflexed at base, spreading above, rotate or saucer-shaped, outside green or brownish purple, densely pubescent with white or rusty hairs, inner surface pale orange or yellow-brown, glabrous; lobes ovate, 8–13 × 4–7.5 mm, acute. Corona lobes solid except for a shallow sinus along upper and inner margins, laterally compressed, 2–5 × 2–4 mm, quadrate with a truncate base and an obliquely truncate top, the upper margins produced into a pair of teeth extending over the stylar head, yellowish green with a brown or purple projecting rim to 1.5 mm wide around the top and base. Anther wings triangular, 2.5 × 1.5 mm; anther appendages broadly ovate, ± 1 × 1 mm, subacute; corpusculum dark brown, ovoid, 0.5 × 0.2 mm; translator arms ± 0.5 × 0.2 mm, flattened and contorted, broadened distally and clasping the pollinia; pollinia 1.1 × 0.5 mm, flattened, obovate. Stylar head flat or undulate. Follicles fusiform, 8–11 × 1.5 cm, and with 4–6 longitudinal toothed wings or ridges running for all or most of their length, minutely puberulent.

UGANDA. Kigezi District: Nyakagyeme, Apr. 1946, *Purseglove* 2042!; Masaka District: Buddu, *Scott Elliot* 7443!
KENYA. Uasin Gishu District: Kipkarren, 1931, *Brodhurst Hill* 57!; S Elgon, May 1941, *Tweedie* 565!
TANZANIA. Morogoro District: Nguru Mts near Maskati Mission, 10 June 1978, *Thulin & Mhoro* 3152!; Mbeya District: Poroto Mts, Ikuyu, 10 Feb. 1979, *Cribb et al.* 11375!; Songea District: 22 km SE of Songea, 26 Mar. 1956, *Milne-Redhead & Taylor* 9341!
DISTR. U 2, 4; **K** 3/5; **T** 6–8; Zambia, Zimbabwe, Lesotho and eastern parts of South Africa
HAB. Open grassland and *Brachystegia* or mixed deciduous woodland; 1000–2100 m. Flowering mostly November to February in southern Tanzania, also recorded in May and June further north

SYN. *Pachycarpus viridiflorus* E. Mey., Comm Pl. Afr. Austr.: 214 (1838). Type: South Africa, Uitenhage Division, N side of the Zuurberg Mountains, 2500–3000', 30 Oct. 1829, *Drège* s.n. (B†, holo.; K!, iso.)
    Xysmalobium viridiflorum (E. Mey.) D. Dietr., Syn. Pl. 2: 903 (1840)
    Gomphocarpus viridiflorus (E. Mey.) Decne. in DC., Prodr. 8: 561 (1844)
    Asclepias rubicunda Schltr. in J.B. 33: 336 (Nov. 1895). Types: Uganda, Masaka District: Buddu, *Scott Elliot* 7443 (K!, lecto., designated by Goyder in K.B. 64: 381 (2009); BM!, isolecto.)
    A. dregeana Schltr. in J.B. 33: 337 (Nov. 1895), *nom. nov.* for *Gomphocarpus marginatus* sensu Schltr., *non* E. Mey
    A. calceolus S. Moore in J.B. 41: 312, 338 (1903). Type: South Africa, N of Johannesburg, *Rand* 966 (BM!, holo.; K!, iso.)
    A. dregeana Schltr. var. *calceolus* (S. Moore) N.E. Br. in Fl. Cap. 4(1): 697 (1908)
    A. dregeana Schltr. var. *sordida* N.E. Br. in Fl. Cap. 4(1): 697 (1908). Type: South Africa, Transkei, Kentani, *Pegler* 655 (K!, holo.; PRE, iso.)
    Pachycarpus fulvus (N.E. Br.) Bullock in K.B. 8: 334 (1953); U.K.W.F. ed. 2: 178 (1994)
    Asclepias viridiflora (E. Mey.) Goyder in K.B. 52: 247 (1997), *non A. viridiflora* Raf. (1808)

10. **Asclepias palustris** (*K. Schum.*) *Schltr.* in J.B. 33: 336 (1895); N.E. Br. in F.T.A. 4(1): 349 (1902). Types: Angola, Malanje, *von Mechow* 401 (K!, lecto., designated by Goyder in K.B. 56: 134 (2001); B†)

Perennial herb with annual stems arising from a vertical rootstock with globose to fusiform lateral tubers; stems erect, usually single, 0.1–0.4(–0.6) m long, generally unbranched, densely pubescent with spreading white hairs. Leaves subsessile or with petiole to 3 mm long, pubescent; lamina linear to lanceolate, 3–7(–8) × (0.1–)0.4–1.3 cm, acute, base rounded to truncate, margins stiffly pubescent, venation prominent with secondary veins at ± 45° to the midrib, indumentum of spreading white hairs on upper surface and on main veins below. Inflorescences extra-axillary with 4–10 flowers in a nodding umbel; peduncles 1–6(–13) cm long, lengthening markedly in fruit, erect, densely spreading-pubescent; bracts filiform or occasionally linear-lanceolate, 0.2–1 cm long, pubescent; pedicels 1–1.5(–2) cm long, pubescent. Calyx lanceolate to broadly triangular, lobes purple, 3–5(–6) × 1–1.5 mm, acute, densely pubescent. Corolla campanulate or occasionally reflexed, white or greenish, tinged with dull pink especially towards tip outside, densely pubescent towards the apex outside, glabrous or minutely papillate at base within; lobes ovate, 5–8 × 3–5 mm, subacute. Corona lobes cucullate but with a rounded tip making the lobe appear subglobose, ± 3 mm long, about half as tall as the column, white with purple tip, the upper margins with a pair of erect triangular teeth proximally, otherwise entire to variously dentate, with a laterally flattened tooth ± 0.5 mm long topped with a dense crest of papillae ± filling the cavity of the lobe, the papillae reaching to or slightly exceeding the upper margins. Anther wings triangular, ± 1.5 mm long; anther appendages reniform, ± 0.5 × 1.5 mm, inflexed over apex of stylar head; corpusculum black, ovoid-subcylindrical, ± 0.5 × 0.2 mm; translator arms ± 0.7 mm long, flattened and geniculate; pollinia obtriangular, 1.2 × 0.5 mm, flattened. Fruiting pedicel not contorted; follicle erect, narrowly fusiform, ± 5 × 0.5 cm, with an attenuate apex, not inflated, smooth, densely pubescent. Seeds ovate, ± 0.4 × 0.3 cm, flattened, with an inflated and somewhat convoluted rim and a verrucose disc. Fig. 96, p. 393.

TANZANIA. Ufipa District: Mbisi, 16 Jan. 1950, *Bullock* 2239!; Mbeya District: Mbeya Peak, 12 km from turn-off, 29 Nov. 1994, *Goyder et al.* 3849!; Songea District: Matengo Hills, Lupembe Hill, 10 Jan. 1956, *Milne-Redhead & Taylor* 8176!
DISTR. T 4, 7, 8; tropical Africa from Nigeria to Uganda in the north, and Angola to Zimbabwe in the south
HAB. Montane or seasonally waterlogged grassland, occasionally in open woodland; 1200–2600 m

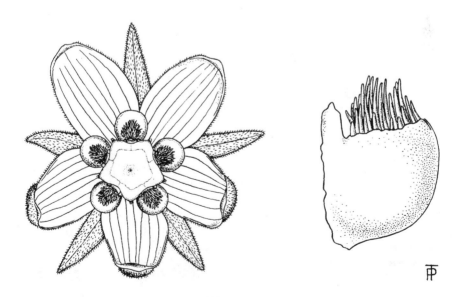

FIG. 96. *ASCLEPIAS PALUSTRIS* — **1**, flower, × 4; **2**, corona lobe, lateral view, × 12. From *Milne-Redhead* 3079. From K.B. 56: 137, drawn by P. Taylor.

SYN. *Gomphocarpus cristatus* Decne. in Ann. Sci. Nat., Sér. 2, 9: 325, t. 11D, Fig. 3 & 4 (1838), *non*
   Asclepias cristata S. Moore in J.B. 50: 343 (1912). Type: Angola, Benguela Plateau, *da Silva*
   s.n. (P! holo., K photo.)
   G. paluster K. Schum. in E.J. 17: 127 (1893), as *palustris*
   Asclepias cristata S. Moore in J.B. 50: 343 (1912), *non Gomphocarpus cristatus* Decne. (1838).
   Types: Angola, Kubango, Kapembe, *Gossweiler* 2288 (BM!, lecto., designated by Goyder
   in K.B. 56: 134 (2001)
   Trachycalymma cristatum (Decne.) Bullock in K.B. 8: 349 (1953) & in F.W.T.A. ed. 2, 2: 92
   (1963); Goyder in K.B. 56: 134 (2001)
   [*Asclepias kyimbilae* Schltr. in sched., based on *Stolz* 502 (K, WAG)]

11. **Asclepias buchwaldii** (*Schltr. & K. Schum.*) *De Wild.* in Ann. Mus. Congo sér. 5,
1: 185 (1904). Type: Tanzania, Lushoto District: Usambara, Mombo, *Buchwald* 375
(B†, holo.); Iringa District, upper slopes of Image Mountain, *Goyder et al.* 3924 (K!,
neo., designated by Goyder in K.B. 56: 146 (2001); DSM!, PRE!, isoneo.)

Perennial herb with annual stems arising from a slender, vertical, fleshy tuber,
occasionally also with fusiform lateral tubers; stems erect, usually single, 0.15–0.6 m
long, simple or occasionally branched below, minutely pubescent with spreading
white hairs. Leaves sessile or with petiole to 2 mm long; lamina somewhat fleshy,
ovate to narrowly lanceolate, 2.5–7(–10) × (0.4–)0.8–2.2 cm, acute to attenuate,
base rounded or rarely subcuneate, margins softly pubescent, venation prominent
in dried material with numerous secondary veins at 45°–90° to the midrib, glabrous
or with an indumentum of minute white hairs on both surfaces. Inflorescences
extra-axillary with 3–6 flowers in a nodding umbel; peduncles erect, 1–4(–6) cm
long, densely spreading-pubescent; bracts filiform, 0.1–0.6 cm long, pubescent;
pedicels 0.7–2 cm long, densely pubescent, extending slightly in fruit. Calyx lobes
dull purple, lanceolate, 2–5 × 0.5–1 mm, acute, densely pubescent. Corolla
campanulate or occasionally reflexed, outer face dull purple with paler margins,
pubescent, green or brownish within and papillose at least towards margins and
apex; lobes oblong, 5–7 × 2–3 mm, subacute. Corona lobes violet with darker
margins, paler within, slightly fleshy, adnate to column for 2–3.5 mm and reaching
the base of the anthers, adnate part forming a pair of vertical wings with slightly
inrolled papillate margins, extending basally into a recurved erect tongue 4–5 × 1–2
mm reaching ± to top of gynostegium, margins inrolled, papillate, apex truncate or
weakly lobes, minutely but densely papillose, inrolled or not, cavity of the lobe
without a tooth. Anther wings vertical, 0.7–1 mm; anther appendages subreniform,
± 1 × 1.5 mm, inflexed over apex of stylar head; corpusculum black, ovoid, ± 0.3 ×
0.1 mm; translator arms ± 0.8 mm long, flattened and geniculate; pollinia obovate,
0.8 × 0.4 mm, flattened. Fruiting pedicel not contorted, erect; follicles narrowly
fusiform, ± 15 × 0.5 cm, with an attenuate apex and base, not inflated, smooth,
densely pubescent. Seeds not seen.

TANZANIA. Ufipa District: 10 km W of Mkowe on road to Chapota, 21 Nov. 1994, *Goyder et al.*
   3778!; Iringa District: Mt Image, 9 Dec. 1994, *Goyder et al.* 3924!; Songea District: 32 km E of
   Songea by R. Mkukire, 8 Feb. 1956, *Milne-Redhead & Taylor* 8719!
DISTR. T 3–4, 7, 8; adjacent regions of Congo-Kinshasa, Burundi, Malawi and Zambia
HAB. Grassland or *Brachystegia* woodland, usually on steep rocky hillsides; 900–2400 m

SYN. *Gomphocarpus buchwaldii* Schltr. & K. Schum. in E.J. 33: 324 (1903)
   Asclepias affinis De Wild. in Ann. Mus. Congo sér. 5, 1: 184 (1904), *non Gomphocarpus affinis*
   Schltr. in E.J. 20, Beibl. 51: 27 (1895), *nec Asclepias affinis* (Schltr.) Schltr. in J.B. 34: 455
   (1896). Type: Congo-Kinshasa, Vieux Kasongo, *Dewèvre* 952 bis (BR!, holo.)
   A. buchwaldii (Schltr. & K. Schum.) De Wild. var. *angustifolia* De Wild. in Ann. Mus.
   Congo sér. 5, 1: 185 (1904). Type: Congo-Kinshasa, Vieux Kasongo, *Dewèvre* 952 (BR!,
   holo.)
   Trachycalymma buchwaldii (Schltr. & K. Schum.) Goyder in K.B. 56: 146 (2001)

12. **Asclepias schumanniana** *Hiern*, Cat. Afr. Pl. Welw. 1: 686 (1898). Types: Angola, Pungo Andongo, Pedra de Cazella, *Welwitsch* 4169 (K!, lecto., designated by Goyder in K.B. 56: 152 (2001); B†, BM!, C, P, isolecto.)

Perennial herb with a single annual stem arising from a subglobose tuber; stems erect, 0.15–0.4 m long, unbranched, pubescent with spreading white hairs. Leaves with petiole to 3 mm long, pubescent; lamina linear or occasionally linear-lanceolate over most of the range, 4–8 × 0.2–0.3(–0.7) cm (Angolan material to 1.3 mm wide and mostly elliptic), acute, base cuneate, venation only visible on wider leaves, secondary veins at ± 45° to the midrib, indumentum of spreading white hairs on both surfaces. Inflorescences terminal or extra-axillary with 2–4(–5) flowers in a suberect umbel; peduncles erect, (1–)2–5 cm long, densely spreading-pubescent; bracts filiform, 0.5–1.3 cm long, pubescent; pedicels 1–1.5 cm long, pubescent. Calyx lobes linear-lanceolate, 6–9 × 1 mm, acute, pubescent. Corolla campanulate, white or pale pink with mauve or purple veins within, sparsely pubescent towards the apex outside, glabrous within; lobes ovate-oblong, ± 10 × 4 mm, subacute. Corona purple with pale margins, lobes fleshy, 5–7 mm long, adnate to the column to the base of the anther wings, somewhat cucullate above and tapering gradually into an erect tongue reaching to the top of the staminal column or slightly higher, proximal margins mostly not reaching the base of the anther wings and with a pair of auricles proximally, with a dense band of papillae across the cavity at the extreme base only, exposed part of tongue glabrous to minutely papillate; interstaminal corona lobes consisting of a truncate or emarginate lobule ± 0.5 × 0.5 mm situated between the auricles of the principal lobes. Anther wings ± 2 mm long, triangular; anther appendages ± 1 × 1 mm, ovate, inflexed over apex of stylar head; corpusculum black, ovoid-subcylindrical, ± 0.3 × 0.1 mm; translator arms ± 0.4 mm long, flattened and geniculate; pollinia flattened, 1 × 0.3 mm, obtriangular. Fruit not seen.

TANZANIA. Bukoba District: Nshamba, Sept.-Oct. 1935, *Gillman* 544!; Buha District: 10 km from Manyovu on Kasulu road, 22 Nov. 1962, *Verdcourt* 3417!; Mpanda District: Sitebi Hill, 9 Feb. 1996, *Congdon* 450!
DISTR. **T** 1, 4; Congo-Kinshasa, Rwanda, Burundi, Angola
HAB. Valley grassland or on rocky hillsides; 1500–1600 m

SYN. *Gomphocarpus amoenus* K. Schum. in E.J. 17: 124 (1893), *non Asclepias amoena* L. (1753), *nec* Brongn. (1831), *nec* Hemsl. (1881). Type as for *A. schumanniana* Hiern
    Asclepias gossweileri S. Moore in J.B. 49: 155 (1911). Type: Angola, Camona, Ambaka, *Gossweiler* 4557 (BM!, holo.)
    Trachycalymma amoenum (K. Schum.) Goyder in K.B. 56: 152 (2001)

NOTE. This species can be distinguished most reliably by the coronal papillae which are restricted to the extreme base of the lobe so the tongue appears glabrous. The shorter pedicels, fewer flowers, colouring of the corona and the narrower leaves also serve to distinguish this species from *T. pulchellum* except in Angola, where material of *T. amoenum* has much broader leaves than elsewhere in its range.

13. **Asclepias foliosa** (*K. Schum.*) *Hiern* in Cat. Afr. Pl. Welw. 1: 686 (1898); N.E. Br. in F.T.A. 4(1): 349 (1902). Types: Congo-Kinshasa, Mukenge, *Pogge* 1130 (B†, holo.; K!, lecto., designated by Goyder in K.B. 56: 153 (2001))

Perennial herb with 1–3(–5) annual stems arising from one or more subglobose or napiform tubers; stems erect, 0.1–0.4 m long, generally unbranched, pubescent with spreading white hairs. Leaves with petiole to 3 mm long, pubescent; lamina linear to elliptic, (1.5–)4–9(–10) × 0.2–1.5 cm, acute, base cuneate, venation only visible on wider leaves, secondary veins at ± 45° to the midrib, glabrous or with indumentum of spreading white hairs on both surfaces. Inflorescences terminal or extra-axillary with 2–4(–6) nodding flowers per umbel; peduncles erect, 1–15 cm

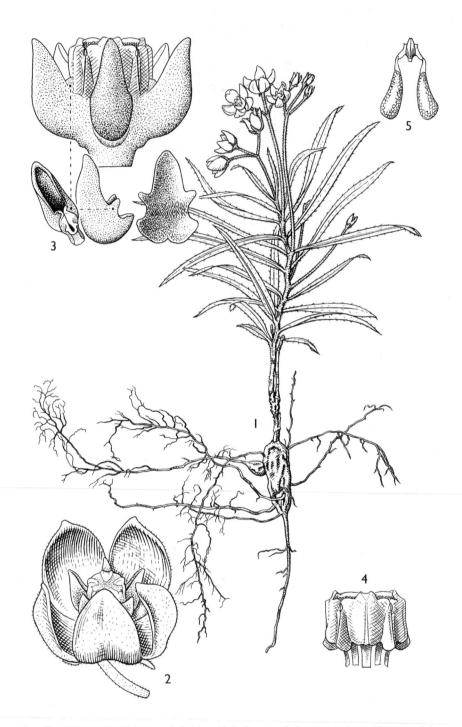

Fig. 97. *ASCLEPIAS FOLIOSA* — **1**, habit, × $^2/_3$; **2**, flower, × 3; **3**, gynostegium with corona lobes, × 7; **4**, gynostegium with corona lobes removed, × 7; **5**, pollinarium, × 14. From Milne-Redhead & *Taylor* 8000. From K.B. 56: 156, drawn by D. Erasmus.

long, pubescent; bracts filiform, 0.2–0.4 cm long, pubescent; pedicels 0.5–2 cm long, pubescent. Calyx lobes purple or occasionally green, ovate to lanceolate, 3–5 × 1(–2) mm, acute, pubescent. Corolla white, campanulate, frequently suffused with pale pink or purple but not strongly veined within, sparsely pubescent outside, glabrous within; lobes ovate-oblong, 7–10(–12) × 3–6(–7) mm, subacute. Corona purple with white margins (green with white margins in West Africa), lobes 2.5–4 mm long, fleshy and adnate to the column to the base of the anther wings, cucullate above and tapering gradually into a rounded spreading tongue half to as long as the staminal column, proximal margins mostly reaching about midway along the anther wings and with a pair of erect teeth extending further up the column, with a dense band of papillae across the cavity halfway along the lobe and entirely concealed within it, exposed part of tongue glabrous to minutely papillate; interstaminal corona lobes minute. Anther wings triangular, ± 1.5 mm long; anther appendages ovate, ± 1 × 1 mm, inflexed over apex of stylar head; corpusculum black, ovoid-subcylindrical, ± 0.4 × 0.2 mm; translator arms ± 0.4 mm long, flattened and geniculate; pollinia obtriangular, 1 × 0.3 mm, flattened. Fruiting pedicel not contorted; follicle erect, narrowly fusiform, 20–25 × 0.5 cm, with a stipe to 17 cm long and an attenuate apex, not inflated, smooth, minutely puberulent. Seeds not seen. Fig. 97, p. 396.

TANZANIA. Ufipa District: N of Tatanda Mission, 16 Nov. 1986, *Goldblatt et al.* 8100!; Iringa District: Kisolanda Farm, 3 Jan. 1996, *de Leyser* 305!; Songea District: Songea airfield, 1 Jan. 1956, *Omari* s.n. in *Milne-Redhead & Taylor* 8000!
DISTR. T 4, 7, 8; West Africa, Congo-Kinshasa and the *Brachystegia* belt of southern tropical Africa
HAB. Open, mixed deciduous woodland; (600–)1100–2000 m

SYN. *Gomphocarpus foliosus* K. Schum. in E.J. 17: 126 (1893)
    Asclepias modesta N.E. Br. in F.T.A. 4(1): 348 (1902). Type: Malawi, Namasi, *Cameron* 6 (K!, lecto., designated by Goyder in K.B. 56: 153 (2001))
    A. modesta N.E. Br. var. *foliosa* N.E. Br. in F.T.A. 4(1): 349 (1902). Type: Angola, Huilla, near Lopollo, *Welwitsch* 4174 (K!, holo.; BM!, iso.)
    A. lepida S. Moore in J.B. 50: 344 (1912). Type: Angola, Kubango, near Forte Colui, *Gossweiler* 2176 (BM!, holo.)
    A. minuta A. Chev. in Mém. Soc. Bot. France 8: 272 (1917). Type: Benin, Atacora Mts, Kouende, *Chevalier* 24227 (P!, holo., K, photo.)
    Trachycalymma pulchellum sensu Bullock in F.W.T.A. ed. 2, 2: 92 (1963); Cribb & Leedal, Mt Fl. S. Tanz.: 105, t. 24b (1982), excl. type
    T. foliosum (K. Schum.) Goyder in K.B. 56: 153 (2001)

14. **Asclepias nuttii** *N.E. Br.* in K.B. 1898: 308 (1898). Type: Tanzania, Ufipa District: 'between Lake Tanganyika and Lake Rukwa', *Nutt* s.n. (K!, holo.)

Perennial herb with one or few annual stems arising from a deep-seated vertical tuber; stems ascending or erect, 6–30 cm long, simple or branched below, reddish, minutely pubescent. Leaves subsessile; lamina narrowly linear, 3–10(–12) × 0.1–0.5 cm, attenuate both apically and basally, glabrous or with minute indumentum restricted to margin and midrib. Inflorescences solitary and terminal forming umbels of 5–10 spreading or erect flowers; subsessile or pedunculate, peduncle reddish, to 2 cm long, minutely but densely pubescent; bracts filiform, pubescent; pedicels 0.8–1.5 cm long, minutely pubescent. Sepals lanceolate or narrowly oblong, 3–5 × 0.5–1 mm, acute, pubescent. Corolla campanulate; lobes ovate-oblong to lanceolate, 6–8 × 2–3 mm, acute, cream or white tinged with pink and glabrous except towards the margins adaxially, abaxial face green tinged with purple-brown, glabrous or pubescent towards the tip. Corona lobes arising at the top of a gynostegial stipe ± 1 mm long, purplish pink with white tip, lobes ± 2 mm tall, reaching ± to top of column, subcylindrical. Anther wings 1.5–2 mm long, with a conspicuous notch near the base ± at the point the margins curl under the gynostegium; corpusculum brown, subcylindrical, ± 0.3–0.4 × 0.1 mm; translator arms ± 0.6 mm long, clearly differentiated into a horizontal proximal part and a vertical distal part, junction strongly geniculate; pollinia narrowly oblong/obtriangular, ± 1 × 0.4 mm, somewhat flattened. Stylar head ± flat. Follicles and seeds not seen.

TANZANIA. Ufipa District: 2 km W of Mkowe on road to Chapota, 21 Nov. 1994, *Goyder et al.* 3772! & 10 km from Mumba on way to Sumbawanga, directly opposite prison camp, 18 Dec. 1986, *D. & J. Moyer* 62!; Mbeya District: Vwawa, 9 Jan. 1976, *Leedal* 3313!
DISTR. **T** 4, 7; a single collection from central Malawi. Surprisingly, this species has not been recorded from the adjacent Mbala region of Zambia
HAB. Seasonally waterlogged 'dambo' grassland; 1500–1800 m

SYN. *Stathmostelma nuttii* (N.E. Br.) Bullock in K.B. 8: 55 (1953)

NOTE. This species is distinctive with a single terminal inflorescence per branch. The translator arms are differentiated into proximal and distal sections, as in *Stathmostelma*, but the proximal part is not so conspicuously expanded. The conspicuous notch in the anther wings suggests a possible link to *Asclepias eminens*.

15. **Asclepias inaequalis** *Goyder* in K.B. 64: 387 (2009). Type: Tanzania, Njombe District, edge of Kitulo Plateau near Matamba, *Goyder, Griffiths, Harvey, Kayombo, Mbago & Paton* 3888 (K!, holo.; DSM!, EA!, PRE!, WAG!, iso.)

Perennial herb with one to several annual stems arising from a slender vertical tuber, the tuber in some cases (*Goyder et al.* 3888) as long as 30 cm; stems erect, 0.2–0.5 m long, simple or little-branched, reddish purple, minutely pubescent with indumentum of white hairs. Leaves subsessile; lamina linear, 3–10 × (0.1–)0.3–0.7 cm, mostly attenuate apically, attenuate or rounded at the base, margins somewhat revolute, minutely pubescent with stiff white hairs on both upper and lower surfaces. Inflorescences extra-axillary forming umbels of 3–6 spreading or erect flowers; peduncles reddish, 1–5 cm long, with a dense indumentum of white hairs; bracts filiform, densely pubescent, deciduous; pedicels 0.6–1.2 cm long, pubescent. Sepals linear to narrowly triangular, 3–4 mm long, attenuate, generally reddish with white hairs. Corolla ± rotate to weakly reflexed; lobes ovate, 4–6 × 2–3 mm, subacute, adaxial surface pale green or cream, glabrous or minutely papillose, abaxial face reddish purple, pubescent. Corona lobes attached ± at the base of the staminal column, erect, broadly obovate, 3–4 × 2.5–3 mm, taller than the column and with the tip arching over it, dorsiventrally compressed but somewhat fleshy with a thickened midline and a pair of thin flanges running along the midline on the inner face, and two further fleshy ridges lying horizontally near the base of the inner face, lobes cream or green, frequently suffused with brown or purple along the midline and the apex. Anther wings 1 mm long, margins ± parallel to the column but curved and distorted in their lower halves; corpusculum brown, subcylindrical, ± 0.3 × 0.1 mm; translator arms ± 0.5 mm long, flattened and ± 0.1 mm broad, very weakly geniculate; pollinia oblong, ± 0.7 × 0.25 mm, flattened. Stylar head ± level with top of anthers. Mature follicles and seed not seen, but very immature follicle on *Goyder et al.* 3895 densely pubescent and held erect on apparently elongating peduncle.

TANZANIA. Njombe District: top of Kitulo Plateau, 4 Dec. 1994, *Goyder et al.* 3895!; Rungwe District: N slopes of Mt Rungwe, 9 Nov. 1966, *Gillett* 17671!; Mbeya District: Mbeya Peak, 25 Nov. 1961, *Kerfoot* 3249!
DISTR. **T** 7; endemic to Mbeya Peak, the Kitulo Plateau, Mt Rungwe and the Poroto and Kipengere ranges
HAB. Montane seasonally burnt grassland but occasionally in more disturbed areas, mostly in peaty soils; 2200–2900 m. Flowering with the early rains (October–January).

NOTE. The asymmetrically distorted base to the anther wings in the new species, where the anther wings of adjacent anthers are differentially contorted, appears to be unique among African *Asclepias*. Goyder in K.B. 58: 718 (2003) reported another case of this anther morphology from the very distantly related South American species *Morrenia stuckertiana* (H. Heger) Malme (*Asclepiadeae*: subtribe *Oxypetalinae*).

16. **Asclepias breviantherae** *Goyder* in K.B. 64: 389 (2009). Type: Tanzania, Rungwe District, Poroto Mts E of Kikondo on road to Kitulo, *Goyder, Griffiths, Harvey, Mbago & Paton* 3872 (K!, holo.; DSM!, EA!, PRE!, WAG!, iso.)

Perennial herb with one to several annual stems arising from a napiform tuber; stems prostrate to erect, 6–12(–25) cm long, simple or little-branched, reddish, minutely pubescent with indumentum of white hairs. Leaves subsessile; lamina linear to narrowly lanceolate or oblong, 3–7 × 0.2–1 cm, mostly attenuate apically, and narrowing abruptly at the base, margins not revolute, minutely pubescent with stiff white hairs on both upper and lower surfaces, or indumentum restricted to principal veins. Inflorescences extra-axillary forming umbels of 2–7 spreading or erect flowers; peduncles reddish, 0.5–4 cm long, with a dense indumentum of white or rusty hairs; bracts filiform, pubescent; pedicels 0.3–1 cm long, pubescent. Sepals narrowly oblong to lanceolate, (2–)2.5–3 × 0.7–1 mm, attenuate, generally reddish with white hairs. Corolla campanulate; lobes oblong to narrowly ovate, (3–)5–6 × 2–2.5 mm, subacute, adaxial surface pale green or cream with pinkish tinge, glabrous or minutely papillose, abaxial face reddish purple, pubescent. Gynostegium with stipe 0.7–1.5 mm long. Corona lobes attached to and obscuring the gynostegial stipe, ± 1.5 mm tall, spreading from the column then curved upwards, somewhat pouched with a thickened fleshy midline and two short rounded auricles apically, cream tinged with red. Anther wings 1 mm long but parallel grooved part (guide rails) only 0.2 mm long, the remainder flared and curved below the head of the gynostegium; corpusculum brown, ovoid-subcylindrical, ± 0.25 × 0.1 mm; translator arms ± 0.4 mm long, flattened but very slender (± 0.05 mm) and strongly geniculate, with distinct proximal and distal sections; pollinia triangular in outline, 0.3–0.4 × 0.3 mm, somewhat flattened. Stylar head flat or domed. Follicles (*Leedal* 5301; *Pawek* 9330) held erect on elongated peduncle, fusiform, to ± 9 cm long, slender, minutely pubescent. Seeds ovoid in outline, ± 3 × 2 mm, with one plane face and one strongly convex, sparsely verrucose and with a narrow marginal rim; coma ± 1.5 cm long.

a. subsp. **breviantherae**

Sepals 2.5–3 mm long. Corolla lobes 5–6 × 2–2.5 mm, oblong to narrowly ovate. Gynostegium with stipe ± 1–1.5 mm long, raising the anthers above the level of the corona. Stylar head domed, extending ± 1 mm beyond top of anthers but obscured beneath the conspicuous membranous anther appendages.

TANZANIA. Rungwe District: Poroto ridge road, 11 Nov. 1966, *Gillett* 17742!; Njombe District: slopes of Matamba Mountains, 22 Nov. 1986, *Goldblatt et al.* 8195!; Mbeya District: summit of Mbeya Peak, 11 Feb. 1978, *Leedal* 4916!
DISTR. **T** 7 (Kitulo, Poroto and Mbeya massifs); Nyika Plateau of northern Malawi
HAB. Thin peaty soil over rock in seasonally burnt montane grassland; (2100–)2400–2800 m

NOTE. The reduced fleshy corona is reminiscent of that shown by *Asclepias minor*. The translator arms are very slender and articulated, perhaps indicating a transitional relationship to *Stathmostelma*. The stipitate gynostegium has very short guide rails between adjacent anthers, the anther wings then curving beneath the stylar head. Characters to distinguish this species from *A. mtorwiensis* are listed under that species.

b. subsp. **minor** *Goyder* in K.B. 64: 391 (2009). Type: Tanzania, Njombe District, Kitulo [Elton] Plateau, Ndumbe River, *Richards* 7685 (K, holo.)

Sepals ± 2 mm long. Corolla lobes ± 3 mm long, ovate. Gynostegial stipe ± 0.7 mm long, the corona therefore reaching the top of the staminal column. Stylar head ± flat, not protruding from the top of the staminal column.

TANZANIA. Njombe District: Kitulo [Elton] Plateau, Ndumbe River, 11 Jan. 1957, *Richards* 7685!; Mbeya District, Kitulo [Elton] Plateau, 24 Jan. 1961, *Richards* 14153B! & 1 Dec. 1963, *Richards* 18510!

DISTR. **T** 7; endemic to the Kitulo Plateau
HAB. Montane grassland; 2100–2400 m

NOTE. These three collections from the Kitulo Plateau, two from the Mbeya side and one from the Njombe side have a very different facies to collections of the type subspecies. However, the principal differences amount essentially to smaller flowers with a shorter gynostegial stipe and a less domed stylar head. As the corona is unchanged, this gives the flower a very different appearance, as the gynostegium is largely obscured by the corona lobes.

These collections are reported from slightly lower altitude than most records of subsp. *breviantherae*, and the specific habitat (shallow soil over rock) of that subspecies is not mentioned. However, Mary Richards' labels, if they are to be believed, imply that on one occasion she collected both subspecies together (*Richards* 14153A & B), at an altitude of ± 2100 m. It is always possible that this was not the case and that the larger flowered plants (*Richards* 14153A) were collected at a higher elevation than the small-flowered ones (*Richards* 14153B), perhaps supplementing collections made earlier in the day, in the belief that this was the same taxon.

Given the somewhat superficial nature of the floral variation, and the suggestion of altitudinal separation, it seems most appropriate to recognise these taxa at the rank of subspecies.

17. **Asclepias mtorwiensis** *Goyder* in K.B. 64: 391 (2009). Type: Tanzania, Njombe District: slopes of Mtorwi Mt above Ndumbi Forest Reserve, *Brummitt, Congdon & Mwasumbi* 18142 (K!, holo.)

Perennial herb with a single annual stem arising from a tuberous rootstock (not seen); stems erect, 8–12 cm long, simple, reddish, minutely pubescent with indumentum of white hairs. Leaves subsessile, semi-succulent; lamina narrowly linear, 5–10 × 0.2–0.3 cm, attenuate both apically and basally, margins revolute, glabrous. Inflorescences extra-axillary forming umbels of 5–9 spreading or erect flowers; peduncles reddish, 0.5–1 cm long, minutely pubescent; bracts filiform, pubescent; pedicels 0.5–0.7 cm long, pubescent. Sepals narrowly triangular, ± 1.5 mm long, glabrous. Corolla campanulate or spreading, perhaps even weakly reflexed; lobes ovate-oblong, ± 3 × 2 mm, subacute, adaxial surface cream with reddish tinge, glabrous, abaxial face reddish green, glabrous. Corona lobes arising from the base of the staminal column, ± 1.5 mm tall and reaching ± the top of the column, pouched and somewhat fleshy, with three short rounded teeth apically, and two rounded lobes on the lateral margins, cream. Anther wings 0.4 mm long; corpusculum brown, ovoid-subcylindrical, ± 0.25 × 0.1 mm; translator arms ± 0.2 mm long, very slender, not geniculate; pollinia oblong-elliptic, ± 0.3 × 0.2 mm, somewhat flattened. Stylar head ± flat. Follicles and seeds not seen.

TANZANIA. Njombe District: Kipengere mountains, slopes of Mtorwi Mt above Ndumbi Forest Reserve, 23 Nov. 1986, *Brummitt et al.* 18142!
DISTR. **T** 7; known only from the type
HAB. Open rocky hillside; ± 2700 m

NOTE. The pouched, fleshy corona is similar to that of *A. breviantherae*, but could also be interpreted as a reduced form of *Asclepias amabilis*. Other differences from *A. breviantherae* include the glabrous leaves and corolla, the gynostegium which is less stipitate, the different form of the anther wings, which are 0.4 mm long and form guide rails for their entire length, rather than for only 0.2 mm as in *A. breviantherae* where they then become flared and curved beneath the head of the gynostegium. The translator arms are filiform and not conspicuously geniculate.

18. **Asclepias alpestris** (*K. Schum.*) *Goyder* in K.B. 64: 392 (2009). Type: Tanzania, Njombe District: Yawuanga Mt, Ukinga [Kinga] range, *Goetze* 1234 (K!, lecto., designated by Goyder in K.B. 64: 392 (2009); B†, BR!, E!, isolecto.)

Perennial herb with one to many annual stems arising from a stout horizontal or vertical tuber; stems ascending or erect, 6–12(–20) cm long, simple or branched below, reddish, minutely pubescent. Leaves subsessile or shortly petiolate; lamina narrowly linear, 3–8 × 0.1–0.4 cm, attenuate both apically and basally, glabrous or with minute indumentum restricted to margin and midrib. Inflorescences extra-axillary forming umbels of 3–6 spreading or erect flowers; peduncles reddish, 0.3–3 cm long, minutely but densely pubescent; bracts filiform, pubescent; pedicels 0.4–1 cm long, minutely pubescent. Sepals lanceolate or narrowly oblong, 2–4 × ± 0.5 mm, acute, pubescent. Corolla rotate; lobes ovate-oblong, 3–4 × 2 mm, subacute, cream or white and glabrous or minutely papillate adaxially, abaxial face streaked with reddish purple above, glabrous. Corona lobes arising at the base of the staminal column, reddish,± 4 mm long, outer face ovate near base but drawn out into a long-attenuate tip, inner face with an erect, cream ligule ± 1 mm long, reaching ± the top of the staminal column, the tip attenuate and recurved. Anther wings 0.5 mm long; corpusculum brown, subcylindrical, ± 0.3 × 0.1 mm; translator arms ± 0.1 mm long, very slender but broadening distally to form a clasping overlap with the pollinia; pollinia narrowly oblong, ± 0.4 × 0.1 mm, somewhat flattened. Stylar head ± flat. Follicles erect, narrowly fusiform, 7–10 × 0.6 cm, minutely pubescent, only one of the pair developing. Seeds not seen.

TANZANIA. Njombe District: top of Kitulo Plateau S of Matamba, 2 Dec. 1994, *Goyder et al.* 3892! & slopes of Matamba Mt, 22 Nov. 1986, *Goldblatt et al.* 8197! & Kitulo Plateau, crystalline ridges, 12 Dec. 1989, *Lovett et al.* 3673!
DISTR. **T** 7; endemic to the upper Kitulo Plateau and Kipengere mountains
HAB. Montane grassland on shallow soil around rocky outcrops; 2600–2900 m

SYN. *Schizoglossum alpestre* K. Schum. in E.J. 30: 384 (1901)

NOTE. Kupicha (1984: 669) excluded this species from the *Schizoglossum* complex despite the well-developed ligule on the inner face of the corona lobe – normally a good indicator for *Schizoglossum*. The pollinia, however, lack a transparent germination zone, which would be expected in that genus. Vegetatively and in other floral characters, this species appears closely allied to the *Asclepias minor/breviantherae* group.

19. **Asclepias randii** *S. Moore* in J.B. 40: 255 (1902). Type: Zimbabwe, Harare [Salisbury], *Rand* 194 (BM!, holo.; K! (fragment and drawing), SRGH!, iso.)

Perennial herb with one to many annual stems arising from a vertical napiform tuber; stems ascending or erect, 6–20 cm long, simple or branched below, reddish, pubescent with spreading white hairs to ± 1 mm long. Leaves subsessile; lamina narrowly linear, 2.5–10 × 0.1–0.5 cm, attenuate both apically and basally, with sparse indumentum of spreading white hairs especially on the margins and midrib. Inflorescences extra-axillary forming umbels of 3–5 erect flowers; peduncles 1–3 cm long, densely pubescent; bracts filiform, pubescent; pedicels 0.7–1 cm long, densely pubescent. Sepals lanceolate or narrowly oblong, 3–4 × 0.5–1 mm, acute, densely spreading-pubescent. Corolla rotate to partially reflexed; lobes oblong to broadly ovate, 4–6 × 2.5–3.5 mm, acute, adaxial surface greenish white tinged with red or brown, glabrous, abaxial face green tinged brown, spreading-pubescent with white hairs. Corona lobes arising at the base of the staminal column, (4–)5–8 mm long, ± cucullate basally with upper margins reaching top of column, the mid-line drawn out into a long-attenuate tip, acute or rounded apically, off-white with a reddish fleshy midline, or the entire lobe mottled red, cavity and mid-line of lobe papillose. Anther wings 1.3–1.5 mm long; corpusculum brown, subcylindrical, ± 0.3 × 0.1 mm, with narrow translucent flanges down the sides; translator arms ± 0.3 mm long, geniculate, broadening distally to form a clasping overlap with the pollinia;

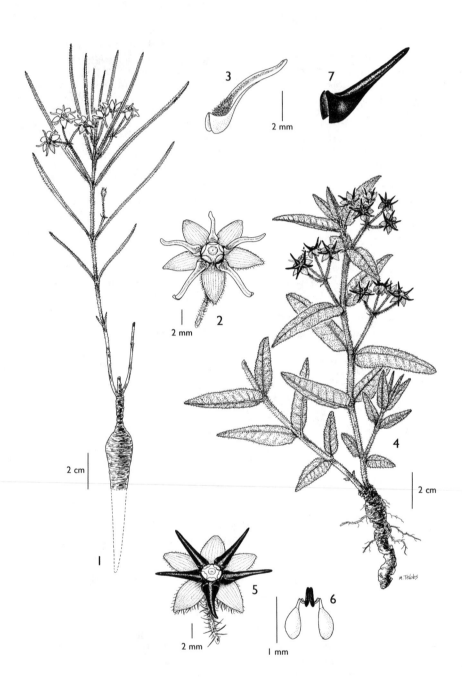

Fig. 98. *ASCLEPIAS RANDII* — **1**, habit; **2**, flower; **3**, corona lobe. *ASCLEPIAS GRANDIRANDII* — **4**, habit; **5**, flower; **6**, pollinarium; **7**, corona lobe. 1–3 from *Goyder et al.* 3854; 4–7 from *Robson* 373. From K.B. 64: 393, drawn by M. Tebbs.

pollinia obtriangular, ± 0.7 × 0.5 mm, somewhat flattened. Stylar head ± flat. Follicles erect, narrowly fusiform, 5–8 × 0.6 cm, minutely but densely pubescent, one or both of the pair developing, the peduncle lengthening in fruit. Seeds ± 5 × 4 mm, ± flattened, verrucose and with a slightly inflated marginal rim ± 0.5 mm wide. Fig. 98/1–3, p. 402.

TANZANIA. Mbeya District: Mbeya Peak, 15 km from turn-off, 29 Nov. 1994, *Goyder et al.* 3854!; Njombe District: Njombe–Igawa road, 4 Feb. 1961, *Richards* 14224! & Livingstone Mountains, on ridge 1.5 km S of Msalaba Mt, 23 Nov. 1992, *Gereau et al.* 5139!
DISTR. **T** 7; S and W Congo-Kinshasa, Angola, Zambia, N Malawi (Nyika Plateau), Zimbabwe, N Namibia
HAB. Montane grassland and seasonally waterlogged 'dambo' grasslands; (1000–)1500–2800 m

20. **Asclepias grandirandii** *Goyder* in K.B. 64: 394 (2009). Type: Malawi, Nyika Plateau, near Chelinda CDC Camp, *Robson* 373 (K!, holo.)

Perennial herb with one to many annual stems arising from a vertical napiform tuber; stems prostrate to ascending or erect, 10–30 cm long, simple or branched below, green, conspicuously pubescent with spreading white hairs. Leaves subsessile; lamina oblong, triangular-oblong or elliptic, 3–5 × 0.5–2 cm, apex acute, base rounded, truncate or weakly cordate, with sparse indumentum of spreading white hairs especially on the margins and midrib. Inflorescences extra-axillary forming umbels of 3–5 erect flowers; peduncles 1–3 cm long, densely pubescent; bracts filiform, pubescent; pedicels 1–2 cm long, densely pubescent. Sepals linear, lanceolate or narrowly triangular, 3–6 × 0.5–2 mm, acute, densely spreading-pubescent. Corolla partially reflexed; lobes oblong to broadly ovate, 5–8 × 3–4 mm, acute, adaxial surface greenish white, glabrous, abaxial face green tinged brown or purple, spreading-pubescent with white hairs. Corona lobes arising at the base of the staminal column, 5–7 mm long, ± cucullate basally with upper margins reaching top of column, the mid-line drawn out into a long-attenuate tip, acute or rounded apically, off-white with a reddish fleshy midline, or the entire lobe mottled red, cavity and mid-line of lobe papillose. Anther wings (1.4–)1.5–2 mm long; corpusculum brown, subcylindrical, 0.35–0.4 × 0.1 mm, with narrow translucent flanges down the sides; translator arms ± 0.5 mm long, geniculate and broadening distally to form a clasping overlap with the pollinia; pollinia obtriangular, ± 0.9 × 0.5 mm, somewhat flattened. Stylar head ± flat. Follicles (immature) erect, fusiform, 6 cm long, densely pubescent; seeds not seen. Fig. 98/4–7, p. 402.

TANZANIA. Ufipa District: Chapota, 4 Dec. 1949, *Bullock* 2033!
DISTR. **T** 4; SE Congo-Kinshasa (Mitumba range between Kolwezi and Kalemia), NE Zambia, Malawi (Nyika and Vipya Plateaus)
HAB. Seasonally burnt upland grassland or woodland; 1400–2200 m

NOTE. This is essentially a broad-leaved version of *Asclepias randii* with longer pedicels and slightly larger flowers. It is at least partially sympatric with *A. randii*, and although the differences seem slight, the overall appearance of the plant is very different – the leaves tend to be shorter and broader in this species than in *A. randii*, and the leaf base is rounded to truncate or weakly cordate, but never attenuate as in the latter species. The geographic range of this species is somewhat narrower than *A. randii*. Pubescence is conspicuous, but the 1 mm long hairs of *A. randii* were not observed.

21. **Asclepias pygmaea** *N.E. Br.* in K.B. 1895: 255 (1895). Type: Tanzania, Rungwe/Njombe District: lower plateau N of Lake Malawi [Lake Nyassa], *Thomson s.n.* (K!, holo.)

Perennial herb with one to several annual stems arising from a vertical napiform tuber; stems ascending or erect, 6–20 cm long, simple or branched below, reddish, minutely pubescent with short white hairs. Leaves subsessile; lamina narrowly linear,

3–6 × 0.1–0.2 cm, attenuate both apically and basally, with sparse indumentum of very short white hairs mostly on the margins and midrib. Inflorescences extra-axillary forming umbels of 3–5 erect flowers; peduncles 1–4 cm long, minutely but densely pubescent; bracts filiform, pubescent; pedicels 0.5–1.3 cm long, minutely pubescent. Sepals lanceolate or narrowly oblong, 2–3 × 0.5 mm, acute, densely spreading-pubescent. Corolla somewhat reflexed; lobes oblong to broadly ovate, 3.5–4 × 1.5–2.5 mm, acute, adaxial surface greenish white tinged with pink, glabrous or minutely papillate, abaxial face green or purplish, spreading-pubescent with white hairs. Corona lobes arising from the top of a minute stipe ± 0.5 mm tall at the base of the staminal column, 3–3.5 mm long, ± cucullate basally with upper margins reaching top of column, the mid-line drawn out into a long-attenuate tip, acute or rounded apically, off-white with a reddish fleshy midline, or the entire lobe mottled red, cavity and mid-line of lobe papillose. Anther wings 1–1.5 mm long; corpusculum brown, subcylindrical, ± 0.3 × 0.1 mm, with short translucent flanges down the sides; translator arms ± 0.3 mm long, proximal part spreading from the corpusculum, differentially thickened, distal part of uniform thickness, curving down abruptly towards pollinia; pollinia obtriangular, ± 0.7 × 0.3 mm, strongly flattened. Stylar head ± flat. Follicles and seeds not seen.

TANZANIA. Iringa District: Sao, 20 July 1933, *Greenway* 3428! & Mufindi, 28 Sep. 1968, *Harris & Paget-Wilkes* 2368!; Njombe District: Njombe–Iringa, Aug. 1965, *Beecher* 97!
DISTR. **T** 7; Nyika Plateau in Malawi
HAB. Montane grassland; 1800–2500 m

NOTE. *Asclepias pygmaea* can be easily confused with *A. randii*. *A. pygmaea* tends to have a more erect habit than *A. randii*, and stem and leaf indumentum is of short curled hairs not the stiff spreading hairs of *A. randii*. The flowers are slightly smaller and the corona has a less drawn-out apex, but are otherwise very similar. However, the pollinarium differs in having clearly differentiated proximal and distal parts to the translator arms, very slender and ± terete distally. The translator arms of *A. randii* are flattened, and broaden gradually towards the pollinia.

74. **MARGARETTA**

Oliv. in Trans. Linn. Soc., Bot. 29: 111, t. 76 (1875); Goyder in K.B. 60: 87–94 (2005)

Perennial herbs with annual stems arising from a slender to stout vertical tuber; latex milky; stems usually simple or branched, erect or ascending. Leaves opposite, subsessile; lamina linear to narrowly oblong. Inflorescences terminal and extra-axillary, umbels erect, borne on long or short peduncles. Calyx lobed to the base. Corolla deeply lobed, apex of lobes often revolute, much reduced and smaller and less conspicuous than the petaloid corona. Corona of 5 petaloid lobes arising above the base of the staminal column; erect and somewhat fleshy, the claw cucullate at least at the base, the inner margins generally produced into a pair of inward- or upward-pointing teeth, with or without a tooth arising from the middle of the cavity, the limb expanded into a flattened oblong to orbicular lobe with entire to dentate margins. Corpusculum ovoid to subcylindrical, black; translator arms short, broad and differentially thickened; pollinia pendant in the anther cells. Stylar head flat. Fruiting pedicel not contorted; follicles single, erect, ovoid to fusiform. Seeds flattened, ovate with a verrucose disc and a narrow inflated rim; coma of silky hairs.

A single, widely distributed species across tropical Africa with eight, regionally localised subspecies.

Margaretta rosea *Oliv.* in Trans. Linn. Soc., Bot. 29: 111, t. 76 (1875); Bullock in F.W.T.A. ed. 2, 2: 91 (1963); U.K.W.F. ed. 2: 179, t. 70 (1994). Type: Uganda, Bunyoro [Unyoro], *Speke & Grant* 531 (K!, holo.)

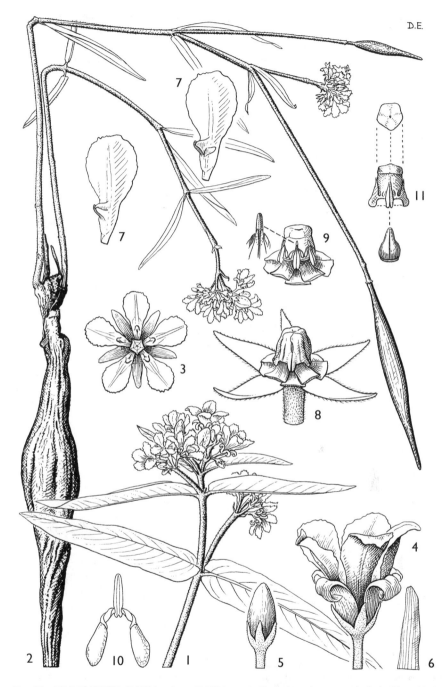

FIG. 99. *MARGARETTA ROSEA* subsp. *ROSEA* — **1**, flowering shoot, × ²/₃; **2**, habit, with tuberous roostock and follicles, × ²/₃; **3**, flower from above, × 1; **4**, flower from the side, × 2; **5**, flower bud, × 2; **6**, corolla lobe, × 2; **7**, corona lobes, × 2; **8**, flower with both corolla and corona lobes removed to expose gynostegium, × 3; **9**, as 8 but with stamens partially removed to show position of pollinaria, × 3; **10**, pollinarium, × 14; **11**, gynoecium showing paired ovaries, the stylar head and apex of stylar head from above, × 3. 1 from *Richards* 195; 2-11 from *Bullock* 1867. Drawn by D. Erasmus.

Perennial herbs with annual shoots arising from a carrot-shaped tuber; stems erect, to ± 60 cm. Leaves lanceolate to narrowly oblong, to ± 19 × 2.3 cm, subcordate at the base, sparsely pubescent on both faces. Inflorescences umbelliform, with 3–10 flowers; peduncles to 7.5 cm long, erect, densely pubescent; pedicels to 1.7 cm long, pubescent. Calyx lobes 2.5–6 mm long, pubescent. Corolla lobed almost to the base; lobes lanceolate, 4–10 × 1–3.5 mm, often revolute apically, glabrous adaxially, abaxial face pubescent. Corona highly variable, petaloid and often brightly coloured, the basal claw with two marginal teeth, a third central tooth present or absent; limb expanded into an entire or toothed dorsiventrally flattened lobe. Fruit narrowly ovoid-fusiform, 4–8 cm long, ± 0.5 cm across. Seeds ovate, ± 5 × 2 mm; coma ± 2 cm long.

Very closely allied to *Stathmostelma*, with which it is almost certainly congeneric in any of the likely future generic realignments. Patterns of morphological variation are strongly geographic. None of the characters are absolutely constant, and there are always a few specimens in which one or other of the diagnostic characters is absent. The shape of the corona limb is particularly variable in subsp. *cornetii*. Nevertheless, few individuals are hard to place and the forms can be successfully keyed out. As the forms generally replace each other geographically, subspecific rank is the most appropriate rank at which to recognise these taxa formally.

In using the following key to subspecies, it may be helpful to consult Fig. 100, p. 407.

1. Corona with 3 teeth on the claw, 2 lateral ones and a
   third centrally 2
   Corona with 2 teeth on the claw; central tooth absent
   or vestigial 4
2. Corona mauve; claw of corona lobes generally ± same
   width as limb; SE Tanzania (Fig. 100.3) . . . .   c. subsp. *whytei*
   Corona red, orange or yellow; claw of corona lobes
   distinctly narrower than limb 3
3. Margins of limb mostly deeply incised all round;
   western Kenya, Uganda and NW Tanzania (Fig. 100.4) a. subsp. *rosea*
   Margins of limb entire or shallowly and irregularly
   toothed or lobed; SW Tanzania (Fig. 100.5) . . . . . .   b. subsp. *corallina*
4. Corona predominantly orange or yellow; N and E
   Tanzania, SE Kenya 5
   Corona violet, purple, magenta or white; SW Tanzania 6
5. Leaves mostly less than 6 cm long, ovate or lanceolate;
   Kilimanjaro region . . . .   e. subsp. *kilimanjarica*
   Leaves mostly 7–12 cm long, narrowly lanceolate;
   Usambaras and Morogoro region (Fig. 100.2) . . . .   d. subsp. *bidens*
6. Limb of corona ± orbicular, narrowing abruptly into
   the claw; SW Tanzania (Fig. 100.1) . . . .   f. subsp. *orbicularis*
   Limb of corona oblong or obovate, tapering gradually
   into the claw; SW Tanzania . . . .   g. subsp. *cornetii*

a. subsp. **rosea**

Corolla pubescent abaxially; adaxial face glabrous to papillose. Corona with 3 teeth on the claw, these teeth frequently further incised or divided in material from Kenya W of the rift valley, claw narrower than the limb, margins of limb mostly deeply incised all round, corona very variable in colour – red, orange or yellow. Figs. 99 & 100/4, pp. 405 & 407.

UGANDA. Karamoja District: Amaler, at foot of Mt Debasien, Jan. 1936, *Eggeling* 2538!; Ankole District: near Rwashamaire [Lwazamaire], 15 Aug. 1938, *A.S. Thomas* 2373A!; Teso District: Serere, Apr. 1931, *Chandler* 631!

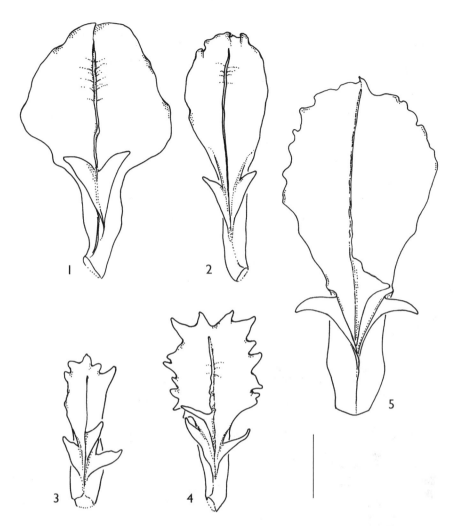

Fig. 100. *MARGARETTA ROSEA* — variation in corona lobes — **1**, subsp. *orbicularis*; **2**, subsp. *bidens*; **3**, subsp. *whytei*; **4**, subsp. *rosea*; **5**, subsp. *corallina*. Scale bar: 3 mm. 1 from *Stolz* 171; 2 from *Volkens* 220; 3 from *Agnew* 50; 4 from *Langdale-Brown* 2064; 5 from *Lea* LV6. Adapted from K.B. 51: 720, drawn by S. Dawson.

KENYA. Trans-Nzoia District: S Cherangani, 18 Feb. 1958, *Symes* 280!; North Kavirondo District: Kakamega Forest, near Forest Station, 12 Oct. 1981, *Gilbert & Mesfin* 6641!; Kisumu-Londiani District: Tinderet Forest Reserve, 2 July 1949, *Maas Geesteranus* 5324!
TANZANIA. Mpanda District: Mahali Mountains, Utahya, 21 Aug. 1958, *Newbould & Jefford* 1702!
DISTR. U 1–4; **K** 3, 5; **T** 4; Central African Republic, NE Congo-Kinshasa, Rwanda, Burundi
HAB. Burnt, sometimes seasonally flooded grassland; (450–)700–2000 m

b. subsp. **corallina** *Goyder* in K.B. 60: 90 (2005). Type: Tanzania, Ufipa District: 4 km W of Mkundi on road to Kamwanga, 22 Nov. 1994, *Goyder, Griffiths, Harvey, Kamwela & Paton* 3791 (K!, holo.; BR!, DSM!, EA!, PRE!, WAG!, iso.)

Corolla pubescent abaxially; adaxial face glabrous to papillose. Corona with 3 teeth on the claw, claw narrower than the limb, margins of limb shallowly toothed, corona red or deep pink. Fig. 100/5.

TANZANIA. Ufipa District: Mbisi Forest Reserve, 6 Nov. 1987, *Ruffo & Kisena* 2617! & 1 km S of Mpanda road on Sumbawanga–Mbala highway, 27 Oct. 1992, *Gereau et al.* 4916! & Mumba Mission, SE of Sumbawanga, 24 Nov. 1994, *Goyder et al.* 3806!
DISTR. **T** 4; restricted to the Ufipa Plateau and areas surrounding Sumbawanga
HAB. Burnt montane grasslands, sometimes on black peaty soils; 1800–2300 m

c. subsp. **whytei** (*K. Schum.*) *Mwanyambo* in K.B. 51: 726 (1996). Type: Malawi, Mulanje, *Whyte* 106 (K!, lecto., designated by Goyder in K.B. 60: 91 (2005); B†, isolecto.)

Corolla pubescent abaxially in southern populations, glabrous in more northerly populations; adaxial face glabrous to papillose. Corona with 3 teeth on the claw, claw of corona lobes ± same width as limb, corona mauve, generally smaller than in other subspecies. Fig. 100/3, p. 407.

TANZANIA. Songea District: 40 km W of Songea by Mbamba Bay Road, 13 Jan. 1956, *Milne-Redhead & Taylor* 8327!
DISTR. **T** 8; S Malawi, Mozambique and adjacent parts of Zimbabwe
HAB. Open *Brachystegia/Uapaca* woodland; 300–1500 m

SYN. *M. whytei* K. Schum. in P.O.A. C: 323 (1895)

d. subsp. **bidens** *Bullock* in K.B. 7: 413 (1952). Type: Tanzania, Lushoto District: Usambara, Mlalo, *Holst* 159 (B†, holo.; K!, iso.)

Leaves mostly 7–12 cm long, narrowly lanceolate. Corolla glabrous. Corona with 2 teeth on the claw, claw of corona lobes narrower than the limb, margins of limb sometimes obscurely toothed, corona predominantly orange or yellow. Fig. 100/2, p. 407.

TANZANIA. Handeni District: 32 km S of Handeni, 5 Apr. 1953, *Drummond & Hemsley* 2036!; Morogoro District: Turiani, 26 Mar. 1953, *Drummond & Hemsley* 1804!; Iringa District: Udzungwa National Park, Camp 210, 23 Sep. 2001, *Luke et al.* 7705!
DISTR. **T** 3, 6, 7; not known elsewhere
HAB. Open woodland and wooded grassland, sometimes on black soils; (50–)500–1500 m

SYN. *Margaretta holstii* K. Schum. in E.J. 17: 133 (1893). Type as for *M. rosea* subsp. *bidens*

e. subsp. **kilimanjarica** *Goyder* in K.B. 60: 91 (2005). Type: Kenya, Laitokitok [Loitokitok], *Napier* 451 (K!, holo.; EA!, iso.)

Leaves less than 6 cm long, ovate or lanceolate. Corolla glabrous or pubescent. Corona with 2 teeth on the claw, claw of corona lobes narrower than the limb, margins of limb entire, corona predominantly orange or yellow, apparently deeper in colour than subsp. *bidens*.

KENYA. Masai District: Laitokitok, Kenya/Tanzania borders, 10 Oct. 1930, *Napier* 451!
TANZANIA. Arusha District: Ngurdoto National Park, Lesokonoi, 27 Feb. 1968, *Richards* 23119! & Arusha National Park, Sakila swamp near Ngurdoto Crater, 14 Dec. 1968, *Richards* 23348!; Moshi District: Oldeani, 10 Nov. 1932, *Geilinger* 3509!
DISTR. **K** 6; **T** 2; Kilimanjaro region only
HAB. Burnt, seasonally waterlogged grassland; ± 1500 m

SYN. *M.* sp. A of U.K.W.F. ed. 2: 179 (1994)

f. subsp. **orbicularis** (*N.E. Br.*) *Goyder* in K.B. 60: 91 (2005). Type: Malawi, North Nyassa, Elephant Marsh, *Scott* s.n. (K!, lecto., designated by Goyder in K.B. 60: 91 (2005))

Corolla glabrous to sparsely pubescent. Corona with 2 teeth on the claw, limb of corona ± orbicular, narrowing abruptly into the claw, corona violet, purple, magenta or white. Fig. 100/1, p. 407.

TANZANIA. Rungwe District: Kyimbila, 2 Apr. 1907, *Stolz* 171! & Kaparogwe Falls, 2 Feb. 1970, *Fuller* 32! & Kiwira, 8 Dec. 1932, *Davies* 782!
DISTR. **T** 7; SE Congo-Kinshasa, N and C Zambia and Malawi

HAB. Burnt, sometimes seasonally waterlogged grassland; 400–1600 m

SYN. *M. orbicularis* N.E. Br. in K.B. 1895: 256 (1895)
    M. pulchella Schltr. in R.E. Fries, Wiss. Ergebn. Schwed. Rhod. Kongo Exped., 1: 266, f. 32,
    t. 18, f. 3 (1916). Type: Zambia, Bwana Mkubwa, *Fries* 491 (UPS!, holo.)

    g. subsp. **cornetii** (*Dewèvre*) *Goyder* in K.B. 60: 92 (2005). Type: Congo-Kinshasa, Katanga, *Cornet* s.n. (BR!, holo.)

Corolla densely pubescent abaxially. Corona with 2 teeth on the claw, violet, purple, magenta or white, limb of corona oblong or obovate, tapering gradually into the claw.

TANZANIA. Ufipa District: Mchata Mountain, 32 km S of Sumbawanga, 27 Nov. 1958, *Napper* 1124!
DISTR. **T** 4; SE Congo-Kinshasa and Zambia
HAB. Burnt grassland; 1000–1800 m

SYN. *M. cornetii* Dewèvre in B.S.B.B. 34(2): 90 (1895)
    M. cornetii Dewèvre var. *pallida* De Wild., Etudes Fl. Katanga 1: 108 (1903). Type: Congo-
    Kinshasa, Lukafu, *Verdick* 133 (BR!, holo.)
    M. verdickii De Wild. in Ann. Mus. Congo sér 5, 1: 183 (1904). Type: Congo-Kinshasa,
    Lukafu, *Verdick* 148bis (BR!, holo.)
    M. decipiens Schltr. in R.E. Fries, Wiss. Ergebn. Schwed. Rhod. Kongo Exped. 1: 265 (1916).
    Type: Zambia, Kamindas, near Lake Bangweulu, *Fries* 987 (UPS!, holo.)

75. **STATHMOSTELMA**

K. Schum. in E.J. 17: 129 (1893) & in E. & P. Pf. 4(2): 239 (1895); Goyder in K.B.
53: 577–616 (1998)

Perennial herbs with annual stems arising from a slender to stout vertical tuber; latex milky; stems 1 to many, simple or branched, erect or ascending, glabrous or pubescent with short, spreading hairs. Leaves opposite, sessile or petiolate. Inflorescences terminal, extra-axillary or both, umbels erect, borne on long or short peduncles. Calyx lobed to the base. Corolla rotate to strongly reflexed, occasionally broadly campanulate. Corolline corona absent. Gynostegium sessile or with stipe to 3 mm long. Gynostegial corona of 5 lobes arising above the base of the staminal column; erect and somewhat fleshy, frequently weakly pouched towards the base and concave-cucullate for at least some of its length, the inner apical margins generally produced into a pair of inward- or upward-pointing teeth, with or without a tooth arising from the middle of the cavity, glabrous except along the midline of the cavity towards the base of the lobe. Anther wings curved, with or without a contorted basal tail. Corpusculum ovoid to subcylindrical or occasionally subglobose, black; translator arms with a membranous, convex and somewhat contorted proximal part, and a filiform distal part broadening abruptly into a short clasping overlap with the pollinium; pollinia flattened, oblong, pendant in the anther cells. Stylar head flat. Fruiting pedicel not contorted; follicles single, erect, ovoid to fusiform, beaked, occasionally inflated but generally not, smooth or with weak longitudinal ridges, glabrous to densely pubescent. Seeds flattened, ovate, with a verrucose disc and a narrow inflated rim; coma of silky hairs.

17 somewhat critical taxa in 13 species. Distributed mostly in eastern parts of tropical Africa, growing mostly at middle altitudes from Ethiopia in the north to Mozambique and Zimbabwe in the south. Very closely allied to *Margaretta*.

1. Leaves at least 4 cm wide . . . . 1. *S. spectabile*
   Leaves not more than 3 cm wide 2

2. Anther wings with contorted basal tails; rootstock a
   narrowly cylindrical tuber; inflorescences mostly terminal   11. *S. pauciflorum*
   Anther wings without contorted basal tails; rootstock a
   napiform to globose tuber; inflorescences mostly extra-
   axillary 3
3. Corona lobes white or greenish, frequently speckled
   purple 4
   Corona lobes red, orange or yellow 6
4. Corona lobes with a laterally flattened tooth within the
   cavity . . . . 10. *S. diversifolium*
   Corona lobes without a tooth within the cavity 5
5. Gynostegium sessile, corona lobes arising at the base of
   the staminal column; anther wing tails adnate to
   column at base; corolla lobes less than 15 mm long . . .   9. *S. angustatum*
   Gynostegium stipitate, corona lobes arising 1–3 mm
   above base of the staminal column; anther wing tails
   proud of column at base; corolla lobes generally
   longer than 20 mm . . . .   8. *S. gigantiflorum*
6. Corolla red or pink 7
   Corolla green or yellow 9
7. Anther wings 1.5 mm long; corona lobes lacking a
   tooth within the cavity; winged part of translator
   arms ± 0.3 mm long . . . .   3. *S. stipitatum*
   Anther wings at least 2 mm long; corona lobes with a
   well-developed tooth in the cavity; winged part of
   translator arms at least 1 mm long 8
8. Inner apical margins of corona lobes produced into a
   pair of erect teeth angled away from the column; tooth
   within cavity of lobe 3–4 mm long . . . .   4. *S. rhacodes*
   Inner apical margins of corona lobes produced into a
   pair of teeth angled upwards or towards the column;
   tooth within cavity of lobe to 2.5 mm long . . . .   2. *S. pedunculatum*
9. Corolla lobes at least 14 mm long; plant generally little
   branched . . . .   5. *S. fornicatum*
   Corolla lobes less than 12 mm long; plant usually well
   branched or with many stems from the base 10
10. Corolla yellow; corona lobes only marginally longer than
   column; leaves minutely pubescent . . . .   6. *S. welwitschii*
   Corolla green; corona lobes 1.5 times height of column;
   leaves densely pubescent . . . .   7. *S. propinquum*

   1. **Stathmostelma spectabile** (*N.E. Br.*) *Schltr.* in E.J. 51: 138 (1913). Type: Malawi, *Buchanan* 553 (K!, lecto., designated by Bullock in K.B. 15: 196 (1961))

   Perennial herb with several annual stems arising from a slender or stout vertical tuber to 2 m long; stems erect, 0.5–1.5 m long, generally unbranched, densely pubescent with short white hairs. Leaves with petiole 5–20(–30) mm long, pubescent; lamina oblong or elliptic to ovate, 10–24 × 4–12 cm, apex acute or obtuse, base rounded to truncate, occasionally somewhat cuneate, both upper and lower surfaces with a dense indumentum of spreading white hairs. Inflorescences extra-axillary with 4–12(–15) flowers in an erect umbel; peduncles erect, (1–)3–10 cm long, densely spreading-pubescent; pedicels 2–5 cm long, pubescent. Sepals linear-lanceolate, (4–)8–16 × 1–2 mm, acute, densely pubescent. Corolla rotate to broadly campanulate, glabrous or pubescent at least towards tip on outer surface, minutely papillose within; lobes ovate to oblong or oblanceolate, 15–20(–30) × 6–10 mm, acute or obtuse. Gynostegium shortly stipitate. Corona lobes arising just above the base of the staminal column and adnate to

FIG. 101. *STATHMOSTELMA SPECTABILE* subsp. *SPECTABILE* — **1**, flowering shoot, × $^2/_3$;
2, follicles, × $^2/_3$; **3**, flower one corona lobe displaced, × 3; **4**, half-flower, × 3; **5**, gynoecium,
× 4; **6**, stamens from within, showing the pollinaria in situ, × 4; **7**, pollinarium, × 10. 1 from
Bax 184; 2–7 from *Burtt* 971. Drawn by D. Erasmus.

it to the base of the anthers, erect and somewhat fleshy, the free part longer than the staminal column, 8–12 × 2–4 mm, weakly pouched towards the base and concave-cucullate for most of its length, apex rounded, the inner apical margins produced into a pair of inward-pointing teeth 2–3 mm long, with or without a tooth arising from the middle of the cavity, glabrous except along the midline of the cavity towards the base of the lobe. Anther wings 3–4 mm long, curved; anther appendages 2–3 mm long, semicircular and inflexed over apex of stylar head; corpusculum black, ovoid, 0.4–0.5 × 0.2 mm; translator arms with a membranous, obovate, convex and somewhat contorted proximal part 2–2.5 × 0.6–1 mm, and a filiform distal part ± 1.5 mm long broadening abruptly into a short clasping overlap with the pollinium; pollinia oblong, ± 1.5 × 0.4 mm, flattened. Stylar head white or green, flat. Fruiting pedicel not contorted; follicle erect, ovoid to fusiform, beaked, inflated or not, smooth or with weak longitudinal ridges, densely pubescent. Seeds ± 5 × 3 mm; coma ± 4 cm long. Fig. 101, p. 411.

a. subsp. **spectabile**

Bracts filiform. Sepals linear-lanceolate to lanceolate, (4–)8–16 × 1–2(–5) mm, acute. Corolla red or orange; lobes weakly to strongly replicate. Corona lobes red, orange or yellow, sometimes paler at the margins, with a slender or robust tooth to 3 mm long in the cavity. Follicle fusiform, 10–14 × 1.5–2 cm, not inflated, smooth or with weak longitudinal ridges. Fig. 101.

TANZANIA. Shinyanga District: Kizumbi Hill, 10 Jan. 1932, *B.D. Burtt* 3515!; Mbulu District: Mt Hanang, 4 Feb. 1946, *Greenway* 7566!; Njombe District: Livingstone Mountains, Ngolo River Valley, 20 Jan. 1991, *Gereau & Kayombo* 3697!
DISTR. T 1, 2, 4, 5, 7; Zambia, Malawi and Zimbabwe
HAB. Seasonally waterlogged grassland and *Brachystegia* or mixed deciduous woodland, often in rocky ground or on limestone; 1000–1900 m. Flowering mostly in December and January

SYN. *Asclepias spectabilis* N.E. Br. in K.B. 1895: 254 (1895) & in F.T.A. 4(1): 325 (1902)
    Stathmostelma odoratum K. Schum. in E.J. 28: 457 (1900). Type: Tanzania, Iringa District: near Lula [Sula], *Goetze* 498 (B†, holo.)
    S. pachycladum K. Schum. in E.J. 28: 458 (1900). Type: Tanzania, Iringa District: near Kigonsive, Jan. 1899, *Goetze* 531 (B†, holo.)
    Asclepias odorata (K. Schum.) N.E. Br. in F.T.A. 4(1): 324 (1902)
    A. pachyclada (K. Schum.) N.E. Br. in F.T.A. 4(1): 325 (1902); T.T.C.L.: 63 (1949)
    Stathmostelma macropetalum Schltr. & K. Schum. in E.J. 33: 325 (1903). Type: Tanzania, Kilimandjaro, near Ndala, *Trotha* 179 (B†, holo.)
    Asclepias macropetala (Schltr. & K. Schum.) N.E. Br. in F.T.A. 4(1): 616 (1904)

b. subsp. **frommii** (*Schltr.*) *Goyder* in K.B. 53: 585 (1988). Types: Tanzania, *Fromm* 89 (B†, syn.) & *Fromm* 195 (B†, syn.); Ufipa District, Lake Sundu, *Richards* 13599 (K!, neo., designated by Goyder in K.B. 53: 585 (1998); BR!, isoneo.)

Bracts lanceolate to ovate. Sepals ovate to broadly ovate, 9–16 × 4–11 mm, acute, green or maroon. Corolla yellow, orange or red; lobes not replicate. Corona lobes yellow or orange, occasionally maroon; tooth within cavity sometimes reduced or absent. Follicle ovoid, ± 8 × 3 cm, shortly beaked, somewhat inflated or not, smooth.

TANZANIA. Ufipa District: near Mpui on Sumbawanga road, 14 Dec. 1956, *Richards* 7268! & Ufipa Plateau, 1 km S of Mutimbwa, 10 Jan. 1987, *Moyer & Sanane* 149!; Mbeya District: Songwe valley, 8 Jan. 1976, *Leedal* 3306!
DISTR. T 4, 7; Congo-Kinshasa (Shaba) and N Zambia
HAB. Open *Brachystegia* woodland or bushland, frequently on limestone or on termitaria; 1200–1900 m. Flowering mostly in December and January

SYN. *Stathmostelma frommii* Schltr. in E.J. 51: 139 (1913)

NOTE. The only character which can be relied upon to separate the two subspecies is the shape of the sepals, which are broadly ovate in subsp. *frommii* and linear-lanceolate in subsp. *spectabile*. The follicle is commonly inflated in subsp. *frommii* and more slender in the type subspecies, but from the limited material available this difference appears inconsistent. Flower colour does not correlate with other characters and cannot be used to differentiate the taxa.

2. **Stathmostelma pedunculatum** (*Decne.*) *K. Schum.* in E.J. 17: 132 (1893); F.P.U.: 120, t. 6 (1962); U.K.W.F. ed. 2: 178 (1994); Goyder in Fl. Eth. 4(1): 130 (2003). Type: Ethiopia, between Adowa and Gondar, 1840, *Quartin-Dillon* s.n. (P!, lecto., designated by Goyder in K.B. 53: 586 (1998), K!, photo)

Perennial herb with 1–4 annual stems arising from a slender, vertical, napiform tuber; stems erect or ascending, 0.1–0.5(–1) m long, simple or occasionally branched below, minutely pubescent with short white hairs. Leaves sessile or with petiole to 5 mm long; lamina linear to lanceolate, (4–)7–15(–19) × 0.2–1.3 cm, acute, base truncate or rounded, minutely pubescent with short, stiff hairs on both upper and lower surfaces. Inflorescences terminal or extra-axillary with 2–4(–7) flowers in an erect umbel; peduncles erect, (6–)8–20(–29) cm long, the lower inflorescences with markedly longer peduncles than the upper ones, thus raising all the flowers to approximately the same level, minutely pubescent; bracts caducous; pedicels 2–5 cm long, pubescent. Sepals lanceolate to ovate, 2–6 × 1–2 mm, acute, usually tinged red towards tip, densely pubescent. Corolla pink to orange-red, rotate or slightly reflexed, pubescent at least towards tip on outer surface, minutely papillose towards the base within; lobes oblong to obovate, 8–17 × 4–6 mm, acute, plane. Gynostegium with stipe 1–3 mm long. Corona lobes pink, red or orange with orange or yellow teeth, erect, 4–7 × 1.5–2.5 mm, 1–2 times as tall as column, weakly pouched towards the base on each side of the mid-line, appearing concave-cucullate for most of its length but in fact solid except for the proximal and upper margins, apex rounded distally, the inner apical margins produced into a pair of inward- or upward-pointing teeth 1–2 mm long and with a tooth to 2.5 mm long arising from the middle of the cavity, glabrous. Anther wings 2–3 mm long, curved; anther appendages semicircular, 1–1.5 mm long, and inflexed over apex of stylar head; corpusculum black, ovoid to subcylindrical, ± 0.4 × 0.1–0.2 mm; translator arms with a membranous, obovate, convex and somewhat contorted proximal part ± 1.4 × 0.6 mm, and a filiform distal part ± 0.6–1 mm long broadening abruptly into a short clasping overlap with the pollinium; pollinia oblong or obovate, ± 1 × 0.4 mm, flattened. Stylar head flat. Fruiting pedicel not contorted; follicle erect, fusiform, 8–12.5 × 0.5–1 cm, beaked, not inflated, smooth, densely pubescent. Seeds ± 5 × 3 mm.

UGANDA. Karamoja District: Kokumongole, 28 May 1939, *A.S. Thomas* 2865!; Ankole District: Mbarara, 6 Nov. 1954, *Lind* 489!; Busoga District: lower slopes of Bugiri Hill, 19 Apr. 1953, *G.H.S. Wood* 703!
KENYA. West Suk District: Kongelai escarpment, July 1961, *Lucas* 190!; Nairobi District: Thika road grasslands 6 km N of Nairobi, 5 Mar. 1951, *Verdcourt* 422!; Kwale District: Mwele Mdogo forest, Shimba Hills, 4 Feb. 1953, *Drummond & Hemsley* 1103!
TANZANIA. Mwanza District: Geita Gold Mine, Nov. 1967, *Procter* 3767A!; Mbulu District: Kitingi, 15 Jan. 1965, *Hukui* 13!; Tanga District: Basi Forest Reserve, 7 Nov. 1964, *Mgaza* 653!
DISTR. **U** 1–4; **K** 1?, 2, 4, 6, 7; **T** 1–5; Cameroon, Sudan, Ethiopia, Congo-Kinshasa (Kivu) and Mozambique
HAB. Grassland, often in seasonally waterlogged grassland; 0–2200 m

SYN. *Gomphocarpus pedunculatus* Decne. in DC., Prodr. 8: 558 (1844)
　　[*Pachycarpus corniculatus* Hochst. in Flora 27: 101 (1844), *nomen nudum*]
　　Asclepias macrantha Oliv. in Trans. Linn. Soc., Bot. 29: 111, t. 75 (1875); N.E. Br. in F.T.A. 4(1): 340 (1902). Type: Ethiopia, Sana District, near Gadding Gale, 23 June 1840, *Schimper* s.n. (K!, holo.; P!, iso. (2 sheets))
　　Gomphocarpus longipes Oliv. in Trans. Linn. Soc., Bot. 29: t. 75 (1875), in error (see note, op. cit. p. 111)
　　Stathmostelma globuliflorum K. Schum. in P.O.A. C: 322 (1895). Type: Tanzania/Kenya, Nyika, near Kiyombe, *Volkens* 87 (B†, holo.; K!, iso.)
　　Asclepias uvirensis S. Moore in J.B. 48: 256 (1910). Type: Congo-Kinshasa, Uvira, shore of Lake Tanganyika, *Kässner* 3162 (BM!, holo.)
　　Stathmostelma macranthum (Oliv.) Schltr. in N.B.G.B. 9: 27 (1924), quoad syn.
　　Asclepias pedunculata (Decne.) Dandy in F.P.S. 2: 401 (1952)

NOTE. In coastal regions of Kenya and Tanzania, and as far up as Nairobi, corona lobes are up to ± 5 mm long excluding the acute teeth, ± as tall as the column and flat-topped. Elsewhere in inland areas they may be 2 times the height of the column, the teeth commonly oblong and obtuse, and angled upwards, the main part of lobe to 7 mm long.

Kirrika 183 from **K** 4, with lemon yellow flowers, may be a pale-flowered form of this species rather than *S. welwitschii*.

3. **Stathmostelma stipitatum** *Goyder*, **sp. nov.** *S. incarnato* similis sed alis antherarum brevioribus 1.5 mm tantum (nec ± 2 mm) longis, stipite gynostegii longiore ± 2 mm (nec 0.5 mm tantum) longo, dentibus lateralibus prominentibus loborum coronae carentibus et parte proximali brachiorum translatorum pollinarii alata breviore 0.3 mm tantum (nec 0.6 mm) longa ornatis differt. Typus: Tanzania, Tabora District: 43 km along the Ipole–Rungwa road, *Bidgood, Leliyo & Vollesen* 7688 (K!, holo.)

Perennial herb with single annual stem arising from underground rootstock; tuber not seen; stems erect, ± 0.6 m long, simple, subglabrous. Leaves subsessile or with petiole to ± 2 mm long; lamina linear, 7–12.5 × 0.1–0.3 cm, attenuate, base narrowly cuneate, tapering gently into the petiole, glabrous. Inflorescences terminal and extra-axillary at upper node(s) with 2–5 flowers in an erect umbel; peduncles erect, 1.5–5 cm long, minutely pubescent; bracts caducous; pedicels 1.5–5 cm long, minutely pubescent, tinged reddish. Sepals ovate-lanceolate, ± 2 × 0.8 mm, acute, usually tinged red towards tip, glabrous. Corolla dark brown, rotate or slightly reflexed, but drying reddish purple, glabrous on outer surface, minutely papillose within; lobes obovate, 5–6 × 3 mm, acute, plane or with somewhat revolute margins. Gynostegium with stipe ± 3 mm long. Corona lobes purple, erect and ± as tall as column, ± 2 mm long, apex rounded distally, the inner apical margins produced into a pair of short, acute teeth directed towards the column. Anther wings 1.5 mm long, triangular, not curved; anther appendages 1–1.5 mm long, semicircular and inflexed over apex of stylar head; corpusculum black, ovoid-subcylindrical, ± 0.25 × 0.1 mm; translator arms with a winged proximal part ± 3 mm long, and a geniculate distal part 0.4 mm long; pollinia obovate, ± 0.8 × 0.3 mm, flattened. Stylar head flat. Fruiting pedicel not contorted; follicle erect, fusiform, ± 11–13 × 0.5 cm, beaked, not inflated, smooth, glabrous. Seeds not seen.

TANZANIA. Tabora District: 43 km along the Ipole–Rungwa road, 7 Feb. 2009, *Bidgood, Leliyo & Vollesen* 7688!
DISTR. **T** 4; known only from the type
HAB. Seasonally waterlogged grassland; ± 1000 m

NOTE. Most closely allied to the Angolan species *S. incarnatum* K. Schum., but differs in the shorter anther wings (1.5 mm rather than ± 2 mm), the longer gynostegial stipe (± 3 mm rather than 0.5 mm), the corona lobes lacking prominent lateral teeth, and the pollinarium with a smaller proximal winged part to the translator arms (0.3 mm long, rather than 0.6 mm).

4. **Stathmostelma rhacodes** *K. Schum.* in E.J. 17: 131, t. 6 D-F (1893); Blundell, Wild Fl. E. Afr.: t. 517 (1987); U.K.W.F. ed. 2: 178, t. 69 (1994); Goyder in Fl. Eth. 4(1): 130 (2003). Type: E.J. 17: t. 6, fig. D-F (1893) (lecto., designated by Goyder in K.B. 53: 590 (1998))

Perennial herb with a single annual stem arising from a vertical, napiform tuber; stems erect or ascending, 0.1–0.6 m long, simple or occasionally branched below, minutely pubescent with short white hairs. Leaves sessile or with petiole to 5 mm long; lamina linear, 5–15 × 0.1–1 cm, acute, base truncate or rounded, minutely pubescent with short, stiff hairs on both upper and lower surfaces. Inflorescences extra-axillary with 2–4 flowers in an erect umbel; peduncles erect, 1.5–15(–25) cm long, the lower inflorescences with markedly longer peduncles than the upper ones, thus raising all the flowers to approximately the same level, minutely pubescent;

bracts caducous; pedicels 1.5–5 cm long, pubescent. Sepals lanceolate to ovate, 3–6 × 1–2 mm, acute, usually tinged red towards tip, pubescent. Corolla pink to orange-red, rotate or slightly reflexed, glabrous on outer surface, minutely papillose towards the base within; lobes oblong to obovate, 9–15 × 4–6 mm, acute, plane. Gynostegium with stipe 1–3 mm long. Corona lobes erect and arching over head of column, red or orange with orange or yellow teeth, at least twice as tall as column including teeth, main part of lobe glabrous and shiny, 4–7 × ± 2 mm, weakly pouched towards the base on each side of the mid-line, appearing concave-cucullate for most of its length but in fact solid except for the papillose proximal and upper margins, apex rounded distally, the inner apical margins produced abruptly into a pair of teeth, teeth erect or angled away from column, ovate to broadly oblong, 2–3 × 1–2 mm, acute or obliquely truncate apically, often somewhat constricted at the base, a third tooth to 3–4 mm long arising from the middle of the cavity, curved upwards or inwards. Anther wings 2–3 mm long, curved; anther appendages 1–1.5 mm long, semicircular and inflexed over apex of stylar head; corpusculum black, subcylindrical, ± 0.4 × 0.1 mm; translator arms with a membranous, obovate, convex and somewhat contorted proximal part 1.2–1.4 × 0.4 mm, and a filiform distal part 0.5–0.8 mm long broadening abruptly into a clasping overlap with the pollinium; pollinia oblong or obovate, ± 0.8 × 0.3 mm, flattened. Stylar head flat. Fruiting pedicel not contorted; follicle erect, fusiform, 10–12 × 0.5 cm, beaked, not inflated, smooth, minutely pubescent when young, glabrous at maturity. Seeds ± 8 × 4 mm.

UGANDA. Bunyoro District: N of Kafu River on Gulu Road, 16 Feb. 1969, *Lye* 2000!; Teso District: Serere, May 1932, *Chandler* 539! (excl. fruiting part); Mengo District: Kakenzi, Kampala, 25 Apr. 1945, *Greenway & Thomas* 7360!
KENYA. Uasin Gishu District: Sergoit Rock, 10 Mar. 1964, *Polhill* 2402!; Nairobi District: Kassani, Thika road, 15 June 1952, *Kirrika* 187!; Kisumu-Londiani District: near Londiani, 15 Apr. 1973, *Hansen* 948!
TANZANIA. Bukoba District: Minziro Forest Reserve, 14 Apr. 1957, *Gane* 112!; North Mara District: Shirati–Utegi road, 23 May 1957, *Gane* 121!
DISTR. U 2–4; K 2–6; T 1; South Sudan
HAB. Grassland, often where seasonally waterlogged; 900–2300 m

SYN. *Asclepias rhacodes* (K. Schum.) N.E. Br. in F.T.A. 4(1): 342 (1902)
     Stathmostelma pedunculatum sensu Schltr. in J.B. 33: 335 (1895), *non* (Decne.) K. Schum.
     S. macranthum sensu Schltr. in N.B.G.B. 9: 27 (1924), *non* (Oliv.) Schltr.
     Asclepias macrantha sensu Bullock in K.B. 1933: 79 (1933), *non* Oliv.

NOTE. Very similar to the upland form of *S. pedunculatum*, differing principally in the corona teeth but perhaps also in the subglabrous follicle.

5. **Stathmostelma fornicatum** (*N.E. Br.*) *Bullock* in K.B. 8: 55 (1953). Type: Malawi(?), Nyika Plateau, *McClounie* 81 (K!, holo.)

Perennial herb to ± 0.6 m tall, with a single annual stem arising from a large napiform tuber; stems erect or ascending, generally branched below, glabrous or minutely pubescent with short white hairs. Leaves with petiole 2–5(–20) mm long; lamina linear, 8–20 × 0.3–1.3 cm, acute, base truncate or rounded, minutely pubescent with short, stiff hairs particularly on the margins and midrib. Inflorescences terminal or extra-axillary with 2–4(–5) flowers in an erect umbel; peduncles 2–14(–18) cm long, the lower inflorescences with markedly longer peduncles than the upper ones, erect, minutely pubescent; bracts caducous; pedicels 2–5 cm long, minutely pubescent. Sepals lanceolate to ovate, 4–7 × 1–3.5 mm, acute or subacute, glabrous or pubescent with ciliate margins, green or tinged purple towards apex. Corolla rotate to campanulate, outer face greenish, chrome-yellow within; lobes lanceolate to ovate, 14–20 × 4–7 mm, acute or subacute, plane or somewhat replicate, glabrous on outer surface, minutely papillose within. Gynostegium with stipe 1–2 mm long. Corona lobes erect, yellow or orange, ± 1.5

times as tall as column, (5–)7–8 × ± 3 mm, appearing concave-cucullate for most of their length but in fact solid except for the papillose proximal and upper margins, ± triangular in section above, apex rounded distally, the inner apical margins extending over the head of the column as a pair of broadly oblong, acute or obliquely truncate teeth 2–3 × 2–3 mm, cavity sinus with or without a well developed central tooth. Anther wings 3–5 × 1 mm, curving somewhat under the anthers; anther appendages ± 2 mm long, semicircular and inflexed over apex of stylar head; corpusculum black, subcylindrical, 0.5–0.6 × 0.1–0.2 mm; translator arms with a membranous, obovate, convex and somewhat contorted proximal part 2–3.5 × 0.7–1 mm, and a filiform distal part 1–1.4 mm long broadening abruptly into a clasping overlap with the pollinium; pollinia oblong, 1–1.3 × 0.3–0.4 mm, flattened. Stylar head flat. Fruiting pedicel not contorted; follicle erect, fusiform, ± 11 × 0.6 cm, beaked, not inflated, smooth, minutely pubescent when young, glabrous at maturity. Seeds ± 5 × 3 mm.

SYN. *Asclepias fornicata* N.E. Br. in K.B. 1906: 250 (1906)

subsp. **tridentatum** *Goyder* in K.B. 53: 593 (1998). Type: Tanzania, Mbeya/Njombe District: Chimala escarpment on Mbeya–Iringa road, *Brummitt & Polhill* 13667C (K!, holo.; EA!, iso.)

Corona with a strongly developed conical or laterally flattened triangular tooth ± 2 mm long arising towards the top of the cavity sinus and pointing upwards or inwards between the two lateral coronal teeth.

TANZANIA. Mpwapwa District: Kongwa grassland, 22 Jan. 1950, *Anderson* 581!; Mbeya/Njombe
    Districts: Chimala escarpment on Mbeya–Iringa road, 9 Jan. 1975, *Brummitt & Polhill* 13667C!
DISTR. **T** 5, 7; not known elsewhere
HAB. Growing among grass in *Brachystegia* woodland; 1200–1800 m

6. **Stathmostelma welwitschii** *Britt. & Rendle* in Trans. Linn. Soc., Bot. ser. 2, Bot. 4: 28 (1894); Goyder in Fl. Eth. 4(1): 131 (2003). Type: Angola, Pungo Andongo, near Pedras de Guinga, *Welwitsch* 4168 (BM!, holo.; K!, iso.)

Perennial herb with 1 to many annual stems arising from a long, slender, vertical, napiform tuber; stems erect or ascending, 0.2–1 m long, simple or branched below, minutely pubescent with short white hairs. Leaves sessile or with petiole to 10 mm long; lamina linear to lanceolate, 7–16 × 0.2–1.2(–1.6) cm, acute, base truncate or rounded, minutely pubescent with short, stiff hairs on both upper and lower surfaces. Inflorescences extra-axillary with 2–7 flowers in an erect umbel; peduncles erect, (4–)7–14(–20) cm long, the lower inflorescences with markedly longer peduncles than the upper ones, thus raising all the flowers to approximately the same level, minutely pubescent; bracts caducous; pedicels 1.5–4 cm long, minutely pubescent. Sepals lanceolate to ovate, 3–6 × 1–2 mm, acute, usually tinged red towards tip, pubescent. Corolla yellow, rotate or slightly reflexed, pubescent at least towards tip on outer surface, minutely papillose towards the base within; lobes 9–12 × 6 mm, oblong to obovate, acute, plane. Gynostegium with stipe 1 mm long. Corona lobes erect, yellow or orange, 4–6 × 2–4 mm, slightly longer than column, appearing concave-cucullate for most of their length but in fact solid except for the proximal and upper margins, apex rounded distally, the inner apical margins produced into a pair of teeth 1–2 mm long and with a rounded or acute tooth arising from the middle of the cavity, glabrous. Anther wings 2–3 mm long, curved; anther appendages 1–1.5 mm long, semicircular and inflexed over apex of stylar head; corpusculum black, ovoid to subcylindrical, ± 0.3 × 0.1 mm; translator arms with a membranous, obovate, convex and somewhat contorted proximal part ± 1.6 × 0.6 mm, and a filiform distal part ± 0.6–1 mm long broadening abruptly into a short clasping overlap with the pollinium; pollinia oblong or obovate, ± 1 × 0.3 mm, flattened. Stylar head flat. Fruiting pedicel not contorted; follicle erect, fusiform, 5–10 × 0.5–1 cm, beaked, not inflated, smooth, densely pubescent. Seeds ± 5 × 3 mm.

NOTE. *Stathmostelma welwitschii* can be readily distinguished from *S. pedunculatum* by the more branched growth form and the flowers which are yellow rather than red. Occasional specimens lack the tooth in the cavity of the corona lobe and can not therefore be assigned to the two varieties below.

a. var. **welwitschii**

Corona with slender acute tooth in cavity.

TANZANIA. Ufipa District: Kasisiwue Plain, 14 Dec. 1958, *Richards* 10347! & Chala–Kisi, 17 Jan. 1950, *Bullock* 2251!
DISTR. **T** 4; Congo-Kinshasa, Ethiopia, Angola and Zambia
HAB. Seasonally waterlogged grassland; 1000–1650 m

SYN. *Stathmostelma laurentianum* Dewèvre in B.S.B.B. 34(2): 87 (1895). Type: Congo-Kinshasa, *Laurent* s.n. (BR!, holo.)
    Asclepias welwitschii (Britt. & Rendle) N.E. Br. in F.T.A. 4(1): 341 (1902)
    A. laurentiana (Dewèvre) N.E. Br. in F.T.A. 4(1): 342 (1902)
    [*Stathmostelma chironioides* De Wild. & Durand in Bull. Herb. Boiss. sér. 2, 1: 829 (≡ De Wild. & Durand, Pl. Gilletianae Congol. 89) (1901), *nom. nud., non Gomphocarpus chironioides* Decne. in DC., Prodr. 8: 562 (1844)]

b. var. **bagshawei** (*S. Moore*) *Goyder* in K.B. 53: 599 (1998). Type: South Sudan, Bari, *Bagshawe* 1640 (BM!, holo.; K!, iso.)

Corona with short, rounded tooth in cavity.

UGANDA. West Nile District: W of Oleiba on border with Congo-Kinshasa, 2 Aug. 1953, *Chancellor* 87! & Koboko, May 1938, *Hazel* 468! & Adumi, Apr. 1941, *Eggeling* 4261!
TANZANIA. Mwanza District: Mwanza, 1 Feb. 1933, *Rounce* 247!
DISTR. **U** 1; **T** 1; South Sudan and N Congo-Kinshasa
HAB. Seasonally waterlogged grassland; 800–1400 m

SYN. *Asclepias bagshawei* S. Moore in J.B. 46: 296 (1908)

7. **Stathmostelma propinquum** (*N.E. Br.*) *Schltr.* in E.J. 51: 134 (1913). Type: Tanzania, Kilimanjaro, Apr. 1893, *C.S. Smith* s.n. (K!, holo.)

Similar in most respects to *S. welwitschii*, but leaves densely pubescent with white hairs, corolla green and corona lobes overtopping gynostegium by about half their length.

TANZANIA. Mbulu District: Tarangire, 5 Jan. 1959, *Mwinijuma* s.n. in *Mahinda* 425!; Pare District: near Kamori, Oct. 1927, *Haarer* 868!; Mpwapwa District: Kongwa, 22 Jan. 1950, *Anderson* 581!
DISTR. **T** 2, 3, 5; not known elsewhere
HAB. Habitat not recorded; the *Haarer* collection was made at 900 m altitude

SYN. *Asclepias propinqua* N.E. Br. in K.B. 1895: 254 (1895) & in F.T.A. 4(1): 343 (1902)

NOTE. Doubtfully distinct from *S. welwitschii*. I have maintained this species as the material is inadequate to make a more informed judgement of its affinities. It is quite possible that further collections will suggest its inclusion within *S. welwitschii*, but as a geographically isolated population with the differences noted above it may deserve recognition at infraspecific rank.

8. **Stathmostelma gigantiflorum** *K. Schum.* in E.J. 17: 129 & t. 6 A-C (1893). Type: E.J. 17: t. 6, fig. A-C (1893) (lecto., designated by Goyder in K.B. 53: 599 (1998))

Perennial herb with a single branched annual stem arising from a stout napiform or globose tuber; stem erect or ascending, 0.3–1 m long, branched from near the ground, minutely pubescent with short white hairs. Leaves mostly subsessile, occasionally with petiole to 10 mm long; lamina linear or occasionally lanceolate, 10–25 × 0.2–1(–2) cm, acute, base cuneate, indumentum mostly restricted to margins and midvein. Inflorescences terminal and extra-axillary with 3–6 flowers in an erect umbel; peduncles erect, 2–17 cm long, the lower inflorescences generally with longer peduncles than the upper ones, minutely pubescent; bracts filiform, pubescent; pedicels 2–5 cm long, minutely pubescent. Sepals triangular to broadly ovate, 6–12(–14) × 2–3(–9) mm, acute, subglabrous to densely pubescent. Corolla rotate to slightly reflexed, subglabrous or pubescent on outer surface, minutely papillose within, white or green outside, white, green or pink and occasionally speckled purple within; lobes ovate to elliptic, (15–)20–30 × 7–15 mm, acute, somewhat replicate. Gynostegium with stipe ± 3 mm long. Corona lobes erect, white or green, frequently speckled with purple towards upper margin, 8–15 × 4–7 mm, mostly overtopping column by at least half their length, concave-cucullate for most of their length, weakly pouched either side of the midline towards the base, apex rounded to attenuate distally, the inner apical margins produced into a pair of upward- or inward-pointing acute teeth, inner margins of lobe papillose, cavity without a median tooth. Anther wings 4–6 mm long, curved, the basal tails free from the column; anther appendages ± 1.5 mm long, semicircular and inflexed over apex of stylar head; corpusculum black, obovoid to subcylindrical, ± 0.7 × 0.3 mm; translator arms with a membranous, obovate, convex and somewhat contorted proximal part ± 3.5 × 1.5 mm, and a filiform distal part ± 1 mm long broadening abruptly into a clasping overlap with the pollinium; pollinia oblong, ± 1.5 × 0.5 mm, flattened. Stylar head flat. Fruiting pedicel not contorted; follicle erect, narrowly fusiform, 10–16 × 0.5–0.8 cm, with a basal stipe, not inflated, smooth, minutely but densely pubescent. Seeds ± 3 × 2 mm, verrucose.

Kenya. Embu District: Kiambere, 24 Nov. 1951, *Kirrika* 133!; Fort Hall District: near Moboloni Rock, 7 Dec. 1952, *Bally* 8380!; Machakos District: 10 km S of Kiboko along track past Rangeland Research Station, 5 Dec. 1992, *Harvey & Vollesen* 64!
Tanzania. Ufipa District: 1 km S of Mutimbwa, 10 Jan. 1987, *Moyer & Sanane* 151!; Kondoa District: Kolo, 25 km N of Kondoa, 11 Jan. 1962, *Polhill & Paulo* 1138!; Mbeya District: Ruaha National Park, Isiki River, 14 Dec. 1972, *Bjørnstad* 2075!
Distr. **K** 4; **T** 1, 2, 4, 5, 7; E Zambia and central Malawi
Hab. Seasonally waterlogged grassland on black clay soils; 500–1800 m

Syn. *Stathmostelma bicolor* K. Schum. in E.J. 28: 457 (1900), *non Asclepias bicolor* Moench (1794). Type: Tanzania, Iringa District, Uhehe, Muhinde, *Goetze* 523 (B†, holo.; K!, iso.)
 Asclepias gigantiflora (K. Schum.) N.E. Br. in F.T.A. 4(1): 326 (1902)
 A. muhindensis N.E. Br. in F.T.A. 4(1): 344 (1902). Type as for *Stathmostelma bicolor*
 Stathmostelma nomadacridum Bullock in K.B. 8: 53 (1953). Type: Tanzania, Ufipa District, Milepa, *Michelmore* 1438 (K!, holo.)
 S. praetermissum Bullock in K.B. 8: 347 (1953); U.K.W.F. ed. 2: 178 (1994). Type: Kenya, Fort Hall District: near Mabaloni [Moboloni] rock, *Bally* 8380 (K!, holo.)

Note. The broadly ovate sepals of *S. gigantiflorum* sensu Bullock (1953a) contrast with the lanceolate sepals of other collections but as the other differences listed by Bullock can be observed elsewhere in the complex I have chosen to interpret the two collections as extreme variants within a variable species. Collections from Kenya and southern Tanzania have distinctive facies recognised by Bullock (1953a, 1953b) when he described *S. praetermissum* and *S. nomadacridum*. However, the characters which combine to give these appearances (principally the form of the apex and teeth of the corona lobes) break down in the intervening territory. I have reluctantly come to the conclusion that, as it is only possible to divide the variation on completely arbitrary grounds, the species must be redefined in the broad sense of N.E. Brown (1902) with no infraspecific taxa recognised. *S. bicolor* (*A. muhindensis*) appears, from the limited material available (a single flower from the type), to be a poorly developed specimen of *S. gigantiflorum* and I have therefore included it in the synonymy of this species. *Bidgood et al.* 2295 from Chunya district in southern Tanzania probably also belongs here but has exceptionally broad leaves and smaller corona lobes than in other material.

9. **Stathmostelma angustatum** *K. Schum.* in E.J. 17: 132 (1893); Goyder in Fl. Eth. 4(1): 131 (2003). Type: Ethiopia, Tigray, Sana, Walcha mountain plain, *Schimper* 1589 (B†, holo.; BM!, K!, iso.)

Perennial herb with a single branched annual stem arising from a slender to stout napiform tuber; stem erect or ascending, 15–25 cm long, branched from near the ground, minutely pubescent with short white hairs. Leaves with petiole to 5 mm long; lamina linear, 4–18 × 0.1–0.9 cm, acute, base cuneate, indumentum mostly restricted to margins and midvein. Inflorescences extra-axillary with 3–4 flowers in an erect umbel; peduncles 1.5–8 cm long, the lower inflorescences generally with longer peduncles than the upper ones, erect, minutely pubescent; bracts filiform, pubescent; pedicels 1.5–3 cm long, minutely pubescent. Sepals triangular to broadly ovate, 4–8 × 2–4 mm, acute, subglabrous to densely pubescent. Corolla rotate to slightly reflexed, subglabrous on outer surface, minutely papillose within; lobes ovate, 12–14 × 5–6 mm, acute. Gynostegium sessile. Corona lobes erect, white or green speckled with purple above on the outer surface, 6–10 × 4–6 mm, 1–1.5 times as long as column, concave-cucullate for most of their length, weakly pouched either side of the midline towards the base, apex rounded distally, the inner apical margins produced into a pair of inward-pointing acute teeth, inner margins of lobe papillose, cavity without a median tooth. Anther wings 3–4 mm long, curved, the basal tails curled under the anthers; anther appendages ± 1.5 mm long, semicircular and inflexed over apex of stylar head; corpusculum black, obovoid to subcylindrical, 0.6 × 0.2 mm; translator arms with a membranous, obovate, convex and somewhat contorted proximal part 2–3 × 0.6–0.7 mm, and a filiform distal part ± 1 mm long broadening abruptly into a clasping overlap with the pollinium; pollinia oblong, ± 1 × 0.3 mm, flattened. Stylar head flat. Fruiting pedicel not contorted; follicle erect, narrowly fusiform, 12–14 × 0.5–1 cm, with a basal stipe, not inflated, smooth, minutely pubescent. Seeds ± 5 × 2–3 mm.

SYN. [*Gomphocarpus angustatus* Hochst. in Pl. Schimp. Abyss. 1589, *nom. nud.*]
     Asclepias angustata (K. Schum.) N.E. Br. in F.T.A. 4(1): 343 (1902)

subsp. **vomeriforme** (*S. Moore*) *Goyder* in K.B. 53: 603 (1998). Type: Uganda, West Nile District: Madi, near Nimule, *Bagshawe* 1612 (BM!, holo.)

Corolla green or purplish brown, lobes revolute.

UGANDA. West Nile District: Madi, 17 May 1907, *Bagshawe* 1612! ; Karamoja District: Kokumongole, 28 May 1939, *A.S. Thomas* 2878! & Napak, 27 May 1940, *A.S. Thomas* 3550!
TANZANIA. Maswa/Mwanza Districts: Duma, 14 Dec. 1960, *Turner s.n.* in EA 12897!
DISTR. U 1; T 1; a single collection from South Sudan
HAB. Open grassland; ± 1200 m

SYN. *Asclepias vomeriformis* S. Moore in J.B. 46: 305 (1908)
     Stathmostelma thomasii Bullock in K.B. 8: 53 (1953). Type: Uganda, Karamoja District, Kokumongole, *A.S. Thomas* 2878 (cited as 2876) (K!, holo.)

NOTE. This species differs from *S. gigantiflorum* in its smaller flowers, a sessile gynostegium, and anther tails curled under column. The Tanzanian collection (*Turner* s.n. in EA 12897) is more robust than the other material but does not have the gynostegial stipe of *S. gigantiflorum*.

10. **Stathmostelma diversifolium** *Goyder* in K.B. 53: 604 (1998); Goyder in Fl. Eth. 4(1): 131 (2003). Type: Ethiopia, Sidamo, old airfield ± 15 km NNE of Yavello, *Gilbert & Jefford* 4554 (K!, holo.; ETH, iso.)

Perennial herb with a single annual stem arising from a long, slender or more usually stout, napiform tuber; stem erect or ascending, 6–30 cm long, branched from near the ground, minutely pubescent with short white hairs. Leaves with petiole 1–10 mm long; lamina linear to lanceolate or oblong, 4–10 × 0.1–3 cm, acute, base

FIG. 102. *STATHMOSTELMA DIVERSIFOLIUM* — **1**, habit, × ¹/₂; **2–4**, leaves, × ¹/₂; **5**, flower, × 3;
6, pollinarium, showing only one of the two translator arms and pollinia, front and rear
views, × 14; **7**, follicle, × ²/₃. 1 from *Gilbert & Jefford* 4554; 2 & 6 from *Friis et al.* 2741; 3 from
Luke 3013; 4 from *Ritchie s.n.* in *Bally* 8472; 5 from *Brown* 194; 7 from *van Someren s.n.* in EA
13078. From K.B. 53: 607, drawn by E. Papadopoulos.

cuneate or truncate, minutely pubescent with short, stiff hairs on both upper and lower surfaces. Inflorescences extra-axillary with up to 4 flowers in an erect umbel; peduncles erect, 0.5–2(–6) cm long, the lower inflorescences with or without longer peduncles than the upper ones, minutely pubescent; bracts caducous; pedicels 1–3 cm long, minutely pubescent. Sepals triangular, 2–7 × 1–2 mm, acute, densely pubescent. Corolla reflexed, dull yellow-green to white or pink, pubescent at least towards tip on outer surface, glabrous within; lobes strongly revolute, ovate, 7–15 × 3–4(–8) mm, acute. Gynostegium with stipe 1–3 mm long. Corona lobes erect, white speckled with purple above on the outer surface, 4–7(–10) × 3–5(–7) mm, 1–2 times as long as column, concave-cucullate for most of their length, weakly pouched either side of the midline towards the base, apex rounded distally, the inner apical margins produced into a pair of acute teeth and a laterally flattened, entire or bifid tooth 1.5–3(–4) × 0.6–1(–1.5) mm arising from the middle of the cavity, inner margins of lobe papillose. Anther wings 2.5–4 mm long, curved; anther appendages ± 1.5 mm long, semicircular and inflexed over apex of stylar head; corpusculum black, obovoid to subcylindrical, 0.5–0.6 × 0.2 mm; translator arms with a membranous, obovate, convex and somewhat contorted proximal part 1.6–2 × 0.6–1 mm, and a filiform distal part 1–1.6 mm long broadening abruptly into a short clasping overlap with the pollinium; pollinia oblong, 1–1.6 × 0.3–0.5 mm, flattened. Stylar head flat. Fruiting pedicel not contorted; follicle erect, ovoid, 5–7 × 1.5–2 cm, with an attenuate beak, not inflated, with 1 or more lines of weakly developed teeth along the dorsal surface, densely pubescent. Fig. 102, p. 420.

KENYA. Northern Frontier District: Huri Hills, 1994, *A. Brown* 194!; Machakos District: 7 km S of Emali, 15 Dec. 1991, *Luke* 3013!; Masai District: Lone Tree area, Ngong Hills, Dec. 1938, *Ritchie s.n.* in *Bally* 8472!
TANZANIA. Musoma District: Serengeti National Park, *Turner s.n.* in EA 12898!; Masai District: Nasera Rock, Ol Doinyo Gol Mts, 12 Nov. 1982, *Vincent* 35!
DISTR. **K** 1, 4, 6; **T** 1, 2; southern Ethiopia
HAB. Open grassland; 1100–1800 m

SYN. *Stathmostelma propinquum* sensu U.K.W.F. ed. 2: 179 (1994), *non* (N.E. Br.) Schltr.

11. **Stathmostelma pauciflorum** (*Klotzsch*) *K. Schum.* in E.J. 17: 132 (1893). Type: Mozambique, Rios de Sena, *Peters s.n.* (B†, holo.; K!, iso.)

Perennial herb with a single annual stem arising from a narrowly cylindrical vertical tuber to 30 cm or more in length; stems erect, 0.2–1 m long, generally unbranched, pubescent with short white hairs. Leaves with petiole 0–5(–10) mm long; lamina narrowly linear to linear-lanceolate, (3–)6–14 × 0.1–0.6(–1.3) cm, acute, cuneate or occasionally truncate, sparsely pubescent with stiff white hairs particularly on margins and midrib below. Inflorescences terminal, sometimes also extra-axillary, with 3–7(–12) flowers in an erect umbel; peduncles erect, (5–)7–30 cm long, minutely spreading-pubescent; bracts filiform, pubescent, caducous; pedicels 1.5–3 cm long, pubescent and frequently purplish. Sepals purplish, ovate-oblong, 2–4 × 1 mm, acute, densely pubescent. Corolla usually strongly reflexed, red or orange (yellow in parts of southern Congo-Kinshasa), glabrous on outer surface, minutely papillose within; lobes oblong-ovate, (6–)7–10 × 3–4 mm, acute, somewhat replicate. Gynostegium sessile. Corona lobes arising at the base of the staminal column, erect and somewhat fleshy, orange or yellow, 7–12(–14) × 2 mm, concave-cucullate in the lower half with the inner margins produced into a pair of inward-pointing falcate teeth ± 2 mm long arching over the column, the upper half forming an erect or spreading, flattened tongue with a rounded to acute apex, no tooth present within cavity of lobe. Anther wings ± 2 mm long with an additional basal tail curled under the anthers; anther appendages ± 1 mm long; corpusculum black, subglobose, 0.3 × 0.2 mm; translator arms with a membranous, oblanceolate, convex and somewhat contorted proximal part ± 1 × 0.2 mm, and a more slender distal part

± 0.5 mm long broadening abruptly into a short clasping overlap with the pollinium; pollinia ± 0.8 × 0.3 mm, flattened, oblong. Stylar head flat. Fruiting pedicel not contorted; follicle erect, fusiform, ± 8 × 0.5 cm, not inflated, smooth, glabrous.

KENYA. Mombasa District: Mombasa, July 1932, *Napier* 2268!; Kwale District: near Mwele, July 1936, *Dale* 3544! & Marenji F.R., 21 June 1994, *Luke & Gray* 4030!
TANZANIA. Arusha District: Sakila, SE of Ngurdoto Crater, 20 Mar. 1968, *Greenway & Kanuri* 13217!; Kilwa District: Selous Game Reserve, Kingupira, 11 Feb. 1971, *Ludanga* 1235!; Songea District: W of Gumbiro, 26 Jan. 1956, *Milne-Redhead & Taylor* 8437!
DISTR. **K** 7; **T** 2, 3, 6–8; S Congo-Kinshasa, Zambia, Malawi, Mozambique and Zimbabwe
HAB. Seasonally waterlogged grassland; sea level–1500 m

SYN. *Gomphocarpus pauciflorus* Klotzsch in Peters, Reise Mossamb., Bot.: 276 (1861)
     Stathmostelma reflexum Britt. & Rendle in Trans. Linn. Soc., ser. 2, 4: 27 & t. 6, figs 4–6 (1894). Type: Malawi, Mlanje, Oct. 1891, *Whyte* s.n. (BM!, holo.)
     Asclepias reflexa (Britt. & Rendle) Britt. & Rendle in Trans. Linn. Soc., ser. 2, 4: 28 (1894); N.E. Br. in F.T.A. 4(1): 344 (1902)
     A. reflexa (Britt. & Rendle) Britt. & Rendle var. *longicauda* S. Moore in J.B. 47: 219 (1909). Type: Zambia, Katanino [Katenina] Hills, *Kassner* 2417 (BM!, lecto., designated by Goyder in K.B. 53: 606 (1998); K!, isolecto.)

76. GOMPHOCARPUS

R. Br., On the Asclepiadeae: 26 (1810); Goyder & Nicholas in K.B. 56: 769–836 (2001)

Gomphocarpus sect. *Eugomphocarpus* Decne., *nom. inval.*, in DC., Prodr. 8: 557 (1844), in part
Gomphocarpus subsect. *Leiocalymma* K. Schum. in E. & P. Pf. 4(2): 236 (1895), in part

Shrubby or pyrophytic perennial herbs; stems erect, branched or not, frequently woody below; root system mostly a fibrous tap root, sometimes somewhat thickened and woody, but never tuberous; latex white, copious. Leaves generally in opposite pairs, occasionally subopposite or in whorls of 3 or 4. Inflorescences extra-axillary with flowers in a nodding umbel. Sepals free. Corolla lobes united only at the base, spreading or reflexed, abaxial face glabrous or pubescent, occasionally tomentose, adaxial face glabrous or minutely papillose, occasionally pubescent, commonly with a line of long white hairs along the right margin. Corolline corona absent. Gynostegial corona of 5 minute interstaminal lobules near the base of the anther wings and 5 well-developed cucullate staminal lobes attached above the base of the staminal column, with or without teeth or processes on the upper margins and mostly lacking a tooth in the cavity. Margins of anther wings straight, slightly convex or slightly sinuous; corpusculum ovoid to subcylindrical, brown or black, with or without translucent flanges up the sides; translator arms flattened and geniculate or differentially thickened; pollinia flattened, oblong or oblanceolate to obovate, occasionally subtriangular, pendant in the anther cells. Stylar head flat. Fruiting pedicel mostly contorted to hold follicle erect, occasionally straight or weakly sinuous. Follicles generally single by abortion, variable in shape from subcylindrical to ovoid or globose, strongly inflated in some taxa, weakly or not at all in others, surface frequently ornamented with pubescent filiform processes. Seeds generally ovate, with one convex and one concave face, verrucose, without a marginal rim, or in the *G. glaucophyllus* group with a convoluted margin, or rarely (*G. rivularis*) smooth, and with an inflated rim, with a coma of silky hairs.

25 taxa in 20 species. Native to drier parts of the African continent and contiguous parts of Arabia, Sinai, Israel and Jordan.

Key to species to be used in conjunction with Figs. 103 & 104, pp. 424 & 425.

1. Leaves broad, at least at base, ovate to lanceolate, or broadly ovate to suborbicular; base of lamina usually somewhat cordate 2
   Leaves linear to narrowly elliptic 5
2. Leaf margin minutely scabrid 3
   Leaf margin smooth 4
3. Corolla lobes to 8 mm long; corona lobes 2–4 mm long, upper margin ± level with top of column and straight or only weakly toothed . . . .   8. *G. swynnertonii*
   Corolla lobes at least 13 mm long; corona lobes 7–9 mm long, upper margin higher than top of column, strongly toothed at around middle . . . .   9. *G. munonquensis*
4. Corolla lobes at least 14 mm long; upper margin of corona lobes considerably higher than head of column; bracts at least 10 mm long, commonly elliptic or oblanceolate; corona lobes mostly 7–12 mm long   11. *G. praticola*
   Corolla lobes to 12 mm long; upper margin of corona lobes ± level with head of column; bracts to 10 mm long, linear to filiform; corona lobes 4–6 mm long . .   10. *G. glaucophyllus*
5. Upper margin of corona lobes lacking a tooth at the proximal end (Fig. 103/6–11); abaxial face of corolla tomentose or pubescent 6
   Upper margin of corona lobes with a tooth at the proximal end (Fig. 103/1–5); abaxial face of corolla glabrous or occasionally pubescent 7
6. Corona lobes oblong, taller than the column; follicles generally subglobose at the base, narrowing abruptly into an attenuate beak; corolla mostly white or cream   6. *G. integer*
   Corona lobes quadrate, ± as tall as the column; follicles more slender, narrowing gradually into the attenuate beak; corolla yellow . . . .   7. *G. stenophyllus*
7. Upper margin of corona lobe without a distinct notch at base of proximal tooth (Fig. 103/1–2) 8
   Upper margin of corona lobe with a distinct notch at base of proximal tooth (Fig. 103/3–5) 9
8. Follicle ovoid below, drawn out into a clearly beaked apex; corona lobe oblong, with well developed proximal tooth; characteristic habit generally much branched below . . . .   1. *G. fruticosus*
   Follicle subglobose, beak absent or minimal; upper margin of corona lobe sloping downwards away from column, proximal tooth weakly developed; characteristic habit with one principal stem branched above . . . .   2. *G. physocarpus*
9. Follicles narrowed gradually into an attenuate beak; stems branched below, not notably stout; occurring in open rocky ground . . . .   5. *G. phillipsiae*
   Follicle subglobose or somewhat ovoid, always without a beak; stem stout, erect, branched above the base; occuring on seasonally flooded plains 10
10. Corona lobes shorter than the staminal column, upper margin denticulate with a vertical cut about half way along separating the proximal tooth; corolla white or pink . . . .   3. *G. semilunatus*
    Corona lobes equalling the column, the upper margin entire, and rounded distally, forming an acute angle at junction with proximal tooth; corolla yellow-green   4. *G. kaessneri*

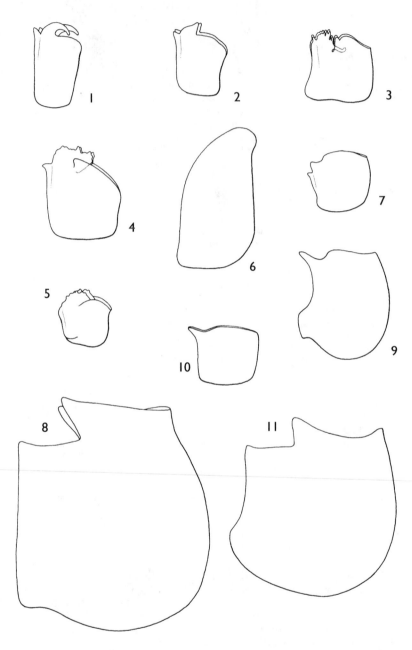

FIG. 103. *GOMPHOCARPUS* corona lobes — corona lobes viewed laterally, with the staminal column (not illustrated) to the left, × 8. **1**, *G. fruticosus*, from *Seydel* 3703; **2**, *G. physocarpus*, from *Phillipson* 958; **3**, *G. semilunatus*, from *Goyder et al.* 3710; **4**, *G. kaessneri*, from *Richards* 23476; **5**, *G. phillipsiae*, from *Hepper & Field* 5036; **6**, *G. integer*, from *Goyder et al.* 3713; **7**, *G. stenophyllus*, from *Tweedie* 2278; **8**, *G. praticola*, from *Sanane* 1415; **9**, *G. glaucophyllus*, from *Pawek* 3959; **10**, *G. swynnertonii*, from *Goyder et al.* 3805; **11**, *G. munonquensis*, from *Congdon* 257. Adapted from K.B. 56: 775, drawn by E. Papadopoulos.

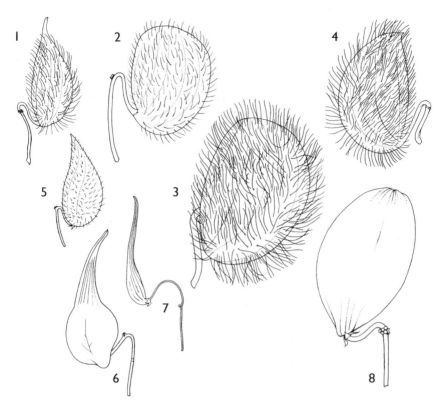

FIG. 104. *GOMPHOCARPUS*— examples of follicles — × ²/₃. **1**, *G. fruticosus* subsp. *fruticosus*, from
Herman 238; **2**, *G. physocarpus*, from *Strey* 4859; **3**, *G. semilunatus*, from *Goyder et al.* 3710;
4, *G. kaessneri*, from *Richards* 13434; **5**, *G. phillipsiae*, from *Goyder et al.* 4064; **6**, *G. integer*, from
Goyder et al. 3715; **7**, *G. stenophyllus*, from *Lind* 3153; **8**, *G. glaucophyllus*, from *Phillips* 4605.
Adapted from K.B. 56: 777, drawn by E. Papadopoulos.

1. **Gomphocarpus fruticosus** (*L.*) *W.T. Aiton*, Hort. Kew. ed. 2, 2: 80 (1811); Goyder
in Fl. Eth. 4(1): 121 (2003) & in Fl. Somalia 3: 145 (2006). Types: Herb. Linn. 310.33
(LINN!, lecto., designated by Wijnands in Bot. Commelins (1983)); South Africa,
Northern Cape, Bloeddrif, *Hardy* 2562 (K!, epi., designated by Goyder & Nicholas in
K.B. 56: 782 (2001); PRE!, WIND!, isoepi.)

Shrubby perennial herb 0.5–1.5(–3) m tall, arising from a tap root; stems erect,
much branched from above the base, woody below, densely spreading-pubescent,
sometimes shortly tomentose on young shoots. Leaves opposite; petiole 1–10 mm
long, pubescent; lamina coriaceous, linear or linear-lanceolate, (2.5–)4–12 ×
(0.2–)0.3–0.8(–1.3) cm, apex acute or attenuate, mucronate, base narrowly to
broadly cuneate, margins smooth, plane or somewhat revolute in northern part of
the range, sparsely to densely pubescent with soft white hairs on the midrib and
margins. Inflorescences extra-axillary with 4–7(–12) flowers in a nodding umbel;
peduncles 1.5–3(–4) cm long, pubescent; bracts filiform, deciduous; pedicels 1–2.5 cm
long, pubescent. Sepals lanceolate or triangular, 2–5 × 0.6–1.3 mm, attenuate,
pubescent. Corolla reflexed, glabrous outside, minutely papillate and frequently with
long white hairs along the right margin within; lobes ovate, 5–8 × 3–5 mm, subacute.
Corona lobes attached 1–1.5 mm above base of staminal column, quadrate, 2–4 ×
1.5–3 mm, laterally compressed, complicate, ± as tall as the column, the upper

margins entire, proximal margins produced into a pair of falcate teeth 1–1.5 mm long, pointing back along the upper margins of the lobe or curved down into the cavity, cavity without teeth or projections. Anther wings 1.5–2 mm long, margins not curved; corpusculum brown, subcylindrical, 0.3 × 0.1–0.15 mm; translator arms ± 0.3–0.4 × 0.1 mm, flattened; pollinia oblong or oblanceolate in outline, 1.2–1.3 × 0.3 mm, flattened. Stylar head flat. Fruiting pedicel contorted to hold follicle erect; follicle ovoid, 4–7 × 1.5–2.5 cm, tapering gradually or abruptly into an attenuate beak, strongly or weakly inflated, pubescent, with or without pubescent filiform processes. Seeds ovate, 3.5–5 × 2 mm, with one convex and one concave face, verrucose; coma ± 3 cm long. Figs. 103/1 & 104/1, pp. 424 & 425.

SYN. *Asclepias fruticosa* L., Sp. Pl. 216 (1753); N.E. Br. in F.T.A. 4(1): 330 (1902) & in Fl. Cap. 4(1): 691 (1908)

    A. glabra Mill., Gard. Dict. ed. 8: no. 12 (1768). Type: '*Asclepias glabra* Mill. Dict. no. 12' (BM!, holo.)

    A. salicifolia Salisb., Prodr.: 150 (1796). Type as for *G. fruticosus* (L.) W.T. Aiton

    ?*A. angustifolia* Schweigger, Enum. Pl. Hort. Regiom.: 13 (1812). Type: 'e horto Berolinensi' (KBG†, holo.)

    ?*Gomphocarpus angustifolius* (Schweigger) Link, Enum. Hort. Berol; 1: 251 (1821)

    G. cornutus Decne. in Ann. Sci. Nat., Bot., sér. 2, 9: 324 (1838). Type: Madagascar, *Bojer* s.n. (P!, holo.)

    G. frutescens E. Mey., Comm Pl. Afr. Austr.: 202 (1838) – error for *fruticosus*

    G. crinitus G. Bertol. in Mem. Reale Accad. Sci. Ist. Bologna 3: 253 & t. 20, fig. 1 (1851). Type: Mozambique, Inhambane, Dec. 1848, *Fornasini* s.n. (BOLO!, holo.; K!, photo.)

    ?*G. arachnoideus* E. Fourn. in Bull. Soc. Bot. France 14: 250 (1867). Type: cult. Paris from seeds ex Mexico (P, holo.)

    Asclepias cornuta (Decne.) Cordem., Fl. Réunion: 482 (1895)

    A. crinita (G. Bertol.) N.E. Br. in F.T.A. 4(1): 352 (1902)

NOTE. Goyder & Nicholas in K.B. 56: 782 (2001) recognised five regional subspecies across Africa and Arabia, of which only subsp. *flavidus* occurs within the F.T.E.A. area.

    Bullock's circumscription of *G. fruticosus* was very much wider than that used here. The specimens cited in F.W.T.A. ed. 2, 2: 92 (1963) are now referred to *G. abyssinicus*.

subsp. **flavidus** (*N.E. Br.*) *Goyder* in K.B. 51: 798 (1996); Goyder in Fl. Eth. 4(1): 121 (2003). Type: Somalia, Darsi, Qar Goliis [Darsa, Golis Range], Mar. 1895, *Cole* s.n. (K!, lecto., designated by Goyder in K.B. 51: 798 (1996))

Stems, inflorescence axis and calyx mostly tinged with purple. Corolla yellow or greenish yellow. Corona brown or purple; upper margin generally curved distally, not strongly angled. Follicles sparsely covered with filiform processes 0.2–0.4 cm long.

UGANDA. Karamoja District: near Moroto, Feb. 1936, *Eggeling* 2887! & Moroto, Kasineri Estate, June 1971, *J. Wilson* 2055! & Napau Pass Road, 16 Nov. 1953, *Dale* U828!

KENYA. Northern Frontier District: Moyale, 17 Dec. 1952, *Woodhouse* s.n. in *Gillett* 15083!; West Suk District: Moribus, May 1932, *Napier* 2039!; Baringo District: 55 km from Nakuru to Lake Bogoria, 2 Apr. 1977, *Hooper & Townsend* 1621!

DISTR. **U** 1; **K** 1–3; South Sudan, Ethiopia and Somalia

HAB. Disturbed areas in sandy or stony soil, commonly in roadside gravel; 800–1600 m

SYN. *Asclepias flavida* N.E. Br. in K.B. 1895: 255 (1895) & in F.T.A. 4(1): 331 (1902); T.T.C.L.: 63 (1949)

    Gomphocarpus fruticosus sensu U.K.W.F. ed. 2: 177, t. 67 (1994) & sensu Blundell, Wild Fl. E. Afr.: t. 221 (1987), *non* (L.) W.T. Aiton

NOTE. The identity of collections from southern Tanzania (**T** 4 & 7 – e.g. *Goyder et al.* 3846) remains uncertain. These are the plants featured in Cribb & Leedal, Mt Fl. S. Tanz.: 103, t. 23b (1982). Flowers are coloured as in subsp. *flavidus* but the flower size is small and follicles are more typical of subsp. *fruticosus*. Some specimens even tend towards *G. physocarpus*. It is possible that these collections represent an introduced taxon but further field observations are needed.

2. **Gomphocarpus physocarpus** E. *Mey.*, Comm Pl. Afr. Austr. 202 (1838); U.K.W.F. ed. 2: 177, t. 67 (1994). Type: '*Gomphocarpus physocarpus* EM. a.'–South Africa, by stream near Glenfilling, alt. 500 ft, *Drège* s.n. (K!, lecto. (specimen in *Herb. Benthamianum*), designated by Goyder in K.B. 53: 418 (1998b); BM!, E!, K!, TCD!, isolecto.)

Shrubby perennial herb to 2.5 m tall arising from a tap root; stems generally single, branching above, woody below, upper parts pubescent with spreading white hairs. Leaves generally opposite, but occasionally subopposite or crowded into pseudowhorls; petiole 3–12 mm long; lamina narrowly oblong to lanceolate, 4–9(–12) × 0.5–1.5(–2) cm, apex acute, base cuneate, margins smooth, subglabrous or sparsely pubescent with soft white hairs, particularly on the midrib and margins. Inflorescences extra-axillary with 5–12 flowers in a nodding umbel; peduncles ascending, 1.5–3.5 cm long, densely spreading-pubescent to tomentose; bracts deciduous; pedicels (1–)1.5–2(–3) cm long, densely pubescent. Sepals narrowly triangular-attenuate, 3–4 × 1 mm, pubescent. Corolla strongly reflexed, white, glabrous outside, minutely papillate and with long white hairs along the right margin within; lobes ovate, 5–8 × 3–4.5 mm, subacute. Corona lobes attached 1.5–2 mm above base of staminal column, quadrate, 2–3 × 1.5–2(–2.5) mm, laterally compressed, complicate, as tall as the column, with a short erect to slightly recurved tooth ± 0.5 mm long on the proximal upper margin, cavity of the lobe lacking any form of tooth, white, frequently tinged with pink or purple. Anther wings 1.8–2 mm long, margins very slightly sinuous; corpusculum black, ovoid-subcylindrical with lateral translucent flanges, 0.4 × 0.15 mm; translator arms ± 0.3 mm, flattened and geniculate; pollinia oblanceolate in outline, ± 1.6 × 0.4 mm, flattened. Stylar head flat. Fruiting pedicel contorted; follicle globose or subglobose, to 4–7 cm in diameter, slightly depressed on one side, not beaked but occasionally somewhat angled apically, strongly inflated, pubescent, densely covered with filiform processes to ± 1 cm long. Seeds ovate, ± 4.5 × 2 mm, with one convex and one concave face, verrucose; coma ± 3 cm long. Figs. 103/2 & 104/2, pp. 424 & 425.

UGANDA. Kigezi District: Muhavura–Mgahinga saddle, Sep. 1946, *Purseglove* 2133! & 28 Oct. 1929, *Snowden* 1580!
KENYA. Naivasha District: 5 km E of Naivasha on Nairobi road, 5 Nov. 1994, *Goyder et al.* 3711!; North Nyeri District: 10 km S of Naro Moru, 1 Jan. 1995, *Goyder & Masinde* 3945!; Masai District: 12 km from Narok town towards Sotik, 20 Jan. 2002, *Kirika et al.* in GBK 32!
TANZANIA. Masai District: Ngorongoro Crater, N end of Munge swamp on the S side of the crater floor, 8 July 1966, *Greenway & Kanuri* 12552! & E Crater Rim, Lemala, 13 Jan. 1989, *Pocs & Chuwa* 89015/B!
DISTR. **U** 2; **K** 3–6; **T** 2; native to South Africa, Swaziland and southern Mozambique, almost certainly introduced elsewhere in Africa. The species is certainly not native elsewhere in the world.
HAB. Common in seasonally wet pastures and flood plains, also occuring in disturbed areas; 1500–3000 m

SYN. *Gomphocarpus brasiliensis* E. Fourn. in Martius, Fl. Brasil. 6(4): 203, t. 53 (1885). Types: Brazil, towards Petropoli, *Glaziou* 6706 (P!, lecto., designated by Goyder & Nicholas in K.B. 56: 788 (2001); C!, K!, isolecto.)
Asclepias physocarpa (E. Mey.) Schltr. in E.J. 21, Beibl. 54: 8 (1896) & in J.B. 34: 453 (1896)
Gomphocarpus fruticosus (L.) W.T. Aiton forma *brasiliensis* (E. Fourn.) Briq. in Kgl. Sv. Vet. Akad. Handl. 34(7): 21 (1900)
Asclepias brasiliensis (E. Fourn.) Schltr. in Meded. Rijksherb. 29: 12 (1916)
Gomphocarpus semilunatus sensu Blundell, Wild Fl. E. Afr.: t. 223 (1987), *non* A. Rich.

NOTE. This species is very closely allied to *G. semilunatus* and *G. kaessneri*, sharing the characteristic growth form of a principal upright stem and many side shoots rather than the more bushy habit of the *G. fruticosus* group. The *G. physocarpus* group also has similar preferences for seasonally flooded meadows. The species of this group can generally be distinguished by the form of the corona, as shown in the key and Fig. 103/2–4. The calyx of *G. physocarpus* is also much shorter than in *G. semilunatus*.

In Kenya and Tanzania *G. physocarpus* appears to occur in drier spots (e.g. road verges) than *G. semilunatus*, and is a shorter, more stiffly branched plant. The corona and branching pattern are as in South Africa. The follicle tends to be more beaked than in typical (South African) material and may be of partially hybrid origin with *G. fruticosus*, as also appears to be the case in Australia where Forster (1996) reports hybrid swarms between these two species.

3. **Gomphocarpus semilunatus** A. *Rich.*, Tent. Fl. Abyss. 2: 39 (1851); U.K.W.F. ed. 2: 177, pl. 68 (1994); Goyder in Fl. Eth. 4(1): 123 (2003). Type: Ethiopia, Tigray, *Quartin Dillon* s.n. (P!, holo.; K!, iso.)

Shrubby perennial herb 1–2.5 m tall arising from a tap root; stems erect, stout, woody below, frequently unbranched, densely pubescent with spreading white hairs above. Leaves opposite or in whorls of 4; petiole 2–6 mm long, pubescent; lamina linear or linear-lanceolate, 8–12(–15) × 0.5–1.5(–2.5) cm, apex acute or attenuate, base cuneate or truncate, margins smooth, revolute, sparsely pubescent with soft white hairs, particularly on the midrib and margins. Inflorescences extra-axillary with 5–9 flowers in a nodding umbel; peduncles ascending, 1–4 cm long, densely spreading-pubescent to tomentose; bracts deciduous; pedicels (1–)1.5–2 cm long, densely pubescent, frequently tinged with purple. Sepals lanceolate, 5–9 × 1–2 mm, attenuate, pubescent, purplish. Corolla ± rotate, white or pink, outer surface deeper than the inner, glabrous outside, minutely papillate and frequently with long white hairs along the right margin within; lobes 7–9 × 4–5 mm, ovate, subacute. Corona lobes attached ± 1 mm above base of staminal column, quadrate, 3 × 2.5–3 mm, laterally compressed, complicate, distinctly shorter than the column, the upper margins sloping down distally, denticulate and with a vertical cut about half way along, cavity of the lobe with or without a dorsally flattened, entire or bifid tooth, purple or pink with white upper and inner margins. Anther wings ± 2 mm long, margins slightly convex; anther appendages ± 1 × 0.5 mm, truncate and inflexed over apex of stylar head; corpusculum black, ovoid-subcylindrical, 0.4 × 0.2 mm; translator arms ± 0.4 × 0.15 mm, flattened and geniculate; pollinia oblanceolate in outline, 1.6–1.8 × 0.5–0.6 mm, flattened. Stylar head flat. Fruiting pedicel contorted; follicle subglobose, to 7 × 5 cm, not beaked, generally inflated, pubescent, densely covered with pubescent filiform processes ± 1 cm long. Seeds ovate, ± 3.5 × 1.5 mm, with one convex and one concave face, verrucose; coma ± 3 cm long. Figs. 103/3 & 104/3, pp. 424 & 425.

UGANDA. Ankole District: Mbarara, Mar. 1939, *Purseglove* 655!; Teso District: Serere, Dec. 1931, *Chandler* 367!; Masaka District: Kalungu county, 3 km S of West Mengo border, 5 June 1971, *Lye* 6190!
KENYA. Naivasha District: Yacht Club, ± 5 km S of Naivasha, 5 Nov. 1994, *Goyder et al.* 3710!; Kericho District: Sotik, 15 June 1953, *Verdcourt* 944!; Masai District: Kipleleo, 12 Nov. 1962, *Glover & Samuel* 3403!
TANZANIA. North Mara District: Bwiregi Plantation, 24 May 1957, *Gane* 122!; Mbulu District: Katesh, 14 June 1941, *Hornby & Hornby* 2120!; Mpanda District: Mugombasi, 100 km S of Kigoma, 30 Aug. 1959, *Harley* 9451!
DISTR. U 1–4; **K** 3–6; **T** 1–5; from Nigeria to Ethiopia in the north, then southwards to N Zambia, S Congo-Kinshasa and Angola
HAB. Seasonally flooded alluvial grasslands and waste places; 1300–2000 m. Flowering ± throughout the year

SYN. *Asclepias denticulata* Schltr. in J.B. 33: 334 (1895). Types: Uganda, Toro District: Ruimi [Wimi], *Scott Elliot* 7904 (BM!, lecto., designated by Goyder & Nicholas in K.B. 56: 791 (2001)); K!, isolecto.)
    A. semilunata (A. Rich.) N.E. Br. in F.T.A. 4(1): 328 (1902); T.T.C.L.: 64 (1949)
    Gomphocarpus physocarpus sensu F.P.U.: 118, fig. 67 (1962) & sensu Bullock in F.W.T.A. ed. 2, 2: 92 (1963) & sensu Blundell, Wild Fl. E. Africa: 147 & fig. 758 (1987), *non* E. Mey.

4. **Gomphocarpus kaessneri** (*N.E. Br.*) *Goyder & Nicholas* in K.B. 56: 792 (2001); Blundell, Wild Fl. E. Afr.: t. 348 (1987); U.K.W.F. ed. 2: 177 (1994), as "N.E. Br.". Type: Kenya, Machakos/Masai Districts, Kiu, *Kässner* 664 (K!, holo.; BM!, iso.)

Shrubby annual or perennial herb 1–2 m tall arising from a tap root; stems erect, branched above the base, densely spreading-pubescent above. Leaves mostly opposite, occasionally in whorls of 4 particularly towards base of stem; sessile or with petiole to 5 mm long; lamina narrowly linear to linear-lanceolate, 7–18 × 0.3–1.5 cm, apex attenuate, margins smooth, weakly to strongly revolute, coriaceous, subglabrous to sparsely pubescent with soft white hairs on both surfaces, particularly on the midrib and margins. Inflorescences extra-axillary with 4–7 flowers in a nodding umbel; peduncles 1–2(–3) cm long, pubescent; pedicels 1–4 cm long, variable even within an inflorescence, pubescent. Sepals narrowly triangular, 4–6 × 1.5 mm, acute, pubescent. Corolla reflexed, greenish yellow, marked with brown outside, glabrous or sparsely pubescent outside, minutely papillate and with white hairs along the right margin within; lobes ovate, 6–10 × 4–5.5 mm, acute or subacute. Corona lobes attached 1.5–2 mm above base of staminal column, 3.5–4 × 2–3.5 mm, laterally compressed, complicate, ± as tall as the column, the upper margins acute at margin with the proximal teeth, rounded distally, proximal margins produced into a pair of broad, usually cristate teeth ± 1.5–2 mm long, pointing back within the upper margins of the lobe, cavity without teeth or projections, brown or purple except for paler inner margin and tooth. Anther wings 2 mm long, margins ± straight; corpusculum brown, ovoid, 0.5 × 0.2 mm; translator arms ± 0.4 × 0.1 mm, flattened, geniculate; pollinia oblong in outline, ± 1.5 × 0.4 mm, flattened. Stylar head flat. Fruiting pedicel contorted; follicle ovoid, 4–7 × 2–4 cm, rounded apically and without a beak, inflated, densely pubescent and with pubescent filiform processes to ± 10 mm long. Seeds ovate, ± 5 × 2 mm, with one convex and one concave face, verrucose; coma ± 3 cm long. Figs. 103/4 & 104/4, pp. 424 & 425.

KENYA. Nairobi District: Thika Road House, 10 Dec. 1950, *Verdcourt* 396!; Masai District: between Magadi and Olorgesaile, 24 Oct. 1955, *Milne-Redhead & Taylor* 7016! & Kijabe–Narok road, 18 June 1960, *Greenway* 9691!
TANZANIA. Musoma District: headwaters of the Seronera River, 9 June 1962, *Greenway & Turner* 10711!; Mbulu, 15 July 1943, *Greenway* 6783!; Lushoto District: Mkomazi, 22 Apr. 1934, *Greenway* 3931!
DISTR. **K** 4, 6; **T** 1–5, 7; N Zambia
HAB. Seasonally flooded alluvial grasslands and waste places – perhaps in somewhat drier climatic zones than *G. semilunatus*; 600–2000 m

SYN. *Asclepias kaessneri* N.E. Br. in J.B. 41: 362 (1903)

NOTE. The growth form of this species is similar to that of both *G. semilunatus* and *G. physocarpus*, but coronally *G. kaessneri* may be confused with *G. phillipsiae*. It can be readily distinguished from *G. phillipsiae*, however, by its more robust growth form and branching pattern, the more exserted cristate coronal tooth, and the inflated ovoid follicle which lacks a beak. The habitat, seasonally flooded plains, also contrasts with the fire-prone rocky grassland in which *G. phillipsiae* occurs.
   Strongly bicoloured flowers with yellow-green corolla and purple corona lobes appear to be diagnostic for this species – both *G. semilunatus* and *G. physocarpus* have a white corolla with pinkish corona lobes.

5. **Gomphocarpus phillipsiae** (*N.E. Br.*) *Goyder* in K.B. 51: 798 (1996); Goyder in Fl. Eth. 4(1): 124 (2003) & in Fl. Somalia 3: 146 (2006). Type: Somalia, *Phillips* s.n. (K!, holo.)

Shrubby perennial herb (0.2–)0.5–1.5 m tall arising from a tap root; stems erect, much branched from above the base, woody below, slender, densely spreading-pubescent. Leaves opposite or in whorls of 4; sessile or with petiole ± 1 mm long; lamina coriaceous, narrowly linear, 2.5–5(–10) × 0.1–0.2(–0.5) cm, apex acute, mucronate, margins smooth, strongly revolute, sparsely pubescent with soft white hairs on the midrib and margins. Inflorescences extra-axillary with 4–7 flowers in a nodding umbel; peduncles 0.5–2(–3) cm long, pubescent; bracts filiform, deciduous; pedicels ± 1.5 cm long, pubescent. Sepals triangular, 3–5 × 1–2 mm, acute, pubescent. Corolla reflexed, white or cream, frequently tinged pink outside, glabrous outside, minutely papillate and with long white hairs along the right margin within; lobes ovate, 5–7 × 3–5 mm, subacute. Corona lobes attached ± 1 mm above base of staminal column, cream or green, quadrate, 2–3 × 2–2.5 mm, laterally compressed, complicate, ± as tall as the column, the upper margins rounded and with a distinct notch at base of proximal teeth, proximal margins produced into a pair of broad, often denticulate teeth 1–1.5 mm long, pointing back along the upper margins of the lobe and curved down into the cavity, cavity without teeth or projections, teeth and margins frequently pink. Anther wings ± 1.5 mm long, margins not curved; corpusculum brown, subcylindrical, 0.3–0.5 × 0.1–0.15 mm; translator arms ± 0.3–0.4 × 0.1 mm, flattened; pollinia oblong or oblanceolate in outline, 1.2–1.3 × 0.3 mm, flattened. Stylar head flat. Fruiting pedicel contorted; follicle ovoid, 4–7 × 1.5–2.5 cm, tapering gradually into an attenuate beak, inflated, pubescent and with pubescent filiform processes (2–)5–7 mm long. Seeds ovate, 3.5–5 × 2 mm, with one convex and one concave face, verrucose; coma ± 3 cm long. Figs. 103/5 & 104/5, pp. 424 & 425.

UGANDA. Karamoja District: Napak, 27 May 1940, *A.S. Thomas* 3571! & Moroto Mt, Sep. 1958, *J. Wilson* 504!; Kigezi District: lower slopes of Mt Muhavura, 17 Oct. 1929, *Snowden* 1521!
KENYA. Northern Frontier District: Marsabit–Isiolo road, 14 May 1970, *Magogo* 1365!; Nakuru District: Sitoton forest camp, 10 Jan. 1946, *Bally* 4733!; North Nyeri District: 10 km S of Naro Moru, 1 Jan. 1995, *Goyder & Masinde* 3944!
TANZANIA. Masai District: Ngorongoro Crater wall, 16 Jan. 1989, *Pócs & Chuwa* 89024/A!; Arusha District: Arusha National Park, above warden's house, 30 Dec. 1969, *Richards* 25020!; Lushoto District: Sangarowe–Monga, 24 Oct. 1929, *Greenway* 1785!
DISTR. U 1–4; K 1–6; T 1–3, 6; E Congo-Kinshasa, Rwanda, Eritrea, Ethiopia and Somalia
HAB. Open rocky ground and disturbed areas; 1200–2200(–3000) m. Flowering ± throughout the year

SYN. *Gomphocarpus fruticosus* (L.) W.T. Aiton var. *angustissimus* Engl., Hochgebirgsl. Trop. Afr.: 341 (1892). Type: Eritrea, Keren, *Steudel* 747 (K!, lecto., designated by Goyder & Nicholas in K.B. 56: 794 (2001); B†, isolecto.)
Asclepias fruticosa L. var. *angustissima* (Engl.) Schltr. in J.B. 33: 335 (1895)
A. phillipsiae N.E. Br. in K.B. 1895: 219 (1895) & in F.T.A. 4(1): 332 (1902); T.T.C.L.: 63 (1949)

6. **Gomphocarpus integer** (*N.E. Br.*) *Bullock* in K.B. 7: 408 (1952); Blundell, Wild Fl. E. Afr.: t. 222 (1987); U.K.W.F. ed. 2: 177 (1994); Goyder in Fl. Eth. 4(1): 126 (2003) & in Fl. Somalia 3: 146, fig. 100 (2006). Type: Somalia, *Phillips* s.n. (K!, lecto., designated by Bullock in K.B. 7: 409 (1952))

Shrubby perennial herb 0.5–2.5 m tall arising from a tap root; stems erect, much branched from base, slender, glabrous below, tomentose above. Leaves opposite; sessile or with petiole to 2 mm long; lamina narrowly linear, 3–10(–16) × 0.1–0.2 cm, apex acute, mucronate, margins smooth, strongly revolute, minutely tomentose on both surfaces. Inflorescences extra-axillary with 4–8(–12) flowers in a nodding umbel; peduncles 0.5–1.5 cm long, glabrous to tomentose; bracts filiform, deciduous; pedicels 1.5–2.5 cm long, glabrous to tomentose. Sepals ovate, 2–2.5 × 1–2 mm, acute, pubescent or tomentose. Corolla reflexed, white, cream or green, frequently tinged pink outside, tomentose outside, glabrous to densely papillate or pubescent and with long white hairs along the right margin within; lobes ovate, 5–9 × 3–6 mm, subacute. Corona lobes attached 0.5–1 mm above base of staminal column, white or green flushed with purple, oblong, (3–)4–7 × 1.5–3(–4) mm, laterally compressed, complicate, overtopping the column by (1–)2–3(–4) mm, rounded or truncate, entire, lacking any form of tooth on either the margins or within the cavity. Anther wings 1.5–2 mm long, margins not curved; corpusculum brown, subcylindrical, 0.4 × 0.1–0.2 mm; translator arms ± 0.3–0.4 × 0.1 mm, flattened; pollinia oblong or oblanceolate in outline, 1–1.2 × 0.3–0.4 mm, flattened. Stylar head flat. Fruiting pedicel contorted; follicles 4–7 × 1–2 cm, subglobose at the base, narrowing abruptly into an attenuate beak (narrowing gradually in southern populations), glabrous to tomentose, usually smooth but occasionally with pubescent filiform processes to ± 4 mm long. Seeds ovate, ± 4 × 2 mm, with one convex and one concave face, verrucose; coma ± 3 cm long. Figs. 103/6 & 104/6, pp. 424 & 425.

KENYA. Northern Frontier District: Huri Hills, 25 Feb. 1963, *Bally* 12540!; Nairobi District: Athi River, by Namanga turn-off, 8 Nov. 1994, *Goyder et al.* 3713!; Kajiado District: 3 km SW of Selengai Game Post, 17 Dec. 1969, *Kibue* 126!
TANZANIA. Masai District: Windy Gap road, Ngorongoro Crater wall, 5 July 1966, *Greenway & Kanuri* 12533! & Engari Nanyuki Springs, 5 June 1962, *Greenway & Turner* 10705!
DISTR. **K** 1, 3–6; **T** 2; Ethiopia, Djibouti, Somalia
HAB. Open rocky ground and disturbed areas; 1200–2200(–3000) m. Flowering ± throughout the year

SYN. *Asclepias integra* N.E. Br. in K.B. 1895: 219 (1895) & in F.T.A. 4(1): 334 (1902); T.T.C.L.: 63 (1949)
   A. negrii Chiov. in Ann. Bot., Roma 10: 393 (1912). Type: Ethiopia, Galla Arussi, *Negri* 920 (FT!, holo.)
   A. litocarpa Chiov., Miss. Biol. Borana, Racc. Bot.: 164 (1939). Type: Ethiopia, Borana, between Mega and Dubuluk, *Cufodontis* 618 (FT!, holo.).

7. **Gomphocarpus stenophyllus** *Oliv.* in Trans. Linn. Soc., Bot. 29: 110, t. 119 (1875), *non* Oliv. in Trans. Linn. Soc., Bot. Ser. 2, Bot. 2: 342 (1887), *nec Asclepias stenophylla* A. Gray (1877); Blundell, Wild Fl. E. Afr.: t. 224 (1987); U.K.W.F. ed. 2: 177, t. 67 (1994); Goyder in Fl. Eth. 4(1): 126 (2003). Type: Tanzania, Tabora [Kazeh District], *Speke & Grant* s.n. (K!, holo.)

Shrubby perennial herb 0.5–2.5 m tall arising from a tap root; stems erect, much branched from base, slender, glabrous below, tomentose above. Leaves opposite; sessile or with petiole to 2 mm long; lamina narrowly linear, 5–12 × ± 0.1 cm, apex acute, mucronate, margins smooth, strongly revolute, minutely tomentose below. Inflorescences extra-axillary with (5–)8–12 flowers in a nodding umbel; peduncles slender to filiform, 1–4 cm long, minutely pubescent; bracts filiform, deciduous; pedicels filiform, 1.5–3 cm long, minutely pubescent. Sepals ovate, 1.5–2 × 1 mm, acute, pubescent. Corolla reflexed, yellow, occasionally tinged pink outside, pubescent outside, glabrous to densely papillate and with long white hairs along the right margin within; lobes ovate, 5–6(–7) × 3–4 mm, subacute. Corona lobes attached 1 mm above base of staminal column, laterally compressed, complicate, 2–3 × 2–2.5 mm, ± as tall as the column, quadrate, proximal angles of upper margins acute, sometimes produced into short teeth pointing towards stylar head, upper margins weakly toothed or angled around the mid-point, lacking any form of tooth within the cavity, yellow or green. Anther wings 1–1.2 mm long, triangular but with a curved margin; corpusculum brown, ovoid, ± 0.2 × 0.1 mm; translator arms ± 0.2 mm long, flattened, geniculate; pollinia oblanceolate in outline, ± 0.6 × 0.3 mm, flattened. Stylar head flat. Fruiting pedicel ± straight; follicles lanceolate in outline, 4–5 × 0.5–1 cm, tapering gradually into an attenuate beak, glabrous to tomentose, usually smooth but occasionally with verrucose longitudinal ridges. Seeds ovate, ± 4 × 2 mm, with one convex and one concave face, verrucose; coma ± 3 cm long. Figs. 103/7 & 104/7, pp. 424 & 425.

KENYA. Northern Frontier District: Lowaweregoi, 15 Dec. 1958, *Newbould* 3229!; Laikipia District: Nyahururu–Rumuruti road, 20 Feb. 1996, *Goyder et al.* 4048!; Machakos District: foot of Lukenya Plateau on Nairobi–Mombasa road, 28 Aug. 1959, *Verdcourt* 2363!
TANZANIA. Mbulu District: Dongobesh, 15 Mar. 1967, *Carmichael* 1353!; Singida, 16 Feb. 1957, *Smith* 1382!; Iringa District: Igurusiri Hill, 22 Apr. 1991, *Bidgood, de Leyser & Vollesen* 2220!
DISTR. **K** 1, 3, 4, 6; **T** 1, 2, 4, 5, 7; semi-arid regions of southern Ethiopia, Kenya and Tanzania.
HAB. Open rocky ground and disturbed areas; 1200–2200(–3000) m. Flowering ± throughout the year

Syn. *Asclepias leucocarpa* Schltr. in J.B. 33: 335 (1895); N.E. Br. in F.T.A. 4(1): 337 (1902);
    T.T.C.L.: 63 (1949). Type: Kenya, Nakuru/Masai District: Mau, *Scott Elliot* 6882 (BM!,
    holo.; K!, iso.)
    Gomphocarpus rostratus (N.E. Br.) Bullock in K.B. 7: 410 (1952), pro parte excl. type

Note. *Gomphocarpus stenophyllus* is florally very similar to *G. integer* but the corona is as broad as
    tall, ± as tall as the column, rather than much taller than wide and significantly overtopping
    the column in *G. integer*. Follicles in *G. stenophyllus* are slender, without the swollen base
    characteristic of *G. integer* which narrows abruptly into a beak.

8. **Gomphocarpus swynnertonii** (*S. Moore*) *Goyder & Nicholas* in K.B. 56: 810
(2001). Type: Zimbabwe, Chimanimani [Melsetter], *Swynnerton* 6092 (BM!, lecto.,
designated by Goyder & Nicholas in K.B. 56: 810 (2001))

Perennial herb arising from stout woody taproot; stems erect, few to many, 10–40
cm long, unbranched, with a dense indumentum of spreading white hairs, or
indumentum occasionally restricted to weak vertical bands. Leaves opposite, sessile
or subsessile, ovate, 1.5–5.5(–8) × 0.7–3 cm, acute, slightly cordate at the base,
margins slightly revolute at least when dry, glabrous except for the minutely
scabrid marginal vein. Inflorescences extra-axillary with 6–13 flowers in a nodding
umbel; peduncles 2–4.5 cm long, generally longer than the subtending leaf,
pubescent; bracts 5–10 mm long, filiform or occasionally broader, ciliate; pedicels
10–23 mm long, slender, pubescent. Calyx lobes ovate or oblong, 3–6 × 1.5–2 mm,
apex acute to rounded, margins ciliate or not. Corolla united at the base for
0.5–1 mm, yellow or greenish yellow, sometimes tinged brownish purple, glabrous
or minutely papillate towards base; lobes spreading to reflexed, ovate or elliptic,
(4–)5–8 × (2–)3–5 mm, acute. Corona lobes attached ± 1 mm above base of
staminal column, laterally compressed, complicate above, solid below line from
point of attachment to column to distal end of upper margin, quadrate, 2–4 ×
2–3 mm, upper margins ± level with top of column, straight or sometimes weakly
toothed at about the middle, produced into a pair of teeth at the proximal end ±
reaching head of column, lower margin spreading or ascending. Anther wings ± 2
× 0.5 mm; anther appendages ± 1 × 1 mm, broadly ovate, obtuse, inflexed over apex
of stylar head; corpusculum black, subcylindrical, 0.4 mm long with narrow
translucent flanges running up sides from attachment of translator arms; translator
arms ± 0.3 mm long, broadened gradually to attachment of pollinium; pollinia
obovate, ± 1.2 × 0.5 mm. Follicles fusiform or lanceolate in outline, 6–9 × 1.5–2 cm,
pubescent, usually only one of the pair developing; fruiting pedicel contorted to
hold follicle erect. Seeds suborbicular, 5–6 × 4–5 mm, with a convoluted margin
and a darker verrucose disc; coma ± 3 cm long. Figs. 103/10, p. 424.

Tanzania. Ufipa District: Mumba Mission, SE of Sumbawanga, 24 Nov. 1994, *Goyder et al.* 3805!;
    Njombe District: Ludewa area, Itimbo, 6 Nov. 1989, *Mwasumbi et al.* 13411!; Songea District,
    Matengo, Miyao, near Rest Camp, 20 Nov. 1956, *Semsei* 2613!
Distr. T 4, 7, 8; from Angola in the west, through the Shaba plateaus of Congo-Kinshasa, to
    the highlands of S Tanzania, Malawi, Zimbabwe and neighbouring territories
Hab. Montane, fire-prone grassland and open *Protea* bushland; (1000–)1600–2400 m

Syn. *Asclepias swynnertonii* S. Moore in J.L.S. 40: 142 (1911)
    A. nyikana Schltr. in E.J. 51: 138 (1913). Type: Tanzania, Rungwe District: Kyimbila, *Stolz*
    105 (B†, holo.; BM!, K!, iso.)
    [*A. stolzii* Schltr. in sched. – based on *Stolz* 105 (type of *A. nyikana*)]

9. **Gomphocarpus munonquensis** (*S. Moore*) *Goyder & Nicholas* in K.B. 56: 812
(2001). Type: Angola, Munonque, *Gossweiler* 3534 (BM!, holo.)

Erect perennial herb arising from stout woody rootstock; stems erect, few, 30–50 cm long, unbranched with a dense indumentum of spreading white hairs. Leaves opposite, coriaceous, ovate, 4–6 × 1.5–3 cm, obtuse or rounded, slightly cordate at the base, marginal vein scabrid, midrib with spreading white hairs, remainder of lamina glabrous. Inflorescences extra-axillary with 6–9 flowers in a nodding umbel; peduncles 1.5–5 cm long, pubescent; bracts filiform or lanceolate, 0.7–1.5 cm long, to 3 mm wide, ciliate; pedicels 2.5–4.5 cm long, pubescent. Calyx lobes triangular or oblong, 6–8 × 2–3 mm, acute or obtuse, sparsely pubescent, ciliate. Corolla united at base for 2–3 mm, greenish yellow, purplish at base, glabrous or minutely papillate towards base; lobes strongly reflexed, ovate or elliptic, 13–17 × 7–10 mm, acute. Corona lobes similar to *G. swynnertonii* but measuring 7–9 × 6–7 mm, the upper margin higher than the staminal column and more strongly toothed, outer margin more strongly curved than in *G. swynnertonii*. Anther wings narrowly triangular but with a convex outer margin, 3–3.5 × 1.5 mm; anther appendages broadly ovate, ± 1.5 × 1.5 mm, inflexed over apex of stylar head; corpusculum black, ovoid, 0.5 × 0.2 mm, with translucent flange 0.05 mm wide running up sides from attachment of translator arms; translator arms ± 0.7 × 0.15 mm, flattened and ± uniform in width; pollinia oblanceolate, ± 1.6 × 0.6 mm. Follicles not seen. Figs. 103/11, p. 424.

TANZANIA. Mpanda District: Kampisa, Sitebi, 22 Dec. 1997, *Congdon* 501!; Kigoma District: Kasye Forest, 1 Mar. 1990, *Congdon* 257!
DISTR. **T** 4; Angola (two collections)
HAB. Grassland or open woodland; 900–1350 m

SYN. *Asclepias munonquensis* S. Moore in J.B. 50: 341 (1912)

NOTE. This species is essentially a large and robust version of *Gomphocarpus swynnertonii* differing most noticably in the more strongly developed corona lobes. Such distribution and habitat details as exist suggest that it occurs at lower altitudes than *G. swynnertonii*, perhaps growing in woodland rather than montane grassland.

10. **Gomphocarpus glaucophyllus** *Schltr.* in E.J. 18, Beibl. 45: 19 (1894). Type: South Africa, Mpumalanga, near Barberton, *Galpin* 663 (B†, holo.; K!, NH!, PRE!, SRGH!, iso.)

Perennial herb arising from stout woody rhizomatous rootstock; stems ascending or erect, 1–4, robust and somewhat fleshy, unbranched, 0.3–1 m long, glabrous and commonly glaucous. Leaves opposite, sessile or subsessile, ovate to lanceolate, (4–)7–12 × (1.3–)3–6 cm, acute, cordate at the base, margins smooth, glabrous, generally glaucous with a waxy bloom on both surfaces; veins prominent, sometimes reddish. Inflorescences extra-axillary with (6–)9–15 flowers in a nodding umbel; peduncles 1.5–6 cm long, mostly $\frac{1}{2}$–$\frac{2}{3}$ length of subtending leaf, glabrous and generally glaucous; bracts filiform to linear, 5–10 × 0.5–1 mm, margins ciliate or not; pedicels 2–3.5 cm long, glabrous, glaucous or not. Calyx lobes oblong to broadly ovate, 7–10 × 2–4 mm, apex obtuse to acute, glabrous, green or glaucous. Corolla united at the base for 1–2 mm, greenish yellow, sometimes purplish on outer face, minutely papillate particularly towards base; lobes strongly or weakly reflexed, ovate, 9–12 × 5–8 mm, acute. Corona lobes attached ± 1.5 mm above base of staminal column, similar to *G. swynnertonii* but measuring 4–6 × 3–6 mm, upper margin ± level with head of staminal column. Anther wings triangular, 2–2.5 × 0.5–1 mm, outer margin straight or weakly concave; anther appendages broadly ovate, ± 1 × 1 mm, obtuse, inflexed over apex of stylar head; corpusculum black, ovoid-subcylindrical, ± 0.4 × 0.15 mm, with translucent flanges running up sides from attachment of translator arms; translator arms ± 0.5 × 0.15 mm, flattened and ± uniform in width but differentially thickened – inner margin thicker near corpusculum, outer margin thicker near pollinium; pollinia oblanceolate, ± 1.2 ×

FIG. 105. *GOMPHOCARPUS PRATICOLA* — **1**, rootstock and lower stem, × ²/₃; **2**, flowering shoot, × ²/₃; **3**, pollinarium, × 14. 1 from *Bullock* 1877; 2 from *Goyder et al.* 3812; 3 from *Sanane* 1415. From K.B. 56: 817, drawn by E. Papadopoulos.

0.4 mm. Follicles fusiform, 5.5–10 × 1.5–3.5 cm, with occasional longitudinal wings and an acute apex or oblong-elliptic in outline and somewhat inflated with an obtuse apex, glabrous, glaucous, usually only one of the pair developing; fruiting peduncle contorted to hold follicle erect. Seeds suborbicular, ± 6 × 5 mm, with a slightly convoluted, pale and minutely pubescent margin surrounding a verrucose pubescent disc; coma ivory, ± 4 cm long. Figs. 103/9 & 104/8, pp. 424 & 425.

UGANDA. Kigezi District: N Rukiga, Shumba Hill, Aug. 1949, *Purseglove* 3094!; Ankole, 6 Sep. 1905, *Dawe* 378!
TANZANIA. Mbulu District: Pienaars Heights or Dauar, between Babati and Bereko, 7 Jan. 1962, *Polhill & Paulo* 1088!; Iringa District:, Uchindire, below Mufindi Scarp, Oct. 1964, *Procter* 2660!; Songea District: Matengo Hills, Mtama, 9 Jan. 1956, *Milne-Redhead & Taylor* 8161!
DISTR. U 2; T 2, 5, 7, 8; Rwanda, Burundi, Zambia, Malawi, Zimbabwe, Swaziland and South Africa
HAB. Grassland or open *Brachystegia* or mixed decidous woodland; 700–2400 m. Flowering mostly October to December

SYN. *Asclepias glaucophylla* (Schltr.) Schltr. in J.B. 34: 455 (1896)
  A. lilacina Weim. in Bot. Notis. 1935: 374 (1935). Type: Zimbabwe, between Nyanga and Rusape, *Fries, Norlindh & Weimarck* 3062 (LD!, holo.; BM!, iso.)

11. **Gomphocarpus praticola** (*S. Moore*) *Goyder & Nicholas* in K.B. 56: 815 (2001), as *praticolus*. Type: Angola, Kuelai, near R. Chipumba ['Sera Pinto, Bie, Munonque' in LISC], *Gossweiler* 3532 (BM!, holo., LISC!, iso.)

Perennial herb arising from a stout woody rhizomatous roostock; stems ascending or erect, 1–3, robust and somewhat fleshy, unbranched, (0.4–)0.6–1 m long, glabrous or rarely with two bands of spreading white hairs along upper parts of stem, glaucous or green. Leaves opposite, held erect, sessile, oblong to elliptic or rarely lanceolate, (5–)6–12(–16) × (1.5–)2–5.5 cm, apex acute or obtuse, base cordate, margins smooth, glabrous, green or glaucous on both surfaces; veins conspicous or obscure. Inflorescences extra-axillary with 6–15 flowers in a nodding umbel; peduncles 1.5–4.7(–7) cm long, usually much shorter than the subtending leaf, glabrous or softly pubescent; bracts variable in size even within an inflorescence, filiform to elliptic or oblanceolate, 10–35 × 0.2–10(–15) mm, acute or obtuse, green or glaucous, glabrous but sometimes ciliate towards tip; pedicels commonly purple, 2.5–5(18.5) cm long, glabrous or with soft spreading hairs. Calyx lobes elliptic to obovate, rarely linear, 5–17(–27) × (2–)4–10 mm, apex obtuse or rarely acute, glabrous, green, glaucous, purple or brown. Corolla united at base for ± 3 mm, green or yellow-green with reddish purple markings outside, papillate towards base; lobes strongly reflexed, ovate, (14–)16–35 × 7–15 mm, acute. Corona lobes attached 1–2 mm above base of the staminal column, laterally compressed but with the half of the lobe furthest from the column subcylindrical and solid except for the rim, outer margin of lobe drawn in apically towards column, brown or purple with greenish yellow upper margin, (5–)7–12 × (4–)6–9 mm, upper margins generally well above head of staminal column, conspicuously toothed at proximal end and half way along upper margin. Anther wings 3–3.5 × 1 mm, outer margin ± straight or weakly curved; anther appendages broadly ovate, 1–1.5 × 1–1.5 mm, obtuse, inflexed above over apex of stylar head; corpusculum black, ovoid, 0.5 × 0.2 mm, with broad translucent wings extending up sides; translator arms ± 0.7 × 0.15 mm, flattened and generally broader distally; pollinia obovate to oblanceolate, 1–1.6 × 0.5 mm. Follicles fusiform, ± 8 × 2 cm, inflated, usually only one of the pair developing, glabrous, glaucous; fruiting pedicel contorted to hold follicle erect. Seeds suborbicular, ± 7 × 4.5 mm, with a slightly convoluted, pale and minutely pubescent margin surrounding a verrucose pubescent disc; coma ivory, ± 4 cm long. Figs. 103/8 & 105, pp. 424 & 434.

TANZANIA. Ufipa District: 2 km from Sumbawanga on Mbisi road, 25 Nov. 1994, *Goyder et al.* 3812! & Ufipa Plateau, 10 km from Mumba towards Sumbawanga, 22 Dec. 1986, *Moyer & Sanane* 82! & Malonje, 21 Nov. 1949, *Bullock* 1877!

DISTR. **T** 4; contiguous upland areas of Zambia, N Malawi, extending westwards through S Congo-Kinshasa into Angola

HAB. Grassland or open *Brachystegia/Uapaca* woodland on sandy or rocky ground; 1500–2400 m

SYN. *Asclepias praticola* S. Moore in J.B. 50: 341 (1912)
    A. katangensis S. Moore in J.B. 50: 340 (1912), *non A. katangensis* De Wild. (1904). Type: Congo-Kinshasa, Katanga, Lovoi R., *Kässner* 3353 (BM!, holo.)
    A. friesii Schltr. in Fries, Wiss. Ergebn. Schwed. Rhod.-Kongo-Exped. 1: 267 (1916). Type: Zambia, Mtali, in marsh N of L. Benguela, *Fries* 1177a (UPS!, holo.; K!, photo)
    A. moorei De Wild. in Contr. Fl. Katanga: 156 (1921). Type as for *A. katangensis* S. Moore

77. **PACHYCARPUS**

E. Mey., Comm Pl. Afr. Austr. 209 (1838); Goyder in K.B. 53: 335–374 (1998) & in F.P.A. 58: 96–103 (2003)

Gomphocarpus sect. *Pachycarpus* (E. Mey.) Decne. in DC., Prodr. 8: 562 (1844); K. Schum. in E. & P. Pf. 4(2): 236 (1895), *pro parte*

Perennial herbs with annual stems arising from a tuberous rootstock; latex milky; stems 1 to many, simple or branched below, erect or ascending, pubescent with short spreading hairs. Leaves opposite, petiolate; lamina generally with prominent secondary veins anastomosing to form an undulating submarginal vein, and an indumentum of scabrid hairs particularly along the margins. Inflorescences terminal and extra-axillary, umbels nodding or erect, sessile or more commonly pedunculate. Calyx lobed to the base. Corolla rotate to campanulate (reflexed in some southern African species). Corolline corona absent. Gynostegial corona of 5 lobes arising near the base of the staminal column (at the top of a gynostegial stipe in *P. medusonema*), spreading or ascending, generally dorsiventrally flattened and usually with a pair of triangular or quadrate appendages arising near the base of the upper surface, but sometimes appearing ± cucullate or pouched, and generally with an apical tongue; upper parts sometimes minutely verrucose or papillate, the remainder of the lobe generally glabrous. Staminal column stout, anther wings vertical or spreading; anther appendages inflexed over stylar head; corpusculum ovoid, occasionally narrower; translator arms flattened and geniculate, frequently with a short clasping overlap with the pollinia; pollinia flattened, sometimes only weakly, and usually broadest distally. Stylar head truncate, ± level with head of staminal column or slightly raised above it. Follicles single by abortion, held ± erect by contortion of the fruiting pedicel, weakly to strongly inflated, frequently with longitudinal wings or ridges, minutely puberulent or glabrous. Seeds flattened, rarely ovoid, generally differentiated into a verrucose disc and an inflated or convoluted rim; coma of silky hairs.

    37 species in tropical and subtropical Africa. Growing mostly in montane grassland and open woodland in the mountains of eastern Africa, and the grasslands of southern Africa from Limpopo Province to the Eastern Cape.

1. Corolla rotate to rotate-campanulate 2
   Corolla campanulate to subglobose 4
2. Peduncles mostly at least 4 cm long; roots a fascicle of horizontal fusiform, fleshy tubers; follicles often strongly inflated, not ornamented with longitudinal wings 3
   Peduncles absent or shorter than 3 cm long; rootstock generally a vertical napiform tuber; follicles generally with 4–6 longitudinal wings distally, never strongly inflated . . . . *3. P. concolor*

3. Corona lobes widest near base, ± as tall as column
    proximally; anther-wing suture parallel to axis of column   1. *P. lineolatus*
   Corona lobes narrower at the base than above, $^1/_2$–$^2/_3$
    height of column proximally; anther-wings angled to
    give a truncate-conical appearance to the column . . . .   2. *P. bisacculatus*
4. Corolla lobes less than 8 mm long 5
   Corolla lobes at least 9 mm long 6
5. Corona lobes with subulate apical tooth 2–3 mm long
    arching over head of column . . . .   10. *P. firmus*
   Corona lobes without an apical tooth . . . .   9. *P. petherickianus*
6. Corona lobes shorter than or ± as long as column 7
   Corona lobes at least 1.5 × as long as column 8
7. Corolla and corona white or cream, sometimes tinged
    pink; corona with an erect triangular tooth arising from
    middle of inner face . . . .   8. *P. goetzei*
   Corolla and corona green or brownish green; corona
    with no tooth on inner face . . . .   11. *P. pachyglossus*
8. Corolla green or brownish green; corona lobes broadly
    ovate . . . .   12. *P. richardsiae*
   Corolla white, cream or reddish purple; corona lobes
    oblong to spatulate 9
9. Basal part of corona lobes spreading outwards from
    staminal column . . . .   4. *P. spurius*
   Basal part of corona lobes ascending or erect 10
10. Corona lobes with single dorsally flattened tongue in
    middle of inner face . . . .   6. *P. eximius*
   Corona lobes with pair of teeth on inner face 11
11. Teeth on inner face of corona lobes arising from margins   5. *P. grantii*
   Teeth on inner face of corona lobes arising from mid-
    line of lobes . . . .   7. *P. distinctus*

1. **Pachycarpus lineolatus** (*Decne.*) *Bullock* in K.B. 8: 333 (1953). Type: Angola, *da Silva* s.n. (P!, holo.; K!, photo.)

Perennial herb with annual stems arising from a fascicle of fleshy fusiform tubers; stems erect or ascending, 0.4–1.5 m long, one to many, usually unbranched, but sometimes branched below, densely pubescent with spreading white hairs. Leaves with petiole (2–)5–13 mm long, pubescent; lamina narrowly lanceolate to broadly ovate, elliptic or oblong, 4–12(–14) × 1–6(–9) cm, apex obtuse or slightly retuse to acute, base cordate, truncate or rounded, margins scabrid, venation prominent with numerous parallel secondary veins at right angles to the midrib, indumentum of slightly scabrid spreading white hairs on both upper and lower surfaces. Inflorescences terminal and extra-axillary with 4–12 flowers in a nodding umbel; peduncles erect or ascending, (2–)4–17 cm long, densely spreading-pubescent; bracts filiform, (2–)5–14 mm long, pubescent; pedicels 1–3(–4) cm long, pubescent, commonly tinged with purple. Calyx lobes lanceolate, (4–)6–11 × (1–)2–3 mm, acute, densely pubescent, green or tinged with purple particularly towards the apex and margins. Corolla rotate or saucer-shaped, underside cream or brownish purple, sometimes with one or more longitudinal purple stripes, sparsely to densely pubescent, upper surface cream or pale green, often marked with a network of purple or chocolate-brown lines in the distal half, minutely papillate and with short hairs towards the apex and margins; lobes ovate, 8–17 × (4–)6–9(–11) mm, acute. Corona lobes arising from the base of the staminal column, solid, fleshy, 7–8 × (1–)3–4 mm, narrowing gradually towards the apex, spreading or ascending from the staminal column, with a pair of triangular or quadrate vertical plates arising from the upper surface, 4–5 mm high at the proximal end reaching to the top of the column, the proximal margins resting against the column, the upper margins meeting to form an enclosed chamber, white or cream with a reddish purple patch on the underside of the fleshy basal plate, often restricted to an area just below the tip, upper parts minutely verrucose or papillate, the remainder of the lobe generally glabrous. Staminal column ± 6 mm long, fertile part forming a cylinder ± 3 × 4–6 mm; anther wings vertical, triangular, 2–3.5 × 1 mm, extending below the fertile part for ± 1.5 mm; anther appendages ovate or semicircular in outline ± 1.5 × 1 mm, and inflexed over stylar head; corpusculum black, ovoid, 0.5–0.6 × 0.2–0.3 mm; translator arms 0.4–0.7 × 0.15–0.2 mm, flattened, with a further 0.2–0.3 mm clasping overlap with the pollinium; pollinia irregularly clavate in outline, 0.8–1.2 × 0.4–0.7 mm, flattened. Stylar head flat, white or green. Follicles and seeds as in *P. bisacculatus*. Fig. 106/1–8, p. 439.

UGANDA. Kigezi District: Ruzhumbura, May 1946, *Purseglove* 2056!; Teso District, Serere, May 1932, *Chandler* 535!; Masaka District: Makole, 21 Apr. 1935, *A.S. Thomas* 1258!
KENYA. West Suk District: Kapenguria, May 1932, *Napier* 2019!; Meru District, Kathiri Hill, Tigania, Nov. 1955, *Adamson* 522!; Teita District: Mbololo Hill, 14 Feb. 1953, *Bally* 8577!
TANZANIA. Mbulu District: near Shesheda, NE slopes of Mt Hanang, 7 Feb. 1946, *Greenway* 7629!; Ufipa District, 10 km W of Mkowe on road to Chapota, 21 Nov. 1994, *Goyder et al.* 3779!; Mbeya District: 8 km NE of Tunduma, 10 Jan. 1975, *Brummitt & Polhill* 13686!
DISTR. U 2–4; K 2–4, 7; T 1–8; widespread from Ivory Coast to South Sudan and S to Angola and Malawi
HAB. Open deciduous woodland or seasonally waterlogged grassland; 700–2500 m

SYN. *Gomphocarpus lineolatus* Decne. in Ann. Sci. Nat. Bot. Sér. 2, 9: 326 (1838)
    Asclepias lineolata (Decne.) Schltr. in J.B. 33: 336 (1895)
    A. schweinfurthii N.E. Br. in K.B. 1895: 253 (1895) & in F.T.A. 4(1): 323 (1902). Type: South Sudan, Jur, *Schweinfurth* 1960 (K!, holo.)
    A. conspicua N.E. Br. in K.B. 1895: 253 (1895) & in F.T.A. 4(1): 324 (1902). Type: Zambia, Fwambo, *Carson* 12 (K!, holo.)
    ?*Calotropis busseana* K. Schum. in E.J. 33: 323 (1903); N.E. Br. in F.T.A. 4(1): 615 (1904). Type: Tanzania, Lushoto District: Usambara, *Busse* 341 (B†, holo.)
    Asclepias browniana S. Moore in J.B. 47: 217 (1909). Type: Congo-Kinshasa, Lake Moero, *Kässner* 2806 (BM!, lecto., designated by Goyder in K.B. 53: 345 (1998))
    Pachycarpus schweinfurthii (N.E. Br.) Bullock in K.B. 8: 330 (1953) & in F.W.T.A. ed. 2, 2: 93 (1963); U.K.W.F. ed. 2: 178, t. 69 (1994)
    P. mildbraedii Bullock in K.B. 8: 334 (1953). Type: Cameroon, Bouar, *Mildbraed* 9391 (K!, holo.)

2. **Pachycarpus bisacculatus** (*Oliv.*) *Goyder* in K.B. 51: 798 (1996) & in Fl. Eth. 4(1): 127 (2003). Type: Tanzania, Kilimanjaro, *Johnston* s.n. (K, holo.)

Perennial herb with annual stems arising from a fascicle of fleshy fusiform tubers; stems erect or ascending, 0.3–1.5 m long, usually single and unbranched, but sometimes branched below, densely pubescent with spreading white hairs. Leaves with petiole (3–)5–13 mm long, pubescent; lamina lanceolate to ovate or oblong, 6–12 × 2–7 cm, apex obtuse or slightly retuse to acute, base cordate or truncate, margins scabrid, venation prominent with numerous parallel secondary veins at right angles to the midrib, indumentum of slightly scabrid spreading white hairs on both upper and lower surfaces. Inflorescences terminal and extra-axillary with 4–8(–14) flowers in a secund umbel; peduncles erect or ascending, (2–)4–17 cm long, densely spreading-pubescent; bracts filiform, 5–12 mm long, pubescent; pedicels (1–)1.5–3 cm long, pubescent, tinged with purple. Calyx lobes lanceolate, 6–10 × 2–3 mm, acute, densely pubescent, purplish. Corolla rotate or saucer-shaped, underside brownish purple, densely pubescent, upper surface pale green with a network of purple or chocolate brown lines particularly in the distal half, minutely papillate and with short hairs towards the apex and margins; lobes ovate, 10–15 × 7–10 mm, acute. Corona lobes 5–6 × 4–6 mm, arising from the base of the staminal column and adnate to it to the base of the anthers, cream or more usually pink with a purple tip, spreading and strongly complicate at extreme base with proximal margins parallel and erect

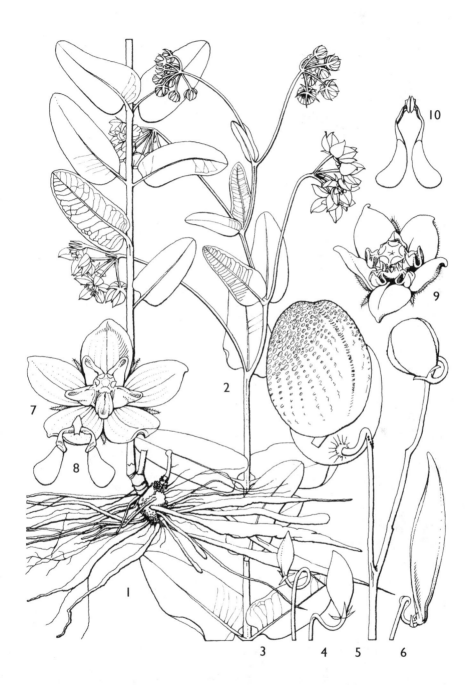

Fig. 106. *PACHYCARPUS LINEOLATUS* — **1** & **2**, habit, × ½; **3–6**, variation in follicle age and degree of inflation, × ½; **7**, flower, × 1½; **8**, pollinarium, × 10. *PACHYCARPUS BISACCULATUS* × **9**, flower, × 1½; **10**, pollinarium, × 10. 1, 7 & 8 from *Goyder et al.* 3779; 3, 4 & 6 from *Dummer* 385; 5 from *Bullock* 2277; 9 & 10 from *Keay* in FHI 25897. From K.B. 53: 348, drawn by E. Papadopoulos.

reaching to the base or middle of the anthers, distal half or two thirds forming an erect or incurved cone with involute margins, flanges between the base of this apical cone and the proximal margins of the lobes sinuous and dilated to form a pair of shallow pouches, the upper and inward-facing parts of the lobes minutely but densely papillate, the outer faces glabrous. Staminal column 6–8 mm long, fertile part forming a cylinder 3–4 × 4–6 mm; anther wings triangular, 3 × 1.5 mm, not extending below fertile part of column, giving the column a truncate conical appearance; anther appendages 2.5–3 × 1–1.5 mm, semicircular in outline and inflexed over apex of stylar head; corpusculum black, ovoid, 0.5 × 0.2 mm; translator arms ± 0.8–1 × 0.15 mm, flattened and geniculate, with a further 0.3–0.4 mm clasping overlap with the pollinium; pollinia irregularly clavate in outline, 1.4–1.6 × 0.7 mm, flattened. Stylar head flat, white or green. Fruiting pedicel contorted to hold follicle erect; follicle ovoid or fusiform, to 12 × 7 cm, generally inflated, without wings or processes, pubescent. Seeds (immature) ovate, 4.5 × 3 mm, with a verrucose disc and a narrow inflated rim; coma ± 3 cm long. Fig. 106/9–10, p. 439.

UGANDA. Karamoja District: Iriri, June 1957, *J. Wilson* 331!; Bunyoro, June 1907, *E. Brown* 400!; Teso District: km 85 on Lira–Soroti road, 17 May 1961, *Lind* 3031!
KENYA. North Kavirindo District: Webuye [Broderick Falls], Apr. 1939, *Tweedie* 451!; Kwale District: Shimba Hills, Mwele Mdogo Forest, 4 Feb. 1953, *Drummond & Hemsley* 1102! & Shimba Hills National Park 'Elephant Lookout', 27 Dec. 1980, *Gilbert* 6015!
TANZANIA. Ufipa District: Lake Sundu, 23 Nov. 1960, *Richards* 13597!; Mbeya District: Songwe, 12 Feb. 1979, *Cribb et al.* 11422!; Songea District: 3 km W of Gumbiro, 26 Jan. 1956, *Milne-Redhead & Taylor* 8549!
DISTR. U 1–4; K 3, 5, 7; T 1, 2, 4, 6–8; widespread from Guinea Bissau to South Sudan and Ethiopia and S to Angola and Mozambique
HAB. Seasonally waterlogged grassland and *Brachystegia* or mixed deciduous woodland; 100–2500 m

SYN. *Gomphocarpus bisacculatus* Oliv. in Trans. Linn. Soc., Bot. Ser. 2, Bot. 2: 341 (1887)
    Asclepias lineolata sensu Hutchinson & Dalziel in F.W.T.A. 2(1): 56 (1931), *non* Decne.
    Pachycarpus lineolatus sensu Bullock in K.B. 8: 333 (1953) & in F.W.T.A. ed. 2, 2: 93 (1963) & sensu Blundell, Wild Fl. E. Afr.: t. 571 (1987) & sensu U.K.W.F. ed. 2: 178 (1994), *non* (Decne.) Bullock
    P. bullockii Cavaco in Bull. Mus. Hist. Nat. (Paris), Sér. 2, 29: 514 (1958). Type: Angola, NE of Lunda, near Cassai stream, *Gossweiler* 13891 (P!, holo.; K!, iso.)

NOTE. Field observations in Zambia and Tanzania by Bullock and Milne-Redhead in K.B. 8: 333 (1953) show that the flowering periods of this species (referred to as *P. lineolatus*) and the closely related *P. lineolatus* (referred to as *P. schweinfurthii*) differ by about six weeks.

3. **Pachycarpus concolor** E. Mey., Comm Pl. Afr. Austr. 210 (1838); N.E. Br. in F.T.A. 4(1): 378 (1902). Type: South Africa, Eastern Cape, between Chalumna R. and Kachu (Yellowwood) R., *Drège* s.n. (K!, lecto., designated by D.M.N.Smith in South African. J. Bot. 54: 411 (1988)); MEL, P!, iso.)

Perennial herb with annual stems arising from a stout, cylindrical, vertical tuber; stems procumbent to erect even in the same population, 0.2–1 m long, one to several, usually unbranched but sometimes branched below, densely pubescent with spreading white or rusty hairs. Leaves spreading, petiole 1–10 mm long, pubescent; lamina linear to narrowly lanceolate or triangular [or lanceolate to ovate outside FTEA area], 5–16 × 0.6–4 cm, widest at or just above the base then tapering gradually into the acute or subacute apex, base shortly cuneate or truncate, margins scabrid, plane or crisped, slightly scabrid with spreading white or rusty hairs on both upper and lower surfaces. Inflorescences terminal and extra-axillary with (1–)2–7 scented flowers in an erect umbel, sessile or with peduncles to 3 cm long, densely spreading-pubescent; bracts filiform, 4–14 mm long, pubescent; pedicels 1–3 cm long, pubescent. Calyx lobes subequal to strongly unequal, longest ovate to lanceolate, 12–16 × 4–8 mm, acute, pubescent, sometimes reddish towards tip. Corolla rotate or broadly campanulate, underside tinged reddish brown, densely pubescent, upper surface white, cream or pale green, with or without purple markings, glabrous or minutely papillate; lobes broadly ovate, 11–15(–18) × 8–13 mm, subacute. Corona lobes dorsiventrally flattened, arising from the base of the staminal column; basal part spreading, rhomboid or continuous with the distal part, ± 3 × 3–6 mm, with an erect pair of pyramidal or deltoid obtuse teeth 2–3.5(–4.5) × 1.5–3.5 mm arising from the middle of this part and reaching ± the top of the staminal column; distal part lingulate or slightly spatulate, 4–9(–15) × 1.5–2.5 mm, rounded or truncate, erect or incurved over the head of the staminal column, glabrous except for a minutely but densely papillate central strip on upper side, entire lobe white or cream, or dull purple at the base with a green tongue. Staminal column truncate-conical, 3–4 × 7–8(–9) mm; anther wings triangular, 3–4 × 1.5–3 mm, appendages broadly ovate, 1.5–2.5 × 1.5–2.5 mm, acute and inflexed over apex of stylar head; corpusculum black, broadly ovoid or ovoid-conical, 0.8 × 0.4–0.6 mm; translator arms 0.8–1 mm long, flattened, ± triangular and weakly geniculate, 0.4 mm wide at attachment to pollinia and with a clasping overlap of ± 0.2 mm with the pollinium on the ventral face; pollinia elliptic in outline, 1.2–1.8 × 0.7–0.8 mm, somewhat flattened. Stylar head flat. Fruiting pedicel contorted to hold follicle erect; follicle fusiform, to 12 × 3 cm, not generally inflated, with 6 narrow longitudinal wings or ridges in the upper half, minutely pubescent. Seeds ovate, 5 × 3 mm, and slightly convex on both faces, pubescent, disc not clearly differentiated from rim; rim with transverse ridges; coma ± 4 cm long.

SYN. *Xysmalobium concolor* (E. Mey.) D. Dietr., Synop. Pl. 2: 902 (1840)
 Gomphocarpus concolor (E. Mey.) Decne. in DC., Prodr. 8: 563 (1844)
 Asclepias concolor (E. Mey.) Schltr. in E.J. 21, Beibl. 54: 6 (1896) & in J.B. 1896: 452 (1896)

 subsp. **rhinophyllus** (*K. Schum.*) *Goyder* in F.P.A. 58: 101 (2003). Type: Kenya, Victoria Nyanza, *Fischer* 399 (B†, holo.; K!, iso.)

 Leaves linear to triangular; inflorescences shortly pedunculate.

KENYA. North Nyeri District: Zawadi Estate, 7 km on Nyeri–Kiganjo road, May 1974, *Faden et al.* 74/574!; Masai District: Oldebesi Lemoko, 28 Apr. 1961, *Glover et al.* 815!; Masai/Machakos Districts: Chyulu Hills, near Half Crater, 22 Oct. 1990, *Luke* 2455B!
TANZANIA. Moshi District: Poverty Gulch Farm, Kilimanjaro, 20 Jan. 1994, *Grimshaw* 94/80! & Simba Farm, W Kilimanjaro, 8 Dec. 1968, *Bigger* 2361!; Lushoto District: West Usambara, 3 km from Mgwashi, 1 Apr. 1979, *Tanner* 208!
DISTR. **K** 3–6; **T** 2, 3; not known elsewhere
HAB. Open burnt grassland; 1450–2100 m

SYN. *Gomphocarpus scaber* K. Schum. in E.J. 17: 124 (1893), *non* Harv. (1863). Type: Kenya, "Victoria Nyanza", *Fischer* 399 (B†, holo.; K iso.)
 G. rhinophyllus K. Schum. in P.O.A. C: 322 (Aug. 1895). Type as for *G. scaber* K. Schum.
 G. trachyphyllus K. Schum. in E. & P. Pf 4(2): 237 (Oct. 1895). Type as for *G. scaber* K. Schum.
 Pachycarpus rhinophyllus (K. Schum.) N.E. Br. in F.T.A. 4(1): 377 (1902); Blundell, Wild Fl. E. Afr.: t. 481 (1987); U.K.W.F. ed. 2: 178 (1994)

4. **Pachycarpus spurius** (*N.E. Br.*) *Bullock* in K.B. 8: 338 (1953). Type: Malawi, Shire Highlands, *Buchanan* 451 (K, holo.)

Perennial herb with 1–5 annual stems arising from a long cylindrical vertical rootstock, sometimes with a few fusiform laterals; stems erect, 0.8–2 m long, unbranched, densely pubescent with spreading white hairs. Leaves with petiole 3–10(–15) mm long, pubescent; lamina lanceolate to broadly ovate, 6–13 × 3–8.5 cm, apex acute or obtuse, base rounded and slightly cordate, margins scabrid, venation prominent with numerous parallel secondary veins, pubescent on both upper and lower surfaces with minute scabrid hairs. Inflorescences terminal and extra-axillary with 4–9 flowers in a nodding or secund umbel; peduncles 2–4.5 cm long, densely pubescent; bracts filiform, ± 10 mm long, pubescent; pedicels 1.5–2.3(–3) cm long, densely pubescent. Calyx lobes ovate to lanceolate, 6–11(–14) × 3–4(–7) mm, acute, pubescent. Corolla globose-campanulate, glabrous or minutely papillate within, pubescent outside, reddish purple or occasionally cream outside, white, cream or purple within, often with darker veins; lobes oblong ovate or obovate, (11–)14–22 × (6–)8–14 mm, acute or obtuse, recurved at the tip. Staminal column stipitate; corona lobes arising from the base of the stipe, dorsiventrally flattened, spreading below, erect or slightly recurved above, spatulate, 10–20 × 4–8 mm, the dilated apex acute to truncate or emarginate, entire or variously toothed. Anther wings triangular, 3–4 × 1–1.5 mm, convex, appendages broadly ovate, 1–2 × 1.5 mm, acute and inflexed over apex of stylar head; corpusculum black, ovoid, 0.8–1 × 0.6 mm; translator arms 0.6–0.8 × 0.1–0.2 mm, flattened; pollinia elliptic or oblanceolate, 1.2 × 0.5–0.6 mm, flattened. Stylar head flat, 3–4 mm diameter, forming a raised circular cushion ± 1 mm above top of anthers. Peduncle not elongating in fruit; fruiting pedicel contorted; follicle ovoid, ± 7 × 3–3.5 cm, somewhat inflated, with 4 or 6 broad longitudinal apically toothed wings, pubescent, not erect. Mature seeds not seen.

Tanzania. Ufipa District: 10 km W of Mkowe on Chapota road, 21 Nov. 1994, *Goyder et al.* 3781!; Mbeya District: Loleza Mt behind Mbeya town, 4 Feb. 1979, *Cribb et al.* 11280!; Songea District: Matengo Hills, 8 km N of Miyau, 1 Mar. 1956, *Milne-Redhead & Taylor* 8923!
Distr. T 4, 5, 7, 8; montane regions of S Congo-Kinshasa, Malawi
Hab. Montane grassland and *Brachystegia* or *Protea* bushland woodland, sometimes in disturbed sites; (1200–)1500–2100 m

Syn. *Xysmalobium spurium* N.E. Br. in K.B. 1895: 251 (1895)
     ?*X. dolichoglossum* K. Schum. in E.J. 28: 456 (1900). Type: Tanzania, Iringa/Mbeta District: Uhehe, Uchungwe (?Udzungwa) Mt, *Goetze* 638 (B†, holo.)
     Schizoglossum spurium (N.E. Br.) N.E. Br. in F.T.A. 4(1): 367 (1902)
     ?*S. dolichoglossum* (K. Schum.) N.E. Br. in F.T.A. 4(1): 367 (1902)
     S. debeersianum K. Schum. in E.J. 33: 323 (1903). Type: Congo-Kinshasa, Haut Marungu, *De Beers* 108 (BR, holo.)

Note. The corona lobes of *Stolz* 553 from Kyimbila in southern Tanzania are unusual in possessing a pair of triangular teeth adnate to the gynostegial stalk. However, these are not attached to the free part of the corona and in all other respects the specimens match *P. spurius.*

5. **Pachycarpus grantii** (*Oliv.*) *Bullock* in K.B. 8: 336 (1953). Type: Tanzania, Bukoba District: Karagwe, *Speke & Grant* 216 (K!, holo.)

Perennial herb with annual stems arising from a slender vertical tuberous rootstock or from white, fleshy roots; stems erect, 0.15 to 1.3 m tall, usually single and unbranched but sometimes branching below and forming small clumps, densely pubescent to villous with spreading white hairs. Leaves with petiole 2–10 mm long, pubescent; lamina lanceolate to ovate or oblong-ovate, rarely broadly ovate, 4.5–10 × 2–5(–6.5) cm, apex acute to obtuse, base rounded or truncate, usually slightly cordate, margins scabrid, venation prominent with numerous parallel secondary veins, indumentum of spreading white hairs on both upper and lower surfaces. Inflorescences terminal and extra-axillary with 5–18 flowers in a secund umbel; peduncles 1.5–6 cm long, elongating in fruit, ascending or erect, densely spreading-pubescent or lanate; bracts filiform, 4–10 mm long, pubescent; pedicels 0.5–2 cm long, densely pubescent. Calyx lobes lanceolate to ovate, 5–10(–13) × 2–5 mm, acute, densely pubescent. Corolla white, cream, dull pink or maroon, campanulate, pubescent on outer face at least above, glabrous within; lobes ovate or oblong, 10–15 × 4–7 mm, apex obtuse or subacute, recurved. Corona lobes arising from the base of the staminal column, maroon or brown, ascending or erect in lower half, erect or somewhat spreading above, oblong or obovate, 7–12 × 2–4.5 mm, dorsiventrally flattened, at least twice length of staminal column, apex obtuse to acute or rarely slightly retuse, lower margins inflexed and produced into a pair of fleshy teeth ± level with top of anthers. Anther wings narrowly triangular, 2–3 × 0.5–1 mm, with widest point just above base or linear oblong; anther appendages ovate, ± 1.5 × 1 mm, acute, erect; corpusculum dark brown, subcylindrical or ovoid, 0.5–0.7 × 0.1–0.2 mm; translator arms 0.2–0.4 × 0.1 mm, flattened; pollinia obovate to oblanceolate, 0.6–1.1 × 0.3–0.4 mm. Stylar head flat, 2–3 mm diameter, forming a raised pentagonal cushion 1–2 mm above top of anthers. Fruiting pedicel contorted to hold follicle erect; follicle narrowly ovoid or fusiform, to 12 × 2.5 cm, with 6 longitudinal wings or ovoid and inflated with no wings (Congo-Kinshasa), pubescent.

SYN. *Schizoglossum grantii* Oliv. in Trans. Linn. Soc., Bot. London 29: 109, t. 74 (1875); N.E. Br. in F.T.A. 4(1): 370 (1902)
    Gomphocarpus grantii (Oliv.) Schltr. in J.B. 33: 269 (1895)
    Asclepias grantii (Oliv.) Schltr. in J.B. 33: 335 (1895)
    Pachycarpus eximius sensu U.K.W.F. ed. 2: 178, t. 68 (1994), *non* (Schltr.) Bullock

NOTE. *Pachycarpus grantii, P. eximius* and *P. distinctus* form a complex of closely allied taxa requiring further investigation.

a. subsp. **grantii**

Robust plant with stems to 1.3 m tall. Leaves generally ovate or oblong ovate, apex obtuse. Corolla white or cream. Corona lobes generally spreading in upper half. Anther wings triangular; corpusculum ovoid; translator arms ± 0.2 mm long; pollinia oblanceolate. Fig. 107/1–4, p. 444.

UGANDA. Karamoja District: Napak, 27 May 1940, *A.S. Thomas* 3590!; Bunyoro District: Bunyoro Ranch, 9 Apr. 1961, *Lind* 2969!; Teso District: Serere, May 1932, *Chandler* 531!
KENYA. Trans-Nzoia District: E Mt Elgon, 13 July 1956, *Irwin* 287! & Mt Elgon, Mar. 1931, *E & C Lugard* 570!
TANZANIA. Bukoba District: Buhamila [Buhamira], Oct. 1931, *Haarer* 2324! & Nshamba, Sept.–Oct. 1935, *Gillman* 545!
DISTR. U 1–4; **K** 3, 5; **T** 1; Congo-Kinshasa, South Sudan
HAB. Wooded grassland; 1000–2000 m

b. subsp. **marroninus** *Goyder* in K.B. 53: 359 (1998). Type: Tanzania, Rungwe District: Poroto Mts, E of Kimondo, *Goyder et al.* 3874 (K!, holo.; DSM!, EA!, PRE!, iso.)

Stems to 40 cm tall, mostly less. Indumentum shorter. Leaves mostly lanceolate, acute, but sometimes ovate and obtuse. Corolla mostly dull pink or maroon, occasionally cream. Corona lobes erect. Anther wings oblong or at most weakly triangular; corpusculum subcylindrical; translator arms ± 0.4 mm long; pollinia obovate. Fig. 107/5–7, p. 444.

TANZANIA. Mbeya District: Mbeya Peak, 13 Dec. 1962, *Richards* 17031!; Rungwe District: Poroto Mts, E of Kimondo, 1 Dec. 1994, *Goyder et al.* 3874! & edge of Kitulo Plateau near Matamba, 2 Dec. 1994, *Goyder et al.* 3886!
DISTR. **T** 7; endemic to the higher plateaux of the southern highlands of Tanzania
HAB. Short montane grassland; 1800–2700 m

NOTE. Differences from the type subspecies are largely a matter of degree, but add up to give a distinctive appearance. As these differences correlate with a geographical separation, it seems appropriate to recognise the two taxa at subspecific rank. It is difficult to link fruiting material with flowering material with certainty, but it appears that follicles of the subsp. *grantii* are shorter than those of subsp. *marroninus* and the wings are much broader distally than in the uniformly narrow wings of the latter subspecies.

6. **Pachycarpus eximius** (*Schltr.*) *Bullock* in K.B. 8:337 (1953). Type: Uganda, Ruwenzori, Virunga, *Scott Elliot* 7627 (BM!, holo.; K!, iso.)

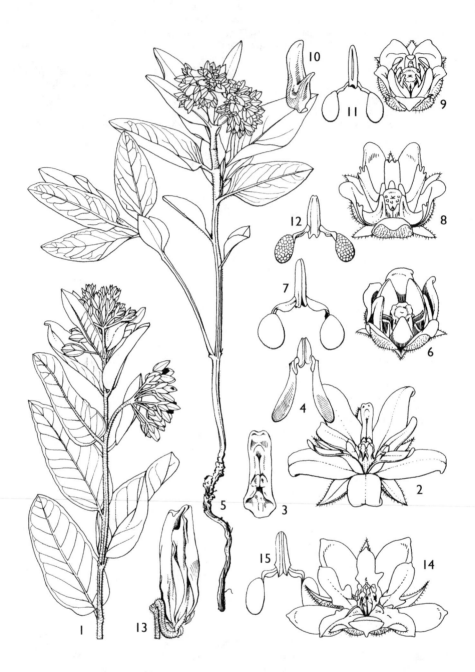

Fig. 107. *PACHYCARPUS GRANTII* subsp. *GRANTII* — **1**, flowering shoot, × ¹/₂; **2**, flower, × 2;
3, corona lobe, × 3; **4**, pollinarium, × 18. *PACHYCARPUS GRANTII* subsp. *MARRONINUS* —
5, habit, × ¹/₂; **6**, flower, × 2; **7**, pollinarium, × 18. *PACHYCARPUS EXIMIUS* — **8** & **9**, flowers,
× 2; **10**, corona lobe, × 3; **11** & **12**, pollinaria, × 18; **13**, follicle, × 1. *PACHYCARPUS
DISTINCTUS* — **14**, flower, × 2; **15**, pollinarium, × 18. 1–4 from *Lugard & Lugard* 570; 5–7
from *Goyder et al.* 3874; 8 & 12 from *Tweedie* 3010; 9–11 from *Bally* 7841; 13 from *Burtt* 2719;
14 & 15 from *Richards* 18035. From K.B. 53: 360, drawn by E. Papadopoulos.

Perennial herb similar in most respects to *P. grantii* with 1–5 annual stems arising from a tuberous rootstock. Inflorescences terminal and extra-axillary with 5–12 flowers in a secund umbel. Calyx lobes often reddish. Corolla white, cream or dull pink. Corona lobes entire or variously toothed at the apex, slightly fleshy towards the base with a narrow dorsally flattened tongue on the inner face, adnate to the corona lobe below, free above from about the top of the staminal column, entire, emarginate or deeply 2–lobed. Follicles with broad, toothed, longitudinal wings. Fig. 107/8–13, p. 444.

UGANDA. Ankole District: Bunyaruguru, Sep. 1938, *Purseglove* 413!; Masaka District: 1 km S of Mutondo, 18 May 1971, *Lye* 6120! & Villa Maria, Masaka, 22 Apr. 1935, *A.S. Thomas* 1266!
KENYA. Elgeyo District: Cherangani Hills, 10 km SW of Kapsowar on road to Kitale, 16 Apr. 1975, *Hepper & Field* 5038!; North Nyeri/Meru Districts: Nanyuki–Meru, 23 Jan. 1932, *van Someren* 1716!; Kericho District: Sotik, 13 June 1950, *Bally* 7841!
TANZANIA. Mbulu District: Katish to Nangwa, Mt Hanang, 4 Apr. 1946, *Greenway* 7570!; Kigoma District: Gombe Stream Reserve, Rutanga Valley, 21 Jan. 1964, *Pirozynski* 266!; Songea District: 44 km N of Hanga Bridge on Njombe Road, 30 June 1956, *Milne-Redhead & Taylor* 10935!
DISTR. U 2, 4; K 2–5; T 1, 2, 4, 5, 8; E Congo-Kinshasa, Rwanda, Burundi and South Sudan
HAB. Grassland; 850–2500 m

SYN. *Asclepias eximia* Schltr. in J.B. 33: 335 (1895)
   Schizoglossum eximium (Schltr.) N.E. Br. in F.T.A. 4(1): 370 (1902)
   Pachycarpus grantii sensu U.K.W.F. ed. 2: 178 (1994), *non* (Oliv.) Bullock
   P. sp. A, Malaisse in Fl. Rwanda 3: 106 (1985)

NOTE. A variant with deeply lobed corona lobes occurs sporadically in Rwanda, Burundi and southern Tanzania and may deserve taxonomic recognition (= *Pachycarpus* sp. A of Fl. Rwanda).

7. **Pachycarpus distinctus** (*N.E. Br.*) *Bullock* in K.B. 8: 335 (1953). Type: Tanzania, Njombe/Songea District: mountains E of Lake Malawi [Nyassa], *Johnson* s.n. (K!, holo.)

Perennial herb similar in most respects to *P. grantii*. Leaf lamina lanceolate to oblong, 4–11 × 1–3.5 cm. Corolla white or reddish purple. Corona lobes spreading or ascending at the base, ascending or erect above, spatulate or obtriangular, 5–13 × 1.5–4.5 mm, entire, emarginate or toothed at the apex, margins sometimes undulate or toothed, with or without a pair of auricles towards the base, inner face with a pair of erect triangular plates along the midline of the lobes between the basal auricles. Anther wings 1.5–2 mm long. Stylar head not exserted from top of column. Follicles not seen. Fig. 107/14–15, p. 444.

TANZANIA. Njombe District: Livingstone Mts, Madunda Mission, 2 Feb. 1961, May 1953, *Eggeling* 6538!; Songea/Njombe escarpment, 26 Mar. 1963, *Richards* 18035!; Kilwa District: Madaba, 4 Mar. 1986, *de Leyser* 99!
DISTR. T 7, 8; endemic to the uplands of SE Tanzania
HAB. Tall grassland and *Brachystegia* woodland; 1000–2000 m

SYN. *Margaretta distincta* N.E. Br. in K.B. 1895: 255 (1895)
   Schizoglossum scyphostigma K. Schum. in E.J. 30: 384 (1901); N.E. Br. in F.T.A. 4(1): 371 (1902). Type: Tanzania, Njombe District: Livingstone Range, Yawulanda Mt, *Goetze* 850 (B†, holo.; BR!, K!, iso.)
   S. distinctum (N.E. Br.) N.E. Br. in F.T.A. 4(1): 371 (1902)

8. **Pachycarpus goetzei** (*K. Schum.*) *Bullock* in K.B. 8: 338 (1953). Types: Tanzania, Iringa District: Uhehe, Kipundi Mts, *Goetze* 674 (B†, syn.); Iringa District: near Gumbira, *Goetze* 706 (B†, syn.); Mbeya District: Kitulo–Matamba road, *Bidgood & Congdon* 154 (K!, neo., designated by Goyder in K.B. 53: 363 (1998); EA!, TPRI, isoneo.)

Perennial herb similar in most respects to *P. eximius* with one to several annual stems, rootstock not seen; stems 0.5 to 1.3 m tall, erect, unbranched. Leaves lanceolate to oblong, (4.5–)7–12 × 1.5–4.5 cm, acute. Inflorescences with 5–18 sweetly scented flowers in a secund umbel; pedicels 1–1.5 cm long, densely pubescent. Calyx often reddish towards tip. Corolla white or cream, often tinged pink; lobes 9–10 × 5–7 mm. Corona lobes ascending or erect with a spreading or slightly recurved tip, cream or white, oblong, 4–5 × 3–5 mm, ± as long as the staminal column, broadened slightly towards the truncate to 3–toothed apex, inner face with an erect triangular tooth in centre of the lobe ± 2 mm long ± reaching the top of the anthers. Anther wings ± 2 × 0.5 mm; corpusculum dark brown, ovoid, 0.8 × 0.4 mm; translator arms 0.5 × 0.2 mm, flattened, obtriangular and with a short clasping overlap with the pollinia; pollinia elliptic, 1 × 0.4 mm. Fruiting pedicel contorted or not; follicle narrowly ovoid, 6–9 × 2–2.5 cm, with 4 broad longitudinal wings, pubescent. Fig. 108/7–8, p. 447.

TANZANIA. Ufipa District: Tatanda Mision, 25 Feb. 1994, *Bidgood, Mbago & Vollesen* 2460!; Mbeya District: Poroto Mts, Ikuyu, 8 km from Irambo, 10 Feb. 1979, *Cribb et al.* 11381!; Rungwe District: Makete, above Ndumbi valley, 17 Jan. 1988, *Lovett & Congdon* 2890!
DISTR. **T** 4, 7; adjacent parts of Zambia and Malawi
HAB. Upland grassland and *Brachystegia* woodland; 1600–2300 m

SYN. *Schizoglossum goetzei* K. Schum. in E.J. 28: 455 (1900); N.E. Br. in F.T.A. 4(1): 371 (1902)
    S. simulans N.E. Br. in F.T.A. 4(1): 369 (1902). Type: Malawi, Nyika Plateau, *Whyte* 108 (K!, lecto., designated by Goyder in K.B. 53: 363 (1998))

9. **Pachycarpus petherickianus** (*Oliv.*) *Goyder* in K.B. 51: 798 (1996) & in Fl. Eth. 4(1): 128 (2003). Type: Trans. Linn. Soc., Bot. London 29: t. 118B (1875) (lecto., designated by Goyder in K.B. 51: 798 (1996))

Perennial herb with annual stems arising from a tuberous rootstock; stems erect, (15–)30–60 cm tall, densely pubescent with spreading white hairs. Leaves with petiole 2–6 mm long, pubescent; lamina linear-lanceolate to ovate, rarely broadly ovate, (3–)5–8 × (0.5–)1–4 cm, apex acute to obtuse, base rounded or truncate, usually slightly cordate, margins scabrid, venation prominent with numerous parallel secondary veins, indumentum of spreading white hairs on both upper and lower surfaces. Inflorescences terminal and extra-axillary with 5–10(–25) flowers in a secund umbel, ascending or erect, subsessile or with peduncles to ± 4 cm long, densely spreading-pubescent or lanate; bracts filiform, 4–6 mm long, pubescent; pedicels 0.5–0.8(–1.6) cm long, densely pubescent. Calyx lobes narrowly lanceolate to ovate, 3–8 × 1–2 mm, acute, densely pubescent. Corolla yellowish green or white, occasionally flushed pink, campanulate, pubescent on outer face at least above, glabrous within; lobes ovate or oblong, 5–8 × 2–4.5 mm, apex subacute or acute, recurved. Corona lobes arising from the base of the staminal column, ascending or erect, green or white, purple in Ethiopian material, slightly fleshy, oblong, 4–5 × 2–3 mm, dorsiventrally flattened or somewhat convex, ± 1.5 times height of gynostegium, apex 3–toothed, the middle tooth sometimes suppressed, margins inflexed for ± ²⁄₃ of their length to form a pair of fleshy plates. Anther wings narrowly triangular, 2 × 0.5 mm; anther appendages ovate, ± 1 × 0.5 mm, acute, erect; corpusculum brown, ovoid, 0.3–0.6 × 0.1–0.2 mm; translator arms 0.2–0.4 × 0.1 mm, flattened; pollinia elliptic, 0.5 × 0.2–0.3 mm. Stylar head flat, 2–3 mm diameter, forming a raised pentagonal cushion 1–2 mm above top of anthers. Fruiting pedicel contorted to hold follicle erect; follicle narrowly ovoid, ± 5 × 1.5 cm, with 4 weakly toothed longitudinal wings, pubescent. Seeds ovate, ± 7 × 5.5 mm, with a verrucose disc and a weakly convoluted rim. Fig. 108/11–15, p. 447.

UGANDA. West Nile District: Kijomoro [Kajamoro], 12 June 1936, *A.S. Thomas* 2013!; Ankole District: Sanga Rest Camp, Oct. 1932, *Eggeling* 974!; Teso District: Serere, Dec. 1931, *Chandler* 140!

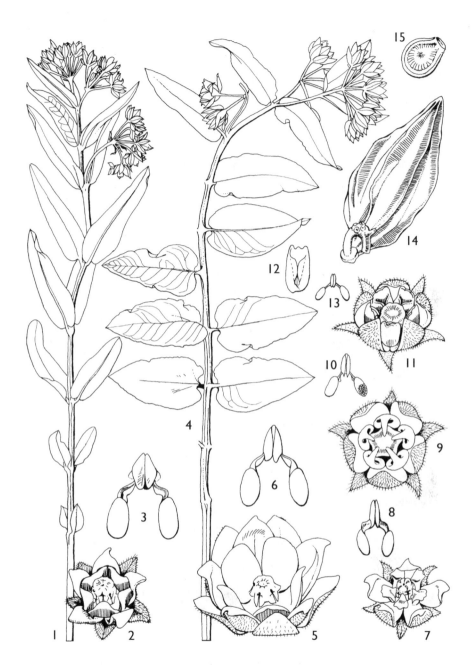

FIG. 108. *PACHYCARPUS PACHYGLOSSUS* — **1**, flowering shoot, × ¹/₂; **2**, flower, × 1¹/₂; **3**, pollinarium, × 10. *PACHYCARPUS RICHARDSIAE* — **4**, flowering shoot, × ¹/₂; **5**, flower, × 1¹/₂; **6**, pollinarium, × 10. *PACHYCARPUS GOETZEI* — **7**, flower, × 1¹/₂; **8**, pollinarium, × 10. *PACHYCARPUS FIRMUS* — **9**, flower, × 3; **10**, pollinarium, × 10. *PACHYCARPUS PETHERICKIANUS* — **11**, flower, × 3; **12**, corona lobe, × 3; **13**, pollinarium, × 10; **14**, follicle, × 1; **15**, seed (without coma), × 2. 1–3 from *Goyder et al.* 3819; 4 from *Richards* 17534; 5 & 6 from *Cribb & Grey-Wilson* 10566; 7 & 8 from *Leedal* 5395; 9 & 10 from *Milne-Redhead & Taylor* 8922; 11–13 from *Tweedie* 3804; 14 & 15 from *Milne-Redhead* 5011A. From K.B. 53: 366, drawn by E. Papadopoulos.

KENYA. West Suk District: Keringet–Lodwar road, May 1970, *Tweedie* 3804!; South Kavirondo
  District: Bukuria, Sep. 1933, *Napier* 5235! & Kuja River, July 1934, *Napier* 6631!
TANZANIA. Bukoba District: near Kamachumu, Sep. 1931, *Haarer* 2164! & Kabirizi, Oct. 1931,
  Haarer 2323!
DISTR. U 1–4; K 2, 5; T 1; Nigeria, Cameroon, Central African Republic, Congo-Kinshasa,
  Burundi, South Sudan and Ethiopia
HAB. Grassland, sometimes in bushland or riparian woodland; 1000–2000 m

SYN. *Schizoglossum petherickianum* Oliv. in Trans. Linn. Soc., Bot. London 29: 109 (1875); N.E. Br.
    in F.T.A. 4(1): 368 (1902)
  S. petherickianum Oliv. var. *cordatum* S. Moore in J.L.S. 37: 184 (1905). Type: Uganda, Ankole
    District, Irunga, *Bagshawe* 394B (BM!, holo.)
  S. cordatum (S. Moore) S. Moore in J.B. 45: 330 (1907)
  Pachycarpus schumannii Chiov. in Ann. Bot. (Rome) 10: 394 (1912). Type: Ethiopia, Galla
    Arussi, *Negri* 1063 (FT!, lecto., designated by Goyder in K.B. 53: 364 (1998); K!, photo)
  Schizoglossum thorbecki Schltr. in E.J. 51: 132 (1913). Types: Cameroon, *Thorbecke* 249, 290;
    Ledermann 1482, 2664, 2814 (B†, syn.)
  Pachycarpus terminans Fiori in Nuovo Giorn. Bot. Ital., n.s., 47: 35 (1940). Type: Ethiopia,
    Sidamo, Wendo [Uondo], June 1937, *Saccardo* s.n. (FT!, holo.; K!, photo)

10. **Pachycarpus firmus** (*N.E. Br.*) *Goyder* in K.B. 53: 367 (1998). Type: Angola,
Huilla, near Lopollo, *Welwitsch* 4191 (K!, holo.; BM!, iso.)

Perennial herb with one to several annual stems arising from a vertical tuberous
rootstock; stems erect, 0.3–0.8 m tall, densely pubescent. Leaves with petiole 2–6 mm
long, pubescent; lamina lanceolate to ovate or oblong-ovate, 4–8 × 2–4 cm, apex
acute to obtuse or rounded, base rounded or truncate, usually slightly cordate,
margins scabrid, venation prominent with numerous parallel secondary veins,
indumentum of spreading white hairs on both upper and lower surfaces.
Inflorescences terminal and extra-axillary with 5–20 flowers in a globose umbel;
peduncles 1–4 cm long, densely spreading-pubescent, not elongating significantly
in fruit; bracts filiform, 4–10 mm long, pubescent; pedicels 0.3–0.8 cm long, densely
pubescent. Calyx lobes lanceolate to ovate, 4–7 × 1–2 mm, acute, densely pubescent,
often purple towards tip. Corolla campanulate, pubescent on outer face at least
above, glabrous within, white or pink, marked with deeper pink or maroon; lobes
ovate, 5–8 × 3–4 mm, acute, not recurved. Corona lobes arising from the base of the
staminal column, erect, white or purple, ± quadrate, 2–3.5 × 2–3.5 mm,
dorsiventrally flattened and slightly convex and about as tall as the column, with a
subulate apical tooth 2–3 mm long arching over the stylar head, inner face of the
lobe with margins inflexed and produced into a pair of fleshy teeth ± level with top
of anthers. Anther wings ± 1.5 × 0.5 mm; anther appendages ovate, ± 1 × 0.5 mm,
acute, inflexed over stylar head; corpusculum dark brown, subcylindrical or ovoid,
± 0.5 × 0.2 mm; translator arms ± 0.2 mm long; pollinia elliptic, 0.6 × 0.2 mm. Stylar
head flat, 2–3 mm diameter, forming a raised pentagonal cushion 1–2 mm above
top of anthers. Fruiting pedicel contorted to hold follicle erect; follicle narrowly
ovoid, ± 6 × 1.5 cm, with 4–6 longitudinal, weakly toothed wings, pubescent. Seeds
subovoid, ± 4.5 × 2 mm, strongly concave and with verrucose or reticulate
sculpturing. Fig. 108/9–10, p. 447.

TANZANIA. Iringa District: Mufindi, Ngwazi grasslands, 10 Feb. 1492, *Lovett* 1492! &
  Lugoda–Ngwazi road, 28 March 2006, *Bidgood, Darbyshire et al.* 5154!; Songea District:
  Matengo Hills, 1 Mar. 1956 (fl.), *Milne-Redhead & Taylor* 8922! & 21 May 1956 (fr.), *Milne-
  Redhead & Taylor* 8922A!
DISTR. T 7, 8; Congo-Kinshasa, Angola and Malawi
HAB. Upland grassland or *Brachystegia* woodland; 1500–2000 m

SYN. *Schizoglossum firmum* N.E. Br. in K.B. 1895: 252 (1895) & in F.T.A. 4(1): 368 (1902)
  Asclepias firma (N.E. Br.) Hiern, Cat. Afr. Pl. Welw. 1: 684 (1898)

11. **Pachycarpus pachyglossus** *Goyder* in K.B. 53: 368 (1998). Type: Tanzania, Ufipa District: Mbisi Forest Reserve, 10 km from Sumbawanga, *Goyder et al.* 3819 (K!, holo.; DSM!, PRE!, iso.)

Perennial herb with 1–3 annual stems arising from a vertical tuberous rootstock; stems erect, 0.2–0.5 m tall, usually unbranched but sometimes branching below, densely pubescent with spreading white hairs. Leaves with petiole 1–5 mm long, pubescent; lamina lanceolate to ovate or oblong-ovate, 4.5–10 × 1.5–4 cm, apex acute, obtuse or rounded, base rounded or truncate, usually slightly cordate, margins scabrid, venation prominent with numerous parallel secondary veins, indumentum of spreading white hairs on both upper and lower surfaces. Inflorescences terminal and extra-axillary with 4–9 unscented flowers in a secund umbel; peduncles ascending or erect, 1–5 cm long, densely spreading-pubescent, not elongating noticably in fruit; bracts lanceolate or filiform, 5–11 mm long, pubescent; pedicels 1–1.5 cm long, densely pubescent. Calyx lobes ovate, 7–11 × 4–7 mm, acute, densely pubescent. Corolla campanulate, pubescent on outer face, glabrous within, green or brownish green; lobes 9–10 × 5–6 mm, ovate or oblong, apex subacute, strongly recurved at the tip. Corona lobes arising from the base of the staminal column, green or brown, erect, broadly oblong, 4–5 × 5–8 mm, dorsiventrally flattened but slightly convex and distinctly fleshy above, as tall as the staminal column or shorter, truncate or slightly retuse, with no teeth or lobes on the inner face. Gynostegium shortly stipitate; anther wings triangular, 3 × 1 mm; anther appendages broadly ovate, ± 2 × 2 mm, inflexed over stylar head; corpusculum dark brown, broadly ovoid-rhomboid, 0.8–1 × 0.8 mm; translator arms 0.5–0.6 × 0.2–0.3 mm, flattened and geniculate distally, with a short clasping overlap with the pollinia; pollinia obovate, 1.1 × 0.5–0.6 mm, not stongly compressed. Stylar head flat, 3–4 mm diameter, forming a raised circular cushion ± 1 mm above top of anthers. Fruiting pedicel contorted; follicle narrowly ovoid, ± 9 × 2.5 cm, with 6 longitudinal toothed ridges, pubescent. Seeds not seen. Fig. 108/1–3, p. 447.

TANZANIA. Ufipa District: Mbisi Forest Reserve, 25 Nov. 1994, *Goyder et al.* 3819 & 7 Nov. 1987, *Ruffo & Kisena* 2786!; mountain above Malonje Farm, 14 Dec. 1956, *Richards* 7282!
DISTR. **T** 4; endemic to the mountains around Sumbawanga
HAB. Montane grassland; 1800–2400 m

12. **Pachycarpus richardsiae** *Goyder* in K.B. 53: 369 (1998). Type: Tanzania, Mbeya Mt, *Napper* 1694 (K!, holo.; EA!, iso.)

Perennial herb with annual stems; rootstock unknown; stems erect, 0.3–1 m tall, usually unbranched, densely pubescent with spreading white hairs. Leaves with petiole 1–5 mm long, pubescent; lamina lanceolate to ovate, 4.5–7 × 1–4.5 cm, apex acute, base rounded or truncate, usually slightly cordate, margins scabrid, venation prominent with numerous parallel secondary veins, indumentum of spreading white hairs on both upper and lower surfaces. Inflorescences terminal and extra-axillary with 4–9 flowers in a secund umbel; peduncles ascending or erect, 1–2 cm long, densely spreading-pubescent; bracts filiform, 5–9 mm long, pubescent; pedicels 1–1.5 cm long, densely pubescent. Calyx lobes ovate, 10–15 × 4–6 mm, acute, densely pubescent. Corolla green, campanulate, pubescent on outer face, glabrous within; lobes ovate, 12–15 × 5–8 mm, apex subacute, recurved at the tip. Corona lobes arising from the base of the staminal column, erect, green or maroon, broadly obovate, 8–11 × 5–9 mm, dorsiventrally flattened, about twice as long as the staminal column, rounded or truncate, with no teeth or lobes on the inner face. Anther wings triangular, 3–3.5 × 1–1.5 mm; anther appendages ovate, ± 3 × 2 mm, acute, inflexed over stylar head; corpusculum dark brown, ovoid, 1–1.2 × 0.6 mm; translator arms

0.5–0.6 × 0.2–0.3 mm, flattened, with a short clasping overlap with the pollinia; pollinia obovate, 1.3 × 0.5–0.6 mm, not strongly compressed. Stylar head flat, 3–4 mm diameter, forming a raised circular cushion ± 1 mm above top of anthers. Follicles not seen. Fig. 108/4–6, p. 447.

Tanzania. Mbeya District: Loleza Mt, 11 Dec. 1977, *Nicholson* 283! & Mbeya range, on way to Mbeya Peak, 6 Feb. 1976, *Cribb et al.* 10566!; Mbeya/Njombe Districts: Kitulo Plateau, 28 Dec. 1975, *Leedal* 3307!
Distr. **T** 7; endemic to the mountains above Mbeya and the Kitulo Plateau
Hab. Montane grassland; 1800–2100 m

78. XYSMALOBIUM

R. Br. in Mem. Wern. Soc. 1: 38 (1809)

Stems simple, 1 or more, arising from a tuberous rootstock; latex white. Leaves opposite, with lateral veins ± parallel and spreading to almost a right angle with the midrib. Inflorescences umbelliform, with erect peduncles. Corolla lobed almost to the base. Corolline corona absent. Gynostegial corona lobes fleshy, generally not strongly compressed either dorsally or laterally. Anthers raised on a variously developed gynostegial stalk formed from the anther filaments; anther appendages membranous. Pollinia attached terminally to the winged translator arms. Follicles inflated or not. Seeds flattened, ovate, with a coma of silky hairs.

An unsatisfactory assemblage of ± 30 species in tropical and southern Africa. Within the Flora region, *X. undulatum* seems only distantly related to the remaining species.

1. Corolla generally bearded at the tip; stems stout; leaf
   margins scabrid to the touch . . . .   1. *X. undulatum*
   Corolla lobes not bearded apically; stems slender; leaf
   margins not scabrid 2
2. Corolla strongly reflexed 3
   Corolla spreading or campanulate, not strongly reflexed 4
3. Corona lobes with flattened, suberect outer face, inner
   face with fleshy projection . . . .   3. *X. stocksii*
   Corona lobes subglobose with fleshy peg or tooth extending
   up side of staminal column . . . .   2. *X. heudelotianum*
4. Gynostegium with conspicuous stipe ± 2 mm long; anther
   wings 0.5–1 mm long, tapering gradually into stipe . . .   7. *X. patulum*
   Gynostegium sessile or subsessile; anther wings (1–)2–3 mm
   long 5
5. Anther wings vertical, extending well-below the anther
   sacs; fleshy apical part of the corona lobes ± horizontal
   in the cavities bounded by anther wing tails . . . .   6. *X. kaessneri*
   Anther wings wider at base than above, without basal
   extensions; corona lobes ± erect 6
6. Anther wings distinctly flared at base, outer margin curved   4. *X. fraternum*
   Anther wings narrowly to broadly triangular, outer margin
   straight . . . .   5. *X.* sp. A

1. **Xysmalobium undulatum** (*L.*) *W.T. Aiton*, Hort. Kew. ed. 2, 2: 79 (1811); Cribb & Leedal, Mt Fl. S. Tanz.: 106, t. 24c (1982); U.K.W.F. ed. 2: 176, pl. 67 (1994); Goyder in Fl. Eth. 4(1): 131, fig. 140.19 (2003). Type: "Apocynum Afric. lapathi folio" in Commelin, Hort. Med. Amstelaed. Pl. Rar.: 16, t. 16 (1706), icono., designated by Wijnands in Bot. Commelins: 48 (1983)

D.E.

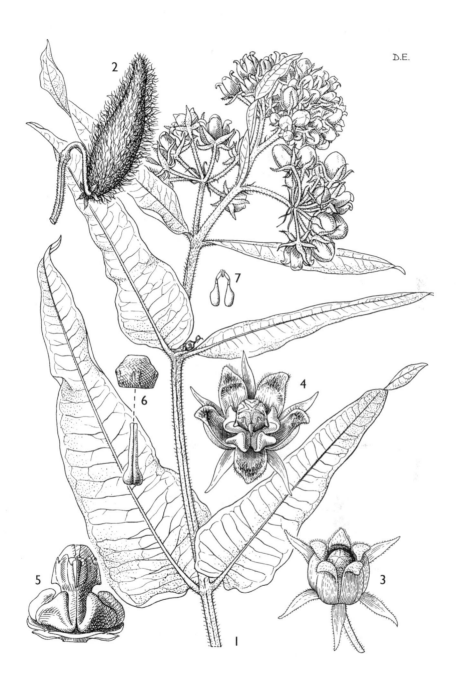

Fig. 109. *XYSMALOBIUM UNDULATUM* — **1**, habit, × ³/₄; **2**, fruit, × ³/₄; **3 & 4**, flowers, × 1¹/₂; **5**, gynostegium and corona, × 3; **6**, gynoecium showing paired ovaries and stylar head, × 3; **7**, pollinarium, × 7. 1 from *Milne-Redhead & Taylor* 8231; 2 from *Milne-Redhead & Taylor* 8231A; 3–7 from *Milne-Redhead & Taylor* 1087. Reproduced with permission, from the Flora of Ethiopia and Eritrea 4, 1; drawn by D. Erasmus.

Robust perennial herb with thick, carrot-like, tuberous rootstock; stems erect, one to several, (0.2–)0.5–1.5 m tall, unbranched, renewed annually, stout at least at the base, sparsely to densely pubescent with spreading white hairs, the base of the stem often glabrescent. Leaves with petiole 1–5(–10) mm; lamina very variable in shape, linear or linear-lanceolate to ovate or ovate-lanceolate, 7.5–19(–27) cm × 0.5–8.5(–12) cm, apex acute or acuminate, base cordate, somewhat hastate, truncate or broadly rounded, margins crispate, undulate or smooth, often slightly revolute, tips of lateral veins anastomosing to form a wavy submarginal vein, with sparse to dense spreading white hairs, particularly on principal veins and near margins. Inflorescences umbelliform, 12–26-flowered, densely pubescent; peduncles arising laterally from the upper nodes, robust, 12–45(–90) mm long, up to 3 mm thick; bracts linear, 3–4(–11) mm long; pedicels slender, (8–)14–18 mm. Sepals ovate to lanceolate, 4–8(–11) × 1–2.5 mm, about half as long to as long as the corolla, apex acute or attenuate, pubescent. Corolla cream or greenish white, often tinged with brown or pink outside, divided almost to the base; lobes spreading, ovate, 5–10 × 2.5–6 mm, lower ²⁄₃ glabrous, somewhat concave, upper ¹⁄₃ variably pubescent, usually lanate but occasionally glabrous or subglabrous, the apex acute, recurved. Corona very fleshy, ± tetrahedral in shape and attached by one of the apices to the base of the gynostegium, or stalk narrow for most of its length and widening abruptly to the ± triangular outer face, the outer face 2–4 × 2–3.5 mm. Gynostegium raised on a distinct stalk formed from the anther filaments; stalk 1–3(–4) mm long, usually only partially obscured by the corona lobes but lobes occasionally overlapping base of anthers when gynostegial stalk short. Anther wings jutting out abruptly from column, anthers 2–3 mm tall, appendages membranous, ovate or ± circular, 1–2 mm, rounded or acute; pollinia slender pear-shaped to short club-shaped, 0.75–1.5 mm long, attached by the tip to the translator arm; translator arms narrowly winged in upper part, 0.5–1 mm long; corpusculum dark brown, ovoid, 0.5 × 0.25 mm. Stylar head ± level with top of anthers, obscured by the anther appendages. Fruit a single follicle, 5–8 cm long, 1.5–2.5 cm wide, somewhat inflated or not, lanceolate or ovate in outline with an obtuse or attenuate apex, the entire surface covered with soft hairy prickles. Fig. 109, p. 451.

UGANDA. Kigezi District: Kigezi, Dec. 1956, *Lind* 2077A!; Buganda District: Kampala–Bombo road, May 1929, *Liebenberg* 825!
KENYA. North Nyeri District: Cole's Mill, Jan. 1922, *Fries & Fries* 1120!; Machakos District: Chyulu Mts, above Kibwezi, May 1975, *Friis & Hansen* 2707! & Chyulu North, May 1938, *Bally* 7929!
TANZANIA. Ufipa District: Malonje Plateau, Old Sumbawanga road, Dec. 1961, *Richards* 15814!; Iringa District: 80 km NE of Iringa, Mt Image, Mar. 1962, *Polhill & Paulo* 1685!; Songea District: Matengo Hills, valley of R. Halau, 3 km S of Miyau, Jan. 1956, *Milne-Redhead & Taylor* 8231!
DISTR. U 2, 4; K 3, 4, 6; T 1, 2, 4, 6–8; Ethiopia, Angola, Zambia, Malawi, Zimbabwe and South Africa
HAB. Fairly common on road or stream banks and in seasonally waterlogged depressions, occasionally in montane grassland; 1000–2400 m

SYN. *Asclepias undulata* L., Sp. Pl.: 214 (1753).
    Xysmalobium angolense Scott Elliot in J.B. 28: 365 (1890). Type: Angola, Huilla, Catumba, *Welwitsch* 4171 (BM!, syn.; K!, isosyn.) & near Huilla and Humpata, *Welwitsch* 4170 (BM!, syn.; K!, isosyn.)
    Woodia trilobata Schltr. in J.B. 33: 337 (1895). Type: Kenya, Nandi District: Nandi Range, *Scott Elliot* 6877 (BM!, holo.; K! iso.)
    Xysmalobium trilobatum (Schltr.) N.E. Br. in F.T.A. 4(1): 306 (1902)
    X. dispar N.E. Br. in F.T.A. 4(1): 307 (1902). Types: Malawi, Manganja Hills, near Mt Soche, 8 Mar. 1862, *Kirk* s.n. (K!, syn.); Malawi, Namasi, *Cameron* 4 (K!, syn.); Zimbabwe, Leshumo Valley, *Holub* 669 (K!, syn.) & 816 (K!, syn.)
    X. barbigerum N.E. Br. in F.T.A. 4(1): 307 (1902). Type: Angola, Amboella, at the mouth of the R. Kuebe, *Baum* 332 (?B, holo.; K!, iso.)
    Asclepias leuchotricha Schltr. in Baum, Kunune-Samb. Expd. 342 (1903). Type: as for *X. barbigerum*
    Xysmalobium leuchotrichum (Schltr.) N.E. Br. in F.T.A. 4(1): 615 (1904)

Note. This species is very variable, with distinctive local forms. One such variant occurs around Lake Victoria and has shorter corolla lobes and a less stipitate gynostegium than typical material, and linear-lanceolate leaves. Another small-flowered form is found in the Southern Highlands of Tanzania and the Nyika Plateau of neighbouring Malawi, but the leaves in this variant are much broader.

2. **Xysmalobium heudelotianum** *Decne.* in DC., Prodr. 8: 520 (1844); Bullock in F.W.T.A. ed. 2, 2: 93 (1963); Goyder in Fl. Eth. 4(1): 132 (2003). Type: "Senegambia", *Heudelot* s.n. (P, holo.)

Erect or occasionally decumbent perennial with an erect carrot-shaped tuber 3–5 × 3 cm; stems one or few, simple or branched, (10–)20–60 cm, pubescent with short curled hairs. Leaves erect or spreading; petiole 1–2(–5) mm long; lamina ovate to oblong- or linear-lanceolate, rarely linear or obovate, (1.5–)4–8(–17) × 0.2–2(–3) cm, apex obtuse with a short mucro or acute, rarely retuse, the base usually truncate or slightly cordate, occasionally rounded or subcuneate, margins smooth or undulate, secondary and often tertiary venation prominent, often raised, reticulate, with a glabrous or pubescent marginal vein, remainder of lamina glabrous or with scattered reddish hairs below. Umbels sessile or very rarely pedunculate, extra-axillary with 4–10(–15) flowers at few or often at many of the nodes, all umbels apparently on one side of the stem; bracts often deciduous, filiform or narrowly triangular, 3–6 mm long, pubescent at least on the margin; pedicels 4–15(–18) mm long, pubescent. Calyx lobes reflexed slightly by the corolla, triangular, ± 3 × 1 mm, acute, pubescent. Corolla cream or white to reddish brown, lobed almost to the base, reflexed; lobes ovate, 4–6(–7) × 2–3.5 mm, acute, glabrous but minutely papillate at base on inner surface, outer surface glabrous or with scattered reddish hairs in the upper half. Corona lobes green or brownish, on short spreading stalks arising near the base of the column, subglobose, 1–2 mm diameter with a smooth or papillate tooth or beak pointing up the side of the anthers and reaching from $^1/_2$ to the full height of the column. Gynostegium 3–4 mm long, gynostegial stalk up to 1 mm long. Anther wings narrowly triangular giving the column a subcylindric or slightly conical appearance; appendages ovate, ± 1 mm long, acute, inflexed over the stylar head. Stylar head truncate, raised slightly above the tip of the anthers; corpusculum dark brown, ovoid, 0.2–0.3 mm long; translator arms weakly geniculate at the middle, proximal half winged on both sides or just the lower one, the winged part (0.2–)0.3–0.5 × 0.1–0.2 mm, translucent, distal part (0.2–)0.4 mm long, slender, terete or sometimes with a narrow translucent wing; pollinia oblanceolate, 0.8–1.2 × 0.2–0.4 mm. Follicles erect, 11–19 cm long, 1 cm wide, narrowly fusiform, long attenuate at both apex and base, smooth and minutely pubescent. Fig. 110, p. 454.

UGANDA. Kigezi District: Katete, St. Francis' Hospital, 11 Jan. 1958, *J.M. Wright* 216!; Teso District: Serere, Dec. 1931, *Chandler* 252!; Masaka District: Kasera, Oct. 1925, *Maitland* 867!
KENYA. West Suk District: Kapenguria, May 1932, *Napier* 1974!; Trans-Nzoia District: Kitale, May 1965, *Tweedie* 3041!; Nairobi, Rowallen Camp, 31 May 1980, *Gilbert* 5955!
TANZANIA. Biharamulo District: 50 km along Kahama road from Biharamulo, 24 Nov. 1962, *Verdcourt* 3452!; Ufipa District: Sumbawanga, 29 Nov. 1954, *Richards* 3410!; Rungwe District: Kyimbila, 17 Dec. 1913, *Stolz* 2377!
DISTR. U 2–4; K 2–6; T 1, 4, 7; Gambia to Nigeria, Cameroon, Congo-Kinshasa, Burundi, South Sudan, Zambia, Malawi, Mozambique and Zimbabwe
HAB. Grassland, occasionally in mixed deciduous woodland; 1200–2000 m

SYN. *Gomphocarpus rubioides* Kotschy & Peyr., Pl. Tinn.: 29, t. 13B (1867). Type: South Sudan, Djur, Dec. 1863, *Heuglin* 29 (W!, holo.).
    Xysmalobium reticulatum N.E. Br. in K.B. 1895: 251 (1895) & in F.T.A. 4(1): 303 (1902); U.K.W.F. ed. 2: 176 (1994). Type: Malawi, Shire Highlands, *Buchanan* s.n. (K!, holo.)
    Schizoglossum truncatulum K. Schum. in P.O.A. C: 322 (1895). Type: Uganda, District unclear, Mpororo, Kanjana Mts, *Stuhlmann* 2096 (B†, holo.; K!, iso. (fragment))
    Xysmalobium schumannianum S. Moore in J.B. 39: 259 (1901). Type: Kenya, Machakos, 1896, *Hinde* s.n. (BM!, holo.; K!, iso. (fragment))

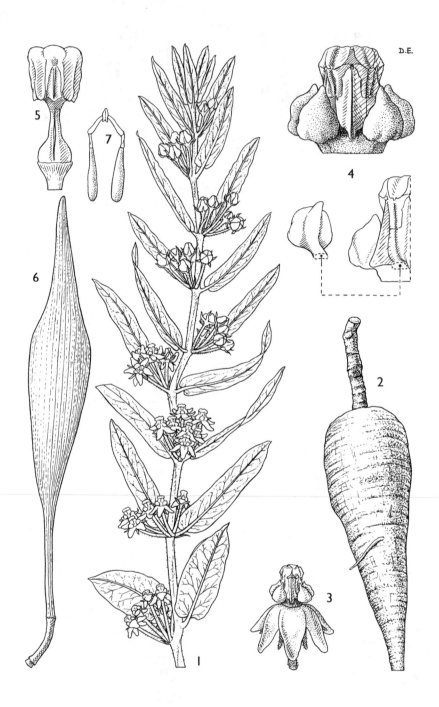

Fig. 110. *XYSMALOBIUM HEUDELOTIANUM* — **1**, flowering stem, × 1; **2**, tuber, × 1; **3**, flower, × 3; **4**, gynostegium and corona, and with one corona lobe detached, × 8; **5**, gynoecium showing paired ovaries and stylar head, × 8; **6**, follicle, × 1; **7**, pollinarium, × 16. 1 & 3–7 from *Richards* 3410; 2 from *Dalziel* 693. Drawn by D. Erasmus.

NOTE. This species is very variable in leaf shape. Two collections from western Uganda are unusual in having broadly oblong or obovate leaves with a retuse apex and material from the Chyulu Hills in Kenya has unusually long leaves. However, in other characters these collections are typical of *X. heudelotianum* and, therefore, have not been recognised taxonomically.

[*Saxymolbium heudelotianum* (Decne.) Bullock, mentioned in U.K.W.F. ed. 2: 176 (1994) as a synonym, is a manuscript name, never validly published.]

3. **Xysmalobium stocksii** *N.E. Br.* in K.B. 1913: 302 (1913); U.K.W.F. ed. 2: 176 (1994). Type: Mozambique, Ibo area, *Stocks* s.n. (K!, holo.)

Erect perennial herb with a globose tuber 2–3 cm diameter, 2–3 cm below soil surface; stems one or two, 20–50 cm, pubescent with short curled hairs. Leaves erect or spreading, petiolate or sessile, petiole up to ± 1 cm long; lamina linear, 5.5–11(–19) × 0.2–0.8(–1.2) cm, apex acute, tapering gradually at the base, veins not visible above, inconspicuous or clearly reticulate below, glabrous. Umbels sessile, terminal or extra axillary with 3–8 flowers at each of the upper 1–6 nodes; bracts often deciduous, linear, 3–8 mm long, pubescent at least on the margins; pedicels 5–17 mm long, densely pubescent on one side. Calyx lobes ovate to lanceolate, 2–4 × 0.5–1 mm, acute, reflexed by the corolla. Corolla reflexed, globose in bud, lobed almost to the base; lobes ovate, 6–7 × 3–3.5 mm, acute, the underside glabrous or with few scattered reddish hairs towards the tip, upper surface minutely papillate or with white hairs especially near the base of the lobe. Corona lobes suberect, reaching from ¹/₂ to about equalling the height of the column, outer face flattened, ovate or obovate, 2–4 × ± 2 mm, obtuse or rounded at the apex, the inner face with a ± pyramidal swelling ± 1.5 mm long in the upper half. Gynostegium ± 4 mm long. Anther wings narrowly triangular, not spreading markedly at the base, the column appearing ± cylindrical; appendages ovate, ± 2 × 1 mm, acute, inflexed over the top of the stylar head; corpusculum black, ovoid, ± 0.3 mm long; translator arms geniculate, proximal half winged on lower side, the winged part obovate, 0.4–0.5 × 0.2–0.3 mm, distal part 0.4–0.5 mm long, slender, terete; pollinium oblanceolate, 1.1–1.3 × ± 0.3 mm. Follicles not seen.

KENYA. Machakos/Masai Districts: Chyulu Hills/plains, near Half Crater Camp, 23 Dec. 2000, *Luke & Luke* 7198A!; Kwale District: Shimba Hills, 26 Mar. 1968, *Magogo & Glover* 481! & near Longomagandi, 29 Apr. 1992, *Luke* 3098B!
TANZANIA. Songea District: 80 km along Tunduru road, 30 Jan. 1991, *Bidgood, Abdallah & Vollesen* 1320! & 12 km W of Songea near Kimarampaka stream, 20 Jan. 1956, *Milne-Redhead & Taylor* 8379! & 1.5 km E of Songea, 4 Feb. 1956, *Milne-Redhead & Taylor* 8599!
DISTR. K 4/6, 7; T 8; Angola, Zambia and Mozambique
HAB. Burnt grassland and degraded *Brachystegia* woodland; 300–1400 m

SYN. *Saxymolbium* sp. A of U.K.W.F.: 374 (1974)

4. **Xysmalobium fraternum** *N.E. Br.* in K.B. 1895: 252 (1895) & in F.T.A. 4(1): 305 (1902). Type: Malawi, Shire Highlands, near Blantyre, *Last* s.n. (K!, holo.)

Erect perennial herb with a subglobose to carrot-shaped tuber; stems usually single, 30–55 cm tall, pubescent with short curled hairs. Leaves petiolate or rarely subsessile, petiole to 8 mm long; lamina often oblong or ovate in lowest leaves, other leaves obovate, 2.5–6.5 × 1.5–3.5 cm, rounded to obtuse or subacute, occasionally slightly retuse, shortly mucronate or not, cuneate at the base, sometimes narrowly so, veins prominent below, reticulate, glabrescent, young leaves with scattered rusty hairs below or on both surfaces. Umbels sessile, extra-axillary, with 4–8 flowers at few or many nodes; bracts linear, 3–4 mm long, acute; pedicels 5–19 mm long, slender, pubescent. Calyx lobes lanceolate, 3–4 × 1–1.5 mm, acute, reflexed by the corolla. Corolla white or yellowish green, campanulate or rotate-campanulate, globose in bud, lobed almost to the base; lobes ovate to oblong, 5 × 2–3 mm, acute, the underside glabrous or with scattered reddish hairs on the upper half, upper surface minutely papillate or with white hairs near the base of the lobe. Corona lobes about as high as the column, spreading outwards for ± 2 mm then arching upwards and inwards to the head of the gynostegium, the outer face broadly ovate, ± 4 × 5 mm with a subacute apex, the inner face with a pubescent or glabrous ridge running down the inner face from the fleshy apex. Gynostegium 3–4 mm high. Anther wings ± triangular, the base spreading outwards from the column; appendages ovate, 1.5 × 1.5 mm, subacute, inflexed over the apex of the stylar head; corpusculum black, ovoid, 0.4–0.5 mm long; translator arms 0.4–0.8 mm long, flattened ± uniformly along the entire length; pollinia triangular, 0.6–0.8 × 0.4–0.6 mm. Tip of stylar head flat, level with or raised slightly above anther tips. Follicles erect, ± 25 cm long, smooth, pubescent, the lower half forming a narrow 'stipe', the upper half narrowly fusiform.

TANZANIA. Ufipa District: Tatanda Mission, 22 Feb. 1994, *Bidgood, Mbago & Vollesen* 2390!; Songea District: 19 km W of Songea near R. Likuyu, 30 Dec. 1955, *Milne-Redhead & Taylor* 7951! & Songea airfield, 1 Jan. 1956, *Omari* s.n. in *Milne-Readhead & Taylor* 7951a!
DISTR. **T** 4, 8; Zambia, Malawi and Mozambique
HAB. *Brachystegia* woodland; 900–1500 m

5. **Xysmalobium** sp. A

Erect herb with stems up to 60 cm arising from a globose tuber. Leaves with petiole up to 9 mm long; lamina oblanceolate to obovate or elliptic, 4–7.5 × 1.4–3.5 cm, lower leaves sometimes ovate or orbicular, apex rounded to acute, usually with a short mucro, base cuneate or rounded, margins commonly undulate at least when dry, secondary and tertiary venation prominent, with scattered reddish hairs particulary on the lower surface, glabrescent. Umbels sessile at upper nodes with up to 6 sweetly scented flowers; bracts filiform, up to 4 mm long, often caducous; pedicels 6–11 mm long, pubescent with spreading hairs. Calyx lobes triangular, 3.5–5 × 1–1.3 mm, apex acute, sparsely pubescent. Corolla white; lobes apparently spreading or perhaps slightly reflexed, ± ovate, 6 × 4 mm, acute, shallowly concave, glabrous but very shortly papillose on inner surface, outer surface usually with scattered reddish hairs towards the tip. Corona lobes reddish brown, ± 3.5 mm long with a stalk ± 1 mm long spreading outwards from the base of the staminal column and a fleshy head 2.5–3 × 2.5–3 mm, the outer face flattened, ± orbicular, the inner face swollen into a conical or pyramidal fleshy structure with a rounded apex, the top of the lobes ± level with the tip of the stamens. Staminal column ± 3 mm long, appearing conical due to the broadly triangular anther wings; anther appendages ovate, ± 1 × 0.5 mm, acute, inflexed over the stylar head; pollinaria with corpusculum 0.2–0.3 mm long; translator arms ± 0.8 mm long, winged in the proximal half, the wings up to 0.2 mm wide, the distal half slender; pollinia obovate-oblong, ± 1 × 0.4 mm. Top of stylar head about level with the top of the anthers, obscured by the anther appendages. Follicles not seen.

TANZANIA. Mpanda District: near Kabungu, 12 Dec. 1958, *Anstey* 4! & 60 km from Ikola on Mpanda road, 8 Nov. 1959, *Richards* 11731!; Iringa District: Ruaha National Park, Magangwe air strip, 17 Dec. 1972, *Bjørnstad* 2157!
DISTR. **T** 4, 7; known only from these three collections
HAB. *Brachystegia* woodland; 1000–1400 m

6. **Xysmalobium kaessneri** *S. Moore* in J.B. 46: 295 (1908). Type: Zambia, Sangolo Spruit, *Kässner* 2104 (BM!, holo.; K!, iso.)

Erect perennial herb arising from a vertical, woody, napiform tuber; stems usually single, 40–70 cm tall, densely pubescent with short curled hairs. Leaves sessile or with petiole up to 1 cm long; lamina oblanceolate, obovate, oblong or elliptic, 3.5–7(–9) × (0.7–)1–4.5 cm, lower leaves occasionally broadly ovate or suborbicular, apex rounded, retuse or, particularly in upper leaves, acute, base rounded to cuneate, margins smooth or slightly crisped, veins prominent especially on lower surface, reticulate, marginal vein present, with white hairs mainly on the midrib and major veins. Umbels sessile, extra-axillary, with 5–8 flowers at each of the upper and middle nodes; bracts linear, 7–9 mm long, acute, moderately to densely pubescent. Corolla campanulate, divided ± to the base, green tinged with purple outside, paler within; lobes ovate, ± 7 × 3.5 mm, acute, outer face pubescent, inner surface minutely pubescent. Corona lobes shorter than the staminal column, basal part of each lobe forming a fleshy claw ± 1 × 0.5 mm spreading upwards and outwards from the base of the column, distal part spatulate, very fleshy, spreading ± horizontally from the cavity between adjacent pairs of anther wings, 2–3 × ± 2 × 1 mm, basal margin often produced into a small backward-pointing tongue. Gynostegium ± 3.5 mm high, cylindrical. Anther wings ± vertical; appendages ovate, 2.5 × 2 mm, acute, inflexed over the stylar head. Pollinaria robust; corpusculum black, ovate, ± 0.5 × 0.4 mm; translator arms dark brown, ± 0.7 × 0.3 mm flattened, the distal end ± as wide as the proximal end of the pollinium; pollinia oblong-obtriangular, 0.7 × 0.4 mm. Apex of stigma head ± 3 mm diameter, flat, ± level with the top of the column and obscured by the anther appendages. Follicles erect, narrowly fusiform, up to ± 30 cm long, 2 cm diameter at the widest point, finely rusty-pubescent, only one of the pair developing.

TANZANIA. Mbeya District: Ipota area, rock gorge leading to Mbaliza River near Mshewe village, 26 Jan. 1990, *Lovett et al.* 3980! & between Mshewe and Muvwa villages, 4 Mar. 1990, *Lovett & Kayombo* 4314!
DISTR. **T** 7; Congo-Kinshasa, Zambia, Malawi and Zimbabwe
HAB. Mixed *Brachystegia* woodland; 1200–1600 m

7. **Xysmalobium patulum** S. *Moore* in J.B. 47: 215 (1909). Type: Zambia, Katanino [Katinina] Hills, *Kässner* s.n., mounted on same sheet as 2167 (BM!, holo.)

Erect perennial herb with short napiform tuber; stems one or two, ± 35 cm tall, pubescent with short curled hairs. Leaves subsessile to shortly petiolate; petiole to ± 0.3 cm long; lamina narrowly oblong to oblanceolate, 2.5–6 × 0.6–1.3 cm, apex obtuse to subacute, tapering gradually at the base; veins prominent, reticulate, glabrous except for occasional hairs on the marginal vein. Umbels sessile, extra-axillary, with 5–9 flowers at each of the middle and upper nodes; bracts linear, 2–4 mm long, acute, pubescent; pedicels 5–8 mm long, pubescent. Calyx lobes ovate, ± 2.5 × 1 mm, acute, pubescent. Corolla green tinged with purple, campanulate, lobed almost to the base; lobes ovate, 5 × 2 mm, acute, pubescent with scattered hairs on lower surface, upper surface minutely papillate especially towards the base. Corona lobes green, reaching ± to the base of the anthers, fleshy but somewhat flattened dorsally, spreading outwards from the base of the staminal column for ± 0.5 mm then becoming erect for ± 2 mm, the outer face obovate, rounded at the apex, thickened in the upper half, sometimes toothed at the basal end of the thickening. Gynostegium ± 2.5 mm tall. Filaments forming an obconic stipe, widest point of anther wings ± 0.5 mm from the top of the column, narrowing gradually below into the stipe; appendages ovate, ± 1 × 0.7 mm, acute, inflexed over the stylar head; corpusculum black, ovoid, 0.4 × 0.3 mm; translator arms ± 0.3 × 0.1 mm, slightly contorted and differentially thickened; pollinia triangular, 0.6 × 0.3 mm, widest around the middle. Apex of stylar head flat, ± level with the anther tips. Follicles not seen.

TANZANIA. Mbeya District: Mbosi circular road, 12 Jan. 1961, *Richards* 13886!; Rungwe District: Kyimbila, *Stolz* 510! & 511!
DISTR. **T** 7; S Congo-Kinshasa and NW Zambia
HAB. *Brachystegia* woodland; ± 1500 m

NOTE. The type differs slightly from other material in the attitude of the basal part of the corona lobes, which is more erect, and the upper part, which is held in a more horizontal position than in other collections of this species.

79. **GLOSSOSTELMA**

Schltr. in J.B. 33: 321 (1895); Goyder in K.B. 50: 527–555 (1995)

Slender or robust, erect, perennial herbs with a short, often stout, vertical rhizome and several fusiform, tuberous, lateral roots; stems usually solitary, occasionally more than one, unbranched, with a line of pubescence running along the stem and alternating at the nodes, or rarely glabrous. Leaves usually petiolate, the petiole weakly channelled. Inflorescences umbelliform, with or without a peduncle; umbels extra-axillary, arising terminally and laterally at the upper 1–4(–6) nodes. Corolla medium to large, campanulate, lobed almost to the base. Corolline corona absent. Gynostegial corona of 5 lobes, often fleshy, sometimes dorsally flattened, united very briefly at the base and attached to the gynostegium near the base of the anthers; gynostegium usually with a conspicuous stalk below the attachment of corona lobes. Stylar head truncate, 5-angled, generally ± level with the top of the anthers and largely obscured by the anther appendages; corpusculum ovoid, brown or black; translator arms flattened, broadened and geniculate near the attachment of the pollinia, held ± at right angles to the axis of the corpusculum; pollinia pendulous, flattened, falcate-oblong to obtriangular. Fruiting pedicel not contorted; follicles erect, lanceolate or ovate-lanceolate in outline, with an attenuate apex, smooth, glabrous or pubescent, usually only one of the pair developing. Seeds flattened with one face convex, ovate to suborbicular in outline with a narrow convoluted rim surrounding a verrucose disc, with a coma of silky hairs.

12 or 13 species, distributed mostly in a narrow belt of southern tropical Africa from Angola and Congo-Kinshasa to Mozambique and southern Tanzania. Generally found in the 'wetter miombo woodland' vegetation of White (1983) or adjacent regions of afromontane vegetation.

1. Leaves linear; corona lobes dorsally compressed for their
   entire length . . . . 1. *G. carsonii*
   Leaves lanceolate to spatulate; corona solid and fleshy at
   least in part 2
2. Corona lobes at least 7 mm long, overtopping the head of
   the column for at least half their length 3
   Corona lobes to 5 mm long, not overtopping the head of
   the column 4
3. Tip of corona lobes solid, clavate, without an apical flap . . 4. *G. ceciliae*
   Tip of corona lobes dorsally flattened with a downward-
   pointing apical flap on inner face . . . . 5. *G. cabrae*
4. Corona lobes spreading, formed of short fleshy outward-
   pointing pegs at base of anther wings; pollinia broadest
   distally; gynostegial stalk inconspicuous, < 1 mm long . . . 6. *G. brevilobum*
   Corona lobes erect, reaching ± to the top of the staminal
   column; pollinia broadest at attachment of translator
   arms; gynostegial stalk conspicuous, at least 1.5 mm long 5
5. Corolla lobes distinctly obovate, apex rounded or obtuse,
   green or maroon with reddish brown, yellow or white
   markings within; corona subglobose, lacking an apical peg 2. *G. spathulatum*
   Corolla lobes ovate, apex acute or subacute, green or
   cream with no internal markings; corona lobes with short
   slender upward-pointing peg at top of inner face . . . . . . 3. *G. mbisiense*

1. **Glossostelma carsonii** (*N.E. Br.*) *Bullock* in K.B. 7: 415 (1952). Type: Zambia, Tanganyika Plateau, Fife Station, *Carson* s.n. (K!, holo.)

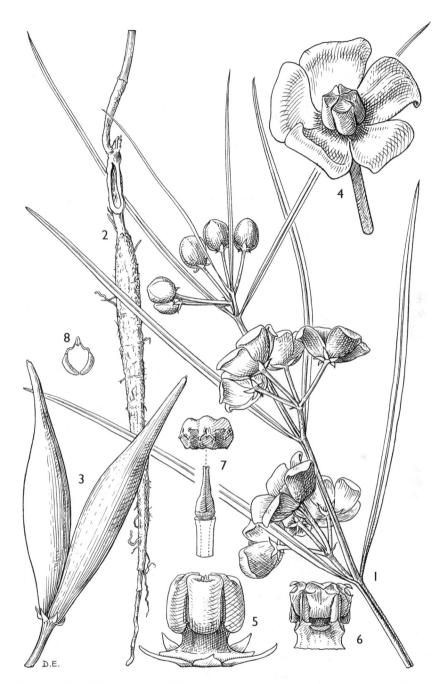

FIG. 111. *GLOSSOSTELMA CARSONII* — **1**, flowering shoot, × 1; **2**, rootstock, × 1 (note laterals appear to have been lost from this rootstock); **3**, paired follicles, × 1; **4** flower, × 2; **5**, gynostegium with corona, × 4; **6**, gynostegium with corona removed, × 4; **7**, gynoecium showing paired ovaries and stylar head, × 4; **8**, pollinarium, × 7. 1 from *Milne-Redhead & Taylor* 7914; 2 from *Milne-Redhead & Taylor* 7914A; 3 from *Cecil* 48A; 4–8 from *Milne-Redhead & Taylor* 1056. From K.B. 50: 534, drawn by D. Erasmus.

Slender erect herb; stems 25–85 cm long. Leaves linear or occasionally narrowly linear-lanceolate, lowest leaves present at anthesis 1.4–6 × 0.1–0.2(–0.4) cm, upper leaves 5.5–20 × 0.1–0.7(–0.9) cm, apex acute, tapering gradually into the sessile or subsessile base, glabrous, margins smooth; lateral veins visible only in the widest leaves. Umbels with 3–5(–9) spreading or erect flowers; peduncles (4–)16–43(–72) mm long, or occasionally absent, pubescent on one side, not lengthening in fruit; bracts filiform, 2–6 mm long, often deciduous; pedicels 10–17(–21) mm long, pubescent on one side. Calyx lobes ovate-triangular, 3–4 × 1–2 mm, acute. Corolla very variable in colour, usually yellowish green or brown with purple stippling, the stippling denser outside, but sometimes entire corolla cream or maroon, campanulate, lobed to 0.5–1 mm from the base, the lobes generally meeting at the margins when fresh; lobes obovate or obovate-oblong, 12–17 × 4.5–11 mm, apex subacute but appearing truncate as tip recurved, glabrous except for the minutely verrucate-papillate apical region and margins. Corona lobes erect, creamy white, orange-yellow or occasionally suffused with purple, oblong or obovate-oblong, (2–)4–6(–7) × 1–2.5 mm, attached (0–)1–2 mm above the base of the gynostegium, dorsally compressed, thin, apex rounded, subacute or, more commonly, variously toothed, the apical part erect or inflexed over the column for up to half its length, rarely shorter than column, the upper margins reflexed or not, basal part of the lobes with incurved margins, glabrous or minutely verrucate-papillate. Anther appendages broadly ovate, 1–2 mm, obtuse, sometimes toothed at margin; corpusculum ± 0.3 mm long; translator arms ± 0.3 mm long; pollinia sickle-shaped, ± 1 × 0.3 mm, tapering towards the base and attached apically to the translator arm. Follicles lanceolate, 11–13 × ± 1.1 cm, apex attenuate, glabrous or white-pubescent. Fig. 111, p. 459.

TANZANIA. Rufiji District: Ngulakula Forest Reserve, Dec. 1968, *Ngoundai* 168!; Mbeya District: Matambwa road to Chimala, Dec. 1963, *Richards* 18541!; Songea airfield, Feb. 1956, *Milne-Readhead & Taylor* 7914A!
DISTR. T 1, 2, 4–8; Congo-Kinshasa, Angola, Zambia, Malawi, Zimbabwe and Mozambique
HAB. *Uapaca-Protea* bushland, *Brachystegia* woodland and montane grassland; 1000–2100 m

SYN. *Xysmalobium carsonii* N.E. Br. in K.B. 1895: 250 (Oct. 1895)
　　Gomphocarpus chlorojodina K. Schum. in E.J. 30: 383 (1901). Type: Tanzania, Mbeya/Rungwe District: Unyiha [Unyika] Plateau, Umalila, *Goetze* 1360 (B†, holo.; K!, iso.)
　　Schizoglossum carsonii (N.E. Br.) N.E. Br. in F.T.A. 4(1): 366 (1902); Weim. in Bot. Notis. 1935(22): 393 (1935)
　　S. chlorojodinum (K. Schum.) N.E. Br. in F.T.A. 4(1): 366 (1902)
　　S. kassneri S. Moore in J.B. 50: 362 (1912). Type: Congo-Kinshasa, Kasai District: Lubi R., *Kässner* 3303 (BM!, holo.)
　　Asclepias carsonii (N.E. Br.) Schltr. in R.E. Fries, Wiss. Ergebn. Schwed. Rhod.-Congo-Exped. 1: 267 (1916)

2. **Glossostelma spathulatum** (*K. Schum.*) *Bullock* in K.B. 7: 414 (1952); Cribb & Leedal, Mt Fl. S. Tanz.: 103, photo 1c (1982). Type: Angola, Cuango, *von Mechow* 539a (K!, lecto., designated by Goyder in K.B. 50: 537 (1995); B†, isolecto.)

Erect herb to 50 cm. Leaves petiolate or occasionally sessile, petiole (2–)5–15(–20) mm long; lamina spreading or ascending, spatulate, obovate, oblanceolate, lanceolate or oblong, lowest leaves present at anthesis 2–3.5 × 0.3–1.2 cm, upper leaves 2.5–8(–10) × (1.1–)2–4(–4.8) cm, apex obtuse, rounded or retuse, apiculate, the apices of lower leaves generally more acute than those of upper leaves, tapering gradually or abruptly at the base into the petiole, glabrous or sparsely hairy beneath; midrib channelled on upper surface, lateral veins clearly visible but neither impressed nor raised, numerous, parallel to each other and at ± 90° to the midrib. Umbels with 2–4(–8) flowers; peduncles pubescent at least on one side, 2–20(–35) mm long; bracts deciduous, narrowly triangular, often of two types, one 2–4 mm long, filiform, the other 5–7(–12) × 1–1.5 mm, margins ciliate; pedicels 11–25(–31) mm long,

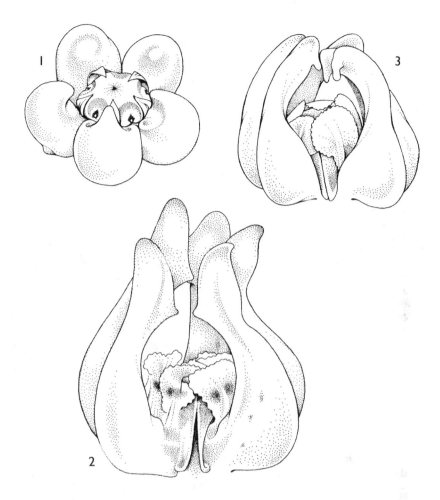

FIG. 112. *GLOSSOSTELMA SPATHULATUM* — **1**, gynostegium with corona, × 4. *GLOSSOSTELMA CECILIAE* — **2**, "Mbala type" gynostegium and corona, × 4; **3**, "general type" gynostegium and corona, × 4. From K.B. 50: 541, drawn by E. Papadopoulos.

pubescent on one side. Calyx lobes ovate, 6–10 × (2.5–)4–5(–6.5) mm, acute or subacute. Corolla green or maroon with reddish brown, yellow or white markings within, campanulate, lobed to 1–2 mm from the base; lobes obovate, 13–20(–25) × 8–14 mm, apex rounded or obtuse, recurved, glabrous. Corona lobes white or yellowish, subglobose, ± 3 × 2 mm, reaching the top of the column or exceeding it slightly, with a flattened inner face and three short, inward-pointing teeth near the top, attached to the staminal column 3–5 mm above the base of the gynostegium. Anther appendages broadly ovate, ± 1.5 × 2 mm; corpusculum ± 0.5 mm long; translator arms ± 0.5 mm long; pollinia falcate-oblong, 0.7–1 × ± 0.5 mm. Follicles ovate-lanceolate, ± 7 × 2 cm, puberulent. Seeds suborbicular, ± 7 × 6 mm. Fig. 112/1.

TANZANIA. Tabora District: Urambo, 12 km SW of station, 26 Jan. 1951, *Welch* 53!; Iringa District: 25 km SE of Iringa on Dabaga road, 8 Feb. 1962, *Polhill & Paulo* 1392!; Songea District: Matengo hills, Miyau, 1 Mar. 1956, *Milne-Redhead & Taylor* 8795!
DISTR. **T** 4, 7, 8; Congo-Kinshasa, Angola, Zambia, Malawi, Mozambique and Zimbabwe
HAB. Montane grassland, open *Parinari* woodland and disturbed miombo; 1100–1900 m

Schizoglossum spathulatum K. Schum. in E.J. 17: 120 (1893) & in E. & P. Pf. 4(2): 233, 234, f. 68 A-D (1895)

Gomphocarpus spathulatus (K. Schum.) Schltr. in J.B. 33: 269 (1895)

Xysmalobium bellum N.E. Br. in K.B. 1895: 69 (1895) & F.T.A. 4(1): 311 (1902). Types: Malawi, *Buchanan* 603 (K!, syn.) & Blantyre, *Buchanan* 43 (K!, syn.; E!, isosyn.) & Manganja Hills, *Kirk* s.n. (K!, syn.); Zambia, Fwambo, *Carson* 62 (K!, syn.)

X. spathulatum (K. Schum.) N.E. Br. in F.T.A. 4(1): 312 (1902)

3. **Glossostelma mbisiense** *Goyder* in K.B. 50: 543 (1995). Type: Tanzania, Ufipa District: Mbisi [Mbizi] Forest, *Sanane* 1426 (K!, holo.)

Erect herb; stems 15–35 cm long. Leaves petiolate or subsessile; petiole 0–6 mm long; lamina slightly fleshy, lanceolate- to ovate-oblong, (3.5–)5–7 × 1–2.1(–2.6) cm, apex acute or subacute, apiculate, glabrous, lateral veins obscure or clearly visible, neither impressed nor raised, parallel to each other and at ± 90° to the midrib. Umbels with 2–6 flowers; peduncles 1–7 mm long, pubescent on one side; bracts filiform, 3–5 mm long; pedicels 10–18 mm long, pubescent. Calyx lobes oblong or lanceolate to triangular, 5–7 × 1.5 mm. Corolla green or cream; lobes ovate or oblong, 13–20 × 4–7 mm, apex acute or subacute, sometimes recurved. Corona lobes attached to the gynostegium 1.5–2 mm above the base of the corolla, 1.5–2 × 1–1.5 mm, swollen at the base, with a slender horn at the top of the inner face, projecting upwards or inwards for up to 1 mm, the tip of the corona lobe ± level with the top of the column or sometimes slightly shorter. Stylar head projecting slightly above the top of the anthers; corpusculum black, ovoid, ± 0.6 mm long; translator arms ± 0.8 mm long, slender, flattened, contorted distally; pollinia broadly oblong but rounded distally, ± 0.7 × 0.5 mm. Follicles (very immature) densely rufous-pubescent; mature follicles not seen.

TANZANIA. Ufipa District: Mbisi Forest Reserve, 4 km SE of Wipanga, 2 Dec. 1993, *Gereau et al.* 5233! & 21–26 Nov. 1958, *Napper* 1046! & 20 Nov. 1987, *Ruffo & Kisena* 2628!
DISTR. T 4; endemic to Mbisi Forest and the Sumbawanga area
HAB. Montane grassland; 2000–2500 m. Flowering November to December

NOTE. At first sight this species could easily be confused with an extreme form of *G. spathulatum* but the corona lobes differ structurally - the lateral teeth are absent from the inner face and the peg at the top is a much more pronounced structure than the upper tooth of *G. spathulatum* and is oriented quite differently. Differences in both the shape and colouring of the corolla lobes correlate with the coronal morphology and there is also a temporal difference in flowering.

4. **Glossostelma ceciliae** (*N.E. Br.*) *Goyder* in K.B. 50: 547 (1995). Type: Zimbabwe, Mashonaland, at Harare [Salisbury], *Cecil* 60 (K!, holo.)

Robust erect herb; stems 20–90 cm long. Leaves petiolate or sessile, petiole (0–)3–6 mm long; lamina narrowly oblong or oblanceolate, spreading or ascending, lowest leaves present at anthesis 2.5–3.5 × 0.7–0.9 cm, upper leaves (3.5–)5–8.5 × 0.9–2.1 cm, apex acute, obtuse or rounded, apiculate, rounded at the base or more commonly tapering gradually into the petiole, glabrous except on the margins; midrib channelled on upper surface, lateral veins clearly visible but neither impressed nor raised, numerous, parallel to each other and at ± 90° to the midrib. Umbels with (1–)3–5 flowers; peduncles pubescent, 1–12 mm long or absent; bracts deciduous, often of two types, one filiform and 5–6 mm long, the other narrowly triangular and 5–9 × 1 mm, margins ciliate; pedicels 16–25(–30) mm, pubescent on one side. Calyx lobes ovate or triangular, 6–10 × 2–4 mm, acute. Corolla green or cream within, tinged brown or reddish brown outside, campanulate, lobed to 2–3 mm from the base; lobes obovate, 23–28 × 9–19 mm, apex acute or obtuse, recurved, glabrous. Corona lobes green or white, attached to the staminal column 2–4 mm above the base of the gynostegium, falcate, 7–10 × 3–4 mm, base slightly swollen, laterally compressed but not flattened above, the upper half arched over the head of the column, dilated slightly at the truncate, clavate apex. Anther appendages broadly ovate, 2–3 × 2–3 mm; corpusculum ovoid, ± 0.3 mm long; translator arms ± 0.5 mm long; pollinia falcate-oblong, 1.25 × 0.5 mm. Follicles lanceolate to long-fusiform, 11–24 × 1–1.3 cm, attenuate at both ends, puberulent. Seeds ovate, ± 5 × 4 mm, convex on one face, the disc verruculose, the rim convoluted. Fig. 112/2 & 3, p. 461.

TANZANIA. Ufipa District: Namwele, Itala Hills, Oct. 1965, *Richards* 20575!; Iringa District: Image Forest Reserve, Nov. 1959, *Procter* 1527! & Ipogoro–M'kawa track, Sao Hill, Dec. 1961, *Richards* 15566!
DISTR. **T** 4, 7; Congo-Kinshasa, Angola, Zambia and Zimbabwe
HAB. *Brachystegia* woodland or seasonally burnt grassland; 1500–2000 m

SYN. *Xysmalobium ceciliae* N.E. Br. in F.T.A. 4(1): 310 (1902)

5. **Glossostelma cabrae** (*De Wild.*) *Goyder* in K.B. 50: 548 (1995). Type: Congo-Kinshasa, Tawa Valley, *Cabra-Michel* 52 (BR!, holo.; K!, photo.)

Erect herb to 50 cm. Leaves with petiole (0–)1–5 mm long; lamina spreading or ascending, oblong, obovate-oblong or rarely linear, lowest leaves present at anthesis 1.5 × 0.2 cm, upper leaves 3.5–8(–20) × (0.4–)1.5–3.1 cm, apex obtuse, rounded or retuse, apiculate, tapering gradually at the base, glabrous or with short, sparse hairs beneath, midrib channelled on upper surface, lateral veins clearly visible but neither impressed nor raised, numerous, parallel to each other and at ± 90° to the midrib. Umbels with 2–4 flowers; peduncles pubescent, 1–12 mm long, or absent; bracts deciduous, filiform, 5–6 mm long; pedicels 18–25 mm long, pubescent on one side. Calyx lobes triangular, 5–6 × 1–1.5 mm, acute. Corolla green or cream with reddish brown markings within, campanulate, lobed to 2–3 mm from the base; lobes recurved, obovate, 23–26(–28) × 9–14(–20) mm, apex acute or obtuse, glabrous except on the margins. Corona lobes green, attached to the staminal column 3–6 mm above the base of the gynostegium, 9–11 mm long, 2 mm wide at the base, the lower half of the lobe somewhat compressed laterally, falcate, arched over the head of the column and narrowed into the base of the erect, dorsally flattened, slightly fleshy, spatulate upper half, the apex rounded or truncate, with one or occasionally two narrowly triangular downward pointing flaps up to 2 mm long on the ventral face. Anther appendages broadly ovate, ± 1.5 × 2 mm; corpusculum ± 0.3 mm long; translator arms ± 0.5 mm long; pollinia falcate-oblong, ± 1 × 0.25 mm. Follicles not seen.

TANZANIA. Rungwe District: Kyimbila District: *Stolz* 191!
DISTR. **T** 7; Congo-Kinshasa, Angola, Zambia and Mozambique
HAB. Deciduous (miombo) woodland or open grassland, 1200–1500 m. Flowering November to December

SYN. *Asclepias cabrae* De Wild. in Ann. Mus. Congo, sér. V, 1: 185 (1904)
      Xysmalobium speciosum S. Moore in J.B. 47: 216 (1909). Type: Zambia, Lisanga Spruit, *Kässner* 2144 (BM!, holo.; K!, iso.)

6. **Glossostelma brevilobum** *Goyder* in K.B. 50: 551 (1995). Type: Malawi, Nkhata Bay District: Vipya Plateau, 40 km SW of Mzuzu, *Pawek* 5954 (K!, holo.; CAH, MAL!, MO, UC, iso.)

Erect herb with stems 12–40 cm long. Leaves sessile or petiolate, petiole to 5 mm long; lamina spatulate, obovate or narrowly oblong to ovate, lowest leaves present at anthesis 1.3–2 × 0.4–1 cm, upper leaves (2–)3–7 × (0.5–)1–3.6 cm, apex rounded, obtuse or subacute, apiculate, the apices of lower leaves generally more acute than those of upper leaves, rounded at the base or tapering gradually into the petiole, glabrous or sparsely hairy beneath; midrib channelled on upper surface, lateral veins clearly visible but neither impressed nor raised, numerous, parallel to each other and at ± 90° to the midrib. Umbels with up to 9 flowers; peduncles pubescent, ± 1 mm long; pedicels 8–13 mm long, pubescent on one side. Calyx lobes ovate or triangular, 2–3 × 1–1.5 mm, acute. Corolla green or brown, campanulate, lobed to 1 mm from the base; lobes ovate, 5–8 × (2–)4–5 mm, apex acute or subacute, glabrous. Corona lobes attached near the base of the gynostegium, ± 1 × 0.5 mm, forming fleshy outward-pointing pegs briefly united into an annulus at the base. Anther appendages broadly ovate, ± 1.5 × 2 mm; corpusculum brown, ovoid, 0.5–0.8 mm long; translator arms 0.4–0.7 mm long; pollinia obtriangular, ± 0.75 × 0.5–0.6 mm Follicles not seen.

TANZANIA. Iringa District: Mufindi, Ngwazi grassland, 3 Feb. 1987, *Lovett* 1424!
DISTR. **T** 7; known from just four localities in Congo-Kinshasa, Burundi, Tanzania and Malawi
HAB. Montane grassland, 1500–1800 m. Flowering November to February

NOTE. *Richards* 10216 from Kito Mountain in SW Tanzania has the initial appearance of narrow-leaved forms of *G. brevilobum* but is anomalous in a number of characters. The rootstock is a simple, slender napiform tuber rather than the typical *Glossostelma* type with fusiform lateral tuberous roots; the corona lobes are fleshy but ovate rather than the reduced pegs of *G. brevilobum*; the translator arms are unusually short and broad and the apex of the stylar head is rostrate. The form of the stylar head is most unusual in this part of the family, while the rootstock and corona lobes may indicate a link to *Xysmalobium.*

80. **PLEUROSTELMA**

Baill., Hist. Pl. 10: 266 (1890); Bullock in K.B. 10: 612 (1956)

Microstephanus N.E. Br. in K.B. 1895: 249 (1895) & in F.T.A. 4(1): 288 (1902)
Podostelma K. Schum. in E.J. 17: 133 (1893)

Perennial herbs or small shrubs with twining stems; latex clear, sparse. Leaves semi-succulent, opposite. Calyx of 5 free sepals. Corolla-tube campanulate; lobes narrowly triangular, contorted in bud. Corolline corona absent. Gynostegial corona of vestigial staminal lobes linked by minute interstaminal pockets around the base of the column. Staminal column raised slightly above the base of the corolla-tube; filaments short; anthers oblong, erect with rounded backs, the horny anther wings incurved toward the centre of the flower; anther appendages sub-erect, membranous. Pollinaria each with 2 pollinia suspended vertically from the short caudicles. Stylar head produced beyond the anthers into a beak. Follicles lanceolate-acuminate, smooth.

Two species native to NE tropical Africa, Madagascar and the Mascarene Islands.

Pleurostelma cernuum (*Decne.*) *Bullock* in K.B. 10: 612 (1956); Liede-Schumann in Fl. Somalia 3: 153, fig. 107 (2006). Type: Tanzania, Pemba, *Bojer* s.n. (P, holo.)

Woody herb or shrub, prostrate or twining to 2 m; stems glabrous, pubescent along one line with curved white hairs or occasionally pubescent all round. Leaves somewhat fleshy, the petiole (2–)4–8(–13) mm long, pubescent along the upper surface; leaf-blade oblong or elliptic to ovate or occasionally rhomboid, 10–40(–55) × 4–15(–25) mm, apex acute, obtuse or rounded, commonly with a short apiculus, base generally acute, but occasionally obtuse or rounded, margins flat or revolute, glabrous or pubescent; lateral veins generally in two pairs beneath, forming an acute angle with the midrib and ending almost parallel with the margin. Inflorescences axillary, 1–4(–6)-flowered, umbelliform, glabrous to densely pubescent; peduncles to 7 mm, or absent; bracts subulate, 0.5–1 mm; pedicels 2–6 mm. Sepals ovate, acute, glabrous or pubescent, hyaline, at least at the margin. Corolla white or cream, 4–8 mm long, glabrous outside; tube campanulate, ± 2 mm, enclosing the staminal column,

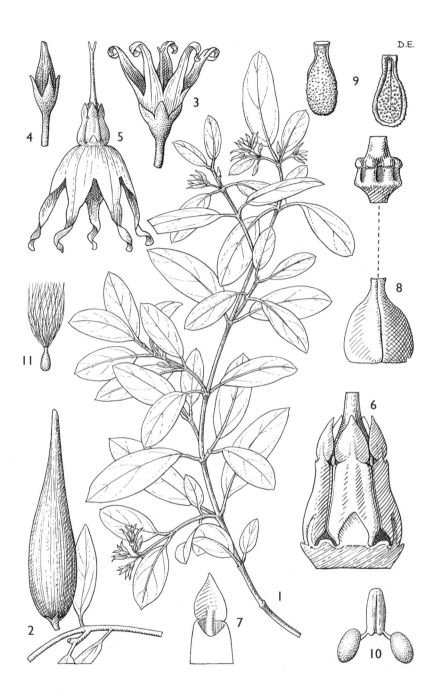

FIG. 113. *PLEUROSTELMA CERNUUM* — **1**, habit, × 1; **2**, fruit, × 1; **3**, flower, × 8; **4**, flower bud, × 8; **5**, flower with corolla folded back to expose gynostegium, × 10; **6**, gynostegium including base of stylar head appendage, × 36; **7**, stamen from within, × 36; **8**, gynoecium, showing paired ovaries and base of stylar head, × 36; **9**, seed, dorsal and ventral views, × 4; **10**, pollinarium, × 40; **11**, seed with coma × 1.5. 1, 3–10 from *Hildebrandt* 178; 2 from *Faulkner* 707. Drawn by D. Erasmus.

sparsely pubescent within; lobes contorted, 2–6 mm, glabrous, tapering gradually to a blunt, slightly offset apex, with small coronal pockets formed at the sinus of 2 lobes within the tube. Staminal column ± 1 mm, forming the base of a narrow cone continued upward by the stylar head appendage. Gynostegial corona inconspicuous, formed of 5 triangular ridges barely projecting from the lower half of each anther and linked by minute pockets at the base of the column, alternating with the anthers. Anther appendages white, ovate, acute and erect-connivent around the base of the apical part of the style; pollen carriers dark reddish brown, ± 0.25 mm; pollinia ovoid, slightly smaller than the carrier. Stylar head projecting beyond the staminal column as a subulate beak, ± 2 mm long, sub-entire or deeply bifid. Follicles reddish brown, 40–60 mm long, ± 10 mm wide, usually only one of the pair developing. Seeds reddish brown, ± 4 × 1.5 mm, crowned with an ivory plume of hairs. Fig. 113, p. 465.

KENYA. Kwale District: Diani Beach, 24 Aug. 1959, *Napper* 1273!; Mombasa District: Shanzu Beach, June 1959, *Ossent* 309A!; Kilifi District: Vipingo, 17 Dec. 1953, *Verdcourt* 1082!
TANZANIA. Tanga District: Amboni Estate, 16 July 1955, *Faulkner* 1625! & coast near Bomalandani [Bomandani], 5 Aug. 1953, *Drummond & Hemsley* 3658!; Uzaramo District: Dar es Salaam, Msasani Beach, 26 Apr. 1988, *Bidgood, Congdon & Vollesen* 1260!; Zanzibar: Mjini District, 23 Nov. 1999, *Fakih & Abdalla* 486!
DISTR. **K** 7; **T** 3, 6, 8; **Z**; **P**; Somalia, Mozambique; Madagascar and the Mascarenes
HAB. Bare ground or coral thickets on sand dunes or coral rock near the coast; sea level to 100 m

SYN. *Astephanus cernuus* Decne. in Ann. Sci. Nat., sér. 2, 9: 342 (1838)
　　Pleurostelma grevei Baill., Hist. Pl. 10: 266 (1891), in adnot.
　　Microstephanus cernuus (Decne.) N.E. Br. in K.B. 1895: 249 (1895) & in F.T.A. 4(1): 288 (1902); T.T.C.L.: 67 (1949)

NOTE. A form of this species with larger corolla, a more shrubby habit and more fleshy leaves occurs around Lamu and in southern Somalia.

81. SCHIZOSTEPHANUS

G.P. 2: 762 (1876); Liede in J.L.S. 114: 81 (1994)

Scandent shrubs or twiners with succulent stems; latex clear or yellowish, not milky. Leaves opposite. Inflorescences extra-axillary, many-flowered, with the flowers scattered along an indeterminate axis. Corolla rotate, the lobes fused basally. Corolline corona absent. Gynostegial corona fused for most of its length into a tube surrounding and obscuring the gynostegium. Gynostegium stipitate. Anther connectives with apical appendages. Pollinaria with a pair of pendent pollinia. Apex of stylar head flat. Follicles paired, or single by abortion, winged longitudinally. Seeds flattened, ovate, with a coma of silky hairs.

Two species in tropical Africa. According to evidence presented by Liede in Ann. Missouri Bot. Gard. 88: 657–668 (2001) its affinities lie most closely with *Cynanchum* and *Pentarrhinum*.

Schizostephanus alatus *K. Schum.* in E.J. 17: 139 (1893); Liede in Fl. Eth. 4(1): 136 (2003); Liede-Schumann in Fl. Somalia 3: 160, fig. 112 (2006). Types: Ethiopia, Mai-Mezano, Djeladjeranne, *Schimper* 1687 (K!, lectotype, designated here; B†, isolecto.); Tanzania, Mwanza District, Kagehi, *Fischer* 391 (B†, paralecto.)

Scandent shrub or twiner to 8 m; stems thick and fleshy, glabrous except on very young shoots. Leaves with petiole 2–6 cm long, glabrous; lamina broadly ovate to suborbicular, 5–10(–12) × 4–6(–10) cm, apex acute to shortly attenuate, base weakly to strongly cordate, glabrous except on principal veins beneath. Inflorescences axes to 10(–13) cm long, somewhat fleshy, subglabrous or minutely pubescent; pedicels 4–6 mm long, minutely pubescent. Calyx lobes ovate, 0.5–1 mm long, rounded to

Fig. 114. *SCHIZOSTEPHANUS ALATUS* — **1**, flowering shoot, × ²/₃; **2**, inflorescence, × 2; **3**, paired follicles from above, single follicle from beneath, × ²/₃; **4**, flower, × 3; **5**, gynostegium largely obscured by the cupular corona, × 10; **6**, stipitate gynostegium, × 14; **7**, pollinarium, × 20. All from *Greenway* 10727. Drawn by M. Tebbs.

subacute, minutely pubescent. Corolla greenish yellow or the lower half reddish purple, united at the base, rotate; lobes oblong or oblong-ovate, 2.5–4 × 1–1.5 mm, truncate and somewhat twisted apically, pubescent with long slender hairs in basal half of lobes adaxially, abaxial face glabrous. Gynostegium raised on a stipe 1–1.5 mm long. Corona forming a fluted cylinder surrounding and obscuring the gynostegium, ± 2 mm long with longitudinal ridges on the inner face, divided apically into ten erect, acute lobes 0.5–1 mm long arising in both staminal and interstaminal positions. Anther wings 0.3–4 mm long. Apex of stylar head not exserted from staminal column. Pollinaria with corpusculum ± 0.3 mm × 0.2 mm with thin lateral wings; translator arms ± 0.15 mm long, flattened and geniculate; pollinia ovoid, ± 0.4 mm long, attached apically to translator arms. Follicles mostly in widely divergent pairs, fusiform, 5–6 cm long, with three longitudinal wings ± 0.5 cm wide running the entire length of the follicle, smooth, glabrous. Seeds flattened, ± 6 × 3 mm, smooth and with a marginal rim ± 0.5 mm wide; coma 1.5–2 cm long. Fig. 114, p. 467.

UGANDA. Mengo District: Kiwatule, Kyandondo, 2 May 1929, *Liebenberg* 814!
KENYA. Northern Frontier District: Dandu, 18 June 1952, *Gillett* 13453!; Masai District: Tsavo National Park, Mzima Springs, 17 Jan. 1961, *Greenway* 9758!; Kwale District: Diani Forest, 2 Oct. 1990, *Luke* 2454!
TANZANIA. Musoma District: Campi ya Mawi, 18 June 1962, *Greenway* 10727!; Mbulu District: Mbagayo, Lake Manyara National Park, 12 Nov. 1963, *Greenway & Kirrika* 11012!; Ulanga District: Mahenge, Issongo, 3 Feb. 1932, *Schlieben* 1715!
DISTR. U 4; K 1, 4, 6, 7; T 1, 2, 6; from Somalia, Ethiopia, Sudan and Kivu in E Congo southwards through Malawi and Zimbabwe to the NE of South Africa
HAB. Mostly on granite or limestone outcrops in dry *Acacia* bushland or degraded forest patches; sea level to 1500 m

SYN. *Cynanchum validum* N.E. Br. in F.T.A. 4(1): 398 (1903); T.T.C.L.: 65 (1949); U.K.W.F. ed. 2: 180 (1994), *non C. alatum* Wight & Arn. (1834); Malaisse in Fl. Rwanda 3: 100 (1985); U.K.W.F. ed. 2: 180 (1994). Type as for *Schizostephanus alatus*

82. PENTARRHINUM

E. Mey., Comm Pl. Afr. Austr. 199 (1838); Liede & Nicholas in K.B. 47: 475 (1992); Liede in Ann. Missouri Bot. Gard. 83: 283 (1996) & in Syst. Bot. 22: 347 (1997)

Twining herbs with perennial rootstock and slender annual stems; latex white. Leaves opposite, herbaceous, petiolate, cordiform, colleters conspicuous at the base of the lamina; small leaves on short axillary shoots frequently giving the appearance of stipules. Inflorescences extra-axillary, cymose, pedunculate. Corolla lobes fused basally, recurved. Corolline corona absent. Gynostegial corona generally consisting of a short tube and prominent free staminal parts, flat or laterally conduplicate, apically either conduplicate or provided with a horn-like ornament projecting towards the gynostegium; if corona fused for much of its length, then with vertical 'suture marks'. Gynostegium sessile. Anther connectives with apical appendages. Pollinia pendulous. Stylar head flat. Follicles mostly single, ellipsoid, rather thick-walled, with or without protruberances of variable length and density. Seeds flattened, ovate, papillate or pubescent, with or without a marginal wing; with a coma of silky hairs.

1. Corona fused for at least half its length, enclosing and largely obscuring the gynostegium . . . . 5. *P. gonoloboides*
   Corona fused only at the base 2
2. Corona lobes clearly 3-dimensional, either solid or with inrolled apex and margins, with an apical or subapical inward-pointing tooth 3
   Corona lobes laminar, either entire or divided, but lacking an inward-pointing apical or subapical tooth 4

3. Corona yellow, papillose, fleshy, expanded abruptly into a flat-
        topped apex with a conspicuous inward-pointing horn ... 1. *P. insipidum*
    Corona white, smooth, thin and with inrolled margins and
        apex forming a tooth on the inner face .... 2. *P. abyssinicum*
4. Corona lobes obtriangular, trifid; *Acacia-Commiphora* bush 4. *P. somaliense*
    Corona lobes rounded, entire; wet forests .... 3. *P. ledermannii*

1. **Pentarrhinum insipidum** *E. Mey.*, Comm Pl. Afr. Austr. 200 (1838); F.P.S. 2: 412
(1952); Malaisse in Fl. Rwanda 3: 108 (1985); U.K.W.F. ed. 2: 179 (1994); Goyder &
Liede in Fl. Eth. 4(1): 142 (2003); Liede-Schumann in Fl. Somalia 3: 158 (2006).
Types: South Africa, Uitenhage, Enon, *Drège* 2220 (B†, holo.; K!, lecto., designated by
Liede & Nicholas in K.B. 47: 484 (1992))

Slender perennial twining to 8 m; stems arising at intervals from a long, horizontal
rhizome, minutely pubescent; latex white. Abbreviated leafy shoots often present in
leaf axils. Leaves with petiole 1–5 cm long, minutely pubescent; lamina ovate to
broadly triangular, 3–8 × 2–6 cm, apex acute or attenuate, base strongly cordate,
minutely pubescent. Inflorescences extra-axillary, simple, very rarely branched
distally, initially ± umbelliform, eventually with flowers scattered along an axis;
peduncle 2–8 cm long, minutely pubescent; rhachis to ± 2 cm; bracts linear,
0.5–1.5 mm long, pubescent; pedicels 10–20 mm long, minutely pubescent. Sepals
oblong or lanceolate, 1–2 mm long, rounded or acute, subglabrous to pubescent.
Flower buds subglobose. Corolla yellow-green or white, frequently tinged brown or
purple, divided almost to the base; lobes ovate to oblong, strongly reflexed, 2.5–4.5
× 1–3 mm, glabrous but with ciliate margins. Corona yellow, sometimes brown or
purple apically and perhaps becoming darker with age, divided almost to the base;
lobes ascending, ± 2 mm long, slightly taller than the staminal column, fleshy and
subcylindrical below, broadening abruptly to form a flattened apex with a well-
developed tooth or horn pointing towards the gynostegium, papillose. Gynostegium
subsessile. Anther wings ± 0.7 mm long; corpusculum rhomboid, ± 0.2 mm long;
translator arms ± 0.2 mm long; pollinia ± 0.5 mm long, attached apically to translator
arms. Follicles occurring singly or in pairs reflexed alongside the pedicel, lanceolate
in outline, 5–8 × 1.5 cm, tapering to a weakly clavate tip, with few to many irregular
winged processes. Seeds ovate, ± 6 × 4 mm, verrucose and with a crenate margin;
coma ± 3 cm long. Fig. 115, p. 470.

UGANDA. Ankole District: Isingiro, May 1939, *Purseglove* 700!; Teso District: Serere, Apr. 1932,
    Chandler 663!; Mengo District: near Busana, Bugerere, Apr. 1932, *Eggeling* 648!
KENYA. Northern Frontier District: Moyale, 8 July 1952, *Gillett* 13538!; Nairobi, Riverside, Aug.
    1945, *Bally* 4622!; Kilifi District: Gede National Monument, 14 Aug. 1991, *S.A. Robertson* 6721!
TANZANIA. Musoma District: NE side of Serengeti Central Plains, 4 June 1962, *Greenway &*
    Turner 10685!; Lushoto District: Lwengera River, 6 km ENE of Korogwe, 27 June 1953,
    Drummond & Hemsley 3062!; Iringa District: near Msembe lower air strip, 1 Apr. 1970,
    Greenway & Kanuri 14243!
DISTR. U 1–4; K 1, 2, 4–7; T 1–3, 5–7; widely distributed from Sudan, Eritrea and Somalia to
    Namibia and South Africa
HAB. Open grassland, *Acacia-Commiphora* bushland and *Brachystegia* woodland, occasionally also
    found in forest; sea level to 1800 m

2. **Pentarrhinum abyssinicum** *Decne.* in DC., Prodr. 8: 553 (1844); U.K.W.F. ed. 2:
179 (1994); Goyder & Liede in Fl. Eth. 4(1): 142 (2003). Type: Delessert, Ic. Sel. 5:
t. 80, 1846 (lecto., designated by Liede & Nicholas in K.B. 47: 481 (1992))

Slender perennial twining to 5 m, stems simple or branched, minutely pubescent;
latex white; rootstock not known; abbreviated leafy shoots often present in leaf axils.
Leaves with petiole 1.5–5 cm long, minutely pubescent; lamina ovate to broadly
triangular, 3–10 × 1.5–6 cm, apex acute or attenuate, base weakly to strongly cordate,

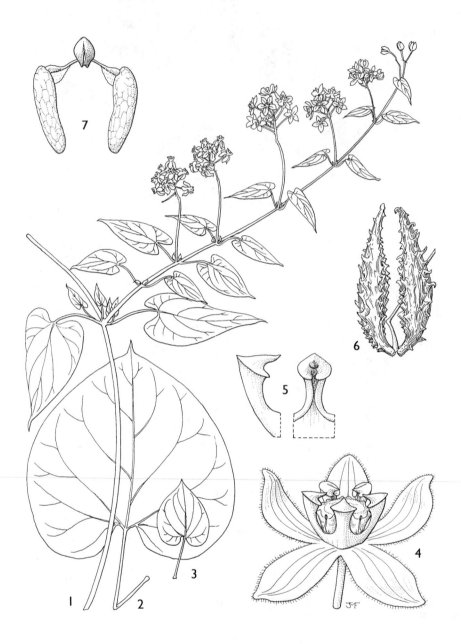

FIG. 115. *PENTARRHINUM INSIPIDUM* — **1**, habit, × ¹/₂; **2 & 3**, leaves, × ¹/₂; **4**, flower, × 4; **5**, corona lobe, lateral and ventral views, × 1; **6**, paired follicles, × ¹/₂; **7**, pollinarium, × 50. 1, 3 & 4 from *Drummond & Hemsley* 3062; 2 from *Polhill & Paulo* 726; 3 from *Milne-Redhead & Taylor* 11310; 6 from *Bally* 9110; 7 from *Pope* 86. From K.B. 47: 485, drawn by M. Fothergill.

shortly pubescent or glabrous. Inflorescences extra-axillary, simple or irregularly branched, umbelliform (most forest forms) or the flowers scattered singly or in pairs along the inflorescence axes (more open habitats); peduncle 1–6 cm long, minutely pubescent or subglabrous; rhachis to ± 3 cm; bracts linear, ± 1 mm long, glabrous or pubescent; pedicels 3–20 mm long, glabrous or minutely pubescent. Sepals triangular, 1–3 mm long, acute, minutely pubescent. Flower buds ovoid or subglobose. Corolla greenish white or pale yellow, sometimes with purple veins, divided almost to the base; lobes reflexed or not, ovate to oblong, 2–6 × 1–2 mm, frequently with somewhat revolute margins, entirely glabrous or pubescent on margins and abaxial face. Corona white, divided almost to the base; lobes spreading or erect, thin or somewhat fleshy, 1.5–2.5 mm long, slightly longer than the staminal column, the lateral and apical margins inrolled forming a weakly to strongly developed apical tooth on the inner face. Gynostegium subsessile. Anther wings 0.5–1 mm long; corpusculum ovoid, ± 0.2 mm long; translator arms ± 0.2 mm long; pollinia ± 0.5 mm long, attached apically to translator arms. Follicles occurring singly or in pairs, lanceolate in outline, 5–6 × 1 cm, tapering to a weakly clavate tip, smooth or with irregular soft processes. Mature seeds not seen.

a. subsp. **abyssinicum**

Corolla less than 10 mm diameter, lobes fully reflexed at anthesis, margins generally glabrous; mostly in grassland or open woodland.

KENYA. Machakos District: Mua Hills Road, May 1958, *Verdcourt & Baring* 2155!; Masai District: Ngong hills, 14 July 1957, *Verdcourt* 1798! & 10 km along Athi–Kajiado road, May 1960, *Archer* 56!
TANZANIA. Masai District: Serengeti National Park, Lagarja Lake, 16 Mar. 1970, *Richards* 25652!; Ufipa District: Tatanda Mission, 24 Feb. 1994, *Bidgood et al.* 2440!; Iringa District: Matanana, 120 km S of Iringa on Great N Road, 27 Mar. 1962, *Polhill & Paulo* 1884!
DISTR. **K** 4–6; **T** 1, 2, 4, 5, 7; from Cameroon and Eritrea to Namibia and Zimbabwe
HAB. Grassland, open bushland or deciduous woodland, less often in forest; 1100–2200 m

NOTE. A number of collections from forest habitats are difficult to place with confidence (e.g. *Bidgood et al.* 4147 from Kigoma; *Grimshaw* 93/377 from Kilimanjaro). These collections have the small flowers of this subspecies but approach *P. ledermannii* in general appearance, and the corona appears intermediate between *P. ledermannii* and *P. abyssinicum* – the inrolled apex and margins of the corona are less well-developed than is typical for *P. abyssinicum*. The length of the anther wings is also very variable.

b. subsp. **angolense** (*N.E. Br.*) *Liede & Nicholas* in K.B. 47: 482 (1992); Burrows in Pl. Nyika Plateau: 73 (2005). Type: Angola, Icolo e Bengo District, near Lagoa de Foto, *Welwitsch* 4240 (K!, lecto., designated by Liede & Nicholas in K.B. 47: 482 (1992), BM!, isolecto.)

Corolla more than 10 mm diameter, lobes fully not reflexed at anthesis, margins ciliate; occuring in wet forest.

KENYA. Nandi District: 6 km from Kapsabet on road to Eldoret, 13 Oct. 1981, *Gilbert & Mesfin* 6716! ; '2ⁿᵈ day's march from Nandi', 1898, *Whyte* s.n.!; Kiambu District: near Limuru Girls High School, 20 Aug. 1961, *Polhill* 457!
TANZANIA. Lushoto District: W Usambaras, Jaegertal Valley, 21 June 1953, *Drummond & Hemsley* 2970!; Ufipa District: Kito Mt on Zambian border, 21 Apr. 1961, *Richards* 15047!; Iringa District, Udzungwa Mountain National Park, 26 May 2002, *Luke et al.* 8472!
DISTR. **K** 3–5; **T** 3, 4, 7; Cameroon, Angola, Namibia, Congo-Kinshasa, Zambia and Malawi
HAB. Wet forest; 1400–2200 m

SYN. *Pentarrhinum abyssinicum* Decne. var. *angolense* N.E. Br. in F.T.A. 4(1): 379 (1902)
   P. insipidum sensu Hiern, Cat. Afr. Pl. Welw. 1(3): 687 (1898) & sensu Bullock in F.W.T.A. ed. 2, 2: 90 (1963), *non* E. Mey.
   Pentarrhinum sp. A of U.K.W.F. ed. 2: 179 (1994)

3. **Pentarrhinum ledermannii** (*Schltr.*) *Goyder & Liede* in K.B. 63: 466 (2009). Types: Cameroon, *Ledermann* 5757, 5892 & 5931a (B†, syn.); neotype: Burundi, Bubanza, Mugomero (Rugazi), *Reekmans* 10069 (K!, neo., designated by Liede in Ann. Missouri Bot. Gard. 83: 320 (1996))

Twining perennial to 5 m, latex white; stems branched, pubescent; abbreviated leafy shoots present in leaf axils. Leaves with petiole 2–6 cm long, sparsely pubescent; lamina ovate, to 12 × 8 cm, apex attenuate, base deeply cordate with rounded auricles, subglabrous or sparsely pubescent on both surfaces. Inflorescences extra-axillary, subumbelliform, with 8–15 flowers opening ± simultaneously; peduncle 1.5–6 cm long, pubescent; bracts 1–3 mm long, linear, pubescent; pedicels 15 mm long, slender, pubescent. Sepals linear to lanceolate, 1–3 mm long, pubescent. Flower buds subglobose. Corolla somewhat reflexed, united only at the base; lobes white marked with longitudinal red veins, ovate, 3–4 × 1.5 mm, glabrous on both surfaces. Corona white, united into a tube ± 0.5 mm long, shorter than the staminal column; staminal parts of the tube each extending for a further 2 mm into a broad, fleshy rounded lobe; adaxial tooth or appendage absent. Gynostegium shortly stipitate. Anther wings ± 0.7 mm long; corpusculum ovoid, ± 0.2 mm long; translator arms ± 0.1 mm long, flattened; pollinia attached subapically to translator arms, ovoid, ± 0.3 × 0.1 mm. Stylar head obscured by anther appendages. Mature follicles not seen.

KENYA. Machakos/Masai Districts: Chyulu Hills, main forest N Camp 3, 24 March 2008, *Luke & Collins* 12055B!
TANZANIA. Morogoro District: Uluguru South Catchment Forest Reserve, W slopes on path from N'gungulu village to Lukwangule plateau, 5 Feb. 2001, *Jannerup & Mhoro* 402! & Uluguru, 13 Feb. 1933, *Schlieben* 3436 (reported by Liede 1996: 322)
DISTR. **K** 4/6; **T** 6; scattered records from Cameroon, Congo-Kinshasa and Burundi
HAB. Moist forest; 2000–2200 m

SYN. *Cynanchum ledermannii* Schltr. in E.J. 51: 140 (1913)
    Pentarrhinum abyssinicum Decne. subsp. *ijimense* Goyder in Cheek *et al.*, Plants Mt Oku & Ijim Ridge: 92 (2000). Type: Cameroon, Boyo Division, Ijim, *Cheek* 9943 (K!, holo.; BR, EA, MO, MSTR, P, PRE, WAG, YA, iso.)

NOTE. The thick-walled and somewhat warty young follicles on *Maitland* 1394 and other Cameroonian collections confirm the suspicions voiced by Liede (1997) that *Cynanchum ledermannii* might be better placed in *Pentarrhinum*. Very close to *P. abyssinicum* from which it can be distinguished by the stipitate gynostegium and the lack of an apical inward pointing tooth on the corona lobe. The Tanzanian collection examined is somewhat larger in most dimensions than the West African material.

4. **Pentarrhinum somaliense** (*N.E. Br.*) *Liede* in Syst. Bot. 22: 368 (1997); Goyder & Liede in Fl. Eth. 4(1): 142 (2003); Liede-Schumann in Fl. Somalia 3: 158 (2006). Type: Ethiopia, Harerge region, Bubi, [Somalia, Boobi], *James & Thrupp* s.n. (K!, holo.)

Erect or scandent perennial to ± 40 cm in our area, twining to 4 m elsewhere, stems simple or branched, minutely pubescent; latex white; rootstock reported to be a deep-seated horizontal rhizome; abbreviated leafy shoots not present in leaf axils. Leaves with petiole 1.5–5.5 cm long, minutely pubescent; lamina oblong to broadly triangular, 3–8 × 1.5–5 cm, apex acute or subacute, base truncate to weakly cordate, glabrescent, any indumentum restricted mostly to margins and principle veins. Inflorescences extra-axillary, simple or irregularly branched, the flowers scattered singly or in pairs along the inflorescence axes; peduncle 1–3 cm long, minutely pubescent or subglabrous; rhachis to 10 cm long but usually much shorter; bracts linear to triangular, 1–4 mm long, glabrous or pubescent; pedicels 2–5 mm long, glabrous or minutely pubescent. Sepals oblong, 2–3 × 0.7–1 mm, acute, glabrous. Flower buds obconical, truncate. Corolla greenish white or pale yellow, divided almost to the base; lobes suberect, oblanceolate or oblong, 3–5 × 1 mm, frequently with somewhat revolute margins, glabrous. Corona white, ageing to purple, divided almost to the base; lobes erect, obtriangular, 2–3 mm long ± as tall as the staminal column, dorsiventrally compressed, with three conspicuous teeth or lobes apically. Gynostegium clearly stipitate. Anther wings ± 0.3 mm long but with anther sacs extending below; corpusculum narrowly ovoid to subcylindrical, 0.2–0.3 mm long; translator arms ± 0.4 mm long, slender and curved around the margins of the pentagonal stylar head; pollinia ± 0.5 mm long, attached apically to translator arms. Follicles occurring singly, lanceolate in outline, 5–6 × 1 cm, tapering to a weakly clavate tip, smooth or with irregular soft processes. Mature seeds not seen.

KENYA. Northern Frontier District: El Wak, 26 May 1952, *Gillett* 13340! & 13 km N of Isiolo on road to Marsabit, 2 Nov. 1978, *Gilbert et al.* 5314!; North Nyeri District: Ngare Ndare Farm, 19 Apr. 1981, *Gilbert* 6093!
DISTR. **K** 1, 4; South Sudan, S Ethiopia and Somalia
HAB. Open *Acacia-Commiphora* bushland; 300–1500 m

SYN. *Schizostephanus somaliensis* N.E. Br. in K.B. 1895: 250 (1895)
    Cynanchum trifurcatum Schltr. in Bull. Herb. Boiss. 4: 448 (1896). Type as for *Pentarrhinum somaliense*
    C. dentatum K. Schum. in Ann. Reale Inst. Bot. Roma 7: 39 (1898). Type: Ethiopia, Harerge region, Ogaden, 'Somalia, inter Sassaber et Cabaden', *Riva* 844 (FT, holo.)
    C. somaliense (N.E. Br.) N.E. Br. in F.T.A. 4(1): 398 (1903); U.K.W.F. ed. 2: 180 (1994)

NOTE. The two Ugandan collections cited by Liede in Ann. Missouri Bot. Gard. 83: 340 (1996) do not belong to this species but to *P. insipidum*, and the Tanzanian collection is a mixed collection – the vegetative parts are probably of *P. insipidum*, while the flowers belong to *Schizostephanus alatus*.

5. **Pentarrhinum gonoloboides** (*Schltr.*) *Liede* in Syst. Bot. 22: 368 (1997); Goyder & Liede in Fl. Eth. 4(1): 142 (2003). Types: Rwanda, *Schlechter* 1617 (B†, holo.); Kenya, Nakuru District: Doboti, near Mau Forest Reserve on track to Nairagie Ngare, *Glover, Gwynne & Samuel* 1492 (K!, neo., designated by Liede in Ann. Missouri Bot. Gard. 83: 317 (1996)); EA!, FT, K!, isoneo.)

Twining perennial to 10 m, stems minutely pubescent when young, becoming glabrous with age; latex white; rootstock not known; abbreviated leafy shoots not present in leaf axils. Leaves with petiole 1.5–3(–8) cm long, glabrous or minutely pubescent; lamina ovate to oblong, 5–11 × 2–8 cm, apex attenuate, base weakly to strongly cordate, veins prominent and raised on lower surface, glabrous. Inflorescences extra-axillary, single or occasionally paired, the flowers scattered singly or in pairs along the inflorescence axes; peduncle 1–3 cm long, minutely rusty-pubescent; rhachis to 3 cm long; bracts linear, ± 3 mm long, minutely pubescent; pedicels 1–2 cm long, minutely rusty-pubescent. Sepals ovate to broadly ovate, 2–4 × 1–2 mm, acute, rusty-pubescent. Flower buds subglobose. Corolla greenish white with longitudinal brown or purple stripes, divided almost to the base; lobes spreading or somewhat reflexed, ovate or oblong, 4–5 × 2.5–3.5 mm, frequently with somewhat revolute margins, glabrous. Corona white or cream, forming a pentagonal cup 1.5–2 mm high enclosing the gynostegium and about as tall as it, corona divided for about half its length into 5 truncate or weakly to strongly three-toothed lobes opposite the stamens. Gynostegium stipitate. Anther wings ± 0.2 mm long; corpusculum ovoid, 0.4–0.5 mm long; translator arms 0.6–0.7 mm long, slender and curved around the margins of the pentagonal stylar head; pollinia ± 0.4 × 0.2 mm, attached subapically to translator arms. Follicles occurring singly or in pairs, lanceolate in outline, 5–11 × 1–1.5 cm, with irregular warty ornamentation. Seeds ovate, 9–11 × 4–6 mm, narrowly winged; coma 20–25 mm long.

UGANDA. Mbale District: Mt Elgon, Sasa trail, 20 Apr. 1997, *Wesche* 1256!

Kenya. Nakuru District: Eburru Forest Reserve, 18 July 2002, *Luke et al.* 8967!; Elgeyo District: Embobut Forest, Cherangani Hills, Jan. 1971, *Tweedie* 3912!; North Nyeri District: Aberdares National Park road, Nyeri, no date?, *Agnew* 7179

Tanzania. Moshi District: forest above Kilimanjaro Timbers, 25 June 1994, *Grimshaw* 94/595!

Distr. **U** 3; **K** 3, 4, 6; **T** 2; Rwanda and southern Ethiopia

Hab. Montane forest; 2500–3200 m

Syn. *Cynanchum gonoloboides* Schltr. in Z.A.E. 2: 543 (1913)
    C. sp. A of U.K.W.F. ed. 2: 180 (1994)

83. CYNANCHUM

L., Sp. Pl. 212 (1753); Liede in E.J. 114: 503 (1993) & in Ann. Missouri Bot. Gard. 83: 283 (1996)

Sarcostemma R. Br., Prodr. 463 (1810); B.R. Adams & R.W.K. Holland in Cact. Succ. J. (U.S.) 50: 107, 166 (1978)
Bunburia Harv., Gen. S. Afr. Pl. 416 (1838)
Colostephanus Harv., Gen. S. Afr. Pl. 417 (1838)
Cyathella Decne. in Ann. Sci. Nat. Bot. sér. 2, 9: 332 (1838)
Cynoctonum E. Mey., Comm Pl. Afr. Austr. 215 (1838), *non* J.F. Gmel. (1791)
Endotropis Endl., Gen. Pl. 591 (1838), *non* Raf. (1825) *nec* Raf. (1838)
Sarcocyphula Harv., Thes. Cap. 2: 58 (1863)
Perianthostelma Baill., Hist. Pl. 10: 247 (1890)
Flanagania Schltr. in E.J. 18, Beibl. 45: 10 (1894)

(synonymy relating to mainland Africa only)

Perennial plants, leafy or leafless; stems herbaceous or succulent, twining or not; latex white or very occasionally clear or yellowish. Leaves when present (i.e. not reduced to scales), opposite, with colleters at the base of the lamina; abbreviated short leafy shoots present in the leaf axils of several species giving the appearance of stipules. Inflorescences extra-axillary, sessile or pedunculate, umbelliform or clustered along an axis that develops with age. Corolla fused basally, glabrous or occasionally pubescent. Corolline corona absent. Gynostegial corona in one or two whorls; inner whorl where present of free, fleshy, staminal lobes; outer whorl a variously developed tube of partially or completely fused staminal and interstaminal parts surrounding the gynostegium. Gynostegium sessile or stipitate. Pollinia pendant in the anther cells. Stylar head ± flat, conical, or with rostrate apical appendage. Follicles generally developing singly by abortion. Seeds flattened, winged or not, smooth or sculptured, glabrous or pubescent; with a coma of white hairs.

About 300 mostly Old-World species distributed across tropical and subtropical regions, with centres of diversity in Madagascar and southern China.

1. Stems and leaves absent at time of flowering; inflorescence emerging direct from ground ...   6. *C. praecox* (p. 481)
    Stems clearly present at time of flowering; inflorescence borne on leafy or leafless shoots 2
2. Leaves reduced to minute scales; stems semi-succulent 3
    Leaves well-developed; stems not succulent 7
3. Corona in a single series forming a tube around the gynostegium; fleshy staminal lobes absent   1. *C. gerrardii* (p. 476)
    Corona in two series – a short tube or annulus around the base of the gynostegium, and discrete fleshy lobes on the backs of the anthers (former *Sarcostemma* spp.) 4

4. Corolla cream, yellow or green, rotate, the lobes spreading; plants never rhizomatous or stoloniferous, although occasionally rooting at the tip; stems erect, scrambling or twining to several metres . . . . 5. *C. viminale* (p. 479)

Corolla purple-pink or reddish brown, lobes erect or suberect; plants with longer stems arching and rooting at the nodes, or with creeping subterranean rhizomes; stems never twining, generally less than 1 m tall 5

5. Stems arising singly from a creeping subterranean rhizome, slender, 2–3 mm in diameter; corolla purple-pink, lobes 2–3 mm long when dry . . . . 2. *C. vanlessenii* (p. 477)

Stems not arising from a subterranean rhizome, or if rhizomatous then stems numerous and congested, generally much more robust than above; corolla reddish brown, lobes 3–6 mm long when dry 6

6. Stems very regularly and stiffly branched, internodes short, 1–3 cm long and of ± uniform length on a branch . . . . 3. *C. stoloniferum* (p. 477)

Branching pattern less regular than above; at least some internodes much longer than 3 cm . . . . . 4. *C. resiliens* (p. 478)

7. Corona united into a tube for most or all of its length; irregularly toothed apically or with short lobes 8

Corona united into a tube for $\frac{1}{2}$ its length or less; lobes conspicuous and well-defined 12

8. Stylar head developed into long-rostrate appendage 11. *C. hastifolium* (p. 485)

Stylar head without long-rostrate appendage 9

9. Leaf base truncate or rounded; corona ± twice as long as broad . . . . 7. *C. altiscandens* (p. 482)

Leaf base cordate or truncate (rounded in *falcatum*); corona shorter than or only slightly longer than wide 10

10. Corona cup-shaped, about equalling the gynostegium; *Acacia-Commiphora* bushland . . . . 8. *C. falcatum* (p. 483)

Corona urceolate, twice as long as gynostegium; forest margins 11

11. Corona white, cream or greenish; leaves with red or purple veins beneath; southern Tanzania . . 9. *C. rungweense* (p. 483)

Corona reddish; leaf veins green; Uganda . . . . . . 10. *C. longipes* (p. 485)

12. Leaves semi-succulent, glabrous or subglabrous, weakly to strongly hastate 13

Leaves thin, sparsely to densely pubescent, ovate to oblong, never hastate 14

13. Corona papillose; tubular part of corona generally taller than broad . . . . 11. *C. hastifolium* (p. 485)

Corona not papillose; tubular part of corona broader than tall . . . . 12. *C. crassiantherae* (p. 486)

14. Corolla 1–2 mm long; flowers globose in bud . . . 15. *C. schistoglossum* (p. 490)

Corolla at least 5 mm long; flowers conical in bud 15

15. Leaves tomentose beneath, oblong to oblong-elliptic; staminal parts of the corona with an adaxial ligule . . . . 13. *C. abyssinicum* (p. 488)

Leaves sparsely pubescent beneath, broadly ovate; staminal parts of the corona without an adaxial ligule . . . . 14. *C. polyanthum* (p. 489)

1. **Cynanchum gerrardii** (*Harv.*) *Liede* in Taxon 40: 117 (1991) & in Fl. Eth. 4(1): 145 (2003); Liede-Schumann in Fl. Somalia 3: 156 (2006). Type: South Africa, KwaZulu-Natal, Tugela, *Gerrard* 1321 (TCD, holo.; BM!, iso.)

Slightly succulent twining perennial to 4 m with white latex; stems much branched, corky below, glaucous, glabrescent. Leaves reduced to frequently subopposite scales to ± 1 mm long. Inflorescences extra-axillary, sessile, forming a subumbelliform cluster of flowers along a congested rachis, 2–4 flowers open at one time; rhachis to 5 mm long; bracts broadly deltoid, to 0.5 mm long, pubescent; pedicels 2–5 mm long, glabrous. Sepals ovate, 0.3–0.7 mm long, glabrous or pubescent. Flower buds globose, ovoid or subcylindrical, 1–2.5 mm long, lobes not contorted in bud. Corolla rotate, united only at the base; lobes green, oblong or ovate, 1.5–2.5 × 0.7–1.5 mm, glabrous. Corona white, united into a tube 0.5–1 mm long ± as tall as the staminal column; staminal parts of the tube very variable in length, extending for a further 0.1–1 mm with a linear apical part; adaxial appendage absent, but staminal part of tube thicker than interstaminal ones; interstaminal parts of the tube truncate to acute, occasionally as long as interstaminal parts. Anther wings 0.2–0.4 mm long; corpusculum ovoid, ± 0.15 mm long; translator arms ± 0.1 mm long, flattened; pollinia attached subapically to translator arms, ovoid, ± 0.4 × 0.15 mm. Stylar head obscured by anther appendages. Follicles occurring singly, fusiform, 8–12 × 0.6–0.8 cm, tapering to a subacute tip. Seeds ovate-convex, 5–6 × 2–3 mm, without a marginal rim; coma 2–2.5 cm long.

SYN. *Cynanchum aphyllum* sensu R. Br. in Mem. Wern. Nat. Hist. Soc. 1: 51 (1810), *non* (Thunb.) Schltr.
  Sarcocyphula gerrardii Harv., Thes. Cap. 2: 58, t. 191 (1863)
  Cynanchum sarcostemmatoides K. Schum. in P.O.A. C: 323 (Aug. 1895); T.T.C.L.: 65 (1949) & *C. sarcostemmoides* K. Schum. in E. & P. Pf. 4(2): 252 (Oct. 1895), orthographic variant. Types: Tanzania, Tanga District: Amboni, *Holst* 2706 (K!, lecto., designated by Liede in E.J. 114: 515 (1993)
  C. edule Jum. & Perr. in Compt. Rend. 152: 1016 (1911), & Rev. Gen. Bot. 22: 258 (1911). Type: not known – synonymy after Liede in E.J. 114: 515 (1993)
  C. tetrapterum sensu Bullock in K.B. 10: 624 (1956) & sensu Malaisse in Fl. Rwanda 3: 100 (1985) & sensu U.K.W.F. ed. 2: 179 (1994), *non Sarcostemma tetrapterum* Turcz. (1848) *nec Cynanchum tetrapterum* (Turcz.) Bullock (1956) [i.e. excl. type which = *C. viminale* sensu lato]

NOTE. The confusion surrounding the names *Cynanchum aphyllum* (Thunb.) Schltr. and *C. tetrapterum* (Turcz.) Bullock is discussed by Liede in Taxon 40: 113–117 (1991). Lectotypifications place these names under the synonymy of *C. viminale*, not *C. gerrardii*.

a. subsp. **gerrardii**

Sepals 0.3–0.5 mm long. Flower buds 1–1.5(–2) mm long. Corona ± as long as gynostegium, fused for most of its length, staminal parts sometimes extended as linear or narrowly oblong teeth.

UGANDA. Karamoja District: Rupa road W of Moroto, Jan. 1959, *Tweedie* 1772!
KENYA. Machakos District: Kibwezi hill, 23 Nov. 1979, *Gatheri et al.* 79/131!; Masai District: SW of Ngong Hills, 4 Aug. 1968, *Gillett* 18689!; Teita District: 1 km W of Mwatate, Teita Hills, 11 Jan. 1995, *Goyder & Masinde* 3964!
TANZANIA. Tanga District: Pangani road, 10 Oct. 1954, *Faulkner* 1577!; Morogoro/Rufiji District: Selous Game Reserve, 10 km SE of Behobeho, 7 June 1977, *Vollesen* in MRC 4645!; Iringa District: foot of Kitonga Gorge 6 km W of Mahenge village, 9 Jan. 1989, *C.M. Taylor et al.* 8468!
DISTR. U 1, ?2; K 1–7; T 1–3, 5–7; Z; widespread in eastern and southern Africa, Arabia, Madagascar and the Mascarene islands
HAB. Dry bushland; sea level to ± 1300 m

NOTE. Corona form is very variable, generally enclosing but not obscuring the gynostegium. However, the staminal (occasionally also the interstaminal) parts of the corona are more strongly developed in plants from K 7 and T 3 than elsewhere, in some cases blurring the distinction with subsp. *lenewtonii*.

b. subsp. **lenewtonii** (*Liede*) *Goyder* in K.B. 63: 465 (2008). Type: Kenya, Northern Frontier District: Moyale, *Gillett* 14031 (K, holo.; B, iso.)

Sepals 0.5–0.7 mm long. Flower buds 2–2.5 mm long. Corona taller than gynostegium, fused for ± half of its length, both staminal and interstaminal parts extending as toothed lobes, at least the staminal parts narrowing into a linear tip.

KENYA. Northern Frontier District: Moyale, 15 km along Wajir road, 6 Nov. 1952, *Gillett* 14154!; Teita District: Voi, 11 May 1931, *Napier* 1064!; Sagala, Oct. 1938, *Joanna* in *Bally* 8900!
DISTR. **K** 1, 7; S Ethiopia and N Somalia
HAB. *Acacia-Commiphora* bushland; 600–1300 m

SYN. *Cynanchum lenewtonii* Liede in K.B. 49: 119 (1994) & in Fl. Eth. 4(1): 145 (2003)

2. **Cynanchum vanlessenii** (*Lavranos*) *Goyder* in K.B. 63: 471 (2008). Type: Yemen, Audhali Plateau, 2 km S of Al Madhan village, *Lavranos & van Lessen* 1832 (K!, holo., in spirit; PRE, iso.)

Strongly rhizomatous succulent perennial with white latex; rhizome horizontal, buried 4–10 cm underground; stems arising singly, erect to ± 20 cm or sprawling to 0.5 m, never twining, slender but 2–3 mm thick, glaucous, glabrescent. Leaves reduced to minute scales. Inflorescences sessile on very short lateral shoots to 2 mm long, umbelliform, 2–8 flowers open at one time; bracts to 0.5 mm long, glabrous; pedicels 3–5 mm long, minutely pubescent. Sepals ovate, 0.6–1 mm long, glabrous. Flower buds ovoid, 2–3 mm long, lobes not contorted in bud. Corolla purple-pink, united only at the base; lobes erect or suberect, ovate, 2–3 × 1.5–2 mm, margins somewhat replicate, glabrous. Corona gynostegial, in two series, pale pink or white; outer corona forming a shallow fluted tube, to ± 1 mm long; inner corona lobes staminal, attached to the back of the anthers, and ± reaching top of the gynostegium, ovoid, fleshy, ± 1 mm long. Anther wings 0.5–0.6 mm long; corpusculum subcylindrical, ± 0.2 mm long; translator arms slender, ± 0.1 mm long; pollinia oblanceolate, ± 0.3 × 0.1 mm. Stylar head obscured by anther appendages. Follicles occurring singly, fusiform, 3.5 × 0.5 cm, tapering to an attenuate apex. Seeds not seen.

KENYA. Northern Frontier District: Baragoi, N Merti Plateau, 1953, *Hennings* s.n. in *Bally* 9708!; Machakos District: Athi River, 29 Nov. 1960, *Archer* 206!; Masai District: Kajiado district 1 km E of marble quarry, 20 Nov. 1992, *Newton et al.* 4153!
TANZANIA. Maswa District: Naabi Hill, 11 Apr. 1961, *Greenway & Turner* 10027!; Arusha District: Ngare Nanyuki, 40 km NE of Arusha, Nov. 1965, *Beesley* 183! & 17 km W of Ngare Nanyuki, 14 Dec. 2000, *Bruyns* 8699 (number used for two separate collections)!
DISTR. **K** 1, 3–6; **T** 1, 2; S Ethiopia; Saudi Arabia and Yemen
HAB. Rocky, frequently volcanic soil, or in grassland on black soil; 1400–1900 m

SYN. *Sarcostemma vanlessenii* Lavranos in Nat. Cact. Succ. J. (U.K.) 29: 35 (1974); Liede in Fl. Eth. 4(1): 148 (2003)
    S. subterranea B.R. Adams & R.W.K. Holland in Cact. Succ. J. (U.S.) 50: 109 (1978); U.K.W.F. ed. 2: 180 (1994). Type: Kenya, Athi River, *Archer* 206 (K!, holo.; EA!, iso.)

NOTE. *Cynanchum vanlessenii*, *C. stoloniferum* and *C. resiliens* appear closely allied. Flowers of *C. vanlessenii* are 2–3 mm long in herbarium material; others ± 4 mm The latter two species frequently have arching stems that root at the nodes. But note that *C. viminale* subsp. *viminale* also reportedly roots at the tips and has relatively small flowers – perhaps this indicated a closer relationship to these three species than to the other subspecies currently placed under *C. viminale*.

3. **Cynanchum stoloniferum** (*B.R. Adams & R.W.K. Holland*) *Goyder* in K.B. 63: 471 (2008). Type: Kenya, Naivasha District: Mt Longonot, *B.R. Adams* 138 (EA!, holo.)

Succulent perennial with white latex; rhizomatous or not; stems tufted and congested, erect initially, but longer stems arching and rooting in contact with the soil, very regularly and stiffly branched; internodes short, 1–3 cm long and of ± uniform length on a branch, glabrescent. Leaves reduced to minute scales. Inflorescences at the tips of short lateral shoots 2–10 mm long, sessile, umbelliform, with 7–9 flowers; bracts ± 0.5 mm long, pubescent; pedicels 3–5 mm long, pubescent. Sepals ovate-triangular, ± 1 mm long, pubescent. Flower buds ovoid, ± 4 mm long, lobes not contorted in bud. Corolla yellowish or greenish brown, united only at the base, lobes held erect or suberect, ovate-oblong, 4–5 × 2–2.5 mm, strongly replicate, glabrous. Corona gynostegial, in two series, white; outer corona forming a fluted tube to ± 1.5 mm long, staminal parts slightly longer than interstaminal sections; inner corona lobes staminal, attached to the back of the anthers, and ± reaching top of the gynostegium, fleshy, ovoid, 1–1.5 mm long. Anther wings ± 0.7 mm long; corpusculum ovoid-subcylindrical, ± 0.25 mm long; translator arms slender, ± 0.15 mm long; pollinia oblanceolate, ± 0.3 × 0.1 mm. Stylar head obscured by anther appendages. Follicles occurring singly, fusiform, 4–6.5 × 0.5 cm, tapering to an attenuate apex. Seeds brown, discoid, ± 6 × 3 mm, flattened; coma white, 2–3 cm long.

KENYA. Northern Frontier District: Seya River, 30 km SE of Maralal, 25 Feb. 1974, *Bally & Carter* 16560!; Uasin Gishu District: Eldoret, 13 Apr. 1951, *G.R. Williams* 121!; Nairobi District: Langata, 24 Feb. 1969, *Napper* 1887!
TANZANIA. Masai District: Oldiang'arangor, East Serengeti, Nov. 1962, *Newbould* 6268!
DISTR. **K** 1, 3, 4, 7; **T** 2; ?Somalia (see Note)
HAB. Rocky ground; 1400–2100 m

SYN. *Sarcostemma stoloniferum* B.R. Adams & R.W.K. Holland in Cact. Succ. J. (U.S.) 50: 110 (1978); U.K.W.F. ed. 2: 180, t. 70 (1994), as *stolonifera*

NOTE. One collection from northern Somalia (*Simmons* B33) may also be referred to this species.

4. **Cynanchum resiliens** (*B.R. Adams & R.W.K. Holland*) *Goyder* in K.B. 63: 471 (2008). Type: Kenya, Teita District: near Buchuma station, 111 km from Mombasa on road to Nairobi, *Adams* 147 (EA!, holo.; K!, iso.)

Similar to the preceding species, with long arching stems rooting at the nodes, but less regularly branched and internodes on main stems generally 5–10 cm long; flowering lateral shoots up to 30 mm long. Corolla brown or purple. Outer corona ± 1 mm tall.

KENYA. Kwale District: between Samburu and Mackinnon Road, 30 Aug. 1953, *Drummond & Hemsley* 4060!; Teita District: between Buchuma and Voi, km 120 from Mombasa, 22 Sep. 1961, *Verdcourt* 3227!; Tana River District: pump station on Tana River N of Galole, 20 Dec. 1964, *Gillett* 16473!
DISTR. **K** 7; not known elsewhere
HAB. *Acacia-Commiphora* bushland, not apparently associated with rocky ground; 40–500 m

SYN. *Sarcostemma resiliens* B.R. Adams & R.W.K. Holland in Cact. Succ. J. (U.S.) 50: 166 (1978)

NOTE. Essentially a low altitude form of *C. stoloniferum* with longer internodes and a shorter outer corona – perhaps deserving of subspecific rank under *C. stoloniferum*. *Richards* 24941 from the Arusha District of Tanzania appears intermediate between the two, having the longer internodes of *C. resiliens*, but occurs at the higher elevations prefered by *C. stoloniferum*. Material from eastern Zimbabwe at K determined as *C. resiliens* and with a similar growth form to the East African material is excluded from this species here, as it has conspicuously tailed bases to the anther wings, suggesting a closer affinity to the widespread *C. viminale* subsp. *suberosum* than to *C. resiliens*. The corolla lobes of Zimbabwean material are more spreading than in typical *C. resiliens*, and the ecology is anomalous also, as the plant is found only on bare rock slabs of inselbergs or granite domes and at high altitude.

5. **Cynanchum viminale** (*L.*) *L.*, Mantissa Pl. 2: 392 (1771); Liede in Fl. Eth. 4(1): 148 (2003); Liede-Schumann in Fl. Somalia 3: 157 (2006). Type: Alpino, De Plantis Aegyptium 190, t. 53, lectotype, designated by Liede & Meve in J.L.S. 112: 2 (1993); "somewhere along the northern East African coast", *Bassi* s.n. in Herb. Linn. 308.1 (LINN!), epitype, designated by Liede & Meve in J.L.S. 118: 47 (1995)

An extremely variable succulent perennial with white latex; lacking rhizomes or stolons, but occasionally some stems rooting at the tips; stems erect, scrambling or twining to several metres, glabrous or minutely pubescent. Leaves reduced to minute scales. Inflorescences umbelliform. Sepals ± 1 mm long. Corolla cream yellow or green, united only at the base; lobes spreading, ovate-oblong, 4–6 × 2–2.5 mm, glabrous. Corona gynostegial, in two series, white; outer corona forming a fluted tube; inner corona lobes staminal, attached to the back of the anthers, and ± reaching top of the gynostegium, ovoid, fleshy. Anther wings 0.5–0.7 mm long; corpusculum ovoid-subcylindrical, ± 0.25 mm long; translator arms slender, ± 0.15 mm long; pollinia oblanceolate, ± 0.3 × 0.1 mm. Stylar head obscured by anther appendages. Follicles occurring singly, fusiform, tapering to an attenuate apex. Seeds discoid, flattened.

SYN. *Euphorbia viminalis* L., Sp. Pl. 452 (1753)
    Cynanchum aphyllum L., Syst. Nat. ed. 12, 3: 235 (1768), *nom. illeg.* [based on *Euphorbia viminalis* L.]
    Sarcostemma viminale (L.) R. Br., Prodr. Fl. Nov. Holland. 464 (1810); T.T.C.L.: 67 (1949); F.P.U.: 121 (1962); Bullock in F.W.T.A. ed. 2, 2: 93 (1963); U.K.W.F. ed. 2: 180 (1994)
    S. 'taxon 3' of B.R. Adams & R.W.K. Holland in Cact. Succ. J. (U.S.) 50: 108, 168 (1978)

1. Plants twining or scrambling to several metres; main stem thicker than other stems (but not generally apparent in herbarium material) 2
   Plants erect or scrambing, but never twining, and generally less than 1 m in height; lacking an obvious main stem 3
2. Flowers mostly in sessile axillary clusters at the nodes; anther wings with basal tails; main stem not conspicuously lenticellate .... b. subsp. **suberosum**
   Flowers mostly in clusters at the ends of slender lateral shoots; anther wings without basal tails; main stem with corky lenticels .... c. subsp. **odontolepis**
3. Stems generally slender, 5 mm or less in diameter; inflorescences sessile at tips of long lateral shoots; corolla lobes 3–5 mm long when dry .... a. subsp. **viminale**
   Stems robust, up to 10 mm across; inflorescences sessile or stipitate; corolla lobes 5–6 mm long when dry 4
4. Inflorescences sessile; gynostegium shorter than broad   e. subsp. **crassicaule**
   Inflorescences stipitate; gynostegium as tall as broad .   d. subsp. **stipitaceum**

a. subsp. **viminale**

Stems numerous, to 2 m, erect or arching but never twining, irregularly branched, glabrous; lacking an obvious main stem. Inflorescences at the tips of long lateral shoots; pedicels 5–9 mm long, subglabrous to pubescent. Corolla lobes 3–5 × 2 mm Outer corona ± 1 mm long; inner corona lobes ± 1 mm long. Anther wings 0.5–0.7 mm long, the margins not extended basally to form tail-like appendages.

KENYA. Teita District: Tsavo National Park, East, Voi Gate camp site, 7 Dec. 1966, *Greenway & Kanuri* 12679!; Kwale District: Mackinnon Road, 96 km from Mombasa, 21 Jan. 1975, *Adams* 146!; Tana River District: Galole, Nov. 1964, *Makin* s.n. in EA 13047!

TANZANIA. Pare/Lushoto Districts: SE of Mkomazi village, 4 Apr. 1988, *Pócs* 88051!
DISTR. **K** 7; **T** 3; Madagascar and the Indian Ocean islands
HAB. Sandy soil or over rock in *Commiphora* bushland; 0–500 m

b. subsp. **suberosum** (*Meve & Liede*) *Goyder* in K.B. 63: 472 (2008). Type: South Africa, Mpumalanga, Komatipoort, *Kirk* 97 (K!, holo.)

Plants scrambling for several metres, with one or occasionally more main stems, much branched, the branching frequently opposite and regular, glabrous. Inflorescences mostly sessile at the nodes, but in addition some frequently as at the tips of lateral shoots; pedicels 5–9 mm long, minutely pubescent. Corolla lobes ± 5 × 2 mm. Gynostegium (including corona) ± as tall as broad. Outer corona ± 1 mm long; inner corona lobes ± 1 mm long. Anther wings ± 0.7 mm long, the margins extended basally to form short tail-like appendages.

UGANDA. Ankole District: Mbarara, Jan. 1948, *Purseglove* 2579!; Teso District: Serere, Aug.–Sep. 1932, *Chandler* 944!; Mengo District: Lake Victoria shore, Kaazi, 1958, *Lind* 2563!
KENYA. North Nyeri District: 25 km NNW of Nanyuki, 2 Jan. 1995, *Goyder & Masinde* 3946!; Masai District: Olorgesailie [Ologosailie] National Museums of Kenya site, 26 Oct. 2000, *S.A.L. Smith et al.* 14!; Teita District: edge of Mwatate on Bura road, 14 Feb. 1996, *Goyder et al.* 4031!
TANZANIA. Mbulu District: upper slopes of the Msasa River Gorge, Lake Manyara National Park, 5 Jan. 1965, *Greenway & Kanuri* 12022!; Pangani District: Bushiri Estate, 21 June 1950, *Faulkner* 612!; Uzaramo District: Dar es Salaam, Bongoyo Island, 2 July 1969, *Mwasumbi* 10544!
DISTR. **U** 1–4; **K** 1–7; **T** 1–3, 5, 6; **Z**; widespread across tropical and southern Africa
HAB. Dry sandy soils or rocky ground, often in association with *Acacia* or *Commiphora*; sea level–2100 m

SYN. *Sarcostemma viminale* sensu U.K.W.F. ed. 2: t. 70 (1994)
  S. viminale (L.) R. Br. subsp. *suberosum* Meve & Liede in J.L.S. 120: 35 (1996); Liede in Fl. Eth. 4(1): 148 (2003)
  S. 'taxon 2', *S. viminale*, sensu B.R. Adams & R.W.K. Holland in Cact. Succ. J. (U.S.) 50: 108, 167 (1978)

NOTE. This is the most common and widespread form of the species in Africa.

c. subsp. **odontolepis** (*Balf. f.*) *Goyder* in K.B. 63: 471 (2008). Type: Mauritius, Rodrigues Island, *Balfour* s.n. (E, holo.)

Similar to subsp. *suberosum* with a single main stem, but flowering branches very slender. Inflorescences generally at tips of lateral branches; pedicels 7–10 mm long, glabrous.

KENYA. Teita District: Rukinga Ranch, 20 km E of Voi, 25 Apr. 1975, *Bally* 16663!; Kilifi District: Cha Simba, 22 km SW of Kilifi on Kaloleni Road, 19 Jan. 1975, *B.R. Adams* 144!; Lamu District: proposed Boni Forest Reserve, 7.7 km W of Marereni and 3 km N of road, 29 Nov. 1988, *Robertson & Luke* 5611!
TANZANIA. Tanga District: Sawa, 18 Aug. 1956, *Faulkner* 1908!; Rufiji District: Mafia Island, Kilindoni, 14 Sep. 1937, *Greenway* 5253!; Zanzibar: Unguja Ukoo, 5 Feb. 1929, *Greenway* 1352!
DISTR. **K** ?4, 7; **T** 3, 6; **Z**, **P**; E Ethiopia and Rodrigues island
HAB. Dry coastal forest or woodland, more rarely in *Acacia-Commiphora* bushland; 0–200(–800) m

SYN. *Sarcostemma odontolepis* Balf. f. in J.L.S. 16: 17 (1877)
  S. 'taxon 4' of B.R. Adams & R.W.K. Holland in Cact. Succ. J. (U.S.) 50: 108, 168 (1978)
  S. viminale (L.) R. Br. subsp. *odontolepis* (Balf. f.) Meve & Liede in J.L.S. 120: 31 (1996); Liede in Fl. Eth. 4(1): 149 (2003)

NOTE. Several specimens from near Nairobi may be this taxon, or perhaps a form of subsp. *suberosum.*

d. subsp. **stipitaceum** (*Forssk.*) *Meve & Liede* in Novon 15: 323 (2005); Liede-Schumann in Fl. Somalia 3: 158 (2006). Type: Yemen, Milhan, *Forsskal* s.n. (C, holo.)

Stems numerous, tangled, thick, erect or sprawling to ± 2 m, never twining, irregularly branched, young shoots minutely tomentose, glaucous; lacking an obvious main stem. Inflorescences terminal and at the nodes, on short lateral shoots; pedicels 5–8 mm long, densely pubescent. Corolla lobes 5–6 × 2.5 mm. Gynostegium (including corona) ± as tall as broad. Outer corona ± 1 mm long; inner corona lobes ± 1.5 mm long. Anther wings 0.7 mm long, the margins not extended basally to form tail-like appendages.

UGANDA. Karamoja District: Moroto, 2 Jan. 1937, *A.S. Thomas* 2140! & Lodoketimit [Lodoketiminit], 19 July 1959, *Kerfoot* 1311!; Acholi District: Paimol, Apr. 1943, *Purseglove* 1534!
KENYA. Northern Frontier District: Dandu, 2 Apr. 1952, *Gillett* 12673!; Turkana District: 16 km NE of Lorengipe, 9 Apr. 1954, *Hemming* 258!; South Kavirondo District: Kanam, Homa Mountain foothills, Nov. 1934, *Turner* 6621!
TANZANIA. Shinyanga District: Usule, 1938, *Koritschoner* 1815!; Masai District: Olduvai Gorge, 26 June 1989, *Chuwa* 2801!; Kilosa District: 4 km N of Mbuyuni on track to Malolo, 20 May 1990, *Carter et al.* 2367!
DISTR. **U** 1; **K** 1, 2, 4–7; **T** 1–3, 5, 6, 8; Sudan, South Sudan, Somalia, Eritrea, Ethiopia, and Angola; Arabian Peninsula
HAB. Dry sandy soils or rock outcrops, often in *Acacia-* or *Commiphora*-dominated vegetation; sea level to 1500 m

SYN. *Asclepias stipitacea* Forssk., Fl. Aegypt. 50 (1775)
    Sarcostemma stipitaceum (Forssk.) Schult. in Roem. & Schult., Syst. Veg. 6(2): 116 (1820)
    S. andongense Hiern, Cat. Afr. Pl. Welw. 689 (1898); T.T.C.L.: 67 (1949); U.K.W.F. ed. 2: 180 (1994), as "*andogense*". Type: Angola, Pungo Andongo, Praesidium, *Welwitsch* 4257 (K!, lectotype, designated here; BM!, G, isolecto.)
    S. 'taxon 5' of B.R. Adams & R.W.K. Holland in Cact. Succ. J. (U.S.) 50: 108, 169 (1978)
    S. viminale (L.) R. Br. subsp. *stipitaceum* (Forssk.) Meve & Liede in J.L.S. 120: 32 (1996); Liede in Fl. Eth. 4(1): 148 (2003)

NOTE. Material from the Jos Plateau in Nigeria may also belong to this taxon.

e. subsp. **crassicaule** *Liede & Meve* in Novon 15: 320 (2005); Liede-Schumann in Fl. Somalia 3: 158 (2006). Type: Kenya, Teita District: SW side of Mt Kasigau, *Goyder et al.* 4029 (K!, holo.; EA!, UBT!, iso.)

As subsp. *stipitatum* but with stems even thicker (to ± 12 mm diameter when fresh), sparsely glabrescent. Inflorescences sessile. Gynostegium distinctly broader than tall.

KENYA. Northern Frontier District: Ol Lolokwe, 15 Apr. 1977, *Gilbert* 5371!; Turkana District: Lokitaung Gorge, 30 Oct. 1977, *Carter & Stannard* 125!; Machakos District: Lukenya Hill, Aug. 1939, *Bally* 119!
DISTR. **K** 1–4, 7; extreme S of Somalia
HAB. Mostly restricted to exposed rock slabs of inselbergs; 500–1600 m

SYN. *Sarcostemma andongense* sensu Blundell, Wild Fl. E. Africa: pl. 225 (1987)

NOTE. Liede-Schumann & Meve (Novon 15: 323, 2005) cite *Dawkins* 647 from Uganda under this taxon, but the Kew duplicate of this material is clearly referable to subsp. *stipitaceum*. I have also excluded *Grimshaw* 94/13 from Kilimanjaro from my concept of subsp. *crassicaule*, as although the inflorescences are sessile rather than stipitate, the gynostegium and habitat are anomalous. Two collections from SE Tanzania (*Bidgood et al.* 1897; *Richards* 17925) have sessile inflorescences with broad gynostegia, but are more slender in habit than is typical in subsp. *crassicaule*. Their minutely tailed anther wings (cf. subsp. *suberosum*) may indicate hybrid origin.

6. **Cynanchum praecox** *S. Moore* in J.B. 40: 256 (1902); Bullock in F.W.T.A. ed. 2, 2: 89 (1963); Burrows in Pl. Nyika Plateau: 70 (2005). Type: Zimbabwe, Harare, valley of Mazoe River, *Rand* 512 (BM!, holo.; K!, iso.)

Low pyrophytic perennial from slender vertical rootstock which sometimes clearly arises from a horizontal rhizome, latex white; leafy shoots developing after flowering, stems pubescent with white hairs; abbreviated leafy shoots not present in leaf axils. Leaves sessile; lamina linear to ovate, 2.5–6 × 0.5–1.2 cm, apex acute, base cuneate, glabrous or sparsely pubescent. Inflorescences forming a subumbelliform head of 5–25(–50) flowers at ground level, irregularly branched; bracts ovate, 1–2 mm long, pubescent; pedicels 5–13(–20) mm long, pubescent. Sepals ovate or oblong, 1–1.5 × 0.3–0.5 mm, pubescent. Flower buds conical, ± 3 mm long, lobes contorted in bud. Corolla rotate, united only at the base; lobes olive-green, yellow-brown or brownish purple, oblong, 3–5 × 1 mm, somewhat twisted distally, glabrous. Corona white or cream, united into a fluted tube 1–1.5 mm long ± as tall as the staminal column, glabrous or papillose; staminal parts of the tube each extending for a further 1–1.5 mm into an oblong lobe with a rounded apex, channelled on the inner face, and continuing along the inner face of the tube as paired longitudinal ridges; adaxial appendage absent; interstaminal parts of the tube ending in a broad, recurved lobe ± 0.5 mm long. Anther wings ± 0.7 mm long; corpusculum ovoid, ± 0.2 mm long; translator arms ± 0.1 mm long, flattened; pollinia attached laterally to translator arms, ovoid, ± 0.2 mm long,. Stylar head conical, shortly exserted from anther appendages. Follicles (immature) occurring singly, pubescent.

TANZANIA. Bukoba District: Bugene, 21 July 1947, *Ford* 190!; Ufipa District: Mbisi Forest Reserve, 25 Nov. 1994, *Goyder et al.* 3828!; Mbeya District: Mbogo Mt, 7 Nov. 1966, *Gillett* 17641!
DISTR. **T** 1, 4, 7; scattered localities from Sierra Leone to Zimbabwe
HAB. Burnt upland grassland; 1500–2300 m

SYN. *Cynanchum pygmaeum* Schltr. in E.J. 51: 140 (1913). Types: Cameroon, *Ledermann* 2226 & 2230 (B†, syn.); neotype: Cameroon, Bamenda, 5 km from Kumbo on Oku road, *Hepper* 2011 (K!, neo., designated by Liede in Ann. Missouri Bot. Gard. 83: 331 (1996))

7. **Cynanchum altiscandens** *K. Schum.* in Abh. Königl. Akad. Wiss. Berlin 64 (1894); Malaisse in Fl. Rwanda 3: 98 (1985); U.K.W.F. ed. 2: 179 (1994); Liede in Fl. Eth. 4(1): 146 (2003). Type: Tanzania, Lushoto District: Usambara, Kwa Mshusa, *Holst* 9078 (B†, holo.; K!, lecto., designated by Liede in Ann. Missouri Bot. Gard. 83: 296 (1996))

Slender twining perennial to several metres with white or occasionally clear latex (*Tanner* 1821); stems branched, glabrous or pubescent; abbreviated leafy shoots usually present in leaf axils. Leaves with petiole 0.5–1.5(–2.5) cm long, pubescent; lamina ovate to elliptic or oblong, 2–4.5 × 1–3 cm, apex acute to attenuate, base truncate or rounded, glabrous to sparsely pubescent on both surfaces. Inflorescences extra-axillary, forming a subumbelliform cluster of flowers along a rachis which develops with age, 5–12 flowers open at one time; peduncle 2–10 mm long, pubescent; rhachis to 10 mm long; bracts ± 1 mm long, pubescent; pedicels 5–8 mm long, glabrous or pubescent. Sepals ovate, 1–1.5 × 0.5–1 mm, glabrous or pubescent. Flower buds ovoid-conical, 3–4 mm long, lobes weakly contorted in bud. Corolla rotate, united only at the base; lobes yellow-green, frequently tinged with purple, oblong, 3–4.5 × ± 1 mm, weakly twisted distally, glabrous. Corona white, united into a fluted tube 3–3.5 mm long, exceeding the staminal column and completely obscuring it; mouth of tube irregularly dentate; adaxial appendages absent. Anther wings ± 0.6 mm long; corpusculum rhomboid, ± 0.3 mm long; translator arms ± 0.2 mm long, flattened; pollinia attached subapically to translator arms, ovoid to oblongoid, ± 0.4 × 0.15 mm. Stylar head obscured by anther appendages. Follicles occurring singly, pendulous, lanceolate in outline, 5.5–6.5 × 1 cm, tapering to a long-attenuate tip, longitudinally ridged, brown, glabrous. Seeds brown, ovate-convex, ± 5 × 2.5 mm, with longitudinal ridges, with a paler marginal rim; coma ± 3 cm long.

UGANDA. Ankole District: Lukiri, 11 Jan. 1989, *Rwaburindore* 2752!; Masaka District: Kalisizo, Aug. 1945, *Purseglove* 1772!; Mubende District: 0.5 km W of Kasanda trading centre, 10 Aug. 1974, *Katende* 2249!
KENYA. Naivasha District: Crater Lake, Nov. 1958, *Newbould* 3634!; North Nyeri District: Naro Moru River Lodge, 23 Oct. 1992, *Harvey et al.* 1!; Masai District: Uaso Nyiro tributary 16 km from Narok, 17 Feb. 1957, *Verdcourt* 1765!
TANZANIA. Masai District: Empakaai [Embagai] Crater, 2 Dec. 1988, *Pócs & Chuwa* 88281/J!; Arusha District: Arusha National Park, 6 Feb. 1971, *Vesey-Fitzgerald* 6967!; Lushoto District: Mkuzi, 6 km NE of Lushoto, 14 Apr. 1953, *Drummond & Hemsley* 2106!

DISTR. U 2–4; **K** 1–4, 6; **T** 1–3, 7; E Congo-Kinshasa, Eritrea and Ethiopia
HAB. Forest margins and disturbed vegetation; 1200–2300 m

SYN. *Cynanchum mensense* K. Schum. in E. & P. Pf. (4)2: 253 (1895). Type: Eritrea, Gheleb, *Schweinfurth* 1505 (B†, holo.; M, lecto., designated by Liede in Ann. Missouri Bot. Gard. 83: 298 (1996))

8. **Cynanchum falcatum** *Hutch. & E.A. Bruce* in K.B. 1941: 145 (1941); Liede in Fl. Eth. 4(1): 147 (2003); Liede-Schumann in Fl. Somalia 3: 156 (2006). Type: Somalia, Woqooyi Galbeed, *Gillett* 4114 (K!, holo.)

Slender twining perennial to several metres, latex white; stems branched, glabrescent, young shoots pubescent with spreading hairs; abbreviated leafy shoots not present in leaf axils. Leaves with petiole 0.2–10 cm long, subglabrous to pubescent; lamina linear, oblong or ovate, 2–4 × 0.2–1.6 cm, apex obtuse to acuminate, base rounded, truncate or weakly cordate, sparsely pubescent on both surfaces. Inflorescences extra-axillary, umbelliform with 5–12 flowers; peduncle 0–5 mm long, pubescent; bracts triangular, ± 1 mm long, pubescent; pedicels 2–5 mm long, densely pubescent. Sepals ovate or triangular, ± 1 × 0.5 mm, densely pubescent. Flower buds ovoid, ± 2 mm long, lobes not contorted in bud. Corolla spreading to campanulate, united at the base; lobes yellow-green, oblong, 2–3 × 1 mm, glabrous. Corona white, united into a tube 1.5–2 mm long ± as tall as the staminal column; staminal parts of the tube each extending briefly as a broadly triangular, somewhat recurved lobe; adaxial appendage absent; interstaminal parts of the tube shorter than staminal parts, bifid, truncate or ovate. Anthers trapezoid, wings ± 0.7 mm long; corpusculum ovoid, ± 0.2 mm long; translator arms ± 0.1 mm long, cylindrical; pollinia attached subapically to translator arms, oblong in outline, ± 0.5 × 0.1 mm. Stylar head obscured by anther appendages. Follicles occurring singly, pendulous, brown, lanceolate in outline, 5–6 × 0.4 cm, tapering gradually to a long-attenuate tip, somewhat angled in section, pubescent. Seeds brown, ovate, ± 6 × 4 mm, smooth with weakly defined radiating ridges, with a paler marginal rim; coma ± 2 cm long.

KENYA. Northern Frontier District: Huri Hills, 25 Feb. 1963, *Bally* 12526!
DISTR. **K** 1; Ethiopia and Somalia
HAB. *Acacia-Commiphora* bushland; ± 1000 m?

9. **Cynanchum rungweense** *Bullock* in K.B. 10: 622 (1956); Cribb & Leedal, Mt Fl. S. Tanz.: 105, t. 24a (1982); Burrows in Pl. Nyika Plateau: 70 (2005). Type: Tanzania, Mbeya District: below Mporoto, Inkuyu, *St. Clair-Thompson* 846 (K!, holo.)

Slender twining perennial to 6 m with white latex; stems branched, glabrous or pubescent; abbreviated leafy shoots usually present in leaf axils. Leaves with petiole 0.5–3.5 cm long, sparsely pubescent; lamina oblong, 3.5–7 × 1–3 cm, apex apiculate, base deeply cordate with rounded auricles, dark green above, glaucous beneath with red-purple veins, sparsely pubescent adaxially, glabrous beneath. Inflorescences extra-axillary, forming a subumbelliform cluster of flowers along a simple or branched rachis which develops with age, 6–9 flowers open at one time; peduncle 6–35 mm long, glabrous or pubescent; bracts minute; pedicels 10–20 mm long, glabrous or pubescent. Sepals ovate, ± 2 × 1–1.5 mm, glabrous or pubescent. Flower buds ovoid, 5–7 mm long, lobes not contorted in bud. Corolla rotate, united only at the base; lobes yellow-green, frequently tinged with purple, oblong, 6–7 × 2–2.5 mm, glabrous. Corona white or cream at the base, yellow-green above, united into an urceolate tube 4–5 mm long, exceeding the staminal column and completely obscuring it; mouth of tube irregularly dentate; adaxial appendages absent. Anther wings ± 1 mm long; corpusculum obovoid, ± 0.3 mm long; translator arms ± 0.5 mm long, flattened and geniculate; pollinia attached apically to translator arms, narrowly obovoid, ± 0.6 × 0.25 mm. Stylar head obscured by anther appendages. Follicles not seen. Fig. 116, p. 484.

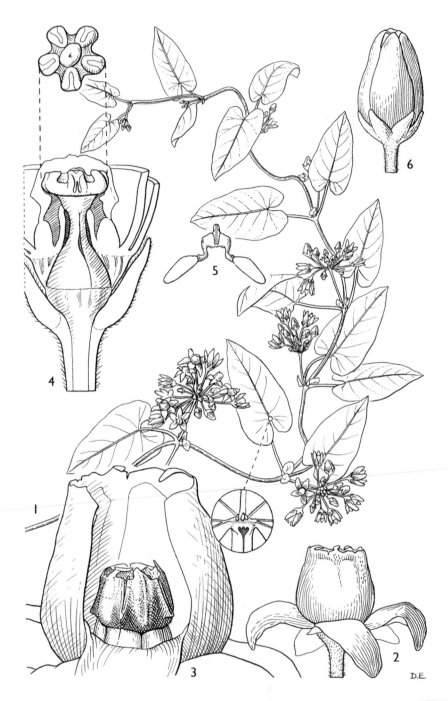

FIG. 116. *CYNANCHUM RUNGWEENSE* — **1**, habit, × ²/₃; **2**, flower with tubular corona compeletely obscuring the gynostegium, × 5; **3**, flower with corona partially deflected to expose gynostegium, × 11; **4**, flower with stamens half cut away to expose gynoecium of two ovaries and a stylar head, the latter also viewed from above, × 11; **5**, pollinarium, × 18; **6**, flower bud, × 8. 1 from *Richards* 6737; 2–5 from *Richards* 6807; 6 from *Stolz* 1983. Drawn by D. Erasmus.

TANZANIA. Sumbawanga District: Mbisi Mts, 28 Feb. 1994, *Bidgood et al.* 2525!; Rungwe District: Poroto Mts, 2 km E of Kimondo on road to Kitulo, 1 Dec. 1994, *Goyder et al.* 3870!; Mbeya District: World's End View, Ipinda, 6 Feb. 1976, *Cribb et al.* 10576!
DISTR. **T** 4, 7; Nyika and Vipya plateaus of N Zambia and Malawi
HAB. Montane forest; 1700–2600 m

10. **Cynanchum longipes** *N.E. Br.* in K.B. 1897: 273 (1897); Bullock in F.W.T.A. ed. 2, 2: 89 (1963). Type: Nigeria, Lagos, Papalayito, *Millen* 48 (K!, holo.)

Twining perennial, latex white; stems branched, subglabrous or pubescent with spreading white hairs; abbreviated leafy shoots present in leaf axils. Leaves with petiole 3–4 cm long, sparsely pubescent; lamina ovate, 5–6 × 2–3.5 cm, apex attenuate, base cordate with rounded auricles, sparsely pubescent on both surfaces. Inflorescences extra-axillary, subumbelliform, with 5–9 flowers opening ± simultaneously; peduncle ± 5 mm long, pubescent; bracts ± 1 mm long, triangular; pedicels 10–15 mm long, slender, pubescent. Sepals broadly ovate or triangular, 1–1.5 × 1 mm, sparsely pubescent. Flower buds ovoid-subglobose, 3–5 mm long, lobes not contorted in bud. Corolla rotate, united only at the base; lobes greenish yellow abaxially, yellow adaxially, ovate, ± 4 × 1.5 mm, somewhat twisted distally, glabrous on both surfaces. Corona reddish purple, united into a tube 3–4 mm long, taller than the staminal column; staminal parts of the tube each extending for a further 0.5–1 mm into a broadly triangular or tri-dentate lobe; adaxial appendage absent; interstaminal parts of the tube shorter than the staminal lobes and more slender. Anther wings ± 1 mm long; corpusculum ovoid, 0.2–0.3 mm long; translator arms ± 0.2 mm long, flattened; pollinia attached subapically to translator arms, oblong in outline, ± 0.6 × 0.2 mm. Stylar head obscured by anther appendages. Follicles occurring singly, fusiform, ± 9 × 0.7 cm.

UGANDA. Mengo District: Kajansi Forest, Entebbe–Kampala Road, July 1935, *Chandler* 1267!
DISTR. **U** 4; widely distributed across West Africa but with just a single record from Congo-Kinshasa
HAB. Forest margins; ± 1150 m

11. **Cynanchum hastifolium** *K. Schum.* in E. & P. Pf. 4(2): 253 (May 1895); U.K.W.F. ed. 2: 179 (1994); Liede-Schumann in Fl. Somalia 3: 157 (2006). Type: Ethiopia, Tigray, near Djeladjeranne, *Schimper* 1690 (K!, lecto., designated by Cufodontis in B.J.B.B. 30: 705 (1960), herbarium not indicated, & Liede-Schumann & Meve in Novon 16: 372 (2006), as 'holo.'; P, isolecto.)

Low twining perennial with white latex; stems to 4 m long, branched, woody below, pale yellow-brown, glabrous or puberulent; abbreviated leafy shoots generally present in leaf axils. Leaves semisucculent, with petiole 0.5–2.5 cm long, glabrous or glabrescent; lamina weakly to strongly hastate, (0.6–)1.5–4(–5.5) × 0.6–2(–3.5) cm, apex rounded to acute or attenuate, frequently shortly mucronate, base truncate to cordate with rounded auricles, subglabrous or sparsely pubescent on both surfaces. Inflorescences extra-axillary, sessile, umbelliform with 3–16 flowers; bracts ± 1 mm long, pubescent; pedicels 3–9 mm long, sparsely to densely pubescent. Sepals linear, ± 2–3 × 0.5 mm, pubescent. Flower ovoid-conical, buds 3–7 mm long, weakly to strongly contorted apically. Corolla rotate, united only at the base; lobes green, lanceolate but appearing linear, 4–7 × 0.8–1.5 mm, margins revolute, somewhat twisted distally, glabrous. Corona united into a fluted tube enclosing and freqently obscuring the staminal column, glabrous or papillose; staminal parts of the tube each extending further as a linear-filiform appendage equalling or longer than the interstaminal parts, channelled on the inner face, and continuing into the inner face of the tube as paired longitudinal ridges; interstaminal parts of the tube longitudinally folded below and ending in a tooth with erect or recurved apex. Anther wings ± 1 mm long; corpusculum ovoid, ± 0.2 mm long; translator arms ±

0.2 mm long, flattened; pollinia attached subapically to translator arms, ovoid, ± 0.4 × 0.2 mm. Stylar head obscured by anther appendages or forming a subcylindrical appendage exserted well beyond the corona. Follicles occurring singly, lanceolate in outline, 6–11 × ± 1.5 cm, tapering to a long-attenuate tip, rounded in section, longitudinally grooved, brown. Seeds brown, ovate, ± 7 × 4 mm, papillose and pubescent, with a broad, denticulate marginal rim; coma 2–2.5 cm long.

a. subsp. **hastifolium**

Flower buds ovoid-conical, 3–4 mm long, weakly contorted apically. Corolla lobes 4–5 mm long. Corona white, glabrous or minutely papillose, united for $^1/_2$–$^2/_3$ of its length into a tube (1–)1.5–2 mm long; free staminal lobes 1–1.5(–2) mm long; free interstaminal lobes 0.5–1(–1.5) mm long. Stylar head generally obscured by anther appendages, but occasionally extending as a subcylindrical appendage ± as long as staminal corona lobes.

UGANDA. Karamoja District: Lodoketeminit, 16 July 1959, *Kerfoot* 1326!
KENYA. Northern Frontier District: Furrole, 14 Sep. 1952, *Gillett* 13863!; Machakos District: Mbirikani–Emali, 12 km from Merueshi, 15 Dec. 1991, *Luke* 3010!
TANZANIA. Shinyanga, *Koritschoner* 2084!; Mbulu District: Lake Manyara National Park, Mbagaya River, 2 Mar. 1964, *Greenway & Kanuri* 11282!
DISTR. **U** 1; **K** 1–4, 6; **T** 1, 2; Mali, Eritrea, Ethiopia, Somalia; reported also from scattered localities in West Africa by Liede (1996)
HAB. Dry *Acacia* bushland; 700–1200 m

SYN. *Cynanchum hastifolium* N.E. Br. in K.B. 1895: 257 (Oct. 1895); U.K.W.F. ed. 2: 179 (1994)
   C. clavidens N.E. Br. in K.B. 1895: 256 (Oct. 1895). Type: Ethiopia, Harar, Bubi ['Somalia, Boobi'], Mar. 1885, *James & Thrupp* s.n. (K!, holo.)
   C. macinense A. Chev. in Explor. Bot. Afrique Occ. Franç. 1: 435 (1920), *nom. nud.* based on: Mali, Macina, pays de Habés, Koboro-Kendé to Kanikombolé, *Chevalier* 24861 (P, holo.)
   C. clavidens N.E. Br. subsp. *hastifolium* (N.E. Br.) Liede in Ann. Missouri Bot. Gard. 83: 306 (1996)
   C. hastifolium N.E. Br. subsp. *clavidens* (N.E. Br.) Liede in Fl. Somalia 3: 157 (March 2006), *comb. illegit.*
   C. hastifolium K. Schum. subsp. *clavidens* (N.E. Br.) Liede in Novon 16: 372 (Nov. 2006)

b. subsp. **longirostrum** Goyder in K.B. 63: 464 (2008). Type: Kenya, Kwale District: 3 km NW of Mackinnon Road, *Goyder & Masinde* 3959 (K!, holo.; EA!, iso.)

Flower buds conical, 4–7 mm long, weakly to strongly contorted apically. Corolla lobes 5–7 mm long. Corona pale green at the base, white above, papillose, united for at least $^3/_4$ of its length into a tube 2–2.7 mm long; free staminal lobes 0.5–1 mm; free interstaminal lobes 0.5–1 mm long. Stylar head exserted from top of gynostegium and the corona as a slender subcylindrical appendage 2–3 mm long, apically bifid or subentire. Fig. 117, p. 487.

KENYA. Machakos District: 3 km NW of Hunters Lodge, 11 Feb. 1996, *Goyder et al.* 4005!; Teita District: 15 km N of Voi on Nairobi road, 12 Feb. 1996, *Goyder et al.* 4015!; Kwale District: Mackinnon Road, Taru Desert, Oct. 1965, *Tweedie* 3199!
TANZANIA. Moshi District: 3 km S of Himo–Korogwe road junction, 11 May 1990, *Carter et al.* 2234!; Kondoa District: between Mondo and Goma, 20 km E of Kondoa, 20 Jan. 1962, *Polhill & Paulo* 1239!
DISTR. **K** 1, 4, 7; **T** 2, 5; three collections from S Somalia and one from the Ogaden
HAB. *Acacia-Commiphora* bushland; 200–1100 m

SYN. *Cynanchum clavidens* N.E. Br. subsp. *clavidens* sensu Liede in Ann. Missouri Bot. Gard. 83: 306 (1996) & Fl. Eth. 4(1): 147 (2003), *non* N.E. Br. in K.B. 1895: 256 (1895)
   C. hastifolium K. Schum. subsp. *clavidens* sensu Liede in Fl. Somalia 3: 157 (2006) & in Novon 16: 372 (2006), *non* N.E. Br. in K.B. 1895: 256 (1895)

12. **Cynanchum crassiantherae** *Liede* in Ann. Missouri Bot. Gard. 83: 307 (1996); Liede-Schumann in Fl. Somalia 3: 156 (2006). Type: Somalia, Shaabeellaha Dhexe, 10–12 km N of Adale on road to Haji Ali, *Gillett & Hemming* 24513 (K!, holo.; EA!, iso.)

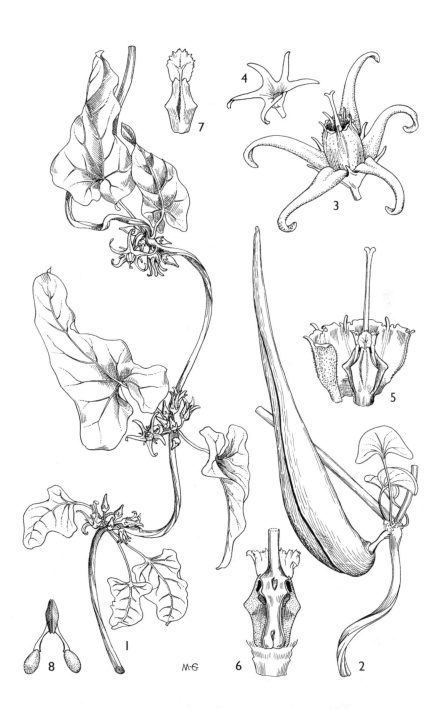

FIG. 117. *CYNANCHUM HASTIFOLIUM* subsp. *LONGIROSTRUM* — **1**, flowering shoot, × 1; **2**, fruiting branch, × 1; **3**, flower, × 4; **4**, calyx, × 4; **5**, corona opened out to expose gynostegium and long-emergent stylar head appendage, × 6; **6**, gynostegium with stamens partially cut away to expose gynoecium, × 6; **7**, stamen, × 8; **8**, pollinarium, × 20. All from *Polhill & Paulo* 1239. Drawn by M. Grierson.

Glabrous twining perennial to 2 m, latex colour not recorded; abbreviated leafy shoots present in leaf axils. Leaves semisucculent, petiole 1.5–3 cm long; lamina broadly triangular, 4–5 × 4–6 cm, apex rounded and apiculate, base deeply cordate with rounded auricles. Inflorescences extra-axillary, umbelliform; peduncle to ± 4 mm long; pedicels 5–8 mm long. Sepals narrowly triangular, 1.5–2 × 0.5 mm,. Flower buds conical, lobes not contorted in bud. Corolla rotate, united only at the base, lobes yellow-green, oblong, 3–4 × 1–1.5 mm,. Corona white, united into a shallow braodly campanulate tube ± 1 mm long ± as tall as the staminal column; staminal parts of the tube each extending for a further 1.5 mm as a suberect linear tooth; interstaminal parts of the tube developed as a broadly triangular lobe ± 0.8 mm long and 1.5 mm broad at the base, with a rounded, somewhat reflexed apex. Anther wings ± 0.5 mm long; corpusculum ovoid, ± 0.2 mm long; translator arms ± 0.1 mm long, flattened; pollinia attached subapically to translator arms, oblong in outline, ± 0.3 × 0.1 mm. Stylar head obscured by anther appendages. Fruit and seen unknown.

KENYA. Lamu District: near Sankuri Camp, 30 July 2006, *Festo et al.* 2777!
DISTR. **K** 7; central and southern Somalia
HAB. *Acacia-Commiphora* bushland and coastal dunes; near sea level

NOTE. The Kenyan collection is much more robust vegetatively than Somali material, but florally appears identical.

13. **Cynanchum abyssinicum** *Decne.* in DC., Prodr. 8: 548 (1844); U.K.W.F. ed. 2: 179 (1994); Liede in Fl. Eth. 4(1): 145 (2003). Type: Ethiopia, *Quartin-Dillon* s.n. (G, holo.)

Slender twining perennial to 4 m with white latex; stems branched, woody below, pubescent with spreading or somewhat reflexed white hairs; abbreviated leafy shoots not present in leaf axils. Leaves with petiole 0.5–2 cm long, pubescent; lamina narrowly oblong to oblong-elliptic, 3–7(–10) × 0.8–2(–3.5) cm, apex acute, frequently shortly mucronate, base cordate with rounded auricles, sparsely pubescent above, tomentose beneath. Inflorescences extra-axillary, forming a subumbelliform cluster of flowers along a rachis which develops with age, 3–6 flowers open at one time; peduncle 5–25 mm long, pubescent; rhachis to 10 mm long; bracts 1–2 mm long, filiform, pubescent; pedicels 7–10(–13) mm long, pubescent. Sepals linear, 1.5–4 × 0.5 mm, densely pubescent. Flower buds conical, 4–7 mm long, lobes contorted in bud. Corolla rotate, united only at the base; lobes yellow-green to brown or reddish purple, oblong, 5–7 × 1–1.5 mm, somewhat twisted distally, pubescent adaxially, sparsely pubescent abaxially. Corona white, united into a fluted tube 1.5–2.5 mm long ± as tall as the staminal column; staminal parts of the tube each extending for a further 2.5–4 mm into a linear-attenuate lobe which narrows gradually into a long filiform apex; adaxial appendage arising on the inner face of the tube, slightly shorter than the outer lobe, narrowly triangular to filiform; interstaminal parts of the tube ending in a broad, tridentate or irregularly lobed tooth 0.5–1 mm long. Anther wings 0.5–1 mm long, margins parallel; corpusculum ovoid, ± 0.4 mm long; translator arms ± 0.2 mm long, flattened; pollinia attached laterally to translator arms, ovoid, ± 0.5 × 0.25 mm. Stylar head obscured by anther appendages. Follicles occurring singly, pendulous, brown, lanceolate in outline, 6–7 × 1 cm, tapering to a weakly clavate tip, somewhat angled in section, longitudinally grooved,. Seeds brown, ovate-convex, ± 5 × 3 mm, smooth, with a paler marginal rim; coma ± 3 cm long.

UGANDA. Kigezi District: Rubaya, July 1946, *Purseglove* 2097! & Kachwekano Farm, Dec. 1949, *Purseglove* 3159!
KENYA. Naivasha District: S Kinangop, 22 July 1961, *Polhill* 433!; Laikipia District: 15 km N of Nyahururu, 23 Aug. 1981, *Gilbert* 6345!; Londiani District: Tinderet Forest Reserve, 22 June 1949, *Maas Geesteranus* 5161!

TANZANIA. Tanga District: Marungu, July 1893, *Volkens* 641!; Iringa District: Mufindi, Lugoda Tea Estate, 9 May 1968, *Renvoize & Abdullah* 2049! & Mufindi, Muisenga Lake, 17 Mar. 1962, *Polhill & Paulo* 1792!
DISTR. U 2, 3; **K** 3–7; **T** 2, 3, 7; Eritrea, Ethiopia and Congo-Kinshasa
HAB. Bushland and forest margins; 1600–2600 m

SYN. [*Cynanchum abyssinicum* Decne. var. *tomentosum* Oliv. in Trans. Linn. Soc., Bot. 2: 342 (1887); T.T.C.L.: 65 (1949); *nomen nudum*, based on: Tanzania, Kilimanjaro, Moshi, *Johnston* 177 (K!, lecto., designated by Liede in Ann. Missouri Bot. Gard. 83: 287 (1996))]
Vincetoxicum abyssinicum (Decne.) Kuntze, Rev. Gen. Pl. 2: 424 (1891)
V. holstii K. Schum. in E.J. 17: 135 (1893). Type: Tanzania, Lushoto District: Usambara, Mlalo, *Holst* 507 (B†, holo.; K!, lecto., designated by Liede in Ann. Missouri Bot. Gard. 83: 287 (1996))
Cynanchum holstii (K. Schum.) K. Schum. in E. & P. Pf. 4(2): 253 (1895)

14. **Cynanchum polyanthum** *K. Schum.* in E. & P. Pf. 4(2): 253 (1895); F.P.S. 2: 407 (1952). Type: Congo-Kinshasa, Monbuttu country near Munza, *Schweinfurth* 3345 (B†, holo.; K!, lecto., designated by Liede in Ann. Missouri Bot. Gard. 83: 329 (1996); P, isolecto.).

Slender twining perennial to 4 m with white latex; stems branched, pubescent with spreading hairs; abbreviated leafy shoots not present in leaf axils. Leaves with petiole 2.5–4 cm long, pubescent; lamina broadly ovate, 3.5–11 × 2–8.5 cm, apex acuminate, base deeply cordate with rounded auricles, sparsely pubescent on both surfaces. Inflorescences extra-axillary, congested, with 2 or more short branches, subumbelliform clusters of flowers along a rachis which develops with age, 5–10 flowers open at one time; peduncle 10–40 mm long, pubescent; rhachis to 10 mm long; bracts filiform, 1–2.5 mm long, pubescent; pedicels 15–25(–35) mm long, slender, pubescent. Sepals ovate or triangular, 1–1.5 × 0.5 mm, densely pubescent. Flower buds conical, 5–6 mm long, lobes contorted in bud. Corolla rotate, united only at the base; lobes yellow-green, frequently flushed with reddish purple, oblong, 5–8 × 1–1.5 mm, somewhat twisted distally, glabrous but minutely papillose adaxially. Corona white or greenish cream, united into a tube 1.5–2 mm long ± as tall as the staminal column; staminal parts of the tube each extending for a further 2 mm into a linear-attenuate lobe with no adaxial appendage; interstaminal parts of the tube ending in a broad, truncate, emarginate or lobed tooth ± 0.5 mm long. Anthers trapezoid, wings ± 1 mm long; corpusculum ovoid, ± 0.4 mm long; translator arms ± 0.1 mm long, flattened; pollinia attached subapically to translator arms, oblong in outline, ± 0.3 × 0.1 mm. Stylar head obscured by anther appendages. Follicles occurring singly, pendulous, brown, lanceolate in outline, 8–9 × 0.8–1 cm, tapering gradually to a weakly clavate tip, somewhat angled in section, pubescent. Seeds brown, ovate, ± 6 × 4 mm, smooth with weakly defined radiating ridges, with a paler marginal rim; coma ± 1.5 cm long.

UGANDA. Busoga District: Musumu swamp 16 km S of Kamuli, 27 May 1953, *G.H.S. Wood* 760!; Mengo District: Kawempe, 19 Oct. 1971, *Synnott* 688!; Masaka District: Masaka–Bukakata road, 11 Oct. 1953, *Drummond & Hemsley* 4736!
DISTR. U 3, 4; Cameroon, Gabon, Congo-Kinshasa, South Sudan and Angola
HAB. Bushland and forest margins; 1000–1200 m

SYN. *Vincetoxicum polyanthum* K. Schum. in E.J. 17: 136 (1893), *non* Kuntze (1891)
Cynanchum obscurum K. Schum. in E. & P. Pf. 4(2): 253 (1895). Type: Angola, Cuanza Norte, Golungo Alto, ad Dumeta in Sobato de Mussengue, *Welwitsch* 4222 (K!, holo.; BM iso.)
C. welwitschii Schltr. & Rendle in J.B. 34: 99 (1896). Based on *Cynanchum obscurum*
Periploca batesii Wernham in J.B. 54: 228 (1916). Type: Cameroon, Bitye, *Bates* 643 (BM!, holo.)

15. **Cynanchum schistoglossum** *Schltr.* in J.B. 33: 271 (Sep. 1895); Malaisse in Fl. Rwanda 3: 98 (1985). Type: South Africa, KwaZulu-Natal, Stanger, Phoenix, *Schlechter* 7090 (B, neo., designated by Liede in Ann. Missouri Bot. Gard. 83: 337 (1996); AMD, BM!, isoneo.)

Slender twining perennial to 4 m, latex white; stems branched, pubescent with spreading white hairs; abbreviated leafy shoots frequently present in leaf axils giving the appearance of broadly ovate stipules. Leaves with petiole 1–2.5 cm long, densely pubescent; lamina ovate, lanceolate or oblong, 3.5–5(–7.5) × 1.5–3(–4) cm, apex acute to attenuate, base weakly to strongly cordate, pubescent on both surfaces. Inflorescences extra-axillary, forming a subumbelliform cluster of flowers along a rachis which develops with age, 3–10 flowers open at one time; peduncle 3–6 mm long, pubescent; rhachis 1–2(–15) mm long; bracts triangular, ± 1 mm long, pubescent; pedicels 3–8 mm long, slender, pubescent. Sepals ovate or lanceolate, 0.5–1(–1.5) × ± 0.5 mm, pubescent. Flower buds globose, 1–1.5 mm long, lobes not contorted in bud. Corolla rotate or campanulate, united only at the base; lobes yellowish green or cream, ovate, 1.5(–2) × 0.7–1 mm, glabrous. Corona white, united into a fluted tube 0.5–1 mm long, ± as tall as the staminal column; staminal parts of the tube each extending for a further 0.7–1 mm into an oblong or attenuate lobe, generally with angular shoulders at the base of the lobe, adaxial appendage absent; interstaminal parts not or only barely extending beyond the mouth of the corona tube, thinner than the staminal parts. Anther wings ± 0.4 mm long; corpusculum ovoid, ± 0.15 mm long; translator arms ± 0.1 mm long, flattened; pollinia attached laterally to translator arms, ovoid, ± 0.2 mm long. Stylar head obscured by anther appendages. Follicles occurring singly, brown, narrowly lanceolate in outline, 4.5–7 × 0.5 cm, tapering gradually to an obtuse or slightly clavate tip, smooth or longitudinally ridged, pubescent. Seeds brown, ovate, 4.5–6 × 3.5–4 mm, smooth, with a paler marginal rim; coma 1.5–2 cm long.

UGANDA. Ankole District: Mitoma, Mar. 1939, *Purseglove* 600!; Teso District: Serere, Dec. 1931, *Chandler* 262!; Masaka District: 1 km E of Lugulama, Koki County, 16 May 1971, *Lye* 6091!
KENYA. North Nyeri District: Nyeri, 19 Dec. 1921, *R.E. & T.C.E. Fries* 139!; South Kavirondo District: Marongo Ridge, 19 km SW of Kisii, 13 Nov. 1974, *Vuyk & Breteler* 239!
TANZANIA. Korogwe District: Kisarake, near Mnyuzi Railway Station, 30 Apr. 1971, *Semsei* 4239!; Ufipa District: near Sakalilo, 25 May 1951, *Bullock* 3896!; Iringa District: near Kisanga, 19 Aug. 1970, *Thulin & Mhoro* 765!
DISTR. U 2–4; K 4, 5; T 2–4, 6, 7; Congo-Kinshasa, Rwanda, Burundi, Angola, Zambia, Malawi, Mozambique, Zimbabwe, Botswana, Namibia and South Africa
HAB. Twining through grasses and bushland, margins of riverine areas; 700–1600 m

SYN. *Cynanchum brevidens* N.E. Br. in K.B. 1895: 257 (Oct. 1895). Type: "Congo", Sep. 1863, *Burton* s.n. (BM, holo.)
  C. brevidens N.E. Br. var. *zambesiaca* N.E. Br. in K.B. 1895: 257 (Oct. 1895). Type: Mozambique, Expedition Island, July 1838, *Kirk* s.n. (K!, holo.)
  C. vagum N.E. Br. in K.B. 1895: 257 (Oct. 1895). Type: Congo-Kinshasa, Stanley Pool, *Hens* 77 (K!, holo.)
  C. minutiflorum K. Schum. in B.S.B.B. 37: 123 (1898), *nom. illeg.* Type as for *C. vagum*
  C. dewevrei De Wild. & T. Durand in Ann. Mus. Congo, Ser. 1, Bot. Ser. 2, 1(2): 42 (1900). Type: Congo-Kinshasa, Mwanana Toumbwè, *Dewèvre* 904 (BR, lecto., designated by Liede in Ann. Missouri Bot. Gard. 83: 337 (1996))

NOTE. The duplicate of *Chandler* 599 in EA is atypical in that the corona approaches that of the West and Central African species *C. adalinae* (K. Schum.) K. Schum., with unusually long interstaminal lobes arising from the top of the fused part if the corona. Flowers on the sheet at K are more typical of the species. Other characters present on both specimens, such as the prominent pseudo-stipules and the elongated inflorescence axis, are consistent with placement in *C. schistoglossum* rather than *C. adalinae*.

84. **BLYTTIA**

Arn. in Mag. Zool. & Bot. 2: 420 (1838), *non Blyttia* Fries (1839); D.V. Field &
J.R.I. Wood in K.B. 38: 215–220 (1983)

Subshrubs or slender woody twiners, stiffly branched; latex clear. Leaves opposite
on new growth or appearing whorled or clustered due to severe contraction of lateral
shoots. Inflorescences alternate and extra-axillary, but often appearing to be
opposite and axillary due to contraction of short lateral shoots, few-flowered, ±
umbelliform; flowers minute. Corolline corona absent. Gynostegial corona of five
distinct fleshy staminal lobes arising from the base of the gynostegium and never
exceeding it. Gynostegium sessile. Anther connectives with apical appendages.
Pollinaria with a pair of pendant pollinia. Stylar head flat or conical. Follicles single,
thin-walled. Seeds flattened, smooth, winged, with a coma of silky hairs.

Two species in East Africa and Arabia.

Blyttia fruticulosa (*Decne.*) *D.V. Field* in K.B. 38: 216 (1983); U.K.W.F. ed. 2: 180
(1994); Liede in Fl. Eth. 4(1): 138 (2003); Liede-Schumann in Fl. Somalia 3: 152, fig.
105 (2006). Type: Saudi Arabia, Jeddah, *Schimper* 816 (P, holo.; K!, iso.)

FIG. 118. *BLYTTIA FRUTICULOSA* — **1**, habit, × 1; **2–4**, leaves showing variation in shape, × 2;
5 & 6, flowers, × 8; **7 & 8** gynostegia with corona, × 16; **9**, pollinarium, × 32; **10**, follicle, ×
1. 1 & 2 from *Trott* 1554; 3 from *Mathew* 6274; 4, 6 & 8 from *Gillett et al.* 22554; 5, 7 & 9 from
Collenette 642; 10 from *Schimper* 816. Adapted from K.B. 38: 217, drawn by D.V. Field.

Stiffly branched subshrubs 0.3–0.8 m high, or scandent to 2 m, young stems grey-green and minutely pubescent, older stems greyish brown, somewhat lenticellate and with prominent nodes. Leaves with petiole 2–5(–8) mm long, minutely pubescent; lamina linear to narrowly oblong or elliptic, 1–3(–5) × 0.2–0.5(–1) cm, rounded to acute apically, generally rounded at the base, minutely sparsely pubescent initially but becoming glabrous with age. Inflorescences very slender, subglabrous; peduncle 5–12 mm long; pedicels 2–4 mm long. Calyx lobes ovate to triangular, ± 0.5 mm long, acute, minutely pubescent. Corolla yellow-green, united at the base; lobes ovate, 1–1.5 × 0.5–0.8 mm, acute, subglabrous to minutely pubescent adaxially, glabrous abaxially. Gynostegium sessile, cylindrical. Corona lobes erect, fleshy, ± 0.5 mm long, rounded apically, reaching ± half-way up the anthers. Anther wings ± 0.6 mm long. Apex of stylar head not exserted from staminal column. Follicles fusiform, 3–5.5 cm, with a long-attenuate beak, smooth, glabrous. Seeds ± 6 × 2.5 mm, flattened, minutely verrucose and with a marginal rim ± 0.2 mm wide; coma 1.5–2 cm long. Fig. 118, p. 491.

KENYA. Northern Frontier District: Yabichu, near Ramu, 23 May 1952, *Gillett* 13296!; Turkana District: 2 km SW of of Lorukumu on the Lodwar–Luiya road, 8 Nov. 1977, *Carter & Stannard* 245!; Teita District: 1.5 km from junction with Nairobi–Mombasa road on eastern turnoff to Voi, 21 Apr. 1974, *Faden & Faden* 74/471!
TANZANIA. Masai District: 57 km S of Kikuletwa, 7 Jan. 1965, *Leippert* 5383!
DISTR. **K** 1–3, 4, 6, 7; **T** 2; Ethiopia, Djibouti, Somalia and Arabia
HAB. Open *Acacia-Commiphora* bushland; 150–1100 m
USES. *Mathew* 6274 reports that at wedding ceremonies near Lokori, Lake Turkana, the bride wears a necklace made from stems of this plant

SYN. *Cynanchum fruticulosum* Decne. in Ann. Sc. Nat., sér. 2, 9: 332 (June 1838)
   Blyttia arabica Arn. in Mag. Zool. & Bot. 2: 420 (1838). Type as above
   Vincetoxicum fruticulosum (Decne.) Decne. in DC., Prodr. 8: 525 (1844)
   Cynanchum defoliascens K. Schum. in Ann. Inst. Bot. Roma 7: 38 (1898); N.E. Br. in F.T.A. 4(1): 400 (1903); U.K.W.F. ed. 2: 179 (1994). Type: Somalia, near Mandah, *Robecchi-Brichetti* 110 (FT, holo.)

85. DIPLOSTIGMA

K. Schum. in P.O.A. C: 324 (1895)

Subshrubs or twiners from a woody rootstock; latex clear. Leaves opposite. Inflorescences extra-axillary, flowers small. Corolline corona absent. Gynostegial corona consisting of two whorls: a ring of connate staminal and interstaminal parts arising from the base of the gynostegium and appressed to the corolla, and a whorl of staminal lobes radiating from the back of the anthers. Gynostegium sessile. Anther connectives with apical appendages; pollinia pendant, attached subapically to the translator arms. Apex of stylar head domed-conical. Follicles almost always single. Seeds flattened, ovate, with a coma of silky hairs.

A monotypic genus in East Africa.

Diplostigma canescens *K. Schum.* in P.O.A. C: 324 (1895) & in E. & P. Pf. 4(2): 257 (1895); N.E. Br. in F.T.A. 4(1): 389 (1903); U.K.W.F. ed. 2: 180 (1994); Liede in Fl. Eth. 4(1): 139 (2003). Type: Tanzania, *Fischer* 232 (B†, holo.)

Plant erect forming a low subshrub at first, shoots twining in contact with other vegetation and scrambling to 2 m; stems grey-green with minute white hairs, thickened and occasionally somewhat corky below. Leaves with petiole 1–6 mm long, pubescent; lamina linear to oblong or elliptic, 1–3 × 0.2–0.8 cm, apex rounded to subacute, base rounded or truncate, minutely pubescent both above and below.

FIG. 119. *DIPLOSTIGMA CANESCENS* — **1**, habit, × 1; **2**, follicle, × 1; **3**, flower, × 8; **4**, gynostegium and corona, × 16; **5**, gynoecium showing paired ovaries and stylar head, × 20; **6**, pollinarium, × 36. 1 from *Geilinger* 4747; 2 from *Gillett* 13832; 3–6 from *Greenway* 4464. Drawn by D. Erasmus.

Inflorescences extra-axillary, sessile with 1–5 flowers, generally only 1 or 2 open at one time; pedicels 0.5–2 mm long, pubescent. Calyx lobes triangular to ovate, 1–1.5 mm long, apex acute, pubescent, green or tinged with purple. Corolla green, rotate or very shallowly campanulate, lobed for $^2/_3$–$^3/_4$ of its length; lobes ovate to triangular, 1.5–3 × 1–1.7 mm, acute, minutely pubescent at least towards the margins adaxially, abaxial face glabrous. Gynostegium 1–1.5 mm high. Corona yellow-green, basal ring forming shallow pouches below and between the more prominent inner staminal lobes; staminal lobes projecting radially from the gynostegium, sub-conical, ± 0.75 mm long. Anther wings ± 0.3 mm long; corpusculum brown, ovoid-subcylindrical, 0.15–0.2 mm long,. Apex of stylar head domed. Follicles narrowly lanceolate in outline, 5–6 × ± 1 cm, tapering gradually into a long drawn out beak, glabrous. Seeds ovate, 5–6 × 3–3.5 mm, with a marginal rim ± 0.5 mm wide and a sparsely verrucose disk; coma 2.5 cm long. Fig. 119, p. 493.

UGANDA. Karamoja District: Turkana Escarpment, Apr. 1960, *J. Wilson* 924!
KENYA. Northern Frontier District: near Kom, 25 Dec. 1976, *Powys* 281!; Machakos District: 60 km NW of Voi on Nairobi road, 11 Feb. 1996, *Goyder et al.* 4012!; Masai District: km 45 on Magadi–Nairobi road, 28 May 1995, *Vollesen* 95/216!
TANZANIA. Mbulu District: 20 km N of Mto wa Mbu, 24 Jan. 1965, *Greenway & Kanuri* 12070!; Masai District: Engaruka Valley, 20 Feb. 1971, *Richards & Arasululu* 26642!; Lushoto District: Mkomazi Game Reserve, 5 km E of Kamakota Hill, 13 June 1996, *Abdallah et al.* 96/203!
DISTR. **U** 1; **K** 1–4, 6, 7; **T** 2, 3; S Ethiopia
HAB. Dry bushland, on gravel or rocky soils; 50–1350 m

86. PENTATROPIS

Wight & Arn., Contr. Bot. India: 52 (1834); D.V. Field & J.R.I. Wood in K.B. 38: 215–220 (1983); Liede in J.L.S. 114: 81–98 (1994)

Strobopetalum N.E. Br. in K.B. 1894: 335 (1894)

Small twiners; latex sparse, clear. Leaves opposite. Inflorescences extra-axillary. Corolla pubescent adaxially. Corolline corona absent. Gynostegial corona of free staminal parts. Gynostegium sessile. Anther wings extended along the base of the corona lobes; connectives with apical appendages; pollinaria with a pair of pendant pollinia. Apex of stylar head flat or conical. Follicles normally one per flower. Seeds flattened, ovate, with a coma of silky hairs.

Three species from semi-arid regions of Africa, Arabia, Madagascar and S Asia.

Pentatropis nivalis (*J.F. Gmel.*) *D.V. Field & J.R.I. Wood* in K.B. 38: 215 (1983); U.K.W.F. ed. 2: 181 (1994); Liede in Fl. Eth. 4(1): 137 (2003); Goyder in Fl. Somalia 3: 151, fig. 104 (2006). Type: Yemen, *Forsskål* 272 (C, holo., IDC microfiche II: III. 1, 2)

Slender herbaceous perennial twining to 5 m, but mostly much less; stems minutely pubescent, older stems somewhat lenticellate. Leaves slightly succulent, petiole 2–10 mm long, minutely pubescent; lamina narrowly oblong to elliptic or ovate, 1.5–4.5 × 0.5–3 cm, obtuse to subacute with a short apiculus, rounded at the base, glabrous or very sparsely pubescent. Inflorescences extra-axillary, sessile or very rarely with peduncle to 1 cm long, with 2–3(–8) flowers open at the same time; pedicels 10–15 mm long, very slender, glabrous. Calyx lobes ± 1 mm long, triangular, acute, glabrous. Corolla broadly conical in bud, fused at the base for ±

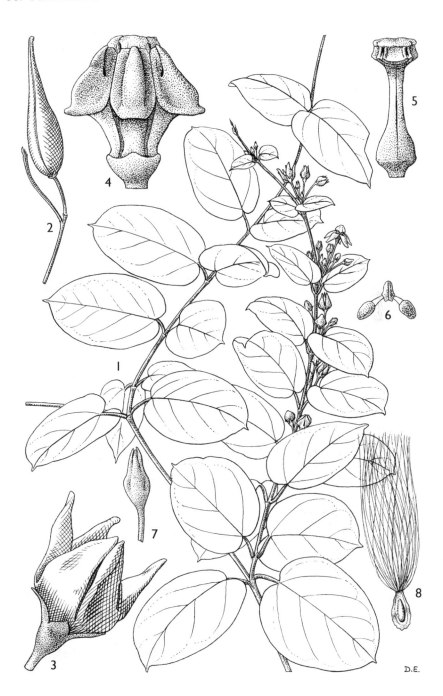

FIG. 120. *PENTATROPIS NIVALIS* — **1**, flowering shoot, × 1; **2**, follicle, × 1; **3**, flower, × 6; **4**, gynostegium with corona, × 10²⁄₃; **5**, gynoecium showing paired ovaries and stylar head, × 10²⁄₃; **6**, pollinarium, × 20; **7**, flower bud, × 6; **8**, seed with coma, × 2. 1 & 2 from *Faulkner* 1859; 3–7 from *Faulkner* 734; 8 from *Wilson* 586. Drawn by D. Erasmus.

1 mm; lobes cream or green, frequently tinged with brown or purple, triangular, 4–6 × 2–3 mm, densely pubescent adaxially at least in lower half, glabrous abaxially. Gynostegium 1.5–2 mm long. Corona lobes ± as tall as staminal column, fleshy, adnate to back of anthers except for the free tip. Anther wings ± 1.5 mm long, the margins strongly curved below, radiating from the column basally and extending beneath the corona lobes. Stylar head not projecting beyond anthers. Follicles narrowly lanceolate in outline, 4.5–6 cm long, strongly beaked, smooth, glabrous. Seeds ovate, ± 5 × 3 mm, with a pale marginal wing ± 0.5 mm wide; coma ± 1 cm long. Fig. 120, p. 495.

UGANDA. Karamoja District: near Kangole, Aug. 1958, *J. Wilson* 586!
KENYA. Northern Frontier District: S end of Huri Hills, 25 Feb. 1963, *Bally* 12527!; Turkana District: by R. Nangolemaret, 30 km NW of Lomoru Itae, 2 Nov. 1977, *Carter & Stannard* 147!; Kilifi District: Kaya Kauma, 14 Apr. 1989, *Robertson & Luke* 5682!
TANZANIA. Mbulu District: Lake Manyara National Park, mouth of Msasa R., 12 Jan. 1965, *Greenway & Kanuri* 12041!; Lushoto District: Mkomazi River on road between Mkomazi and Buiko, 30 Apr. 1953, *Drummond & Hemsley* 2310!; Uzaramo District: Dar es Salaam, Oyster Bay, 2 June 1969, *Mwasumbi* 10521!
DISTR. U 1; **K** 1–3, 6, 7; **T** 1–3, 6, 8; widely distributed across the southern margins of the Sahara from Senegal to Ethiopia and Eritrea, Somalia, and from the Arabian Peninsula to Afghanistan and NW India; Madagascar
HAB. Scrambling over shrubs in dry bushland, frequently in seasonally flooded ground, or in riverine forests or thickets; sea level to 900 m

SYN. *Asclepias nivea* Forssk., Fl. Aegypt.-Arab.: 51 (1775), *non* L. (1753). Type as for *P. nivalis*
    A. nivalis J.F. Gmel. in Linnaeus, Syst. Nat. ed. 13, 2(1): 444 (1791)
    [*Asclepias* Forssk. in Roem. & Schult., Syst. Veg. 6: 85 (1820), cited in Index Kewensis as *Asclepias forskohlei* Roem. & Schult. Based on *A. nivea* Forssk.]
    Pentatropis spiralis sensu Decne. in Ann. Sc. Nat., sér. 2, 9: 327 (1838); F.P.S. 412 (1952); Bullock in F.W.T.A. ed. 2, 2: 91 (1963), *non Asclepias spiralis* Forssk.
    P. senegalensis Decne. in Ann. Sc. Nat., sér. 2, 9: 328 (1838). Type: 'Senegambia', *?Heudelot* 530 in *Leprieur* (P, holo.; K!, iso.)
    P. madagascariensis Decne. in DC., Prodr. 8: 536 (1844). Type: Madagascar, *Bojer* s.n. (P, holo.)
    Tylophora cirrosa Aschers. in Schweinf., Beitr. Fl. Aethiop.: 132 (1867). Types: Eritrea, Dahlak Island, *Ehrenberg* s.n. (B†, syn.) and near Togodele, Schoho, *Ehrenberg* s.n. (B†, syn.)
    Pentatropis spiralis Decne. var. *longepetiolata* Engl., Hochgebirgsfl. Trop. Afr.: 343 (1892). Types: Ethiopia, Gondar, near Senka-Berr, 1863, *Schimper* s.n. (?P, syn.); Eritrea, Schoho, *Petit* s.n. (?P, syn.); Anadehr, *Schimper* 596 (?P, syn.); Massawa, *Steudner* 752 (B†, syn.)
    Pentarrhinum fasciculatum K. Schum. in P.O.A. C: 323 (1895). Type: Tanzania, Pangani, *Stuhlmann* I: 809 (B†, holo.; K!, iso.)
    Pentatropis hoyoides K. Schum. in Ann. Inst. Bot. Roma 7: 40 (1898); N.E. Br. in F.T.A. 4(1): 381 (1902). Types: Ethiopia/Somalia, "Ogaden & Webi", *Robecchi-Bricchetti* 321, 333 (FT, syn.)
    P. cynanchoides N.E. Br. in F.T.A. 4(1): 380 (1902). Types as for *Tylophora cirrosa* Aschers.
    P. cynanchoides N.E. Br. var. *senegalensis* (Decne.) N.E. Br. in F.T.A. 4(1): 381 (1902)
    P. fasciculatus (K. Schum.) N.E. Br. in F.T.A. 4(1): 381 (1902)
    P. rigida Chiov., Result. Sci. Miss. Stef.-Paoli, Coll. Bot. 1: 114 (1916). Types: Somalia, Mogadishu, Amarr Gegeb, *Paoli* 21, 130 (FT, syn.)
    P. nivalis (J.F. Gmel.) D.V. Field & J.R.I. Wood subsp. *madagascariensis* (Decne.) Liede & Meve in Adansonia 23: 348 (2001)

NOTE. Madagascan material has been treated as a separate subspecies, subsp. *madagascariensis*, on the basis of its generally smaller leaves, but some Somali collections blur this distinction.

87. **TYLOPHORA**

R. Br., Prodr. 460 (1810)

Sphaerocodon Benth. in G.P. 2: 772 (1876)
Tylophoropsis N.E. Br. in Gard. Chron. ser. 3, 16: 244 (1894)

Erect or twining shrubs, vines or pyrophytic herbs; latex clear. Leaves opposite. Inflorescences extra-axillary, frequently lax and with many flower clusters along an extended axis, more rarely few-flowered. Corolla deeply five-lobed, normally adaxially with indumentum. Corolline corona absent. Gynostegial corona consisting of five discrete lobes adnate to the staminal column below the anther wings. Anther connectives with apical appendages; pollinia ± horizontal or pendant, rarely erect, globose to ovoid or cylindrical, minute. Stylar head flat or capitate. Follicles usually occuring singly, smooth, wingless; seeds flattened, smooth, winged, with a coma of silky hairs.

About 100 species in the Old World tropics and subtropics.

1. Pyrophytic herbs with erect or scrambling stems to 60 cm; pollinia subcylindrical, erect, translator arms attached basally .... 1. *T. caffra* (p. 499)
   Slender or robust twiners, usually to several metres; pollinia generally ovoid to sub-orbicular and mostly pendant, translator arms attached laterally or apically 2
2. Stems robust, at least 5 mm diameter; leaves fleshy, elliptic to orbicular, mostly at least 10 cm long and wide .... 2. *T. cameroonica* (p. 500)
   Stems more slender, to 3 mm diameter; leaves thin to at most semisucculent, distinctly longer than wide, if orbicular then less than 10 cm long 3
3. Leaves semisucculent; corolla green with minute velvety indumentum adaxially .... 3. *T. anomala* (p. 501)
   Leaves not succulent; corolla pink or purple, if green, then glabrous or with slender hairs adaxially–never densely velvety-pubescent 4
4. Inflorescences 2 to many times as long as subtending leaf, with 5–20+ sessile clusters of flowers distributed at regular intervals along the inflorescence axes; leaves deeply cordate at base .... 12. *T. sylvatica* (p. 508)
   Not as above 5
5. Mature leaves 10–16 cm long, mostly oblong or obovate, base deeply cordate to somewhat truncate; corolla at least 1 cm across, lobes broad, 3–4 mm wide and with a clearly displaced apex .... 4. *T. conspicua* (p. 502)
   Mature leaves mostly less than 10 cm long; corolla not as above 6
6. Inflorescences sessile 7
   Inflorescences pedunculate 8
7. Leaves elliptic; pedicels 15–25 mm long ...... 11. *T. stenoloba* (p. 507)
   Leaves linear, with or without additional basal lobes; pedicels 5–7 mm long .... 18. *T. tridactylata* (p. 510)

8. Whole plant glabrous; leaves 3.5–6 cm wide ..   5. *T. apiculata* (p. 503)
   At least the young stems minutely pubescent; if
   glabrous or subglabrous, then leaves less than
   3 cm wide 9
9. Corona narrowing abruptly into a free apical
   appendage 10
   Corona lacking a free apical appendage,
   corona lobes adnate to staminal column for
   their entire length 12
10. Corona lobes erect, corolla green or cream;
    upland Kenya ....   8. *T. lugardiae* (p. 504)
    Corona lobes spreading from column; corolla
    maroon; southern Tanzania 11
11. Appendage to corona 0.5–0.8 mm long;
    southern Highlands ....   6. *T. iringensis* (p. 503)
    Appendage to corona less than 0.3 mm long;
    Ulugurus ....   7. *T.* sp. A (p. 504)
12. Apex of stylar head conical, exserted from top
    of anthers ....   9. *T. fleckii* (p. 505)
    Apex of stylar head flat, not exserted from top
    of anthers 13
13. Leaves linear or linear-lanceolate, less than
    3 mm wide 14
    Leaves ovate, oblong or triangular, broader than
    3 mm 15
14. Leaves linear except where broadened into a
    somewhat rhomboid base; anther wings
    extended basally into falcate tails; Usambaras .   19. *T.* sp. E (p. 512)
    Leaves linear lanceolate, lacking a rhomboid
    base and tapering gradually towards the
    apex; anther wings without falcate tails
    basally; Ulugurus and Udzungwa ....   15. *T. gracillima* (p. 509)
15. Leaves less than 2 cm long, if longer (up to 4 cm)
    then broadly ovate with crisped or undulate
    margins 16
    Leaves at least 2 cm long, usually much larger,
    margins never crisped or undulate 17
16. Anther wings 0.1 mm long; leaves mostly shorter
    than 1.5 cm ....   15. *T. gracillima* (p. 509)
    Anther wings 0.3–0.4 mm long; leaves to 4 cm
    long ....   16. *T.* sp. C (p. 510)
17. Leaf base cuneate; flowers small at tips of very
    open branched inflorescences – corolla less
    than 2 mm long ....   17. *T.* sp. D (p. 510)
    Leaf base rounded, truncate or weakly cordate;
    flowers more than 2 mm long 18
18. Corolla lobes 4–9 mm long, attenuate and
    somewhat twisted at apex ....   10. *T. heterophylla* (p. 505)
    Corolla lobes less than 4 mm long, if longer,
    then apex rounded or subacute 19
19. Corona lobes less than 0.6 mm long;
    inflorescences very delicate; Kenya, Uganda
    and coastal or lake regions of Tanzania 20
    Corona lobes at least 0.8 mm long; inflorescences
    more robust; mountains of **T** 6 (Ulugurus) and
    T 7 21

20. Anther wings extended basally into falcate tails;
    K 7 (Taita Hills) .... 14. *T.* sp. B (p. 509)
    Anther wings without tails .... 13. *T. tenuipedunculata* (p. 509)
21. Leaves broadly ovate; anther wings ± 0.4 mm
    long; S Ulugurus .... 20. *T.* sp. F (p. 512)
    Leaves lanceolate-oblong; anther wings ± 1 mm
    long; Poroto Mts .... 21. *T.* sp. G (p. 513)

Tylophora sp. C of U.K.W.F. ed. 2: 181 (1994) – not clear which taxon this refers to, possibly *fleckii* or *heterophylla* – which are both shrubby and variable!

1. **Tylophora caffra** *Meisn.* in J.B. 2: 542 (1846). Type: South Africa, 'Port Natal', *Krauss* 85 (BM!, holo.)

Perennial herb with one or more annual erect or trailing stems to ± 60 cm arising from a tuberous rootstock; stems minutely but densely pubescent. Leaves herbaceous; petiole 1–8 mm long, densely pubescent; lamina generally oblong-elliptic, sometimes ovate-oblong or obovate-oblong, 2–9 × (0.7–)1–4.5 cm, apex rounded to subacute, frequently shortly apiculate, base rounded to truncate, frequently slightly asymmetric, indumentum sometimes restricted to main veins, subglabrous to sparsely pubescent adaxially, denser and more prominent abaxially. Inflorescences extra-axillary, solitary at each node, sparsely to densely pubescent; peduncles (to 1ˢᵗ cluster of flowers) 0.1–4 cm long; bracts filiform to narrowly oblong, 1–5 mm long, densely pubescent; pedicels 3–13 mm long, densely pubescent. Calyx lobes oblong, 2–7 × 0.5–2 mm, pubescent. Corolla green tinged with red or purple outside, deep reddish purple within, the centre occasionally pale, globose in bud, campanulate when open, generally united for half to ²⁄₃ of its length, rarely more deeply divided, extremely variable in size, 5–14 mm long, glabrous or sparsely pubescent at least towards the tip abaxially, subglabrous or sparsely pubescent within; lobes oblong to broadly triangular, 2–6 × 1.5–6 mm, apex rounded, but margins sometimes somewhat revolute making the apex appear acute. Gynostegium 1.5 mm long. Corona forming a cream, papillose cylinder around the staminal column; free lobes maroon or purple, fleshy, suborbicular, 0.3–0.4 mm long, radiating from the top of the column just below the base of the anthers. Anther wings falcate, 0.3–0.4 mm; pollinaria minute, corpusculum ovoid-subcylindrical, ± 0.1 mm long; translator arms slender, terete, somewhat curved, ± 0.1 mm long, attached to the pollinia basally; pollinia ± 0.25 mm long, ovoid-subcylindrical, erect. Apex of stylar head flat. Follicles mostly occuring singly, ovoid, 4–6 × 1.5–2 cm, smooth, with thick fleshy walls. Seeds 6–10 × 5–7 mm, oblong to ovate, winged; coma 0.7–2 cm long.

UGANDA. Karamoja District: Mt Moroto, Mar. 1959, *J. Wilson* 744!; Ankole District: Kyabagenyi Hill, 29 Apr. 1946, *A.S. Thomas* 4482!; Teso District: Serere, July 1932, *Chandler* 814!
KENYA. Mt Elgon, 19 Apr. 1931, *E & C Lugard* 618!
TANZANIA. Musoma District: junction of Kleins Camp–Tabora and Wogakuria roads, 20 Dec. 1964, *Greenway & Turner* 11787!; Masai District: Essimingor Forest Reserve, 2 km S of water intake for Makuyuni Village, 13 Apr. 2000, *Gereau et al.* 6483!; Ufipa District: Tatanda Mission, 22 Feb. 1994, *Bidgood et al.* 2377!
DISTR. U 1–4; K 3/5; T 1, 2, 4, 6, 7; widespread across tropical and subtropical Africa from Sierra Leone to South Sudan, to N Namibia and N South Africa
HAB. Mixed deciduous woodland, frequently on rocky slopes, or seasonally burnt grassland; 1000–2400 m

SYN. *Sphaerocodon obtusifolium* Benth. in Hooker's Ic. Pl. 12: 78, t. 1190 (1876); N.E. Br. in F.T.A. 4(1): 412 (1903); Fl. Rwanda 3: 116 (1985). Type: Malawi, Shire River near Miramballa, *Kirk* s.n. (K!, holo.)
    S. natalense Benth. in Hooker's Ic. Pl. 12: 79 (1876). Type: South Africa, KwaZulu-Natal, *Gerrard* 1797 (K!, holo.)

Vincetoxicum caffrum (Meisn.) Kuntze, Rev. Gen. Pl. 2: 424 (1891)

Gongronema welwitschii K. Schum. in E.J. 17: 145 (1893). Type: Angola, *Welwitsch* 4196 (BM!, holo.; K!, LISC, iso.)

?*Sphaerocodon platypodum* K. Schum. in De Wild. in Ann. Mus. Congo ser. 4, 1: 225 (Jan. 1903). Type: Congo-Kinshasa, Lukafu, *Verdick* 411 (BR, holo.)

S. melananthum N.E. Br. in F.T.A. 4(1): 412 (March 1903). Type: Angola, Cuando Cubango, Rio Cuito, 'below the River Longa, 10 Dec. 1899', *Baum* 526 (K!, holo.)

Gymnema melananthum K. Schum. in Warb., Kunene–Sambesi Exped.: 344 (May 1903). Type: Angola, Cuando Cubango, Rio Cuito, 'Unwelt des Kuito, 11 Dec. 1899', *Baum* 526 (B†, holo.; K!, iso.)

Sphaerocodon caffrum (Meisn.) Schltr. in J.B. 33: 339 (1909); Bullock in F.W.T.A. ed. 2, 2: 97 (1963); U.K.W.F. ed. 2: 181 (1994)

S. angolense S. Moore in J.B. 47: 219 (1909). Type: Angola, Kutchi, *Gossweiler* 4124 (BM!, holo.; K!, iso.)

NOTE. A polymorphic species occurring across tropical Africa. Plants from open habitats tend to have smaller leaves and reduced corolla dimensions in comparison with plants from more wooded environments. They are frequently also stiffly erect, whereas woodland plants often have a more sprawling habit. The most distinctive variant of this erect, small-flowered form occurs in southern Angola, Namibia and western Zambia, probably associated with Kalahari sands, and was described as *Sphaerocodon melananthum*. However, dimensions of the functional parts of the gynostegium – the pollinaria and anther wings – are consistent with other material of *Tylophora caffra*, and this variant is treated here as an extreme form of *T. caffra*, deserving recognition at varietal rank at most.

2. **Tylophora cameroonica** *N.E. Br.* in K.B. 1895: 258 (1895) & in F.T.A. 4(1): 407 (1903); Bullock in F.W.T.A. ed. 2, 2: 96 (1963). Type: Cameroon, Rio del Rey, *Johnston* s.n. (K!, holo.)

Robust vine to several metres; stems at least 5 mm in diameter, glabrous or minutely tomentose with golden hairs. Leaves semi-succulent; petiole 1–3 cm long, glabrous or pubescent; lamina elliptic to suborbicular, 8–16 × (4–)5–13 cm, apex rounded and with a short apiculus, base cordate, completely glabrous or with indumentum of minute hairs on margins and on principal veins beneath. Inflorescences extra-axillary, with 1–several much-branched and generally congested (in the Flora area) panicles at each node, inflorescence axes subglabrous to minutely tomentose; peduncles (to 1st branching point) 2–4.5 cm long; bracts triangular, minute; pedicels 3–5 mm long, minutely tomentose with golden hairs. Calyx lobes broadly triangular, ± 1 mm long, minutely tomentose with golden hairs. Corolla pink or reddish purple, rotate, united at the base, sparsely pubescent with white hairs on upper surface, glabrous beneath; corolla lobes ovate, ± 2–3 × 2 mm, apex rounded. Staminal corona lobes maroon or purple, fleshy, ± 0.5 mm long, shorter than the gynostegium and radiating from it. Anther wings 0.2 mm long. Pollinaria minute, corpusculum ovoid-subcylindrical, ± 0.05 mm long; translator arms slender, terete, somewhat curved, ± 0.05 mm long; pollinia ovoid, ± 0.1 mm long. Stylar head flat. Follicles mostly occuring singly, to ± 8 × 1 cm, narrowly lanceolate in outline, smooth. Seeds oblong to ovate, 6–7 × 3 mm, winged and with a minutely pubescent convex face and a glabrous concave face; coma 1.5 cm long.

UGANDA. Toro District: near Crater Valley Resort, on road to Kanyanchu, 15 Aug. 1997, *Eilu* 176!; Kigezi District: Kisoro [cult. Kiwenda, Namalonge, W Mengo, 22 Feb. 1968], *Ferreira & Serwadda* in NC 230!; Mengo District: Budo, near Kampala, Apr. 1939, *Chandler* 2789!

DISTR. U 2, 4; scattered localities from Ivory Coast to Cameroon and E Congo-Kinshasa

HAB. Moist forest; 1200–1300 m

USES. Cultivated as an ornamental in parts of Uganda

SYN. *Tylophora glauca* Bullock in K.B. 9: 585 (1955) & in F.W.T.A. ed. 2, 2: 96 (1963). Type: Nigeria, Asaba District: Obomkpa village, *Onochie* 33430 (K!, holo.)

NOTE. Most collections of this species are robust climbers with minutely golden-tomentose stems and inflorescences – characters used by Bullock (K.B. 9: 585 (1955)) in describing *Tylophora glauca*. Collections of *T. cameroonica* sensu Bullock have more diffuse, subglabrous inflorescences and glabrous stems. However, these characters are inconsistent even in the specimens Bullock cited from Uganda. While extreme forms are indeed distinctive, many specimens are more difficult to place. Variation in inflorescence architecture parallels that in *T. anomala*, so the more slender collections from Cameroon (*T. cameroonica* sensu stricto) are regarded here as merely a distinctive form within a more broadly defined species.

3. **Tylophora anomala** *N.E. Br.* in Fl. Cap. 4(1): 766 (March 1908); Meve in Syst. Geogr. Pl. 68: 257 (1999); Burrows in Pl. Nyika Plateau 75 (2005). Types: South Africa, KwaZulu-Natal, 'Buck-bush, near Durban,' *Gerrard* 1320 ex herb. *McKen* 4 (K!, lecto., designated here; TCD, isolecto.); 'Buck-bush, Umgeni,' *Gerrard* 1320 (BM!, K!, TCD, paralecto.)

Twining herbaceous vine to several metres; stems to 3 mm diameter, glabrous or pubescent. Leaves semi-succulent, leathery and frequently with prominent venation when dry; petiole 0.2–2 cm long, glabrous or pubescent; lamina ovate to oblong or suborbicular, (3–)4–13(–17) × (1.2–)2–7.5 cm, apex rounded to acute, frequently apiculate or mucronate, base weakly to strongly cordate, occasionally rounded or truncate, generally with a pair of colleters, completely glabrous or with indumentum of minute hairs on principal veins. Inflorescences extra-axillary, sessile or pedunculate, or with both at the same node, axis of pedunculate inflorescences frequently continuing beyond the first cluster of flowers, forming further clusters at intervals, axes generally simple or occasionally branched; peduncles (to 1st cluster of flowers) to 7 cm long, glabrous or minutely pubescent; bracts minute, triangular, ciliate; pedicels 3–11(–20) mm long, glabrous or pubescent. Calyx lobes narrowly to broadly triangular, (0.5–)1–1.5 mm long, pubescent or more rarely glabrous. Corolla green or cream, sometimes tinged with purple, generally rotate, united at the base, but in occasional plants united to ± half-way, resulting in campanulate or even urceolate corollas, densely velvety-pubescent with white hairs on upper surface, glabrous beneath; corolla lobes triangular to ovate, 2–3 × 1–2 mm, apex rounded, margins strongly revolute (most Tanzanian collections) or ± plane (most Kenyan material). Staminal corona lobes fleshy, erect, 0.8–1.2 mm high, almost as tall as the gynostegium and free from it for much of their length, oblong or ovate in outline with a truncate apex, slightly flattened dorsoventrally. Anther wings 0.2 mm long. Pollinaria minute, corpusculum 0.15–0.2 mm long, subcylindrical; translator arms slender, terete, somewhat curved, 0.1–0.17 mm long; pollinia ovoid, ± 0.2 mm long, attached to the translator arms apically. Stylar head flat. Follicles mostly occuring singly, narrowly lanceolate in outline, to ± 12 × 1 cm, smooth or slightly winged. Seeds oblong to ovate, 6–10 × 3.5–5 mm, winged, glabrous or pubescent; coma 2.5–3.5 cm long.

UGANDA. Ankole District: Bushenyi, Nyakolokwe, 11 Sep. 2000, *Rwaburindore* 4982!; Mbale District: Bugishu, *Maitland* 1245!
KENYA. Trans Nzoia District: Hoey's Bridge, Nzoia River, Aug. 1971, *Tweedie* 4109!; N Kavirondo District: near Maragoli, Oct. 1986, *Agesa* s.n. in *Boppre* 87–120!; Machakos/Masai District: Chyulu Hills, above Ol Donyo Wuas Camp, 28 Oct. 1989, *Luke* 1964!
TANZANIA. Mpanda District: Kungwe-Mahali Peninsula, below Kungwe Mt, 11 Sep. 1959, *Harley* 9600!; Njombe District: 10 km SE of Njombe on road to Songea, 8 Nov. 1987, *Mwasumbi et al.* 13523!; Songea District: Luhila, 20 Sep. 1956, *Semsei* 2483!
DISTR. U 2, 3; K 1, 3–7; T 2, 4, 6–8; widely distributed across eastern Africa to S Mozambique and NE South Africa; also in Cameroon and Bioko
HAB. Moist riverine forest or thicket; (0–)1400–2200 m

SYN. ?*Pentarrhinum coriaceum* Schltr. in J.B. 32: 357 (1894). Type: South Africa, KwaZulu-Natal, without locality, *Gerrard & McKen* s.n. (B†, holo.). But see note below

Cynanchum chirindense S. Moore in J.B. 46: 305 (Oct. 1908). Type: Zimbabwe, Chirinda Forest, *Swynnerton* 137 (BM!, holo.; K!, iso.)

C. papillosum H.Weim. in Bot. Notis. 1935: 398 (1935). Type: Zimbabwe, Inyanga, above Pungwe River, *Fries, Norlindh & Weimarck* 3879 (LD, holo.)

Tylophora sp. B of U.K.W.F. ed. 2: 181 (1994)

T. urceolata Meve in K.B. 51: 585 (1996). Type: Cameroon, Mt Cameroon, *Meve* 901 (K!, holo.; LBG, MSUN, iso.)

NOTE. Meve (K.B. 1999) suggested that *Pentarrhinum coriaceum* Schltr. might be conspecific with *T. anomala*, although Schlechter's diagnosis makes no mention of the conspicuous indumentum on the adaxial face of the corolla, which throws some doubt on this view. Were *P. coriaceum* proved to be conspecific, its publication predates not only *T. anomala*, but also *T. coriacea* Marais (a nomen novum for *T. laevigata* Decne.).

The inflorescence of *Tylophora anomala* is hugely variable in its architecture, varying from sessile subumbelliform clusters of flowers, through typical *Tylophora* pedunculate zig-zag inflorescences with clusters of flowers on the angles of a single shoot, to many-branched inflorescences in a single leaf axil. Local forms are distinctive (sp. B of U.K.W.F., for example with its sessile subumbelliform inflorescences), but are not given formal taxonomic recognition here as they do not correlate with other characters, and intermediates occur between the main inflorescence types. *T. urceolata* Meve, described from Mt Cameroon, is regarded here as an occasional variant with an inflorescence reduced to a single flower. Urceolate corollas occur sporadically throughout the range of the species and are not associated elsewhere with such reduced inflorescences.

4. **Tylophora conspicua** *N.E. Br.* in K.B. 1895: 257 (1895) & in F.T.A. 4(1): 405 (1903); Bullock in F.W.T.A. ed. 2, 2: 96 (1963). Types: Angola, Golungo Alto, near Sobato Bumba, *Welwitsch* 4214 (K!, lecto., designated by Bullock in K.B. 9: 583 (1954, publ. 1955); BM!, LISC, isolecto.)

Twining herbaceous vine to several metres; stems glabrous or minutely pubescent. Leaves not succulent; petiole 1.5–5 cm long, minutely pubescent; lamina elliptic to slightly obovate or oblong, 10–13(–16) × 6(–10) cm, apex mostly attenuate, base ± truncate to deeply cordate, glabrous. Inflorescences extra-axillary, with several spiral clusters of flowers arising at the angles of a weakly zig-zag axis, inflorescence axes subglabrous or minutely pubescent; peduncles (to 1st branching point) 3–5.5 cm long; bracts broadly triangular, pubescent; pedicels 5–9 mm long, minutely pubescent. Calyx lobes 1–1.5 mm long, ovate or triangular, apex acute or rounded, minutely pubescent, frequently with ciliate margins. Corolla dark reddish purple, rotate, united at the base for ± 2 mm, glabrous; lobes oblong, 3–4 × 3–4 mm, apex displaced, rounded. Staminal corona lobes maroon or purple, fleshy, subovoid, 0.5–0.8 mm long, adnate to the staminal column and reaching to the base of the anthers. Anther wings 0.2 mm long. Pollinaria minute, corpusculum ovoid-subcylindrical, ± 0.05 mm long; translator arms slender, terete, ± 0.05 mm long; pollinia ovoid, ± 0.1 mm long. Stylar head flat. Follicles occuring singly, narrowly lanceolate in outline, 7–9 × 1 cm, smooth. Seeds ovate, 8 × 5 mm, irregularly winged, verrucose; coma 1.5 cm long.

UGANDA. Mengo District: Entebbe Botanic Gardens, Oct. 1935, *Chandler* 1434!

KENYA. Kwale District: Shimba Hills N.R., near Risley's Ridge, 2 July 2001, *Luke & Pakia* 7458! & Miongoni Forest, 26 May 1990, *Luke & Robertson* 2309!

TANZANIA. Moshi District: by Njoro R, 8 km W of Moshi, 3 Nov. 1955, *Milne-Redhead & Taylor* 7035!; Lushoto/Tanga Districts: Amani Zigi Forest Reserve, Kisiwani village, 20 Jan. 1999, *Sallu* 186!; Iringa District: Kilombero West Scarp Forest Reserve, 11 March 1996, *Frimodt-Möller et al.* in NG663!

DISTR. U 4; K 7; T 2, 3, 6, 7; from Liberia to Kenya and Angola to Zimbabwe

HAB. Moist forest; 0–1800 m

NOTE. *Drummond & Hemsley* 1943 from Lusunguru Forest, Morogoro District, has the characteristic leaf-shape of this species – and other characters of the species such as the minutely pubescent inflorescence axes, but the flowers are much smaller and resemble those of *T. apiculata*.

5. **Tylophora apiculata** *K. Schum.* in P.O.A. C: 325 (1895), *non* Schltr. (1907); N.E. Br. in F.T.A. 4(1): 407 (1903); T.T.C.L.: 69 (1949). Type: Tanzania, Pangani, *Stuhlmann* I: 848 (B†, holo., K!, iso. (fragment, as I: 884))

Twining herbaceous vine to several metres, whole plant glabrous. Leaves not succulent; petiole 1–3 cm long; lamina elliptic to oblong or ovate, 5–12 × 3.5–6 cm, apex rounded and apiculate, or acute to attenuate, base rounded or very weakly cordate. Inflorescences extra-axillary, appearing much branched with several clusters of flowers arising at the angles of a strongly zig-zag axis; peduncles (to 1ˢᵗ branching point) 1–5 cm long; bracts minute; pedicels 5–15 mm long, very slender. Calyx lobes triangular, 0.5–1 mm long, apex acute. Corolla pale pink or mauve, glabrous, rotate; lobes ovate, 1.5–3 mm long, apex rounded. Gynostegium 0.5 mm long. Staminal corona lobes green, fleshy, ± 0.4 mm long, adnate to the column for their entire length but thicker basally and radiating from the staminal column. Anther wings 0.1 mm long. Pollinaria minute, corpusculum ovoid-subcylindrical, ± 0.05 mm long; translator arms slender, ± 0.1 mm long; pollinia ovoid, ± 0.1 mm long,. Stylar head flat. Follicles occuring singly, narrowly lanceolate in outline, 6–7 × 1 cm, smooth. Seeds not seen.

Kenya. Tana River District: Hewani Forest 60, 13 Feb. 1992 (cult. Malindi, 29 Apr. 1995), *S.A. Robertson* 6894!; Kwale District: Gongoni Forest Reserve, 1 June 1990, *Robertson & Luke* 6313! & Buda Mafisini Forest, 12 km WSW of Gazi, 21 Aug. 1953, *Drummond & Hemsley* 3941!
Tanzania. Lushoto/Tanga Districts: Pangani River, Korogwe, 6 Mar. 1954, *Faulkner* 1374!; Rufiji District: Kiwengoma Forest, 12 Feb. 1990, *Clunies-Ross & Sheil* 728!; Lindi District: Litipo Forest Reserve, E side of Lake Lutamba, 25 Feb. 1991, *Bidgood et al.* 1722!
Distr. **K** 7; **T** 3, 6, 8; coastal Mozambique
Hab. Moist coastal or riverine forest; 0–300 m

Note. Very similar to *T. conspicua*, but generally more delicate, with rounded not deeply cordate base to the leaf, a more open, often branched, glabrous inflorescence, with smaller paler flowers and a shorter gynostegium.

6. **Tylophora iringensis** (*Markgr.*) *Goyder* in K.B. 60: 613 (2006). Type: Tanzania, Iringa District: Lupembe, near Ditimi, *Schlieben* 557 (B†, holo.; BM!, iso.)

Twining herbaceous vine to several metres; stems minutely pubescent. Leaves not succulent; petiole 3–6 mm long, pubescent; lamina ovate to lanceolate, 1–2.3 × 0.4–1.5 cm, apex acute, usually apiculate, base rounded, with a pair of minute colleters, sub-glabrous with indumentum restricted to margins and principal veins. Inflorescences extra-axillary, pedunculate, generally with a single cluster of 2–4 flowers; peduncles 4–5 mm long, slender, glabrous; bracts minute, triangular; pedicels 4–5 mm long, glabrous. Calyx lobes ovate or triangular, ± 1 mm long, pubescent. Corolla maroon or purple, rotate, united at the base, upper surface of the lobes densely pubescent with white hairs, glabrous beneath; corolla lobes ovate, 1–2 × 1–1.5 mm, apex acute or subacute. Staminal corona lobes red, fleshy, oblong, spreading horizontally from the gynostegium, with a free, erect, ligulate appendage 0.5–0.8 mm long, about as tall as the column, apex recurved. Anther wings ± 0.15 mm long. Pollinaria minute, corpusculum 0.15 mm long; translator arms slender, terete, ± 0.1 mm long; pollinia globose, ± 0.2 mm, attached to the translator arms apically. Stylar head domed, cream. Follicles occuring singly, narrowly lanceolate in outline, ± 8 × 1 cm, with a long attenuate beak, smooth.

Tanzania. Rungwe District: Livingstone Mts, track N of Isalala River and below ridge-top trail from Bumbigi, 4 Mar. 1991, *Gereau & Kayombo* 4212!; Mufindi District: E Mufindi, 5 Aug. 1933, *Greenway* 3495! & Uhafiwa-Luhega Forest, 2 Aug. 1989, *Kayombo* 787!
Distr. **T** 7; Nyika Plateau in N Malawi – endemic to the northern end of Lake Malawi
Hab. Moist evergreen montane forest; 1300–2100 m

SYN. *Pentarrhinum iringense* Markgr. in N.B.G.B. 11: 404 (1932)
    Tylophoropsis erubescens Liede & Meve in K.B. 49: 752 (1994). Type: Tanzania, Iringa District: Mufindi, forest opposite Lugoda Tea Estate factory entrance, *Renvoize & Abdallah* 2112 (K! (sheet annotated as type by Liede), lecto., designated by Goyder in K.B. 60: 613 (2005, publ. 2006), K!, isolecto.)
    Tylophora erubescens (Liede & Meve) Liede in Taxon 45: 206 (1996); Burrows in Pl. Nyika Plateau 75 (2005)

7. Tylophora sp. A

Twining herbaceous vine, stems minutely pubescent. Leaves not succulent; petiole 4–10 mm long, pubescent at least above; lamina ovate, 1.5–4.5 × 1–3 cm, apex acute or attenuate, shortly apiculate, base rounded, with a pair of minute colleters, sub-glabrous with indumentum restricted to margins and principal veins. Inflorescences extra-axillary, pedunculate, with several clusters of flowers; peduncles 8–15 mm long, slender, glabrous; bracts minute, narrowly triangular; pedicels 3–7 mm long, glabrous. Calyx lobes ovate or triangular, ± 1 mm long, sparsely pubescent. Corolla purple, rotate, united at the base, upper surface of the lobes pubescent at least towards the tip with white hairs, glabrous beneath; lobes ovate, 1.5–2 × 1 mm, apex acute or subacute. Gynostegium ± 0.5 mm long. Staminal corona lobes red, fleshy, obovate, spreading horizontally from the gynostegium, with a free, erect tooth 0.15–0.3 mm long, reaching to the base of the anthers or for their full length. Anther wings 0.15–0.2 mm long; pollinaria minute; corpusculum 0.05 mm long. Stylar head domed. Follicles not seen.

TANZANIA. Morogoro District: Uluguru Mts, Mkumbaku Mt, 14 June 1978, *Thulin & Mhoro* 3198! & Nguru Mts, Divue headwaters, 6 km SSE of Maskati Mission, 12 Feb. 1991, *Manktelow et al.* 91321!
DISTR. T 6; not known elsewhere
HAB. Moist evergreen montane forest; 1700–2100 m

NOTE. Vegetatively this species appears very close to *Tylophora iringensis* and *T. lugardiae* with its minutely pubescent stems, petioles and midveins to the leaves. The inflorescence is more branched than *T. iringensis*. The pubescent corolla lobes suggest affinities with *T. iringensis*, but the much reduced free tooth on the corona is neither dorsiventrally flattened nor recurved. *T. lugardiae* has a similarly reduced tooth, but the oblong corona lobes adhere to the much taller staminal column rather than radiating from it.

8. Tylophora lugardiae *Bullock* in K.B. 1932: 496 (1932); U.K.W.F. ed. 2: 181 (1994); Liede in Fl. Eth. 4(1): 135 (2003). Type: Kenya, Mt Elgon, 22 May 1931, *Lugard* 656 (K!, holo.; EA, iso.)

Twining herbaceous vine to several metres, stems pubescent with spreading white hairs. Leaves not succulent; petiole 0.3–1 cm long, pubescent; lamina ovate to elliptic, 2.5–6 × 1.5–3 cm, apex acute, frequently minutely apiculate, base rounded or obtuse, with minute colleters, glabrous or sub-glabrous with indumentum restricted to margins and principal veins. Inflorescences extra-axillary, pedunculate, with one or more clusters of nodding flowers along the axis; peduncles (to 1st cluster of flowers) 1–1.5 cm long, slender, glabrous; bracts minute, triangular; pedicels 3–11 mm long, glabrous. Calyx lobes ovate or triangular, 1–1.5 mm long, pubescent. Corolla green or cream, rotate, united at the base, glabrous; lobes ovate to triangular, 2.5–3.5 × 1–2 mm, acute. Staminal corona lobes maroon or cream, fleshy, erect, ± 1.1 mm high, ± as tall as the gynostegium and fused to it for much of their length, cuboid, narrowing abruptly into a free, linear appendage. Anther wings 0.25 mm long; pollinaria minute; corpusculum ovoid, ± 0.1 mm long; translator arms slender, terete, ± 0.05 mm long; pollinia suberect, ovoid, ± 0.15 mm long, attached to the translator arms basally. Stylar head flat. Follicles occuring singly, narrowly lanceolate in outline, ± 9.5 × 1 cm, with a long attenuate beak, smooth.

KENYA. Baringo District: Grandich, site of Kabarnet Dam, 5 Jan. 1995, *Goyder & Masinde* 3953!; cult. Nairobi ex Elgeyo District: Elgeyo Escarpment, 2 km N of Tambach, 3 Apr. 1961, *Archer* 303!; Kiambu District: Thiririka, 16 Jan. 1952, *Kirrika* 158!

DISTR. **K** 3, 4; N Ethiopia

HAB. Margins of montane forest, frequently degraded; 2000–2500 m

USES. None recorded

SYN. *Tylophora corollae* Meve & Liede in Edinb. J. Bot. 53: 323 (1996). Type: Ethiopia, 82 km S of Gondar, *J. de Wilde* 7170 (WAG, holo.; BR, K!, iso.)

9. **Tylophora fleckii** (*Schltr.*) *N.E. Br.* in Fl. Cap. 4(1): 766 (1908); Meve & Liede in Edinburgh J. Bot. 53: 325 (1996); Goyder in Fl. Somalia 3: 150, fig. 103 (2006). Type: Namibia, 'Gross-Namaland, Gansberg', *Fleck* 431 (B/Z? holo.)

Shrub twining to 2 m (erect subshrub outside the Flora area), stems minutely but densely pubescent when young. Leaves not succulent; petiole 0.5–1(–2) cm long, sparsely to densely pubescent; lamina ovate to lanceolate-oblong, 1.5–5 × 0.5–3 cm, apex attenuate, base rounded or truncate, generally with a pair of colleters, subglabrous or finely pubescent with indumentum of minute hairs on both upper and lower surfaces. Inflorescences extra-axillary, pedunculate, with a single cluster of flowers; peduncles 1–1.5 cm long, minutely pubescent; bracts minute, triangular, ciliate; pedicels 3–17 mm long, very slender, glabrous or minutely pubescent. Calyx lobes ovate, 1–1.5 mm long, glabrous or sparsely pubescent. Corolla green, cream or reddish purple, rotate, united at the base, puberulent with white hairs on upper surface, glabrous beneath; lobes ovate, 2–3 × 1–1.5 mm, apex rounded. Staminal corona lobes fleshy, somewhat swollen at the base which is adnate to the corolla tube, tapering above, shorter than the gynostegium and fused to it for their entire length. Anther wings 0.15–0.2 mm long; pollinaria minute; corpusculum ovoid, ± 0.05 mm long; translator arms slender, terete, ± 0.05 mm long; pollinia subglobose, ± 0.1 mm long, attached laterally to the translator arms. Stylar head conical. Follicles occuring singly, lanceolate in outline, 3.5–5 × 1 cm, and tapering into a beak, smooth. Seeds not seen.

KENYA. Marsabit District: Mt Kulal, Gatab, 21 Nov. 1978, *Hepper & Jaeger* 6988!; Teita District: near Mwatate, 17 Apr. 1960, *Verdcourt & Polhill* 2737! & Msau River Valley, 18 May 1985, *Taita Hills Exped.* 594!

TANZANIA. Musoma District: Serengeti, Seronera, 12 Mar. 1962, *Greenway* 10511! & 4 May 1961, *Greenway* 10142; Mbulu District: Ngorongoro Conservation Area, Marera Forest Reserve, 19 Jan. 1989, *Pócs & Chuwa* 89031/N!

DISTR. **K** 1, 4, 6, 7; **T** 1, 2; N Somalia and Namibia

HAB. Remnants of dry forest; 800–1600 m

SYN. *Tylophoropsis fleckii* Schltr. in Bull. Herb. Boiss. 7: 39 (1899)

NOTE. The habit of material from the Flora region is laxer and more scandent than in Somalia and Namibia, and with larger dimensions vegetatively, presumably reflecting the more mesic conditions in our region.

10. **Tylophora heterophylla** *A. Rich.*, Tent. Fl. Abyss. 2: 41 (1850); Liede in Fl. Eth. 4(1): 135 (2003). Type: Ethiopia, Tigray, Uogerat [Ouodgerate], Gumasir [Goumassa mts], *Petit* s.n. (P, holo.)

Twining shrub to several metres; young stems sparsely to densely pubescent, rarely subglabrous. Leaves not succulent; petiole 0.5–1.7(–2.5) cm long, pubescent; lamina ovate to oblong, 3–6 × 1.5–3.5 cm, apex acute, frequently shortly apiculate, base truncate or weakly cordate, with minute colleters, sub-glabrous with indumentum restricted to margins and principal veins, rarely more densely pubescent. Inflorescences extra-axillary, pedunculate, with one or more clusters of

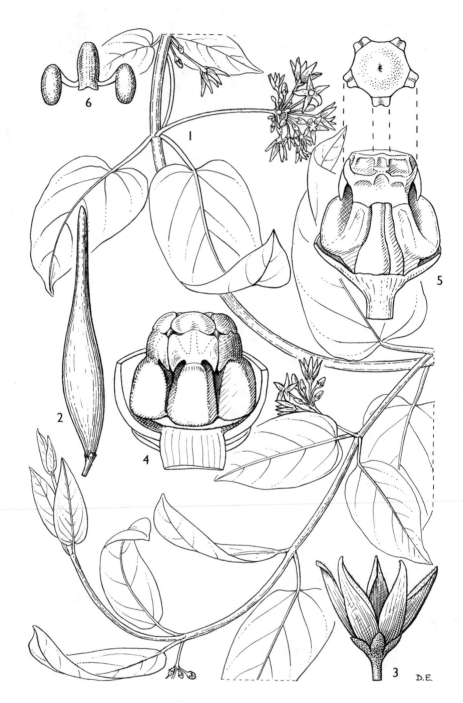

FIG. 121. *TYLOPHORA HETEROPHYLLA* — **1**, flowering shoot, × 1; **2**, follicle, × 1; **3**, flower, × 3; **4**, gynostegium and corona, × 12; **5**, gynostegium with stamens partially cut away to expose paired ovaries and stylar head, the latter also from above, × 12; **6**, pollinarium, × 22. 1 & 3–6 from *Drummond & Hemsley* 1302; 2 from *Linder* 2075. Drawn by D. Erasmus.

flowers along the axis; peduncles (to 1ˢᵗ cluster of flowers) 1–4.5 cm long, glabrous or pubescent, slender; bracts minute, triangular; pedicels 4–8(–10) mm long, glabrous. Calyx lobes ovate or triangular, 1–2 mm long, glabrous or with a pubescent apex, margins ciliate. Corolla crimson, brown or purple, generally rotate, united at the base, upper surface of the lobes minutely but densely pubescent with white hairs at least towards the tips, glabrous beneath; corolla lobes triangular, oblong or ovate, 4–9 × 2–2.5 mm, often extended into a long, spatulate, twisted tip, margins revolute. Staminal corona lobes fleshy, erect, 0.6–1.2 mm high, shorter than the gynostegium and fused to it for much of their length, oblong, brown or purple. Anther wings 0.5 mm long; pollinaria minute, corpusculum subcylindrical, 0.2–0.35 mm long; translator arms slender, terete, ± 0.15 mm long; pollinia ovoid, 0.2–0.3 mm long, attached to the translator arms apically. Stylar head flat, green. Follicles occuring singly, narrowly lanceolate in outline, ± 8 × 1 cm, with a long attenuate beak, smooth. Fig. 121, p. 506.

UGANDA. Kigezi District: Virunga range, between Muhavura and Mgahinga, 24 Oct. 1954, *Stauffer* 606! & Mgahinga, *Eggeling* 1073!
KENYA. Elgeyo District: Cherangani Hills, 6 km NW of Kaibibich, 27 July 1969, *Mabberley & McCall* 65!; Naivasha District: Aberdares National Park, S of Fort Jerusalem, 30 July 1960, *Polhill* 248!; Narok District: Enesambulai valley, 7 Dec. 1969, *Greenway & Kanuri* 13885!
TANZANIA. Arusha District: Mt Meru crater rim, 28 Dec. 1965, *Richards* 20906! & 23 Apr. 1968, *Greenway & Kanuri* 13517!
DISTR. U 2; K 3–6; T 2; Congo-Kinshasa (Kivu), Ethiopia, Eritrea and Yemen
HAB. Montane and riverine forest; 2200–3000 m

SYN. *Vincetoxicum heterophyllum* (A. Rich.) Vatke in Linnaea 40: 212 (1876)
Tylophora yemenensis Defl., Voy. Yemen 165 (1889). Type: Yemen, *Deflers* s.n. (K, iso.)
Gymnema longipedunculata Schweinf. in Höhnel, Reise zum Rudolf & Stephanie-See: 859 (1892), as *longepedunculatum*. Type: Kenya/Ethiopia?, *Höhnel* s.n. (B†, holo.)
Tylophoropsis heterophylla (A. Rich.) N.E. Br. in Gard. Chron. (1894) 2: 245 (1894) & in F.T.A. 4(1): 403 (1903); T.T.C.L.: 69 (1949); Fl. Rwanda 3: 120 (1985); U.K.W.F. ed. 2: 180 (1994)
T. yemenensis (Defl.) N.E. Br. in Gard. Chron. (1894) 2: 244 (1894)
Sphaerocodon longipedunculatum (Schweinf.) K. Schum. in P.O.A. C: 326 (1895), as *longepedunculatum*
Tylophora longipedunculata (Schweinf.) Schltr. in J.B. 33: 338 (1895)
Cynanchum roseum Chiov., Racc. Bot. Miss. Consol. Kenya: 82 (1935). Type: Kenya, *Balbo* 83 (TOM, holo.)

11. **Tylophora stenoloba** (*K. Schum.*) *N.E. Br.* in K.B. 1895: 257 (Oct. 1895) & in F.T.A. 4(1): 409 (1903); T.T.C.L.: 69 (1949). Type: Tanzania, Tanga District: 'Usambara', Doda, *Holst* 2977a (B†, holo.; K!, iso.)

Slender twining herbaceous vine to 3 metres, stems minutely pubescent or subglabrous. Leaves semi-succulent; petiole 2–4(–6) mm long, pubescent adaxially; lamina elliptic, 2–3.5 × 0.7–1.7 cm, apex rounded and with a short apiculus, base rounded, glabrous except for the midrib and basal margins, a minute colleter present at base of lamina. Inflorescences extra-axillary, mostly sessile, rarely (*Vollesen* MRC 3102) with peduncle to 1.3 cm; bracts minute, triangular; pedicels filiform, 1.5–2.5 cm long, glabrous. Calyx lobes narrowly to broadly triangular, 0.5–1 mm long, glabrous. Corolla pink or purple, rotate, united at the base for ± 1 mm; lobes spirally twisted in bud, spreading at anthesis, 6–12 mm long, narrowing from a triangular base into a weakly to strongly twisted filiform limb with a somewhat spatulate apex, pubescent with minute scattered white hairs at least towards the base adaxially, otherwise glabrous. Gynostegium 0.5 mm long. Staminal corona lobes maroon or purple, fleshy, ovoid, ± 0.3 mm long, adnate to the staminal column for most of its length. Anther wings ± 0.1 mm long; pollinaria minute, corpusculum ovoid-subcylindrical, ± 0.05 mm long; translator arms slender, terete, ± 0.05 mm long; pollinia ovoid, ± 0.1 mm long. Stylar head flat. Follicles occurring singly, narrowly lanceolate in outline, 3.5–5 × 0.7–0.8 cm, with a long slender beak, smooth. Seeds ovate, ± 7 × 4 mm, with a paler marginal wing ± 1 mm wide, minutely pubescent on both faces; coma 2 cm long.

KENYA. Tana River District: National Primate Reserve, Mulondi Swamp, 22 Mar. 1990, *Luke et al.* TPR 782!; Kilifi District: Kaya Ribe S, 2 Sep. 1999, *Luke & Luke* 5975!; Kwale District: Duruma River, 20 Mar. 1902, *Kässner* 395!
TANZANIA. Tanga District: Machu, Tanga–Pangani road, 11 Feb. 1955, *Faulkner* 1555!; Uzaramo District: University College Dar es Salaam, 28 Oct. 1969, *Harris* 3539!; Kilwa District: Selous Game Reserve, Kingupira, 16 Dec. 1975, *Vollesen* in MRC 3102!; Zanzibar: Mazizini [Massazine], 21 Oct. 1961, *Faulkner* 2929!
DISTR. **K** 7; **T** 3, 6, 8; **Z**; **P**; coastal Mozambique
HAB. Dry evergreen coastal or riverine forest; 0–300 m
USES. None recorded

SYN. *Astephanus stenolobus* K. Schum. in P.O.A. C: 321 (Aug. 1895)

NOTE. A note by Jan Gillett in the Kew Herbarium, and Diana Polhill's Gazetteer, suggests that the type locality Doda is on the coast between Tanga and Moa, not inland as Holst's label 'Flora von Usambara' implies.

12. **Tylophora sylvatica** *Decne.* in Ann. Sc. Nat. sér. 2, 9: 273 (1838); N.E. Br. in F.T.A. 4(1): 407 (1903); T.T.C.L.: 69 (1949); Bullock in F.W.T.A. ed. 2, 2: 96 (1963); Fl. Rwanda 3: 118 (1985); U.K.W.F. ed. 2: 181 (1994); Liede in Fl. Eth. 4(1): 136 (2003). Type: Senegambia, *Leprieur* s.n. (P, holo.)

Twining herbaceous vine to several metres, stems minutely pubescent when young, becoming glabrescent. Leaves not succulent; petiole 1–3 cm long, pubescent adaxially; lamina ovate to oblong, 3–9 × 1.5–5 cm, apex acute and somewhat attenuate, base cordate, rarely truncate, with minute colleters, glabrous or sub-glabrous with minute indumentum restricted to margins and principal veins. Inflorescences extra-axillary, up to ± 20 cm long with clusters of minute flowers scattered at intervals of 1–3 cm along the simple or branched axes; peduncles (to 1st cluster of flowers) 2–5 cm long, minutely pubescent; bracts minute, narrowly triangular; pedicels 1–5 mm long, minutely pubescent. Calyx lobes ovate or triangular, 0.5–0.8 mm long, pubescent. Corolla reddish brown or dull purple, rotate, united at the base, glabrous; lobes triangular, ovate or oblong, 1.5–2.5 × 1–1.5 mm, acute. Gynostegium 1 mm long. Staminal corona lobes maroon, ± 0.5 mm high, adnate to the staminal filaments to the base of the anthers and lacking a free apical appendage, broadest near the base of the column forming a fleshy subtriangular tubercle. Anther wings 0.25 mm long; pollinaria minute, corpusculum ± 0.1 mm long. Stylar head flat. Follicles mostly occuring singly, if paired then held at 180°, narrowly fusiform, 6–9.5 × ± 0.5 cm, smooth, glabrous. Seeds ± elliptic, 7–9 × 3 mm, flattened, with a narrow marginal rim which is irregularly toothed at the end opposite the 1.5–2 cm long coma.

UGANDA. West Nile District: Maracha, Apr. 1940, *Eggeling* 3873!; Ankole District: Buhweju County, Nyarambu-Rugongo, 4 Nov. 1993, *Rwaburindore* 3626!; Masaka District: Malabigambo forest, 3 km SSW of Katera, 2 Oct. 1953, *Drummond & Hemsley* 4534!
KENYA. North Kavirondo District: Nzoia R near Webuye [Broderick Falls], Mar. 1965, *Tweedie* 3027!; Kisumu-Londiani District: edge of fishing village near Kisumu, Mar. 1958, *Tweedie* 1503!; Masai District: Mara R, 21 Aug. 1961, *Archer* 269!
TANZANIA. Mpanda District: Mahale Mts, Kasiha, Apr. 1982, *Hasegawa* 23!; Ulanga District: Udzungwa Mts National Park, Campsite #3, 5 Feb. 1999, *Massawe* 165!; Lindi District: Nyangao, 23 Oct. 1934, *Schlieben* 5515!
DISTR. **U** 1, 2, 4; **K** 5, 6; **T** 4, 6–8; widely distributed across Central and West Africa, South Sudan, SW Ethiopia, Angola and NW Zambia
HAB. Edge of swamps, riverine or swamp forest, occasionally on open river banks; 200–1800 m

SYN. ?*Tylophora adalinae* K. Schum. in E. & P. Pf. 4(2): 286 (1895). Type not designated

13. **Tylophora tenuipedunculata** *K. Schum.* in E.J. 17: 144 (1893); N.E. Br. in F.T.A. 4(1): 409 (1903). Type: Congo (Brazaville), Louango, Chinchocho district, near Povo Zala, *Soyaux* 163 (B†, holo.; K!, iso.)

Slender twining herbaceous vine to 3 metres, stems subglabrous to uniformly but minutely pubescent. Leaves not succulent; petiole 0.6–2 cm long, pubescent with hairs mostly in groove along adaxial surface, occasionally subglabrous; lamina ovate to oblong or narrowly triangular, 2–6 × 1–2 cm, apex acute or attenuate, base rounded to truncate and with minute colleters, sub-glabrous. Inflorescences extra-axillary, pedunculate, with one or more clusters of flowers along the axis; peduncles (to 1st cluster of flowers) 1.5–3.5 cm long, slender, glabrous; bracts minute, pubescent; pedicels 0.6–2 cm long, glabrous. Calyx lobes triangular, 0.5–1 mm long, glabrous. Corolla pale to deep pink, rotate, united at the base only, sparsely pubescent or subglabrous adaxially, glabrous beneath; lobes ovate to triangular, 1.5–3.5 × 0.5–1.5 mm, apex acute to subacute. Gynostegium 0.6–0.8 mm long. Staminal corona lobes fleshy, ovoid, 0.4–0.6 mm high, reaching the base of the anthers and adnate to the staminal column for their entire length. Anther wings 0.2 mm long; pollinaria minute, corpusculum ± 0.07 mm long; translator arms slender, terete, ± 0.05 mm long; pollinia ovoid, ± 0.15 mm long, attached to the translator arms laterally. Stylar head flat. Follicles narrowly lanceolate in outline, 4–6 × ± 0.5 cm, with a long slender beak, smooth; seeds not seen.

UGANDA. Mengo District: Kampala, Kawanda Hill, Nov. 1937, *Chandler* 1998! & Kyiwaga [Kyewaga] Forest, Nov. 1922, *Maitland* 542!
KENYA. North Kavirondo District: Mlaba [Malaba] Forest, Kabras, Aug. 1965, *Tweedie* 3099!; Kilifi District: Arabuko-Sokoke Forest Reserve, 25 km SW of Malindi, 27 Nov. 1961, *Polhill & Paulo* 863! & 9 Jan. 1995, *Goyder et al.* 3963!
TANZANIA. Handeni District: Kabuku Forest, 6 June 1969, *Faulkner* 4231!; Kigoma District: Kubila Forest, 1 Apr. 1994, *Bidgood & Vollesen* 3025!; Uzaramo District: Pugu Forest Reserve, 8 km S of Pugu, 13 Nov. 1994, *Goyder et al.* 3726!
DISTR. U 4; **K** 5, 7; **T** 3, 4, 6, 8; Cameroon and Congo (Brazzaville)
HAB. Upland forest in the west, dry coastal forest in the east; 50–1500 m

NOTE. Corolla lobes acute/attenuate, mostly with a sparse indumentum on upper surface, at least towards tip. Material from Kenyan and Tanzanian coastal forests generally has shorter corolla lobes and a less rectangular outline to the anther wings as compared with western, upland material. Collections from the Rondo Plateau in southern Tanzania, however, are ± intermediate between these forms, and I have not recognised the entities formally.

14. **Tylophora sp. B**

Similar to *T. tenuipedunculata*, but differs most notably in the free, falcate, basal extensions to the anther wings, ± 0.3 mm long.

KENYA. Teita District: Taita Hills, Ngangao Forest, 1 Dec. 1998, *Luke et al.* 5494!
DISTR. **K** 7; known only from this collection
HAB. Montane forest; ± 1750 m

15. **Tylophora gracillima** *Markgr.* in N.B.G.B. 13: 284 (1936); T.T.C.L.: 69 (1949). Type: Tanzania, Morogoro, Uluguru Mts, *Schlieben* 3067 (B†, holo.; EA! K!, iso.).

Slender herbaceous twiner to ± 3 m, stems uniformly spreading-pubescent. Leaves not succulent; petiole 1–3.5 mm long, pubescent with spreading hairs; lamina linear-lanceolate to broadly ovate, 0.7–1.6 × 0.2–0.8 cm, apex acute or occasionally rounded, minutely apiculate, base rounded to subcuneate and with minute colleters, margins minutely undulate or crisped, minutely pubescent with any indumentum ± restricted to margins and principal veins. Inflorescences slender, extra-axillary, pedunculate, with very open branching; peduncles (to 1st branching point) 0.5–2 cm long, glabrous or minutely spreading pubescent; bracts minute; pedicels 0.3 cm long, glabrous. Calyx lobes ovate to narrowly triangular, 0.5–1 mm long, glabrous or minutely pubescent. Corolla greenish white to deep pink or purple, rotate, united at the base only, glabrous; lobes ovate, 1–2 × 0.7–1 mm, apex rounded to subacute. Gynostegium purple or green, 0.3 mm long. Staminal corona lobes fleshy, ovoid, 0.2–0.3 mm long, radiating from the staminal column and adnate to it to the base of the anthers. Anther wings 0.1 mm long; pollinaria minute, corpusculum ± 0.06 mm long. Stylar head flat. Follicles usually single, occasionally paired and held at ± 180°, narrowly fusiform, 4–5 × ± 0.3 cm, with a long slender beak, smooth; seeds not seen.

TANZANIA. Morogoro District: Uluguru Mts, Tegetero, 20 Mar. 1953, *Drummond & Hemsley* 1698! & Uluguru North Catchment Forest Reserve, NW of Tegetero Mission, 21 Jan. 2001, *Jannerup & Mhoro* 247!; Mufindi District: Lulando Forest Reserve, Fufu forest patch, 3 Aug. 1999, *Kayombo et al.* 2824!
DISTR. T 6, 7; endemic to Uluguru mountains, Udzungwa Escarpment and Lulando forest
HAB. Moist forest; 1100–1800 m

NOTE. One element of *Polhill & Wingfield* 4671 (the other element is *Tylophora* sp. F) appears to be this species, but the leaves are much larger (to 2.7 × 1.2 cm), and the collection was made at higher altitude – 2200 m near the Lukwangule plateau.

16. **Tylophora sp. C**

Similar to *T. gracillima*, but leaves much larger, to 4 × 2.5 cm, margins minutely undulate or crisped, and with much larger anther wings (0.3–0.4 mm).

TANZANIA. Iringa District: Ndunduru Forest Reserve, 7 Oct. 2000, *Luke et al.* 7053! & 10 Sep. 2004, *Luke et al.* 10429!; Udzungwa Scarp Forest Reserve, 12 Dec. 1997, *Frimodt-Möller et al.* in TZ 516! & 16 Dec. 1997, *Frimodt-Möller et al.* in TZ 632!
DISTR. T 7; known only from these collections
HAB. Moist montane forest; 1450–1800 m

17. **Tylophora sp. D**

Similar florally to *T. gracillima*, but the inflorescence is even more branched with several main axes per node, the leaves are much larger: 4–5 × 2–3 cm, ovate with a cuneate base, with plane, not crisped or undulate margins; young stems minutely pubescent in two lines.

TANZANIA. Morogoro District: Uluguru North catchment Forest Reserve, 5.5 km WNW of Tegetero Mission, 15 Jan. 2001, *Jannerup & Mhoro* 0170!
DISTR. T 6; known only from this collection
HAB. Open forest; ± 1850 m

18. **Tylophora tridactylata** *Goyder* in K.B. 63: 467 (2009). Type: Tanzania, Iringa District: Mufindi, Kigogo River, *Polhill & Paulo* 1813 (K!, holo.)

Slender herbaceous twiner to 5 m, stems with pubescence restricted to two lines. Leaves not succulent; petiole 1–8 mm long, glabrous; lamina linear or very rarely narrowly triangular, 2–3.5 × 0.1–0.2 cm, with or without a pair of basal lobes to 1 cm long held at an angle of ± 45° to the midrib, tips to lobes rounded, apex of principal blade rounded to acute, minutely apiculate, base generally cuneate, with minute colleters, minutely pubescent with any indumentum ± restricted to margins and principal veins. Inflorescences extra-axillary, sessile, few-flowered; pedicels 5–7 mm

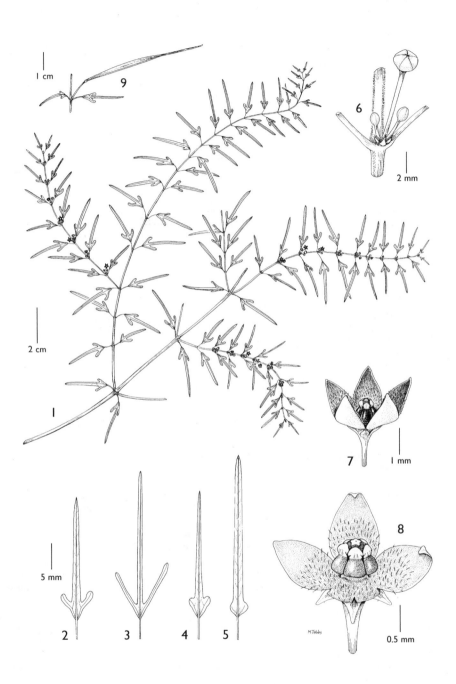

FIG. 122. *TYLOPHORA TRIDACTYLATA* — **1**, habit; **2–5**, variation in leaf morphology; **6**, node with young inflorescence; **7**, flower; **8**, flower with two corolla lobes removed to display gynostegium and corona; **9**, node with single follicle. 1, 2 & 6 from *Polhill & Paulo* 1813; 3 from *Rodgers* 1059; 4 & 9 from *Ndangalasi* 0438; 5 from *Ndangalasi* 0441; 7 & 8 from *Frimodt-Möller et al.* in NG 097. Drawn by M. Tebbs. Reproduced from K.B. 63: 468, Fig. 1.

long, glabrous. Calyx lobes ovate to narrowly triangular, ± 0.5 mm long, glabrous or minutely pubescent. Corolla dark red, pubescent adaxially at least in lower half of lobes, rotate, united at the base only; lobes ovate, 1–1.5 × 0.7–1 mm, apex rounded to subacute. Gynostegium green, ± 0.4 mm long. Staminal corona lobes fleshy, ovoid, ± 0.3 mm long, radiating from the staminal column and adnate to it to the base of the anthers. Anther wings 0.1 mm long; pollinaria minute. Stylar head flat. Follicles single, narrowly fusiform, ± 7 × ± 0.3 cm, with a long slender beak, smooth; seeds not seen. Fig. 122, p. 511.

TANZANIA. Iringa District: Kigogo River, 19 Mar. 1962, *Polhill & Paulo* 1813! & Luhega Forest Reserve, 22 Feb. 1996, *Frimodt-Möller et al.* in NG 097! & Mbawi forest, Udzungwa Scarp Forest Reserve, 24 Feb. 2000, *Ndangalasi* 0438!
DISTR. **T** 6, 7; endemic to the Sanje/Udzungwa region
HAB. Wet forest; 1200–1600 m

NOTE. Differs from *T. gracillima* in the less prominent indumentum of young shoots which is restricted to two lines rather than uniformly distributed around the stem; the leaf lamina is longer, linear rather than lanceolate to ovate, and in addition may have weakly or strongly developed basal lobes; base cuneate; inflorescence sessile not pedunculate and corolla pubescent adaxially.

19. **Tylophora sp. E**

Very similar to *T. tridactylata*, with rhomboid base to linear leaves, but with uniformly pubescent stems and pedunculate branched inflorescences as in *T. gracillima*, and conspicuously falcate anther wings with basal extensions as in sp. B (*Luke et al.* 5494) from Taita Hills, **K** 7.

TANZANIA. Lushoto District: West Usambara Mts, Mkusi Valley between Mkuzi and Kifungilo, 23 Apr. 1953, *Drummond & Hemsley* 2224!
DISTR. **T** 3; known only from this collection
HAB. Wet forest; 1600 m

20. **Tylophora sp. F**

Twining herbaceous vine to several metres, stems minutely pubescent, mostly in two lines, becoming glabrous with age. Leaves not succulent; petiole 0.6–1(–1.5) cm long, pubescent with hairs restricted mostly to groove along adaxial surface; lamina ovate, 2–4(–7) × 1.5–3(–4.5) cm, apex acute or attenuate, base rounded to truncate and with minute colleters, sub-glabrous, secondary veins prominent beneath. Inflorescences extra-axillary, pedunculate and irregularly branched; peduncles (to 1st branching point) 1–3 cm long, slender, glabrous; bracts triangular, ± 1 mm long, pubescent; pedicels 3–5 mm long, glabrous. Calyx lobes broadly ovate, rarely triangular, (1–)2–4 mm long, glabrous or pubescent, margins ciliate. Corolla deep pink or purple, rotate, united at the base for ± 1 mm, upper surface minutely puberulent with white hairs, glabrous beneath; lobes ovate, 2–4 × 2–3.5 mm, apex rounded or subacute, margins somewhat crisped but not revolute. Gynostegium ± 1.2 mm long. Staminal corona lobes fleshy, ovoid, ± 0.9 mm high, reaching the base of the anthers and adnate to the staminal column for their entire length. Anther wings 0.4 mm long; pollinaria minute, corpusculum 0.15 mm long; translator arms slender, terete, ± 0.1 mm long; pollinia subglobose, ± 0.1 mm long, attached to the translator arms laterally. Stylar head flat, green. Follicles not seen.

TANZANIA. Morogoro District: S Uluguru Mts, Lukwangule Plateau, above Chenzema Mission, 13 Mar. 1953, *Drummond & Hemsley* 1532! & 18 Feb. 1932, *Schlieben* 3481! & 25 May 2000, *Jannerup & Mhoro* 0046!

Distr. **T** 6; endemic to the S Uluguru mountains
Hab. Montane moist forest; 1700–2500 m

Note. Florally, this taxon is very similar to *T. umbellata* Schltr., a South African species ± restricted to the Eastern Cape. However, it has broadly ovate rather than lanceolate calyx and corolla lobes, and leaf shape is also different, with sp. F having a rounded or truncate leaf base rather than the cuneate base of *T. umbellata*.

21. **Tylophora sp. G**

Twining herbaceous vine to 2 metres, stems minutely pubescent, mostly in two lines, becoming glabrous with age. Leaves not succulent; petiole 0.5 cm long, pubescent with hairs mostly in groove along adaxial surface; lamina oblong-lanceolate, 3–5 × 1–1.5 cm, apex acute, base rounded to truncate and with minute colleters, sub-glabrous, secondary veins prominent beneath. Inflorescences extra-axillary, pedunculate; peduncles (to 1ˢᵗ cluster of flowers) 2.5 cm long, slender, glabrous; bracts mostly deciduous; pedicels 6–9 mm long, glabrous. Calyx lobes ovate to lanceolate, 2–3 mm long, glabrous, margins ciliate. Corolla maroon, broadly campanulate, united at the base for ± 0.5 mm, minutely puberulent with white hairs towards the tips on both adaxial and abaxial faces; lobes ovate, 3–4.5 × 2–3 mm, apex rounded or subacute, margins somewhat crisped but not revolute. Gynostegium ± 1.5 mm long. Staminal corona lobes fleshy, ovoid, ± 0.9 mm high, reaching the lower $\frac{1}{3}$ of the anthers and adnate to the staminal column for their entire length. Anther wings 1 mm long; pollinaria minute, corpusculum 0.3 mm long. Stylar head flat, green. Follicles not seen.

Tanzania. Rungwe District: Poroto Mts, E of Kimondo on road to Kitulo, 1 Dec. 1994, *Goyder et al.* 3873!
Distr. **T** 7; not known elsewhere
Hab. Scrambling over bush at edge of montane grassland; 2600 m

Note. Similar to the previous species, but leaves are much narrower and the anther wings and corpusculum much longer.

INDEX TO APOCYNACEAE

var. *concolor* Krug & Urb., 385
Asclepias daemia Forssk., 366
Asclepias denticulata Schltr., 428
Asclepias dregeana Schltr., 392
 var. *calceolus* (S. Moore) N.E. Br., 392
 var. *sordida* N.E. Br., 393
Asclepias echinata Roxb., 367
Asclepias edentata *Goyder*, 389, 391
Asclepias eximia Schltr., 445
Asclepias firma (N.E. Br.) Hiern, 448
Asclepias flavida N.E. Br., 426
Asclepias fluviatilis A. Chev., 372
Asclepias foliosa (*K. Schum.*) *Hiern*, 395
Asclepias fornicata N.E. Br., 416
Asclepias forskohlei Roem. & Schult., 496
Asclepias friesii Schltr., 436
Asclepias fruticosa L., 426
 var. *angustissima* (Engl.) Schltr., 430
Asclepias fulva *N.E. Br.*, 391
Asclepias gigantea L., 370
Asclepias gigantiflora (K. Schum.) N.E. Br., 418
Asclepias glaberrima (Oliv.) Schltr., 372
Asclepias glabra Forssk., 366
Asclepias glabra Miller, 426
Asclepias glaucophylla (Schltr.) Schltr., 435
Asclepias gossweileri S. Moore, 395
Asclepias grandirandii *Goyder*, 403
Asclepias grantii (Oliv.) Schltr., 443
Asclepias inaequalis *Goyder*, 398
Asclepias integra N.E. Br., 431
Asclepias kaessneri N.E. Br., 429
Asclepias katangensis S. Moore, 436
Asclepias kyimbilae Schltr., 394
Asclepias laniflora Del., 372
Asclepias laniflora Forssk., 372
Asclepias laurentiana (Dewèvre) N.E. Br., 417
Asclepias lepida S. Moore, 397
Asclepias leuchotricha Schltr., 452
Asclepias leucocarpa Schltr., 432
Asclepias lilacina Weim., 435
Asclepias lineolata (Decne.) Schltr., 438
Asclepias lineolata sensu auct., 440
Asclepias litocarpa Chiov., 431
Asclepias longissima (*K. Schum.*) *N.E. Br.*, 386, 388
Asclepias macrantha Oliv., 413, 415
Asclepias macrantha sensu auct., 415
Asclepias macropetala (Schltr. & K. Schum.) N.E. Br., 412
Asclepias margaritacea Schult., 385
Asclepias minuta A. Chev., 397
Asclepias modesta N.E. Br., 397
 var. *foliosa* N.E. Br., 397
Asclepias moorei De Wild., 436
Asclepias mtorwiensis *Goyder*, 400, 399
Asclepias muhindensis N.E. Br., 418
Asclepias munonquensis S. Moore, 433
Asclepias negrii Chiov., 431
Asclepias nivalis J.F. Gmel., 496
Asclepias nivea Forssk., 496
Asclepias nivea L., 385
 var. *curassavica* (L.) O. Kuntze, 385
Asclepias nuttii *N.E. Br.*, 397

Asclepias nyikana Schltr., 432
Asclepias odorata (K. Schum.) N.E. Br., 412
Asclepias pachyclada (K. Schum.) N.E. Br., 412
Asclepias palustris (*K. Schum.*) *Schltr.*, 392
Asclepias pedunculata (Decne.) Dandy, 413
Asclepias phillipsiae N.E. Br., 430
Asclepias physocarpa (E. Mey.) Schltr., 427
Asclepias praticola S. Moore, 436
Asclepias propinqua N.E. Br., 417
Asclepias pseudoamabilis *Goyder*, 391
Asclepias pygmaea *N.E. Br.*, 403
Asclepias randii *S. Moore*, 401, 403, 404
Asclepias reflexa (Britt. & Rendle) Britt. & Rendle, 422
 var. *longicauda* S. Moore, 422
Asclepias rhacodes (K. Schum.) N.E. Br., 415
'Asclepias rostrata' of T.T.C.L., 384
Asclepias rivalis S. Moore, 372
Asclepias rubicunda Schltr., 392
Asclepias salicifolia Salisb., 426
Asclepias scandens P. Beauv., 366
Asclepias schumanniana *Hiern*, 395
Asclepias schweinfurthii N.E. Br., 438
Asclepias semilunata (A. Rich.) N.E. Br. 428
Asclepias spectabilis N.E. Br., 412
Asclepias spiralis Forssk, 496
Asclepias stathmostelmoides *Goyder*, 385, 386, 388
Asclepias stipitacea Forssk., 481
Asclepias stolzii Schltr., 432
Asclepias swynnertonii S. Moore, 432
Asclepias tanganyikensis *E.A. Bruce*, 388
Asclepias undulata L., 452
Asclepias uvirensis S. Moore, 413
Asclepias viridiflora (E. Mey.) Goyder, 392
Asclepias vomeriformis S. Moore, 419
Asclepias welwitschii (Britt. & Rendle) N.E. Br., 417
Aspidoglossum *E. Mey.*, 372
Aspidoglossum angustissimum (*K. Schum.*) *Bullock*, 375
Aspidoglossum breve *Kupicha*, 374
Aspidoglossum connatum (*N.E. Br.*) *Bullock*, 376
Aspidoglossum elliotii (*Schltr.*) *Kupicha*, 380
Aspidoglossum interruptum (*E. Mey.*) *Bullock*, 378
Aspidoglossum kulsii Cuf., 378
Aspidoglossum lanatum (*Weim.*) *Kupicha*, 375
Aspidoglossum masaicum (*N.E. Br.*) *Kupicha*, 378
Aspidoglossum sp. A, 381
Aspidoglossum sp. B, 381
Aspidoglossum sp. B of Kupicha, 381
Aspidoglossum whytei (N.E. Br.) Bullock, 376
Aspidosperma condylocarpon Muell. Arg., 58
Astephanus cernuus Decne., 466
Astephanus stenolobus K. Schum., 508

Baissea *A. DC.*, 101
Baissea alborosea Gilg & Stapf, 102
Baissea angolensis Stapf
 var. *major* Stapf, 104

Caralluma rivae Chiov., 323
Caralluma rosengrenii Vierhapper, 323
Caralluma russeliana (Brongn.) Cufod., 320
Caralluma schweinfurthii A. Berger, 345
Caralluma scutellata Deflers, 328
Caralluma socotrana (Balf. f.) N.E. Br., 323
Caralluma speciosa (N.E. Br.) N.E. Br., 321
Caralluma sprengeri (Schweinf.) N.E. Br., 346
Caralluma subterranea E.A. Bruce & P.R.O. Bally, 351
Caralluma tombuctuensis (A. Chev.) N.E. Br., 320
Caralluma tubiformis E.A. Bruce & P.R.O. Bally, 347
Caralluma turneri *E.A. Bruce*, 314
 subsp. **turneri**, 314
 subsp. **ukambensis** (*P.R.O. Bally*) *L.E. Newton* 314
Caralluma vibratilis E.A. Bruce & P.R.O. Bally, 354
Caralluma wilsonii P.R.O. Bally, 352
Carissa *L.*, 11
Carissa arduina (L.) Lam., 12
Carissa bispinosa (*L.*) *Brenan*, 11
Carissa candolleana Jaub. & Spach., 12
Carissa carandas L., 11
Carissa cornifolia Jaub. & Spach., 12
Carissa bispinosa (L.) Merxm.,
Carissa edulis (Forssk.) Vahl, 12
 var. *major* Stapf, 12
 var. *tomentosa* (A. Rich.) Stapf, 12
Carissa friesiorum (Markgr.) Cufod.
Carissa grandiflora (E. Meyer) A. DC., 11
Carissa longiflora (Stapf) Lawrence, 17
Carissa macrocarpa (Eckl.) A.DC., 11
Carissa oppositifolia (Lam.) Pichon, 17
Carissa schimperi A. DC., 17
Carissa spinarum *L.*, 12
Carissa tetramera (*Sacleux*) *Stapf*, 14
Carissa tomentosa A. Rich., 12
Carissa richardiana Jaub. & Spach., 12
Carpodinus landolphioides (Hallier f.) Stapf, 21
Carvalhoa *K. Schum.*, 52
Carvalhoa campanulata *K. Schum.*, 54
Carvalhoa macrophylla K. Schum., 54
Carvalhoa petiolata K. Schum., 54
Catharanthus roseus (L.) G. Don, 6
Cerbera *L.*, 65
Cerbera manghas *L.*, 67
Ceropegia *L.*, 219
 sect. *Riocreuxia* sensu auct., 216
Ceropegia aberrans Schltr., 242
Ceropegia abyssinica *Decne.*, 250, 241, 252, 254
 var. **abyssinica**, 250
 var. **songeensis** *H. Huber*, 251
Ceropegia achtenii *De Wild.*, 251, 241, 254, 255
 subsp. *adolfii* (Werderm.) Huber, 252
Ceropegia adolfi Werderm., 252
 var. *gracillima* Werderm., 252
Ceropegia affinis *Vatke*, 260, 263
Ceropegia albertina S. Moore, 271
Ceropegia albisepta *Jum. & H. Perrier*, 275
 var. *bruceana* H. Huber, 277

 var. **robynsiana** (*Werderm.*) *H. Huber*, 275, 274
Ceropegia ampliata *E. Mey.*, 287
 var. *oxyloba* H. Huber, 287
Ceropegia anceps S. Moore, 272
Ceropegia angiensis De Wild., 233
Ceropegia angusta N.E. Br., 263
Ceropegia angustiloba De Wild., 283
Ceropegia arabica *H. Huber*, 288, 287, 291
 var. **powysii** (*D.V. Field*) *U. Meve & R.M. Mangelsdorff*, 288
Ceropegia archeri Bally, 269
Ceropegia aristolochioides *Decne.*, 270, 239, 269, 272
 subsp. *albertina* (S. Moore) H. Huber, 271
 subsp. **aristolochioides**, 270
 var. *wittei* Staner, 271
Ceropegia aristolochioides sensu auct., 272
Ceropegia atacorensis A. Chev., 263
Ceropegia bajana Bullock, 263
Ceropegia ballyana *Bullock*, 274, 277
Ceropegia barbigera P.V. Bruyns, 288
Ceropegia baringii P.R.O. Bally, 271
Ceropegia batesii S. Moore, 272
Ceropegia beccariana Martelli, 271
Ceropegia bequaerti De Wild., 251
Ceropegia biddumana K. Schum., 261
Ceropegia boussingaultifolia Dinter, 284
Ceropegia brevirostris P.R.O. Bally & D.V. Field, 268
Ceropegia brosima E.A. Bruce & P.R.O. Bally, 255
Ceropegia brownii Ledger, 287
Ceropegia bulbosa *Roxb.*, 255, 256
Ceropegia burchellii (K. Schum.) H. Huber, 218
 subsp. *burchellii*, 218
 subsp. *profusa* (N.E. Br.) H. Huber, 218
Ceropegia burgeri M.G. Gilbert, 271
Ceropegia butaguensis De Wild., 263
Ceropegia calcarata N.E. Br., 233
Ceropegia campanulata *G. Don*, 256
 var. **pulchella** *H. Huber*, 257
Ceropegia chipiaensis Stopp, 260
Ceropegia chortophylla Werderm., 242
Ceropegia chrysochroma H. Huber, 219
Ceropegia claviloba *Werderm.*, 236, 235, 244
Ceropegia collaricorona Werderm., 246
Ceropegia constricta N.E. Br., 284
Ceropegia copleyae E.A. Bruce & P.R.O. Bally, 281
Ceropegia cordiloba *Werderm.*, 235, 239
Ceropegia crassifolia *Schltr.*, 280
 var. **copleyae** (*E.A. Bruce & P.R.O. Bally*) *H. Huber*, 281
Ceropegia crassula Schltr., 271
Ceropegia criniticaulis Werderm., 233
Ceropegia cufodontis *Chiov.*, 243
Ceropegia cynanchoides Schltr., 263
Ceropegia cyrtoidea Werderm., 268
Ceropegia decumbens P.R.O. Bally, 285
Ceropegia degemensis S. Moore, 272
Ceropegia dentata N.E. Br., 238
Ceropegia denticulata *K. Schum.*, 285
 var. **brownii** (Ledger) P.R.O. Bally, 286
 var. **denticulata**, 286

Echidnopsis virchowii *K. Schum.*, 327
   var. *stellata* (Lavranos) Plowes, 328
Echidnopsis watsonii *P.R.O. Bally*, 332
Echites guineensis Thonn., 99
Echites pubescens Buch.-Ham., 92
Ectadium Benth., 126
Ectadium oblongifolia (Meisn.) B.D. Jacks., 132
Ectadiopsis oblongifolia (Meisn.) Schltr., 132
Ectadiopsis producta (N.E. Br.) Bullock, 134
Ectadiopsis suffruticosa K. Schum., 132
Ectadiopsis oblongifolium Meisn., 132
Edithcolea N.E. Br., 323
Edithcolea grandis *N.E. Br.*, 325
   var. *baylissiana* Lavranos & D.S. Hardy, 325
Edithcolea sordida N.E. Br., 325
Endotropis Endl., 474
Ephippiocarpa Markgr., 50
Ephippiocarpa humilis (Chiov.) Boiteau, 52
Ephippiocarpa orientalis (S. Moore) Markgr., 52
Epistemma *D.V. Field and J.B. Hall*, 166
Epistemma neuerburgii *Eb. Fisch., Venter,*
   Killman & Meve, 166
Euphorbia viminalis L., 479

Flanagania Schltr., 474
Fockea *Endl.*, 186
Fockea angustifolia *K. Schum.*, 186
Fockea dammarana Schltr., 187
Fockea lugardii N.E. Br., 187
Fockea mildbraedii Schltr., 187
Fockea monroi S. Moore, 187
Fockea multiflora *K. Schum.*, 187
Fockea schinzii N.E. Br., 189
Fockea sessiliflora Schltr., 187
Fockea tugelensis N.E. Br., 187
Funtumia *Stapf*, 86
Funtumia africana (*Benth.*) *Stapf*, 86
Funtumia congolana (De Wild.) Jumelle, 88
Funtumia gilletii (De Wild.) Jumelle, 88
Funtumia elastica (*Preuss*) *Stapf*, 88
Funtumia latifolia (Stapf) Stapf, 88

Gabunia odoratissima Stapf, 47
Garcinia sciura Spirlet, 17
Gilgia candida Pax, 359
Glossonema *Decne.*, 358
Glossonema boveanum (Decne.) Decne., 361
Glossonema elliotii Schltr., 359
Glossonema lineare (Fenzl) Decne., 214
Glossonema macrosepalum Chiov., 359
Glossonema revoilii *Franch.*, 359
Glossonema rivaei K. Schum., 359
Glossonema sp. A, 361
Glossostelma *Schltr.*, 458, 382
Glossostelma brevilobum *Goyder*, 463
Glossostelma cabrae (*De Wild*) *Goyder*, 463
Glossostelma carsonii (*N.E. Br.*) *Bullock*, 458
Glossostelma ceciliae (*N.E. Br.*) *Goyder*, 462
Glossostelma mbisiense *Goyder*, 462
Glossostelma spathulatum (*K. Schum.*)
   Bullock, 460, 462
Gomphocarpus *R. Br.*, 422, 382
   sect. *Eugomphocarpus* Decne., 422

   sect. *Pachycarpus* (E. Mey.) Decne., 436
   subsect. *Leiocalymma* K. Schum., 422
Gomphocarpus amabilis (N.E. Br.) Bullock, 389
Gomphocarpus amoenus K. Schum., 395
Gomphocarpus angustatus Hochst., 419
Gomphocarpus angustifolius (Schweigger) Link,
   426
Gomphocarpus arachnoideus E. Fourn., 426
Gomphocarpus bisacculatus Oliv., 440
Gomphocarpus brasiliensis E. Fourn., 427
Gomphocarpus buchwaldii Schltr. & K. Schum.,
   394
Gomphocarpus chlorojodina K. Schum., 460
Gomphocarpus concolor (E. Mey.) Decne., 441
Gomphocarpus cornutus Decne., 426
Gomphocarpus crinitus G. Bertol., 426
Gomphocarpus cristatus Decne., 394
Gomphocarpus foliosus K. Schum., 397
Gomphocarpus frutescens E. Mey., 426
Gomphocarpus fruticosus (*L.*) *W.T. Aiton*,
   425
   forma *brasiliensis* (E. Fourn.) Briq., 427
   subsp. **flavidus** (*N.E. Br.*) *Goyder*, 426
   var. *angustissimus* Engl., 430
Gomphocarpus fruticosus sensu auct., 426
Gomphocarpus glaberrimus Oliv., 372
Gomphocarpus glaucophyllus *Schltr.*, 433
Gomphocarpus grantii (Oliv.) Schltr., 443
Gomphocarpus integer (*N.E. Br.*) *Bullock*,
   430, 432
Gomphocarpus kaessneri (*N.E. Br.*) *Goyder &*
   Nicholas, 428, 427
Gomphocarpus lineolatus Decne., 438
Gomphocarpus longipes Oliv., 413
Gomphocarpus longissimus K. Schum., 386
Gomphocarpus munonquensis (*S. Moore*)
   Goyder & Nicholas, 432
Gomphocarpus paluster K. Schum., 394
Gomphocarpus pauciflorus Klotzsch, 422
Gomphocarpus pedunculatus Decne., 413
Gomphocarpus phillipsiae (*N.E. Br.*) *Goyder*,
   429
Gomphocarpus physocarpus *E. Mey.*, 427,
   426, 429
Gomphocarpus physocarpus sensu auct., 428
Gomphocarpus praticola (*S. Moore*) *Goyder &*
   Nicholas, 435
Gomphocarpus rhinophyllus K. Schum., 441
Gomphocarpus rostratus (N.E. Br.) Bullock, 432
Gomphocarpus rubioides Kotschy & Peyr., 453
Gomphocarpus scaber Harv., 441
Gomphocarpus scaber K. Schum., 441
Gomphocarpus semilunatus *A. Rich.*, 428,
   427, 429
Gomphocarpus semilunatus sensu auct., 427
Gomphocarpus spathulatus (K. Schum.) Schltr.,
   462
Gomphocarpus stenophyllus *Oliv.*, 431
Gomphocarpus swynnertonii (*S. Moore*)
   Goyder & Nicholas, 432, 433
Gomphocarpus tanganyikensis (E.A. Bruce)
   Bullock, 388
Gomphocarpus trachyphyllus K. Schum., 441

var. *angustifolia* (Engl.) Stapf, 30
var. *rotundifolia* Dew., 30
var. *rufa* Stapf, 30
var. *schweinfurthiana* (Hallier f.) Stapf, 29
var. *tubeufii* Busse ex Stapf, 32
Landolphia polyantha K. Schum., 21
Landolphia rogersii Stapf, 19
Landolphia scandens (K. Schum. & Thonn.)
    Didr., 29
    var. *angustifolia* (Engl.) Hallier f, 30
    var. *ferruginea* Hallier f., 29
    var. *petersiana* (Klotzsch) Hallier f., 30
    var. *rigida* Hallier f., 29
    var. *rotundifolia* (Dew.) Hallier f., 30
    var. *schweinfurthiana* Hallier f., 29
    var. *stuhlmanii* Hallier f., 30
    var. *tubeufii* (Stapf) Busse
Landolphia sennensis (Klotzsch) K. Schum., 30
Landolphia stolzii Busse, 22
Landolphia subturbinata Dawe, 22
Landolphia tayloris Stapf, 32
Landolphia turbinata A. Chev., 22
Landolphia ugandensis Stapf, 19
Landolphia watsoniana *Romburgh*, 24
Lanugia variegata (Britt. & Rendle) N.E. Br., 89
Lepistoma Blume, 126
Leposma Blume, 126
Leptadenia *R. Br.*, 210
Leptadenia hastata (*Pers.*) *Decne.*, 210
    subsp. **hastata**, 210
    subsp. **meridionalis** Goyder, 212
Leptadenia lancifolia (Schumach. & Thonn.)
    Decne., 212
    var. *scabra* Decne., 212
Leptopharyngia elegans (Stapf) Boiteau, 46
Lochnera rosea L., 6

Macropelma K. Schum., 135
Macropelma angustifolia K. Schum., 137
Madorius Kuntze, 367
Madorius giganteus (L.) Kuntze, 370
Madorius procerus (Aiton) Kuntze, 368
Mandevilla amoena, 10
Mandevilla laxa (Ruiz & Pavon) R.E.
    Woodson, 10
Mandevilla splendens (Hook.) R.E.
    Woodson, 10
Mandevilla suaveolens Lindl., 10
Margaretta cornetii Dewèvre, 409
    var. *pallida* De Wild., 409
Margaretta decipiens Schltr., 409
Margaretta distincta N.E. Br., 445
Margaretta holstii K. Schum., 408
Margaretta *Oliv.*, 404, 382, 409
Margaretta orbicularis N.E. Br., 409
Margaretta pulchella Schltr., 409
Margaretta rosea Oliv., 404
    subsp. **bidens** *Bullock*, 408
    subsp. **corallina** *Goyder*, 407
    subsp. **cornetii** (*Dewèvre*) *Goyder*, 409
    subsp. **kilimanjarica** *Goyder*, 408
    subsp. **orbicularis** (*N.E. Br.*) *Goyder*, 408
    subsp. **rosea**, 406

    subsp. **whytei** (*K. Schum.*) *Mwanyambo*, 408
M. sp. A, 408
Margaretta verdickii De Wild., 409
Margaretta whytei K. Schum., 408
Marsdenia *R. Br.*, 190
Marsdenia abyssinica (*Hochst.*) *Schltr.*, 204
    forma *complicata* Bullock, 204
Marsdenia angolensis *N.E. Br.*, 195
Marsdenia crinita *Oliv.*, 203
Marsdenia cynanchoides *Schltr.*, 197
Marsdenia exellii *Norman*, 205
Marsdenia faulkneri (*Bullock*) *Omlor*, 202
Marsdenia glabriflora Benth., 196
    var. *orbicularis* N.E. Br., 196
Marsdenia gondarensis Chiov., 195
Marsdenia grandiflora (Decne.) Choux, 205
Marsdenia grandiflora Norman, 205
Marsdenia latifolia (*Benth.*) *K. Schum.*, 196
Marsdenia leonensis Benth., 196
Marsdenia macrantha (*Klotzsch*) *Schltr.*, 198
Marsdenia magniflora P.T. Li, 205
Marsdenia normaniana Omlor, 205
Marsdenia profusa N.E. Br., 196
Marsdenia racemosa K. Schum., 196
Marsdenia rubicunda (*K. Schum.*) *N.E. Br.*, 199
Marsdenia schimperi *Decne.*, 201
Marsdenia spissa S. Moore, 204
Marsdenia stefaninii Chiov., 201
Marsdenia stelostigma *K. Schum.*, 201
Marsdenia sylvestris (*Retz.*) *P.I. Forst.*, 194
Marsdenia taylori *Schltr. & Rendle*, 196
Marsdenia umbellifera K. Schum, 208
Marsdenia zambesiaca Schltr, 199
Mascarenhasia *A. DC.*, 89
Mascarenhasia arborescens *A. DC*, 89
Mascarenhasia elastica K. Schum., 89
Mascarenhasia fischeri K. Schum., 89
Mascarenhasia variegata Britt. & Rendle, 89
Microstephanus N.E. Br., 464
Microstephanus cernuus (Decne.) N.E. Br., 466
Mondia *Skeels*, 143
Mondia ecornuta (*N.E. Br.*) *Bullock*, 143
Mondia whitei (*Hook.f.*) *Skeels*, 145
Monolluma *Plowes*, 321
Monolluma socotrana (*Balf. f.*) *Meve & Liede*,
    323
Morrenia stuckertiana (H. Heger) Malme,
    398
Motandra *A. DC.*, 99
Motandra altissima Stapf, 99
Motandra erlangeri K. Schum., 109
Motandra glabrata Baill., 107
Motandra guineensis (*Thonn.*) *A. DC.*, 99
Motandra viridiflora K. Schum., 105

Neoschumannia *Schltr.*, 214
Neoschumannia cardinea (*S. Moore*) *Meve*, 215
Nerium obesum Forssk., 69
Nerium oleander L., 10
Neurolobium cymosum Baill.,

Odontanthera sensu Mabberley, 358
Odontanthera linearis (Fenzl) Mabberley, 214

Odontanthera reniformis sensu Mabberley, 359
Odontanthera reniformis Wight, 359
Omphalogonus Baill., 122
Omphalogonus calophyllus Baill., 123
Omphalogonus nigritanus N.E. Br., 123
Oncinotis *Benth.*, 105
Oncinotis glabrata (*Baill.*) *Hiern*, 106
Oncinotis hirta *Oliv.*, 107
Oncinotis jespersenii De Wild., 107
Oncinotis melanocephala K. Schum., 104
Oncinotis oblanceolata Engl., 109
Oncinotis obovata De Wild., 109
Oncinotis pontyi Dubard, 107
  var. *breviloba* Pichon, 109
Oncinotis tenuiloba *Stapf*, 109
Oncinotis sp. of T.T.C.L., 109
Orbea *Haw.*, 343
Orbea caudata (*N.E. Br.*) *Bruyns*, 354
  subsp. **caudata**, 354
Orbea denboefii (*Lavranos*) *Bruyns*, 348, 351
Orbea distincta (*E.A. Bruce*) *Bruyns*, 350
Orbea dummeri (*N.E. Br.*) *Bruyns*, 348
Orbea gemugofana (*M.G. Gilbert*) *Bruyns*, 346
Orbea laikipiensis (*M.G. Gilbert*) *Bruyns*, 347
Orbea laticorona, 345
Orbea schweinfurthii (*A. Berger*) *Bruyns*, 345
Orbea semitubiflora (*L.E. Newton*) *Bruyns*, 352
Orbea semota (*N.E. Br.*) *L.C. Leach*, 356
  subsp. *orientalis* Bruyns, 356
Orbea sprengeri (*Schweinf.*) *Bruyns*, 345
  subsp. **sprengeri**, 346
Orbea subterranea (*E.A. Bruce & P.R.O. Bally*) *Bruyns*, 351
Orbea taitica *Bruyns*, 353
Orbea tubiformis (*E.A. Bruce & P.R.O. Bally*) *Bruyns*, 347
Orbea vibratilis (*E.A. Bruce & P.R.O. Bally*) *Bruyns*, 353
Orbea wilsonii (*P.R.O. Bally*) *Bruyns*, 352
Orbeopsis L.C. Leach, 343
Orbeopsis caudata (N.E. Br.) L.C. Leach, 356
Oxypetalum coeruleum (Sweet) Decne, 178
Oxystelma *R. Br.*, 361
Oxystelma alpini Decne., 362
Oxystelma bornouense *R. Br.*, 362
Oxystelma esculentum (L. f.) Schult., 362
Oxystelma esculentum (*L. f.*) *Sm.*, 362
  var. *alpini* (Decne.) N.E. Br., 362
Oxystelma secamone sensu auct., 362
Oxystelma senegalense Decne., 364

Pachycarpus *E. Mey.*, 436, 382
Pachycarpus bisacculatus (*Oliv.*) *Goyder*, 438
Pachycarpus bullockii Cavaco, 440
Pachycarpus concolor *E. Mey.*, 440
  subsp. **rhinophyllus** (*K. Schum.*) *Goyder*, 441
Pachycarpus corniculatus Hochst., 413
Pachycarpus distinctus (*N.E. Br.*) *Bullock*, 445, 443
Pachycarpus eximius (*Schltr.*) *Bullock*, 443
Pachycarpus eximius sensu auct., 443
Pachycarpus firmus (*N.E. Br.*) *Goyder*, 448

Pachycarpus fulvus (N.E. Br.) Bullock, 392
Pachycarpus goetzei (*K. Schum.*) *Bullock*, 445
Pachycarpus grantii (*Oliv.*) *Bullock*, 442, 445
  subsp. **grantii**, 443
  subsp. **marroninus** *Goyder*, 443
Pachycarpus grantii sensu auct., 445
Pachycarpus lineolatus (*Decne.*) *Bullock*, 437, 440
Pachycarpus lineolatus sensu auct., 440
Pachycarpus mildbraedii Bullock, 438
Pachycarpus pachyglossus *Goyder*, 449
Pachycarpus petherickianus (*Oliv.*) *Goyder*, 446
Pachycarpus rhinophyllus (K. Schum.) N.E. Br., 441
Pachycarpus richardsiae *Goyder*, 449
Pachycarpus schumannii Chiov., 448
Pachycarpus schweinfurthii (N.E. Br.) Bullock, 438
Pachycarpus sp. A, 445
Pachycarpus spurius (*N.E. Br.*) *Bullock*, 441
Pachycarpus terminans Fiori, 448
Pachycarpus viridiflorus E. Mey., 392
Pachycymbium L.C. Leach, 343
Pachycymbium baldratii (White & Sloane) M.G. Gilbert, 351
  subsp. *subterraneum* (E.A. Bruce & P.R.O. Bally) M.G. Gilbert, 351
Pachycymbium denboefii (Lavranos) M.G. Gilbert, 350
Pachycymbium distinctum (E.A. Bruce) M.G. Gilbert, 351
Pachycymbium dummeri (N.E. Br.) M.G. Gilbert, 348
Pachycymbium gemugofanum (M.G. Gilbert) M.G. Gilbert, 346
Pachycymbium laikipiense M.G. Gilbert, 347
Pachycymbium laticoronum, 345
Pachycymbium schweinfurthii (A. Berger) M.G. Gilbert, 345
Pachycymbium semitubiflorum (L.E. Newton) M.G. Gilbert, 352
Pachycymbium sprengeri (Schweinf.) M.G. Gilbert, 346
Pachycymbium tubiforme (E.A. Bruce & P.R.O. Bally) M.G. Gilbert, 348
Pachycymbium vibratile (E.A. Bruce & P.R.O. Bally) M.G. Gilbert, 354
Pachycymbium wilsonii (P.R.O. Bally) M.G. Gilbert, 352
Pacouria amoena (Hua) Pichon, 29
Pacouria angustifolia (Engl.) O. Kuntze, 32
Pacouria owariensis (P. Beauv.) Hiern, 22
Pacouria parvifolia (K. Schum.) Hiern, 24
Pacouria petersiana (Klotzsch) S. Moore, 29, 32
  var. schweinfurthiana (Hallier f.) S. Moore, 29
Parquetina *Baill.*, 122
Parquetina calophylla (*Baill.*) *Venter*, 123
Parquetina gabonica Baill., 125
Parquetina nigrescens (*Afzel.*) *Bullock*, 125, 123

Parquetina nigrescens sensu auct., 123
Pentagonanthus Bullock, 146
Pentagonanthus grandiflorus (N.E. Br.)
  Bullock, 150
  subsp. *glabrescens* (Bullock) Bullock, 150
Pentarrhinum *E. Mey.*, 468, 466
Pentarrhinum abyssinicum Decne., 469
  subsp. **abyssinicum**, 471
  subsp. **angolense** (N.E. Br.) Liede &
    Nicholas, 471
  subsp. *ijimense*, 472
  var. *angolense* N.E. Br., 471
Pentarrhinum coriaceum Schltr., 501
Pentarrhinum fasciculatum K. Schum., 496
Pentarrhinum gonoloboides (Schltr.) Liede,
  473
Pentarrhinum insipidum E. Mey., 469, 471,
  473
Pentarrhinum insipidum sensu auct., 471
Pentarrhinum iringense Markgr., 504
Pentarrhinum ledermannii (Schltr.) Goyder
  & Liede, 472, 471
Pentarrhinum somaliense (N.E. Br.) Liede,
  472
Pentarrhinum sp. A., 471
Pentatropis *Wight & Arn.*, 494
Pentatropis cynanchoides N.E. Br., 496
  var. *senegalensis* (Decne.) N.E. Br., 496
Pentatropis fasciculatus (K. Schum.) N.E. Br.,
  496
Pentatropis hoyoides K. Schum., 496
Pentatropis madagascariensis Decne., 496
Pentatropis nivalis (*J.F. Gmel.*) *D.V. Field &*
  J.R.I. Wood, 494
  subsp. *madagascariensis* (Decne.) Liede &
    Meve, 496
Pentatropis rigida Chiov., 496
Pentatropis senegalensis Decne., 496
Pentatropis spiralis sensu auct., 496
  var. *longepetiolata* Engl., 496
Pergularia *L.*, 364
Pergularia africana N.E. Br., 210
Pergularia daemia (*Forssk.*) *Chiov.*, 364
  subsp. **daemia**, 366
  var. *macrantha* Chiov., 367
Pergularia extensa (Jacq.) N.E. Br., 367
Pergularia glabra (Forssk.) Chiov., 367
Pergularia sanguinolenta Lindl., 135
Pergularia tacazzeana Chiov., 210
Perianthostelma Baill., 474
Periploca *L.*, 118
Periploca afzelii G. Don, 125
Periploca batesii Wernham, 489
Periploca calophylla (Baill.) Roberty, 123
Periploca esculenta L. f., 362
Periploca gabonica (Baill.) A. Chev., 125
Periploca latifolia K. Schum., 146
Periploca linearifolia *Quart.-Dill. & A. Rich.*,
  118
  var. *gracilis* Browicz, 120
Periploca linearis Hochst., 120
Periploca nigrescens Afzel., 125
Periploca petersiana Vatke, 199

Periploca preussii K. Schum., 125
Periploca refractifolia Gilli, 120
Periploca secamone L., 174, 362
Periploca sylvestris Retz., 195
Periplocoideae K. Schum., 2
Piaggiaea boranensis Chiov., 71
Piaggiaea demartiniana (Chiov.) Chiov., 71
Picralima *Pierre*, 34
Picralima nitida (*Stapf*) *Th. & H. Durand*, 36
Pleiocarpa *Benth.*, 39
Pleiocarpa bagshawei S. Moore, 41
Pleiocarpa bicarpellata *Stapf*, 39
Pleiocarpa microcarpa Stapf, 41
Pleiocarpa pycnantha (*K. Schum.*) *Stapf*, 39
  var. *tubicina* (Stapf) Pichon, 41
Pleiocarpa tubicina Stapf, 41
Pleiocarpa, sp. near *swynnertonii* of T.T.L.C.,
  41
Pleioceras *Baill.*, 71
Pleioceras orientale *Vollesen*, 73
Pleurostelma *Baill.*, 464
Pleurostelma africanum Schltr., 159
Pleurostelma cernuum (*Decne.*) *Bullock*, 464
Pleurostelma grevei Baill., 466
Pleurostelma Schltr., 157
Plumeria acuminata Ait., 10
Plumeria alba L., 10
Plumeria alba sensu auctt., 10
Plumeria obtusa L., 10
Plumeria rubra L., 10
Podostelma K. Schum., 464
Pseudopectinaria Lavranos, 325
Pseudopectinaria malum Lavranos, 334
Pterygocarpus Hochst., 190
Pterygocarpus abyssinicus Hochst., 204

Raphionacme *Harv.*, 146
Raphionacme abyssinica Chiov., 159
Raphionacme angolensis (Baill.) N.E. Br., 157
Raphionacme bagshawii S. Moore, 155
Raphionacme baguirmiensis A. Chev., 148
Raphionacme bingeri (A. Chev.) Lebrun &
  Stork, 155
Raphionacme borenensis *Venter & M.G.*
  Gilbert, 147
Raphionacme brownii Scott-Elliot, 148
Raphionacme daronii Berhaut, 155
Raphionacme decolor Schltr., 152
Raphionacme denticulata N.E. Br., 157
Raphionacme ernstiana Meve, 153
Raphionacme excisa Schltr., 155
Raphionacme flanaganii *Schltr.*, 149
Raphionacme globosa *K. Schum.*, 149
Raphionacme gossweileri S. Moore, 154
Raphionacme grandiflora *N.E. Br.*, 150
  subsp. *glabrescens* Bullock, 150
Raphionacme jurensis N.E. Br., 155
Raphionacme loandae Schltr. & Rendle, 140
Raphionacme longifolia *N.E. Br.*, 150
Raphionacme longituba *E.A. Bruce*, 152
Raphionacme madiensis *S. Moore*, 153
Raphionacme michelii *De Wild.*, 153
Raphionacme monteiroae (Oliv.) N.E. Br., 140

Traunia albiflora K. Schum., 202
Triodoglossum Bullock, 157
Triodoglossum abyssinicum (Chiov.) Bullock, 159
Tylophora *R. Br.*, 497
Tylophora adalinae K. Schum., 508
Tylophora anomala *N.E. Br.*, 501, 206
Tylophora apiculata *K. Schum.*, 503
Tylophora caffra *Meisn.*, 499
Tylophora cameroonica *N.E. Br.*, 500
Tylophora cirrosa Aschers., 496
Tylophora conspicua *N.E. Br.*, 502
Tylophora corollae Meve & Liede, 505
Tylophora erubescens (Liede & Meve) Liede, 504
Tylophora fleckii (*Schltr.*) *N.E. Br.*, 505
Tylophora glauca Bullock, 500, 501
Tylophora gracillima *Markgr.*, 509, 510, 512
Tylophora heterophylla *A. Rich.*, 505
Tylophora incanum Brunner, 212
Tylophora iringensis (*Markgr.*) *Goyder*, 503
Tylophora longipedunculata (Schweinf.) Schltr., 507
Tylophora lugardiae *Bullock*, 504
Tylophora sp. A, 504
Tylophora sp. B, 509
Tylophora sp. B of U.K.W.F., 502
Tylophora sp. C. of U.K.W.F., 499
Tylophora sp. C, 510
Tylophora sp. D, 510
Tylophora sp. E, 512
Tylophora sp. F, 512
Tylophora sp. G, 513
Tylophora stenoloba (*K. Schum.*) *N.E. Br.*, 507
Tylophora sylvatica *Decne.*, 508
Tylophora tenuipedunculata *K. Schum.*, 509
Tylophora umbellata Schltr., 513
Tylophora urceolata Meve, 502
Tylophora tridactylata *Goyder*, 510
Tylophora yemenensis Defl., 507
Tylophoropsis erubescens Liede & Meve, 504
Tylophoropsis fleckii Schltr., 505
Tylophoropsis heterophylla (A. Rich.) N.E. Br., 507
Tylophoropsis N.E. Br., 497
Tylophoropsis yemenensis (Defl.) N.E. Br., 507

Vahea comorensis Bojer, 34
Vinca minor L., 10
Vincetoxicum abyssinicum (Decne.) Kuntze, 489
Vincetoxicum caffrum (Meisn.) Kuntze, 500
Vincetoxicum fruticulosum (Decne.) Decne., 492
Vincetoxicum heterophyllum (A. Rich.) Vatke, 507
Vincetoxicum holstii K. Schum., 489
Vincetoxicum polyanthum K. Schum., 489
Vincetoxicum polyanthum Kuntze, 489
Virchowia A. Schenk, 325
Virchowia K. Schum., 325
Voacanga *Thouars*, 42
Voacanga africana Stapf, 42
   var. *lutescens* (Stapf) Pichon, 44
   var. *auriculata* Pichon, 44

Voacanga angustifolia K. Schum., 42
Voacanga bequaertii De Willd., 42
Voacanga boehmii K. Schum., 42
Voacanga densiflora Engl., 44
Voacanga dichotoma K. Schum., 42, 47
Voacanga lutescens Stapf, 42
Voacanga obtusa K. Schum., 44
Voacanga schweinfurthii Stapf, 42
Voacanga thouarsii *Roem. & Schult.*, 44
   var. *obtusa* (K. Schum.) Pichon, 44

Willughbeia cordata Klotzsch, 34
Willughbeia petersiana Klotzsch, 30
Willughbeia sennensis Klotzsch, 30
Woodia trilobata Schltr., 452
Wrightia *R. Br.*, 69
Wrightia boranensis (Chiov.) Cufod., 71
Wrightia demartiniana *Chiov.*, 71
Wrightia stuhlmannii K. Schum., 95

Xysmalobium *R. Br.*, 450, 382, 464
Xysmalobium angolense Scott Elliot, 452
Xysmalobium barbigerum N.E. Br., 452
Xysmalobium bellum N.E. Br., 462
Xysmalobium carsonii N.E. Br., 460
Xysmalobium ceciliae N.E. Br., 463
Xysmalobium concolor (E. Mey.) D. Dietr., 441
Xysmalobium dispar N.E. Br., 452
Xysmalobium dolichoglossum K. Schum., 442
Xysmalobium fraternum *N.E. Br.*, 455
Xysmalobium heudelotianum *Decne.*, 453
Xysmalobium kaessneri S. Moore, 456
Xysmalobium leuchotrichum (Schltr.) N.E. Br., 452
Xysmalobium patulum S. Moore, 457
Xysmalobium reticulatum N.E. Br., 453
Xysmalobium schumannianum S. Moore, 453
Xysmalobium sp. A, 456
Xysmalobium spathulatum (K. Schum.) N.E. Br., 462
Xysmalobium speciosum S. Moore, 463
Xysmalobium spurium N.E. Br., 442
Xysmalobium stocksii *N.E. Br.*, 455
Xysmalobium trilobatum (Schltr.) N.E. Br., 452
Xysmalobium undulatum (*L.*) *W.T. Aiton*, 450
Xysmalobium viridiflorum (E. Mey.) D. Dietr., 392

Zuchellia Decne., 146
Zuchellia angolensis Decne., 157
Zygodia Benth., 101
Zygodia kidengensis K. Schum., 105
Zygodia melanocephala (K. Schum.) Stapf, 105
Zygodia myrtifolia Benth., 104
Zygonerion Baill., 75
Zygonerion welwitschii Baill., 78

LIST OF ABBREVIATIONS

A.V.P. = O. Hedberg, Afroalpine Vascular Plants; **B.J.B.B.** = Bulletin du Jardin Botanique de l'Etat, Bruxelles; Bulletin du Jardin Botanique Nationale de Belgique; **B.S.B.B.** = Bulletin de la Société Royale de Botanique de Belgique; **C.F.A.** = Conspectus Florae Angolensis; **E.J.** = A. Engler, Botanische Jahrbücher für Systematik, Pflanzengeschichte und Pflanzengeographie; **E.M.** = A. Engler, Monographieen Afrikanischer Pflanzen-Familien und Gattungen; **E.P.** = A. Engler, Das Pflanzenreich; **E.P.A.** = G. Cufodontis, Enumeratio Plantarum Aethiopiae Spermatophyta; in B.J.B.B. 23, Suppl. (1953) et seq.; **E. & P. Pf.** = A. Engler & K. Prantl, Die Natürlichen Pflanzenfamilien; **F.A.C.** = Flore d'Afrique Centrale (*formerly* F.C.B.); **F.C.B.** = Flore du Congo Belge et du Ruanda-Urundi; Flore du Congo, du Rwanda et du Burundi; **F.E.E.** = Flora of Ethiopia & Eritrea; **F.D.-O.A.** = A. Peter, Flora von Deutsch-Ostafrika; **F.F.N.R.** = F. White, Forest Flora of Northern Rhodesia; **F.P.N.A.** = W. Robyns, Flore des Spermatophytes du Parc National Albert; **F.P.S.** = F.W. Andrews, Flowering Plants of the Anglo-Egyptian Sudan *or* Flowering Plants of the Sudan; **F.P.U.** = E. Lind & A. Tallantire, Some Common Flowering Plants of Uganda; **F.R.** = F. Fedde, Repertorium Specierum Novarum Regni Vegetabilis; **F.S.A.** = Flora of Southern Africa; **F.T.A.** = Flora of Tropical Africa; **F.W.T.A.** = Flora of West Tropical Africa; **F.Z.** = Flora Zambesiaca; **G.F.P.** = J. Hutchinson, The Genera of Flowering Plants; **G.P.** = G. Bentham & J.D. Hooker, Genera Plantarum; **G.T.** = D.M. Napper, Grasses of Tanganyika; **I.G.U.** = K.W. Harker & D.M. Napper, An Illustrated Guide to the Grasses of Uganda; **I.T.U.** = W.J. Eggeling, Indigenous Trees of the Uganda Protectorate; **J.B.** = Journal of Botany; **J.L.S.** = Journal of the Linnean Society of London, Botany; **K.B.** = Kew Bulletin, *or* Bulletin of Miscellaneous Information, Kew; **K.T.S.** = I. Dale & P.J. Greenway, Kenya Trees and Shrubs; **K.T.S.L.** = H.J. Beentje, Kenya Trees, Shrubs and Lianas; **L.T.A.** = E.G. Baker, Leguminosae of Tropical Africa; **N.B.G.B.** = Notizblatt des Botanischen Gartens und Museums zu Berlin-Dahlem; **P.O.A.** = A. Engler, Die Pflanzenwelt Ost-Afrikas und der Nachbargebiete; **R.K.G.** = A.V. Bogdan, A Revised List of Kenya Grasses; **T.S.K.** = E. Battiscombe, Trees and Shrubs of Kenya Colony; **T.T.C.L.** = J.P.M. Brenan, Check-lists of the Forest Trees and Shrubs of the British Empire no. 5, part II, Tanganyika Territory; **U.K.W.F.** = A.D.Q. Agnew (or for ed. 2, A.D.Q. Agnew & S. Agnew), Upland Kenya Wild Flowers; **U.O.P.Z.** = R.O. Williams, Useful and Ornamental Plants in Zanzibar and Pemba; **V.E.** = A. Engler & O. Drude, Die Vegetation der Erde, IX, Pflanzenwelt Afrikas; **W.F.K.** = A.J. Jex-Blake, Some Wild Flowers of Kenya; **Z.A.E.** = Wissenschaftliche Ergebnisse der Deutschen Zentral-Afrika-Expedition 1907–1908, 2 (Botanik).

FAMILIES OF VASCULAR PLANTS REPRESENTED IN
THE FLORA OF TROPICAL EAST AFRICA

The family system used in the Flora has diverged in some respects from that now in use at Kew and the herbaria in East Africa. The accepted family name of a synonym or alternative is indicated by the word "see". Included family names are referred to the one used in the Flora by "in" if in accordance with the current system, and "as" if not. Where two families are included in one fascicle the subsidiary family is referred to the main family by "with".

PUBLISHED PARTS

- Foreword and preface
*Glossary
Index of Collecting Localities

Acanthaceae
    Part 1
    **Part 2
*Actiniopteridaceae
*Adiantaceae
Aizoaceae
Alangiaceae
Alismataceae
*Alliaceae
*Aloaceae
*Amaranthaceae
*Amaryllidaceae
*Anacardiaceae
*Ancistrocladaceae
Anisophyllaceae — as Rhizophoraceae
Annonaceae
*Anthericaceae
Apiaceae — see Umbelliferae
Apocynaceae
    *Part 1
    Part 2
*Aponogetonaceae
Aquifoliaceae
*Araceae
Araliaceae
Arecaceae — see Palmae
*Aristolochiaceae
*Asclepiadaceae — see Apocynaceae
Asparagaceae
*Asphodelaceae
Aspleniaceae
Asteraceae — see Compositae
Avicenniaceae — as Verbenaceae
*Azollaceae

*Balanitaceae
*Balanophoraceae
*Balsaminaceae
Basellaceae
Begoniaceae
Berberidaceae
Bignoniaceae
Bischofiaceae — in Euphorbiaceae
Bixaceae
Blechnaceae
*Bombacaceae
*Boraginaceae
Brassicaceae — see Cruciferae
Brexiaceae
Buddlejaceae — as Loganiaceae
*Burmanniaceae
*Burseraceae
Butomaceae
Buxaceae

Cabombaceae
Cactaceae
Caesalpiniaceae — in Leguminosae
*Callitrichaceae
Campanulaceae
Canellaceae
Cannabaceae
Cannaceae — with Musaceae
Capparaceae
Caprifoliaceae
Caricaceae
Caryophyllaceae
*Casuarinaceae
Cecropiaceae — with Moraceae
*Celastraceae
*Ceratophyllaceae
Chenopodiaceae
Chrysobalanaceae — as Rosaceae

*Palmae
Pandaceae — with Euphorbiaceae
*Pandanaceae
Papaveraceae
Papilionaceae — in Leguminosae
*Parkeriaceae
Passifloraceae
Pedaliaceae
Periplocaceae — see Apocynaceae (Part 2)
Phytolaccaceae
*Piperaceae
Pittosporaceae
Plantaginaceae
Plumbaginaceae
Poaceae — see Gramineae
Podocarpaceae
Podostemaceae
Polemoniaceae — see Cobaeaceae
Polygalaceae
Polygonaceae
*Polypodiaceae
Pontederiaceae
*Portulacaceae
Potamogetonaceae
Primulaceae
*Proteaceae
*Psilotaceae
*Ptaeroxylaceae
*Pteridaceae

*Rafflesiaceae
Ranunculaceae
Resedaceae
Restionaceae
Rhamnaceae
Rhizophoraceae
Rosaceae
Rubiaceae
  Part 1
  *Part 2
  *Part 3
*Ruppiaceae
*Rutaceae

*Salicaceae
Salvadoraceae
*Salviniaceae
Santalaceae
*Sapindaceae
Sapotaceae
*Schizaeaceae

Scrophulariaceae
Scytopetalaceae
Selaginellaceae
Selaginaceae — in Scrophulariaceae
*Simaroubaceae
*Smilacaceae
Solanaceae
Sonneratiaceae
Sphenocleaceae
Strychnaceae — in Loganiaceae
*Surianaceae
Sterculiaceae

Taccaceae
Tamaricaceae
Tecophilaeaceae
Ternstroemiaceae — in Theaceae
Tetragoniaceae — in Aizoaceae
Theaceae
Thelypteridaceae
Thismiaceae — in Burmanniaceae
Thymelaeaceae
*Tiliaceae
Trapaceae
Tribulaceae — in Zygophyllaceae
*Triuridaceae
Turneraceae
Typhaceae

Uapacaceae — in Euphorbiaceae
Ulmaceae
*Umbelliferae
*Urticaceae

Vacciniaceae — in Ericaceae
Valerianaceae
Velloziaceae
*Verbenaceae
*Violaceae
*Viscaceae
*Vitaceae
*Vittariaceae

*Woodsiaceae

*Xyridaceae

*Zannichelliaceae
*Zingiberaceae
*Zosteraceae
*Zygophyllaceae

Editorial adviser, National Museums of Kenya: Quentin Luke
Adviser on Linnaean types: C. Jarvis

Parts of this Flora, unless otherwise indicated, are obtainable from:
Royal Botanic Gardens, Kew, Richmond, Surrey TW9 3AB, England. www.kew.org or www.kewbooks.com

* Only available through CRC Press at:
UK and Rest of World (except North and South America):
CRS Press/ITPS,
Cheriton House, North Way, Andover, Hants SP10 5BE.
e: uk.tandf@thomsonpublishingservices. co.uk

North and South America:
CRC Press,
2000NW Corporate Blvd, Boco Raton, FL 33431-9868, USA.
e: orders@crcpress.com

For availability and expected publication dates please check on our website, www.kew.books.com or email publishing@kew.org

Information on current prices can be found at www.kewbooks.com or www.tandf.co.uk/books/

ROYAL BOTANIC GARDENS

© The Board of Trustees of the Royal Botanic Gardens, Kew 2012

Illustrations copyright © Contributing artists

The authors and illustrators have asserted their rights to be identified as the authors of this work in accordance with the Copyright, Designs and Patents Act 1988.

Great care has been taken to maintain the accuracy of the information contained in this work. However, neither the publisher, the editors nor authors can be held responsible for any consequences arising from use of the information contained herein.

First published in 2012 by
Royal Botanic Gardens, Kew
Richmond, Surrey, TW9 3AB, UK
www.kew.org

ISBN 978 1 84246 396 3

Distributed on behalf of the Royal Botanic Gardens, Kew in North America by the University of Chicago Press, 1427 East 60th Street, Chicago, IL 60637, USA.

British Library Cataloguing in Publication Data
A catalogue record for this book is available from the British Library

Design and typesetting by Margaret Newman,
Kew Publishing, Royal Botanic Gardens, Kew.

Printed in the the USA by The University of Chicago Press

Kew's mission is to inspire and deliver science-based plant conservation worldwide, enhancing the quality of life.

Kew receives half of its running costs from Government through the Department for Environment, Food and Rural Affairs (Defra). All other funding needed to support Kew's vital work comes from members, foundations, donors and commercial activities including book sales.